Fundamentals of Telecommunications

WILEY SERIES IN TELECOMMUNICATIONS AND SIGNAL PROCESSING

John G. Proakis, Editor
Northeastern University

Introduction to Digital Mobil Communications
Yoshihiko Akaiwa
Digital Telephony, 2nd Edition
John Bellamy
Elements of Information Theory
Thomas M. Cover and Joy A. Thomas
Fundamentals of Telecommunications
Roger L. Freeman
Practical Data Communications
Roger L. Freeman
Radio System Design for Telecommunications, 2nd Edition
Roger L. Freeman
Telecommunication System Engineering, 3rd Edition
Roger L. Freeman
Telecommunications Transmission Handbook, 4th Edition
Roger L. Freeman
Introduction to Communications Engineering, 2nd Edition
Robert M. Gagliardi
Optical Communications, 2nd Edition
Robert M. Gagliardi and Sherman Karp
Active Noise Control Systems: Algorithms and DSP Implementations
Sen M. Kuo and Dennis R. Morgan
Mobile Communications Design Fundamentals, 2nd Edition
William C. Y. Lee
Expert System Applications for Telecommunications
Jay Liebowitz
Digital Signal Estimation
Robert J. Mammone, Editor
Digital Communication Receivers: Synchronization, Channel Estimation, and Signal Processing
Heinrich Meyr, Marc Moeneclaey, and Stefan A. Fechtel
Synchronization in Digital Communications, Volume I
Heinrich Meyr and Gerd Ascheid
Business Earth Stations for Telecommunications
Walter L. Morgan and Denis Rouffet
Wireless Information Networks
Kaveh Pahlavan and Allen H. Levesque
Satellite Communications: The First Quarter Century of Service
David W. E. Rees
Fundamentals of Telecommunication Networks
Tarek N. Saadawi, Mostafa Ammar, with Ahmed El Hakeem
Meteor Burst Communications: Theory and Practice
Donald L. Schilling, Editor
Vector Space Projections: A Numerical Approach to Signal and Image Processing, Neural Nets, and Optics
Henry Stark and Yongyi Yang
Signaling in Telecommunication Networks
John G. van Bosse
Telecommunication Circuit Design
Patrick D. van der Puije
Worldwide Telecommunications Guide for the Business Manager
Walter H. Vignault

Fundamentals of Telecommunications

Roger L. Freeman

A Wiley-Interscience Publication
JOHN WILEY & SONS, INC.
New York • Chichester • Weinheim • Brisbane • Singapore • Toronto

The texts extracted from the ITU material have been reproduced with the prior authorization of the Union as the copyright holder. The sole responsibility for selecting extracts for reproduction lies with the author alone and can in no way be attributed to the ITU. The complete volumes of the ITU material from which the texts reproduced here are extracted, can be obtained from

International Telecommunication Union
General Secetariat—Sales and Marketing Service
Place de Nations
CH-1211 GENEVA 20 (Switzerland)
Telephone: +41 22 730 51 11 Telex: 421 000 itu ch Telegram: ITU GENEVE Fax: +41 22 730 51 94 X.400: S = Sales; P = itu; A = Arcom; C = ch Internet: Sales@itu.ch

As referenced, some figures and tables throughout this text are copyrighted by The Institute of Electrical and Electronics Engineers, Inc. The IEEE disclaims any responsibility or liability resulting from the placement and use in this publication. All figures are reprinted with permission.

As referenced, some figures and tables are copyrighted by Bellcore and are reprinted with permission.

This book is printed on acid-free paper. ♾

Published by John Wiley & Sons, Inc.
Published simultaneously in Canada.

For ordering and customer information call, 1-800-CALL-WILEY.

Library of Congress Cataloging-in-Publication Data:

Freeman, Roger L.
Fundamentals of telecommunications / Roger L. Freeman.
p. cm.
"A Wiley-Interscience publication."
Includes index.
ISBN 0-471-29699-6 (cloth : alk. paper)
1. Telecommunication. I. Title.
TK5101.F6595 1999
621.382—dc21 98-4272
CIP

Printed in the United States of America.

10 9 8 7 6 5 4 3 2 1

To Paquita

CONTENTS

PREFACE

This book is an entry-level text on the technology of telecommunications. It has been crafted with the newcomer in mind. The eighteen chapters of text have been prepared for high-school graduates who understand algebra, logarithms, and basic electrical principles such as Ohm's law. However, many users require support in these areas so Appendices A and B review the essentials of electricity and mathematics through logarithms. This material was placed in the appendices so as not to distract from the main theme: the technology of telecommunication systems. Another topic that many in the industry find difficult is the use of decibels and derived units. Appendix C provides the reader with a basic understanding of decibels and their applications. The only mathematics necessary is an understanding of the powers of ten.

To meet my stated objective, whereby this text acts as a tutor for those with no experience in telecommunications, every term and concept is carefully explained. Nearly all terminology can be traced to the latest edition of the IEEE dictionary and/or to the several ITU (International Telecommunication Union) glossaries. Other tools I use are analogies and real-life experiences.

We hear the expression "going back to basics." This book addresses the basics and it is written in such a way that it brings along the novice. The structure of the book is purposeful; later chapters build on earlier material. The book begins with some general concepts in telecommunications: What is connectivity, What do nodes do? From there we move on to the voice network embodied in the public switched telecommunications network (PSTN), digital transmission and networks, an introduction to data communications, followed by enterprise networks. It continues with switching and signaling, the transmission transport, cable television, cellular/PCS, ATM, and network management. CCITT Signaling System No. 7 is a data network used exclusively for signaling. It was located after our generic discussion of data and enterprise networks. The novice would be lost in the explanation of System 7 without a basic understanding of data communications.

I have borrowed heavily from my many enriching years of giving seminars, both at Northeastern University and at the University of Wisconsin—Madison. The advantage of the classroom is that the instructor can stop to reiterate or explain a sticky point. Not so with a book. As a result, I have made every effort to spot those difficult issues, and then give clear explanations. Brevity has been a challenge for me. Telecommunications is developing explosively. My goal has been to hit the high points and leave the details to my other texts.

A major source of reference material has been the International Telecommunication Union (ITU). The ITU had a major reorganization on January 1, 1993. Its two principal subsidiary organizations, CCITT and CCIR, changed their names to ITU Telecommunication Standardization Sector and the ITU Radio Communications Sector, respectively. Reference publications issued prior to January 1993 carry the older title: CCITT and CCIR. Standards issued after that date carry ITU-T for Telecommunication Sector publications and ITU-R for the Radio Communications Sector documents.

ACKNOWLEDGMENTS

Some authors are fortunate to have a cadre of friends who pitch in to help and advise during the preparation of a book. I am one of these privileged people. These friends have stood by me since the publication of my first technical text. In this group are John Lawlor, principal, John Lawlor and Associates of Sharon, MA; Dr. Ron Brown, independent consultant, Melrose, MA; Bill Ostaski, an expert on Internet matters who is based in Beverly Farms, MA; Marshall Cross, president, Megawave Corp., Boylston, MA; and Jerry Brilliant, independent consultant based in Fairfax, VA.

I am grateful to my friends at Motorola in Chandler, AZ, where I learned about mentoring young engineers. In that large group, four names immediately come to mind: Dr. Ernie Woodward, Doug White, Dr. Ali Elahi, and Ken Peterson—all of the Celestri program.

Then there is Milt Crane, an independent consultant in Phoenix, AZ, who is active in local IEEE affairs. Dan Danbeck, program director with Engineering Professional Development, University of Wisconsin–Madison, who provided constructive comments on the book's outline. Ted Myers, of Ameritech Cellular, made helpful suggestions on content. John Bellamy, independent consultant and Prof. John Proakis, series editor and well-known author in his own right, reviewed the outline and gave constructive comments to shorten the book to some reasonable length.

I shall always be indebted to Dr. Don Schilling, professor emeritus, City College of New York and great proponent of CDMA in the PCS and cellular environment. Also, my son, Bob Freeman, major accounts manager for Hispanic America, Axis Communications, for suggestions on book promotion. Bob broke into this business about five years ago. Also, my thanks to Dr. Ted Woo of SCTE for help on CATV; to Fran Drake, program director, University of Wisconsin–Madison, who gave me this book idea in the first place; and Dr. Bob Egri, principal investigator at MaCom Lowell (MA) for suggestions on the radio frequency side.

ROGER L. FREEMAN

Scottsdale, Arizona
November, 1998

Fundamentals of Telecommunications

1

INTRODUCTORY CONCEPTS

1.1 WHAT IS TELECOMMUNICATION?

Many people call *telecommunication* the world's most lucrative industry. If we add cellular and PCS users,[1] there are about 1800 million subscribers to telecommunication services world wide (1999). Annual expenditures on telecommunications may reach 900,000 million dollars in the year 2000.[2]

Prior to divestiture, the *Bell System* was the largest commercial company in the United States even though it could not be found on the Fortune 500 listing of the largest companies. It had the biggest fleet of vehicles, the most employees, and the greatest income. Every retiree with any sense held the safe and dependable Bell stock. In 1982, Western Electric Co., the Bell System manufacturing arm, was number seven on the Fortune 500. However, if one checked the Fortune 100 Utilities, the Bell System was up on the top. Transferring this information to the Fortune 500, again put Bell System as the leader on the list.

We know telecommunication is big business; but what is it? *Webster's* (Ref. 1) calls it *communications at a distance*. The IEEE dictionary (Ref. 2) defines telecommunications as "the transmission of signals over long distance, such as by telegraph, radio or television." Another term we often hear is *electrical communication*. This is a descriptive term, but of somewhat broader scope.

Some take the view that telecommunication deals only with voice telephony, and the typical provider of this service is the local telephone company. We hold with a wider interpretation. Telecommunication encompasses the electrical communication at a distance of voice, data, and image information (e.g., TV and facsimile). These media, therefore, will be major topics of this book. The word *media* (medium, singular) also is used to describe what is transporting telecommunication signals. This is termed transmission media. There are four basic types of medium: (1) wire-pair, (2) coaxial cable, (3) fiber optics, and (4) radio.

1.2 TELECOMMUNICATION WILL TOUCH EVERYBODY

In industrialized nations, the telephone is accepted as a way of life. The telephone is connected to the public switched telecommunications network (PSTN) for local, national,

[1]PCS, personal communication services, is a cellular-radiolike service covering a smaller operational area.
[2]We refrain from using *billion* because it is ambiguous. Its value differs, depending on where you come from.

and international voice communications. These same telephone connections may also carry data and image information (e.g., television).

The personal computer (PC) is beginning to take on a similar role as the telephone, that of being ubiquitous. Of course, as we know, the two are becoming married. In most situations, the PC uses telephone connectivity to obtain internet and e-mail services. Radio adjuncts to the telephone, typically cellular and PCS, are beginning to offer similar services such as data communications (including internet) and facsimile (fax), as well as voice. The popular press calls these adjuncts *wireless*. Can we consider wireless in opposition to *being wired*?

Count the number of devices one has at home that carry out some kind of controlling or alerting function. They also carry out a personal communication service. Among these devices are television remote controls, garage-door openers, VCR and remote radio and CD player controllers, certain types of home security systems, pagers, and cordless telephones. We even take cellular radios for granted.

In some countries, a potential subscriber has to wait months or years for a telephone. Cellular radio, in many cases, provides a way around the problem, where equivalent telephone service can be established in an hour—just enough time to buy a cellular radio in the local store and sign a contract for service.

The PSTN has ever-increasing data communications traffic, where the network is used as a conduit for data. PSTN circuits may be leased or used in a dial-up mode for data connections. Of course, the Internet has given added stimulus to data circuit usage of the PSTN. The PSTN sees facsimile as just another data circuit, usually in the dial-up mode. Conference television traffic adds still another flavor to PSTN traffic and is also a major growth segment.

There is a growing trend for users to bypass the PSTN partially or completely. The use of satellite links in certain situations is one method for PSTN bypass. Another is to lease capacity from some other provider. *Other provider* could be a power company with excess capacity on its microwave or fiber optic system. There are other examples, such as a railroad with extensive rights-of-way that are used by a fiber optic network.

Another possibility is to build a private network using any one or a combination of fiber optics, line-of-sight-microwave, and satellite communications. Some private networks take on the appearance of a mini-PSTN.

1.3 INTRODUCTORY TOPICS IN TELECOMMUNICATIONS

An overall telecommunications network (i.e., the PSTN) consists of local networks interconnected by a long-distance network. The concept is illustrated in Figure 1.1. This is the PSTN, which is open to public correspondence. It is usually regulated by a government authority or may be a government monopoly, although there is a notable trend toward privatization. In the United States the PSTN has been a commercial enterprise since its inception.

1.3.1 End-Users, Nodes, and Connectivities

End-users, as the term tells us, provide the inputs to the network and are recipients of network outputs. The end-user employs what is called an I/O, standing for input/output (device). An I/O may be a PC, computer, telephone instrument, cellular/PCS telephone or combined device, facsimile, or conference TV equipment. It may also be some type

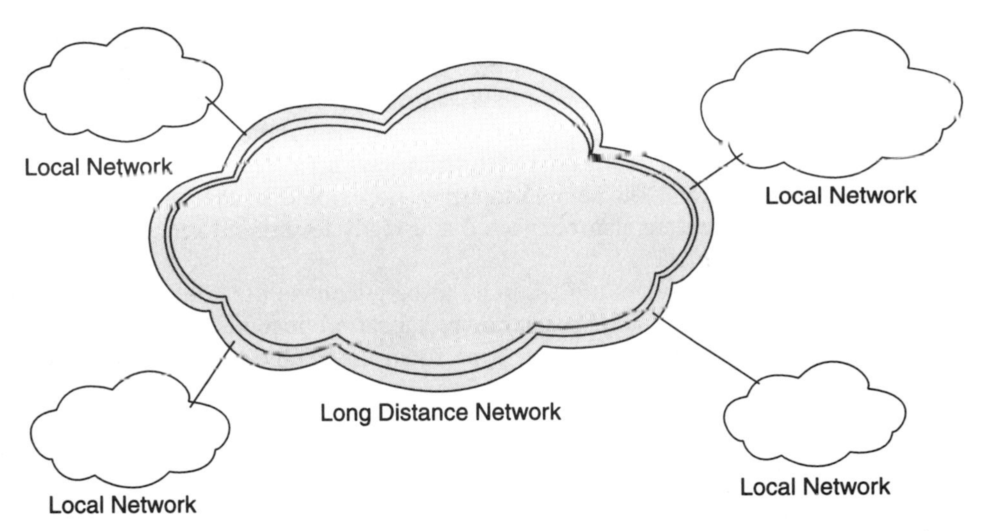

Figure 1.1 The PSTN consists of local networks interconnected by a long-distance network.

of *machine* that provides a stimulus to a coder or receives stimulus from a decoder in, say, some sort of SCADA system.[3]

End-users usually connect to *nodes*. We will call a node a point or junction in a transmission system where lines and trunks meet. A node usually carries out a switching function. In the case of the local area network (LAN), we are stretching the definition. In this case a network interface unit is used, through which one or more end-users may be connected.

A *connectivity* connects an end-user to a node, and from there possibly through other nodes to some final end-user destination with which the initiating end-user wants to communicate. Figure 1.2 illustrates this concept.

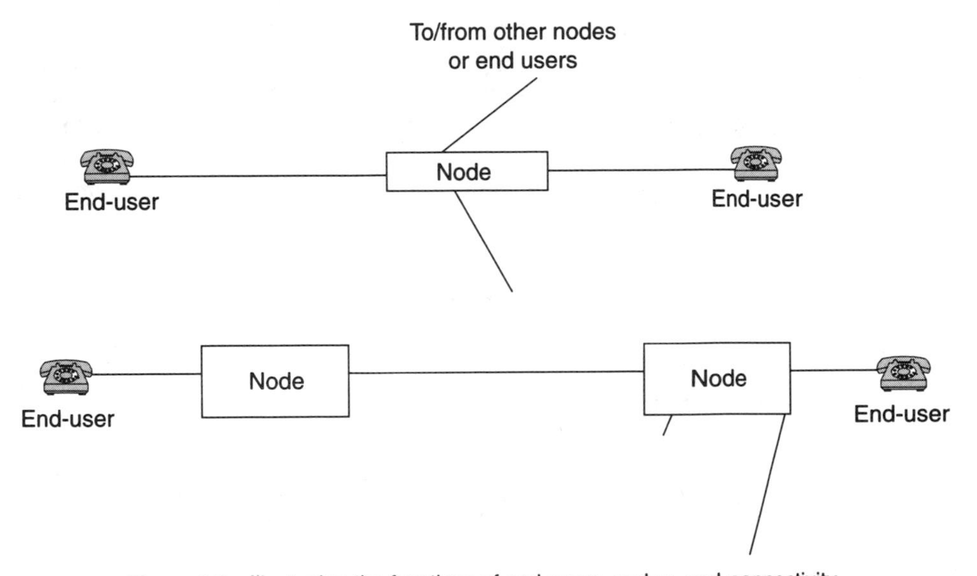

Figure 1.2 Illustrating the functions of end-users, nodes, and connectivity.

[3]SCADA stands for supervisory control and data acquisition.

The IEEE (Ref. 2) defines a *connection* as "an association of channels, switching systems, and other functional units set up to provide means for a transfer of information between two or more points in a telecommunications network." There would seem to be two interpretations of this definition. First, the equipment, both switching and transmission facilities, is available to set up a path from, say, point A to point B. Assume A and B to be user end-points. The second interpretation would be that not only are the circuits available, but they are also connected and ready to pass information or are in the information-passing mode.

At this juncture, the end-users are assumed to be telephone users, and the path that is set up is a speech path (it could, of course, be a data or video path). There are three sequential stages to a telephone call:

1. Call setup;
2. Information exchange; and
3. Call take down.

Call setup is the stage where a circuit is established and activated. The setup is facilitated by *signaling*, which is discussed in Chapter 7.[4] It is initiated by the calling subscriber (user) going *off-hook*. This is a term that derives from the telephony of the early 1900s. It means "the action of taking the telephone instrument out of its cradle." Two little knobs in the cradle pop up, pushed by a spring action, causing an electrical closure. If we turn a light on, we have an electrical closure allowing electrical current to pass. The same thing happens with our telephone set; it now passes current. The current source is a "battery" that resides at the local serving switch. It is connected by the *subscriber loop*. This is just a pair of copper wires connecting the battery and switch out to the subscriber premises and then to the subscriber instrument. The action of current flow alerts the serving exchange that the subscriber requests service. When the current starts to flow, the exchange returns a dial tone, which is audible in the headset (of the subscriber instrument). The calling subscriber (user) now knows that she/he may start dialing digits or pushing buttons on the subscriber instrument. Each button is associated with a digit. There are 10 digits, 0 through 9. Figure 1.3 shows a telephone end instrument connected through a subscriber loop to a local serving exchange. It also shows that all-important *battery* (battery feed bridge), which provides a source of current for the subscriber loop.

If the called subscriber and the calling subscriber are in the same local area, only

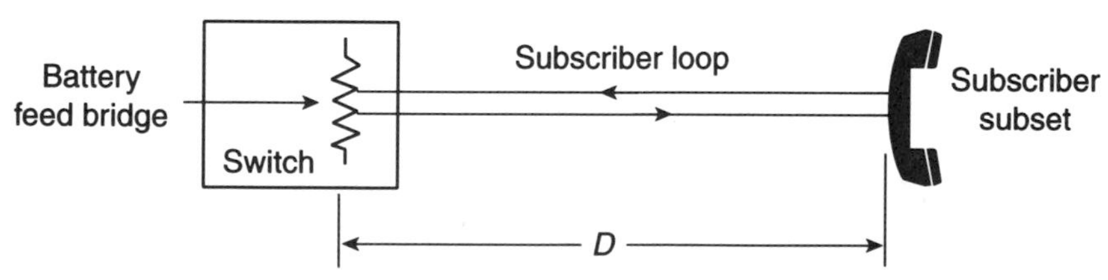

Figure 1.3 A subscriber set is connected to a telephone exchange by a subscriber loop. Note the battery feed in the telephone serving switch. Distance D is the loop length discussed in Section 5.4.

[4]*Signaling* may be defined as the exchange of information specifically concerned with the establishment and control of connections, and the transfer of user-to-user and management information in a circuit-switched (e.g., the PSTN) network.

seven digits need be dialed. These seven digits represent the telephone number of the called subscriber (user). This type of signaling, the dialing of the digits, is called *address signaling*. The digits actuate control circuits in the local switch, allowing a connectivity to be set up. If the calling and called subscribers reside in the serving area of that local switch, no further action need be taken. A connection is made to the called subscriber line and the switch sends a special ringing signal down that loop to the called subscriber, and her/his telephone rings, telling her/him that someone wishes to talk to her/him on the telephone. This audible ringing is called *alerting*, another form of signaling. Once the called subscriber goes off-hook (i.e., takes the telephone out of its cradle), there is activated connectivity, and the call enters the information-passing phase, or phase 2 of the telephone call.

When the call is completed, the telephones at each end are returned to their cradles, breaking the circuit of each subscriber loop. This, of course, is analogous to turning off a light; the current stops flowing. Phase 3 of the telephone call begins. It terminates the call, and the connecting circuit in the switch is taken down and freed-up for another user. Both subscriber loops are now *idle*. If a third user tries to call either subscriber during stages 2 and 3, she/he is returned a *busy-back* by the exchange (serving switch). This is the familiar "busy signal," a tone with a particular cadence. The return of the busy-back is a form of signaling called *call-progress* signaling.

Suppose now that a subscriber wishes to call another telephone subscriber outside the local serving area of her/his switch. The call setup will be similar as before, except that at the calling subscriber serving switch the call will be connected to an outgoing *trunk*. As shown in Figure 1.4, trunks are transmission pathways that interconnect switches. To repeat: subscriber loops connect end-users (subscriber) to a local serving switch; trunks interconnect exchanges or switches.

The IEEE (Ref. 2) defines a *trunk* as "a transmission path between exchanges or central offices." The word *transmission* in the IEEE definition refers to one (or several) transmission media. The medium might be wire-pair cable, fiber optic cable, microwave radio and, stretching the imagination, satellite communications. In the conventional telephone plant, coaxial cable has fallen out of favor as a transmission medium for this application. Of course, in the long-distance plant, satellite communication is

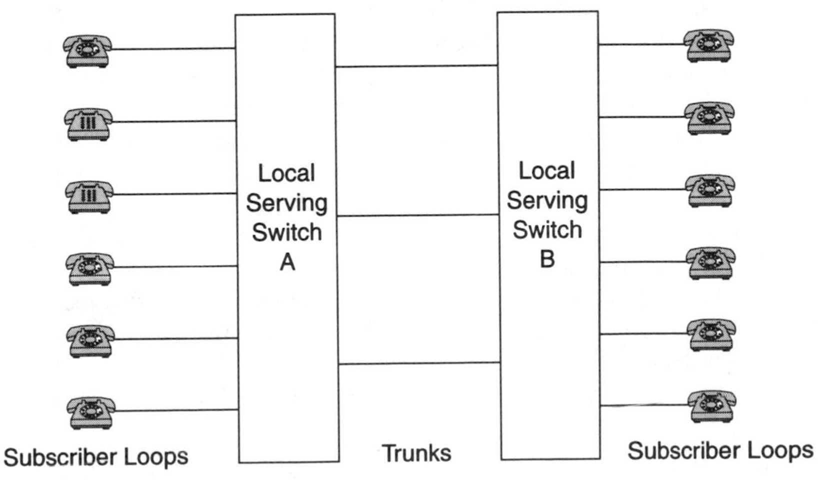

Figure 1.4 Subscriber loops connect telephone subscribers to their local serving exchange; trunks interconnect exchanges (switches).

fairly widely employed, particularly for international service. Our preceding reference was for local service.

1.3.2 Telephone Numbering and Routing

Every subscriber in the world is identified by a number, which is geographically tied to a physical location.[5] This is the *telephone number*. The telephone number, as we used it here, is seven digits long. For example:

$$234 - 5678$$

The last four digits identify the subscriber line; the first three digits (i.e., 234) identify the serving switch (or exchange).

For a moment, let's consider theoretical numbering capacity. The subscriber number, those last four digits, has a theoretical numbering capacity of 10,000. The first telephone number issued could be 0000, the second number, if it were assigned in sequence, would be 0001, the third, 0002, and so on. At the point where the numbers ran out, the last number issued would be 9999.

The first three digits of the preceding example contain the exchange code (or central office code). These three digits identify the exchange or switch. The theoretical maximum capacity is 1000. If again we assign numbers in sequence, the first exchange would have 001, the next 002, then 003, and finally 999. However, particularly in the case of the exchange code, there are blocked numbers. Numbers starting with 0 may not be desirable in North America because 0 is used to dial the operator.

The numbering system for North America (United States, Canada, and Caribbean islands) is governed by the NANP or North American Numbering Plan. It states that central office codes (exchange codes) are in the form NXX where N can be any number from 2 through 9 and X can be any number from 0 through 9. Numbers starting with 0 or 1 are blocked numbers. This cuts the total exchange code capacity to 800 numbers. Inside these 800 numbers there are five blocked numbers such as 555 for directory assistance and 958/959 for local plant test.

When long-distance service becomes involved, we must turn to using still an additional three digits. Colloquially we call these area codes. In the official North American terminology used in the NANP is NPA for numbering plan area, and we call these area codes *NPA codes*. We try to assure that both exchange codes and NPA codes do not cross political/administrative boundaries. What is meant here are state, city, and county boundaries. We have seen exceptions to the county/city rule, but not to the state. For example, the exchange code 443 (in the 508 area code, middle Massachusetts) is exclusively for the use of the town of Sudbury, Massachusetts. Bordering towns, such as Framingham, will not use that number. Of course, that exchange code number is meant for Sudbury's singular central office (local serving switch).

There is similar thinking for NPAs (area codes). In this case it is that these area codes may not cross state boundaries. For instance, 212 is for Manhattan and may not be used for northern New Jersey.

Return now to our example telephone call. Here the calling party wishes to speak

[5]This will change. At least in North America, we expect to have telephone number portability. Thus, whenever one moves to a new location, she/he takes her/his telephone number with them. Will we see a day when telephone numbers are issued at birth, much like social security numbers?

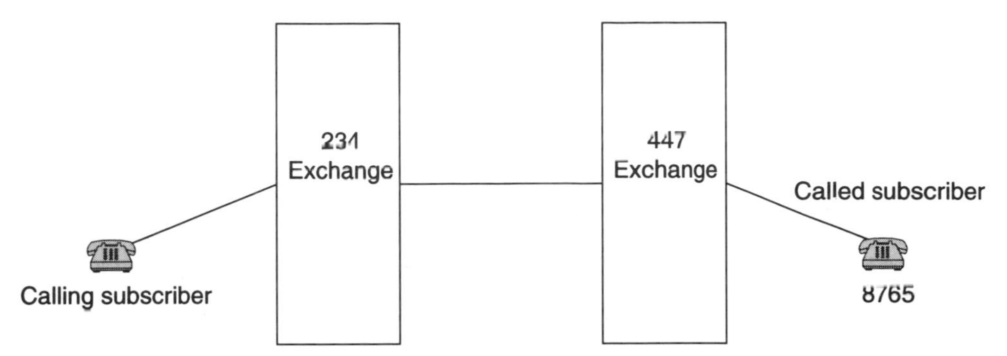

Figure 1.5 Example connectivity subscriber-to-subscriber through two adjacent exchanges.

to a called party that is served by a different exchange (central office).[6] We will assign the digits 234 for the calling party's serving exchange; for the called party's serving exchange we assign the digits 447. This connectivity is shown graphically in Figure 1.5. We described the functions required for the calling party to reach her/his exchange. This is the 234 exchange. It examines the dialed digits of the called subscriber, 447–8765. To route the call, the exchange will only work upon the first three digits. It accesses its local look-up table for the routing to the 447 exchange and takes action upon that information. An appropriate vacant trunk is selected for this route and the signaling for the call advances to the 447 exchange. Here this exchange identifies the dialed number as its own and connects it to the correct subscriber loop, namely, the one matching the 8765 number. Ringing current is applied to the loop to alert the called subscriber. The called subscriber takes her/his telephone off-hook and conversation can begin. Phases 2 and 3 of this telephone call are similar to our previous description.

1.3.3 Use of Tandem Switches in a Local Area Connectivity

Routing through a tandem switch is an important economic expedient for a telephone company or administration. We could call a tandem switch a *traffic concentrator*. Up to now we have discussed direct trunk circuits. To employ a direct trunk circuit, there must be sufficient traffic to justify such a circuit. One reference (Ref. 3) suggests a break point of 20 erlangs.[7] For a connectivity with traffic intensity under 20 erlangs for the busy hour (BH), the traffic should be routed through a tandem (exchange). For traffic intensities over that value, establish a direct route. Direct route and tandem connectivities are illustrated in Figure 1.6.

1.3.4 Busy Hour and Grade of Service

The PSTN is very inefficient. This inefficiency stems from the number of circuits and the revenue received per circuit. The PSTN would approach 100% efficiency if all the circuits were used all the time. The fact is that the PSTN approaches total capacity utilization for only several hours during the working day. After 10 P.M. and before 7 A.M. capacity utilization may be 2% or 3%.

The network is dimensioned (sized) to meet the period of maximum usage demand.

[6]The term *office* or *central office* is commonly used in North America for a switch or an exchange. The terms switch, office, and exchange are synonymous.

[7]The *erlang* is a unit of traffic intensity. One erlang represents one hour of line (circuit) occupancy.

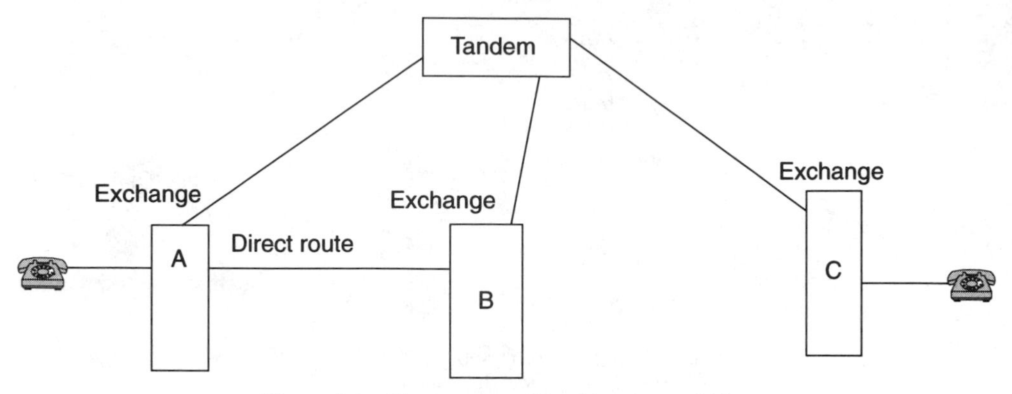

Figure 1.6 Direct route and tandem connectivities.

This period is called the *busy hour* (BH). There are two periods where traffic demand on the PSTN is maximum–one in the morning and one in the afternoon. This is illustrated in Figure 1.7.

Note the two traffic peaks in Figure 1.7. These are caused by business subscribers. If the residential and business curves were combined, the peaks would be much sharper. Also note that the morning peak is somewhat more intense than the afternoon busy hour. In North America (i.e., north of the Rio Grande river), the busy hour BH is between 9:30 A.M. and 10:30 A.M. Because it is more intense than the afternoon high-traffic period, it is called the BH. There are at least four distinct definitions of the busy hour. We quote only one: "That uninterrupted period of 60 minutes during the day when the traffic offered is maximum." Other definitions may be found in (Ref. 4).

BH traffic intensities are used to dimension the number of trunks required on a connectivity as well as the size of (a) switch(es) involved. Now a PSTN company (administration) can improve its revenue versus expenditures by cutting back on the number

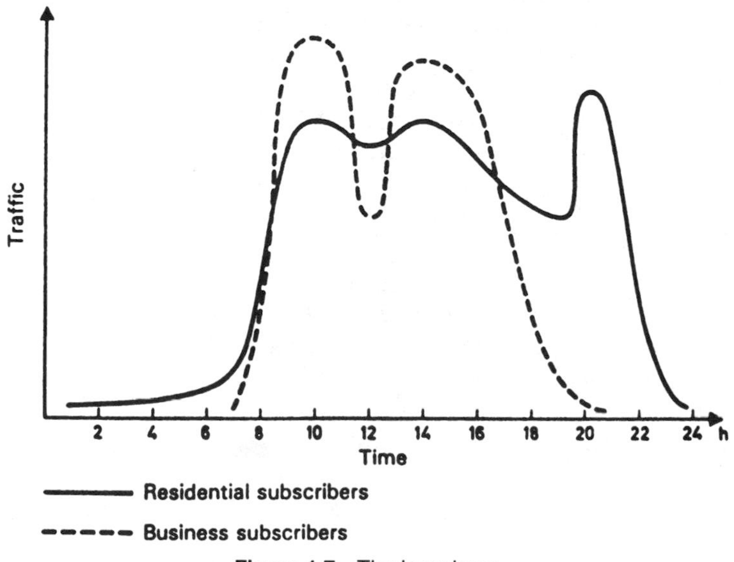

Figure 1.7 The busy hour.

of trunks required and making switches "smaller." Of course, network users will do a lot of complaining about poor service. Let's just suppose the PSTN does just that—cuts back on the number of circuits. Now, during the BH period, a user may dial a number and receive either a voice announcement or a rapid-cadence tone telling the user that all trunks are busy (ATB) and to try again later. From a technical standpoint, the user has encountered *blockage*. This would be due to one of two reasons, or may be due to both causes. These are: insufficient switch capacity and not enough trunks to assign during the BH. There is a more in-depth discussion of the busy hour in Section 4.2.1.

Networks are sized/dimensioned for a traffic load expected during the busy hour. The sizing is based on probability, usually expressed as a decimal or percentage. That probability percentage or decimal is called the *grade of service*. The IEEE (Ref. 2) defines grade of service as "the proportion of total calls, usually during the busy hour, that cannot be completed immediately or served within a prescribed time."

Grade of service and *blocking probability* are synonymous. Blocking probability objectives are usually stated as B = 0.01 or 1%. This means that during the busy hour 1 in 100 calls can be expected to meet blockage.

1.3.5 Simplex, Half-Duplex and Full Duplex

These are operational terms, and they will be used throughout this text. *Simplex* is one-way operation; there is no reply channel provided. Radio and television broadcasting are simplex. Certain types of data circuits might be based on simplex operation.

Half-duplex is a two-way service. It is defined as transmission over a circuit capable of transmitting in either direction, but only in one direction at a time.

Full duplex or just *duplex* defines simultaneous two-way independent transmission on a circuit in both directions. All PSTN-type circuits discussed in this text are considered using full duplex operation unless otherwise specified.

1.3.6 One-Way and Two-Way Circuits

Trunks can be configured for either one-way or two-way operation.[8] A third option is a hybrid, where one-way circuits predominate and a number of two-way circuits are provided for overflow situations. Figure 1.8*a* shows two-way trunk operation. In this case any trunk can be selected for operation in either direction. The insightful reader will observe that there is some fair probability that the same trunk can be selected from either side of the circuit. This is called *double seizure*. It is highly undesirable. One way to reduce this probability is to use normal trunk numbering (from top down) on one side of the circuit (at exchange A in the figure) and to reverse trunk numbering (from the bottom up) at the opposite side of the circuit (exchange B).

Figure 1.8*b* shows one-way trunk operation. The upper trunk group is assigned for the direction from A to B and the lower trunk group for the opposite direction, from exchange B to exchange A. Here there is no possibility of double seizure.

Figure 1.8*c* illustrates a typical hybrid arrangement. The upper trunk group carries traffic from exchange A to exchange B exclusively. The lowest trunk group carries traffic in the opposite direction. The small, middle trunk group contains two-way circuits. Switches are programmed to select from the one-way circuits first, until all these circuits become busy, then they may assign from the two-way circuit pool.

Let us clear up some possible confusion here. Consider the one-way circuit from A

[8]Called *both-way* in the United Kingdom and in CCITT documentation.

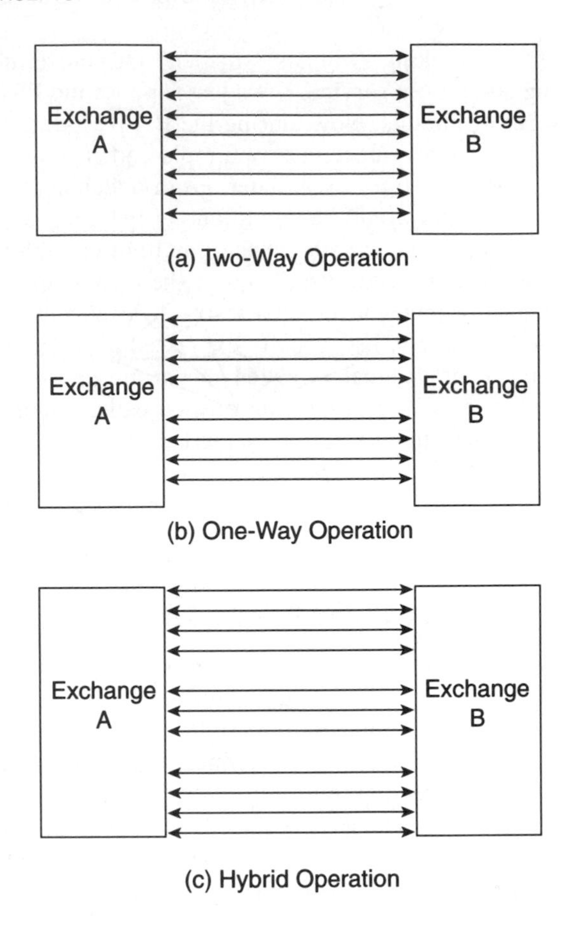

Figure 1.8 Two-way and one-way circuits: two-way operation (*a*), one-way operation (*b*), and a hybrid scheme, a combination of one-way and two-way operation (*c*).

to B, for example. In this case, calls originating at exchange A bound for exchange B in Figure 1.8*b* are assigned to the upper trunk group. Calls originating at exchange B destined for exchange A are assigned from the pool of the lower trunk group. Do not confuse these concepts with two-wire and four-wire operation, discussed in Chapter 4, Section 4.4.

1.3.7 Network Topologies

The IEEE (Ref. 2) defines *topology* as "the interconnection pattern of nodes on a network." We can say that a telecommunication network consists of a group of interconnected nodes or switching centers. There are a number of different ways we can interconnect switches in a telecommunication network.

If every switch in a network is connected to all other switches (or nodes) in the network, we call this "pattern" a *full-mesh* network. Such a network is shown in Figure 1.9*a*. This figure has 8 nodes.[9]

[9]The reader is challenged to redraw the figure adding just one node for a total of nine nodes. Then add a tenth and so on. The increasing complexity becomes very obvious.

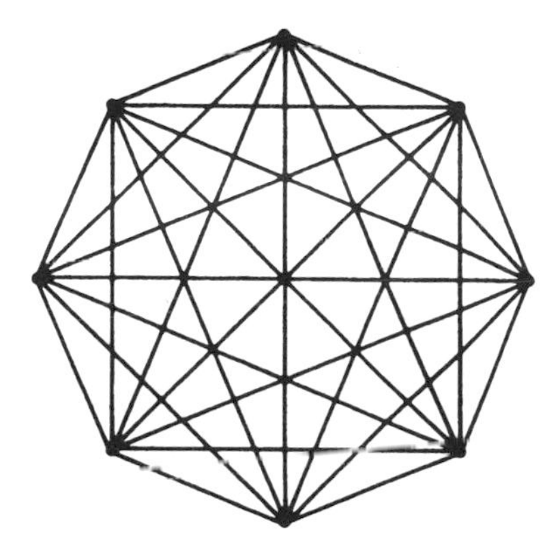

Figure 1.9*a* A full-mesh network connecting eight nodes.

In the 1970s, Madrid (Spain) had 82 switching centers connected in a full-mesh network. A full-mesh network is very survivable because of a plethora of possible alternative routes.

Figure 1.9*b* shows a *star network.* It is probably the least survivable. However, it is one of the most economic nodal patterns both to install and to administer. Figure 1.9*c* shows a *multiple star network.* Of course we are free to modify such networks by adding direct routes. Usually we can apply the 20 erlang rule in such situations. If a certain *traffic relation* has 20 erlangs or more of BH traffic, a direct route is usually justified. The term *traffic relation* simply means the traffic intensity (usually the BH traffic intensity) that can be expected between two known points. For instance, between Albany, NY, and New York City there is a traffic relation.[10] On that relation we would probably expect thousands of erlangs during the busy hour.

Figure 1.9*d* shows a hierarchical network. It is a natural outgrowth of the multiple star network shown in Figure 1.9*c*. The PSTNs of the world universally used a hierarchical network; CCITT recommended such a network for international application. Today there is a trend away from this structure or, at least, there will be a reduction of the number of levels. In Figure 1.9*d* there are five levels. The highest rank or order in the hierarchy

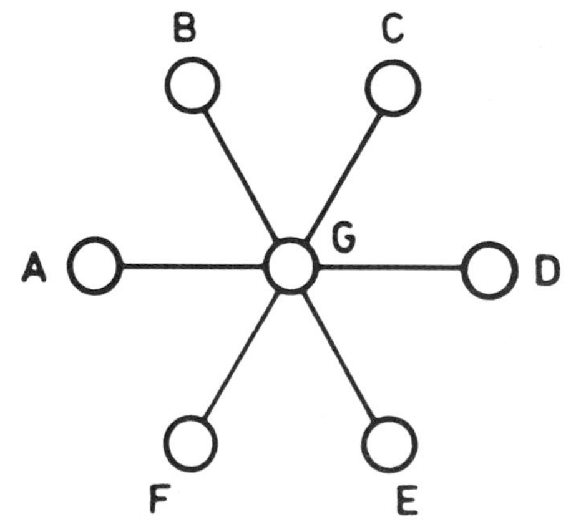

Figure 1.9*b* A star network.

[10]Albany is the capital of the state of New York.

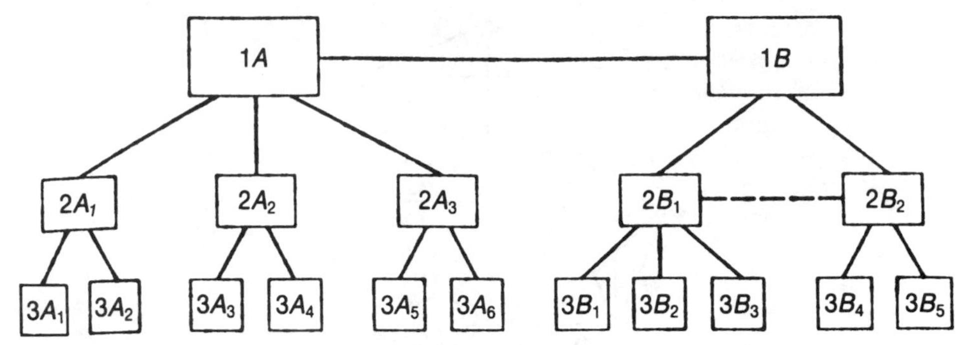

Figure 1.9*c* A higher-order or multiple star network. Note the direct route between $2B_1$ and $2B_2$. There is another direct route between $3A_5$ and $3A_6$.

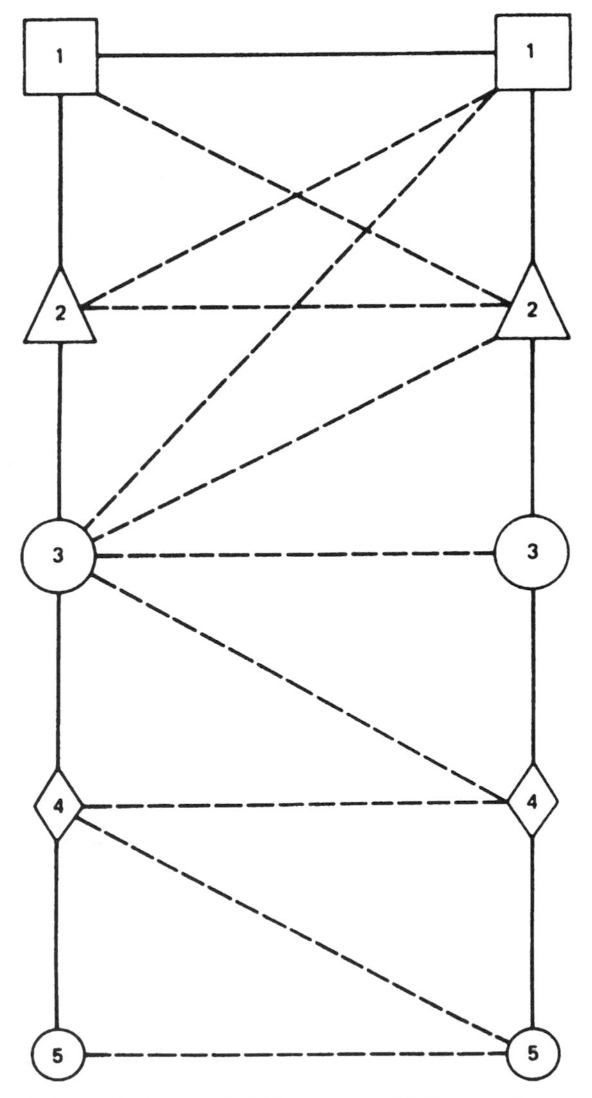

Figure 1.9*d* A typical hierarchical network. This was the AT&T network around 1988. The CCITT-recommended network was very similar.

is the class 1 center (shown as 1 in the figure), and the lowest rank is the class 5 office (shown as 5 in the figure). The class 5 office (switch), often called an *end office*, is the local serving switch, which was discussed previously. Remember that the term *office* is a North American term meaning switching center, node, or switch.

In a typical hierarchical network, *high-usage* (HU) routes may be established, regardless of rank in the hierarchy, if the traffic intensity justifies. A high-usage route or connectivity is the same as a direct route. We tend to use direct route when discussing the local area and we use high-usage routes when discussing a long-distance or toll network.

1.3.7.1. Rules of Conventional Hierarchical Networks. One will note the backbone structure of Figure 1.9*d*. If we remove the high-usage routes (dashed lines in the figure), the backbone structure remains. This backbone is illustrated in Figure 1.10. In the terminology of hierarchical networks, the backbone represents the *final route* from which no *overflow* is permitted.

Let us digress and explain what we mean by overflow. It is defined as that part of the offered traffic that cannot be carried by a switch over a selected trunk group. It is that traffic that met congestion, what was called *blockage* earlier. We also can have overflow of a buffer (a digital memory), where overflow just spills, and is lost.

In the case of a hierarchical network, the overflow can be routed over a different route. It may overflow on to another HU route or to the final route on the backbone (see Figure 1.10).

A hierarchical system of routing leads to simplified switch design. A common expression used when discussing hierarchical routing and multiple star configurations is that lower-rank exchanges *home* on higher-rank exchanges. If a call is destined for an exchange of lower rank in its chain, the call proceeds down the chain. In a similar

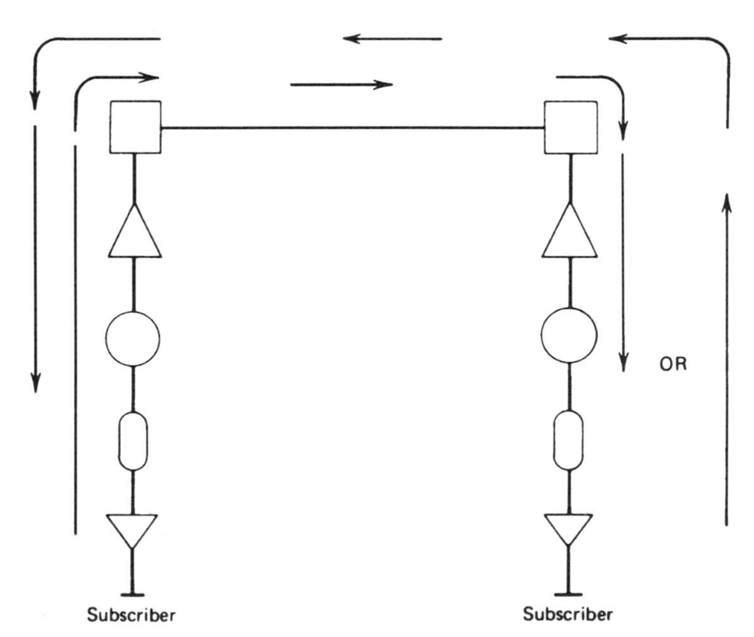

Figure 1.10 The backbone of a hierarchical network. The backbone traces the final route.

manner, if a call is destined for another exchange outside the chain (the opposite side of Figure 1.9*d*), it proceeds up the chain and across. When high-usage routes exist, a call may be routed on a route additional or supplementary to the pure hierarchy, proceeding to the distant transit center and then descending to the destination.[11] Of course, at the highest level in a pure hierarchy, the call crosses from one chain over to the other. In hierarchical networks only the order of each switch in the hierarchy and those additional high usage links (routes) that provide access need be known. In such networks administration is simplified, and storage or routing information is reduced, when compared to the full-mesh type of network, for example.

1.3.7.2. Trend Away from the Hierarchical Structure. There has been a decided trend away from hierarchical routing and network structure. However, there will always be some form of hierarchical structure into the foreseeable future. The change is brought about due to two factors: (1) transmission and (2) switching. Since 1965, transmission techniques have taken leaps forward. Satellite communications allowed direct routes some one-third the way around the world. This was followed by the introduction of fiber optic transmission, providing nearly infinite bandwidth, low loss, and excellent performance properties. These transmission techniques are discussed in Chapter 9.

In the switching domain, the stored program control (SPC) switch had the computer brains to make nearly real-time decisions for routing.[12] This brought about dynamic routing such as AT&T DNHR (dynamic nonhierarchical routing). The advent of CCITT Signaling System No. 7 (Chapter 7), working with high-speed computers, made it possible for optimum routing based on real-time information on the availability of route capacity and shortest routes. Thus the complex network hierarchy started to become obsolete.

Nearly all reference to routing hierarchy disappeared from CCITT in the 1988 Plenary Session (Melbourne) documents. International connectivity is by means of direct/high-usage routes. In fact, CCITT Rec. E.172 (Geneva 10/92) states that "In the ISDN era, it is suggested that the network structure be non-hierarchical, ..."[13] Of course, reference is being made here to the international network.

1.3.8 Variations in Traffic Flow

In networks covering large geographic expanses and even in cases of certain local networks, there may be a variation of the time of day of the BH or in a certain direction of traffic flow. It should be pointed out that the busy hour is tied up with a country's culture. Countries have different working habits and standard business hours vary. In Mexico, for instance, the BH is more skewed toward noon because Mexicans eat lunch later than do people in the United States.

In the United States business traffic peaks during several hours before and several hours after the noon lunch period on weekdays, and social calls peak in early evening. Traffic flow tends to be from suburban living areas to urban center in the morning, and the reverse in the evening.

In national networks covering several time zones, where the difference in local time

[11]A *transit center* or *transit exchange* is a term used in the long-distance network for a tandem exchange. The term *tandem exchange* is reserved for the local network.

[12]SPC stands for stored program control. This simply means a switch that is computer controlled. SPC switches started appearing in 1975.

[13]ISDN stands for Integrated Services Digital Network(s). This is discussed in Section 12.4.

may be appreciable, long-distance traffic tends to be concentrated in a few hours common to BH peaks at both ends. In such cases it is possible to direct traffic so that peaks of traffic in one area (time zone) fall into valleys of traffic of another area. This is called taking advantage of the *noncoincident busy hour*. The network design can be made more optimal if configured to take advantage of these phenomena, particularly in the design of direct routes and overflow routes.

1.4 QUALITY OF SERVICE

Quality of service (QoS) appears at the outset to be an intangible concept. However, it is very tangible for a telephone subscriber unhappy with his or her service. The concept of service quality must be covered early in an all-encompassing text on telecommunications. System designers should never once lose sight of the concept, no matter what segment of the system they may be responsible for. Quality of service means *how happy* the telephone company (or other common carrier) is keeping the customer. For instance, we might find that about half the time a customer dials, the call goes awry or the caller cannot get a dial tone or cannot hear what is being said by the party at the other end. All these have an impact on quality of service. So we begin to find that QoS is an important factor in many areas of the telecommunications business and means different things to different people. In the old days of telegraphy, a rough measure of how well the system was working was the number of service messages received at the switching center. In modern telephony we now talk about *service observation*.

The transmission engineer calls QoS *customer satisfaction*, which is commonly measured by how well the customer can hear the calling party. The unit for measuring how well we can hear a distant party on the telephone is *loudness rating*, measured in decibels (dB). From the network and switching viewpoints, the percentage of lost calls (due to blockage or congestion) during the BH certainly constitutes another measure of service quality. Remember, this item is denominated *grade of service*. One target figure for grade of service is 1 in 100 calls lost during the busy hour. Other elements to be listed under QoS are:

- Delay before receiving dial tone (*dial tone delay*);
- Post dial(ing) delay (time from the completion of dialing the last digit of a number to the first ring-back of the called telephone).[14] This is the primary measure of signaling quality;
- Availability of service tones [e.g., busy tone, telephone out of order, time out, and all trunks busy (ATB)];
- Correctness of billing;
- Reasonable cost of service to the customer;
- Responsiveness to servicing requests;
- Responsiveness and courtesy of operators; and
- Time to installation of a new telephone and, by some, the additional services offered by the telephone company.

[14]*Ring-back* is a call-progress signal telling the calling subscriber that a ringing signal is being applied to the called subscriber's telephone.

One way or another each item, depending on the service quality goal, will have an impact on the design of a telecommunication system.

1.5 STANDARDIZATION IN TELECOMMUNICATIONS

Standardization is vital in telecommunications. A rough analogy—it allows worldwide communication because we all "speak a standard language." Progressing through this book, the reader will find that this is not strictly true. However, a good-faith attempt is made in nearly every case.

There are international, regional, and national standardization agencies. There are at least two international agencies that impact telecommunications. The most encompassing is the International Telecommunication Union (ITU) based in Geneva, Switzerland, which has produced more than 1000 standards. Another is the International Standardization Organization (ISO), which has issued a number of important data communication standards.

Unlike other standardization entities, the ITU is a treaty organization with more treaty signatories than the United Nations. Its General Secretariat produces the *Radio Regulations*. This document set is the only one that is legally binding on the nations that have signed the treaty. In addition, two of the ITU's subsidiary organizations prepare and disseminate documents that are recommendations, reports, or opinions, and are not legally binding on treaty signatories. However they serve as worldwide standards.

The ITU went through a reorganization on January 1, 1993. Prior to that the two important branches were the CCITT, standing for International Consultive Committee for Telephone and Telegraph, and the CCIR, standing for International Consultive Committee for Radio. After the reorganization, the CCITT became the Telecommunication Standardization Sector of the ITU, and the CCIR became the ITU Radiocommunication Sector. The former produces ITU-T Recommendations and the latter produces ITU-R Recommendations. The ITU Radiocommunications Sector essentially prepares the *Radio Regulations* for the General Secretariat.

We note one important regional organization, ETSI, the European Telecommunication Standardization Institute. For example, it is responsible for a principal cellular radio specification—GSM or Ground System Mobile (in the French). Prior to the 1990s, ETSI was the Conference European Post and Telegraph or CEPT. CEPT produced the European version of digital network PCM, previously called CEPT30+2 and now called E-1.

There are numerous national standardization organizations. There is the American National Standards Institute based in New York City that produces a wide range of standards. The Electronics Industries Association (EIA) and the Telecommunication Industry Association (TIA), are both based in Washington, DC, and are associated one with the other. Both are prolific preparers of telecommunication standards. The Institute of Electrical and Electronic Engineers (IEEE) produces the 802 series specifications, which are of particular interest to enterprise networks. There are the Advanced Television Systems Committee (ATSC) standards for video compression, and the Society of Cable Telecommunication Engineers that produce CATV (cable television) standards. Another important group is the Alliance for Telecommunication Industry Solutions. This group prepares standards dealing with the North American digital network. Bellcore (Bell Communications Research) is an excellent source for standards with a North American flavor.

These standards were especially developed for the Regional Bell Operating Companies (RBOCs). There are also a number of *forums*. A forum, in this context, is a group of manufacturers and users that band together to formulate standards. For example, there is the Frame Relay Forum, the ATM Forum, and so on. Often these ad hoc industrial standards are adopted by CCITT, ANSI, and the ISO, among others.

1.6 ORGANIZATION OF THE PSTN IN THE UNITED STATES

Prior to 1984 the PSTN in the United States consisted of the Bell System (part of AT&T) and a number of independent telephone companies such as GTE. A U.S. federal court considered the Bell System/AT&T a monopoly and forced it to divest its interests.

As part of the divestiture, the Modification and Final Judgment (MFJ) called for the separation of *exchange* and *interexchange* telecommunications functions. Exchange services are provided by RBOCs; interexchange services are provided by other than RBOC entities. What this means is that local telephone service may be provided by the RBOCs and long-distance (interexchange) services by non-RBOC entities such as AT&T, Sprint, MCI, and WorldCom.

New service territories called local access and transport areas (LATAs), also referred to as *service areas* by some RBOCs, were created in response to the MFJ exchange-area requirements. LATAs serve the following two basic purposes:

1. They provide a method for delineating the area within which the RBOCs may offer services.
2. They provided a basis for determining how the assets of the former Bell System were to be divided between the RBOCs and AT&T at divestiture.

Appendix B of the MFJ requires each RBOC to offer *equal access* through RBOC end offices (local exchanges) in a LATA to all interexchange carriers (IXCs). All carriers must be provided services that are equal in type, quality, and price to those provided by AT&T.

We define a LEC (local exchange carrier) as a company that provides intraLATA telecommunication within a franchised territory. A LATA defines those areas within which a LEC may offer telecommunication services. Many independent LECs are associated with RBOCs in LATAs and provide exchange access individually or jointly with a RBOC.

1.6.1 Points of Presence

A point of presence (POP) is a location within a LATA that has been designated by an access customer for the connection of its facilities with those of a LEC. Typically, a POP is a location that houses an access customer's switching system or facility node. Consider an "access customer" as an interexchange carrier, such as Sprint or AT&T.

At each POP, the access customer is required to designate a physical point of termination (POT) consistent with technical and operational characteristics specified by the LEC. The POT provides a clear demarcation between the LEC's exchange access functions and the access customer's interexchange functions. The POT generally is a distribution frame or other item of equipment (a cross-connect) at which the LEC's access facilities terminate and where cross-connection, testing, and service verification can occur. A later federal court judgement (1992) required an LEC to provide space for equipment for CAPs (competitive access providers).

REVIEW EXERCISES

1. Define *telecommunications*.
2. Identify *end-users*.
3. What is/are the function(s) of a *node*?
4. Define a *connectivity*.
5. What are the three phases of a telephone call?
6. Describe *on-hook* and *off-hook*.
7. What is the function of the *subscriber loop*?
8. What is the function of the *battery*?
9. Describe *address signaling* and its purpose.
10. Differentiate *trunks* from subscriber loops (subscriber lines).
11. What is the theoretical capacity of a four-digit telephone number? Of a three-digit exchange number?
12. What is the common colloquial name for an *NPA code*?
13. What is the rationale for having a *tandem switch*?
14. Define *grade of service*. What value would we have for an objective grade of service?
15. How can we improve grade of service? Give the downside of this.
16. Give the basic definition of the *busy hour*.
17. Differentiate *simplex*, *half-duplex*, and *full duplex*.
18. What is *double seizure*?
19. On what kind of trunk would double seizure occur?
20. What is a *full-mesh* network? What is a major attribute of a mesh network?
21. What are two major attributes of a *star network*?
22. Define a *traffic relation*.
23. On a hierarchical network, what is *final route*?
24. Give at least three reasons for the trend away from hierarchical routing.
25. List at least six QoS items.
26. List at least one international standardization body, one regional standardization group, and three U.S. standardization organizations.
27. Define a POP and POT.

REFERENCES

1. *Webster's Third International Dictionary*, G&C Merriam Co., Springfield, MA, 1981.
2. *IEEE Standard Dictionary of Electrical and Electronic Terms*, 6th ed., IEEE Std. 100-1996, IEEE, New York, 1996.
3. *Telecommunication Planning*, ITT Laboratories of Spain, Madrid, 1973.
4. R. L. Freeman, *Telecommunication System Engineering*, 3rd ed., Wiley, New York, 1996.

2

SIGNALS CONVEY INTELLIGENCE

2.1 OBJECTIVE

Telecommunication deals with conveying information with electrical signals. This chapter prepares the telecommunication novice with some very basic elements of telecommunications. We are concerned about the transport and delivery of information. The first step introduces the reader to early signaling techniques prior to the middle of the 19th century when Samuel Morse opened the first electrical communication circuit in 1843.

The next step is to introduce the reader to some basic concepts in *electricity*, which are mandatory for an understanding of how telecommunications works from a technical perspective. For an introduction to electricity, the reader should consult Appendix A. After completion of this chapter, the user of this text should have a grasp of electrical communications and its units of measure. Specifically we will introduce an *electrical signal* and how it can carry intelligence. We will differentiate analog and digital transmission with a very first approximation.

Binary digital transmission will then be introduced starting with binary numbers and how they can be very simply represented electrically. We then delve into conducted transmission. That is the transport of an electrical signal on a copper-wire pair, on coaxial cable, and then by light in a fiber optic strand of glass. Radio transmission and the concept of *modulation* will then be introduced.

2.2 SIGNALS IN EVERYDAY LIFE

Prior to the advent of practical electrical communication, human beings have been signaling over a distance in all kinds of ways. The bell in the church tower called people to religious services or "for whom the bell tolls"—the announcement of a death. We knew a priori several things about church bells. We knew approximately when services were to begin, and we knew that a long, slow tolling of the bells announced death. Thus we could distinguish one from the other, namely, a call to religious services or the announcement of death. Let us call lesson 1, *a priori knowledge.*

The Greeks used a relay of signal fires to announce the fall of Troy. They knew a prior that a signal fire in the distance announced victory at Troy. We can assume that no fire meant defeat. The fires were built in a form of relay, where a distant fire was just visible with the naked eye, the sight of which caused the lighting of a second fire, and then a third, fourth, and so on, in a line of fires on nine hills terminating in Queen Clytemenestra's palace in Argos, Greece. It also announced the return of her husband from the battle of Troy.

Human beings communicated with speech, which developed and evolved over thousands of years. This was our principal form of communication. However, it was not exactly "communication at a distance." Speech distance might be measured in feet or meters. At the same time there was visual communication with body language and facial expressions. This form of communication had even more limited distance.

Then there was semaphore, which was very specialized and required considerable training. Semaphore was slow but could achieve some miles of distance using the manual version.

Semaphore consisted of two flags, one in each hand. A flag could assume any one of six positions 45 degrees apart. The two flags then could have six times six, or 36 unique positions. This accommodated the 25-letter alphabet and 10 numbers. The letters i and j became one letter for the 26-letter alphabet.

A similar system used in fixed locations, often called *signal hills* or *telegraph hills*, was made up of a tower with a movable beam mounted on a post. Each end of a beam had a movable indicator or arm that could assume seven distinct positions, 45 degrees apart. With two beams, there were 49 possibilities, easily accommodating the alphabet, ten digits, and punctuation. The origin of this "telegraph" is credited to the French in the very late eighteenth century. It was used for defense purposes linking Toulon to Paris. There were 120 towers some three to six miles apart. It took forty minutes to transmit signals across the system, with about three signals a minute. It was called the Chappe semaphore, named after its inventor. Weather and darkness, of course, were major influences. One form of railroad signals using the signal arm is still in use in some areas today.

The American Indian used smoke signals by day and fires at night. The use of a drum or drums for distance communication was common in Africa.

Electrical telegraph revolutionized distance communications. We use the date of 1843 for its practical inception. It actually has roots well prior to this date. Many of the famous names in the lore of electricity became involved. For example, Hans Christian Oersted of Denmark proposed the needle telegraph in 1819. Gauss and Weber built a 2.3-km (1.4-mile), two-wire telegraph line using a technique known as the galvanometer-mirror device in 1833 in Germany. Then there was the Cooke and Wheatstone five-needle telegraph, which was placed into operation in 1839 in the United Kingdom. It was meant for railroad application and used a code of 20 letters and 10 numerals to meet railway requirements.

It was while the United States Congress in 1837 was considering a petition to authorize a New York to New Orleans Chappe semaphore line that Samuel F. B. Morse argued for the U.S. government to support his electrical telegraph. The government appropriated the money in early 1843. The first operational line was between New York and Baltimore. Within 20 years the telegraph covered the United States from coast to coast. The first phase of electrical communications was completed. It revolutionized our lives. (Ref. 1).

2.3 BASIC CONCEPTS OF ELECTRICITY FOR COMMUNICATIONS

2.3.1 Early Sources of Electrical Current

Rather crude dry-cell batteries were employed in the earlier periods of telegraph as an electrical current source. Their development coincided with the Morse telegraph (ca. 1835–1840). They produced about 1.5 volts (direct current) per cell. To achieve a higher voltage, cells were placed in series. Figure 2.1*a* shows the standard graphic notation for

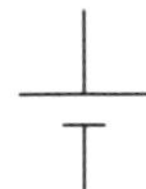

Figure 2.1*a* Graphic notation of a single dry cell.

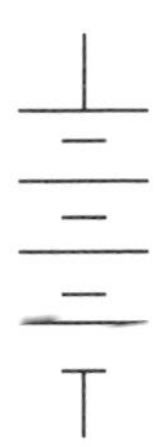

Figure 2.1*b* Graphic notation of a "battery" of dry cells.

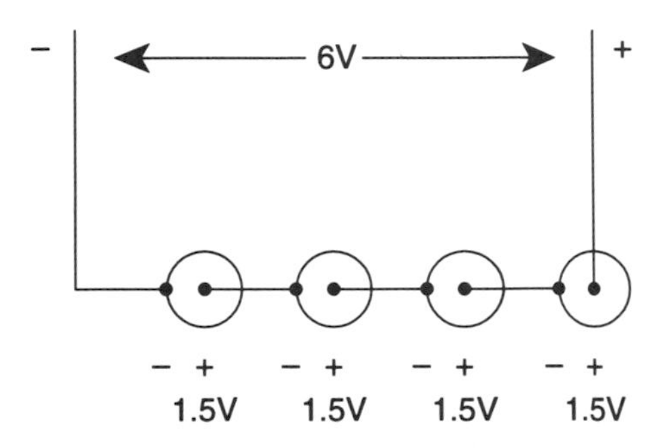

Figure 2.1*c* How dry cells can be connected in *series* to increase voltage.

a cell; Figure 2.1*b* shows the graphic symbol for a battery made up of several cells. A drawing of a battery made up of four cells is illustrated in Figure 2.1*c*. A dry cell stores chemical energy from which, when its positive electrode is connected through some resistive device to the negative electrode, a current will flow. A battery of cells was the simple power source for a telegraph circuit.

2.3.2 Electrical Telegraph: An Early Form of Long-Distance Communications

Let's connect a battery terminal (or electrode) with a *length* of copper wire looping it back to the other electrode. A buzzer or other sound-generating device is inserted into that loop at the farthest end of the wire before looping back; we now have the essentials of a telegraph circuit. This concept is shown in Figure 2.2. The loop has a certain resistance, which is a function of its length and the diameter of the wire. The longer we make the loop, the greater the resistance. As the length increases (the resistance increases), the current in the loop decreases. There will be some point where the current (in amperes) is so low that the buzzer will not work. The maximum loop length can be increased by using wire with a greater diameter. It can be increased still further by using electrical repeaters placed near the maximum length point. Another relay technique involves a human operator. At the far end of the loop an operator copies the message and retransmits it down the next leg of the circuit.

2.3.2.1. Conveying Intelligence over the Electrical Telegraph. This model of a simple telegraph circuit consists of a copper wire loop with a buzzer inserted at the

Figure 2.2 A simple electrical telegraph circuit.

distant end where the wire pair loops around. At the near end, which we may call the transmitting end, there is an electrical switch, which we will call a *key*. The key consists of two electrical contacts, which, when pressed together, make contact, closing the circuit and permitting current to flow. The key is spring-loaded, which keeps it normally in the open position (no current flow).

To convey intelligence, the written word, a code was developed by Morse, consisting of three elements: a *dot*, where the key was held down for a very short period of time; a *dash*, where the key was held down for a longer period of time; and a *space*, where the key was left in the "up" position and no current flowed. By adjusting the period of time of spaces, the receiving operator could discern the separation of characters (A, B, C, . . . Z) and separation of words, where the space interval was longer. Table 2.1 shows

Table 2.1 Two Versions of the Morse Code

	A	B		A	B
A	• —	• —	P	• • • • •	• — — •
Ä		• — • —	Q	• • — •	— — • —
Á		• — — • —	R	• • •	• — •
Å		• — — • —	S	• • •	• • •
B	— • • •	— • • •	T	—	—
C	• • •	— • — •	U	• • —	• • —
CH		— — — —	Ü		• • — —
D	— • •	— • •	V	• • • —	• • • —
E	•	•	W	• — —	• — —
É		• • — • •	X	• — • •	— • • —
F	• — •	• • — •	Y	• • • •	— • — —
G	— — •	— — •	Z	• • • •	— — • •
H	• • • •	• • • •	1	• — — •	• — — — —
I	• •	• •	2	• • — • •	• • — — —
J	— • — •	• — — —	3	• • • — •	• • • — —
K	— • —	— • —	4	• • • • —	• • • • —
L	⸺	• — • •	5	— — —	• • • • •
M	— —	— —	6	• • • • • •	— • • • •
N	— •	— •	7	— — • •	— — • • •
Ñ		— — • — —	8	— • • • •	— — — • •
O	• •	— — —	9	— • • —	— — — — •
Ö		— — — •	0	⸻	— — — — —

Column A is the American Morse Code; Column B is the International Morse Code.

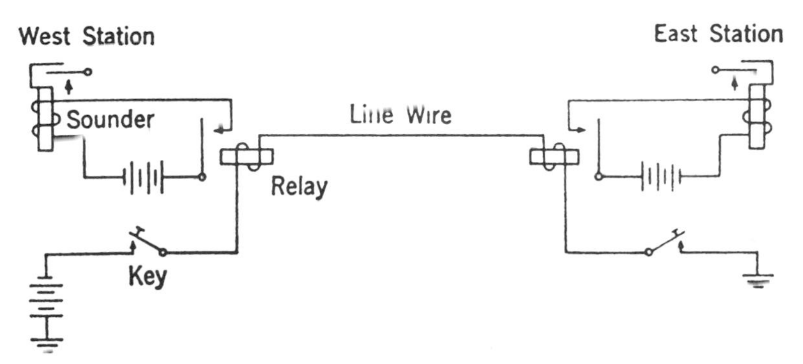

Figure 2.3 A practical elementary telegraph circuit with ground return.

two versions of the Morse code: land-line and international. By land-line, we mean a code used to communicate over land by means of wire conductors. The international Morse code was developed somewhat later, and was used by radio.

A more practical telegraph system is illustrated in Figure 2.3. Note that the figure has just one metallic wire connecting the west station to the east station. The second wire is replaced with *ground*. The earth is a good conductor, and so we use earth, called ground, as the second conductor (or wire). Such a telegraph system is called *single-wire ground return*.

This is a similar circuit as that shown in Figure 2.2. In this case, when both keys are closed, a dc (direct current) circuit is traced from a battery in the west station through the key and relay at that point to the line wire; from there through the relay and key at the east station and back through the earth (ground) to the battery. The relays at each end, in turn, control the local circuits, which include a separate battery and a sounder (e.g., buzzer or other electric sounding device). Opening and closing the key at one end, while the key at the other end is closed, causes both sounders to operate accordingly.

A relay is a switch that is controlled electrically. It consists of wire wrapped around an iron core and a hinged metal strip is normally open. When current flows through the windings (i.e., the wire wrapped around the core) a magnetic field is set up, drawing the hinged metal strip into a closed position, causing current to flow in the secondary circuit. It is a simple open-and-closed device such that when current flows there is a contact closure (the metal strip), and when there is no current through the windings, the circuit is open. Of course, there is a spring on the metal strip holding it open except when current flows.

Twenty years after Morse demonstrated his telegraph on a New York–Baltimore–Washington route, telegraph covered the country from coast to coast. It caused a revolution in communications. When I worked in Ecuador in 1968, single-wire ground return telegraph reached every town in the country. It was the country's principal means of electrical communication (Ref. 2).

2.3.3 What Is Frequency?

To understand more advanced telecommunication concepts, we need a firm knowledge of frequency and related parameters such as band and bandwidth, wavelength, period, and phase. Let us first define frequency and relate it to everyday life.

The IEEE defines *frequency* as "the number of complete cycles of sinusoidal variation per unit time." The time unit we will use is the second. For those readers with a

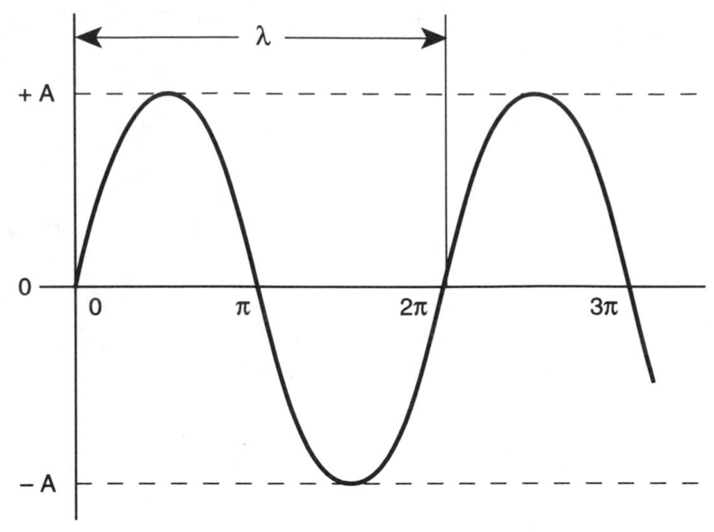

Figure 2.4 A sine wave. Here frequency is the number of times per second that a wave cycle (one peak and one trough) repeats at a given amplitude. In the figure, A is the amplitude and λ is the wavelength, π is π radians or 180°, and 2π is the radian value at 360°.

mathematical bent, if we plot $y = \sin x$, where x is expressed in radians, a "sine wave" is developed, as shown in Figure 2.4.

Figure 2.5 shows two sine waves; the left side illustrates a lower frequency and the right side shows a higher frequency. The *amplitude*, measured in this case as voltage, is the excursion, up or down, at any singular point. Amplitude expresses the intensity at that point. If we spoke of amplitude without qualifying it at some point, it would be the maximum excursion in the negative or positive direction (up or down). In this case it is 6 volts. If it is in the "down" direction, it would be −6 volts, based on Figure 2.5; and in the "up" direction it would be +6 volts.

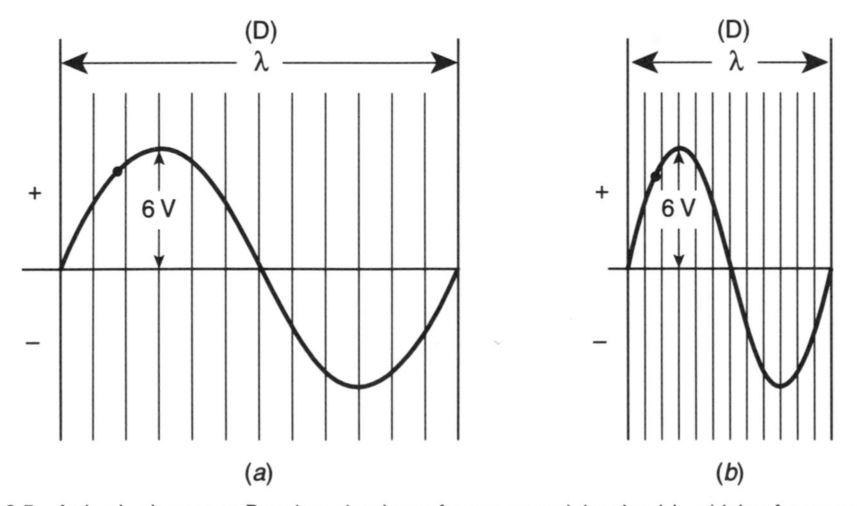

Figure 2.5 A simple sine wave. Drawing *a* is a lower frequency and drawing *b* is a higher frequency. Note that the wavelength is shown traditionally as λ (Greek letter lambda) and that *a* has a longer wavelength than *b*.

Frequency is an important aspect of music. For example, the key of A is 440 Hz and middle C is 263 Hz. Note that the unit of measurement of frequency used to be cycles per second (prior to 1963) and now the unit of measure is hertz (Hz), named for Heinrich Hertz, a German physicist credited with the discovery of radio waves. Simple sine waves can be produced in the laboratory with a *signal generator*, which is an electronic oscillator that can be tuned to different frequencies. An audio signal generator can be tuned to 263 Hz, middle C, and we can hear it if the generator output is connected to a loudspeaker. These are sound frequencies.

When we listen to the radio on the AM broadcast band, we may listen to a talk show on WOR, at a frequency of 710 kHz (kilohertz, equal to 710,000 Hz). On the FM band in the Phoenix, AZ, area, we may tune to a classical music station, KBAQ, at 89.5 MHz (89,500,000 Hz). These are radio frequencies.

Metric prefixes are often used, when appropriate, to express frequency, as illustrated in the preceding paragraph. For example, kilohertz (kHz), megahertz (MHz), and gigahertz (GHz) are used for Hz × 1000, Hz × 1,000,000 and Hz × 1,000,000,000. Accordingly, 38.71 GHz is 38.710,000,000 Hz.

Wavelength is conventionally measured in meters and is represented by the symbol λ. It is defined as the distance between successive peaks or troughs of a sinusoidal wave (i.e., D in Fig. 2.5). Both sound and radio waves each travel with a certain velocity of propagation. Radio waves travel at 186,000 mi/sec in a vacuum, or 3×10^8 m/sec.[1] If we multiply frequency in Hz times the wavelength in meters, we get a constant, the velocity of propagation. In a vacuum (or in free space):

$$F\lambda = 3 \times 10^8 \text{m/sec}, \tag{2.1}$$

where F is measured in Hz and λ is measured in meters (m).

Example 1. The international calling and distress frequency is 500 kHz. What is the equivalent wavelength in meters?

$$\begin{aligned} 500,000\lambda &= 3 \times 10^8 \text{m/sec} \\ \lambda &= 3 \times 10^8 / 5 \times 10^5 \\ &= 600 \text{ m.} \end{aligned}$$

Example 2. A line-of-sight millimeter wave radio link operates at 38.71 GHz.[2] What is the equivalent wavelength at this frequency?

$$\begin{aligned} 38.71 \times 10^9\, \lambda &= 3 \times 10^8 \text{m/sec} \\ \lambda &= 3 \times 10^8 / 38.71 \times 10^9 \\ &= 0.00775 \text{ m or } 7.75 \text{ mm.} \end{aligned}$$

[1]Sound waves travel at 1076 ft/sec (331 m/sec) in air at 0°C and with 1 atmosphere of atmospheric pressure. However, our interest here is in radio waves, not sound waves.

[2]This is termed *millimeter* radio because wavelengths in this region are measured in millimeters (i.e., for frequencies above 30 GHz), rather than in centimeters or meters.

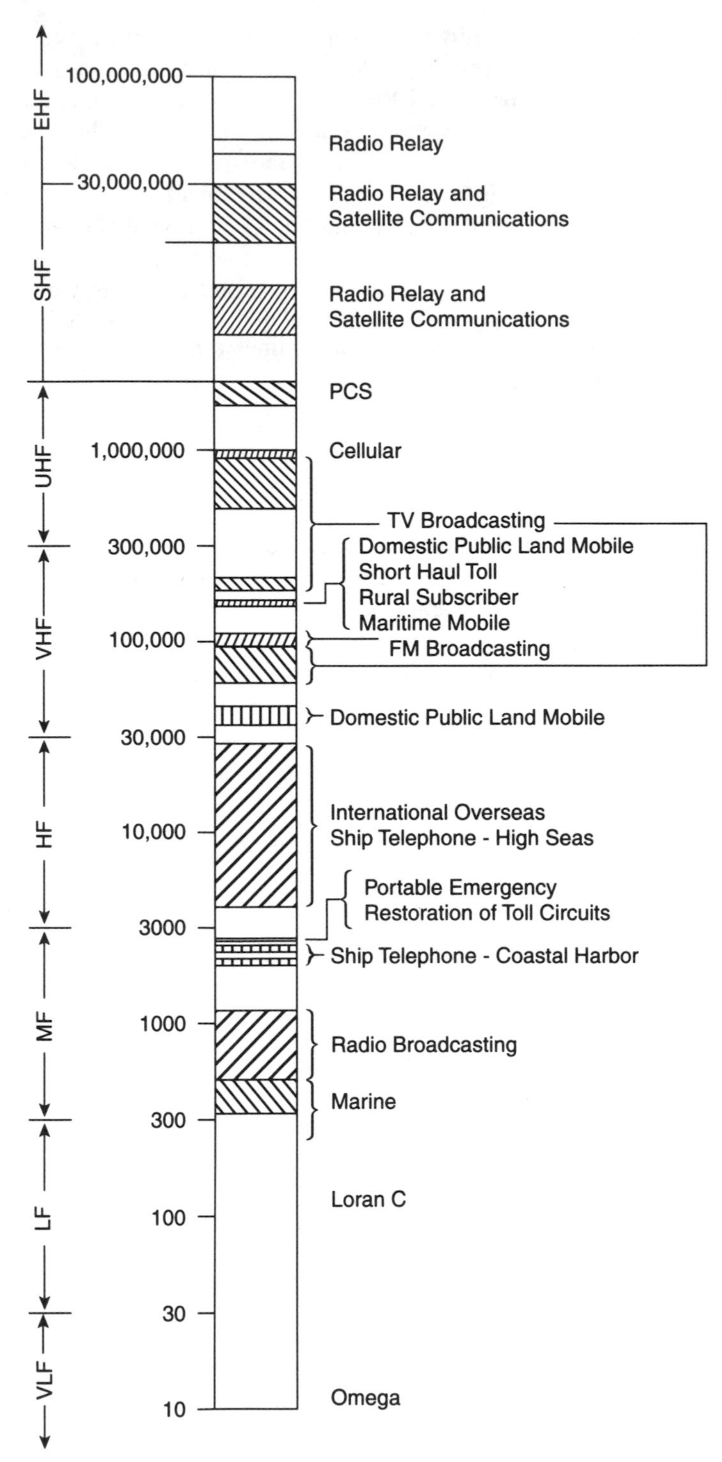

Figure 2.6 The radio frequency spectrum showing some frequency band assignments.

Figure 2.6 is an outline drawing of the radio frequency spectrum from nearly 0 Hz to 100 GHz. The drawing shows several frequency bands assigned to specific services.

2.3.3.1. *Introduction to Phase.* The IEEE defines *phase* as "a relative measurement that describes the temporal relationship between two signals that have the same frequency." We can plot a sine wave (representing a certain frequency) by the method shown in Figure 2.7, where the horizontal lines are continuation of points a, b, c, etc., and the vertical lines a′, b′, c′, and so on, are equally spaced and indicate *angular degrees of rotation*. The intersection of lines a and a′, b and b′, and so forth, indicates points on the sine wave curve.

To illustrate what is meant by *phase relation*, we turn to the construction of a sine wave using a circle, as shown in Figure 2.7. In the figure the horizontal scale (the abscissa) represents time and the vertical scale (the ordinate) represents instantaneous values of current *or* voltage. The complete curve shows values of current (or voltage) for all instants during one complete cycle. It is convenient and customary to divide the time scale into units of *degrees* rather than seconds, considering one complete cycle as being completed always in 360 degrees or units of time (regardless of the actual time taken in seconds). The reason for this convention becomes obvious from the method of constructing the sine wave, as shown in Figure 2.7, where, to plot the complete curve, we take points around the circumference of the circle through 360 *angular degrees*. It needs to be kept in mind that in the sense now used, the *degree* is a measure of *time* in terms of the frequency, not of an angle.

We must understand phase and phase angle because those terms will be used in our discussions of modulation and of certain types of distortion that can limit the rate of transmitting information and/or corrupt a wanted signal (Ref. 2).

An example of two signals of the same frequency, in phase and with different amplitudes, is illustrated in Figure 2.8*a*, and another example of two signals of the same frequency and amplitude, but 180 degrees out of phase is shown in Figure 2.8*b*. Note the use of π in the figure, meaning π radians or 180°, 2π radians or 360°. (See Appendix A, Section A.9.)

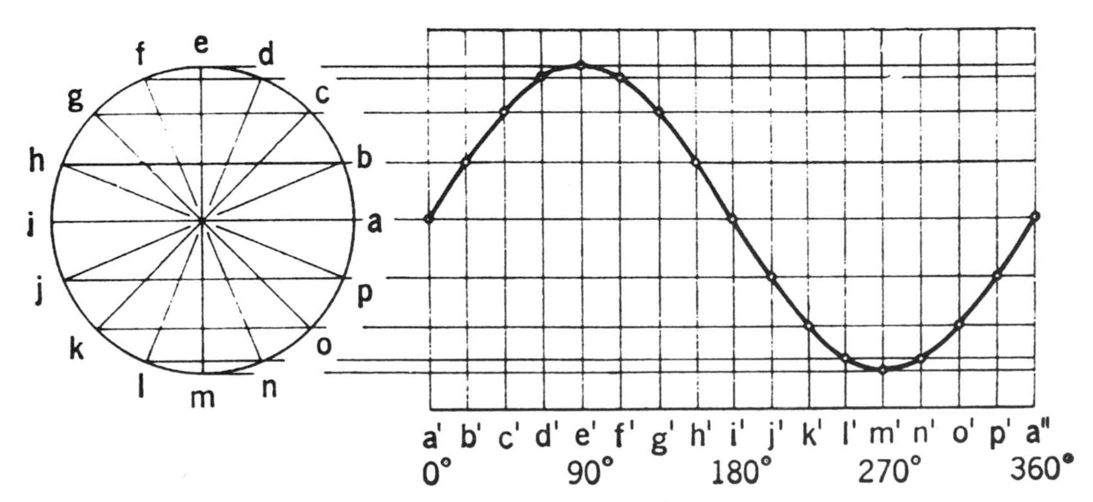

Figure 2.7 Graphical construction of a sine wave.

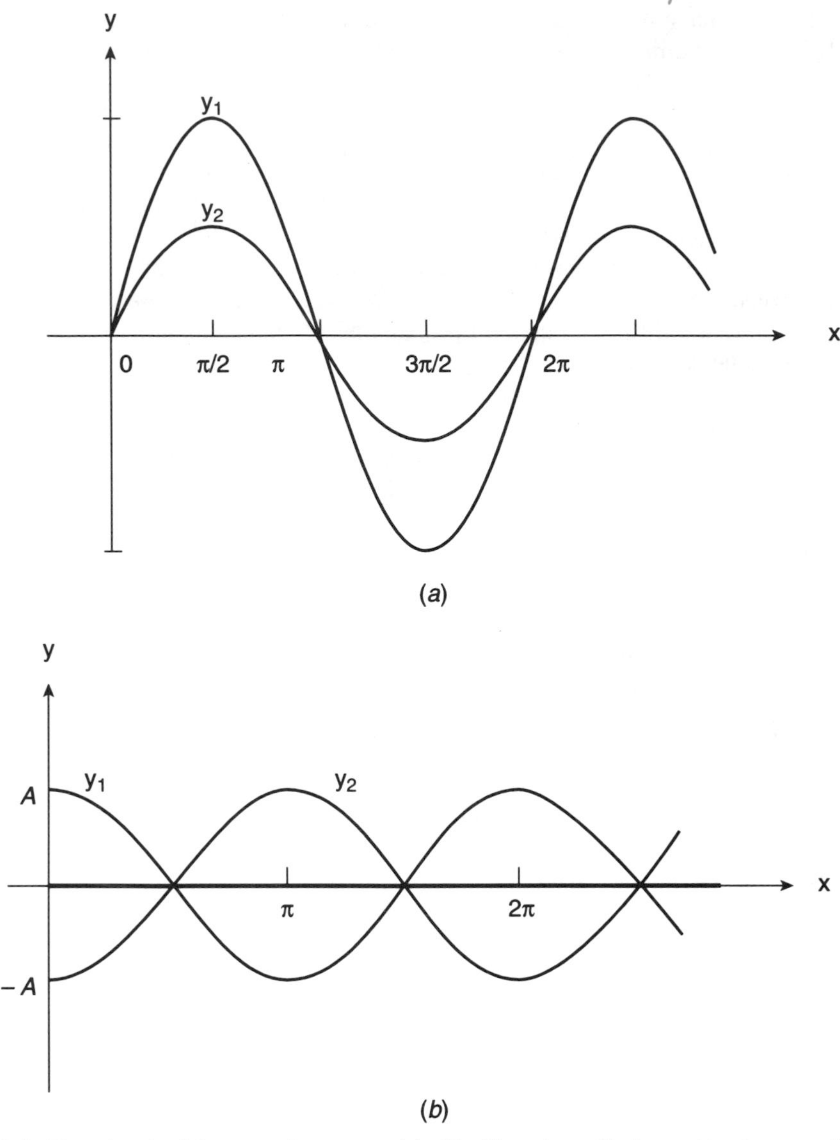

Figure 2.8 Two signals of the same frequency: (*a*) with different amplitudes and in-phase; and (*b*) with the same amplitudes but 180° out of phase.

2.4 ELECTRICAL SIGNALS

2.4.1 Introduction to Transmission

Transmission may be defined as the electrical transfer of a signal, message, or other form of intelligence from one location to another. Traditionally, transmission has been one of the two major disciplines of telecommunication. *Switching* is the other principal specialty. Switching establishes a connection from user X to some distant user Y. Simplistically we can say that transmission is responsible for the transport of the signal from user X to user Y. In the old days of telephony these disciplines were separate, with strong demarcation between one and the other. Not so today. The demarcation line is fast disappearing. For example, under normal circumstances in the PSTN, a switch provides network timing which is vital for digital transmission.

What we have been dealing with so far is *baseband* transmission. This is the transmission of a raw electrical signal described in Section 2.3.2. This type of baseband signal is very similar to the 1s and 0s transmitted electrically from a PC. Another type of baseband signal is the alternating current derived from the mouthpiece of a telephone handset (subset). Here the alternating current is an electrical facsimile of the voice soundwave impinging on the telephone microphone.

Baseband transmission can have severe distance limitations. We will find that the signal can only be transmitted so far before being corrupted one way or another. For example, a voice signal transmitted from a standard telephone set over a fairly heavy copper wire pair (19 gauge) may reach a distant subset earpiece some 30 km or less distant before losing all intelligibility. This is because the signal strength is so very low that it becomes inaudible.

To overcome this distance limitation we may turn to *carrier* or *radio* transmission. Both transmission types involve the generation and conditioning of a radio signal. Carrier transmission usually (not always) implies the use of a conductive medium such as wire pair, coaxial cable, or fiber optic cable to carry a radio or light-derived signal. Radio transmission always implies radiation of the signal in the form of an electromagnetic wave. We listen to the radio or watch television. These are received and displayed or heard as the result of the reception of radio signals.

2.4.2 Modulation

At the transmitting side of a telecommunication link a radio carrier is generated. The carrier is characterized by a frequency, described in Section 2.3.3. This single radio frequency carries no useful information for the user. Useful information may include voice, data, or image (typically facsimile or television). *Modulation* is the process of impinging that useful information on the carrier and *demodulation* is the recovery of that information from the carrier at the distant end near the destination user.

The IEEE defines modulation as "a process whereby certain characteristics of a wave, often called the carrier, are varied or selected in accordance with a modulation function." The *modulating function* is the information baseband described previously.

There are three generic forms of modulation:

1. Amplitude modulation (AM);
2. Frequency modulation (FM); and
3. Phase modulation (PM).

Item 1 (amplitude modulation) is where a carrier is varied in amplitude in accordance with the information baseband signal. In the case of item 2 (frequency modulation), a carrier is varied in frequency in accordance with the baseband signal; and for item 3 (phase modulation), a carrier is varied in its phase in accordance with the information baseband signal.

Figure 2.9 graphically illustrates amplitude, frequency, and phase modulation. The modulating signal is a baseband stream of bits: 1s and 0s. We deal with digital transmission (e.g., 1s and 0s) extensively in Chapters 6 and 10.

Prior to 1960, all transmission systems were *analog*. Today, in the PSTN, all telecommunication systems are *digital*, except for the preponderance of subscriber access lines.

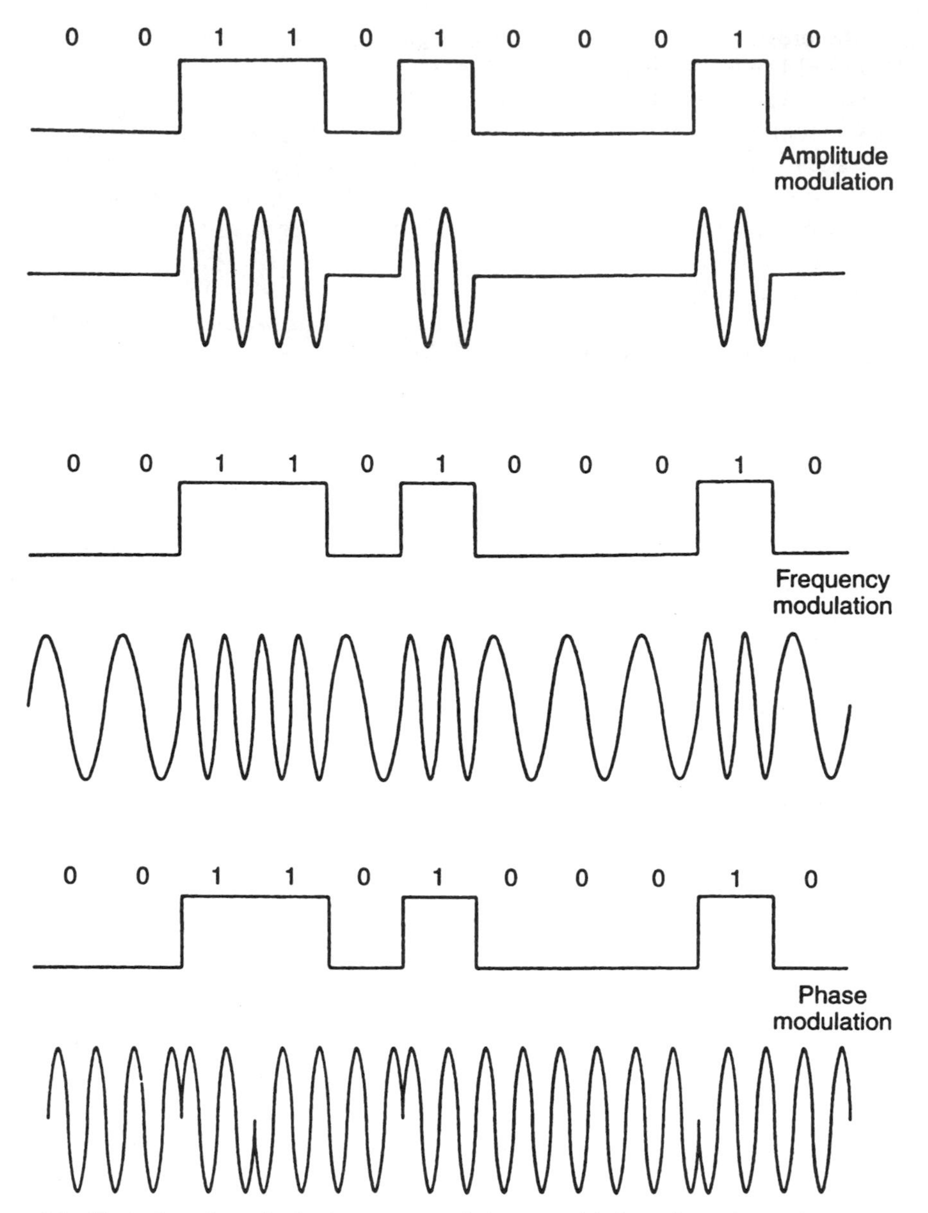

Figure 2.9 Illustration of amplitude, frequency, and phase modulation, where the modulating signal is the binary digital sequence 00110100010, an electrical baseband signal.

These are the subscriber loops described in Chapter 1. Let us now distinguish and define analog and digital transmission.

2.4.2.1. Analog Transmission. *Analog* transmission implies continuity, as contrasted with digital transmission, which is concerned with discrete states. Many signals can be used in either the analog or digital sense, the means of carrying the information being the distinguishing feature. The information content of an analog signal is conveyed by the value or magnitude of some characteristic(s) of the signal such as amplitude, frequency or a phase of a voltage, the amplitude or duration of a pulse, the angular position of a shaft, or the pressure of a fluid. Typical analog transmissions are the signals we hear on AM and FM radio and what we see (and hear) on television. In fact, television is rather unique. The video itself uses amplitude modulation, the sound subcarrier uses frequency modulation, and the color subcarrier employs phase modulation. All are in analog formats.

2.4.2.2. Digital Transmission. The information content of a digital signal is concerned with discrete states of the signal, such as the presence or absence of a voltage (see Section 2.3.2), a contact is the open or closed position, or a hole or no hole in certain positions on a card or paper tape. The signal is given meaning by assigning numerical values or other information to the various combinations of the discrete states of the signal. We will be dealing extensively with digital transmission as the discussion in this text proceeds.

2.4.3 Binary Digital Signals

In Section 2.4.1, we defined a digital waveform as one that displayed discreteness. Suppose we consider the numbers 0 through 9. In one case only integer values are permitted in this range, no in-between values such as 3.761 or 8.07. This is digital, where we can only assign integer values between 0 and 9. These are discrete values. On the other hand, if we can assign any number value between 0 and 9, there could be an infinite number of values such as 7.01648754372100. This, then, is analog. We have continuity, no discreteness.

Consider now how neat it would be if we had only two values in our digital system. Arbitrarily, we'll call them a 1 and a 0. This is indeed a binary system, just two possible values. It makes the work of a decision circuit really easy. Such a circuit has to decide on just one of two possibilities. Look at real life: a light is on or it is off, two values: on and off. A car engine is running or not running, and so on. In our case of interest, we denominate one value a 1 and the other, a 0. We could have a condition where current flows and we'll call that condition a 1; no current flowing we'll call a 0.[3]

Of course, we are defining a *binary* system with a number base of 2. Our day-to-day numbers are based on a decimal number system where the number base is 10.

The basic key in binary digital transmission is the *bit*, which is the smallest unit of information in the binary system of notation. It is the abbreviation of the term *binary digit*. It is a unit of information represented by either a "1" or a "0."

A 1 and a 0 do not carry much information, yet we do use just one binary digit in many applications. One of the four types of telephone signaling is called *supervisory signaling*. The only information necessary in this case is that the line is busy or it is idle. We may assign the idle state a 0 and the busy state a 1. Another application where only a single binary digit is required is in *built-in test equipment* (BITE). In this case, we accept one of two conditions: (1) a circuit, module, or printed circuit board (PCB) is operational or (2) it is not. BITE automates the troubleshooting of electronic equipment.

To increase the information capacity of a binary system is to place several bits (binary digits) contiguously together. For instance, if we have a 2-bit code, there are four possibilities: 00, 01, 10, and 11. A 3-bit code provides eight different binary sequences, each 3 bits long. In this case we have 000, 001, 010, 011, 100, 101, 110, and 111. We could assign letters of the alphabet to each sequence. There are only eight distinct possibilities

[3]The reader with insight will note an ambiguity here. We could reverse the conditions, making the 1 state a 0 and the 0 state a 1. We address this issue in Chapter 10.

so only eight letters can be accommodated. If we turn to a 4-bit code, 16 distinct binary sequences can be developed, each 4 bits long. A 5-bit code will develop 32 distinct sequences, and so on.

As a result, we can state that for a binary code of length n, we will have 2^n different possibilities. The American Standard Code for Information Interchange (ASCII) is a 7-bit code (see Section 10.4), it will then have 2^7 or 128 binary sequence possibilities. When we deal with pulse code modulation (PCM) (Chapter 6), as typically employed on the PSTN, a time slot contains 8 bits. We know that an 8-bit binary code has 256 distinct 8-bit sequences (i.e., $2^8 = 256$).

Consider the following important definitions when dealing with the bit and binary transmission: *Bit rate* is defined as the number of bits (those 1s and 0s) that are transmitted per second. *Bit error rate* (BER) is the number of bit errors measured or expected per unit of time. Commonly the time unit is the second. An error, of course, is where a decision circuit declares a 1 when it was supposed to be a 0, or declares a 0 when it was supposed to be a 1 (Ref. 3).

2.5 INTRODUCTION TO TRANSPORTING ELECTRICAL SIGNALS

To transport electrical signals, a transmission medium is required. There are four types of transmission media:

1. Wire pair;
2. Coaxial cable;
3. Fiber optic cable; and
4. Radio.

2.5.1 Wire Pair

As one might imagine, a wire pair consists of two wires. The wires commonly use a copper conductor, although aluminum conductors have been employed. A basic impairment of wire pair is *loss*. Loss is synonymous with attenuation. Loss can be defined as the dissipation of signal strength as a signal travels along a wire pair, or any other transmission medium, for that matter. Loss or attenuation is usually expressed in decibels (dB). In Appendix C the reader will find a tutorial on decibels and their applications in telecommunications.

Loss causes the signal power to be dissipated as a signal passes along a wire pair. Power is expressed in watts. For this application, the use of milliwatts may be more practical. If we denominate loss with the notation L_{dB}, then:

$$L_{dB} = 10 \log(P_1/P_2), \tag{2.2}$$

where P_1 is the power of the signal where it enters the wire pair, and P_2 is the power level of the signal at the distant end of the wire pair. This is the traditional formula defining the decibel in the power domain (see Appendix C).

Example 1. Suppose a 10-mW (milliwatt) 1000-Hz signal is launched into a wire pair. At the distant end of the wire pair the signal is measured at 0.2 mW. What is the loss in dB on the line for this signal?

$$
\begin{aligned}
L_{\text{dB}} &= 10 \log(10/0.2) \\
&= 10 \log(50) \\
&\approx 17 \text{ dB},
\end{aligned}
$$

All logarithms used in this text are to the base 10. Appendix B provides a review of logarithms and their applications.

The opposite of loss is *gain*. An attenuator is a device placed in a circuit to purposely cause loss. An *amplifier* does just the reverse, it gives a signal *gain*. An amplifier increases a signal's intensity. We will use the following graphic symbol for an attenuator:

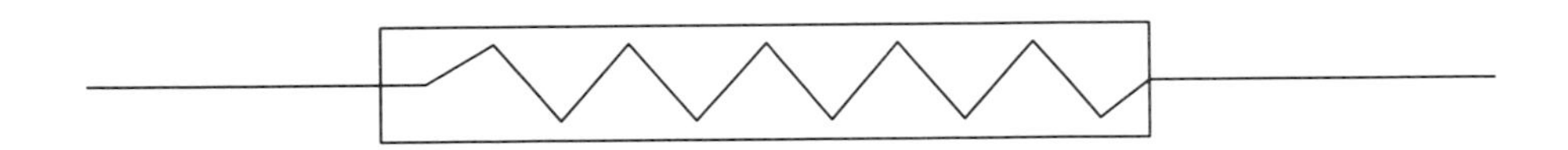

and the following symbol for an amplifier:

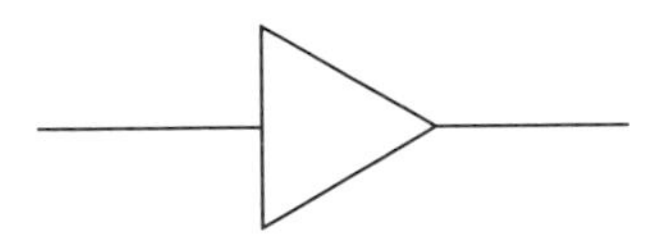

Wire-pair transmission suffers other impairments besides loss. One of these impairments is *crosstalk*. Most of us have heard crosstalk on our telephone line. It appears as another, "foreign" conversation having nothing to do with our telephone call. One basic cause of crosstalk is from other wire pairs sharing the same cable as our line. These other conversations are electrically induced into our line. To mitigate this impairment, physical twists are placed on each wire pair in the cable. Generally there are from 2 to 12 twists per foot of wire pair. From this we get the term *twisted pair*. The figure below shows a section of twisted pair.

Another impairment causes a form of *delay distortion* on the line, which is cumulative and varies directly with the length of the line as well as with the construction of the wire itself. It has little effect on voice transmission, but can place definition restrictions on data rate for digital/data transmission on the pair. The impairment is due to the *capacitance* between one wire and the other of the pair, between each wire and ground, and between each wire and the shield, if a shield is employed. Delay distortion is covered in greater depth in Chapters 6 and 10.

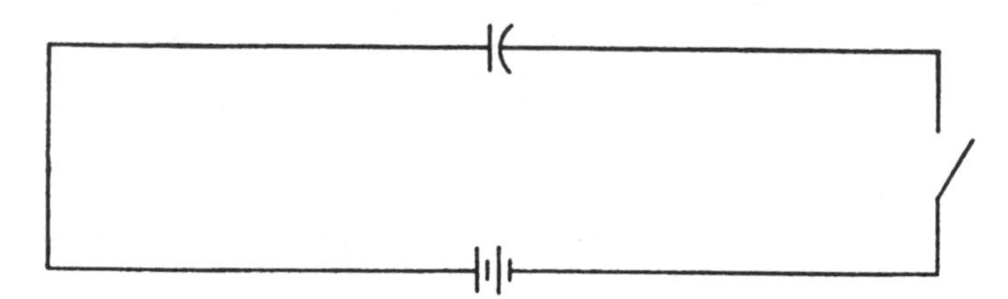

Figure 2.10 A simple capacitive circuit.

2.5.1.1. Capacitance. Direct current circuits are affected by resistance, whereas alternating current (ac) circuits, besides resistance, are affected by the properties of inductance and capacitance. In this subsection, we provide a brief description of capacitance. (Also see Appendix A, Section A.8.)

Capacitance is somewhat analogous to elasticity. While a storage battery stores electricity as another form of energy (i.e., chemical energy), a capacitor stores electricity in its natural state. An analogy of capacitance is a closed tank filled with compressed air. The quantity of air, since air is elastic, depends upon the pressure as well as the size or capacity of the tank. If a capacitor is connected to a direct source of voltage through a switch, as shown in Figure 2.10, and the switch is suddenly closed, there will be a rush of current in the circuit.[4] This current will charge the capacitor to the same voltage value as the battery, but the current will decrease rapidly and become zero when the capacitor is fully charged.

Let us define a capacitor as two conductors separated by an insulator. A *conductor* conducts electricity. Certain conductors conduct electricity better than others. Platinum and gold are very excellent conductors, but very expensive. Copper does not conduct electricity as well as gold and platinum, but is much more cost-effective. An insulator carries out the opposite function of a conductor. It tends to prevent the flow of electricity through it. Some insulators are better than others regarding the conduction of electricity. Air is an excellent insulator. However, we well know that air can pass electricity if the voltage is very high. Consider lightning, for example. Other examples of insulators are bakelite, celluloid, fiber, formica, glass, lucite, mica, paper, rubber, and wood.

The insulated conductors of every circuit, such as our wire pair, have to a greater or lesser degree this property of capacitance. The capacitance of two parallel open wires or a pair of cable conductors of any considerable length is appreciable in practice.

2.5.1.2. Bandwidth of a Twisted Pair. The usable bandwidth of a twisted wire pair varies with the type of wire pair used and its length. Ordinary wire pair used in the PSTN subscriber access plant can support 2 MHz over about 1 mile of length. Special Category-5 twisted pair displays 67 dB loss at 100 MHz over a length of 1000 feet.

2.5.1.3. Bandwidth Defined. The IEEE defines bandwidth as "the range of frequencies within which performance, with respect to some characteristic, falls within specific limits." One such limit is the amplitude of a signal within the band. Here it is commonly defined at the points where the response is 3 dB below the reference value. This 3-dB power bandwidth definition is illustrated graphically in Figure 2.11.

[4]A *capacitor* is a device whose primary purpose is to introduce capacitance into a circuit.

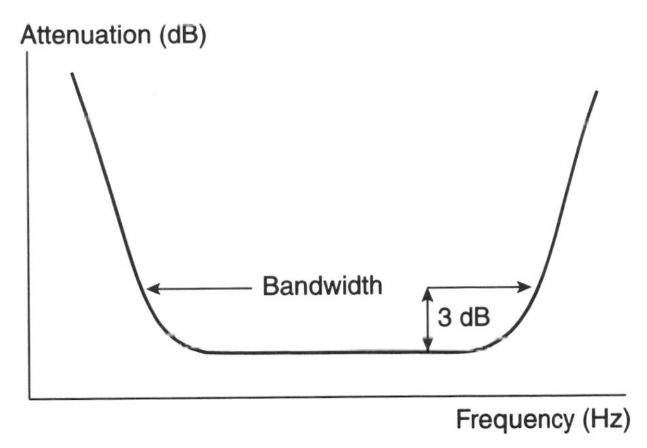

Figure 2.11 The concept of the 3-dB power bandwidth.

2.5.2 Coaxial Cable Transmission

Up to this point we have been discussing two parallel conductors, namely, wire pair. An entirely different configuration of two conductors may be used to advantage where high and very high radio frequencies are involved. This is a *coaxial* configuration. Here the conducting pair consists of a cylindrical tube with a single wire conductor going down its center, as shown in Figure 2.12. In practice, the center conductor is held in place accurately by a surrounding insulating material, which may take the form of a solid core, discs, or beads strung along the axis of the wire or a spirally wrapped string. The nominal *impedance* is 75 ohms, and special cable is available with a 50-ohm impedance.

Impedance can be defined as the combined effect of a circuit's resistance, inductance, and capacitance taken as a single property, and is expressed in *ohms* (Ω) for any given sine wave frequency. Further explanation of impedance will be found in Appendix A.

From about 1953 to 1986 coaxial cable was widely deployed for long-distance, multichannel transmission. Its frequency response was exponential. In other words, its loss increased drastically as frequency was increased. For example, for 0.375-inch coaxial cable, the loss at 100 kHz was about 1 dB and about 12 dB at 10 MHz. Thus, equalization was required. *Equalization* tends to level out the frequency response. With the advent of fiber optic cable, with its much greater bandwidth and comparatively flat frequency response, the use of coaxial cable on long-distance circuits fell out of favor. It is still widely used as an RF (radio frequency) transmission line connecting a radio to its antenna. It is also extensively employed in cable television plants, especially in the "last mile" or "last 100 feet."

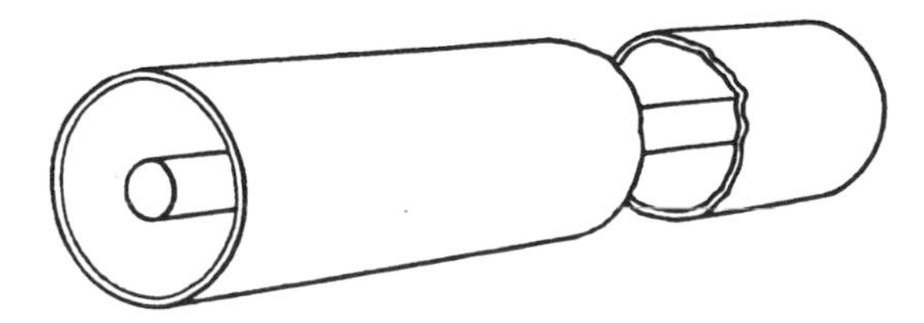

Figure 2.12 A pictorial representation of a coaxial cable section.

2.5.3 Fiber Optic Cable

Fiber optic cable is the favored transmission medium for very wideband terrestrial links, including undersea applications. It is also used for cable television "super trunks." The bandwidth of a fiber optic strand can be measured in terahertz (THz). In fact, the whole usable radio frequency spectrum can be accommodated on just one such strand. Such a strand is about the diameter of a human hair. It can carry one serial bit stream at 10 Gbps (gigabits per second) transmission rate, or by wave division multiplexing (WDM) methods, an aggregate of 100 Gbps or more. Fiber optic transmission will be discussed further in Chapter 9.

Fiber optic systems can be *loss limited* or *dispersion limited.* If a fiber optic link is limited by loss, it means that as the link is extended in distance the signal has dissipated so much that it becomes unusable. The maximum loss that a link can withstand and still operate satisfactorily is a function of the type of fiber, wavelength of the light signal, the bit rate and error rate, signal type (e.g., TV video), power output of the light source (transmitter), and the sensitivity of the light detector (receiver).[5]

Dispersion limited means that a link's length is limited by signal corruption. As a link is lengthened, there may be some point where the bit error rate (BER) becomes unacceptable. This is caused by signal energy of a particular pulse that arrives later than other signal energy of the same pulse. There are several reasons why energy elements of a single light pulse may become delayed, compared with other elements. One may be that certain launched *modes* arrive at the distant end before other modes. Another may be that certain frequencies contained in a light pulse arrive before other frequencies. In either case, delayed power spills into the subsequent bit position, which can confuse the decision circuit. The decision circuit determines whether the pulse represented a 1 or a 0. The higher the bit rate, the worse the situation becomes. Also, the delay increases as a link is extended.

The maximum length of fiber optic links range from 20 miles (32 km) to several hundred miles (km) before requiring a repeater. This length can be extended by the use of amplifiers and/or repeaters, where each amplifier can impart 20 to 40 dB gain. A fiber optic repeater detects, demodulates, and then remodulates a light transmitter. In the process of doing this, the digital signal is regenerated. A regenerator takes a corrupted and distorted digital signal and forms a brand new, nearly perfect digital signal.

A simplified model of a fiber optic link is illustrated in Figure 2.13. In this figure, the driver conditions the electrical baseband signal prior to modulation of the light signal; the optical source is the transmitter where the light signal is generated and modulated; the fiber optic transmission medium consists of a fiber strand, connectors, and splices; the optical detector is the receiver, where the light signal is detected and demodulated; and the output circuit conditions the resulting electrical baseband signal for transmission to the electrical line (Ref. 3). A more detailed discussion of fiber optic systems will be found in Chapter 9.

2.5.4 Radio Transmission

Up to now we have discussed *guided* transmission. The signal is guided or conducted down some sort of a "pipe." The "pipes" we have covered included wire pair, coaxial cable, and fiber optic cable. Radio transmission, on the other hand, is based on radiated emission.

[5]In the world of fiber optics, wavelength is used rather than frequency. We can convert wavelength to frequency using Eq. (2.1). One theory is that fiber optic transmission was developed by physicists who are more accustomed to wavelength than frequency.

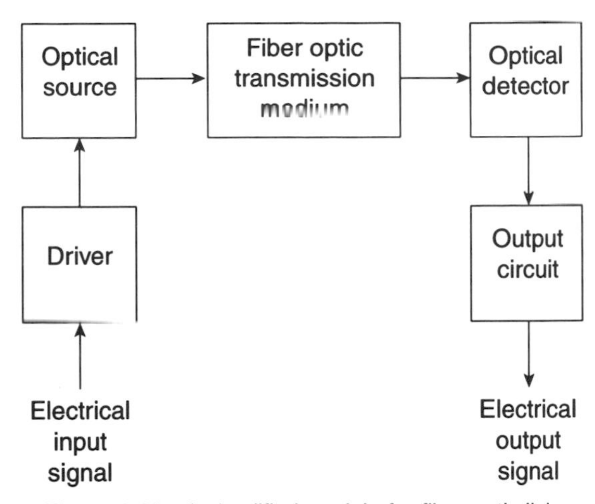

Figure 2.13 A simplified model of a fiber optic link.

The essential elements of any radio system are: (1) a transmitter for generating and modulating a "high-frequency" carrier wave with an information baseband[6]; (2) a transmitting antenna that will radiate the maximum amount of signal energy of the modulated carrier in the desired direction; (3) a receiving antenna that will intercept the maximum amount of the radiated energy after its transmission through space; and (4) a receiver to select the desired carrier wave, amplify the signal, detect it, or separate the signal from the carrier. Although the basic principles are the same in all cases, there are many different designs of radio systems. These differences depend upon the types of signals to be transmitted, type of modulation (AM, FM, or PM or a hybrid), where in the frequency spectrum (see Figure 2.6) in which transmission is to be affected, and licensing restrictions. Figure 2.14 is a generalized model of a radio link.

The information-transport capacity of a radio link depends on many factors. The first factor is the application. The following is a brief list of applications with some relevant RF bandwidths:

- Line-of-sight microwave, depending on the frequency band: 2, 5, 10, 20, 30, 40, or 60 MHz;
- SCADA (system control and data acquisition): up to 12 kHz in the 900-MHz band;
- Satellite communications, geostationary satellites: 500-MHz or 2.5-GHz bandwidths broken down into 36-MHz and 72-MHz segments;
- Cellular radio: 25-MHz bandwidth in the 800/900-MHz band; the 25-MHz band is split into two 12.5-MHz segments for two competitive providers;
- Personal communication services (PCS): 200-MHz band just below 2.0 GHz, broken down into various segments such as licensed and unlicensed users;
- Cellular/PCS by satellite (e.g., Iridium, Globalstar); 10.5-MHz bandwidth in the 1600-MHz band; and
- Local multipoint distribution system (LMDS) in 28/38-GHz bands; 1.2-GHz bandwidth for CATV, Internet, data, and telephony services (Ref. 3).

[6]"High-frequency" takes on the connotation in the context of this book of any signal from 400 MHz to 100 GHz.

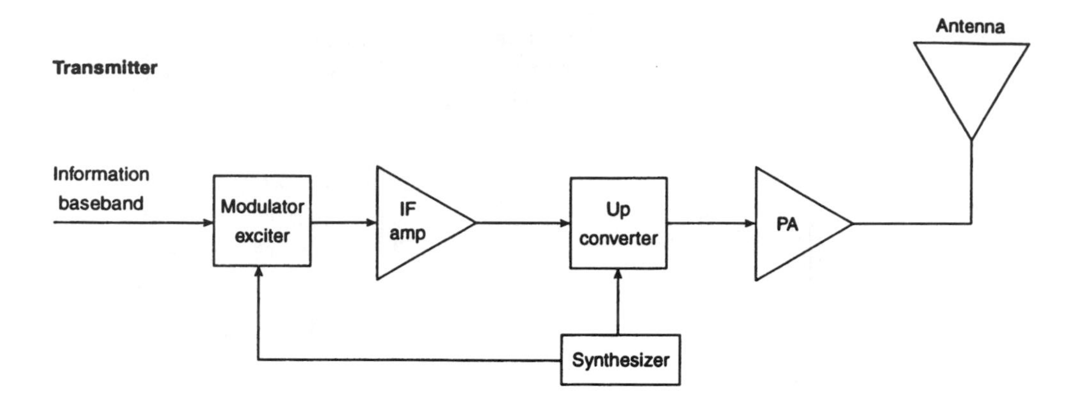

Receiver

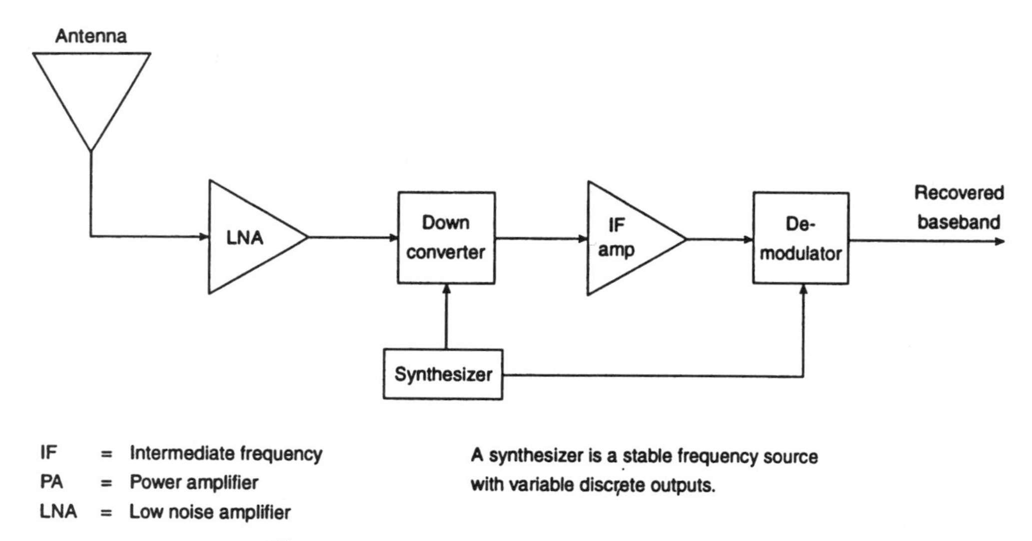

Figure 2.14 A generic model of a typical radio link.

Bandwidth is also determined by the regulating authority (e.g., the FCC in the United States) for a particular service/application. Through *bit packing* techniques, described in Chapter 9, the information-carrying capacity of a unit of bandwidth is considerably greater than 1 bit per Hz of bandwidth. On line-of-sight microwave systems, 5, 6, 7, and 8 bits per Hz of bandwidth are fairly common. Chapter 9 provides a more detailed discussion of radio systems.

REVIEW EXERCISES

1. Name at least four different ways of communicating at a distance prior to the advent of electrical communication.
2. What kind of energy is stored in a battery?
3. How did the old electric telegraph communicate intelligence?

4. What limited the distance we could transmit with electrical telegraph before using a repeater? Give at least two ways we could extend the distance.
5. How could that old-time electrical telegraph operate with just one wire?
6. Name at least four ways we might characterize a "sine wave," either partially or wholly.
7. What is the equivalent wavelength (λ) of 850 MHz? of 7 GHz?
8. What angle (in degrees) is equivalent to $3\pi/2$? $\pi/4$?
9. Give two examples of baseband transmission.
10. Define *modulation*.
11. What are the three generic forms of modulation? What popular device we find in the home utilizes all three types of modulation simultaneously. *Hint:* The answer needs a modifier in front of the word.
12. Differentiate an analog signal from a digital signal.
13. Give at least four applications of a 1-bit code. Use your imagination.
14. What is the total capacity of a 9-bit binary code? The Hollerith code was a 12-bit code. What was its total capacity?
15. Name four different transmission media.
16. What is the opposite of *loss*? What is the most common unit of measurement to express the amount of loss?
17. What is the reason for *twists* in twisted pair?
18. What is the principal cause of data rate limitation on wire pair?
19. What is the principal drawback of using coaxial cable for long-distance transmission?
20. What is the principal, unbeatable advantage of fiber optic cable?
21. Regarding limitation of bit rate and length, a fiber optic cable may be either ______ or ______?
22. Explain *dispersion* (with fiber optic cable).
23. What are some typical services of LMDS?

REFERENCES

1. *From Semaphore to Satellite*, International Telecommunication Union, Geneva, 1965.
2. *Principles of Electricity Applied to Telephone and Telegraph Work*, American Telephone and Telegraph Co., New York, 1961.
3. R. L. Freeman, *Telecommunication Transmission Handbook*, 4th ed., Wiley, New York, 1998.

3

QUALITY OF SERVICE AND TELECOMMUNICATION IMPAIRMENTS

3.1 OBJECTIVE

Quality of service (QoS) was introduced in Section 1.4. In this chapter we will be more definitive in several key areas. There are a number of generic impairments that will directly or indirectly affect quality of service. An understanding of these impairments and their underlying causes is extremely important if one wants to grasp the entire picture of a telecommunication system.

3.2 QUALITY OF SERVICE: VOICE, DATA, AND IMAGE

3.2.1 Introduction to Signal-to-Noise Ratio

Signal-to-noise ratio (S/N or SNR) is the most widely used parameter for measurement of signal quality in the field of transmission. Signal-to-noise ratio expresses in decibels the amount by which signal level exceeds the noise level in a specified bandwidth.

As we review the several types of material to be transmitted on a network, each will require a minimum S/N to satisfy the user or to make a receiving instrument function within certain specified criteria. The following are S/N guidelines at the corresponding receiving devices:

Voice: 40 dB;

Video (TV): 45 dB;

Data: ~15 dB, based upon the modulation type and specified error performance.

To illustrate the concept of S/N, consider Figure 3.1. This oscilloscope presentation shows a nominal analog voice channel (300–3400 Hz) with a 1000-Hz test signal. The vertical scale is signal power measured in dBm (see Appendix C for a tutorial on dBs), and the horizontal scale is frequency, 0 Hz to 3400 Hz. The S/N as illustrated is 10 dB. We can derive this by inspection or by reading the levels on the oscilloscope presentation. The signal level is +15 dBm; the noise is +5 dBm, then:

$$(S/N)_{dB} = \text{level}_{(\text{signal in dBm})} - \text{level}_{(\text{noise in dBm})} \qquad (3.1)$$

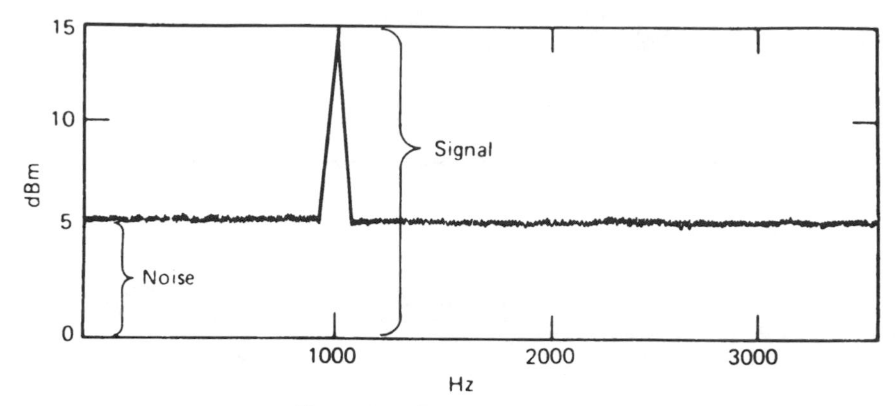

Figure 3.1 Signal-to-noise ratio.

Inserting the values given in the oscilloscope example, we have:

$$\begin{aligned} S/N &= +15 \text{ dBm} - (+5 \text{ dBm}) \\ &= 10 \text{ dB}. \end{aligned}$$

This expression is set up as shown because we are dealing with logarithms (see Appendix B). When multiplying in the domain of logarithms, we add. When dividing, we subtract. We are dividing because on the left side of the equation we have S/N or S divided by N.

Signal-to-noise ratio really has limited use in the PSTN for characterizing speech transmission because of the "spurtiness" of the human voice. We can appreciate that individual talker signal power can fluctuate widely so that the S/N ratio is far from constant during a telephone call and from one telephone call to the next. In lieu of actual voice, we use a test tone to measure level and S/N. A test tone is a single frequency, usually around 800 or 1000 Hz, generated by a signal generator and inserted in the voice channel. The level of the tone (often measured in dBm) can be easily measured with the appropriate test equipment. Such a tone has constant amplitude and no silent intervals, which is typical of voice transmission (Ref. 3).

3.2.2 Voice Transmission

3.2.2.1. Loudness Rating and Its Predecessors. Historically, on telephone connections, the complaint has been that the distant talker's voice was not loud enough at the receiving telephone. "Hearing sufficiently well" on a telephone connection is a subjective matter. This is a major element of QoS. Various methods have been derived over the years to rate telephone connections regarding customer satisfaction.

The underlying cause of low signal level is loss across the network. Any method to measure "hearing sufficiently well" should incorporate intervening losses on a telephone connection. As discussed in Chapter 2, losses are conventionally measured in dB. Thus the unit of measure of "hearing sufficiently well" is the decibel. From the present method of measurement we derive the *loudness rating*, abbreviated LR. It had several predecessors: *reference equivalent* and *corrected reference equivalent*.

3.2.2.2. Reference Equivalent. The reference equivalent value, called the *overall reference equivalent* (ORE), was indicative of how loud a telephone signal is. How loud is a subjective matter. Given a particular voice level, for some listeners it would be satisfactory, others unsatisfactory. The ITU in Geneva brought together a group of telephone users to judge telephone loudness. A test installation was set up made up of two standard telephone subsets, a talker's simulated subscriber loop and a listener's simulated loop. An adjustable attenuating network was placed between the two simulated loops. The test group, on an individual basis, judged level at the receiving telephone earpiece. At a 6-dB setting of the attenuator or less, calls were judged too loud. Better than 99% of the test population judged calls to be satisfactory with an attenuator setting of 16 dB; 80% rated a call satisfactory with an ORE 36 dB or better, and 33.6% of the test population rated calls with an ORE of 40 dB as unsatisfactory, and so on.

Using a similar test setup, standard telephone sets of different telephone administrations (countries) could be rated. The mouthpiece (transmitter) and earpiece (receiver) were rated separately and given a dB value. The dB value was indicative of their working better or worse than the telephones used in the ITU laboratory. The attenuator setting represented the loss in a particular network connection. To calculate overall reference equivalent (ORE) we summed the three dB values (i.e., the transmit reference equivalent of the telephone set, the intervening network losses, and the receive reference equivalent of the same type subset).

In one CCITT recommendation, 97% of all international calls were recommended to have an ORE of 33 dB or better. It was found that with this 33-dB value, less than 10% of users were unsatisfied with the level of the received speech signal.

3.2.2.3. Corrected Reference Equivalent. Because difficulties were encountered in the use of reference equivalents, the ORE was replaced by the *corrected reference equivalent* (CRE) around 1980. The concept and measurement technique of the CRE was essentially the same as RE (reference equivalent) and the dB remained the measurement unit. CRE test scores varied somewhat from its RE counterparts. Less than 5 dB (CRE) was too loud; an optimum connection had an RE value of 9 dB and a range from 7 dB to 11 dB for CRE. For a 30-dB value of CRE, 40% of a test population rated the call excellent, whereas 15% rated it poor or bad.

3.2.2.4. Loudness Rating. Around 1990 the CCITT replaced corrected reference equivalent with *loudness rating*. The method recommended to determine loudness rating eliminates the need for subjective determinations of loudness loss in terms of corrected reference equivalent. The concept of overall loudness loss (OLR) is very similar to the ORE concept used with reference equivalent.

Table 3.1 gives opinion results for various values of OLR in dB. These values are based upon representative laboratory conversation test results for telephone connections in which other characteristics such as circuit noise have little contribution to impairment.

3.2.2.4.1 Determination of Loudness Rating. The designation with notations of loudness rating concept for an international connection is given in Figure 3.2. It is assumed that telephone sensitivity, both for the earpiece and microphone, have been measured. OLR is calculated using the following formula:

$$\mathrm{OLR} = \mathrm{SLR} + \mathrm{CLR} + \mathrm{RLR}. \tag{3.2}$$

Table 3.1 Overall Loudness Rating Opinion Results

Overall Loudness Rating (dB)	Representative Opinion Results[a] Percent "Good plus Excellent"	Percent "Poor plus Bad"
5–15	<90	<1
20	80	4
25	65	10
30	45	20

[a] Based on opinion relationship derived from the transmission quality index (see Annex A, ITU-T Rec. P.11).

Source: ITU-T Rec. P.11, Table 1/P.11, p. 2, Helsinki, 3/93.

The measurement units in Eq. (3.2) are dB.

OLR is defined as the loudness loss between the speaking subscriber's mouth and the listening subscriber's ear via a telephone connection. The send loudness rating (SLR) is defined as the loudness loss between the speaking subscriber's mouth and an electrical interface in the network. The receive loudness rating (RLR) is the loudness loss between an electrical interface in the network and the listening subscriber's ear. The circuit loudness rating (CLR) is the loudness loss between two electrical interfaces in a connection or circuit, each interface terminated by its nominal impedance (Refs. 1, 2).

3.2.3 Data Circuits

Bit error rate (BER) is the underlying QoS parameter for data circuits. BER is *not* subjective; it is readily measurable. Data users are very demanding of network operators regarding BER. If a network did not ever carry data, BER requirements could be much less stringent. CCITT/ITU-T recommends a BER of 1×10^{-6} for at least 80% of a month.[1] Let us assume that these data will be transported on the digital network, typical of a PSTN. Let us further assume that conventional analog modems are not used, and the data is exchanged bit for bit with "channels" on the digital network. Thus, the BER of the data reflects the BER of the underlying digital channel that is acting as its transport. BERs encountered on digital networks in the industrialized/postindustrialized

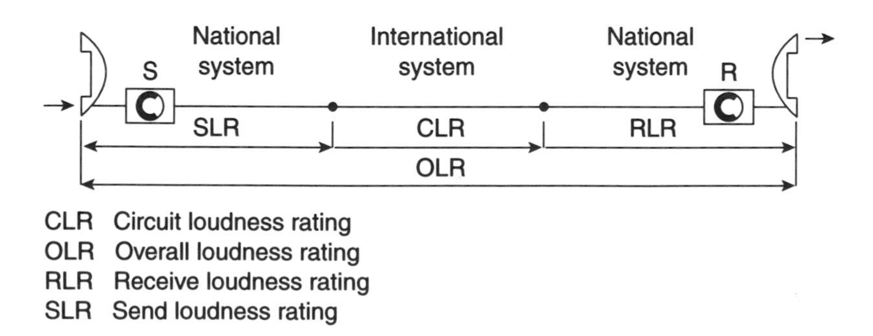

Figure 3.2 Designation of LRs in an international connection.

[1] See CCITT Rec. G.821.

nations are far improved, some attaining an end-to-end BER of 5×10^{-10}. Thus the data being transported can expect a similar BER. The genesis of frame relay, discussed in Chapter 10, is based on the premise that these excellent BERs can be expected.

3.2.4 Video (Television)

Television picture quality is subjective to the viewer. It is based on the S/N of the picture channel. The S/N values derived from two agencies are provided below. The TASO (Television Allocations Study Organization) ratings follow:

TASO PICTURE RATING

Quality	S/N
1. Excellent (no perceptible snow)	45 dB
2. Fine (snow just perceptible)	35 dB
3. Passable (snow definitely perceptible but not objectionable)	29 dB
4. Marginal (snow somewhat objectionable)	25 dB

Snow is the visual perception of high levels of thermal noise typical with poorer S/N values.

CCIR developed a five-point scale for picture quality versus impairment. This scale is shown in the table below:

CCIR FIVE GRADE SCALE

Quality	Impairment
5. Excellent	5. Imperceptible
4. Good	4. Perceptible, but not annoying
3. Fair	3. Slightly annoying
2. Poor	2. Annoying
1. Bad	1. Very annoying

Later CCIR/ITU-R documents steer clear of assigning S/N to such quality scales. In fact, when digital compression of TV is employed, the use of S/N to indicate picture quality is deprecated.

3.3 THREE BASIC IMPAIRMENTS AND HOW THEY AFFECT THE END-USER

There are three basic impairments found in all telecommunication transmission systems. These are:

1. Amplitude (or attenuation) distortion;
2. Phase distortion; and
3. Noise.

3.3.1 Amplitude Distortion

The IEEE defines attenuation distortion (amplitude distortion) as the change in attenuation at any frequency with respect to that of a reference frequency. For the discussion in

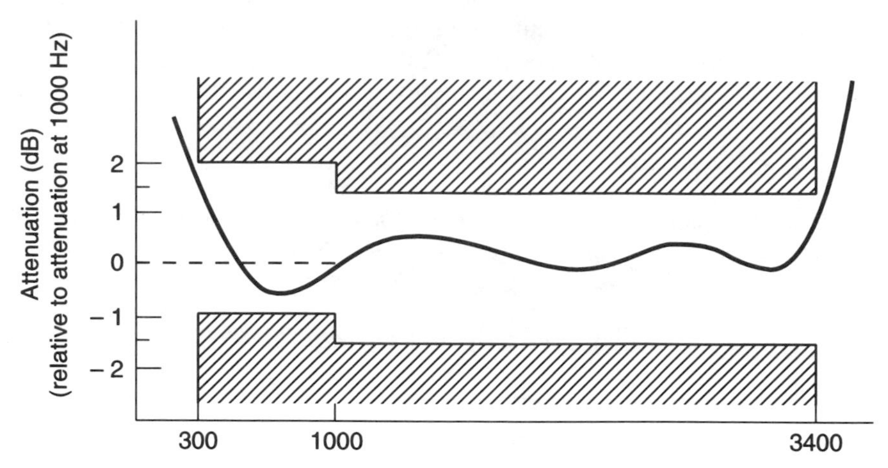

Figure 3.3 Typical attenuation distortion across a voice channel bandpass filter. Cross-hatched areas are response specifications, whereas the wavy line is the measured response.

this section, we'll narrow the subject to the (analog) voice channel. In most cases a user is connected, through his/her metallic subscriber loop, to the local serving exchange. This circuit is analog. Based upon the CCITT definition, the voice channel occupies the band from 300 Hz to 3400 Hz. We call this the *passband*.

Attenuation distortion can be avoided if all frequencies within the passband are subjected to the same loss (or gain). Whatever the transmission medium, however, some frequencies are attenuated more than others. Filters are employed in most active circuits (and in some passive circuits) and are major causes of attenuation distortion. Figure 3.3 is a response curve of a typical bandpass filter with voice channel application.

As stated in our definition, amplitude distortion across the voice channel is measured against a reference frequency. CCITT recommends 800 Hz as the reference; in North America the reference is 1000 Hz.[2] Let us look at some ways attenuation distortion may be stated. For example, one European requirement may state that between 600 Hz and 2800 Hz the level will vary no more than −1 to +2 dB, where the plus sign means more loss and the minus sign means less loss. Thus if an 800-Hz signal at −10 dBm is placed at the input of the channel, we would expect −10 dBm at the output (if there were no overall loss or gain), but at other frequencies we can expect a variation at the output of −1 to +2 dB. For instance, we might measure the level at the output at 2500 Hz at −11.9 dBm and at 1100 Hz at −9 dBm.

When filters or filterlike devices are placed in tandem, attenuation distortion tends to sum.[3] Two identical filters degrade attenuation distortion twice as much as just one filter.

3.3.2 Phase Distortion

We can look at a voice channel as a band-pass filter. A signal takes a finite time to pass through the telecommunication network. This time is a function of the velocity

[2]Test frequencies of 800-Hz and 1000-Hz are not recommended if the analog voice channel terminates into the digital network. In this case CCITT and Bellcore recommend 1020 Hz. The reason for this is explained in Chapter 6.

[3]Any signal-passing device, active or passive, can display filterlike properties. A good example is a subscriber loop, particularly if it has load coils and bridged taps. Load coils and bridged taps are discussed in Chapter 5.

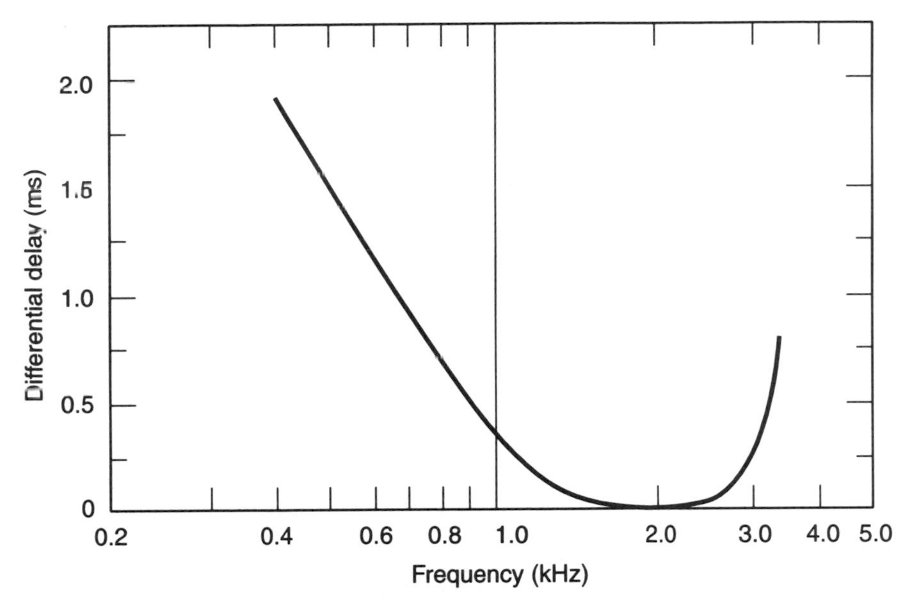

Figure 3.4 Typical differential delay across a voice channel.

of propagation for the medium and, of course, the length of the medium. The value can vary from 10,000 mi/sec (16,000 k/sec) to 186,000 mi/sec (297,600 km/sec). The former value is for heavily loaded subscriber pair cable.[4] This latter value is the velocity of propagation in free space, namely, radio propagation.

The velocity of propagation also tends to vary with frequency because of the electrical characteristics associated with the network. Again, the biggest culprit is filters. Considering the voice channel, therefore, the velocity of propagation tends to increase toward band center and decrease toward band edge. This is illustrated in Figure 3.4.

The finite time it takes a signal to pass through the total extension of the voice channel or through any network is called *delay*. Absolute delay is the delay a signal experiences while passing through the channel end-to-end at a reference frequency. But we have learned that propagation time is different for different frequencies with the wavefront of one frequency arriving before the wavefront of another frequency in the passband. A modulated signal will not be distorted on passing through the channel if the phase shift changes uniformly with frequency, whereas if the phase shift is nonlinear with respect to frequency, the output signal is distorted with respect to frequency.

In essence, we are dealing with phase linearity of a circuit. If the phase–frequency relationship over a passband is not linear, phase distortion will occur in the transmitted signal. Phase distortion is often measured by a parameter called *envelope delay distortion* (EDD). Mathematically, EDD is the derivative of the phase shift with respect to frequency. The maximum variation in the envelope over a band of frequencies is called envelope delay distortion. Therefore EDD is always a difference between the envelope delay at one frequency and that at another frequency of interest in the passband. It should be noted that envelope delay is often defined the same as *group delay*—which is the ratio of change, with angular frequency, of phase shift between two points in the network (Ref. 2).[5]

[4]Wire-pair loading is discussed in Chapter 5.

[5]*Angular frequency* and just the term *frequency* are conceptually the same for this text. Actually, angular frequency is measured in radians per second. There are 2π radians in 1 Hz.

Figure 3.4 shows that absolute delay is minimum around 1700 Hz and 1800 Hz in the voice channel. The figure also shows that around 1700 Hz and 1800 Hz, envelope delay distortion is flattest.[6] It is for this reason that so many data modems use 1700 Hz or 1800 Hz for the characteristic tone frequency, which is modulated by the data. A data modem is a device that takes the raw electrical baseband data signal and makes it compatible for transmission over the voice channel.

This brings up an important point. Phase distortion (or EDD) has little effect on speech communications over the telecommunications network. However, regarding data transmission, phase distortion is the greatest bottleneck for data rate (i.e., the number of bits per second that a channel can support). It has probably more effect on limiting data rate that any other parameter (Ref. 3).

3.3.3 Noise

3.3.3.1. General. Noise, in its broadest definition, consists of any undesired signal in a communication circuit. The subject of noise and noise reduction is probably the most important single consideration in transmission engineering. It is the major limiting factor in overall system performance. For our discussion in this text, noise is broken down into four categories:

1. Thermal noise;
2. Intermodulation noise;
3. Impulse noise; and
4. Crosstalk.

3.3.3.2. Thermal Noise. Thermal noise occurs in all transmission media and all communication equipment, including passive devices such as waveguide. It arises from random electron motion and is characterized by a uniform distribution of energy over the frequency spectrum with a Gaussian distribution of levels.

Gaussian distribution tells us that there is statistical randomness. For those of you who have studied statistics, this means that there is a "normal" distribution with standard deviations. Because of this, we can develop a mathematical relationship to calculate noise levels given certain key parameters.

Every equipment element and the transmission medium itself contributes thermal noise to a communication system if the temperature of that element or medium is above absolute zero on the Kelvin temperature scale. Thermal noise is the factor that sets the lower limit of sensitivity of a receiving system and is often expressed as a temperature, usually given in units referred to absolute zero. These units are called kelvins (K), not degrees.

Thermal noise is a general term referring to noise based on thermal agitations of electrons. The term "white noise" refers to the average uniform spectral distribution of noise energy with respect to frequency. Thermal noise is directly proportional to bandwidth and noise temperature.

We turn to the work of the Austrian scientist, Ludwig Boltzmann, who did landmark work on the random motion of electrons. From Boltzmann's constant, we can write a relationship for the thermal noise level (P_n) in 1 Hz of bandwidth at absolute zero (Kelvin scale) or

[6]"Flattest" means that there is little change in value. The line is flat, not sloping.

$$P_n = -228.6 \text{ dBW per Hz of bandwidth for a perfect receiver at absolute zero.} \quad (3.3a)$$

At room temperature (290 K or 17°C) we have:

$$P_n = -204 \text{ dBW per Hz of bandwidth for a perfect receiver} \quad (3.3b)$$

or

$$= -174 \text{ dBm/Hz of bandwidth for a perfect receiver.}$$

A *perfect receiver* is a receiving device that contributes no thermal noise to the communication channel. Of course, this is an idealistic situation that cannot occur in real life. It does provide us a handy reference, though. The following relationship converts Eq. (3.3b) for a real receiver in a real-life setting.

$$P_n = -204 \text{ dBW/Hz} + NF_{dB} + 10 \log B, \quad (3.4)$$

where B is the bandwidth of the receiver in question. The bandwidth must always be in Hz or converted to Hz.

NF is the noise figure of the receiver. It is an artifice that we use to quantify the amount of thermal noise a receiver (or any other device) injects into a communication channel. The noise figure unit is the dB.

An example of application of Eq. (3.4) might be a receiver with a 3-dB noise figure and a 10-MHz bandwidth. What would be the thermal noise power (level) in dBW of the receiver? Use Eq. (3.4).

$$\begin{aligned} P_n &= -204 \text{ dBW/Hz} + 3 \text{ dB} + 10 \log(10 \times 10^6) \\ &= -204 \text{ dBW/Hz} + 3 \text{ dB} + 70 \text{ dB} \\ &= -131 \text{ dBW.} \end{aligned}$$

3.3.3.3. Intermodulation Noise. Intermodulation (IM) noise is the result of the presence of intermodulation products. If two signals with frequencies F_1 and F_2 are passed through a nonlinear device or medium, the result will contain IM products that are spurious frequency energy components. These components may be present either inside and/or outside the frequency band of interest for a particular device or system. IM products may be produced from harmonics of the desired signal in question, either as products between harmonics, or as one of the basic signals and the harmonic of the other basic signal, or between both signals themselves.[7] The products result when two (or more) signals beat together or "mix." These products can be sums and/or differences. Look at the mixing possibilities when passing F_1 and F_2 through a nonlinear device. The coefficients indicate the first, second, or third harmonics.

- Second-order products $F_1 \pm F_2$;
- Third-order products $2F_1 \pm F_2$; $2F_2 \pm F_1$; and
- Fourth-order products $2F_1 \pm 2F_2$; $3F_1 \pm F_2$ . . . .

[7]A harmonic of a certain frequency F can be 2F (twice the value of F), 3F, 4F, 5F, and so on. It is an integer multiple of the basic frequency.

Devices passing multiple signals simultaneously, such as multichannel radio equipment, develop IM products that are so varied that they resemble white noise. Intermodulation noise may result from a number of causes:

- Improper level setting. If the level of an input to a device is too high, the device is driven into its nonlinear operating region (overdrive).
- Improper alignment causing a device to function nonlinearly.
- Nonlinear envelope delay.
- Device malfunction.

To summarize, IM noise results from either a nonlinearity or a malfunction that has the effect of nonlinearity. The causes(s) of IM noise is (are) different from that of thermal noise. However, its detrimental effects and physical nature can be identical with those of thermal noise, particularly in multichannel systems carrying complex signals.

3.3.3.4. Impulse Noise. Impulse noise is noncontinuous, consisting of irregular pulses or noise spikes of short duration and of relatively high amplitude. These spikes are often called *hits*, and each spike has a broad spectral content (i.e., impulse noise *smears* a broad frequency bandwidth). Impulse noise degrades voice telephony usually only marginally, if at all. However, it may seriously degrade error performance on data or other digital circuits. The causes of impulse noise are lightning, car ignitions, mechanical switches (even light switches), flourescent lights, and so on. Impulse noise will be discussed in more detail in Chapter 10.

3.3.3.5. Crosstalk. Crosstalk is the unwanted coupling between signal paths. There are essentially three causes of crosstalk:

1. Electrical coupling between transmission media, such as between wire pairs on a voice-frequency (VF) cable system and on digital (PCM) cable systems;
2. Poor control of frequency response (i.e., defective filters or poor filter design); and
3. Nonlinear performance in analog frequency division multiplex (FDM) system.

Excessive level may exacerbate crosstalk. By "excessive level" we mean that the level or signal intensity has been adjusted to a point higher than it should be. In telephony and data systems, levels are commonly measured in dBm. In cable television systems levels are measured as voltages over a common impedance (75 Ω). See the discussion of level in Section 3.4.

There are two types of crosstalk:

1. *Intelligible*, where at least four words are intelligible to the listener from extraneous conversation(s) in a seven-second period; and
2. *Unintelligible*, crosstalk resulting from any other form of disturbing effects of one channel on another.

Intelligible crosstalk presents the greatest impairment because of its distraction to the listener. Distraction is considered to be caused either by fear of loss of privacy or primarily by the user of the primary line consciously or unconsciously trying to understand what is being said on the secondary or interfering circuits; this would be true for any interference that is syllabic in nature.

Received crosstalk varies with the volume of the disturbing talker, the loss from the disturbing talker to the point of crosstalk, the coupling loss between the two circuits under consideration, and the loss from the point of crosstalk to the listener. The most important of these factors for this discussion is the coupling loss between the two circuits under consideration. Also, we must not lose sight of the fact that the effects of crosstalk are subjective, and other factors have to be considered when crosstalk impairments are to be measured. Among these factors are the type of people who use the channel, the acuity of listeners, traffic patterns, and operating practices (Ref. 4).

3.4 LEVEL

Level is an important parameter in the telecommunications network, particularly in the analog network or in the analog portion of a network. In the context of this book when we use the word *level*, we mean *signal magnitude or intensity*. Level could be comparative. The output of an amplifier is 30 dB higher than the input. But more commonly, we mean absolute level, and in telephony it is measured in dBm (decibels referenced to 1 milliwatt) and in radio systems we are more apt to use dBW (decibels referenced to 1 watt). Television systems measure levels in voltage, commonly the dBmV (decibels referenced to 1 millivolt).

In the telecommunication network, if levels are too high, amplifiers become overloaded, resulting in increases in intermodulation noise and crosstalk. If levels are too low, customer satisfaction suffers (i.e., loudness rating). In the analog network, level was a major issue; in the digital network, somewhat less so.

System levels are used for engineering a communication system. These are usually taken from a level chart or reference system drawing made by a planning group or as a part of an engineered job. On the chart, a 0 TLP (zero test level point) is established. A TLP is a location in a circuit or system at which a specified test-tone level is expected during alignment. A 0 TLP is a point at which the test-tone level should be 0 dBm. A test tone is a tone produced by an audio signal generator, usually 1020 Hz. Note that these frequencies are inside the standard voice channel which covers the range of 300–3400 Hz. In the digital network, test tones must be applied on the analog side. This will be covered in Chapter 6.

From the 0 TLP other points may be shown using the unit dBr (decibel reference). A minus sign shows that the level is so many decibels below reference and a plus sign, above. The unit dBm0 is an absolute unit of power in dBm referred to the 0 TLP. The dBm can be related to the dBr and dBm0 by the following formula:

$$\mathrm{dBm} = \mathrm{dBm0} + \mathrm{dBr}. \tag{3.5}$$

For instance, a value of −32 dBm at a −22 dBr point corresponds to a reference level of −10 dBm0. A −10-dBm0 signal introduced at the 0-dBr point (0 TLP) has an absolute signal level of −10 dBm (Ref. 5).

3.4.1 Typical Levels

Earlier measurements of speech level used the unit of measure VU, standing for volume unit. For a 1000-Hz sinusoid signal (simple sine wave signal), 0 VU = 0 dBm. When a VU meter is used to measure the level of a voice signal, it is difficult to exactly equate VU and dBm. One of the problems, of course, is that speech transmission is characterized by spurts of signal. However, a good approximation relating VU to dBm is the following formula:

$$\text{Average power of a telephone talker} \approx \text{VU} - 1.4(\text{dBm}). \tag{3.6}$$

In the telecommunication network, telephone channels are often multiplexed at the first serving exchange. When the network was analog, the multiplexers operated in the frequency domain and were called *frequency division multiplexers* (FDM). Voice channel inputs were standardized with a level of either −15 dBm or −16 dBm, and the outputs of demultiplexers were +7 dBm. These levels, of course, were test-tone levels. In industrialized and postindustrialized nations, in nearly every case, multiplexers are digital. These multiplexers have an overload point at about +3.17 dBm0. The digital reference signal is 0 dBm on the analog side using a standard test tone between 1013 Hz and 1022 Hz (Ref. 4).

3.5 ECHO AND SINGING

Echo and singing are two important impairments that impact QoS. Echo is when a talker hears her/his own voice delayed. The annoyance is a function of the delay time (i.e., the time between the launching of a syllable by a talker and when the echo of that syllable is heard by the same talker). It is also a function of the intensity (level) of the echo, but to some lesser extent. Singing is audio feedback. It is an "ear-splitting" howl, much like the howl one gets by placing a public address microphone in front of a loudspeaker. We will discuss causes and cures of echo and singing in Chapter 4.

REVIEW EXERCISES

1. Define *signal-to-noise ratio.*

2. Give signal-to-noise ratio guidelines at a receiving device for the following three media: (1) voice, (2) video-TV, and (3) data. Base the answer on where a typical customer says the signal is very good or excellent.

3. Why do we use a sinusoidal test tone when we measure S/N on a speech channel rather than just the speech signal itself?

4. The noise level of a certain voice channel is measured at −39 dBm and the test-tone signal level is measured at +3 dBm. What is the channel S/N?

5. If we know the loudness rating of a telephone subset earpiece and of the subset mouthpiece, what additional data do we need to determine the overall loudness rating (OLR) of a telephone connection?

6. The BER of an underlying digital circuit is 1×10^{-8}, for data riding on this circuit. What is the best BER we can expect on the data?

7. What are the three basic impairments on a telecommunication transmission channel?

8. Of the three impairments, which one affects data error rate the most and thus limits bit rate?

9. Explain the cause of phase distortion.

10. Name the four types of noise we are likely to encounter in a telecommunication system.

11. What will be the thermal noise level of a receiver with a noise figure of 3 dB and a bandwidth of 1 MHz?

12. Define third-order products based on the mixing of two frequencies F_1 and F_2.

13. Give four causes of impulse noise.

14. Relate VU to dBm for a simple sinusoidal signal to a complex signal such as human voice.

15. Echo as an annoyance to a telephone listener varies with two typical causes. What are they? What is the most important (most annoying)?

REFERENCES

1. *Effect of Transmission Impairments*, ITU-T Rec. P.11, ITU Helsinki, March 1993.
2. *Loudness Ratings on National Systems*, ITU-T Rec. G.121, ITU Helsinki, March, 1993.
3. R. L. Freeman, *Telecommunication System Engineering*, 3rd ed., Wiley, New York, 1996.
4. R. L. Freeman, *Telecommunication Transmission Handbook*, 4th ed., Wiley, New York, 1998.
5. CCITT G. Recommendations Fascicles III.1 and III.2, IXth Plenary Assembly, Melbourne, 1988.

4

TRANSMISSION AND SWITCHING: CORNERSTONES OF A NETWORK

4.1 TRANSMISSION AND SWITCHING DEFINED

The IEEE defines *transmission* as the propagation of a signal, message, or other form of intelligence by any means such as optical fiber, wire, or visual. Our definition is not so broad. Transmission provides the transport of a signal from an end-user source to the destination such that the signal quality at the destination meets certain performance criteria.

Switching selects the route to the desired destination that the transmitted signal travels by the closing of switches either in the space domain or the time domain or some combination(s) of the two.

Prior to 1985, transmission and switching were separate disciplines in telecommunication with a firm dividing line between the two. Switching engineers knew little about transmission, and transmission engineers knew little about switching. As mentioned in Chapter 1, that dividing line today is hazy at best. *Signaling* develops and carries the control information for switches. If a transmission path becomes impaired, signaling becomes ineffectual and the distant-end switch either will not operate or will not function correctly, misrouting the connectivity. Timing, which is so vital for the digital transmission path, derives from the connected switches.

4.2 TRAFFIC INTENSITY DEFINES THE SIZE OF SWITCHES AND THE CAPACITY OF TRANSMISSION LINKS

4.2.1 Traffic Studies

As already mentioned, telephone exchanges (switches) are connected by trunks or junctions.[1] The number of trunks connecting exchange X with exchange Y is the number of voice pairs or their equivalent used in the connection. One of the most important steps in telecommunication system design is to determine the number of trunks required on a route or connection between exchanges. We could say we are *dimensioning* the route. To dimension the route correctly we must have some idea of its usage—that is, how many people will wish to talk at once over the route. The usage of a transmission route

[1]The term *junction* means a trunk in the local area. It is a British term. *Trunk* is used universally in the long-distance plant.

or switch brings us into the realm of traffic engineering; and usage may be defined by two parameters: (1) *calling rate*, or the number of times a route or traffic path is used per unit time period; or more properly defined, "the call intensity per traffic path during the busy hour (BH);" and (2) *holding time*, or "the average duration of occupancy of one or more paths by calls." A *traffic path* is a "channel, time slot, frequency band, line, trunk switch, or circuit over which individual communications pass in sequence." *Carried traffic* is the volume of traffic actually carried by a switch, and *offered traffic* is the volume of traffic offered to a switch. Offered traffic minus carried traffic equals *lost calls*. A lost call is one that does not make it through a switch. A call is "lost" usually because it meets congestion or blockage at that switch.

To dimension a traffic path or size a telephone exchange, we must know the traffic intensity representative of the normal busy season. There are weekly and daily variations in traffic within the busy season. Traffic is random in nature. However, there is a certain consistency we can look for. For one thing, there is usually more traffic on Mondays and Fridays, and a lower volume on Wednesdays. A certain consistency can also be found in the normal workday variation. Across a typical day the variation is such that a one-hour period shows greater usage than any other one-hour period. From the hour of the day with least traffic intensity to the hour of greatest traffic, the variation can exceed 100 : 1. Figure 4.1 shows a typical hour-to-hour traffic variation for a serving switch in the United States. It can be seen that the busiest period, the *busy hour* (BH), is between 10 A.M. and 11 A.M. (The busy hour from the viewpoint of grade of service was introduced in Section 1.3.4). From one workday to the next, originating BH calls can vary as much as 25%. To these fairly "regular" variations, there are also unpredictable peaks caused by stock market or money market activity, weather, natural disaster, international events, sporting events, and so on. Normal traffic growth must also be taken into account. Nevertheless, suitable forecasts of BH traffic can be made. However, before proceeding further in this discussion, consider the following definitions of the busy hour.

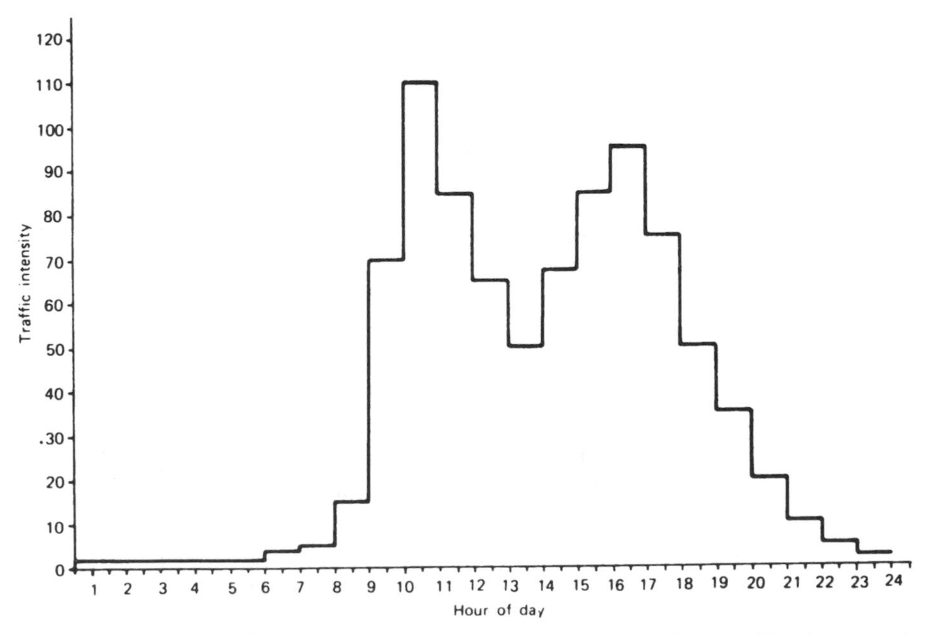

Figure 4.1 Bar chart of traffic intensity over a typical working day (U.S., mixed business, and residential).

1. *Busy hour.* The busy hour refers to the traffic volume or number of call attempts, and is that continuous one-hour period being wholly in the time interval concerned for which this quantity (i.e., traffic volume or call attempts) is greatest.
2. *Average busy season busy hour.* The ABSBH is used for trunk groups and always has a grade of service criterion applied.[2] For example, for the ABSBH load, a call requiring a circuit in a trunk group should encounter *all trunks busy* (ATB) no more than 1% of the time.

Other definitions of the busy hour may be found in Ref. 1.

When dimensioning telephone exchanges and transmission routes, we shall be working with BH traffic levels and care must be used in the definition of the busy hour. Peak traffic loads are of greater concern than average loads for the system planner when dimensioning switching equipment.

Another concern in modern digital switching systems is *call attempts*. We could say that call attempts is synonymous with offered traffic. Even though a call is not *carried* and is turned away, the switch's processor or computer is still exercised. In many instances a switch's capability to route traffic is limited by the peak number of call attempts its processor can handle.

4.2.1.1. Measurement of Telephone Traffic. If we define *telephone traffic* as the aggregate of telephone calls over a group of circuits or trunks with regard to the duration of calls as well as their number, we can say that traffic flow (A):

$$A = C \times T, \tag{4.1}$$

where C designates the number of calls originated during the period of one hour and T is the average *holding time*, usually given in hours. A is a dimensionless unit because we are multiplying calls/hour by hour/call.

Suppose that the average holding time is 2.5 minutes and the calling rate in the BH for a particular day is 237. The traffic flow (A) would then be 237 × 2.5, or 592.5 call-minutes (Cm) or 593.5/60, or about 9.87 call-hours (Ch).

The preferred unit of traffic intensity is the *erlang*, named after the Danish mathematician A. K. Erlang (Copenhagen Telephone Company, 1928). The erlang is a dimensionless unit. One erlang represents a circuit occupied for 1 hour. Considering a group of circuits, traffic intensity in erlangs is the number of call-seconds per second or the number of call-hours per hour. If we knew that a group of 10 circuits had a call intensity of 5 erlangs, we would expect half of the circuits to be busy at the time of measurement.

In the United States the term *unit call* (UC), or its synonymous term, *hundred call-second*, abbreviated ccs, generally is used.[3] These terms express the sum of the number of busy circuits, provided that the busy trunks were observed once every 100 seconds (36 observations in 1 hour) (Ref. 2). The following simple relationship should be kept in mind: 1 erlang = 36 ccs, assuming a 1-hour time-unit interval.

Extensive traffic measurements are made on switching systems because of their numerous traffic-sensitive components. Usual measurements for a component such as a service circuit include call attempts, calls carried, and usage. The typical holding time

[2]*Grade of service* refers to the planned value criterion of probability of blockage of an exchange. This is the point where an exchange just reaches its full capacity to carry traffic. This usually happens during the busy hour.

[3]The first letter "c" in ccs stands for the Roman number 100.

for a common-control element in a switch is considerably shorter than that for a trunk, and short sampling intervals (e.g., 10 seconds) or continuous monitoring are used to measure usage (Ref. 1).

4.2.1.2. Blockage, Lost Calls, and Grade of Service. Let's assume that an isolated telephone exchange serves 5000 subscribers and that no more than 10% of the subscribers wish service simultaneously. Therefore, the exchange is dimensioned with sufficient equipment to complete 500 simultaneous connections. Each connection would be, of course, between any two of the 5000 subscribers. Now let subscriber 501 attempt to originate a call. She/he cannot complete the call because all the connecting equipment is busy, even though the line she/he wishes to reach may be idle. This call from subscriber 501 is termed a *lost call* or *blocked call*. She/he has met blockage. The probability of encountering blockage is an important parameter in traffic engineering of telecommunication systems. If congestion conditions are to be met in a telephone system, we can expect that those conditions will usually be encountered during the BH. A switched is dimensioned (sized) to handle the BH load. But how well? We could, indeed, far overdimension the switch such that it could handle any sort of traffic peaks. However, that is uneconomical. So with a well-designed switch, during the busiest of BHs we can expect moments of congestion such that additional call attempts will meet blockage. *Grade of service* expresses the probability of meeting blockage during the BH and is commonly expressed by the letter "*p*."[4] A typical grade of service is p = 0.01. This means that an average of one call in 100 will be blocked or "lost" during the BH. Grade of service, a term in the erlang formula, is more accurately defined as the *probability of blockage*. It is important to remember that lost calls (blocked calls) refer to calls that fail at *first trial*. We discuss attempts (at dialing) later, that is, the way blocked calls are handled.

We exemplify grade of service by the following problem: If we know that there are 345 seizures (i.e., lines connected for service) and 6 blocked calls (i.e., lost calls) during the BH, what is the grade of service?

$$\begin{aligned} \text{Grade of service} &= \text{Number of lost calls/Number of offered calls} \\ &= 6/(354+6) = 6/360 \\ p &\approx 0.017 \end{aligned} \tag{4.2}$$

The average grade of service for a network may be obtained by adding the grade of service provided by a particular group of trunks or circuits of specified size and carrying a specified traffic intensity. It is the probability that a call offered to the group will find available trunks already occupied on first attempt. This probability depends on a number of factors. The most important of which are (1) the distribution in time and duration of offered traffic (e.g., random or periodic arrival and constant or exponentially distributed holding time), (2) the number of traffic sources [limited or high (infinite)], (3) the availability of trunks in a group to traffic sources (full or restricted availability), and (4) the manner in which lost calls are "handled." Several new concepts are suggested in these four factors. These must be explained before continuing.

4.2.1.2.1 Availability. Switches were previously discussed as devices with lines and trunks, but better terms for describing switches are *inlets* and *outlets*. When a switch

[4]Grade of service was introduced in Section 1.3.4.

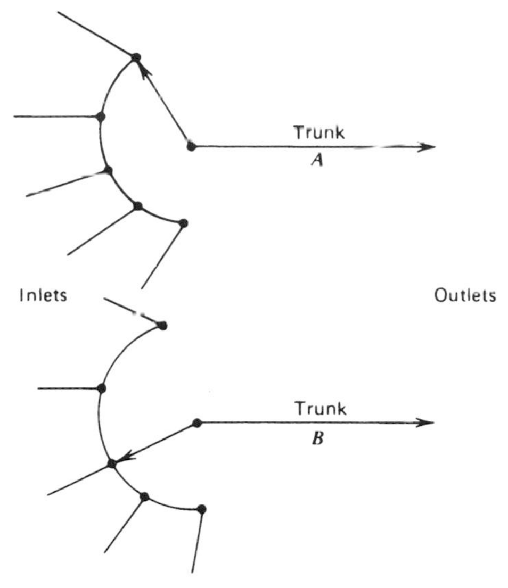

Figure 4.2a An example of a switch with limited availability.

has *full availability*, each inlet has access to *any* outlet. When not all the free outlets in a switching system can be reached by inlets, the switching system is referred to as one with *limited availability*. Examples of switches with limited and full availability are shown in Figures 4.2*a* and 4.2*b*.

Of course, full-availability switching is more desirable than limited availability, but is more expensive for larger switches. Thus full-availability switching is generally found only in small switching configurations and in many new digital switches (see Chapter 6). *Grading* is one method of improving the traffic-handling capabilities of switching configurations with limited availability. Grading is a scheme for interconnecting switching subgroups to make the switching load more uniform.

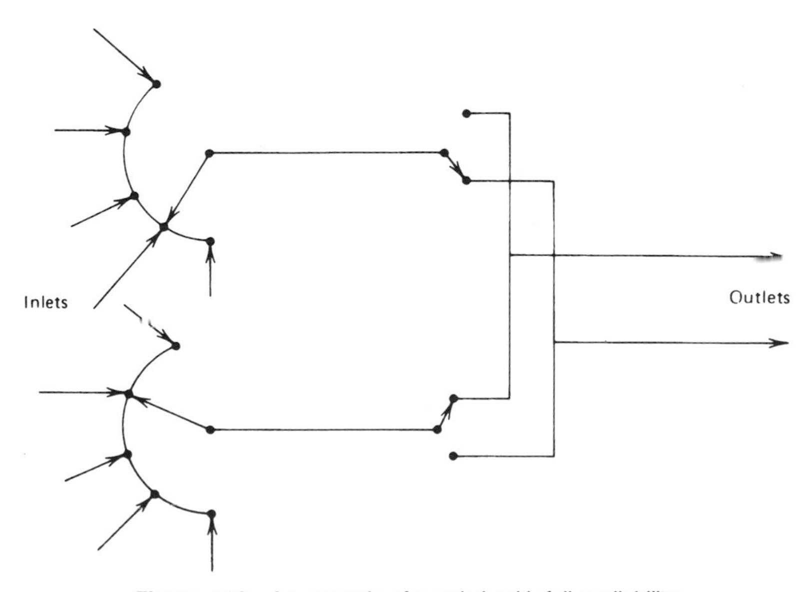

Figure 4.2b An example of a switch with full availability.

4.2.1.2.2 "Handling" of Lost Calls. In conventional telephone traffic theory, three methods are considered for the handling or dispensing of lost calls:

1. Lost calls held (LCH);
2. Lost calls cleared (LCC); and
3. Lost calls delayed (LCD).

The LCH concept assumes that the telephone user will immediately reattempt the call on receipt of a congestion signal and will continue to redial. The user hopes to seize connection equipment or a trunk as soon as switching equipment becomes available for the call to be handled. It is the assumption in the LCH concept that lost calls are held or waiting at the user's telephone. This concept further assumes that such lost calls extend the average holding time theoretically and, in this case, the average holding time is zero, and all the time is waiting time. The principal traffic formula (for conventional analog space division switching) in North America is based on the LCH concept.

The LCC concept, which is primarily used in Europe or those countries that have adopted European practice, assumes that the user will hang up and wait some time interval before reattempting if the user hears the congestion signal on the first attempt. Such calls, it is assumed, disappear from the system. A reattempt (after the delay) is considered as initiating a new call. The Erlang B formula is based on this criterion.

The LCD concept assumes that the user is automatically put in queue (a waiting line or pool). For example, this is done, of course, when an operator is dialed. It is also done on all modern digital switching systems. Such switches are computer-based for the "brains" of the control functions and are called switches with *stored program control* (SPC). The LCD category may be broken down into three subcategories, depending on how the queue or pool of waiting calls is handled. The waiting calls may be handled *last in first out* (LIFO), *first in first out* (FIFO), or *at random*.

4.2.1.2.3 Infinite and Finite Traffic Sources. We can assume that traffic sources are either infinite or finite. For the infinite-traffic-sources case the probability of call arrival is constant and does not depend on the occupancy of the system. It also implies an infinite number of call arrivals, each with an infinitely small holding time. An example of finite traffic sources is when the number of sources offering traffic to a group of trunks is comparatively small in comparison to the number of circuits. We can also say that with a finite number of sources the arrival rate is proportional to the number of sources that are not already engaged in sending a call (Ref. 2).

4.2.1.2.4 Probability-Distribution Curves. Telephone-call originations in any particular area are random in nature. We find that originating calls or call arrivals at an exchange closely fit a family of probability-distribution curves following a Poisson distribution.[5] The Poisson distribution is fundamental in traffic theory.

Most probability distribution curves are two-parameter curves; that is, they may be described by two parameters: mean and variance. The *mean* is a point on the probability-distribution curve where an equal number of events occur to the right of the point as to the left of the point. Mean is synonymous with average (see Figure 4.3).

The second parameter used to describe a distribution curve is the *dispersion*, which tells us how the values or population are dispersed about the center or mean of the

[5] S. D. Poisson was a nineteenth-century French mathematician/physicist specializing in "randomness."

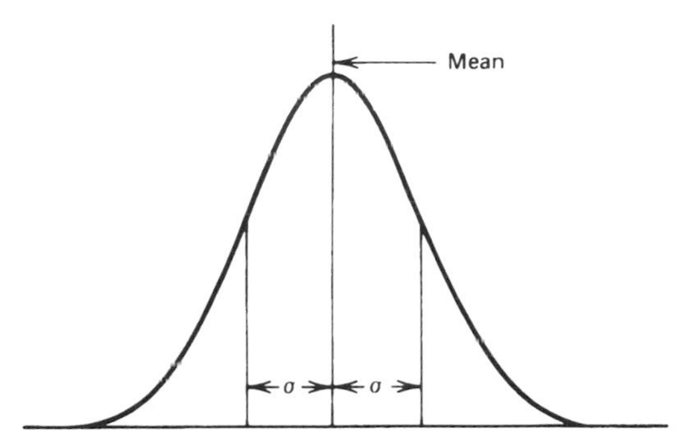

Figure 4.3 A normal distribution curve showing the mean and standard deviation, σ.

curve. There are several measures of dispersion. One is the familiar *standard deviation.* The standard deviation is usually expressed by the Greek letter sigma (σ). For example, 1 σ either side of the mean in Figure 4.3 will contain about 68% of the population or measurements, 2 σ will contain about 95% of the measurements and around 99% of the subject, population, or whatever is being measured. The curve shown in Figure 4.3 is a *normal distribution curve.*

4.2.2 Discussion of the Erlang and Poisson Traffic Formulas

When dimensioning a route, we want to find the optimum number of circuits to serve the route. There are several formulas at our disposal to determine that number of circuits based on the BH traffic load. In Section 4.2.1.2 four factors were discussed that will help us to determine which traffic formula to use given a particular set of circumstances. These factors primarily dealt with (1) call arrivals and holding-time distributions, (2) number of traffic sources, (3) availability (full or limited), and (4) handling of lost calls.

The Erlang B loss formula was/is very widely used outside of the United States. Loss in this context means the probability of encountering blockage at the switch due to congestion or to "all trunks busy" (ATB). The formula expresses grade of service or the probability of finding x channels busy. The other two factors in the Erlang B formula are the mean of the *offered traffic* and the number of trunks or servicing channels available. The formula assumes that:

- Traffic originates from an infinite number of sources;
- Lost calls are cleared assuming a zero holding time;
- The number of trunks or servicing channels is limited; and
- Full availability exists.

The actual Erlang B formula is beyond the scope of this text. For more detailed information, it is recommended that the reader consult Ref. 3. It is far less involved to use traffic tables as found in Table 4.1, which gives trunk-dimensioning information for some specific grades of service, from 0.001 to 0.05 and from 1 to 49 trunks. The table uses traffic-intensity units UC (unit call) and TU (traffic unit), where TU is in erlangs assuming BH conditions and UC is in ccs (cent-call-seconds, meaning 100 call seconds). Remember that 1 erlang = 36 ccs (based on a 1-hour time interval).

To exemplify the use of Table 4.1, suppose a route carried 16.68 erlangs of traffic with a desired grade of service of 0.001; then 30 trunks would be required. If the grade

Table 4.1 Trunk-Loading Capacity, Based on Erlang B Formula, Full Availability

	Grade of Service 1 in 1000		Grade of Service 1 in 500		Grade of Service 1 in 200		Grade of Service 1 in 100		Grade of Service 1 in 50		Grade of Service 1 in 20	
Trunks	UC	TU	UC	TU	UC	TU	UC	TU	UC	TU	UC	TU
1	0.04	0.001	0.07	0.002	0.2	0.005	0.4	0.01	0.7	0.02	1.8	0.05
2	1.8	0.05	2.5	0.07	4	0.11	5.4	0.15	7.9	0.22	14	0.38
3	6.8	0.19	9	0.25	13	0.35	17	0.46	22	0.60	32	0.90
4	16	0.44	19	0.53	25	0.70	31	0.87	39	1.09	55	1.52
5	27	0.76	32	0.90	41	1.13	49	1.36	60	1.66	80	2.22
6	41	1.15	48	1.33	58	1.62	69	1.91	82	2.28	107	2.96
7	57	1.58	65	1.80	78	2.16	90	2.50	106	2.94	135	3.74
8	74	2.05	83	2.31	98	2.73	113	3.13	131	3.63	163	4.54
9	92	2.56	103	2.85	120	3.33	136	3.78	156	4.34	193	5.37
10	111	3.09	123	3.43	143	3.96	161	4.46	183	5.08	224	6.22
11	131	3.65	145	4.02	166	4.61	186	5.16	210	5.84	255	7.08
12	152	4.23	167	4.64	190	5.28	212	5.88	238	6.62	286	7.95
13	174	4.83	190	5.27	215	5.96	238	6.61	267	7.41	318	8.83
14	196	5.45	213	5.92	240	6.66	265	7.35	295	8.20	350	9.73
15	219	6.08	237	6.58	266	7.38	292	8.11	324	9.01	383	10.63
16	242	6.72	261	7.26	292	8.10	319	8.87	354	9.83	415	11.54
17	266	7.38	286	7.95	318	8.83	347	9.65	384	10.66	449	12.46
18	290	8.05	311	8.64	345	9.58	376	10.44	414	11.49	482	13.38
19	314	8.72	337	9.35	372	10.33	404	11.23	444	12.33	515	14.31
20	339	9.41	363	10.07	399	11.09	433	12.03	474	13.18	549	15.25
21	364	10.11	388	10.79	427	11.86	462	12.84	505	14.04	583	16.19
22	389	10.81	415	11.53	455	12.63	491	13.65	536	14.90	617	17.13

23	415	11.52	442	12.27	483	13.42	521	14.47	567	15.76	651	18.08
24	441	12.24	468	13.01	511	14.20	550	15.29	599	16.63	685	19.03
25	467	12.97	495	13.76	540	15.00	580	16.12	630	17.50	720	19.99
26	493	13.70	523	14.52	569	15.80	611	16.96	662	18.38	754	20.94
27	520	14.44	550	15.28	598	16.60	641	17.80	693	19.26	788	21.90
28	546	15.18	578	16.05	627	17.41	671	18.64	725	20.15	823	22.87
29	573	15.93	606	16.83	656	18.22	702	19.49	757	21.04	858	23.83
30	600	16.68	634	17.61	685	19.03	732	20.34	789	21.93	893	24.80
31	628	17.44	662	18.39	715	19.85	763	21.19	822	22.83	928	25.77
32	655	18.20	690	19.18	744	20.68	794	22.05	854	23.73	963	26.75
33	683	18.97	719	19.97	774	21.51	825	22.91	887	24.63	998	27.72
34	711	19.74	747	20.76	804	22.34	856	23.77	919	25.53	1033	28.70
35	739	20.52	776	21.56	834	23.17	887	24.64	951	26.43	1068	29.68
36	767	21.30	805	22.36	864	24.01	918	25.51	984	27.34	1104	30.66
37	795	22.03	834	23.17	895	24.85	950	26.38	1017	28.25	1139	31.64
38	823	22.86	863	23.97	925	25.69	981	27.25	1050	29.17	1175	32.63
39	851	23.65	892	24.78	955	26.53	1013	28.13	1083	30.08	1210	33.61
40	880	24.44	922	25.60	986	27.38	1044	29.01	1116	31.00	1246	34.60
41	909	25.24	951	26.42	1016	28.23	1076	29.89	1149	31.92	1281	35.59
42	937	26.04	981	27.24	1047	29.08	1108	30.77	1182	32.84	1317	36.58
43	966	26.84	1010	28.06	1078	29.94	1140	31.66	1215	33.76	1353	37.57
44	995	27.64	1040	28.88	1109	30.80	1171	32.54	1248	34.68	1388	38.56
45	1024	28.45	1070	29.71	1140	31.66	1203	33.43	1282	35.61	1424	39.55
46	1053	29.26	1099	30.54	1171	32.52	1236	34.32	1315	36.53	1459	40.54
47	1083	30.07	1129	31.37	1202	33.38	1268	35.21	1349	37.46	1495	41.54
48	1111	30.88	1159	32.20	1233	34.25	1300	36.11	1382	38.39	1531	42.54
49	1141	31.69	1189	33.04	1264	35.11	1332	37.00	1415	39.32	1567	43.54

of service were reduced to 0.05, the 30 trunks could carry 24.80 erlangs of traffic. When sizing a route for trunks or an exchange, we often come up with a fractional number of servicing channels or trunks. In this case we would opt for the next highest integer because we cannot install a fraction of a trunk. For instance, if calculations show that a trunk route should have 31.4 trunks, it would be designed for 32 trunks.

The Erlang B formula, based on lost calls cleared, has been standardized by the CCITT (CCITT Rec. Q.87) and has been generally accepted outside the United States. In the United States the Poisson formula is favored. This formula is often called the *Molina formula*. It is based on the LCH concept. Table 4.2 provides trunking sizes for various grades of service deriving from the *P* formula; such tables are sometimes called "*P*" tables (Poisson) and assume full availability. We must remember that the Poisson equation also assumes that traffic originates from a large (infinite) number of independent subscribers or sources (random traffic input), with a limited number of trunks or servicing channels and LCH (Ref. 3).

4.2.3 Waiting Systems (Queueing)

The North American PSTN will be entirely digital by the year 2000. Nearly all digital switches operate under some form of queueing discipline, which many call *waiting systems* because an incoming call is placed in queue and waits its turn for service. These systems are based on our third assumption, namely, LCD. Of course a queue in this case is a pool of callers waiting to be served by a switch. The term *serving time* is the time a call takes to be served from the moment of arrival in the queue to the moment of being served by the switch. For traffic calculations in most telecommunication queueing systems, the mathematics is based on the assumption that call arrivals are random and Poissonian. The traffic engineer is given the parameters of offered traffic, the size of the queue, and a specified grade of service and will determine the number of serving circuits or trunks that are required.

The method by which a waiting call is selected to be served from the pool of waiting calls is called *queue discipline*. The most common discipline is the first-come, first-served discipline, where the call waiting longest in the queue is served first. This can turn out to be costly because of the equipment required to keep order in the queue. Another type is random selection, where the time a call has waited is disregarded and those waiting are selected in random order. There is also the last-come, first-served discipline and bulk service discipline, where batches of waiting calls are admitted, and there are also priority service disciplines, which can be preemptive and nonpreemptive. In queueing systems the grade of service may be defined as the probability of delay. This is expressed as $P(t)$, the probability that a call is not being immediately served and has to wait a period of time greater than t. The average delay on all calls is another parameter that can be used to express grade of service, and the length of queue is yet another.

The probability of delay, the most common index of grade of service for waiting systems when dealing with full availability and a Poissonian call-arrival process (i.e., random arrivals), is calculated using the Erlang C formula, which assumes an infinitely long queue length. A more in-depth coverage of the Erlang C formula, along with Erlang C traffic tables, may be found in Ref. 3.

4.2.4 Dimensioning and Efficiency

By definition, if we were to dimension a route or estimate the required number of servicing channels where the number of trunks (or servicing channels) just equaled the

Table 4.2 Trunk-Loading Capacity, Based on Poisson Formula, Full Availability

	Grade of Service 1 in 1000		Grade of Service 1 in 100		Grade of Service 1 in 50		Grade of Service 1 in 20		Grade of Service 1 in 10	
Trunks	UC	TU	UC	TU	UC	TU	UC	TU	UC	TU
1	0.1	0.003	0.4	0.01	0.7	0.02	1.9	0.05	3.8	0.10
2	1.6	0.05	5.4	0.15	7.9	0.20	12.9	0.35	19.1	0.55
3	6.9	0.20	16	0.45	20	0.55	29.4	0.80	39.6	1.10
4	15	0.40	30	0.85	37	1.05	49	1.35	63	1.75
5	27	0.75	46	1.30	56	1.55	71	1.95	88	2.45
6	40	1.10	64	1.80	76	2.10	94	2.60	113	3.15
7	55	1.55	84	2.35	97	2.70	118	3.25	140	3.90
8	71	1.95	105	2.90	119	3.30	143	3.95	168	4.65
9	88	2.45	126	3.50	142	3.95	169	4.70	195	5.40
10	107	2.95	149	4.15	166	4.60	195	5.40	224	6.20
11	126	3.50	172	4.80	191	5.30	222	6.15	253	7.05
12	145	4.05	195	5.40	216	6.00	249	6.90	282	7.85
13	166	4.60	220	6.10	241	6.70	277	7.70	311	8.65
14	187	5.20	244	6.80	267	7.40	305	8.45	341	9.45
15	208	5.80	269	7.45	293	8.15	333	9.25	370	10.30
16	231	6.40	294	8.15	320	8.90	362	10.05	401	11.15
17	253	7.05	320	8.90	347	9.65	390	10.85	431	11.95
18	276	7.65	346	9.60	374	10.40	419	11.65	462	12.85
19	299	8.30	373	10.35	401	11.15	448	12.45	492	13.65
20	323	8.95	399	11.10	429	11.90	477	13.25	523	14.55
21	346	9.60	426	11.85	458	12.70	507	14.10	554	15.40
22	370	10.30	453	12.60	486	13.50	536	14.90	585	16.25
23	395	10.95	480	13.35	514	14.30	566	15.70	616	17.10
24	419	11.65	507	14.10	542	15.05	596	16.55	647	17.95
25	444	12.35	535	14.85	572	15.90	626	17.40	678	18.35
26	469	13.05	562	15.60	599	16.65	656	18.20	710	19.70

Table 4.2 ***(Continued)***

Trunks	Grade of Service 1 in 1000		Grade of Service 1 in 100		Grade of Service 1 in 50		Grade of Service 1 in 20		Grade of Service 1 in 10	
	UC	TU	UC	TU	UC	TU	UC	TU	UC	TU
27	495	13.75	590	16.40	627	17.40	686	19.05	741	20.60
28	520	14.45	618	17.15	656	18.20	717	19.90	773	21.45
29	545	15.15	647	17.95	685	19.05	747	20.75	805	22.35
30	571	15.85	675	18.75	715	19.85	778	21.60	836	23.20
31	597	16.60	703	19.55	744	20.65	809	22.45	868	24.10
32	624	17.35	732	20.35	773	21.45	840	23.35	900	25.00
33	650	18.05	760	21.10	803	22.30	871	24.20	932	25.90
34	676	18.80	789	21.90	832	23.10	902	25.05	964	26.80
35	703	19.55	818	22.70	862	23.95	933	25.90	996	27.65
36	729	20.25	847	23.55	892	24.80	964	26.80	1028	28.55
37	756	21.00	876	24.35	922	25.60	995	27.65	1060	29.45
38	783	21.75	905	25.15	951	26.40	1026	28.50	1092	30.35
39	810	22.50	935	25.95	982	27.30	1057	29.35	1125	31.25
40	837	23.25	964	26.80	1012	28.10	1088	30.20	1157	32.14
41	865	24.05	993	27.60	1042	28.95	1120	31.10	1190	33.05
42	892	24.80	1023	28.40	1072	29.80	1151	31.95	1222	33.95
43	919	25.55	1052	29.20	1103	30.65	1183	32.85	1255	34.85
44	947	26.30	1082	30.05	1133	31.45	1214	33.70	1287	35.75
45	975	27.10	1112	30.90	1164	32.35	1246	34.60	1320	36.65
46	1003	27.85	1142	31.70	1194	33.15	1277	35.45	1352	37.55
47	1030	28.60	1171	32.55	1225	34.05	1309	36.35	1385	38.45
48	1058	29.40	1201	33.35	1255	34.85	1340	37.20	1417	39.35
49	1086	30.15	1231	34.20	1286	35.70	1372	38.10	1450	40.30
50	1115	30.95	1261	35.05	1317	36.60	1403	38.95	1482	41.15

erlang load, we would attain 100% efficiency. All trunks would be busy with calls all the time or at least for the entire BH. This would not even allow time for call set-up (i.e., making the connection) or for switch processing time. In practice, if we sized our trunks, trunk routes, or switches this way, there would be many unhappy customers.

On the other hand, we do, indeed, want to dimension our routes (and switches) to have a high efficiency and still keep our customers relatively happy. The goal of our previous exercises in traffic engineering was just that. The grade of service is one measure of subscriber satisfaction. As an example, let us assume that between cities X and Y there were 47 trunks on the interconnecting telephone route. The tariffs, from which the telephone company derives revenue, are a function of the erlangs of carried traffic. Suppose we allow $1.00 per erlang-hour. The very upper limit of service on the route is 47 erlangs and the telephone company would earn $47 for the busy hour (much less for all other hours) for that trunk route and the portion of the switches and local plant involved with these calls. As we well know, many of the telephone company's subscribers would be unhappy because they would have to wait excessively to get calls through from X to Y. How, then, do we optimize a trunk route (or serving circuits) and keep the customers as satisfied as possible with service?

Remember from Table 4.1, with an excellent grade of service of 0.001, that we relate grade of service to subscriber satisfaction (one element of quality of service) and that 47 trunks could carry 30.07 erlangs during the busy hour. Assuming the route did carry 30.07 erlangs, let's say at $1.00 per erlang, it would earn $30.07 for that hour. From a revenue viewpoint that would be the best hour of the day. If the grade of service were reduced to 0.01, 47 trunks would bring in $35.21 (i.e., 35.21 erlangs) for the busy hour. Note the improvement in revenue at the cost of reducing grade of service.

Here we are relating efficiency on trunk utilization. Trunks not carrying traffic do not bring in revenue. If we are only using some trunks during the busy hour only minutes a day to cover BH traffic peaks, the remainder of the day they are not used. That is highly inefficient. As we reduce the grade of service, the trunk utilization factor improves. For instance, 47 trunks will only carry 30.07 erlangs with a grade of service of 1 in 1000 (0.001), whereas if we reduce the grade of service to 1 in 20 (0.05), we carry 41.54 erlangs (see Table 4.1). Efficiency has improved notably. Quality of service, as a result, has decreased markedly (Ref. 1).

4.2.4.1. *Alternative Routing.* One method to improve efficiency is to use alternative routing (called *alternate routing* in North America). Suppose we have three serving areas, X, Y, and Z, served by three switches (exchanges), X, Y, and Z, as illustrated in Figure 4.4. Let the grade of service be 0.005 (1 in 200 in Table 4.1). We find that

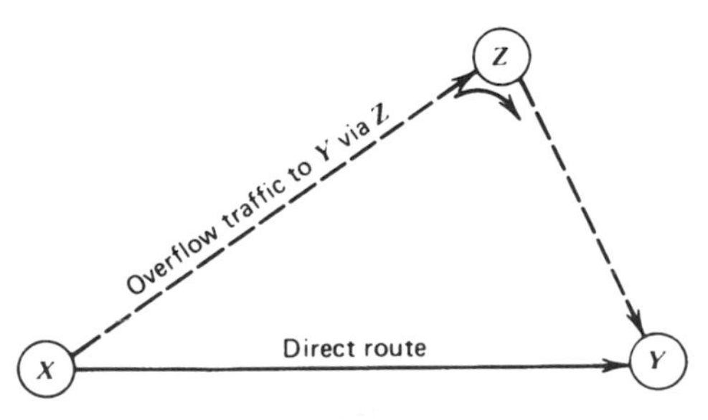

Figure 4.4 Simplified diagram of the alternative (alternate) routing concept. (Solid line represents the direct route, dashed lines represent the alternative route carrying the overflow traffic from X to Y.)

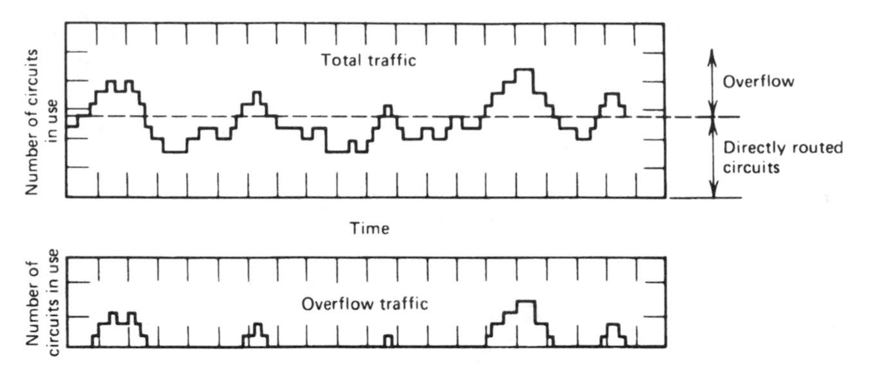

Figure 4.5 Traffic peakedness: The peaks are carried on alternative routes.

it would require 48 trunks to carry 34.25 erlangs of traffic during the BH to meet that grade of service between X and Y. Suppose we reduce the number of trunks between X and Y, still keeping the BH traffic intensity at 34.25 erlangs. We would thereby increase efficiency on the X-Y route at the cost of reducing grade of service. With a modification of the switch at X, we could route traffic bound for Y that met congestion on the X-Y route via switch Z. Then Z would route that traffic on the Z-Y link. Essentially this is alternative routing in its simplest form. Congestion would probably only occur during very short peaking periods in the BH, and chances are that these peaks would not occur simultaneously with peaks of traffic intensity on the Z-Y route. Furthermore, the incremental load on the X-Z-Y route would be very small. The concept of traffic peakedness that would overflow onto the secondary (X-Z-Y) is shown in Figure 4.5.

4.2.4.2. Efficiency versus Circuit Group Size. In the present context a *circuit group* refers to a group of circuits performing a specific function. For instance, all the trunks (circuits) routed from X to Y in Figure 4.4 make up a circuit group irrespective of size. This circuit group should not be confused with the "group" used in transmission-engineering of carrier systems.[6]

If we assume full loading, we find that efficiency improves with circuit group size. From Table 4.1, given a grade of service of 1 in 100, 5 erlangs of traffic require a group with 11 trunks, more than 2 : 1 ratio of trunks to erlangs, and 20 erlangs require 30 trunks, a 3 : 2 ratio. If we extend this to 100 erlangs, 120 trunks are required, a 6 : 5 ratio. Figure 4.6 shows how efficiency improves with group size.

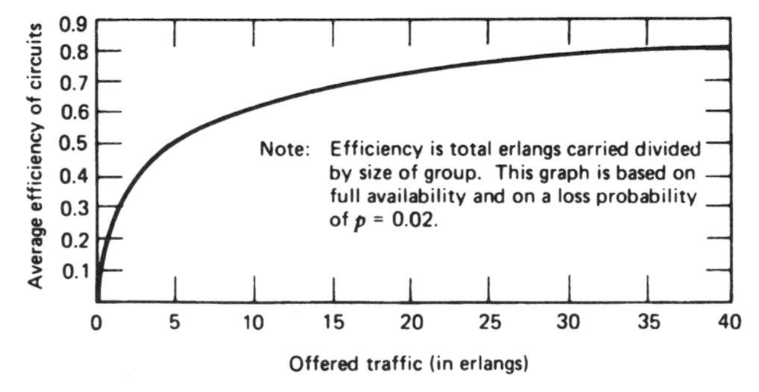

Figure 4.6 Group efficiency increases with size.

[6]*Carrier systems* are frequency division multiplex systems introduced in Section 4.5.

4.2.5 Quantifying Data Traffic

Data traffic usually consists of short, bursty transactions from a few milliseconds duration to several seconds, depending on the data transmission rate (i.e., the number of bits per second). This is particularly true on local area networks (LANs). As the data rate slows down, such as we might find on a wide area network (WAN), transaction time increases, possibly to a minute or so. For these reasons, it is dangerous to apply speech telephony traffic theory and practice to the data environment.

There is an exception here, that is, when a data protocol specifies a permanent virtual circuit (PVC). This is a circuit that is set up in advance for one or several data transactions. One group of traffic engineers has proposed the milli-erlang for LAN and PVC applications. We think this idea bears merit.

4.3 INTRODUCTION TO SWITCHING

In this section our concern is telephone switching, the switching of voice channels. We will deal with some switching concepts and with several specifics. Switching was defined in Section 4.1 in contraposition with transmission.

Actual connectivity is carried out by the switching function. A connectivity may involve more than one switch. As we pointed out in Chapter 1, there are local switches, tandem switches, and *transit* switches. A transit switch is simply a tandem switch that operates in the long-distance or "toll" service.

A local switch has an area of responsibility. We call this its *serving area*. All subscriber loops in a serving area connect to that switch responsible for the area. Many calls in a local area traverse no more than one switch. These are calls to neighbors. Other calls, destined for subscribers outside that serving area, may traverse a tandem switch to another local serving switch if there is no direct route available. If there is a direct route, the tandem is eliminated for that *traffic relation*. It is unnecessary.

Let us define a *traffic relation* as a connectivity between exchange A and B. The routing on calls for that traffic relation is undetermined. Another connotation for the term traffic relation implies that not only would there be a connectivity capability, but also the BH traffic expected on that connectivity.

To carry out these functions, a switch had to have some sort of intelligence. In a manually operated exchange, the intelligence was human, the telephone operator. The operator was replaced by an automatic switch. Prior to the computer age, a switch's intelligence was "hard-wired" and its capabilities were somewhat limited. Today, all modern switches are computer based and have a wide selection of capabilities and services. Our interest here is in the routing of a call. A switch knows how to route a call through the dialed telephone number, as described in Section 1.3.2. There we showed that a basic telephone number consists of seven digits. The last four digits identify the subscriber; the first three digits identify the local serving exchange responsible for that subscriber. The three-digit exchange code is unique inside of an area code. In North America, an area code is a three-digit number identifying a specific geographical area. In many countries, if one wishes to dial a number which is in another area code, an access code is required. In the United States that access code is a 1.

4.3.1 Basic Switching Requirements

Conceptually, consider that a switch has inlets and outlets. Inlets serve incoming calls; outlets serve outgoing calls. A call from a calling subscriber enters an exchange through an inlet. It connects to a called subscriber through an outlet. There are three basic switching requirements:

1. An exchange (switch) must be able to connect any incoming call to one of a multitude of outgoing circuits.
2. It has the ability not only to establish and maintain (or *hold*) a physical connection between a caller and the called party for the duration of the call, but also to be able to disconnect (i.e., "clear") it after call termination.
3. It also has the ability to prevent new calls from intruding into circuits that are already in use. To avoid this, a new call must be diverted to another circuit that is free or it must be temporarily denied access, where the caller will hear a "busy back" (i.e., a tone cadence indicating that the line is busy) or an "all trunks busy" tone cadence signal or voice announcement (i.e., indicating congestion or blockage).

Let's differentiate between local and tandem/transit exchanges. A local exchange connects lines (subscriber loops) to other lines or to trunks. A tandem/transit exchange switches trunks. Local exchanges *concentrate* and *expand*. Tandem and transit exchanges do not.

4.3.2 Concentration and Expansion

Trunks are expensive assets. Ideally there should be one trunk available for every subscriber line (loop). Then there never would be a chance of blockage. Thus whenever a subscriber wished to connect to a distant subscriber, there would be a trunk facility available for that call. Our knowledge of telephone calling habits of subscribers tells us that, during the busy hour, on the order of 30% of subscriber lines will be required to connect to trunks for business customers and some 10% for residential customers. Of course, these values are rough estimates. We would have to apply the appropriate traffic formula based on a grade of service, as described in Section 4.2.1, for refined estimates.

Based on these arguments, a local exchange serving residential customers might have 10,000 lines, and only 1000 trunks would be required. This is concentration. Consider that those 1000 incoming trunks to that exchange must expand out to 10,000 subscribers. This is expansion and it provides all subscribers served by the switch with access to incoming trunks and local switching paths. The concentration/expansion concept of a local serving exchange is illustrated in the following diagram:

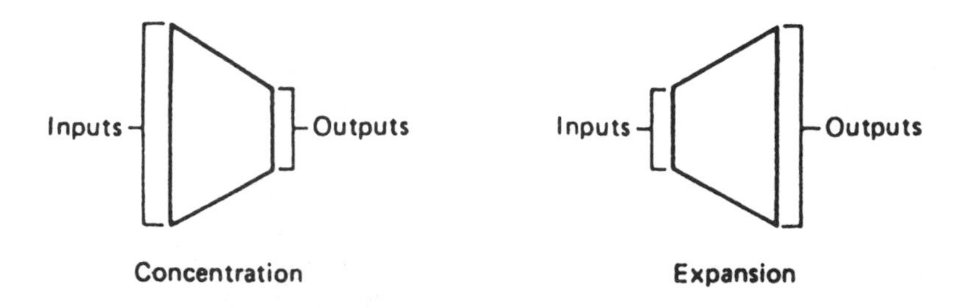

4.3.3 Essential Functions of a Local Switch

As mentioned previously, means are provided in a local switch to connect each subscriber line to any other in the same exchange. In addition, any incoming trunk must be able to connect to any subscriber line and any subscriber to any outgoing trunk.[7] These switching functions are remotely controlled by the calling subscriber, whether she/he is a local subscriber or long-distance subscriber. These remote instructions are transmitted to the switch (exchange) by "off-hook," "on-hook," and dial information.[8] There are eight basic functions that must be carried out by a conventional switch or exchange:

1. Interconnection;
2. Control;
3. Alerting;
4. Attending;
5. Information receiving;
6. Information transmitting;
7. Busy testing; and
8. Supervising.

Consider a typical manual switching center illustrated in Figure 4.7. Here the eight basic functions are carried out for each call. The important *interconnection* function is illustrated by the jacks appearing in front of the operator. There are subscriber-line jacks and jacks for incoming and outgoing trunks.[9] The connection is made by double-ended connecting cords, which can connect subscriber-to-subscriber or subscriber-to-trunk. The cords available are always less than half the number of jacks appearing on the board, because one interconnecting cord occupies two jacks, one on either end. Concentration takes place at this point on a manual exchange. Distribution is also carried out because any cord may be used to complete a connection to any of the terminating jacks. The operator is *alerted* by a lamp becoming lit when there is an incoming call requiring connection. This is the *attending-alerting* function. The operator then assumes the *control* function, determining an idle connecting cord and plugging it into the incoming jack. She/he then determines call destination, continuing her/his control function by plugging the cord into the terminating jack of the called subscriber or proper trunk to terminate her/his portion of control of the incoming call. Of course, before plugging into the terminating jack, she/he carries out a *busy test* function to determine that the called line or trunk is not busy. To alert the called subscriber that there is an incoming call, she/he uses the manual ring-down by connecting the called line to a ringing current source, as illustrated in Figure 4.7.[10]

Other signaling means are used for trunk signaling if the incoming call is destined for another exchange. On such a call the operator performs the information function orally or by dialing the call information to the next exchange in the routing.

The *supervision* function is performed by lamps to show when a call is completed

[7]The statement assumes *full availability*.

[8]*Off-hook* and *on-hook* are defined in Section 1.3.1.

[9]A jack is an electric receptacle. It is a connecting device, ordinarily employed in a fixed location, to which a wire or wires may be attached, and is arranged for the insertion of a plug.

[10]Ring-down is a method of signaling to alert an operator or a distant subscriber. In old-time telephone systems, a magneto was manually turned, generating an alternating current which would ring a bell at the other end. Today, special ringing generators are used.

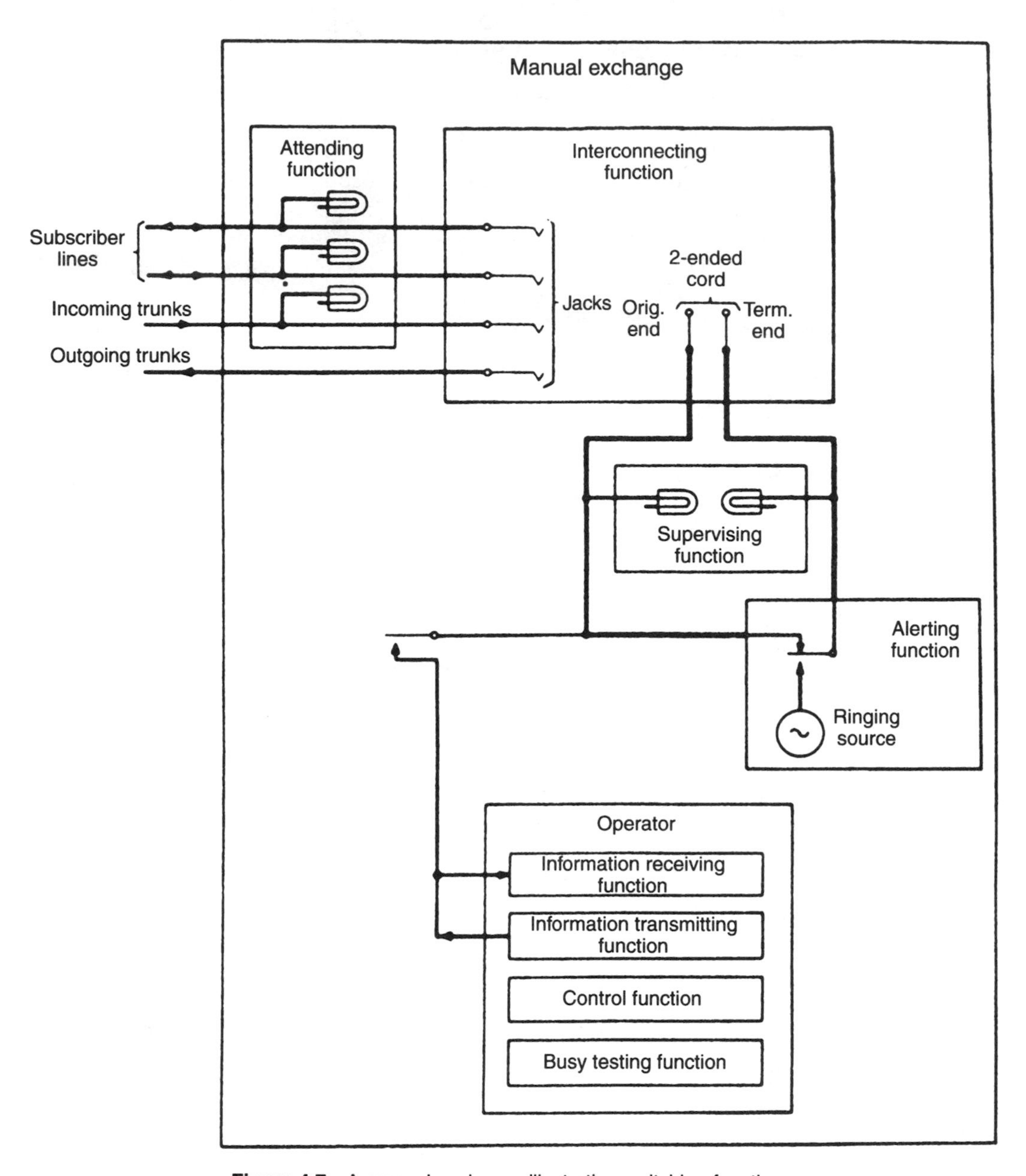

Figure 4.7 A manual exchange illustrating switching functions.

and the call is taken down (i.e., the patch cord can be removed). The operator conducts numerous control functions to set up a call, such as selecting a cord, plugging it into the originating jack of the calling line, connecting her/his headset to determine calling information, selecting (and busy-testing) the called subscriber jack, and then plugging the other end of the cord into the proper terminating jack and alerting the called subscriber by ring-down. Concentration is the ratio of the field of incoming jacks to cord positions. Expansion is the number of cord positions to outgoing (terminating) jacks. The terminating and originating jacks can be interchangeable. The called subscriber at one moment in time can become the calling subscriber at another moment in time. On

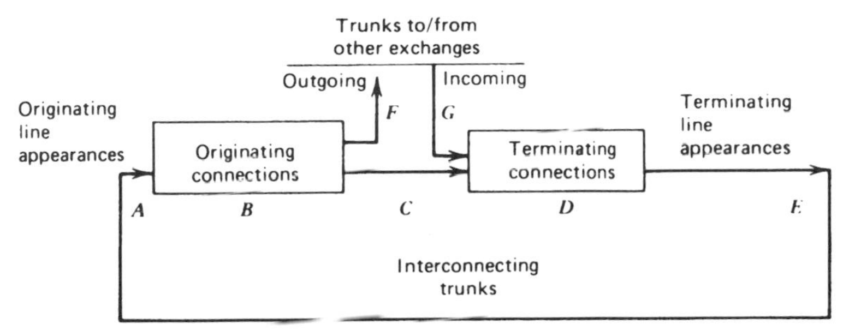

Figure 4.8 Originating and terminating line appearances.

the other hand, incoming and outgoing trunks may be separated. In this case they would be one-way circuits. If not separated, they would be both-way circuits, accepting both incoming and outgoing traffic.

4.3.4 Some Introductory Switching Concepts

All local telephone switches have, as a minimum, three functional elements: (1) concentration, (2) distribution, and (3) expansion. Concentration and expansion were discussed in Section 4.3.3. Viewing a switch another way, we can say that it has *originating line* appearances and *terminating line* appearances. These are illustrated in a simplified conceptual drawing in Figure 4.8, which shows three different call possibilities of a typical local exchange:

1. A call originated by a subscriber who is served by the exchange and bound for a subscriber who is served by the same exchange (route A-B-C-D-E);
2. A call originated by a subscriber who is served by the exchange and bound for a subscriber who is served by another exchange (route A-B-F); and
3. A call originated by a subscriber who is served by another exchange and bound for a subscriber served by the exchange in question (route G-D-E).

Call concentration takes place in B and call expansion at D. Figure 4.9 is simply a redrawing of Figure 4.8 to show the concept of distribution. The distribution stage in switching serves to connect by switching the concentration stage to the expansion stage.

4.3.5 Early Automatic Switching Systems

4.3.5.1. Objective. We summarize several earlier, space division switching systems because of the concepts involved. Once the reader grasps these concepts, the ideas and

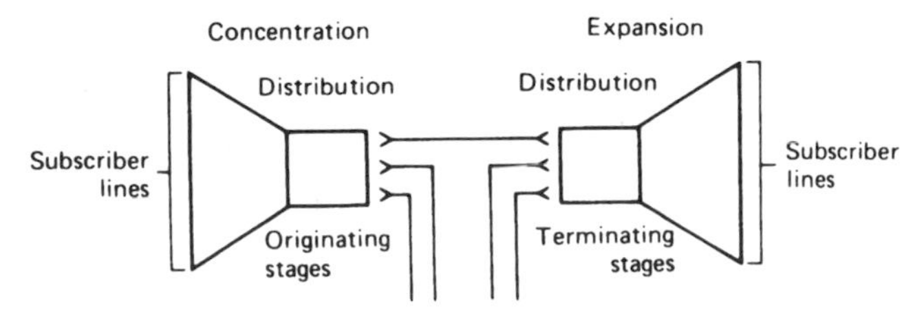

Figure 4.9 The concept of distribution.

notions of digital switching will be much easier to understand. First, the operation of the original step-by-step switch is described. This is followed by a discussion of the crossbar switch.

4.3.5.2. Step-by-Step Switch. The step-by-step (SXS) switch was widely used in the United States prior to 1950, when the crossbar switch tended to replace it. Its application was nearly universal in the United Kingdom, where it was called the *Strowger switch.*

The step-by-step switch has a curious history. Its inventor was Almon B. Strowger, an undertaker in Kansas City. Strowger suspected that he was losing business because the town's telephone operator was directing all requests for funeral services to a competitor, which some say was a boyfriend, and who others say was a relative. We do not know how talented Strowger was as a mortician, but he certainly goes down in history for his electromechanical talents for the invention of the automatic telephone switch. The first "step" switch was installed in Indiana in 1892. They were popular with independent telephone companies, but installation in AT&T's Bell System did not start until 1911.

The step-by-step switch is conveniently based on a stepping relay of 10 levels. In its simplest form, which uses direct progressive control, dial pulses from a subscriber's telephone activate the switch. For example, if a subscriber dials a 3, three pulses from the subscriber subset are transmitted to the switch. The switch then steps to level 3 in the first relay bank. The second relay bank is now connected, waiting for the second dialed digit. It accepts the second digit from the subscriber and steps to it equivalent position and connects to the third relay bank, and so on, for four or seven dialed digits. Assume a certain exchange only serves three-digit numbers. A dialed number happens to be 375 and will be stepped through three sets of banks of 10 steps each. This is conceptually illustrated in Figure 4.10.

4.3.5.3 Crossbar Switch. Crossbar switching dates back to 1938 and reached a peak of installed lines in 1983. Its life had been extended by using stored-program control

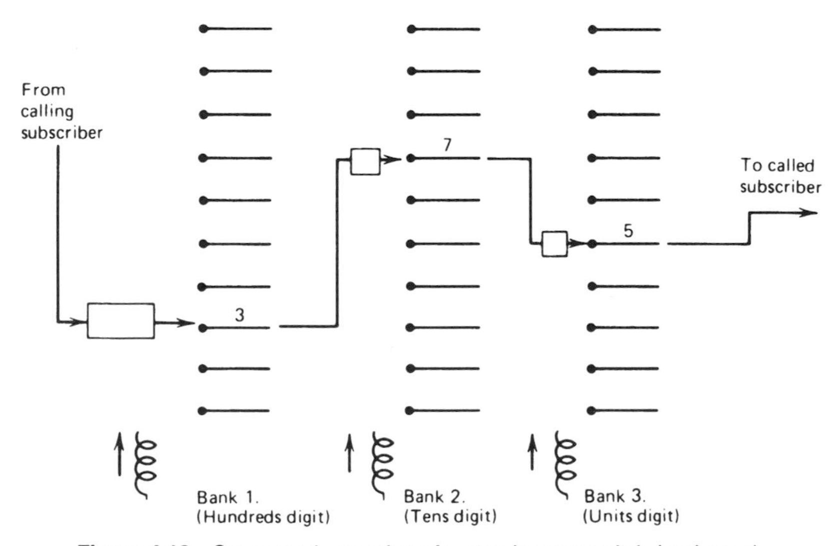

Figure 4.10 Conceptual operation of a step-by-step switch (exchange).

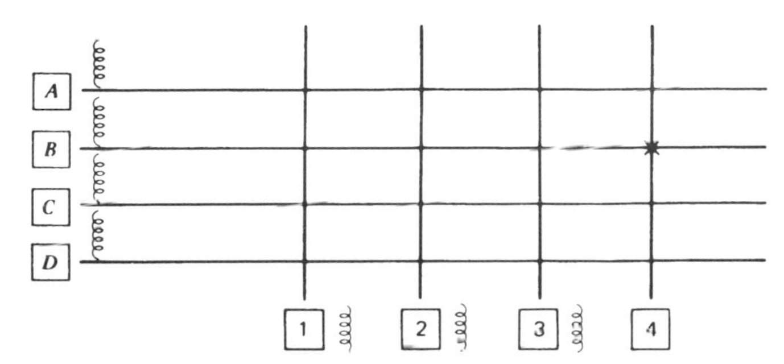

Figure 4.11 The crossbar concept.

(SPC) rather than hard-wire control in the more conventional crossbar configuration.[11] The crossbar is actually a matrix switch used to establish the speech path. An electrical contact is made by actuating a horizontal and vertical relay. Consider the switching mechanism illustrated in Figure 4.11. To make contact at point B_4 on the matrix, horizontal relay B and vertical relay 4 must close to establish the connection. Such closing is usually momentary, but sufficient to cause *latching*. Two forms of latching are found in crossbar practice: mechanical and electrical. The latch keeps the speech path connection until an "on-hook" condition occurs. Once the latching occurs, connection B_4 is "busied out," and the horizontal and vertical relays are freed-up to make other connections for other calls.[12]

4.3.6 Common Control (Hard-Wired)

First, we must distinguish *common control* from *direct progressive control* described in Section 4.3.5.2. With direct progressive control a subscriber dialed a digit, and the first relay bank stepped to the dialed digit; the subscriber dialed a second digit, and the second stepping relay bank actuated, stepping to that digit level, and so on through the entire dialed number. With common control, on the other hand, the dialed number is first stored in a register.[13] These digits are then analyzed and acted upon by a *marker*, which is a hard-wired processor. Once the call setup is complete, the register and marker are free to handle other call setups. The marker was specifically developed for the crossbar switch. Such marker systems are most applicable to specialized crossbar switching matrices of crossbar switches. SPC is a direct descendent of the crossbar common control system and is described in the following text.

4.3.7 Stored Program Control

4.3.7.1 Introduction. SPC is a broad term designating switches where common control is carried out entirely by computer. In some exchanges, this involves a large, powerful computer. In others, two or more minicomputers may carry out the SPC function. Still, with other switches, the basic switch functions are controlled by distributed micro-

[11]SPC, stored program control, simply means that the switch or exchange is computer controlled. Of course, all modern digital switches are computer controlled.

[12]"Busy-out" means that a line or connection is taken out of the pool because it is busy; it is being used, and is not available for others to utilize.

[13]A *register* is a device that receives and stores signals. In this particular case, it receives and stores dialed digits.

processors. Software may be hard-wired on one hand or programmable on the other. There is a natural marriage between a binary digital computer and the switch control functions. In most cases these also work in the binary digital domain. The crossbar markers and registers are typical examples.

The conventional crossbar marker requires about half a second to service a call. Up to 40 expensive markers are required on a large exchange. Strapping points on the marker are available to laboriously reconfigure the exchange for subscriber change, new subscribers, changes in traffic patterns, reconfiguration of existing trunks or their interface, and so on.

Replacing register markers with programmable logic—a computer, if you will—permits one device to carry out the work of 40. A simple input sequence on the keyboard of the computer workstation replaces strapping procedures. System faults are displayed as they occur, and circuit status may be indicated on the screen periodically. Due to the high speed of the computer, postdial delay is reduced. SPC exchanges permit numerous new service offerings, such as conference calls, abbreviated dialing, "camp-on-busy," call forwarding, voice mail, and call waiting.

4.3.7.2 Basic SPC Functions.

There are four basic functional elements of an SPC switching system:

1. Switching matrix;
2. Call store (memory);
3. Program store (memory); and
4. Central processor (computer).

The earlier switching matrices consisted of electromechanical crosspoints, such as a crossbar matrix, reed, correed, or ferreed cross points. Later switching matrices employed solid-state crosspoints.

The *call store* is often referred to as the "scratch-pad" memory. This is temporary storage of incoming call information ready for use, on command from the central processor. It also contains availability and status information of lines, trunks, and service circuits and on internal switch-circuit conditions. Circuit status information is brought to the memory by a method of scanning. All speech circuits are scanned for a busy/idle condition.

The *program store* provides basic instructions to the controller (central processor). In many installations translation information is held in this store (memory), such as directory number (DN) to equivalent number (EN) translation and trunk-signaling information. A simplified functional diagram of a basic SPC system is shown in Figure 4.12.

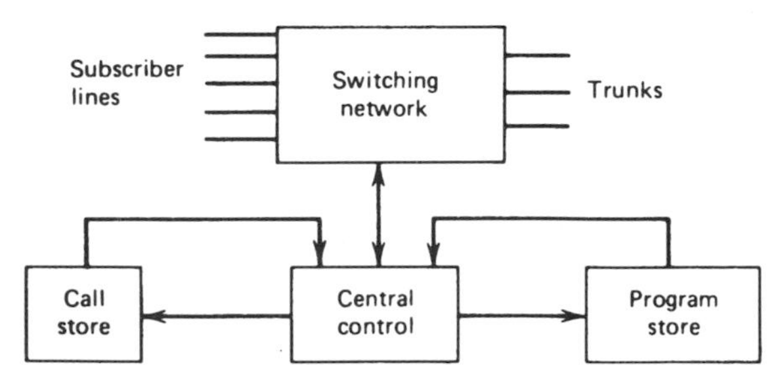

Figure 4.12 A simplified functional diagram of an SPC exchange.

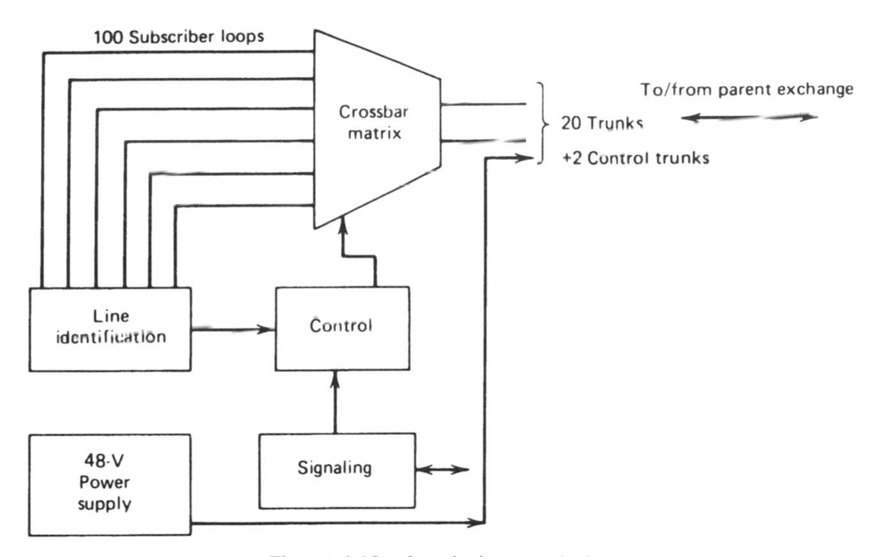

Figure 4.13 A typical concentrator.

4.3.8 Concentrators and Remote Switching

In Chapter 5 we discuss the design of a subscriber loop. There we will find that there are very definite length limitations on subscriber loops. As we look further into subscriber loop design, methods of extending loops still further are described. One way to extend such loops is with a remote concentrator or switch.

The simplest form of extending a switch is to use a concentrator some distance from the switch (exchange). *Concentrators* or *line concentrators* consolidate subscriber loops, are remotely operated, and are a part of the concentration and expansion portion of a switch placed at a remote location. The concentrator may be based on electromechanical facilities or solid-state crosspoints for the concentration matrix. For instance a 10 : 1 concentrator might serve 100 subscriber loops and deliver 10 trunks to the "mother" exchange. A concentrator does no switching whatsoever. All switching is carried out at the controlling or "mother" exchange. A typical line concentrator is illustrated in Figure 4.13, where 100 subscriber loops are consolidated to 20 trunks plus 2 trunks for control from the nearby "mother" exchange. Of course, the ratio of loops to trunks is a key issue, and is based on calling habits and whether the subscribers are predominantly business or residential.

A remote switch, sometimes called a satellite or satellite exchange, originates and terminates calls from the parent exchange. It differs from a concentrator in that local calls (i.e., calls originating and terminating inside the same satellite serving area) are served by the remote switch and do not have to traverse the parent exchange as remote concentrator calls do. A block of telephone numbers is assigned to the satellite serving area and is usually part of the basic number block assigned to the parent exchange. Because of the numbering arrangement, a satellite exchange can discriminate between local calls and calls to be handled by the parent exchange. A satellite exchange can be regarded as a component of the parent exchange that has been dislocated and moved to a distant site. The use of remote switching is very common in rural areas, and the distance a remote switch is from the parent exchange can be as much as 100 miles (160 km). Satellite exchanges range in size from 300 to 2000 lines. Concentrators are cost effective for 300 or fewer subscribers. However, AT&T's SLC-96 can serve 1000 subscribers or more.

4.4 SOME ESSENTIAL CONCEPTS IN TRANSMISSION

4.4.1 Introduction

In this section we discuss two-wire and four-wire transmission and two impairments commonly caused by two-wire-to-four-wire conversion equipment. These impairments are *echo* and *singing*. The second part of this section is an introduction to multiplexing. *Multiplexing* allows two or more communication channels to share the same transmission bearer facility.

4.4.2 Two-Wire and Four-Wire Transmission

4.4.2.1 Two-Wire Transmission. A telephone conversation inherently requires transmission in both directions. When both directions are carried on the same pair of wires, it is called *two-wire transmission*. The telephones in our homes and offices are connected to a local switching center (exchange) by means of two-wire circuits. A more proper definition for transmitting and switching purposes is that when oppositely directed portions of a single telephone conversation occur over the same electrical transmission channel or path, we call this *two-wire operation*.

4.4.2.2 Four-Wire Transmission. Carrier and radio systems require that oppositely directed portions of a single conversation occur over separate transmission channels or paths (or use mutually exclusive time periods). Thus we have two wires for the transmit path and two wires for the receive path, or a total of four wires, for a full-duplex (two-way) telephone conversation. For almost all operational telephone systems, the end instrument (i.e., the telephone subset) is connected to its intervening network on a two-wire basis.

Nearly all long-distance (toll) telephone connections traverse four-wire links. From the near-end user the connection to the long-distance network is two-wire or via a two-wire link. Likewise, the far-end user is also connected to the long-distance (toll) network via a two-wire link. Such a long-distance connection is shown in Figure 4.14. Schematically, the four-wire interconnection is shown as if it were a single-channel wire-line with amplifiers. However, it would more likely be a multichannel multiplexed configuration

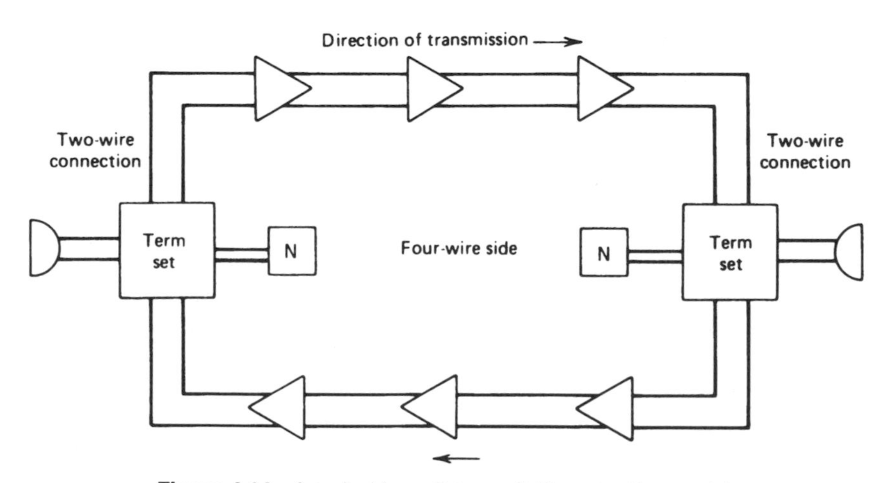

Figure 4.14 A typical long-distance (toll) connection model.

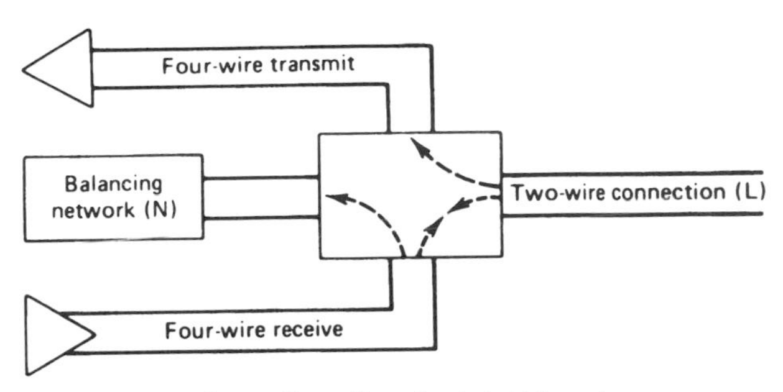

Figure 4.15 Operation of a hybrid transformer.

on wire/fiber optic cable or over radio. Nevertheless, the amplifiers in Figure 4.14 serve to convey the ideas this section considers. As illustrated in Figure 4.14, conversion from two-wire to four-wire operation is carried out by a *hybrid*, which is a four-port, four-winding transformer.

4.4.2.3 *Operation of a Hybrid.* A hybrid, in terms of telephony (at voice frequency), is a transformer with four separate windings. Based on a simplified description, a hybrid may be viewed as a power splitter with four sets of wire-pair connections. A functional block diagram of a hybrid device is shown in Figure 4.15. Two of the wire-pair connections belong to the four-wire path, which consists of a transmit pair and a receive pair. The third pair is a connection to the two-wire link, which is eventually connected to the subscriber subset via one or more switches. The last pair of the four connects the hybrid to a resistance–capacitance balancing network, which electrically balances the hybrid with the two-wire connection to the subscriber subset over the frequency range of the balancing network. *Balancing*, in this context, means matching impedances, that is, the impedance of the two-wire side to the hybrid two-wire port.

Signal energy entering from the two-wire subset connection divides equally. Half of it dissipates (as heat) in the impedance of the four-wire side receive path and the other half goes to the four-wire side transmit path, as illustrated in Figure 4.15. Here the *ideal* situation is that no energy is to be dissipated by the balancing network (i.e., there is a perfect balance or impedance match). The balancing network is supposed to display the characteristic impedance of the two-wire line (subscriber connection) to the hybrid.[14] Signal energy entering from the four-wire side receive path is also split in half in the ideal situation where there is a perfect balance (i.e., a perfect match). Half of the energy is dissipated by the balancing network (N) and half at the two-wire port (L) (see Figure 4.15.)

The reader should note that in the description of a hybrid, in every case, ideally half of the signal energy entering the hybrid is used to advantage and half is dissipated or wasted. Also keep in mind that any passive device inserted in a circuit, such as a hybrid, has an insertion loss. As a rule of thumb, we say that the insertion loss of a hybrid is 0.5 dB. Thus there are two losses here that the reader must not lose sight of:

[14]*Characteristic impedance* is the impedance that the line or port on a device is supposed to display. For most subscriber loops it is 900 Ω with a 2.16 μF capacitor in series at 1000 Hz for 26-gauge wire pair or 600 Ω resistive. The notation for characteristics impedance is Z_o.

Hybrid insertion loss	0.5 dB	
Hybrid dissipation loss	3.0 dB	(half of the power)
	3.5 dB	(total)

As far as this section is concerned, any signal passing through a hybrid suffers a 3.5-dB loss. This is a good design number for gross engineering practice. However, some hybrids used on short subscriber connections purposely have higher losses, as do special resistance-type hybrids.

In Figure 4.15, consider the balancing network (N) and the two-wire side of the hybrid (L). In all probability (L), the two-wire side, will connect to a subscriber through at least one switch. Thus the two-wire port on the hybrid could look into at least 10,000 possible subscriber connections, some short loops, some long loops, and some loops in poor condition. Because of the fixed conditions on the four-wire side, we can generally depend on holding a good impedance match. Our concern under these conditions is the impedance match on the two-wire side, that is, the impedance match between the compromise network (N) and the two-wire side (L). Here the impedance can have high variability from one subscriber loop to another.

We measure the capability of impedance match by *return loss*. In this particular case we call it *balance return loss*:

$$\text{Balance return loss}_{\text{dB}} = 20 \log_{10} \frac{Z_L + Z_N}{Z_L - Z_N}.$$

Let us say, for argument's sake, that we have a perfect match. In other words, the impedance of the two-wire subscriber loop side (L) on this particular call was exactly 900 Ω and the balancing network (N) was 900 Ω. Substitute these numbers in the preceding formula above and we get

$$\text{Balance return loss}_{\text{dB}} = 20 \log \frac{900 + 900}{900 - 900}.$$

Examine the denominator. It is zero. Any number divided by zero is infinity. Thus we have an infinitely high return loss. And this happens when we have a perfect match, an ideal condition. Of course it is seldom realized in real life. In real life we find that the balance return loss for a large population of hybrids connected in service and serving a large population of two-wire users has a median more on the order of 11 dB with a standard deviation of 3 dB (Ref. 4). This is valid for North America. For some other areas of the world, balance return loss median may be lower with a larger standard deviation.

When the return loss becomes low (i.e., there is a poor impedance match), there is a reflection of the speech signal. That is, speech energy from the talker at her/his distant hybrid leaks across from the four-wire receive to the four-wire transmit side (see Figure 4.15). This signal energy is heard by the talker. It is delayed due to the propagation time. This is echo, which can be a major impairment depending on its intensity and amount of time it is delayed. It can also be very disruptive on a data circuit.

We define the cause of echo as any impedance mismatch in the circuit. It is most commonly caused by this mismatch that occurs at the hybrid. Echo that is excessive becomes singing. *Singing* is caused by high positive feedback on the intervening amplifiers (Figure 4.14). Singing on the analog network could take the network down by

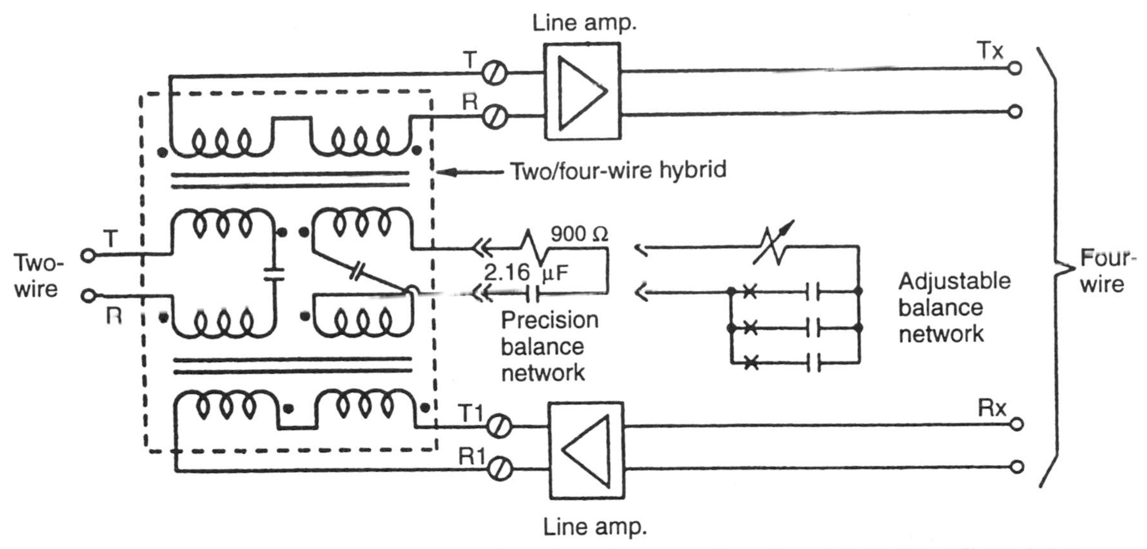

Figure 4.16 Schematic diagram of two-wire to four-wire conversion using a hybrid. (From Figure 5-9, p. 104, Ref. 5. Reprinted with permission of the IEEE Press.)

overloading multiplex equipment. The possibility of singing on the digital network is very low. The control of echo and singing is discussed in Chapter 8. Figure 4.16 is a schematic diagram of a hybrid circuit.

4.5 INTRODUCTION TO MULTIPLEXING

4.5.1 Definition

Multiplexing is used for the transmission of a plurality of information channels over a single transmission medium. An information channel may be a telephone voice channel, data channel, or a channel carrying image information. Our discussion as follows will concentrate on a telephone channel. A *telephone channel* is a channel optimized for carrying voice traffic, in this case the voice of a single telephone user. We will define it as an analog channel as occupying the band of frequencies between 300 Hz and 3400 Hz (CCITT definition). Before launching into our discussion, keep in mind that all multiplex equipment is four-wire equipment. If we look at one side of a circuit, there will be a multiplexer used for transmission and a demultiplexer used for reception.

The number of channels that can be multiplexed on a particular circuit depends on the bandwidth of the transmission medium involved. We might transmit 24 or 48 or 96 channels on a wire pair, depending on the characteristics of that wire pair. Coaxial cable can support many thousands of voice channels; line-of-sight microwave radio is capable of carrying from several hundred to several thousand voice channels. A single fiber optic thread can support literally tens of thousands of channels. A communication satellite transponder can carry from 700 or possibly up to 2000 such voice channels, depending on the transponder's bandwidth.

There are essentially two generic methods of multiplexing information channels:

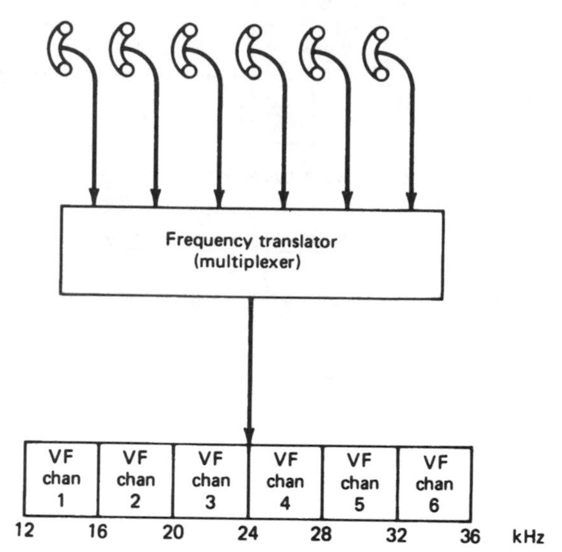

Figure 4.17 The frequency division multiplex concept illustrated.

1. In the frequency domain; we call this frequency division multiplex (FDM); and
2. In the time domain, which we call time division multiplex (TDM).

The concepts of frequency division multiplexing are discussed in this chapter. Time division multiplexing (pulse code modulation) is covered in Chapter 6.

4.5.2 Frequency Division Multiplex

4.5.2.1 Introduction. With FDM the available channel bandwidth is divided into a number of nonoverlapping frequency slots. Each frequency slot or bandwidth segment carries a single information-bearing signal such as a voice channel. We can consider an FDM multiplexer as a frequency translator. At the opposite end of the circuit, a demultiplexer filters and translates the frequency slots back into the original information-bearing channels. In the case of a telephone channel, a frequency slot is conveniently 4 kHz wide, sufficient to accommodate the standard 300 Hz to 3400 kHz voice channel. Figure 4.17 illustrates the basic concept of frequency division multiplex.

In practice, the *frequency translator* (multiplexer) uses single sideband modulation of radio frequency (RF) carriers. A different RF carrier is used for each channel to be multiplexed. This technique is based on mixing or heterodyning the signal to be multiplexed, typically a voice channel, with an RF carrier.

An RF carrier is an unmodulated radio frequency signal of some specified frequency. In theory, because it is not modulated, it has an infinitely small bandwidth. In practice, of course, it does have some measurable bandwidth, although very narrow. Such a carrier derives from a simple frequency source such as an oscillator or a more complex source such as a synthesizer, which can generate a stable output in a range of frequencies.

A simplified block diagram of an FDM link is shown in Figure 4.18.

4.5.2.2 Mixing. The heterodyning or mixing of signals of frequencies A and B is shown as follows. What frequencies may be found at the output of the mixer?

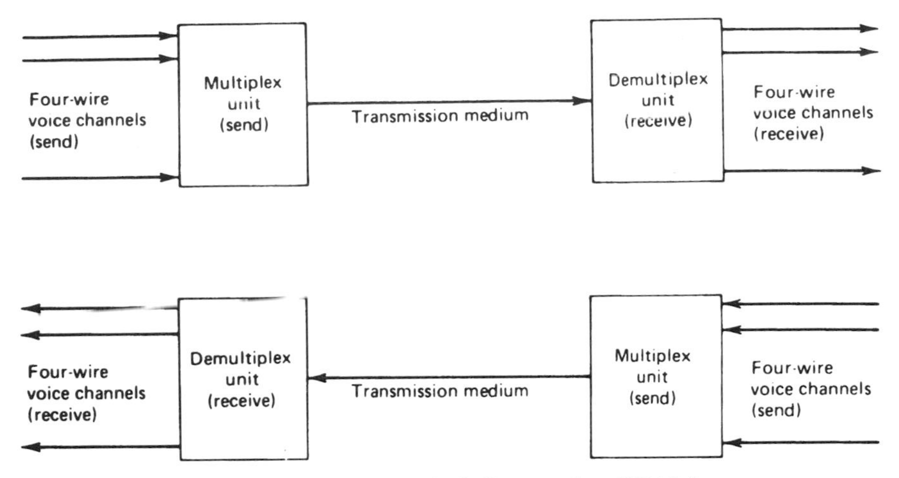

Figure 4.18 Simplified block diagram of an FDM link.

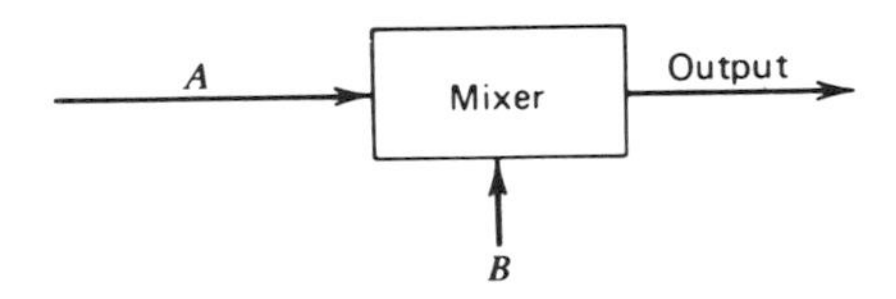

Both the original signals will be present, as well as the signals representing their sum and their difference in the frequency domain. Thus at the output of the illustrated mixer we will have present the signals of frequency A, B, $A + B$, and $A - B$. Such a mixing process is repeated many times in FDM equipment.

Let us now look at the boundaries of the nominal 4-kHz voice channel. These are 300 Hz and 3400 Hz. Let us further consider these frequencies as simple tones of 300 Hz and 3400 Hz. Now consider the following mixer and examine the possibilities at its output:

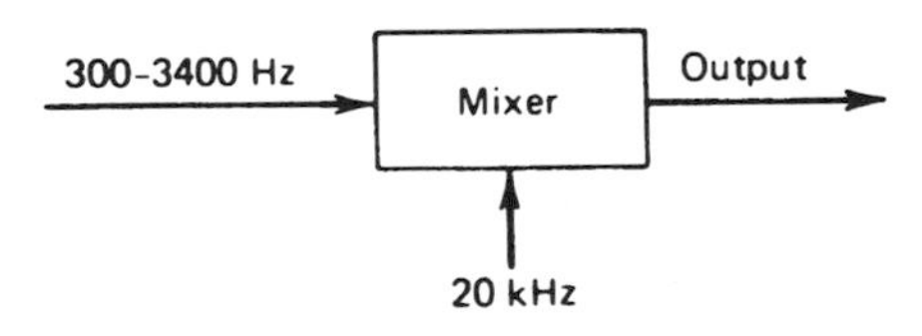

First, the output may be the sum or

$$\begin{array}{r} 20{,}000\ \text{Hz} \\ +\quad 300\ \text{Hz} \\ \hline 20{,}300\ \text{Hz} \end{array} \qquad \begin{array}{r} 20{,}000\ \text{Hz} \\ +\ 3{,}400\ \text{Hz} \\ \hline 23{,}400\ \text{Hz} \end{array}$$

A simple low-pass filter could filter out all frequencies below 20,300 Hz.

Now imagine that instead of two frequencies, we have a continuous spectrum of frequencies between 300 Hz and 3400 Hz (i.e., we have the voice channel). We represent the spectrum as a triangle:

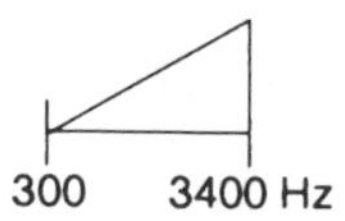

As a result of the mixing process (translation) we have another triangle, as follows:

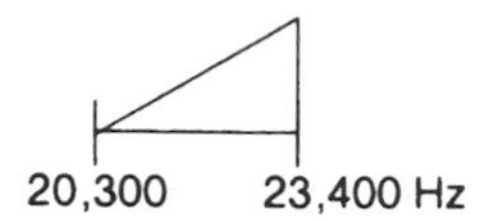

When we take the sum, as we did previously, and filter out all other frequencies, we say we have selected the upper sideband. Therefore we have a triangle facing to the right, and we call this an upright or erect sideband. We can also take the difference, such that

$$\begin{array}{r} 20,000 \text{ Hz} \\ -\quad 300 \text{ Hz} \\ \hline 19,700 \text{ Hz} \end{array} \qquad \begin{array}{r} 20,000 \text{ Hz} \\ -\ 3,400 \text{ Hz} \\ \hline 16,600 \text{ Hz} \end{array}$$

and we see that in the translation (mixing process) we have had an inversion of frequencies. The higher frequencies of the voice channel become the lower frequencies of the translated spectrum, and the lower frequencies of the voice channel become the higher when the difference is taken. We represent this by a right triangle facing the other direction:

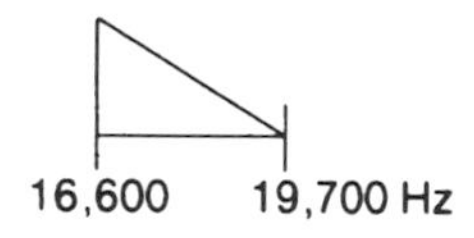

This is called an inverted sideband. To review, when we take the sum, we get an erect sideband. When we take the difference, frequencies invert and we have an inverted sideband represented by a triangle facing left.

Again, this modulation technique is called single-sideband suppressed carrier (SSBSC). It is a type of amplitude modulation (AM). With conventional AM, the modulation produces two sidebands, an upper sideband and a lower sideband, symmetrical on either side of the carrier. Each sideband carries the information signal. If we tune to 870 kHz on the AM dial, 870 kHz is the frequency of the RF carrier, and its sidebands fall on either side, where each sideband is about 7.5 kHz wide. Thus a radio station on the AM dial requires about 15 kHz of spectrum bandwidth.

In our case, there are also two sidebands extending about 3.4 kHz either side of the carrier frequency. Of course, the carrier frequency is the local oscillator frequency, which is suppressed at the output, as is the upper sideband. All that remains is the lower sideband, which contains the voice channel information.

4.5.2.3 CCITT Modulation Plan

4.5.2.3.1 Introduction. A modulation plan sets forth the development of a band of frequencies called the *line frequency* (i.e., ready for transmission on the line or transmission medium). The modulation plan usually is a diagram showing the necessary mixing, local oscillator mixing frequencies, and the sidebands selected by means of the triangles described previously in a step-by-step process from voice channel input to line frequency output. The CCITT has recommended a standardized modulation plan with a common terminology. This allows large telephone networks, on both national and multinational systems, to interconnect. In the following paragraphs the reader is advised to be careful with terminology.

4.5.2.3.2 Formation of the Standard CCITT Group. The standard *group* as defined by the CCITT, occupies the frequency band of 60 kHz to 108 kHz and contains 12 voice channels. Each voice channel is the nominal 4-kHz channel occupying the 300-Hz to 3400-Hz spectrum. The group is formed by mixing each of the 12 voice channels with a particular carrier frequency associated with each channel. Lower sidebands are then selected, and the carrier frequencies and the upper sidebands are suppressed. Figure 4.19 shows the preferred approach to the formation of the standard CCITT group. It should be noted that in the 60-kHz to 108-kHz band, voice channel 1 occupies the highest frequency segment by convention, between 104 kHz and 108 kHz. The layout of the standard group is illustrated in Figure 4.19. Single sideband suppressed carrier (SSBSC) modulation techniques are utilized universally.

4.5.2.3.3 Formation of the Standard CCITT Supergroup. A supergroup contains five standard CCITT groups, equivalent to 60 voice channels. The standard supergroup, before further translation, occupies the frequency band of 312 kHz to 552 kHz. Each of the five groups making up the supergroup is translated in frequency to the supergroup frequency band by mixing with the appropriate carrier frequencies. The carrier frequencies are 420 kHz for group 1, 468 kHz for group 2, 516 kHz for group 3, 564 kHz for group 4, and 612 kHz for group 5. In the mixing process, in each case, the difference is taken (i.e., the lower is selected). This frequency translation process is illustrated in Figure 4.20.

4.5.2.4 Line Frequency. The band of frequencies that the multiplexer applies to the line, whether the line is a radiolink, wire-pair, or fiber optic cable, is called the line frequency. Some texts use the term *high frequency* (HF) for the line frequency. This is not to be confused with HF radio, which is a radio system that operates in the band of 3 MHz to 30 MHz.

The line frequency in this case may be the direct application of a group or supergroup to the line. However, more commonly a final frequency translation stage occurs, particularly on high-density systems.[15] An example of line frequency formation is illustrated in Figure 4.21. This figure shows the makeup of the basic 15-supergroup assembly. Its capacity is (15 × 60) 900 voice channels.

4.5.3 Pilot Tones

Pilot tones are used to control level in FDM systems. They may also be used to actuate maintenance alarms. In Figures 4.20 and 4.21, the pilot tones are indicated by the vertical

[15]"High-density" meaning, in this context, a system carrying a very large number of voice channels.

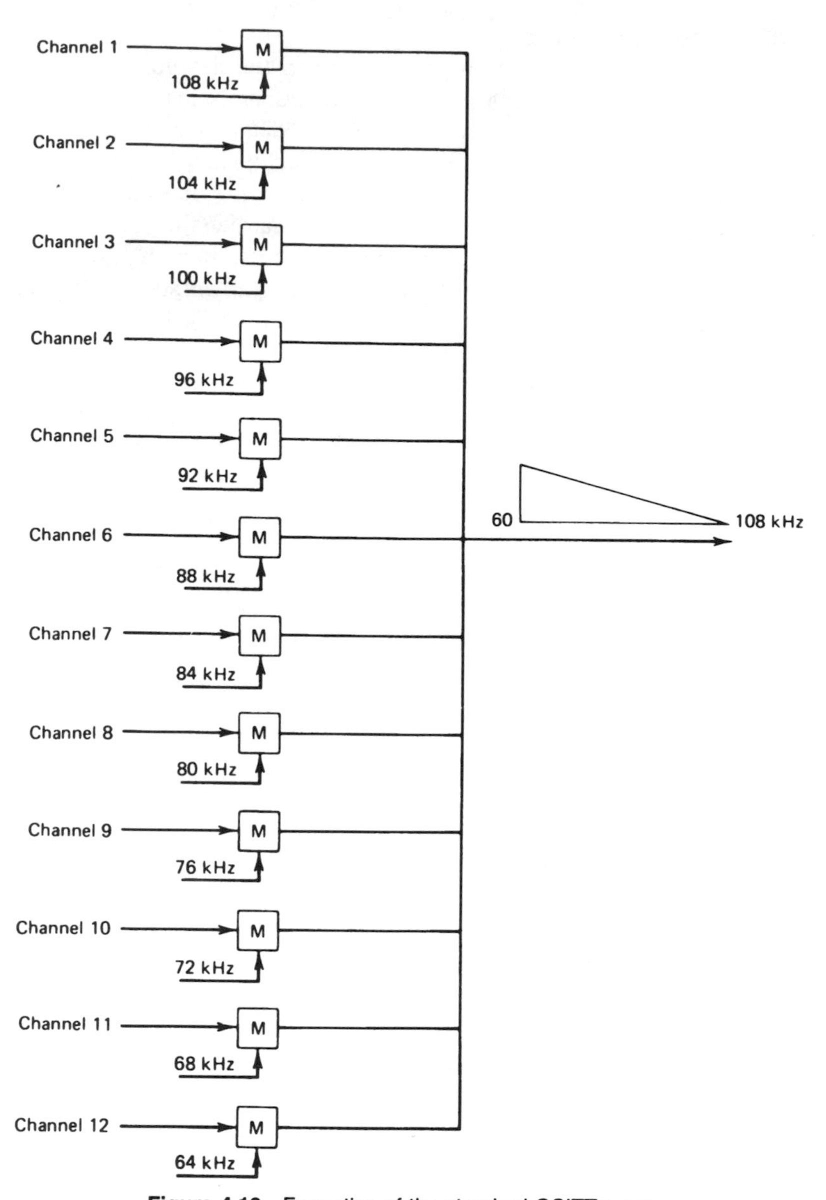

Figure 4.19 Formation of the standard CCITT group.

lines with little triangles on top. For example in the 15-supergroup assembly shown in Figure 4.21, the pilot tone is at 1552 kHz.

A pilot tone provides a comparatively constant amplitude reference for an automatic gain control (AGC) circuit. Frequency division multiplex equipment was designed to carry speech telephony. The nature of speech, particularly its varying amplitude, makes it a poor prospect as a reference for level control. Ideally, simple single-sinusoid (a sine wave signal), constant-amplitude signals with 100% duty cycles provide simple control information for level regulating equipment (i.e., the AGC circuit).[16]

[16]*Duty cycle* refers to how long (in this case) a signal is "on." A 100% duty cycle means the signal is on all the time.

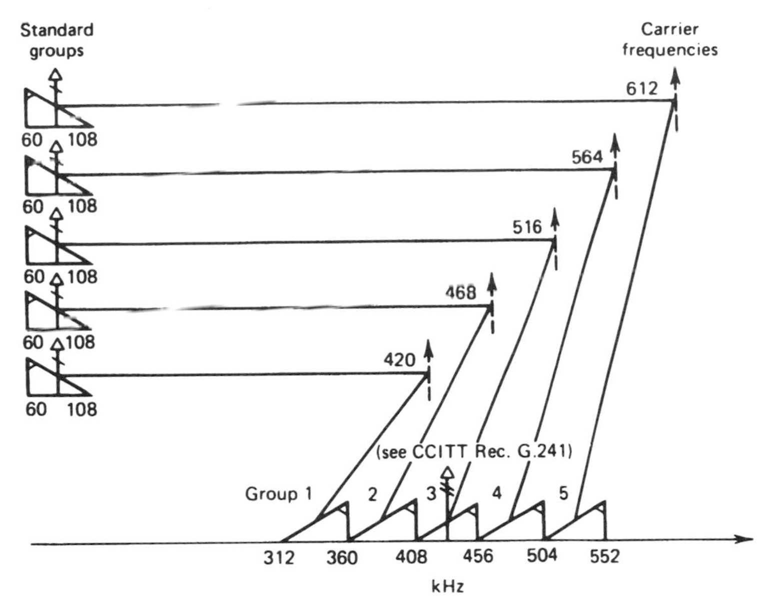

Figure 4.20 Formation of the standard CCITT supergroup. The vertical arrows show the frequencies of the group level regulating pilot tones. (From CCITT Rec. G.233, courtesy of ITU-T Organization, Ref. 6.)

Modern FDM equipments initiate a level-regulating pilot tone on each group on the transmitting side of the circuit. Individual level-regulating pilots are inserted on each supergroup and other frequency configurations. The intent is to maintain the system level within ±0.5 dB.

Pilots are assigned frequencies that are part of the transmitted FDM spectrum yet do not interference with the voice channel operation. They are standardized by CCITT and are usually inserted on a frequency in the guard between voice channels or are residual carriers (i.e., partially suppressed carriers).

4.5.4 Comments on the Employment and Disadvantages of FDM Systems

FDM systems began to be implemented in the 1950s, reaching a peak employment in the 1970s. All long-haul (long-distance) broadband systems, typically line-of-sight microwave, satellite communication, and coaxial cable systems, almost universally used FDM configurations. One transcontinental system in North America, called the L-5 system, carried 10,800 voice channels on each cable pair. (Remember the system is four-wire, requiring two coaxial cables per system.) There were 10 working cable pairs and one pair as spare. The total system had 108,000 voice channel capacity.

Few, if any, new FDM systems are being installed in North America today. FDM is being completely displaced by TDM systems (i.e., digital PCM systems, see Chapter 6). The principal drawback of FDM systems is noise accumulation. At every modulation–demodulation point along a circuit, noise is inserted. Unless the system designer was very careful, there would be so much noise accumulated at the terminal end of the system that the signal was unacceptable and the signal-to-noise ratio was very poor. Noise does not accumulate on digital systems.

We incorporated this section on frequency division multiplex so that the reader would understand the important concepts of FDM. Many will find that frequency division techniques are employed elsewhere such as on satellite communications, cellular, and PCS

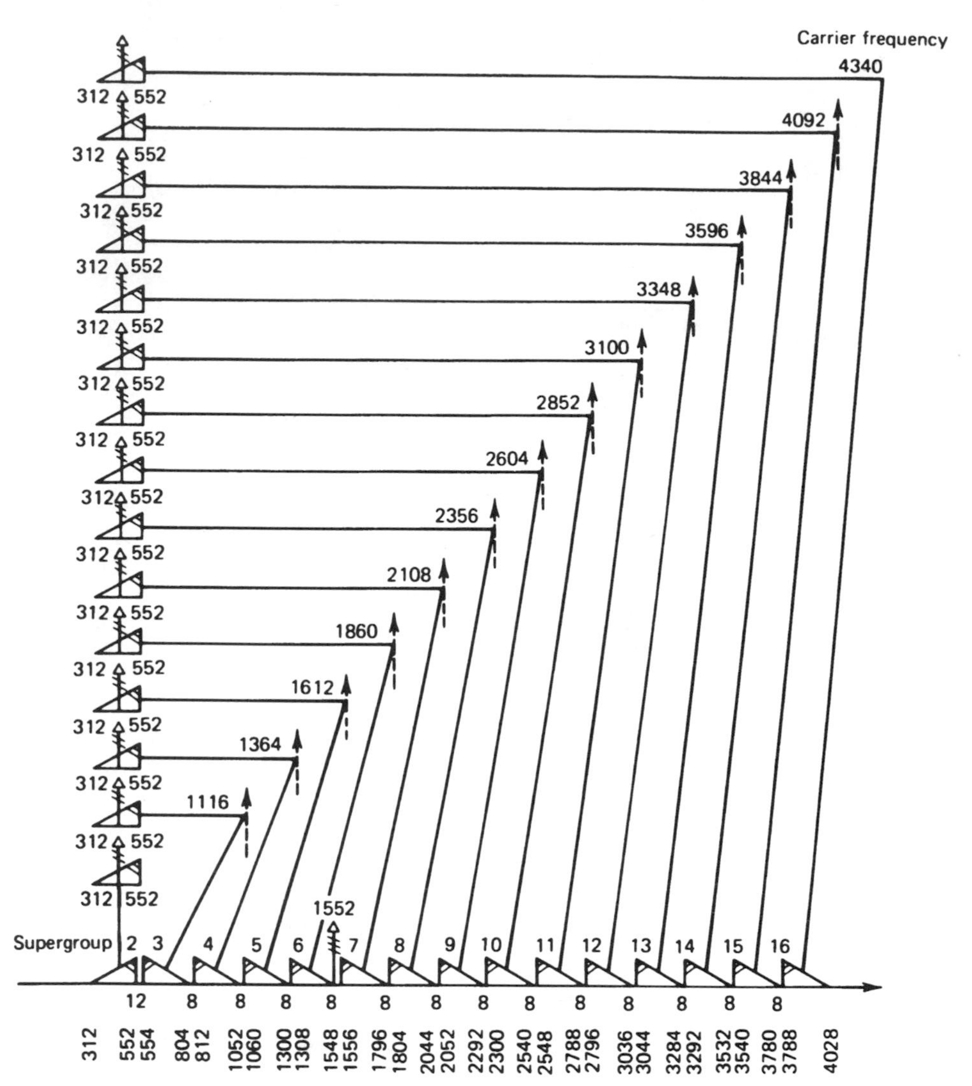

Figure 4.21 Makeup of the basic CCITT 15-supergroup line frequency assembly. (From CCITT Rec. G.233, courtesy of the ITU-T Organization, Geneva, Ref. 6)

systems. Also keep in mind that FDM as described herein is still widely used in emerging nations. However, these networks are also starting to phase out FDM in favor of a digital network based on TDM (PCM). TDM techniques are covered in Chapter 6.

REVIEW EXERCISES

1. Define *switching* in light of *transmission.*
2. Define *signaling.*
3. Define *calling rate* and *holding time.*
4. Define *lost calls* using the terms *offered traffic* and *carried traffic.*

5. Why are *call attempts* so important in the design of modern SPC switches?

6. Suppose the average holding time of a call is 3.1 minutes and the calling rate in the busy hour is 465 on a particular work day. What is the traffic flow?

7. Under normal operating conditions, when can we expect blockage (of calls)?

8. The statistics during the BH for a particular exchange is 5 lost calls in 841 offered calls. What is the grade of service?

9. When dealing with traffic formulas and resulting traffic tables, we have to know how lost calls are "handled." What are the three ways that lost calls may be handled?

10. On a probability-distribution curve, define the *mean*.

11. What percentage of events are encompassed in one standard deviation?

12. What traffic formula (and resulting traffic tables) would be used for modern digital exchanges (we assume such exchanges are SPC based)?

13. Consider one traffic relation that has been designed to meet grade of service objectives. We vary one characteristic—the number of traffic channels. Argue the case for efficiency for 10 circuits versus 49 circuits during the BH. Assume grade of service as 0.01 and $1.00 per erlang.

14. Define *serving time* when dealing with waiting systems (Erlang C).

15. Leaving aside issues of survivability, describe how *alternative routing* can improve grade of service or allow us to reduce the number of trunks on a traffic relation and still meet grade of service objectives.

16. Why is data traffic so different from telephone traffic?

17. Distinguish between a tandem/transit switch and a local serving switch.

18. Why use a *tandem switch* in the first place?

19. What are the three basic requirements of a telephone switch?

20. For a local serving switch with 10,000-line capacity, exemplify concentration ratios (lines/trunks) for a residential area, for an industrial/office area.

21. List six of the eight basic functions of a local switch.

22. Give the three basic functions of a local serving switch.

23. In a very small town, its local serving exchange has only three-digit subscriber numbers. Theoretically, what is the maximum number of subscribers it can serve?

24. How many switch banks will a step-by-step (SXS) switch have if it is to serve up to 10,000 subscribers?

25. What happens to a line that is "busied out?"

26. Give at least three advantages of an SPC exchange when compared with a register-marker crossbar exchange.

27. What are the three basic functional blocks of an SPC exchange (SPC portion only)?

28. Differentiate remote concentrators from remote switches. Give one big advantage each provides.

29. Describe two-wire operation and four-wire operation.

30. What is the function of a *hybrid* (transformer)?

31. Describe the function of the *balancing network* as used with a hybrid.

32. There are two new telephone network impairments we usually can blame on the hybrid. What are they?

33. What is the balance return loss on a particular hybrid connectivity when the balancing network is set for a 900-Ω loop, and this particular loop has only a 300-Ω impedance?

34. What are the two generic methods of multiplexing?

35. A mixer used in an FDM configuration has a local oscillator frequency of 64 kHz and is based on the CCITT modulation plan. After mixing (and filtering), what is the resulting extension of the desired frequency band (actual frequency limits at 3 dB points)?

36. The standard CCITT supergroup consists of _______ groups and occupies the frequency band _______ kHz to _______ kHz? How many standard voice channels does it contain?

37. What are pilot tones and what are their purpose in an FDM link?

REFERENCES

1. R. L. Freeman, *Telecommunication System Engineering*, 3rd ed., Wiley, New York, 1996.
2. R. A. Mina, "The Theory and Reality of Teletraffic Engineering," *Telephony* article series, *Telephony*, Chicago, 1971.
3. R. L. Freeman, *Reference Manual for Telecommunication Engineering*, 2nd ed., Section 1, Wiley, New York, 1994.
4. *Transmission Systems for Communications*, 5th ed., Bell Telephone Laboratories, Holmdel, New Jersey, 1982.
5. W. D. Reeve, *Subscriber Loop Signaling and Transmission Handbook—Analog*, IEEE Press, New York, 1992.
6. *Recommendations Concerning Translating Equipments*, CCITT Rec. G.233, Geneva, 1982.
7. R. L. Freeman, *Telecommunication Transmission Handbook*, 4th ed., Wiley, New York, 1998.

5

TRANSMISSION ASPECTS OF VOICE TELEPHONY

5.1 OBJECTIVE

The goal of this chapter is to provide the reader with a firm foundation of the *analog voice channel*. Obviously, from the term, we are dealing with the transmission of the human voice. Voice is a sound signal. That sound is converted to an electrical signal by the mouthpiece of the subscriber subset. The electrical signal traverses down a subscriber loop to a local serving switch or to a PABX.[1] The local serving switch is the point of connectivity with the PSTN. As the network evolves to an all-digital network, the local serving switch is the point where the analog signal is converted to an equivalent digital signal. Digital transmission and switching are discussed in Chapter 6. However, there are still locations in North America where digital conversion takes place deeper in the PSTN, perhaps at a tandem exchange. Local service, inside the local serving area in these cases, remains analog, and local trunks (junctions) may consist of wire pairs carrying the analog signals. Analog wire-pair trunks are even more prevalent outside of North America.

In this chapter we will define the analog voice channel and describe its more common impairments. The subscriber subset's functions are reviewed, as well as the sound-to-electrical-signal conversion, which takes place in that subset. We then discuss subscriber loop and analog trunk design.

The reader should not lose sight of the fact that more than 60% of the revenue-bearing traffic on the PSTN is voice traffic, with much of the remainder being data traffic in one form or another. It should also be kept in mind that the connectivity to the PSTN, that is, the portion from the user's subset to the local serving exchange, will remain analog until some time in the future. ISDN, of course, is a notable exception.

5.2 DEFINITION OF THE VOICE CHANNEL

The IEEE (Ref. 1) defines a *voice-band channel* as "a channel that is suitable for transmission of speech or analog data and has the maximum usable frequency range of 300 to 3400 Hz." CCITT also defines it in the range of 300 Hz to 3400 Hz. Bell Telephone

[1]PABX stands for private branch exchange. It is found in the office or factory environment, and is used to switch local telephone calls and to connect calls to a nearby local serving exchange. In North America and in many other places a PABX is privately owned.

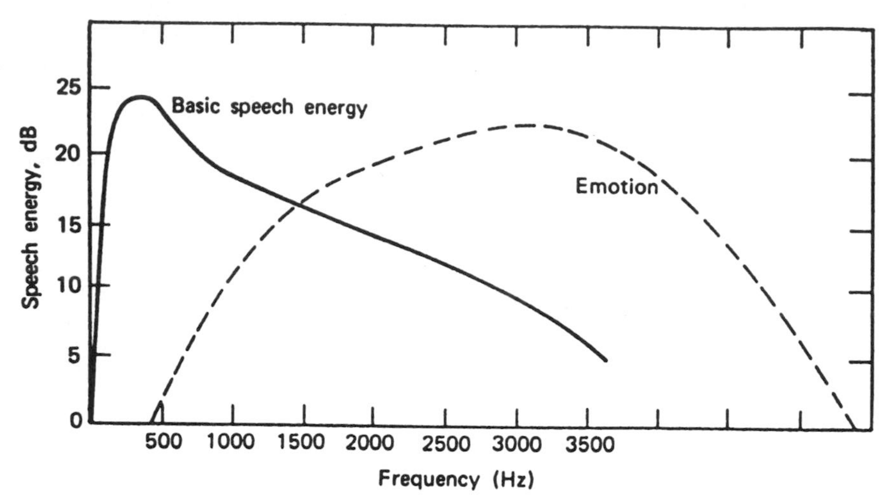

Figure 5.1 Energy and emotion distribution of human speech (From *BSTJ*, July 1931).

Laboratories (Ref. 2) defines it in the range of 200 Hz to 3300 Hz. We remain with the CCITT/IEEE definition.

5.2.1 Human Voice

Human voice communication depends on the voice-generating mechanism of mouth and throat being the initial transmitter, and the acuity of the ear being the receiver. Frequency components of the human voice extend down to some 20 Hz and as high as 32,000 Hz. The lower frequency components carry the voice energy and the higher frequency components carry emotion. Figure 5.1 shows a distribution of energy and emotion of the typical human voice.

The human ear and many devices and components of the telecommunication network tend to constrain this frequency range. Young people can hear sound out to about 18,000 Hz, and as we get older this range diminishes. People in their sixties may not be able to hear sounds above 7000 Hz.

It is not the intent of the PSTN operator to provide high-fidelity communications between telephone users, only intelligible connections. Not only is there the frequency constraint of voice communications brought about by the human ear, but there are also constraints brought about by the subset transmitter (mouthpiece) and receiver (earpiece), and the subscriber loop (depending on its length, condition, and make up). Then purposely the electrical voice signal will then enter a low-pass filter limiting its high-frequency excursion to 3400 Hz. The filter is in the input circuit of the multiplex equipment. Thus we say that the voice channel or VF (voice-frequency) channel occupies the band from 300 Hz to 3400 Hz. Figure 5.2 shows the overall frequency response of a simulated telephone network using the standard 500/2500 North American telephone subset.

5.3 OPERATION OF THE TELEPHONE SUBSET

A telephone subset consists of an earpiece, which we may call the receiver; the mouthpiece, which we may call the transmitter; and some control circuitry in the telephone cradle-stand. Figure 5.3 illustrates a telephone subset connected all the way through its subscriber loop to the local serving exchange. The control circuits are aptly shown.

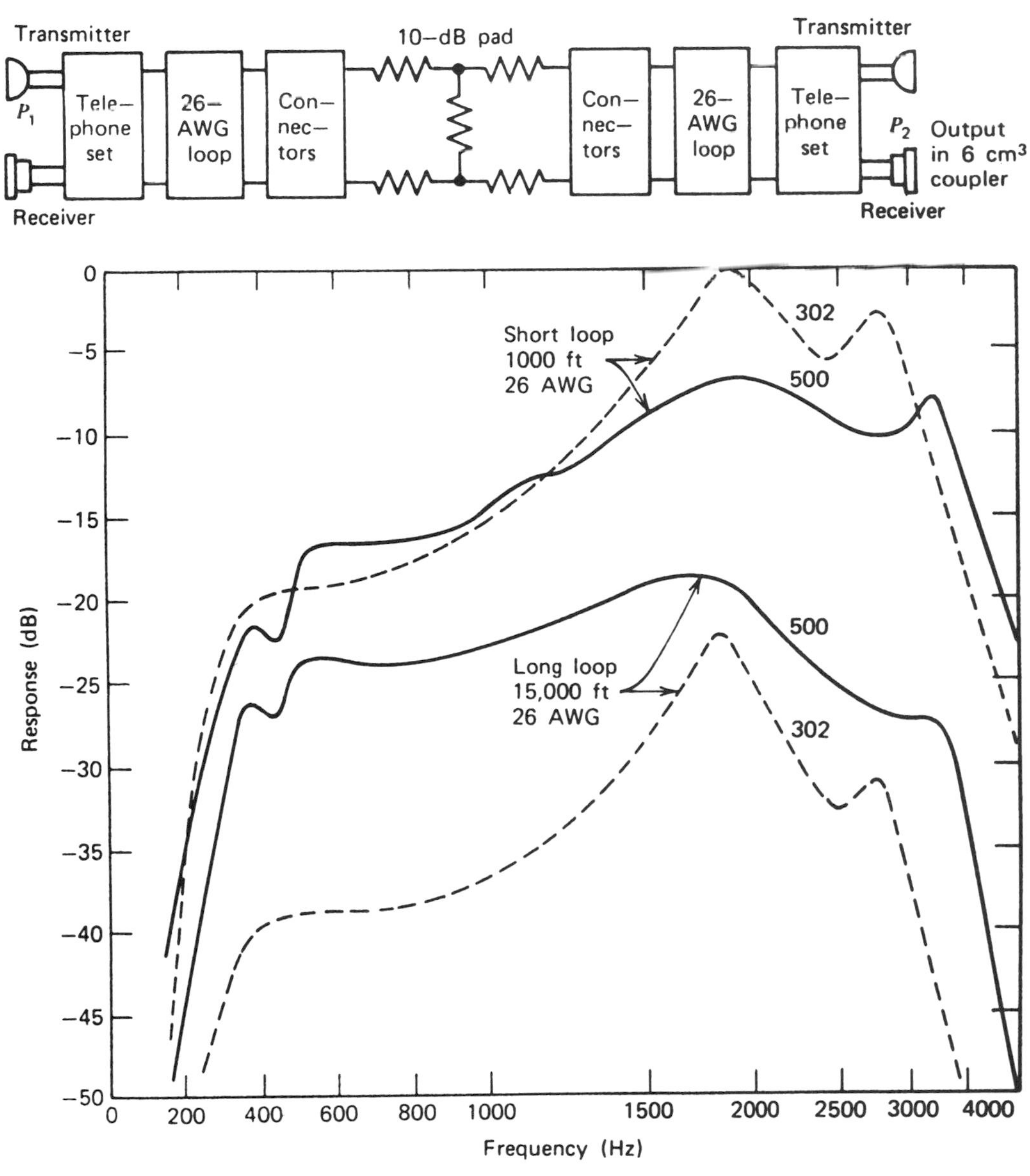

Figure 5.2 Comparison of overall response of the North American 302 and 500/2500 telephone sets. (From W. F. Tuffnell: *500-Type Telephone S*, Bell Labs Record, Sept. 1951, copyright 1951 Bell Telephone Laboratories.) *Note:* the 302 type subset is now obsolete (Ref. 3).

The hook-switch in Figure 5.3 (there are two) are the two little "knobs" that pop up out of the cradle when the subset is lifted from its cradle. When the subset is replaced in its cradle, these knobs are depressed. This illustrates the "on-hook" and "off-hook" functions described earlier. The "dial-switch" represents the function of the telephone dial. It simply closes and opens (i.e., makes contact, breaks contact). If a 1 is dialed, there is a single break-and-make (the loop is opened and the loop is closed). When the "loop is closed," current flows; when open, current stops flowing. If, for instance, a 5 is dialed, there will be five break-and-make operations.

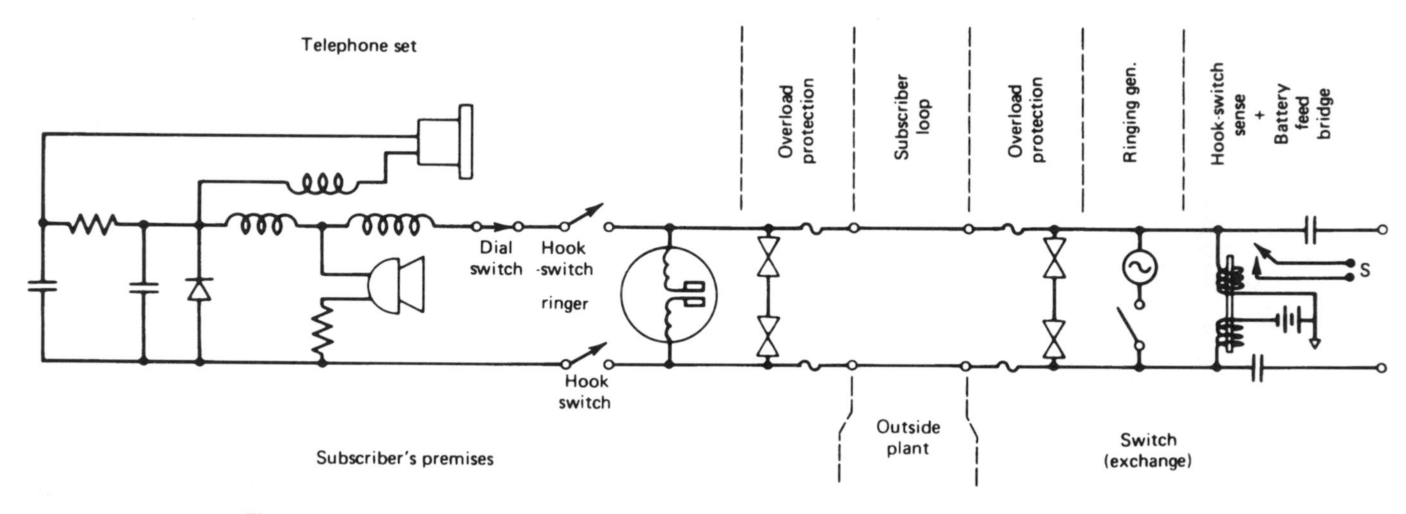

Figure 5.3 The conventional telephone subset. Note the hook-switch and dial-switch "controls."

5.3.1 Subset Mouthpiece or Transmitter

The mouthpiece converts acoustic energy (i.e., sound) into equivalent electric energy by means of a carbon granule transmitter. The transmitter requires a direct current (dc) voltage, a minimum of 3 V to 5 V, across its electrodes. We call this *talk battery*. In modern systems it is supplied over the subscriber loop and derives from a battery source at the local serving switch, as illustrated in Figure 5.3.

Current deriving from the local switch supply flows through the carbon grains or granules, which are contained just below a diaphragm that hold the carbon granules in place. This current flow occurs when the telephone is taken off-hook (i.e., out of its cradle). When sound impinges on the diaphragm of the transmitter, resulting variations of air pressure are transferred from the diaphragm to the carbon granules, and the resistance of the electrical path through the carbon changes in proportion to the pressure. A pulsating direct current in the subscriber loop results. The frequency response of the carbon transmitter peaks between 800 Hz and 1000 Hz. This is illustrated in Figure 5.2.

5.3.2 Telephone Earpiece or Receiver

A typical receiver consists of a diaphragm of magnetic material, often soft iron alloy, placed in a steady magnetic field supplied by a permanent magnet, and a varying magnetic field, caused by the voice currents flowing through the voice coils. Such voice currents are alternating (ac) in nature and originate at the far-end telephone transmitter. These currents cause the magnetic field of the receiver to alternately increase and decrease, making the diaphragm move and respond to the variations. As a result, an acoustic pressure wave is set up, reproducing, more or less exactly, the original sound wave from the distant telephone transmitter. The telephone receiver, as a converter of electrical energy to acoustic energy, has a comparatively low efficiency, on the order of 2% to 3%.

Sidetone is the sound of the talker's voice heard in his own receiver. The sidetone level must be controlled. When the level is high, the natural human reaction is for the talker to lower his voice. Thus by regulating the sidetone, talker levels can be regulated. If too much sidetone is fed back to the receiver, the output level of the transmitter is reduced, owing to the talker lowering his/her voice, thereby reducing the level (voice volume) at the distant receiver, deteriorating performance.

5.4 SUBSCRIBER LOOP DESIGN

5.4.1 Basic Design Considerations

We speak of the telephone subscriber as the user of the subset. As mentioned in Section 1.3, telephone subscribers are connected via a subscriber loop to a local serving switch, which can connect a call to another subscriber served by that same switch or via other switches through the PSTN to a distant called subscriber. The subscriber loop is a wire pair. Present-day commercial telephone service provides for both transmission and reception on the same pair of wires that connect the subscriber to the local serving switch. In other words, it is two-wire operation.

The subscriber loop is a dc loop, in that it is a wire pair supplying a metallic path for the following:[2]

[2]Metallic path is a path that is "metal," usually copper or aluminum. It may be composed of a wire pair or coaxial cable. We could have a *radio* path or a *fiber optic* path.

1. Talk battery;
2. An ac ringing voltage for the bell or other alerting device on the telephone instrument supplied from a special ringing voltage source;
3. Current to flow through the loop when the telephone subset is taken out of its cradle (off-hook), which tells the switch that it requires "access" and causes line seizure at the local serving switch; and
4. The telephone dial that, when operated, makes and breaks the dc current on the closed loop, which indicates to the switching equipment the telephone number of the distant telephone with which communication is desired.[3]

The typical subscriber loop is supplied its battery voltage by means of a battery feed circuit, illustrated in Figure 5.3. Battery voltages have been standardized at −48 V dc. It is a negative voltage, to minimize cathodic reaction, which is a form of corrosion that can be a thermal noise source.

5.4.2 Subscriber Loop Length Limits

It is desirable from an economic standpoint to permit subscriber loop lengths to be as long as possible. Thus the subscriber serving area could become very large. This, in turn, would reduce the number of serving switches required per unit area, affording greater centralization, less land to buy, fewer buildings, simpler maintenance, and so forth. Unfortunately, there are other tradeoffs forcing the urban/suburban telecommunication system designer into smaller serving areas and more switches.

The subscriber loop plant, sometimes called *outside plant,* is the largest single investment that a telecommunication company has. Physically, we can extend a subscriber loop very long distances—5, 10, 20, 50, or even 100 miles. Such loops require expensive conditioning, which we will delve into later in this chapter. It is incumbent on a telecommunication company to optimize costs to have the fewest possible specially conditioned loops.

Two basic criteria, which limit loop length, must be considered when designing a subscriber loop:

1. Attenuation (loss) limits; and
2. Resistance limits.

Attenuation (loss) must be limited to keep within *loudness rating* requirements. Loudness rating was discussed in Section 3.2. If a subscriber loop has too much loss, the telephone user signal level suffers, and the signal cannot be heard well enough. The user may consider the connection unsatisfactory. In North America the maximum loss objective is 8 dB for a subscriber loop. In some other countries that value is 7 dB. Remember that it takes two subscriber loops to make a connection—the subscriber loop of the calling subscriber and the loop of the called subscriber.

The attenuation is referenced to 1000 Hz in North America and 800 Hz elsewhere in the world. In other words, when we measure loss, unless otherwise stated, it is measured at the reference frequency. Loss (attenuation) is a function of the diameter of the copper wire making up the pair and the length of the pair.

[3]Modern telephone subsets almost universally use a touch-tone pad, which transmits a unique two-tone signal for each digit actuated. This type of subscriber address signaling is described in Chapter 7.

Consider this example. We take a reel of 19-gauge (American wire gauge or AWG) copper wire and connect a telephone transmitter at one end.[4] Now extend the reel, laying out the wire along a track or road. At intervals somebody is assigned to talk on the transmitter, and we test the speech level, say, every 5 km. At about 30 km the level of the voice heard on the test receiver is so low that intelligible conversation is impossible. What drops the level of the voice as the wire is extended is the loss, which is a function of length.

Let us digress a moment and discuss *American wire gauge* or AWG. It is a standardized method of measuring wire diameter. Just like the gauge on shotguns, as the AWG number increases, the wire diameter decreases. The following equivalents will give us a basic idea of AWG versus diameter:

American Wire Gauge	Diameter (mm)	Diameter (inches)
19	0.91	0.036
22	0.644	0.025
24	0.511	0.020
26	0.405	0.016
28	0.302	0.012

Signaling limits of a subscriber loop are based on dc resistance. When we go "off-hook" with a telephone, a certain minimum amount of current must flow in the loop to actuate the local serving switch. 20 mA is the generally accepted minimum loop current value in North America. If subscriber loop current is below this value, we have exceeded the signaling limits. Applying Ohm's law, the loop resistance should not exceed 2400 Ω. Budget 400 Ω for the battery feed bridge and we are left with 2000 Ω for the loop itself. We must account for the resistance of the subset wiring. Budget 300 Ω for this. Thus the resistance of the wire itself in the loop must not exceed 1700 Ω.

Once we exceed the signaling limit (the loop resistance, wire only, exceeds 1700 Ω), when the telephone goes off-hook, no dial tone is returned. This just means that there is insufficient loop current to actuate the switch, telling the switch we wish to make a call. When there is sufficient current, the switch, in turn, returns the dial tone. When there is insufficient loop current, we hear nothing. If we cannot effect signaling, the telephone just will not operate. So between the two limiting factors, loss and resistance, resistance is certainly the more important of the criteria.

5.4.3 Designing a Subscriber Loop

Figure 5.4 is a simplified model of a subscriber loop. Distance D in the figure is the length of the loop. As mentioned earlier, D must be limited in length, owing to (1) attenuation of the voice signal on the loop and (2) dc resistance of the loop for signaling.

The maximum loop loss is taken from the national transmission plan.[5] In North America, it is 8 dB measured at 1000 Hz. We will use the maximum resistance value calculated previously, namely, 1700 Ω (wire only).

[4]A 19-gauge copper wire has a diameter of 0.91 mm.

[5]National transmission plan for North America; see Bellcore, *BOC Notes on the LEC Networks*, latest edition (Ref. 5).

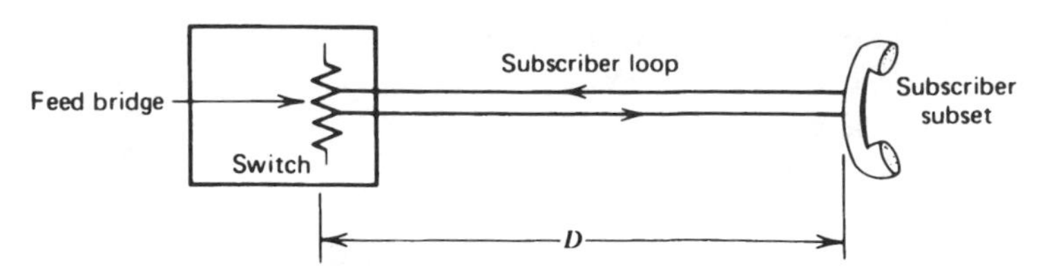

Figure 5.4 Subscriber loop model.

5.4.3.1 Calculating the Resistance Limit. To calculate the dc loop resistance for copper conductors, the following formula is applicable:

$$R_{\mathrm{dc}} = \frac{0.1095}{d^2}, \tag{5.1}$$

where R_{dc} = loop resistance (Ω/mi) and d = diameter of the conductor (in).

If we want a 17-mile loop, allowing 100 Ω per mile of loop (for the 1700-Ω limit), what diameter of copper wire would we need? Apply Eq. (5.1)

$$100 = 0.1095/d^2$$
$$d^2 = 0.1095/100 = 0.001095$$
$$d = 0.0331 \text{ inches} \quad \text{or } 0.84 \text{ mm} \quad \text{or about 19 gauge.}$$

By applying resistance values from Table 5.1, we can calculate the maximum loop length for 1700 Ω maximum signaling resistance. As an example, for a 26-gauge loop,

$$1700/83.5 = 20.359 \text{ kft} \quad \text{or } 20,359 \text{ feet}$$

This, then, is the signaling limit for 26-gauge (copper) subscriber loop. It is *not* the loss (attenuation) limit, or what some call the *transmission limit.*

Another guideline in the design of subscriber loops is the minimum loop current off-hook for effective subset operation. For instance, the Bell System 500-type subset requires at least 20 mA for efficient operation.

5.4.3.2 Calculation of the Loss Limit. For our discussion here, the loss at 1000 Hz of a subscriber loop varies with diameter of the wire and the length of the loop. Table 5.2 gives values of loss (attenuation) per unit length for typical subscriber low-capacitance wire pair.

Table 5.1 Loop Resistance for Various Conductor Gauges

AWG	Ohms/1000 ft of Loop	Ohms/Mile of Loop	Ohms/km of Loop
28	132	697	433
26	83.5	440	268
24	51.9	274	168.5
22	32.4	171	106
19	16.1	85	53

Table 5.2 Loss per Unit Length of Subscriber Wire Pairs

AWG	Loss/1000 ft (dB)	dB/km	dB/mi
28	0.615	2.03	3.25
26	0.51	1.61	2.69
24	0.41	1.27	2.16
22	0.32	1.01	1.69
19	0.21	0.71	1.11
16	0.14	0.46	0.74

Work the following examples based on a maximum loss of 8 dB. Here we are to calculate the maximum loop length for that 8-dB loss. Use simple division with the values in column 2 of Table 5.2. The answers, of course, will be in kilofeet.

$$\text{28 gauge } 8/0.615 = 13.0 \text{ kft}$$
$$\text{26 gauge } 8/0.51 = 15.68 \text{ kft}$$
$$\text{24 gauge } 8/0.41 = 19.51 \text{ kft}$$
$$\text{22 gauge } 8/0.32 = 25.0 \text{ kft}$$
$$\text{19 gauge } 8/0.21 = 38.1 \text{ kft}$$
$$\text{16 gauge } 8/0.14 = 57.14 \text{ kft}$$

Copper is costly. Thus many telecommunication companies employ gauges with diameters no greater than 22 gauge.

5.4.4 Extending the Subscriber Loop

In many situations, subscribers will reside outside of the maximum subscriber loop lengths previously described. There are five generally accepted methods that can be used to extend these maximums. They are:

1. Increasing conductor diameter (covered in the preceding text);
2. Using amplifiers and/or range extenders;[6]
3. Employing inductive loading;
4. Using digital subscriber line (DSL) techniques (covered in Chapter 6); and
5. Employing remote concentrators or switches (see Section 4.3).

Amplifiers in the subscriber loop extend the transmission range. Perhaps better said, they compensate for loop loss. Commonly such amplifiers are set for about 7-dB gain. Care must be used to assure that dc signaling is not lost.

5.4.4.1 Inductive Loading. Inductive loading of a subscriber loop (or metallic VF trunk) tends to reduce the transmission loss at the expense of amplitude-frequency response beyond 3000 Hz to 3400 Hz, depending on the loading technique employed. *Loading* a particular subscriber loop (or metallic pair trunk) consists of inserting series inductances (loading coils) into the loop at fixed distance intervals. Adding load coils tends to:

[6]A range extender increases the battery voltage to either −84 or −96 V dc. In some texts the term *loop extender* is used rather than range extender.

Table 5.3 Code for Load Coil Spacing

Code Letter	Spacing (ft)	Spacing (m)
A	700	213.5
B	3000	915
C	929	283.3
D	4500	1372.5
E	5575	1700.4
F	2787	850
H	6000	1830
X	680	207.4
Y	2130	649.6

- Decrease the velocity of propagation;[7]
- Increase impedance.

Loaded cables are coded according to the spacing of the load coils. The standard code for the spacing of load coils is shown in Table 5.3. Loaded cables typically are designated 19H44, 24B88, and so forth. The first number indicates the wire gauge, the letter is taken from Table 5.3 and is indicative of the spacing, and the third number is the inductance of the load coil in millihenries (mH). For example, 19H66 cable has been widely used in Europe for long-distance operation. Thus this cable has 19-gauge wire pairs with load coils inserted at 1830-m (6000-ft) intervals with coils of 66-mH inductance. The most commonly used spacings are B, D, and H.

Table 5.4 will be useful in calculating the attenuation (loss) of loaded loops for a given length. For example, in 19H88, the last entry in the table, the attenuation per kilometer is 0.26 dB (0.42 dB per statute mile). Thus for our 8-dB loop loss limit, we have 8/02.6, limiting the loop to 30.77 km (19.23 mi).

When determining signaling limits in loop design, add about 15 Ω per load coil as a series resister. In other words, the resistance values of the series load coils must be included in the total loop resistance.

5.4.5 "Cookbook" Design Methods for Subscriber Loops

5.4.5.1 Resistance Design Concept. Resistance design (RD) dates back to the 1960s and has since been revised. It was basic North American practice. Our inclusion of resistance design helps understand the "cookbook" design concept. At the time of its inception, nearly all local serving area switches could handle loops up to 1300 Ω resistance. In virtually every case, if the RD rules were followed there would be compliance with the attenuation limit of 8 dB. The maximum resistance limit defines a perimeter around a local switch which is called the *resistance design boundary*. For subscribers outside this boundary served by the switch, *long-route design* (LRD) rules were imposed. LRD is briefly covered in Section 5.4.5.2.

The following additional terms dealing with RD are based on Ref. 5.

[7]Velocity of propagation is the speed (velocity) that an electrical signal travels down a particular transmission medium.

Table 5.4 Some Properties of Cable Conductors

Diameter (mm)	AWG No.	Mutual Capacitance (nF/km)	Type of Loading	Loop Resistance (Ω/km)	Attenuation at 1000 Hz (dB/km)
0.32	28	40	None	433	2.03
		50	None		2.27
0.40		40	None	277	1.62
		50	H66		1.42
		50	H88		1.24
0.405	26	40	None	270	1.61
		50	None		1.79
		40	H66	273	1.25
		50	H66		1.39
		40	H88	274	1.09
		50	H88		1.21
0.50		40	None	177	1.30
		50	H66	180	0.92
		50	H88	181	0.80
0.511	24	40	None	170	1.27
		50	None		1.42
		40	H66	173	0.79
		50	H66		0.88
		40	H88	174	0.69
		50	H88		0.77
0.60		40	None	123	1.08
		50	None		1.21
		40	H66	126	0.58
		50	H88	127	0.56
0.644	22	40	None	107	1.01
		50	None		1.12
		40	H66	110	0.50
		50	H66		0.56
		40	H88	111	0.44
0.70		40	None	90	0.92
		50	H66		0.48
		40	H88	94	0.37
0.80		40	None	69	0.81
		50	H66	72	0.38
		40	H88	73	0.29
0.90		40	None	55	0.72
0.91	19	40	None	53	0.71
		50	None		0.79
		40	H44	55	0.31
		50	H66	56	0.29
		50	H88	57	0.26

Source: ITT, *Outside Plant*, Telecommunication Planning Documents. (Courtesy of Alcatel.) (Ref. 4)

1. *Resistance design limit* is the maximum value of loop resistance to which the RD method is applicable. The value was set at 1300 Ω primarily to control transmission loss. In the revised resistance design (RRD) plan, this value is increased to 1500 Ω.

2. *Switch supervisory limit* is the conductor loop resistance beyond which the operation of the switch supervisory equipment (loop signaling equipment) is uncertain.

3. *Switch design limit.* With RD procedures, this limit was set at 1300 Ω (in RRD it is increased to 1500 Ω).

4. The *design loop* is the subscriber loop under study for a given distribution area to which the switch design limit is applied to determine conductor sizes (i.e., gauges or diameters). It is normally the longest loop in the cable of interest.

5. The *theoretical design* is the subscriber cable makeup consisting of the two finest (smallest in diameter) standard consecutive gauges necessary in the design loop to meet the switch design limit.

The application of resistance design to subscriber loops begins with three basic steps: (1) determination of the resistance design boundary, (2) determination of the design loop, and (3) selection of the cable gauge(s) to meet design objectives.

The resistance design boundary is applied in medium and high subscriber density areas. LRD procedures are applied in areas of sparser density (e.g., rural areas).

The design loop length is based on local and forecast service requirements. The planned ultimate longest loop length for the project under consideration is the design loop, and the theoretical design and gauge(s) selection are based on it.

The theoretical design is used to determine the wire gauge or combinations of gauges for any loop. If more than one gauge is required, Ref. 6 states that the most economical approach, neglecting existing plant, is the use of the two finest consecutive standard gauges that meet a particular switch design limit. The smaller of the two gauges is usually placed outward from the serving switch because it usually has a larger cross section of pairs. Since the design loop length has been determined, the resistance per kft (or km) for each gauge may be determined from Tables 5.1, 5.2, and 5.4. The theoretical design can now be calculated from the solution of two simultaneous equations.

The following example was taken from Ref. 6. Suppose we wished to design a 32-kft loop with a maximum loop resistance of 1300 Ω. If we were to use 24-gauge copper pair, Table 5.1 shows that we exceed the 1300-Ω limit; if we use 22 gauge, we are under the limit by some amount. Therefore, what combination of the two gauges in series would just give us 1300 Ω? The loop requires five H66 load coils, each of which has a 9-Ω resistance. It should be noted that the 1300-Ω limit value does not include the resistance of the telephone subset.

Let X = the kilofeet value of the length of the 24-gauge pair and Y = the kilofeet value of the length of the 22-gauge pair. Now we can write the first equation:

$$X + Y = 32 \text{ kft.}$$

Table 5.1 shows the resistance of a 24-gauge wire pair as 51.9 Ω/kft, and for a 22-gauge wire pair as 32.4 Ω/kft. We can now write a second simultaneous equation:

$$51.9X + 32.4Y + 5(9) = 1300\ \Omega$$

$$X = 11.2 \text{ kft of 24-gauge cable}$$

$$Y = 32 - X = 20.8 \text{ kft of 22-gauge cable}$$

(See Appendix B for the solution of simultaneous equations.)

We stated earlier that if the resistance design rules are followed, the North American 8-dB objective loss requirement will be met for all loops. However, to ensure that this is the case, these additional rules should be followed:

- Inductive load all loops over 18 kft long.
- Limit the cumulative length of all bridged taps on nonloaded loops to 6 kft or less.

Ref. 6 recommends H88 loading where we know the spacing between load coils is 6000 ft with a spacing tolerance of ±120 ft. Wherever possible it is desirable to take deviations greater than ±120 ft on the short side so that correction may later be applied by normal build-out procedures.

The first load section out from the serving switch is 3000 ft for H66/H88 loading. In the measurement of this length, due consideration should be given to switch wiring so that the combination is equivalent to 3000 ft. It should be remembered that the spacing of this first coil is most critical to achieve acceptable return loss and must be placed as close to the recommended location as physically and economically possible.

5.4.5.2 Long-Route Design (LRD). The long-route design procedure uses several zones corresponding to the resistance of the loop in excess of 1300 Ω. Of course, each subscriber loop must be able to carry out the supervisory signaling function and meet the 8-dB maximum loop attenuation rule (North America). LRD provides for a specific combination of fixed-gain devices (VF repeaters/amplifiers) to meet the supervision and loss criteria.

On most long loops a range extender with gain is employed at the switch. A range extender (loop extender) boosts the standard −48 V by an additional 36 V to 48 V, an amplifier provides a gain of from 3 dB to 6 dB. Inductive loading is H88. Any combination of cable gauges may be used between 19 and 26 gauge.

5.4.6 Current North American Loop Design Rules

There are three subscriber loop design methods in this category: (1) RRD (revised resistance design), (2) MLRD (modified long-route design), and (3) CREG (concentration range extender with gain).

5.4.6.1 Revised Resistance Design (RRD). RRD covers subscriber loops as long as 24 kft. Loop length is broken down into two ranges—from 0 ft to 18,000 ft, where the maximum loop resistance is 1300 Ω, and from 18,000 ft to 24,000 ft, where the maximum loop resistance is 1500 Ω. H88 loading is used on loops longer than 18,000 ft. Two gauge combinations may be employed selected from the following three wire gauges: 22, 24, and 26 gauge.

5.4.6.2 Modified Long-Route Design (MLRD). Loop resistances up to 1500 Ω are served by RRD procedures. The range beyond 1500 Ω is served by MLRD, CREG (see the following subsection), or DLC (digital loop carrier; see Chapter 6).

Under MLRD loop resistances from 1500 Ω to 2000 Ω are placed in the RZ18 category and require 3-dB gain. The loop resistance range from 2000 Ω to 2800 Ω is designated RZ28, and loops in this range require 6-dB gain. New switches have range extenders with gain that automatically switch their gain setting to provide the 3-dB or 6-dB net gain as required. This automatic switching removes the need to maintain and administer transmission zones. From this standpoint, MLRD a single range-extended zone.

5.4.6.3 Concentration with Range Extension and Gain (CREG). The CREG plan is designed for use with finer gauge copper pair cable. It can accomplish this by providing VF amplifier gain behind a stage of switching concentration. It employs H88 loading beyond 1500 Ω. Any two gauges in the combination of 22, 24, or 26 gauge may be employed. Gain and range extension applies only to loops beyond the 1500-Ω demarcation.

5.4.6.4 *Digital Loop Carrier (DLC).* Of the three long-route design techniques, digital loop carrier is the most attractive, especially for facilities greater than 28,000 ft. Many of these DLC systems are based on T1 digital techniques (described in Chapter 6) with specially designed terminals. One such T1 system can serve up to 40 subscriber loops over a single repeatered line consisting of two wire pairs, one for transmission in one direction and one for transmission in the other direction. Such a system is typically used on long routes when relatively high subscriber density is forecast for the planning period and when such feeder routes would require expansion.[8]

Another advantage of DLC is that it can provide improved transmission loss distributions. One such system displays a 1000-Hz transmission loss of 2 dB between a serving switch and a remote terminal regardless of the length of the digital section. The low insertion loss of the digital portion of such systems allows up to 6 dB to be apportioned to the analog subscriber loop distribution plant (Ref. 5).

5.5 DESIGN OF LOCAL AREA WIRE-PAIR TRUNKS (JUNCTIONS)

5.5.1 Introduction

Exchanges in a common local area often are connected in a full-mesh topology (see Section 1.3.7, a definition of *full mesh*). Historically, depending on distance and certain other economic factors, these trunks used VF (analog wire-pair) transmission over cable. In North America, this type of transmission is phasing out in favor of a digital connectivity. However, analog wire-pair transmission still persists in a number of parts of the world, and especially outside of North America. In the United Kingdom the term *junction* is used for trunks serving the local area, whether analog or digital.

There are notably fewer trunks than subscriber lines for which they serve. This is due to the concentration at a local serving switch. The ratio of trunks to subscriber lines varies from 3 to 25. Because there are fewer trunks, more investment can be made on this portion of the plant. Losses are generally kept around 2 dB and return losses are well over 24 dB because of excellent impedance matches. These low trunk insertion losses can be accomplished by several means, such as using larger diameter wire pairs, employing VF amplifiers, and inductive loading.

5.5.2 Inductive Loading of Wire-Pair Trunks (Junctions)

The approach to inductive loading of wire-pair trunks is similar to that for loading subscriber loops. The distance (D) between load coils is all important. The spacing (D) should not vary more than ±2% from the specified spacing.

The first load coil is spaced D/2 from an exchange main frame, where D is the specified distance between load coils (see Table 5.3).[9] Take the case of H loading, for instance. The distance between load coil points is 6000 ft (1830 m), but the first load coil is place at D/2 or 3000 ft (915 m) from the exchange. Then if the exchange is by-passed by some of the pairs, a full-load section exists. This concept is illustrated in Figure 5.5.

[8] *Planning period* refers to telecommunication planning. Here we mean advanced planning for growth. Planning periods may be 5, 10, or 15 years in advance of an installation date.

[9] *Main frame* is a facility, often a frame at a switching center, where all circuits terminate and where they may be cross connected. In other words, this is a location where we can get physical or virtual access to a circuit and where we may reconfigure assets. Main frames in local serving switches often are extremely large, where 10,000 or more subscriber lines terminate.

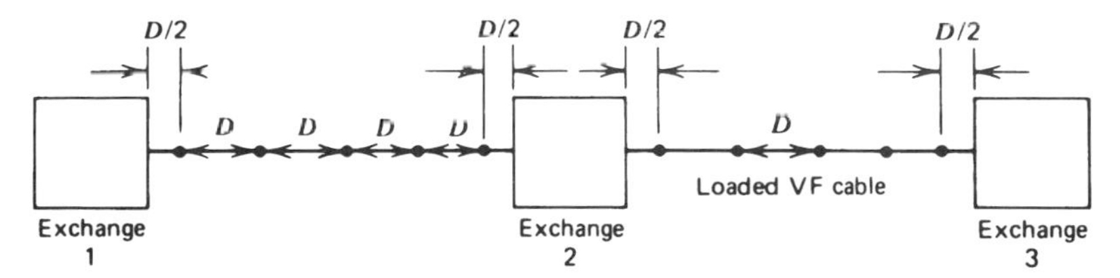

Figure 5.5 Loading of VF trunks (junctions).

Now consider this example. A loaded 500-pair VF trunk cable extends across town. A new switching center is to be installed along the route where 50 pairs are to be dropped and 50 inserted. It would be desirable to establish the new switch midway between load points. At the switch, 450 circuits will by-pass the office (switch). Using this $D/2$ technique, these circuits need no conditioning; they are full-load sections (i.e., $D/2 + D/2 = 1D$, a full-load section). Meanwhile, the 50 circuits entering from each direction are terminated for switching and need conditioning so that each looks electrically like a full-load section. However, the physical distance from the switch out to the first load point is D/2 or, in the case of H loading, 3000 ft or 915 m. To make the load coil distance electrically equivalent to 6000 ft or 1830 m, line build-out (LBO) is used. LBO is described in Section 5.5.2.1.

Suppose that the location of a new switching center was such that it was not halfway, but at some other fractional distance. For the section comprising the shorter distance, LBO is used. For the other, longer run, often a half-load coil is installed at the switching center and LBO is added to trim up the remaining electrical distance.

5.5.2.1 Line Build-Out (LBO). In many instances the first (and last) load coil cannot be placed at a D/2 distance from a switch or the separation between load coils cannot be D within tolerance. The reasons for the inability of an installation crew to meet the siting requirements are varied. Buildings could be in the way; the right-of-way requires a detour; hostile cable ground conditions exist; and so forth. In these cases, we install the load coil at a distance less than D/2 and use LBO (line build out).

Line build-out networks are used to increase the electrical length of a wire-pair cable section. These networks range in complexity from a simple capacitor, which simulates the capacitance of the missing cable length, to artificial cable sections. Network complexity increases as the frequency range over which the network has to operate increases. There is no comparable simple means to shorten the electrical length of a cable section. LBO can also be used for impedance matching (Ref. 7).

5.5.3 Local Trunk (Junction) Design Considerations

The basic considerations in the design of local trunks (junctions) are loss, stability, signaling, noise, and cost. Each are interrelated such that a change in value of one may affect the others. This forces considerable reiteration in the design process, and such designs are often a compromise.

One major goal is to optimize return loss on trunk facilities. This turns out to be a more manageable task than that required in the subscriber distribution plant. In North America the characteristic impedance of local wire trunks in most cases is 900 Ω in series with a 2.16-μF capacitor to match the impedance of the local (end-offices) exchanges. It should be pointed out that some tandem and intertandem trunks connect to 600-Ω tandem switches.

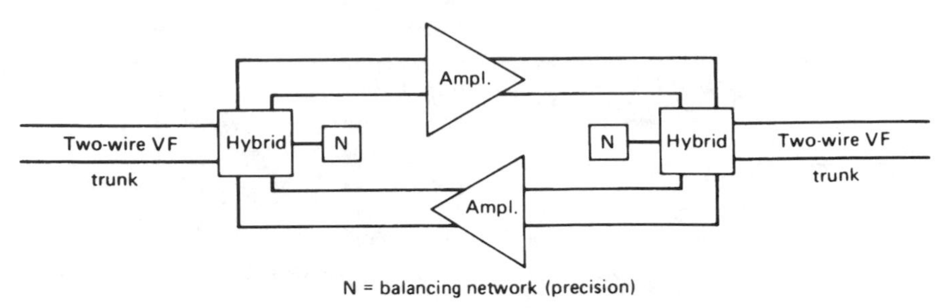

Figure 5.6 Simplified block diagram of a VF repeater.

5.6 VF REPEATERS (AMPLIFIERS)

Voice frequency (VF) repeaters (amplifiers) in telephone terminology imply the use of *uni*directional amplifiers on VF trunks.[10] With one approach on a two-wire trunk, two amplifiers are used on each pair connected by a hybrid at the input and a hybrid at the output. A simplified block diagram is shown in Figure 5.6.

The gain of a VF repeater can be run up as high as 20 dB or 25 dB, and originally they were used at 50-mi intervals on 19-gauge loaded cable in the long-distance (toll) plant. Today they are seldom found on long-distance circuits, but they do have application on local trunk circuits where the gain requirements are considerably less. Trunks using VF repeaters have the repeater's gain adjusted to the equivalent loss of the circuit minus the 4-dB loss to provide the necessary singing margin. In practice, a repeater is installed at each end of the trunk circuit to simplify maintenance and power feeding. Gains may be as high as 6–8 dB.

Another repeater commonly used on two-wire trunks is the negative-impedance repeater. This repeater can provide a gain as high as 12 dB, but 7 or 8 dB is more common in practice. The negative-impedance repeater requires an LBO at each port and is a true, two-way, two-wire repeater. The repeater action is based on regenerative feedback of two amplifiers. The advantage of negative-impedance repeaters is that they are transparent to dc signaling. On the other hand, VF repeaters require a composite arrangement to pass dc signaling. This consists of a transformer by-pass (Ref. 7).

REVIEW EXERCISES

1. Define the voice channel using the band of frequencies it occupies from the CCITT perspective. What is its bandwidth?

2. Where does the sensitivity of a telephone set peak (i.e., at about what frequency) from a North American perspective? from a CCITT perspective?

3. A local serving switch provides a battery voltage source for the subscriber loop. What is the nominal voltage of this battery? Name at least three functions that this emf source provides.

4. A subscriber loop is designed basically on two limiting conditions (impairments). What are they?

[10]VF or *voice frequency* refers to the nominal 4-kHz analog voice channel defined at the beginning of this chapter.

5. What is the North American maximum loss objective for the subscriber loop?
6. What happens when we exceed the "signaling" limit on a subscriber loop?
7. What are the two effects that a load coil has on a subscriber loop or a metallic VF trunk?
8. Using resistance design or revised resistance design, our only concern is resistance. What about loss?
9. What is the *switch design limit* of RD? of RRD?
10. Where will we apply *long-route design* (LRD)?
11. Define a *range extender.*
12. With RRD, on loops greater than 1500 Ω, what expedients do we have using standardized design rules?
13. What kind of inductive loading is used with RRD?
14. Beyond what limit on a subscriber loop is it advisable to employ DLC?
15. For VF trunks, if D is the distance between load coils, why is the first load coil placed at D/2 from the exchange?
16. Define a *main frame.*
17. What does *line build-out* do?
18. What are the two types of VF repeaters that may be used on VF trunks or subscriber loops? What are practical gain values used on these amplifiers?

REFERENCES

1. *IEEE Standard Dictionary of Electrical and Electronics Terms*, 6th ed., IEEE Std-100-1996, IEEE, New York, 1996.
2. *Transmission Systems for Communications*, 5th ed., Bell Telephone Laboratories, Holmdel, NJ, 1982.
3. W. F. Tuffnell, *500-Type Telephone Sets*, Bell Labs Record, Bell Telephone Laboratories, Holmdel, NJ, Sept. 1951.
4. *Outside Plant, Telecommunication Planning Documents*, ITT Laboratories Spain, Madrid, 1973.
5. *BOC Notes on the LEC Networks—1994*, Bellcore, Livingston, NJ, 1994.
6. *Telecommunication Transmission Engineering*, 2nd ed., Vols. 1–3, American Telephone and Telegraph Co., New York, 1977.
7. R. L. Freeman, *Telecommunication Transmission Handbook*, 4th ed., Wiley, New York, 1998.

6

DIGITAL NETWORKS

6.1 INTRODUCTION TO DIGITAL TRANSMISSION

The concept of digital transmission is entirely different from its analog counterpart. With an analog signal there is *continuity*, as contrasted with a digital signal that is concerned with *discrete* states. The information content of an analog signal is conveyed by the value or magnitude of some characteristic(s) of the signal such as amplitude, frequency, or phase of a voltage; the amplitude or duration of a pulse; the angular position of a shaft; or the pressure of a fluid. To extract the information it is necessary to compare the value or magnitude of the signal to a standard. The information content of a digital signal is concerned with discrete states of the signal, such as the presence or absence of a voltage, a contact in an open or closed position, a voltage either positive- or negative-going, or that a light is on or off. The signal is given meaning by assigning numerical values or other information to the various possible combinations of the discrete states of the signal (Ref. 1).

The examples of digital signals just given are all binary. Of course, with a binary signal (a bit), the signal can only take on one of two states. This is very provident for several reasons. First, or course, with a binary system, we can utilize the number base 2 and apply binary arithmetic, if need be. The other good reason is that we can use a decision circuit where there can only two possible conditions. We call those conditions a 1 and a 0.

Key to the principal advantage of digital transmission is the employment of such simple decision circuits. We call them *regenerators*. A corrupted digital signal enters on one side, and a good, clean, nearly perfect square-wave digital signal comes out the other side. Accumulated noise on the corrupted signal stops at the regenerator. This is the principal disadvantage of analog transmission: noise accumulates. Not so with digital transmission.

Let's list some other advantages of binary digital transmission. It is compatible with the integrated circuits (ICs) such as LSI, VLSI, and VHSIC. PCs are digital. A digital signal is more tolerant of noise than its analog counterpart. It remains intelligible under very poor error performance, with bit error rates typically as low as 1×10^{-2} for voice operation. The North American PSTN is one hundred percent digital. The remainder of the industrialized world should reach that goal by the turn of the century.

The digital network is based on pulse code modulation (PCM). The general design of a PCM system was invented by Reeve, an ITT engineer from Standard Telephone Laboratories (STL), in 1937, while visiting a French ITT subsidiary. It did not become a reality until Shockley's (Bell Telephone Laboratories) invention of the transistor. Field trials of PCM systems were evident as early as 1952 in North America.

PCM is a form of time division multiplex. It has revolutionized telecommunications. Even our super-high fidelity compact disk (CD) is based on PCM.

6.1.1 Two Different PCM Standards

As we progress through this chapter, we must keep in mind that there are two quite different PCM standards. On the one hand there is a North American standard, which some call T1; we prefer the term DS1 PCM hierarchy. The other standard is the E1 hierarchy, which we sometimes refer to as the "European" system. Prior to about 1988, E1 was called CEPT30+2, where CEPT stood for Conference European Post and Telegraph (from the French). Japan has sort of a hybrid system. We do not have to travel far in the United States to encounter the E1 hierarchy—just south of the Rio Grande river (Mexico).

6.2 BASIS OF PULSE CODE MODULATION

Let's see how we can develop an equivalent PCM signal from an analog signal, typified by human speech. Our analog system model will be a simple tone, say, 1200 Hz, which we represent by a sine wave. There are three steps in the development of a PCM signal from that analog model:

1. Sampling;
2. Quantization; and
3. Coding.

6.2.1 Sampling

The cornerstone of an explanation of how PCM works is the Nyquist sampling theorem (Ref. 2), which states:

> If a band-limited signal is sampled at regular intervals of time and at a rate equal to or higher than twice the highest significant signal frequency, then the sample contains all the information of the original signal. The original signal may then be reconstructed by use of a low-pass filter.

Consider some examples of the Nyquist sampling theorem from which we derive the sampling rate:

1. The nominal 4-kHz voice channel: sampling rate is 8000 times per second (i.e., 2×4000);
2. A 15-kHz program channel: sampling rate is 30,000 times per second (i.e., $2 \times 15{,}000$);[1]
3. An analog radar product channel 56-kHz wide: sampling rate is 112,000 times per second (i.e., $2 \times 56{,}000$).

[1]A *program channel* is a communication channel that carries radio broadcast material such as music and commentary. It is a facility offered by the PSTN to radio and television broadcasters.

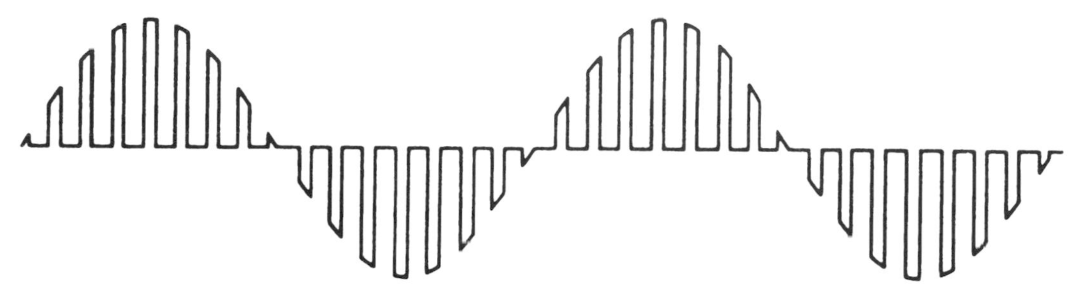

Figure 6.1 A PAM wave as a result of sampling a single sinusoid.

Our interest here, of course, is the nominal 4-kHz voice channel sampled 8000 times a second. By simple division, a sample is taken every 125 μsec, or 1-second/8000.

6.2.1.1 Pulse Amplitude Modulation (PAM) Wave. With several exceptions, practical PCM systems involve time-division multiplexing. Sampling in these cases does not just involve one voice channel but several. In practice, one system (T1) samples 24 voice channels in sequence and another (E1) samples 30 voice channels. The result of the multiple sampling is a pulse amplitude modulation (PAM) wave. A simplified PAM wave is shown in Figure 6.1. In this case, it is a single sinusoid (sine wave). A simplified diagram of the processing involved to derive a multiplex PAM wave is shown in Figure 6.2. For simplicity, only three voice channels are sampled sequentially.

The sampling is done by *gating*. It is just what the term means, a "gate" is opened for a very short period of time, just enough time to obtain a voltage sample. With the North American DS1 (T1) system, 24 voice channels are sampled sequentially and are interleaved to form a PAM-multiplexed wave. The gate is open about 5.2 μs (125/24) for each voice. The full sequence of 24 channels is sampled successively from channel 1 through channel 24 in a 125-μs period. We call this 125-μs period a *frame*, and inside the frame all 24 channels are successively sampled just once.

Another system widely used outside of the United States and Canada is E1, which is a 30-voice-channel system plus an additional two service channels, for a total of 32 channels. By definition, this system must sample 8000 times per second because it is also optimized for voice operation, and thus its frame period is 125 μs. To accommodate the 32 channels, the gate is open 125/32 or about 3.906 μs.

6.2.2 Quantization

A PCM system simply transmits to the distant end the value of a voltage sample at a certain moment in time. Our goal is to assign a binary sequence to each of those voltage samples. For argument's sake we will contain the maximum excursion of the PAM wave to within +1 V to −1 V. In the PAM waveform there could be an infinite number of different values of voltage between +1 V and −1 V. For instance, one value could be −0.3875631 V. To assign a binary sequence to each voltage value, we would have to construct a code of infinite length. So we must limit the number of voltage values between +1 V and −1 V, and the values must be discrete. For example, we could set 20 discrete values between +1 V and −1 V, with each value at a discrete 0.1-V increment.

Because we are working in the binary domain, we select the total number of discrete values to be a binary number multiple (i.e., 2, 4, 8, 16, 32, 64, 128, etc.). This facilitates binary coding. For instance, if there were four values, they would be as follows: 00,

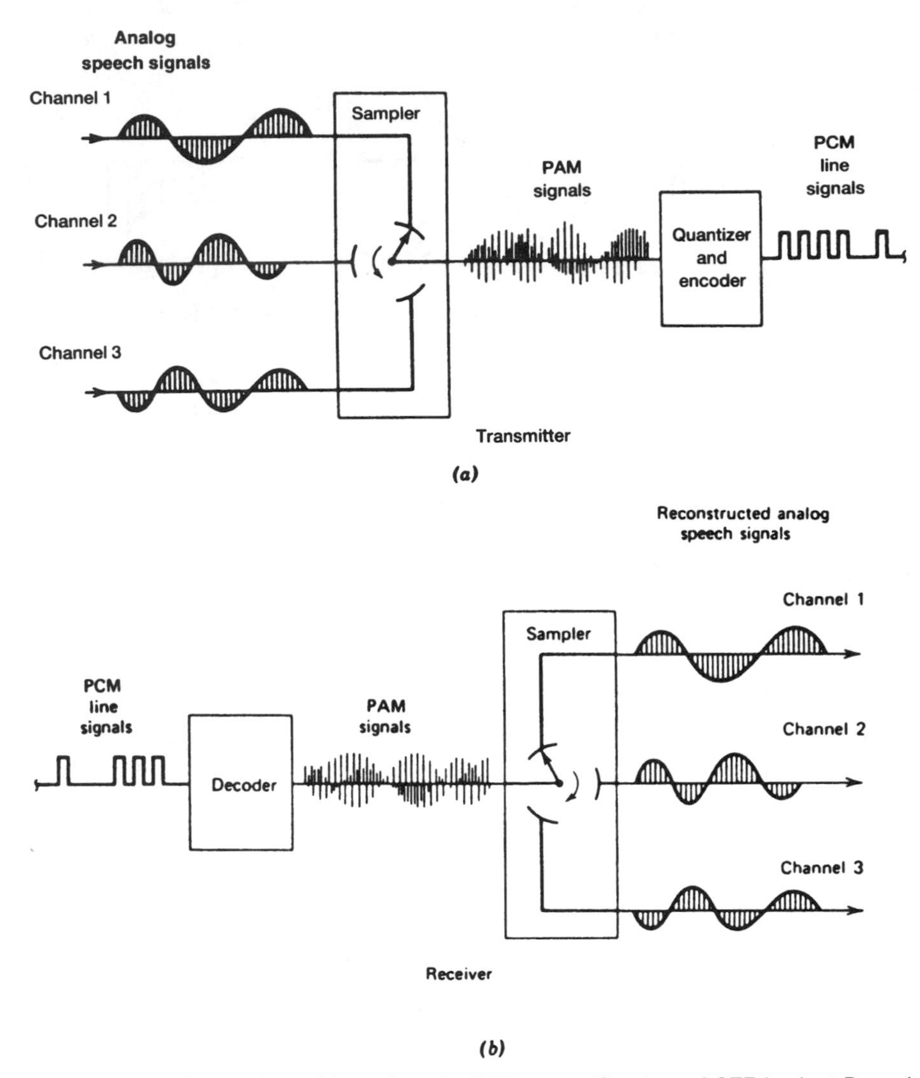

Figure 6.2 A simplified analogy of formation of a PAM wave. (Courtesy of GTE Lenkurt Demodulator, San Carlos, CA.)

01, 10, and 11. This is a 2-bit code. A 3-bit code would yield eight different binary numbers. We find, then, that the number of total possible different binary combinations given a code of n binary symbols (bits) is 2^n. A 7-bit code has 128 different binary combinations (i.e., $2^7 = 128$).

For the quantization process, we want to present to the coder a discrete voltage value. Suppose our quantization steps were on 0.1-V increments and our voltage measure for one sample was 0.37 V. That would have to be rounded off to 0.4 V, the nearest discrete value. Note here that there is a 0.03-V error, the difference between 0.37 V and 0.40 V.

Figure 6.3 shows one cycle of the PAM wave appearing in Figure 6.1, where we use a 4-bit code. In the figure a 4-bit code is used that allows 16 different binary-coded possibilities or levels between +1 V and −1 V. Thus we can assign eight possibilities above the origin and eight possibilities below the origin. These 16 quantum steps are coded as follows:

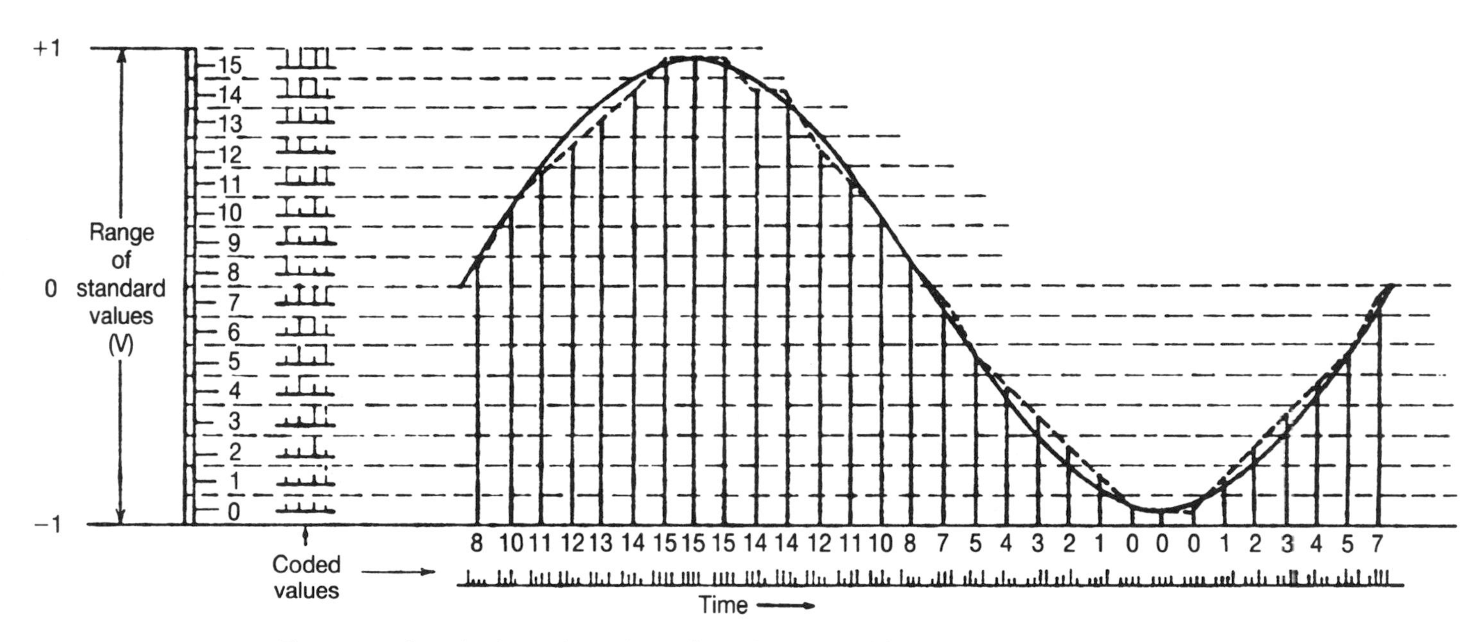

Figure 6.3 Quantization and resulting coding using 16 quantizing steps.

STEP NUMBER	CODE	STEP NUMBER	CODE
0	0000	8	1000
1	0001	9	1001
2	0010	10	1010
3	0011	11	1011
4	0100	12	1100
5	0101	13	1101
6	0110	14	1110
7	0111	15	1111

Examining Figure 6.3 shows that step 12 is used twice. Neither time it is used is it the true value of the impinging sinusoid voltage. It is a rounded-off value. These rounded-off values are shown with the dashed lines in Figure 6.3, which follows the general outline of the sinusoid. The horizontal dashed lines show the point where the quantum changes to the next higher or lower level if the sinusoid curve is above or below that value. Take step 14 in the curve, for example. The curve, dropping from its maximum, is given two values of 14 consecutively. For the first, the curve resides above 14, and for the second, below. That error, in the case of 14, from the quantum value to the true value is called *quantizing distortion*. This distortion is a major source of imperfection in PCM systems.

Let's set this discussion aside for a moment and consider an historical analogy. One of my daughters was having trouble with short division in school. Of course her Dad was there to help. Divide 4 into 13. The answer is 3 with a remainder. Divide 5 into 22 and the answer is 4 with a remainder. Of course she would tell what the remainders were. But for you, the reader, I say let's just throw away the remainders. That which we throw away is the error between the quantized value and the real value. That which we throw away gives rise to quantization distortion.

In Figure 6.3, maintaining the −1, 0, +1 volt relationship, let us double the number of quantum steps from 16 to 32. What improvement would we achieve in quantization distortion? First determine the step increment in millivolts in each case. In the first case the total range of 2000 mV would be divided into 16 steps, or 125 mV/step. The second case would have 2000/32 or 62.5 mV/step. For the 16-step case, the worst quantizing error (distortion) would occur when an input to be quantized was at the half-step level or, in this case, 125/2 or 62.5 mV above or below the nearest quantizing step. For the 32-step case, the worst quantizing error (distortion) would again be at the half-step level, or 62.5/2 or 31.25 mV. Thus the improvement in decibels for doubling the number of quantizing steps is

$$20 \log \frac{62.5}{31.25} = 20 \log 2 \text{ or } 6 \text{ dB (approximately).}$$

This is valid for linear quantization only. Thus increasing the number of quantizing steps for a fixed range of input values reduces quantizing distortion accordingly.

Voice transmission presents a problem. It has a wide dynamic range, on the order of 50 dB. That is the level range from the loudest syllable of the loudest talker to lowest-level syllable of the quietest talker. Using linear quantization, we find it would require 2048 discrete steps to provide any fidelity at all. Since 2048 is 2^{11}, this means we

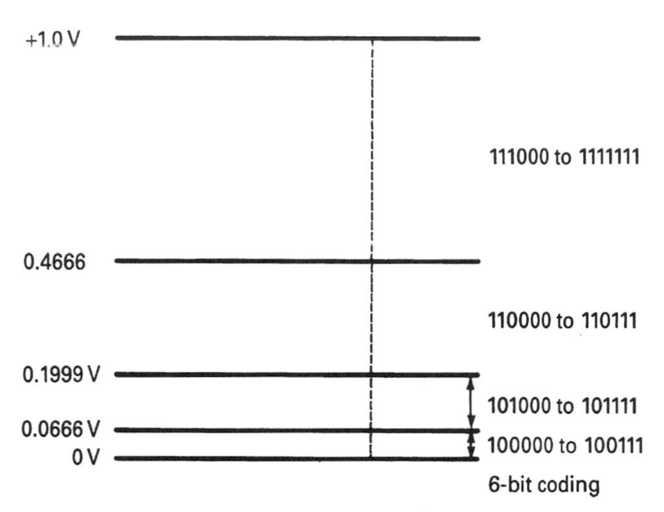

Figure 6.4 Simple graphic representation of compression. Six-bit coding, eight six-bit sequences per segment.

would need an 11-bit code. Such a code sampled 8000 times per second leads to 88,000-bps equivalent voice channel and an 88-kHz bandwidth, assuming 1 bit per Hz. Designers felt this was too great a bit rate/bandwidth.

They turned to an old analog technique of companding. *Companding* derives from two words: compression and expansion. Compression takes place on the transmit side of the circuit; expansion on the receive side. Compression reduces the dynamic range with little loss of fidelity, and expansion returns the signal to its normal condition. This is done by favoring low-level speech over higher-level speech. In other words, more code segments are assigned to speech bursts at low levels than at the higher levels, progressively more as the level reduces. This is shown graphically in Figure 6.4, where eight coded sequences are assigned to each level grouping. The smallest range rises only 0.0666 V from the origin (assigned to 0-V level). The largest extends over 0.5 V, and it is assigned only eight coded sequences.

6.2.3 Coding

Older PCM systems used a 7-bit code, and modern systems use an 8-bit code with its improved quantizing distortion performance. The companding and coding are carried out together, simultaneously. The compression and later expansion functions are logarithmic. A pseudologarithmic curve made up of linear segments imparts finer granularity to low-level signals and less granularity to the higher-level signals. The logarithmic curve follows one of two laws, the A-law and the μ-law (pronounced mu-law). The curve for the A-law may be plotted from the formula:

$$F_A(x) = \left(\frac{A|x|}{1 + \ln(A)} \right) \qquad 0 \leq |x| \leq \frac{1}{A}$$

$$F_A(x) = \left(\frac{1 + \ln|Ax|}{1 + \ln(A)} \right) \qquad \frac{1}{A} \leq |x| \leq 1,$$

where A = 87.6. (Note: The notation ln indicates a logarithm to the natural base called *e*. Its value is 2.7182818340.) The A-law is used with the E1 system. The curve for the μ-law is plotted from the formula:

$$F_\mu(x) = \frac{\ln(1 + \mu|x|)}{\ln(1 + \mu)},$$

where x is the signal input amplitude and μ = 100 for the original North American T1 system (now outdated), and 255 for later North American (DS1) systems and the CCITT 24-channel system (CCITT Rec. G.733). (See Ref. 3).

A common expression used in dealing with the "quality" of a PCM signal is *signal-to-distortion* ratio (S/D, expressed in dB). Parameters A and μ, for the respective companding laws, determine the range over which the signal-to-distortion ratio is comparatively constant, about 26 dB. For A-law companding, an S/D = 37.5 dB can be expected (A = 87.6). And for μ-law companding, we can expect S/D = 37 dB (μ = 255) (Ref. 4).

Turn now to Figure 6.5, which shows the companding curve and resulting coding for the European E1 system. Note that the curve consists of linear piecewise segments, seven above and seven below the origin. The segment just above and the segment just below the origin consist of two linear elements. Counting the collinear elements by the

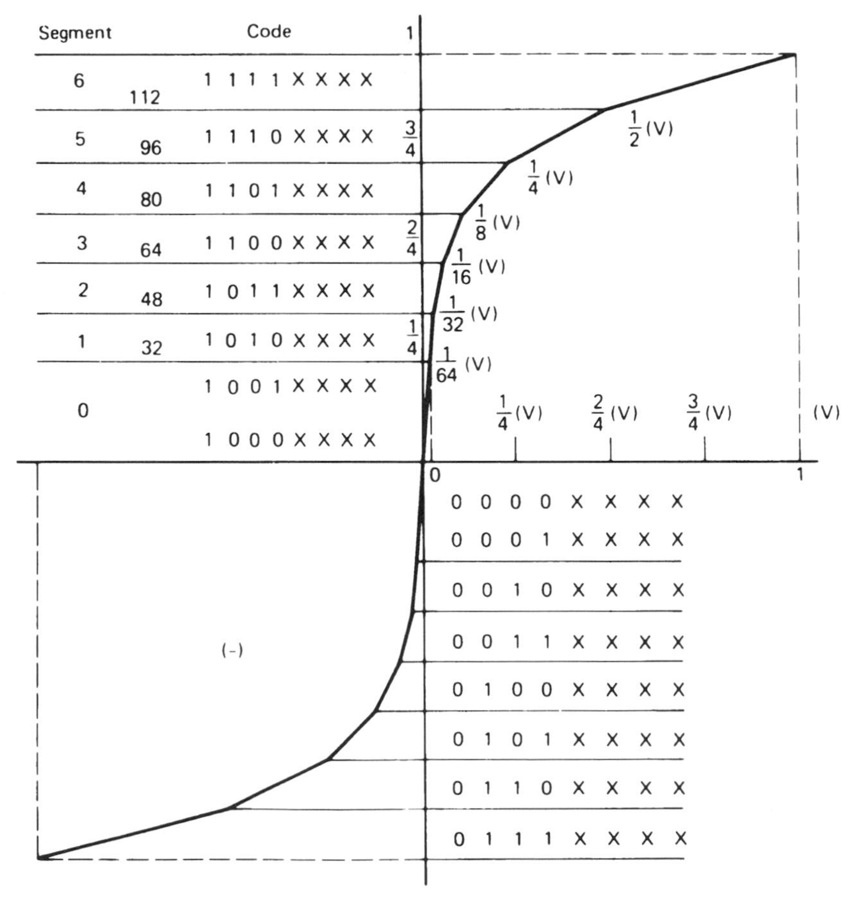

Figure 6.5 The 13-segment approximation of the A-law curve used with E1 PCM equipment.

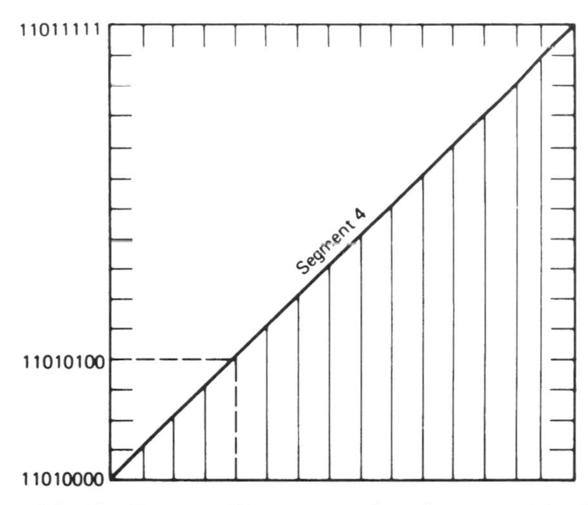

Figure 6.6 The European E1 system, coding of segment 4 (positive).

origin, there are 16 segments. Each segment has 16 8-bit PCM codewords assigned. These are the codewords that identify the voltage level of a sample at some moment in time. Each codeword, often called a PCM "word," consists of 8 bits. The first bit (most significant bit) tells the distant-end receiver if the sample is a positive or negative voltage. Observe that all PCM words above the origin start with a binary 1, and those below the origin start with a binary 0. The next 3 bits in sequence identify the segment. There are eight segments (or collinear equivalents) above the origin and eight below ($2^3 = 8$). The last 4 bits, shown in the figure as XXXX, indicate exactly where in a particular segment that voltage line is located.

Suppose the distant-end received the binary sequence 11010100 in an El system. The first bit indicates that the voltage is positive (i.e., above the origin in Figure 6.5). The next three bits, 101, indicate that the sample is in segment 4 (positive). The last 4 bits, 0100, tell the distant end where it is in that segment as illustrated in Figure 6.6. Note that the 16 steps inside the segment are linear. Figure 6.7 shows an equivalent logarithmic curve for the North American DS1 system.[2] It uses a 15-segment approximation of the logarithmic μ-law curve ($\mu = 255$). The segments cutting the origin are collinear and are counted as one. So, again, we have a total of 16 segments.

The coding process in PCM utilizes straightforward binary codes. Examples of such codes are illustrated in Figure 6.5, and expanded in Figure 6.6 and Figure 6.7.

The North American DS1 (T1) PCM system uses a 15-segment approximation of the logarithmic μ-law ($\mu = 255$), shown in Figure 6.7. The segments cutting the origin are collinear and are counted as one. As can be seen in Figure 6.7, similar to Figure 6.5, the first code element (bit), whether a 1 or a 0, indicates to the distant end whether the sample voltage is positive or negative, above or below the horizontal axis. The next three elements (bits) identify the segment and the last four elements (bits) identify the actual quantum level inside the segment.

6.2.3.1 Concept of Frame. As is illustrated in Figure 6.2, PCM multiplexing is carried out with the sampling process, sampling the analog sources sequentially. These sources may be the nominal 4-kHz voice channels or other information sources that

[2]More popularly referred to as T1.

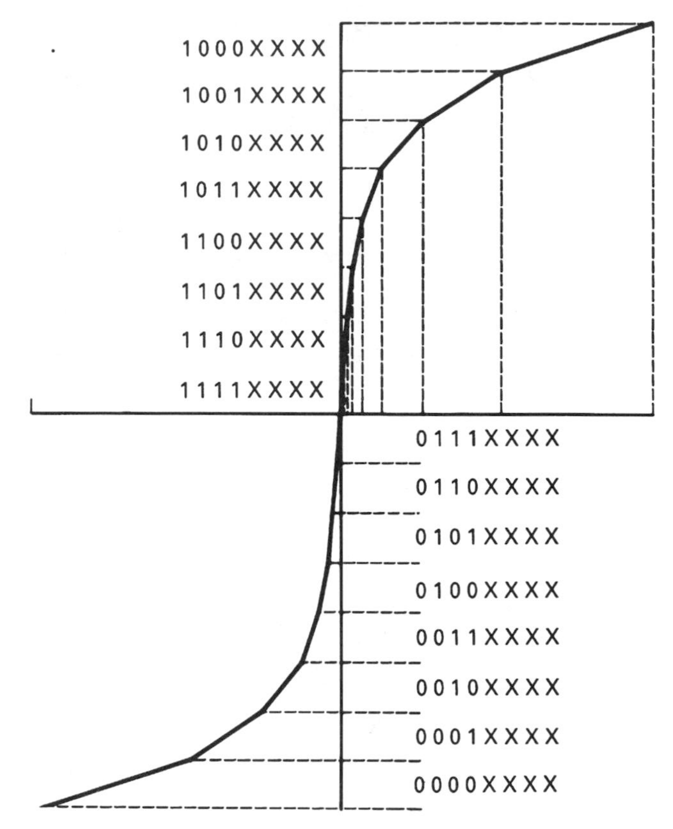

Figure 6.7 Piecewise linear approximation of the μ-law logarithmic curve used with the DS1 format.

have a 4-kHz bandwidth, such as data or freeze-frame video. The final result of the sampling and subsequent quantization and coding is a series of electrical pulses, a serial bit stream of 1s and 0s that requires some identification or indication of the beginning of a sampling sequence. This identification is necessary so that the far-end receiver knows exactly when the sampling sequence starts. Once the receiver receives the "indication," it knows a priori (in the case of DS1) that 24 eight-bit slots follow. It synchronizes the receiver. Such identification is carried out by a *framing bit*, and one full sequence or cycle of samples is called a *frame* in PCM terminology.

Consider the framing structure of the two widely implemented PCM systems: the North American DS1 and the European E1. The North American DS1 system is a 24-channel PCM system using 8-level coding (e.g., $2^8 = 256$ quantizing steps or distinct PCM code words). Supervisory signaling is "in-band" where bit 8 of every sixth frame is "robbed" for supervisory signaling.[3–5] The DS1 format shown in Figure 6.8 has one bit added as a framing bit. (This is that indication to tell the distant end receiver where the frame starts.) It is called the "S" bit. The DS1 frame then consists of:

$$(8 \times 24) + 1 = 193 \text{ bits},$$

[3]"In-band," an unfortunate expression harking back to the analog world.

[4]In the DS1 system it should be noted that in each frame that has bit 8 "robbed," 7-bit coding is used versus 8-bit coding employed on the other five frames.

[5]*Supervisory signaling* is discussed in Chapter 7. All supervisory signaling does is tell us if the channel is busy or idle.

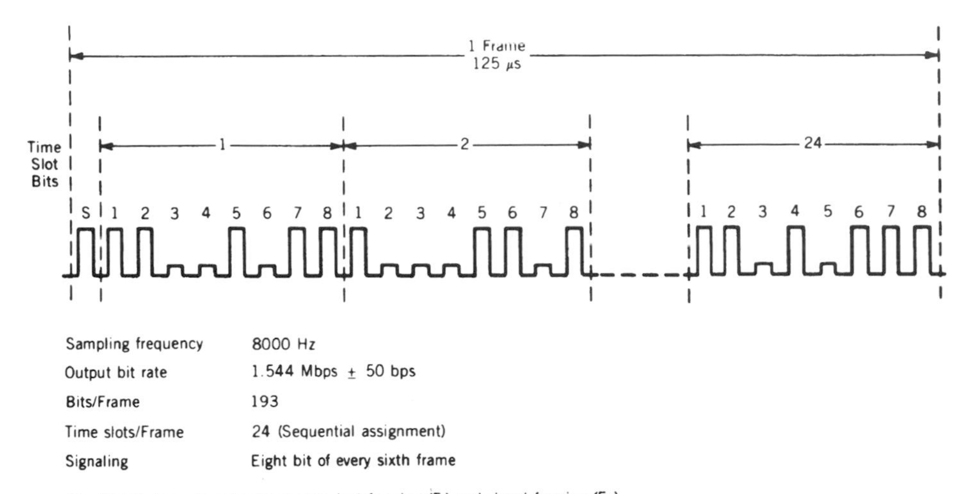

Figure 6.8 DS1 signal format.

making up a full sequence or frame. By definition, 8000 frames are transmitted per second (i.e., 4000 × 2, the Nyquist sampling rate), so the bit rate of DS1 is:

$$193 \times 8000 = 1,544,000 \text{ bps} \qquad \text{or } 1.544 \text{ Mbps.}$$

The E1 European PCM system is a 32-channel system. Of the 32 channels, 30 transmit speech (or data) derived from incoming telephone trunks and the remaining 2 channels transmit synchronization-alignment and signaling information. Each channel is allotted an 8-bit time slot (TS), and we tabulate TS 0 through 31 as follows:

TS	TYPE OF INFORMATION
0	Synchronizing (framing)
1–15	Speech
16	Signaling
17–31	Speech

In TS 0 a synchronizing code or word is transmitted every second frame, occupying digits 2 through 8 as follows:

0011011

In those frames without the synchronizing word, the second bit of TS 0 is frozen at a 1 so that in these frames the synchronizing word cannot be imitated. The remaining bits of TS 0 can be used for the transmission of supervisory information signals [Ref. 16].

Again, E1 in its primary rate format transmits 32 channels of 8-bit time slots. An E1 frame therefore has 8 × 32 = 256 bits. There is no framing bit. Framing alignment is carried out in TS 0. The E1 bit rate to the line is:

$$256 \times 8000 = 2{,}048{,}000 \text{ bps} \qquad \text{or } 2.048 \text{ Mbps}$$

Framing and basic timing should be distinguished. "Framing" ensures that the PCM receiver is aligned regarding the beginning (and end) of a bit sequence or frame; "timing" refers to the synchronization of the receiver clock, specifically, that it is in step with its companion far-end transmit clock. Timing at the receiver is corrected via the incoming "1"-to-"0" and "0"-to-"1" transitions.[6] It is mandatory that long periods of no transitions do not occur. This important point is discussed later in reference to line codes and digit inversion.

6.3 PCM SYSTEM OPERATION

PCM channel banks operate on a four-wire basis. Voice channel inputs and outputs to and from a PCM multiplex channel bank are four-wire, or must be converted to four-wire in the channel bank equipment. Another term commonly used for channel bank is *codec*, which is a contraction for *coder-decoder* even though the equipment carries out more functions than just coding and decoding. A block diagram of a typical codec (PCM channel bank) is shown in Figure 6.9.

A codec accepts 24 or 30 voice channels, depending on the system used; digitizes and multiplexes the information; and delivers a serial bit stream to the line of 1.544

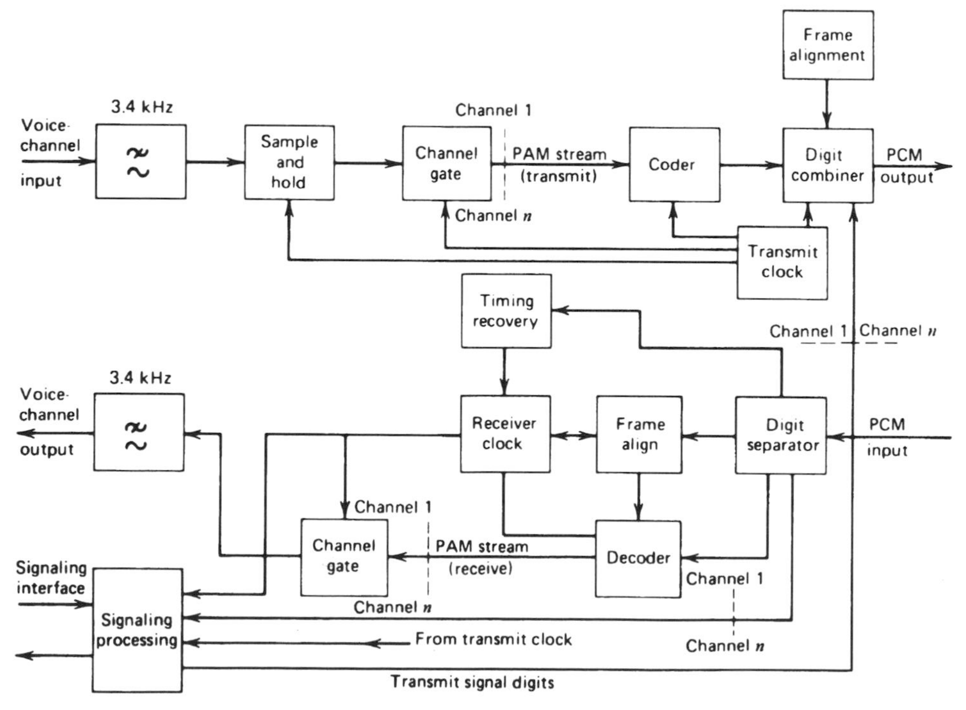

Figure 6.9 Simplified functional block diagram of a PCM codec or channel bank.

[6]A *transition* in this context is a change of electrical state. We often use the term "mark" for a binary 1 and "space" for a binary 0. The terms mark and space come from old-time automatic telegraphy and have been passed on through the data world to the parlance of digital communications technology.

Mbps or 2.048 Mbps. It accepts a serial bit stream at one or the other modulation rate, demultiplexes the digital information, and performs digital-to-analog conversion. Output to the analog telecommunications network is the 24 or 30 nominal 4-kHz voice channels. Figure 6.9 illustrates the processing of a single analog voice channel through a codec. The voice channel to be transmitted is first passed through a 3.4-kHz low-pass filter. The output of the filter is fed to a sampling circuit. The sample of each channel of a set of n channels (n usually equals 24 or 30) is released in turn to the pulse amplitude modulation (PAM) highway. The release of samples is under control of a channel gating pulse derived from the transmit clock. The input to the coder is the PAM highway. The coder accepts a sample of each channel in sequence and then generates the appropriate 8-bit signal character corresponding to each sample presented. The coder output is the basic PCM signal that is fed to the digit combiner where framing-alignment signals are inserted in the appropriate time slots, as well as the necessary supervisory signaling digits corresponding to each channel (European approach), and are placed on a common signaling highway that makes up one equivalent channel of the multiplex serial bit stream transmitted to the line. In North American practice supervisory signaling is carried out somewhat differently, by "bit robbing," as previously mentioned, such as bit 8 in frame 12. Thus each equivalent voice channel carries its own signaling.

On the receive side the codec accepts the serial PCM bit stream, inputting the digit separator, where the signal is regenerated and split, delivering the PCM signal to four locations to carry out the following processing functions: (1) timing recovery, (2) decoding, (3) frame alignment, and (4) signaling (supervisory). Timing recovery keeps the receive clock in synchronism with the far-end transmit clock. The receive clock provides the necessary gating pulses for the receive side of the PCM codec. The frame-alignment circuit senses the presence of the frame-alignment signal at the correct time interval, thus providing the receive terminal with frame alignment. The decoder, under control of the receive clock, decodes the code character signals corresponding to each channel. The output of the decoder is the reconstituted pulses making up a PAM highway. The channel gate accepts the PAM highway, gating the n-channel PAM highway in sequence under control of the receive clock. The output of the channel gate is fed, in turn, to each channel filter, thus enabling the reconstituted analog voice signal to reach the appropriate voice path. Gating pulses extract signaling information in the signaling processor and apply this information to each of the reconstituted voice channels with the supervisory signaling interface as required by the analog telephone system in question.

6.4 LINE CODE

When PCM signals are transmitted to the cable plant, they are in the bipolar mode, as illustrated in Figure 6.10. The marks or 1s have only a 50% duty cycle. There are some advantages to this mode of transmission.

- No dc return is required; thus transformer coupling can be used on the line.
- The power spectrum of the transmitted signal is centered at a frequency equivalent to half the bit rate.

It will be noted in bipolar transmission that the 0s are coded as absence of pulses and 1s are alternately coded as positive and negative pulses, with the alternation taking

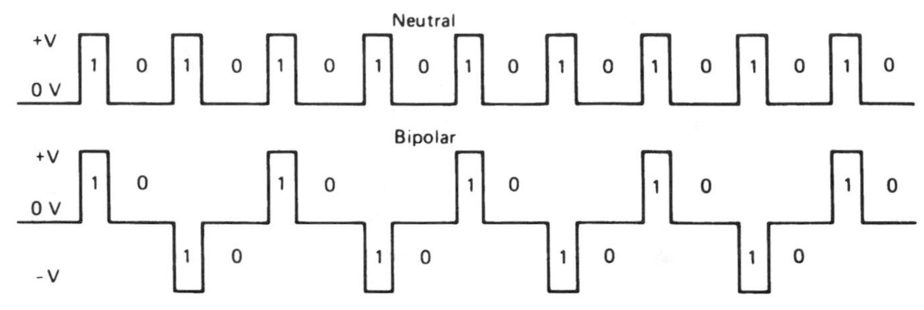

Figure 6.10 Neutral versus bipolar bit streams. The upper diagram illustrates alternating 1s and 0s transmitted in the neutral mode; the lower diagram the equivalent in the bipolar mode, which is also called alternate mark inversion or AMI. Note that in the neutral mode, the 0 state is inactive, 0 volts. Neutral transmission is discussed in Chapter 10.

place at every occurrence of a 1. This mode of transmission is also called *alternate mark inversion* (AMI).

One drawback to straightforward AMI transmission is that when a long string of 0s is transmitted (e.g., no transitions), a timing problem may arise because repeaters and decoders have no way of extracting timing without transitions. The problem can be alleviated by forbidding long strings of 0s. Codes have been developed that are bipolar but with *N* 0s substitution; they are called "B*N*ZS" codes. For instance, a B6ZS code substitutes a particular signal for a string of six 0s. B8ZS is used on subscriber loop carrier and inserts a violation after a string of 8 zeros.

Another such code is the HDB3 code (high-density binary 3), where the 3 indicates substitution for binary sequences with more than three consecutive 0s. With HDB3, the second and third 0s of the string are transmitted unchanged. The fourth 0 is transmitted to the line with the same polarity as the previous mark sent, which is a "violation" of the AMI concept. The first 0 may or may not be modified to a 1, to ensure that the successive violations are of opposite polarity. HDB3 is used with European E series PCM systems and is similar to B3ZS.

6.5 SIGNAL-TO-GAUSSIAN-NOISE RATIO ON PCM REPEATERED LINES

As mentioned earlier, noise accumulation on PCM systems is not a crucial issue. However, this does not mean that Gaussian noise (or crosstalk or impulse noise) is unimportant.[7] Indeed, it will affect error performance expressed as error rate. Errors are cumulative, and as we go down a PCM-repeatered line, the error performance degrades. A decision in error, whether a 1 or a 0, made anywhere in the digital system, is not recoverable. Thus such an incorrect decision made by one regenerative repeater adds to the existing error rate on the line, and errors taking place in subsequent repeaters further down the line add in a cumulative manner, thus deteriorating the received signal.

In a purely binary transmission system, if a 22-dB signal-to-noise ratio is maintained, the system operates nearly error free.[8] In this respect, consider Table 6.1.

As discussed in Section 6.4, PCM, in practice, is transmitted on-line with alternate mark inversion (in the bipolar mode). The marks (1s) have a 50% duty cycle, permitting signal energy concentration at a frequency equivalent to half the transmitted bit rate.

[7]Gaussian noise is the same as thermal noise.

[8]It is against the laws of physics to have a completely error-free system.

Table 6.1 Error Rate of a Binary Transmission System Versus Signal-to-rms-Noise Ratio

Error Rate	S/N (dB)	Error Rate	S/N (dB)
10^{-2}	13.5	10^{-7}	20.3
10^{-3}	16.0	10^{-8}	21.0
10^{-4}	17.5	10^{-9}	21.6
10^{-5}	18.7	10^{-10}	22.0
10^{-6}	19.6	10^{-11}	22.2

Thus it is advisable to add 1 dB or 2 dB to the values shown in Table 6.1 to achieve the desired error performance in a practical system.

6.6 REGENERATIVE REPEATERS

As we are probably aware, pulses passing down a digital transmission line suffer attenuation and are badly distorted by the frequency characteristic of the line. A regenerative repeater amplifies and reconstructs such a badly distorted digital signal and develops a nearly perfect replica of the original at its output. Regenerative repeaters are an essential key to digital transmission in that we could say that the "noise stops at the repeater."

Figure 6.11 is a simplified block diagram of a regenerative repeater and shows typical waveforms corresponding to each functional stage of signal processing. As illustrated in the figure, at the first stage of signal processing is amplification and equalization. With many regenerative repeaters, equalization is a two-step process. The first is a fixed equalizer that compensates for the attenuation-frequency characteristic (attenuation distortion), which is caused by the standard length of transmission line between repeaters (often 6000 ft or 1830 m). The second equalizer is variable and compensates for departures between nominal repeater section length and the actual length as well as loss variations due to temperature. The adjustable equalizer uses automatic line build-

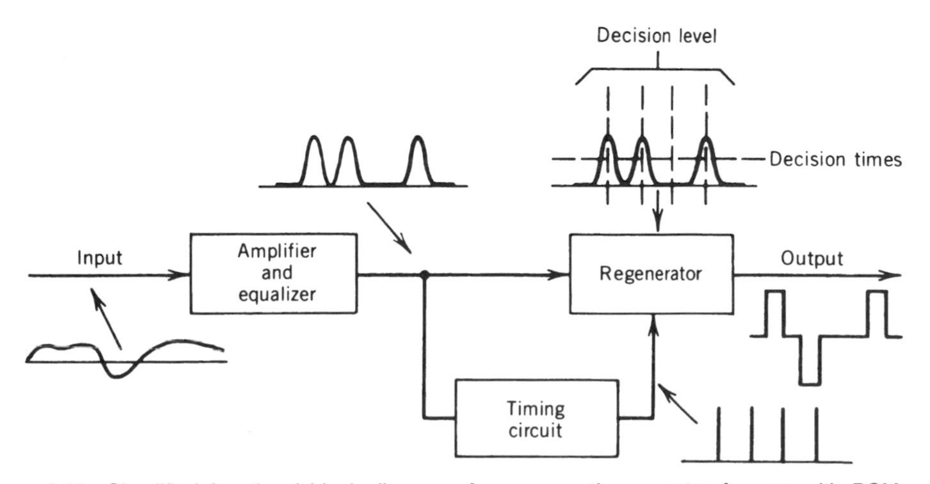

Figure 6.11 Simplified functional block diagram of a regenerative repeater for use with PCM cable systems.

out (ALBO) networks that are automatically adjusted according to characteristics of the received signal.[9]

The signal output of the repeater must be precisely timed to maintain accurate pulse width and space between the pulses. The timing is derived from the incoming bit stream. The incoming signal is rectified and clipped, producing square waves that are applied to the timing extractor, which is a circuit tuned to the timing frequency. The output of the circuit controls a clock-pulse generator that produces an output of narrow pulses that are alternately positive and negative at the zero crossings of the square-wave input.

The narrow positive clock pulses gate the incoming pulses of the regenerator, and the negative pulses are used to run off the regenerator. Thus the combination is used to control the width of the regenerated pulses.

Regenerative repeaters are the major source of timing jitter in a digital transmission system. Jitter is one of the principal impairments in a digital network, giving rise to pulse distortion and intersymbol interference. Jitter is discussed in more detail in Section 6.9.2.

Most regenerative repeaters transmit a bipolar (AMI) waveform (see Figure 6.10). Such signals can have one of three possible states in any instant in time: positive, zero or negative (volts), and are often designated +, 0, −. The threshold circuits are gates to admit the signal at the middle of the pulse interval. For instance, if the signal is positive and exceeds a positive threshold, it is recognized as a positive pulse. If it is negative and exceeds a negative threshold, it is recognized as a negative pulse. If it has a (voltage) value between the positive and negative voltage thresholds, it is recognized as a 0 (no pulse).

When either threshold is exceeded, the regenerator is triggered to produce a pulse of the appropriate duration, polarity, and amplitude. In this manner the distorted input signal is reconstructed as a new output signal for transmission to the next repeater or terminal facility.

6.7 PCM SYSTEM ENHANCEMENTS

6.7.1 Enhancements to DS1

The PCM frame rate is 8000 frames a second. With DS1, each frame has one framing bit. Thus 8000 framing bits are transmitted per second. With modern processor technology, all of the 8000 framing bits are not needed to keep the system frame-aligned. Only one-quarter of the 8000 framing bits per second are actually necessary for framing and the remainder of the bits, 6000 bits per second, can be used for other purposes such as on-line gross error detection and for a maintenance data link. To make good use of these overhead bits, DS1 frames are taken either 12 or 24 at a time. These groupings are called *superframe* and *extended superframe*, respectively. The extended superframe, in particular, provides excellent facilities for on-line error monitoring and troubleshooting.

6.7.2 Enhancements to E1

Remember that timeslot 0 in the E1 format is the synchronization channel, with a channel bit rate of 64 kbps. Only half of these bits are required for synchronization; the remainder, 32 kbps, is available for on-line error monitoring, for a data channel for

[9]*Line buildout* is the adding of capacitance and/or resistance to a transmission line to look "electrically" longer or shorter than it actually is physically.

remote alarms. These remote alarms tell the system operator about the status of the distant PCM terminal.

6.8 HIGHER-ORDER PCM MULTIPLEX SYSTEMS

6.8.1 Introduction

Higher-order PCM multiplex is developed out of several primary multiplex sources. Primary multiplex is typically DS1 in North America and E1 in Europe; some countries have standardized on E1, such as most of Hispanic America. Not only are E1 and DS1 incompatible, the higher-order multiplexes, as one might imagine, are also incompatible. First we introduce *stuffing*, describe some North American higher-level multiplex, and then discuss European multiplexes based on the E1 system.

6.8.2 Stuffing and Justification

Stuffing (justification) is common to all higher-level multiplexers that we describe in the following. Consider the DS2 higher-level multiplex. It derives from an M12 multiplexer, taking inputs from four 24-channel channel banks. The clocks in these channel banks are free running. The transmission rate output of each channel bank is *nominally* 1,544,000 bps. However, there is a tolerance of ±50 ppm (±77 bps). Suppose all four DS1 inputs were operating on the high side of the tolerance or at 1,544,077 bps. The input to the M12 multiplexer is a buffer. It has a finite capacity. Unless bits are read out of the buffer faster than they are coming in, at some time the buffer will overflow. This is highly undesirable. Thus we have bit stuffing.

Stuffing in the output aggregate bit stream means adding extra bits. It allows us to read out of a buffer faster than we write into it.

In Ref. 1 the IEEE defines *stuffing bits* as "bits inserted into a frame to compensate for timing differences in constituent lower rate signals." CCITT uses the term *justification*. Figure 6.12 illustrates the stuffing concept.

6.8.3 North American Higher-Level Multiplex

The North American digital hierarchy is illustrated in Figure 6.13. The higher-level multiplexers are type-coded in such a way that we know the DS levels (e.g., DS1, DS1C,

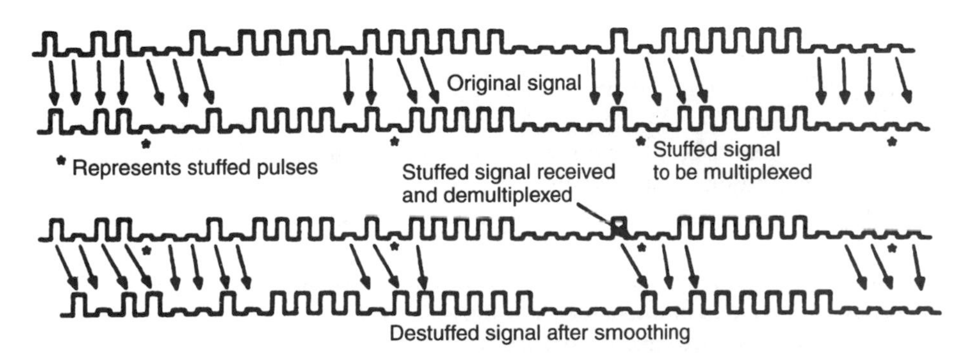

Figure 6.12 Pulse stuffing synchronization. (From Ref. 5, Figure 29-2.)

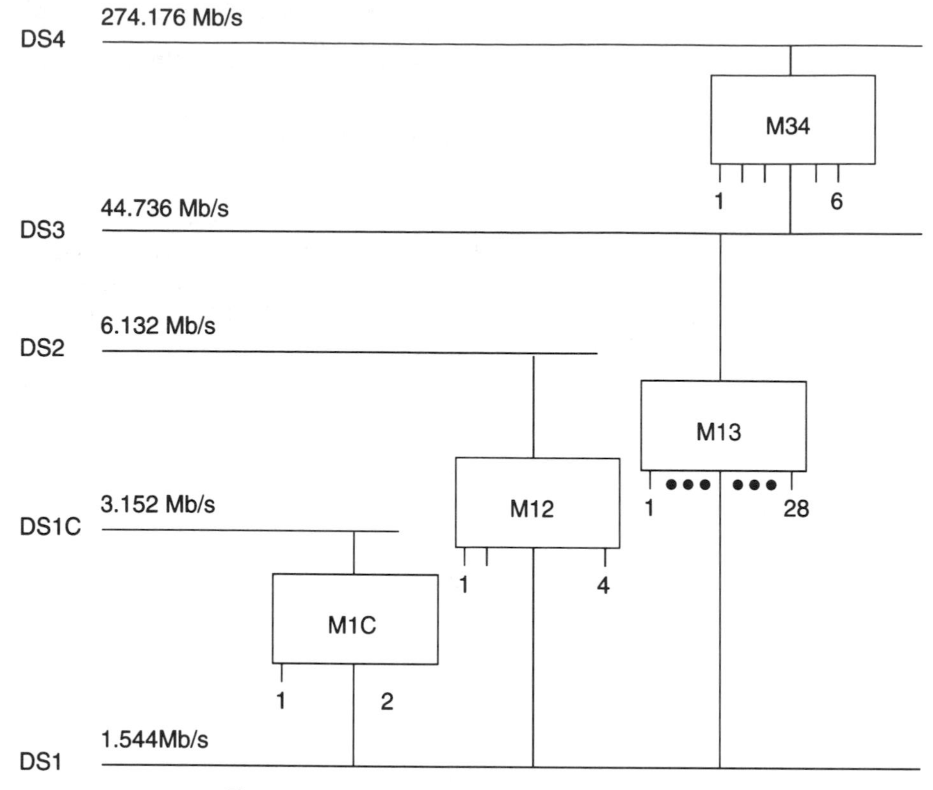

Figure 6.13 The North American Digital Hierarchy.

DS2, DS3, DS4) that are being combined. For instance, the M34 multiplexer takes four DS3 bit streams at its input to form the DS4 bit stream at its output. We describe the operation of the M12 multiplexer because it typifies the series.

The formation of the second-level North American multiplex, DS2, from four DS1 inputs is shown in Figure 6.14. There are four inputs to the M12 multiplexer, each operating at the nominal 1.544 Mbps rate. The output bit rate is 6.312 Mbps. Now multiply 1.544 Mbps by 4 and get 6.176 Mbps. In other words, the output of the M12 multiplexer is operating 136 kbps faster than the aggregate of the four inputs. Some of these extra bits are overhead bits and the remainder are stuff bits. Figure 6.15 shows the makeup of the DS2 frame.

The M12 multiplex frame consists of 1176 bits. The frame is divided into four 294-bit subframes, as illustrated in Figure 6.15. There is a control bit word that is distributed throughout the frame and that begins with an M bit. Thus each subframe begins with an M bit. There are four M bits forming the series 011X, where the fourth bit (X), which may be a 1 or a 0, may be used as an alarm indicator bit. When transmitted as a 1, no alarm condition exists. When it is transmitted as a 0, an alarm is present. The 011 sequence for the first three M bits is used in the receiving circuits to identify the frame.

It is noted in Figure 6.15 that each subframe is made up of six 49-bit blocks. Each block starts with a control bit, which is followed by a 48-bit block of information. Of these 48 bits, 12 bits are taken from each of the four input DS1 signals. These are interleaved sequentially in the 48-bit block. The first bit in the third and sixth block is

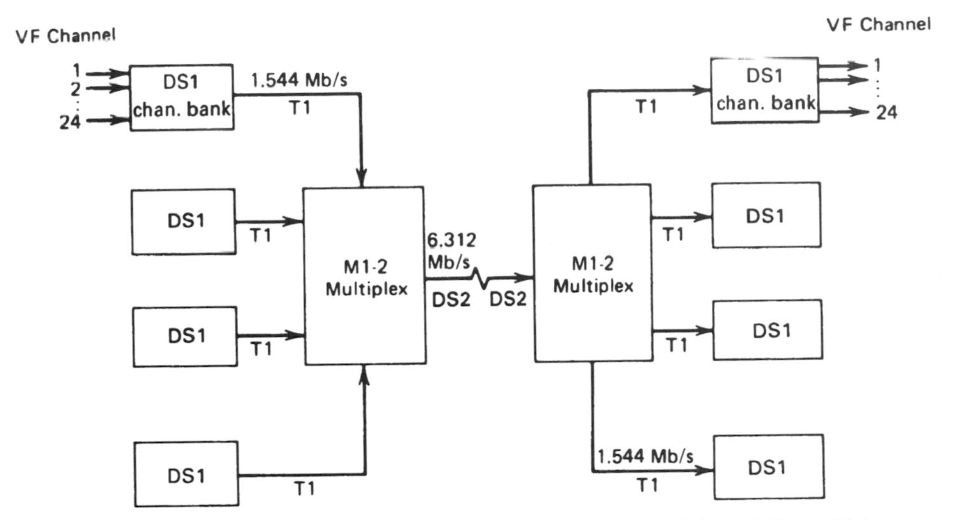

Figure 6.14 The formation of the DS2 signal from four DS1 signals in an M12 multiplexer.

designated an F bit. The F bits are a 0101 ... sequence used to identify the location of the control bit sequence and the start of each block of information bits.

6.8.4 Euopean E1 Digital Hierarchy

The E1 hierarchy is identified in a similar manner as the DS1 hierarchy. E1 (30 voice channels) is the primary multiplex; E2 is the second level and is derived from four E1s. Thus E2 contains 120 equivalent digital voice channels. E3 is the third level and it is derived from four E2 inputs and contains 480 equivalent voice channels. E4 derives from four E3 formations and contains the equivalent of 1920 voice channels. International digital hierarchies are compared in Table 6.2. Table 6.3 provides the basic parameters for the formation of the E2 level in the European digital hierarchy.

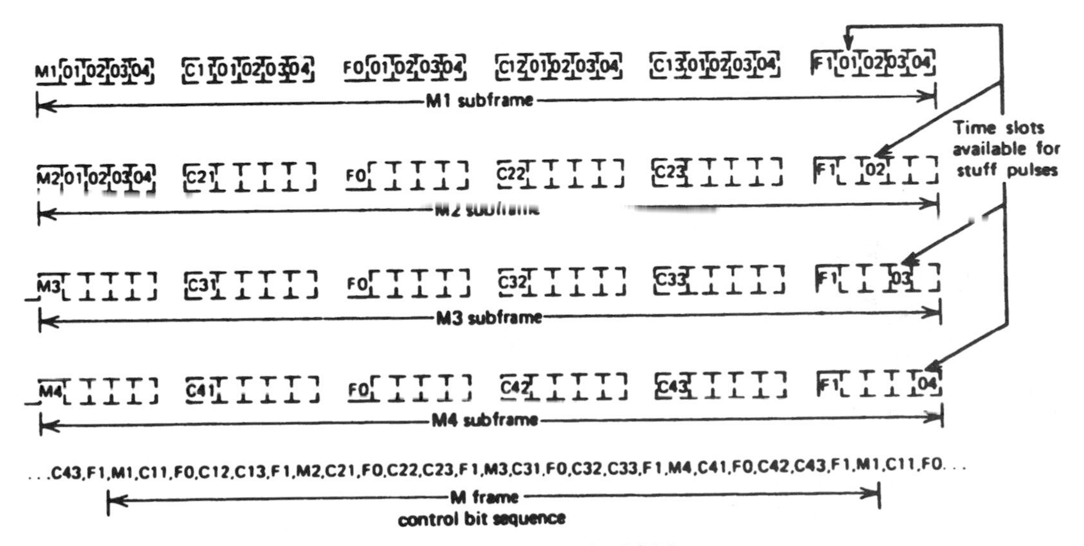

Figure 6.15 Makeup of a DS2 frame.

Table 6.2 Higher-Level PCM Multiplex Comparison

System Type	Level 1	2	3	4	5
North American T/D type	1	2	3	4	
Numer of voice channels	24	96	672	4032	
Line bit rate (Mbps)	1.544	6.312	44.736	274.176	
Japan					
Number of voice channels	24	96	480	1440	5760
Line bit rate (Mbps)	1.544	6.312	32.064	97.728	400.362
Europe					
Number of voice channels	30	120	480	1920	
Line bit rate (Mbps)	2.048	8.448	34.368	139.264	

Source: Ref. 11.

Table 6.3 8448-kbps Digital Multiplexing Frame Structure Using Positive/Zero/Negative Justification

Tributary bit rate (kbit/s) *2048*

Number of tributaries *4*

Frame Structure	Bit Number
	Set I
Frame-alignment signal (11100110)	1 to 8
Bits from tributaries	9 to 264
	Set II
Justification control bits C_{j1} (see Note)	1 to 4
Bits for service functions	5 to 8
Bits from tributaries	9 to 264
	Set III
Justification control bits C_{j2} (see Note)	1 to 4
Spare bits	5 to 8
Bits from tributaries	9 to 264
	Set IV
Justification control bits C_{j3} (see Note)	1 to 4
Bits from tributaries available for negative justification	5 to 8
Bits from tributaries available for positive justification	9 to 12
Bits from tributaries	12 to 264
Frame length	1056 bits
Frame direction	125 μs
Bits per tributary	256 bits
Maximum justification rate per tributary	8 kbps

Note: C_{jn} indicates nth justification control bit of the jth tributary.

Source: Table 1/G.745, CCITT Rec. G.745, p. 437, Fascicle III.4, IXth Plenary Assembly, Melbourne, 1988 (Ref. 6).

CCITT Rec. G.745 (Ref. 6) recommends cyclic bit interleaving in the tributary (i.e., E1 inputs) numbering order and positive/zero/negative justification with two-command control.[10] The justification control signal is distributed and the C_{jin} bits ($n = 1, 2, 3$; see Table 6.3) are used for justification control bits.

Positive justification is indicated by the signal 111, transmitted in each of two consecutive frames. Negative justification is indicated by the signal 000, also transmitted in each of two consecutive frames. No-justification is indicated by the signal 111 in one frame and 000 in the next frame. Bits 5, 6, 7, and 8 in Set IV (Table 6.3) are used for negative justification of tributaries 1, 2, 3, and 4, respectively, and bits 9 to 12 for positive justification of the same tributaries.

Besides, when information from tributaries 1, 2, 3, and 4 is not transmitted, bits 5, 6, 7, and 8 in Set IV are available for transmitting information concerning the type of justification (positive or negative) in frames containing commands of positive justification and intermediate amount of jitter in frames containing commands of negative justification.[11] The maximum amount of justification rate per tributary is shown in Table 6.3.

6.9 LONG-DISTANCE PCM TRANSMISSION

6.9.1 Transmission Limitations

Digital waveforms lend themselves to transmission by wire pair, coaxial cable, fiber-optic cable, and wideband radio media. The PCM multiplex format using an AMI signal was first applied to wire-pair cable (see Section 5.5). Its use on coaxial cable is now deprecated in favor of fiber optic cable. Each transmission medium has limitations brought about by impairments. In one way or another each limitation is a function of the length of a link employing the medium and the transmission rate (i.e., bit rate). We have discussed loss, for example. As loss increases (i.e., between regenerative repeaters), signal-to-noise ratio suffers, directly impacting bit error performance. The following transmission impairments to PCM transmission are covered: jitter, distortion, noise, and crosstalk. The design of long-distance digital links is covered in Chapter 9.

6.9.2 Jitter and Wander

In the context of digital transmission, *jitter* is defined as short-term variation of the sampling instant from its intended position in time or phase. Longer-term variation of the sampling instant is called *wander*. Jitter can cause transmission impairments such as:

- Displacement of the ideal sampling instant. This leads to a degradation in system error performance;
- Slips in timing recovery circuits, manifesting in degraded error performance;
- Distortion of the resulting analog signal after decoding at the receive end of the circuit.

[10]Positive/zero/negative justification. This refers to stuffing to compensate for input channel bit rates that are either too slow, none necessary, or too fast.
[11]*Jitter*; see Section 6.9.2.

Think of jitter as minute random motion of a timing gate. Timing gates are shown in the lower-right-hand corner of Figure 6.11. An analogy is a person with shaky hands trying to adjust a screw with a screwdriver.

The random phase modulation, or *phase jitter*, introduced at each regenerative repeater accumulates in a repeater chain and may lead to crosstalk and distortion on the reconstructed analog signal. In digital switching systems, jitter on the incoming lines is a potential source of slips.[12] Jitter accumulation is a function of the number of regenerative repeaters in tandem. Keep in mind that switches, fiber optic receivers, and digital radios are also regenerative repeaters.

Certainly by reducing the number of regenerative repeaters in tandem, we reduce jitter accordingly. Wire-pair systems transporting PCM at the DS1 rate have repeaters every 6000 ft (1830 m). If we are to reduce jitter, wire pair is not a good candidate for long circuits. On the other hand, fiber optic systems, depending on design and bit rate, have repeaters every 40 miles to 200 miles (64 km to 320 km). This is another reason why fiber optic systems are favored for long-haul application. Line-of-sight microwave radio, strictly for budgeting purposes, may have repeaters every 30 miles (48 km), so it, too, is a candidate for long-haul systems. Satellite radio systems have the potential for the least number of repeaters per unit length.

6.9.3 Distortion

On metallic transmission links, such as coaxial cable and wire-pair cable, line characteristics distort and attenuate the digital signal as it traverses the medium. There are three cable characteristics that create this distortion: (1) loss, (2) amplitude distortion (amplitude-frequency response), and (3) delay distortion. Thus the regenerative repeater must provide amplification and equalization of the incoming digital signal before regeneration. There are also trade-offs between loss and distortion on the one hand and repeater characteristics and repeater section length on the other.

6.9.4 Thermal Noise

As in any electrical communication system, thermal noise, impulse noise, and crosstalk affect system design. Because of the nature of a binary digital system, these impairments need only be considered on a per-repeater-section basis because noise does not accumulate due to the regenerative process carried out at repeaters and nodes. Bit errors do accumulate, and these noise impairments are one of the several causes of errors. One way to limit error accumulation is to specify a stringent bit error rate (BER) requirement for each repeater section. Up to several years ago, repeater sections were specified with a BER of 1×10^{-9}. Today a BER of 1×10^{-10} to 1×10^{-12} is prevalent in the North American network.

It is interesting to note that PCM provides intelligible voice performance for an error rate as low as 1 in 100 (1×10^{-2}). However, the bottom threshold (worst tolerable) BER is one error in one thousand (1×10^{-3}) at system endpoints. This value is required to ensure correct operation of supervisory signaling. The reader should appreciate that such degraded BER values are completely unsuitable for data transmission over the digital network.

[12]*Slips* are a major impairment in digital networks. Slips and slip rate are discussed in Section 12.7.

6.9.5 Crosstalk

Crosstalk is a major impairment in PCM wire-pair systems, particularly when "go" and "return" channels are carried in the same cable sheath. The major offender of single-cable operation is near-end crosstalk (NEXT). When the two directions of transmission are carried in separate cables or use shielded pairs in a common cable, far-end crosstalk (FEXT) becomes dominant.

One characteristic has been found to be a major contributor to poor crosstalk coupling loss. This is the capacitance imbalance between wire pairs. Stringent quality control during cable manufacture is one measure to ensure that minimum balance values are met.

6.10 DIGITAL LOOP CARRIER

Digital subscriber loop carrier is a method of extending the metallic subscriber plant by using one or more DS1 configurations. As an example, the SLC-96 uses four DS1 configurations to derive an equivalent of 96 voice channels.

The digital transmission facility used by a DLC system may be repeatered wire-pair cable, optical fibers, either or both combined with digital multiplexers, or other appropriate media. In Bellcore terminology, the central office termination (COT) is the digital terminal colocated with the local serving switch. The RT is the remote terminal. The RT must provide all of the features to a subscriber loop that the local serving switch normally does, such as supervision, ringing, address signaling, both dial pulse and touch tone, and so on (Ref. 7).

6.10.1 New Versions of DSL

ADSL, or asymmetric digital subscriber line, as described by Bellcore, provides 1.544 Mbps service "downstream," meaning from the local serving switch to the subscriber, out to 18,000 ft (5500 m). In the upstream direction 16 kbps service is furnished. Such service has taken on new life in providing a higher bit rate for Internet customers.

There is an ANSI version of ADSL that can provide 6 Mbps downstream service using a complex digital waveform and devices called automatic equalizers to improve bandwidth characteristics, particularly amplitude and phase distortion. The upstream bit rate can be as high as 640 kbps. Some manufacturers purport to be able to extend this service out to 12,000 kft (3700 m).

6.11 DIGITAL SWITCHING

6.11.1 Advantages and Issues of Digital Switching

There are both economic and technical advantages to digital switching; in this context we refer to PCM switching. The economic advantages of time-division PCM switching include the following:

- There are notably fewer equivalent cross points for a given number of lines and trunks than in a space-division switch.
- A PCM switch is of considerably smaller size.

- It has more common circuitry (i.e., common modules).
- It is easier to achieve full availability within economic constraints.

The technical advantages include the following:

- It is regenerative (i.e., the switch does not distort the signal; in fact, the output signal is "cleaner" than the input).
- It is noise-resistant.
- It is computer-based and thus incorporates all the advantages of SPC.
- The binary message format is compatible with digital computers. It is also compatible with signaling.
- A digital exchange is lossless. There is no insertion loss as a result of a switch inserted in the network.
- It exploits the continuing cost erosion of digital logic and memory; LSI, VLSI, and VHSIC insertion.[13]

Two technical issues may be listed as disadvantages:

1. A digital switch deteriorates error performance of the system. A well-designed switch may only impact network error performance minimally, but it still does it.
2. Switch and network synchronization, and the reduction of wander and jitter, can be gating issues in system design.

6.11.2 Approaches to PCM Switching

6.11.2.1 General. A digital switch's architecture is made up of two elements, called T and S, for time-division switching (T) and space-division switching (S), and can be made up of sequences of T and S. For example, the AT&T No. 4 ESS is a TSSSST switch; No. 3 EAX is an SSTSS; and the classic Northern Telecom DMS-100 is TSTS-folded. Many of these switches (e.g., DMS-100) are still available.

One thing these switches have in common is that they had multiple space (S) stages. This has now changed. Many of the new switches, or enhanced versions of the switches just mentioned, have very large capacities (e.g., 100,000 lines) and are simply TST or STS switches.

We will describe a simple time switch, a space switch, and methods of making up an architecture combining T and S stages. We will show that designing a switch with fairly high line and trunk capacity requires multiple stages. Then we will discuss the "new look" at the time stage.

6.11.2.2 Time Switch. On a conceptual basis, Figure 6.16 shows a time-switch or time-slot interchanger (TSI). A time-slot is the 8-bit PCM word. Remember, it expresses the voltage value of a sample taken at a certain moment in time. Of course, a time-slot consists of 8 bits. A time-slot represents one voice channel, and the time-slot is repeated 8000 times a second (with different binary values of course). DS1 has 24 time slots in a frame, one for each channel. E1 has 32 time slots.

[13]VHSIC stands for very high speed integrated circuit.

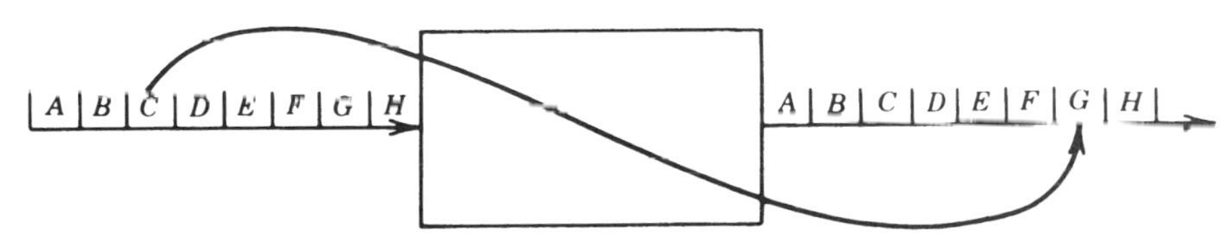

Figure 6.16 A time-division switch, which we call a time-slot interchanger (TSI). Connectivity shown is from user C in the incoming slot C to user G in outgoing slot G.

The time duration of an 8-bit time slot in each case is (125 μsec)/24 = 5.2083 μsec for the DS1 case, and (125 μsec)/32 = 3.906 μsec for the E1 case. Time-slot interchanging involves moving the data contained in each time slot from the incoming bit stream to an outgoing bit stream, but with a different time-slot arrangement in the outgoing stream, in accordance with the destination of each time slot. What is done, of course, is to generate a new frame for transmission at the appropriate switch outlet.

Obviously, to accomplish this, at least one time slot must be stored in memory (write) and then called out of memory in a changed position (read). The operations must be controlled in some manner, and some of these control actions must be kept in memory together with the software managing such actions. Typical control functions are time-slot "idle" or "busy." Now we can identify three of the basic functional blocks of a time swtich:

1. Memory for speech;
2. Memory for control; and
3. Time-slot counter or processor.

These three blocks are shown in Figure 6.17. There are two choices in handling a time switch: (1) sequential write, random read, as illustrated in Figure 6.17*a*, and (2) the

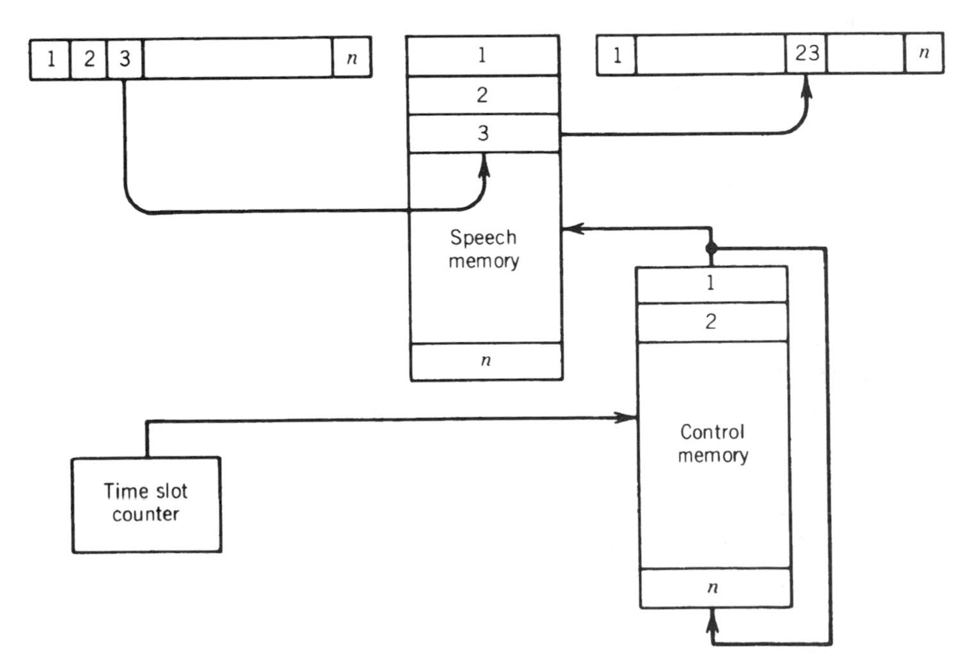

Figure 6.17a Time-slot interchange: time switch (T). Seuqential write, random read.

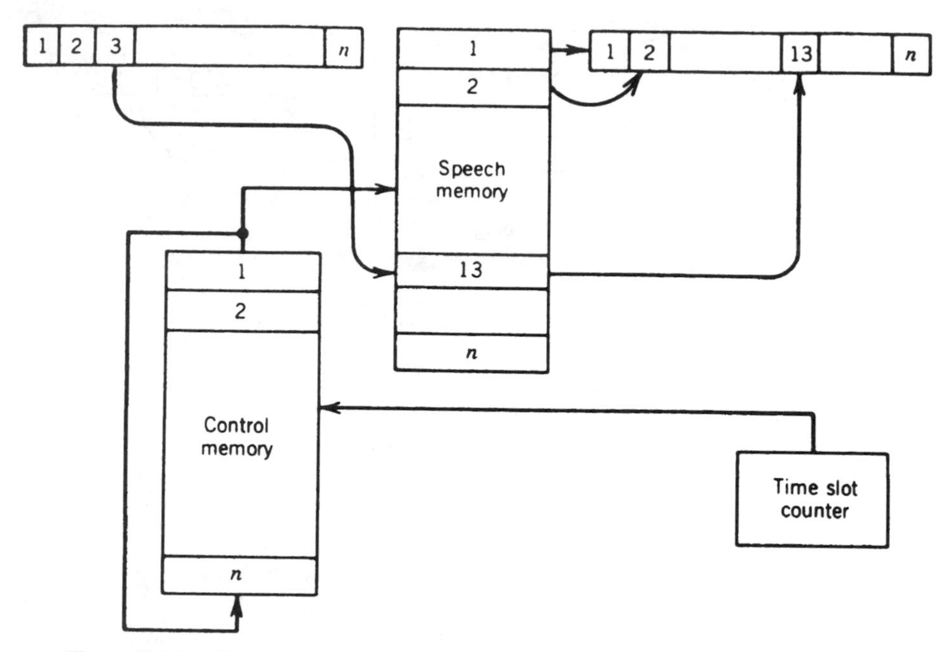

Figure 6.17*b* Time-switch, time-slot interchange (T). Random write, sequential read.

reverse, namely, random write, sequential read as shown in Figure 6.17*b*. In the first case, sequential write, the time slots are written into the speech memory as they appear in the *incoming* bit stream. They are read out of the memory in the correct order for the outgoing bit stream.

For the second case, random write (Figure 6.17*b*), the incoming time slots are written into memory in the order of appearance in the *outgoing* bit stream. This means that the incoming time slots are written into memory in the desired *output* order. The writing of incoming time slots into the speech memory can be controlled by a simple time-slot counter and can be sequential (e.g., in the order in which they appear in the incoming bit stream, as in Figure 6.17*a*). The readout of the speech memory is controlled by the control memory. In this case the readout is random where the time slots are read out in the desired output order. The memory has as many cells as there are time slots. For the DS1 example there would be 24 cells. This time switch, as shown, works well for a single inlet–outlet switch. With just 24 cells, it can handle 23 stations besides the calling subscriber, not an auspicious number.

How can we increase a switch's capacity? Enter the space switch (S). Figure 6.18 affords a simple illustration of this concept. For example, time slot B_1 on the B trunk is moved to the Z trunk into time-slot Z_1; and time-slot C_n is moved to trunk W into time-slot W_n. However, the reader should note that there is no change in the time-slot position.

6.11.2.3 Space Switch. A typical time-division space switch (S) is shown in Figure 6.19. It consists of a cross point matrix made up of logic gates that allow the switching of time slots in a spatial domain. These PCM time slot bit streams are organized by the switch into a pattern determined by the required network connectivity. The matrix consists of a number of input horizontals and output verticals with a logic gate at each crosspoint. The array, as shown in the figure, has M horizontals and N verticals, and

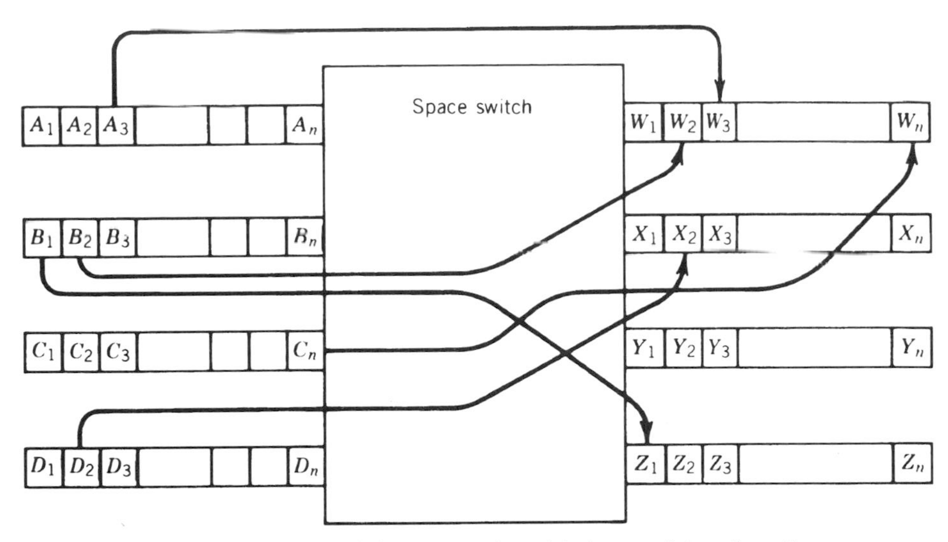

Figure 6.18 Space switch connects time slots in a spatial configuration.

we call it an M × N array. If M = N, the switch is nonblocking; If M > N, the switch concentrates, and if M < N, the switch expands.

Return to Figure 6.19. The array consists of a number of (M) input horizontals and (N) output verticals. For a given time slot, the appropriate logic gate is enabled and the time slot passes from the input horizontal to the desired output vertical. The other

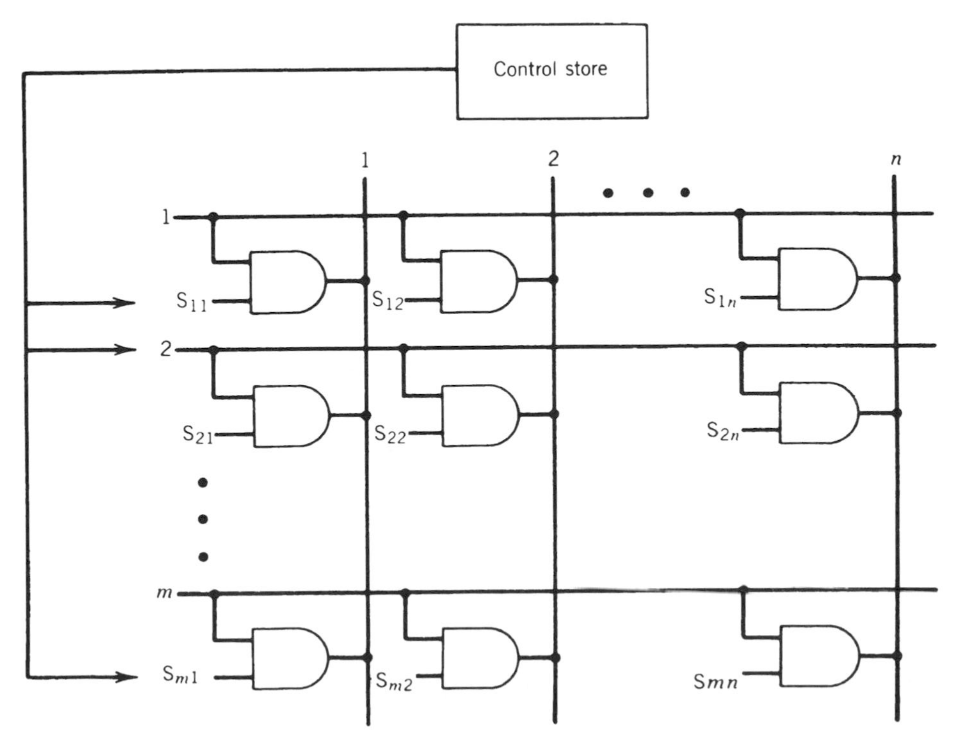

Figure 6.19 Time-division space switch cross point array showing enabling gates.

horizontals, each serving a different serial stream of time slots, can have the same time slot (e.g., a time slot from time slots number 1–24, 1–30, or 1 – n; e.g., time slot 7 on each stream) switched into other verticals enabling their gates. In the next time-slot position (e.g., time slot 8), a completely different path configuration could occur, again allowing time slots from horizontals to be switched to selected verticals. The selection, of course, is a function of how the traffic is to be routed at that moment for calls in progress or being set up.

The space array (cross point matrix) does not switch time slots as does a time switch (time-slot interchanger). This is because the occurrences of time slots are identical on the horizontal and on the vertical. It switches in the space domain, not in the time domain. The control memory in Figure 6.19 enables gates in accordance with its stored information.

If an array has M inputs and N outputs, M and N may be equal or unequal depending on the function of the switch on that portion of the switch. For a tandem or transit switch we would expect M = N. For a local switch requiring concentration and expansion, M and N would be unequal.

If, in Figure 6.19, if it is desired to transmit a signal from input 1 (horizontal) to output 2 (vertical), the gate at the intersection would be activated by placing an enable signal on S_{12} during the desired time-slot period. Then the eight bits of that time slot would pass through the logic gate onto the vertical. In the same time slot, an enable signal on S_{M1} on the Mth horizontal would permit that particular time slot to pass to vertical 1. From this we can see that the maximum capacity of the array during any one time-slot interval measured in simultaneous call connections is the smaller value of M or N. For example, if the array is 20 × 20 and a time-slot interchanger is placed on each input (horizontal) line and the interchanger handles 30 time slots, the array then can serve 20 × 30 = 600 different time slots. The reader should note how the TSI multiplies the call-handling capability of the array when compared with its analog counterpart.

6.11.2.4 Time-Space-Time Switch. Digital switches are composed of time and space switches in any order.[14] We use the letter T to designate a time-switching stage and use S to designate a space switching stage. For instance, a switch that consists of a sequence of a time-switching stage, a space-switching stage, and a time-switching stage is called a TST switch. A switching consisting of a space-switching stage, a time-switching stage, and a space-switching stage is designated an STS switch. There are other combinations of T and S. As we mentioned earlier, the AT&T No. 4 ESS switch is a good example. It is a TSSSST switch.

Figure 6.20 illustrates the time-space-time (TST) concept. The first stage of the switch is the TSI or time stages that interchange time slots (in the time domain) between external incoming digital channels and the subsequent space stage. The space stage provides connectivity between time stages at the input and output. It is a multiplier of call-handling capacity. The multiplier is either the value for M or value for N, whichever is smaller. We also saw earlier that space-stage time slots need not have any relation to either external incoming or outgoing time slots regarding number, numbering, or position. For instance, incoming time-slot 4 can be connected to outgoing time-slot 19 via space network time-slot 8.

If the space stage of a TST switch is nonblocking, blocking in the overall switch occurs if there is no internal space-stage time slot during which the link from the inlet

[14]The order is a switch designer's decision.

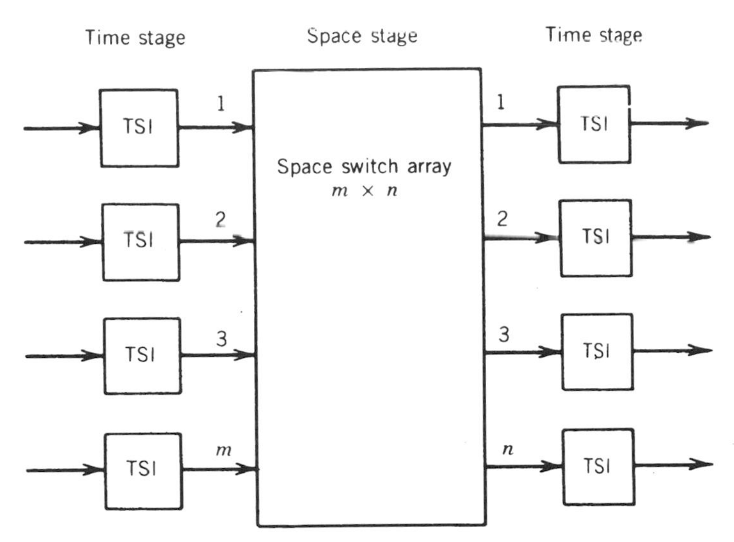

Figure 6.20 A time-space-time (TST) switch. TSI = time-slot interchanger.

time stage and the link to the outlet time stage are both idle. The blocking probability can be minimized if the number of space-stage time slots is large. A TST switch is strictly nonblocking if

$$l = 2c - 1, \tag{6.1}$$

where l is the number of space-stage time slots and c is the number of external TDM time slots (Ref. 3).

6.11.2.5 Space-Time-Space Switch. A space-time-space switch reverses the order architecture of a TST switch. The STS switch consists of a space cross point matrix at the input followed by an array of time-slot interchangers whose ports feed another cross point matrix at the output. Such a switch is shown in Figure 6.21. Consider this operational example with an STS. Suppose that an incoming time-slot 5 on port No. 1 must be connected to an output slot 12 at outgoing port 4. This can be accomplished by time-slot interchanger No. 1, which would switch it to time-slot 12; then the outgoing space stage would place that on outgoing trunk No. 4. Alternatively, time-slot 5 could be placed at the input of TSI No. 4 by the incoming space switch, where it would be switched to time-slot 12, and then out port No. 4.

6.11.2.6 TST Compared with STS. Both TST and STS switches can be designed with identical call-carrying capacities and blocking probabilities. It can be shown that a direct one-to-one mapping exists between time-division and space-division networks (Ref. 3).

The architecture of TST switching is more complex than STS switching with space concentration. The TST switch becomes more cost-effective because time expansion can be achieved at less cost than space expansion. Such expansion is required as link utilization increases because less concentration is acceptable as utilization increases.

It would follow, then, that TST switches have a distinct implementation advantage over STS switches when a large amount of traffic must be handled. Bellamy (Ref. 3) states that for small switches STS is favored due to reduced implementation complexi-

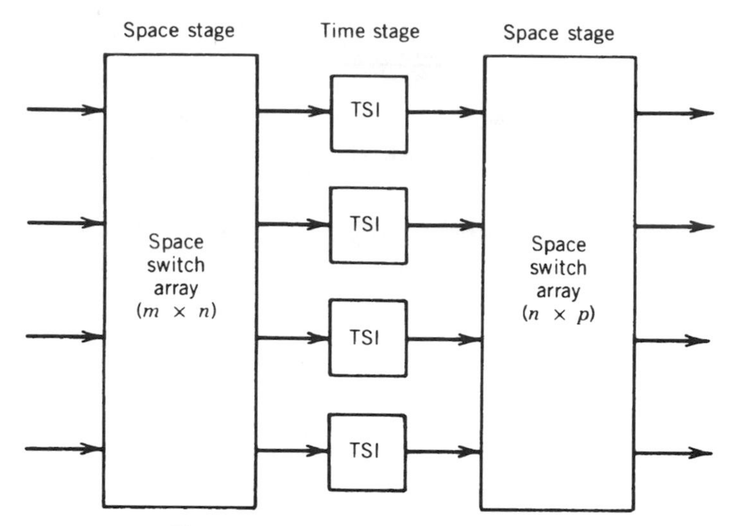

Figure 6.21 A space-time-space switch.

ties. The choice of a particular switch architecture may be more dependent on such factors as modularity, testability, and expandability.

One consideration that generally favors an STS implementation is the relatively simpler control requirements. However, for large switches with heavy traffic loads, the implementation advantage of the TST switch and its derivatives is dominant. A typical large switch is the ATT No. 4 ESS, which has a TSSSST architecture and has the capability of terminating 107,520 trunks with a blocking probability of 0.5% and channel occupancy of 0.7.

6.11.3 Review of Some Digital Switching Concepts

6.11.3.1 Early Ideas and New Concepts. In Section 6.11.2 the reader was probably led to believe that the elemental time-switching stage, the TSI, would have 24 or 30 time-slot capacity to match the North American DS1 rate of the "European" E1 rate, respectively. That means that a manufacturer would have to develop and produce two distinct switches, one to satisfy the North American market and one for the European market. Most switch manufacturers made just one switch with a common internal switching network, the time and space arrays we just discussed. For one thing, they could map five DS1 groups into four E1 groups, the common denominator being 120 DS0/E0 (64-kbps channels). Peripheral modules cleaned up any differences remaining, such as signaling. The "120" is a number used in AT&T's 4ESS. It maps 120 eight-bit time slots into 128 time slots. The eight time slots of the remainder are used for diagnostic and maintenance purposes (Ref. 8).

Another early concept was a common internal bit rate, to be carried on those "highways" we spoke about or on *junctors*.[15] At the points of interface that a switch has with the outside world, it must have 8-bit time slots in DS1 (or high-level multiplex) or E1 (or higher: E2, E3) frames each 125 μs in duration. Inside the switch was another matter. For instance, with Nortel's DMS-100 the incoming 8-bit time slot was mapped into a 10-bit time slot, as shown in Figure 6.22.[16] The example used in the figure is DS1.

[15] *Junctor* is a path connecting switching networks internal to a switch.
[16] Nortel was previously called Northern Telecom.

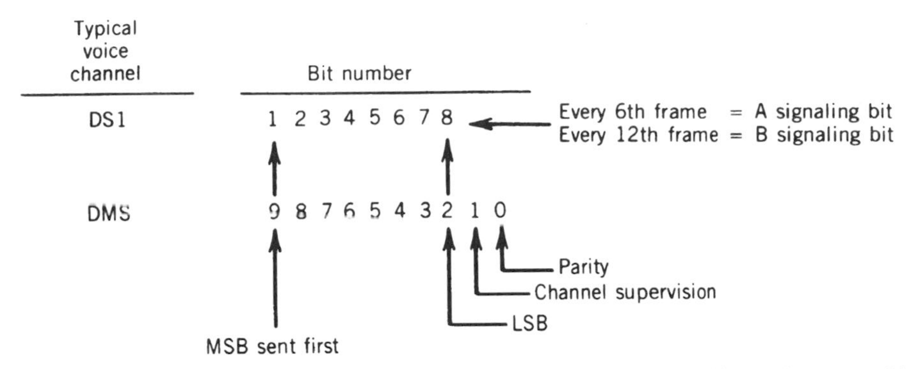

Figure 6.22 Bit mapping in the DMS-100, DS1 to DMS. DMS is the internal bit rate/structure. MSB = most significant bit; LSB = least significant bit.

Note in Figure 6.22 that one bit is a parity bit (bit 0) and the other appended bit (bit 1) carries the supervisory signaling information, telling the switch whether the time slot is idle or busy.[17] Bits 2 through 9 are the bits of the original 8-bit time slot. Because Nortel in their DMS-100 wanted a switch that was simple to convert from E1 to DS1, they built up their internal bit rate to 2.560 Mbps as follows: 10 bits per time slot, 32 time slots × 8000 (the frame rate) or 2.560 Mbps.[18] This now can accommodate E1, all 32 channels. As mentioned, 5 DS1s are easily mapped into 4 E1s and vice versa.

Another popular digital switch is AT&T's 5ESS, which maps each 8-bit time slot into a 16-bit internal PCM word. It actually appends eight additional bits onto the 8-bit PCM word, as shown in Figure 6.23.

6.11.3.2 Higher-Level Multiplex Structures Internal to a Digital Switch. We pictured a simple time-slot interchanger switch with 24 eight-bit time slots to satisfy DS1 requirements. It would meet the needs of 24 subscribers without blocking. There is no reason why we could build a TSI with a DS3 rate. The basic TSI then could handle 672 subscribers (i.e., 672 time slots). If we added a concentrator ahead of it for 4:1 concentration, then the time switch could handle 4 × 672, or 2688 subscribers.

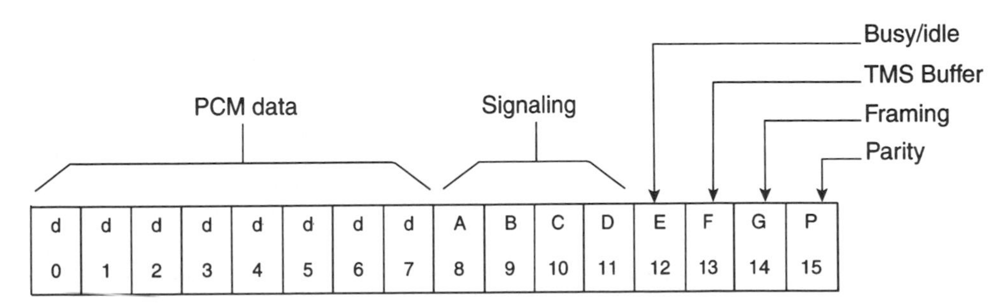

Figure 6.23 The composition of the AT&T 5ESS internal 16-bit time slot.

[17]*Parity bit* is used for error detection. It is a redundant bit appended to an array of bits to make the sum of all the 1 bits (marks) (in the array) always odd or always even.

[18]The 8000 frames per second or frame rate is common on all conventional PCM systems. As the reader will recall from Section 6.2.1, this is the Nyquist sampling rate for the 4-kHz analog voice channel on converting it to a PCM equivalent.

An example of this new thinking is the AT&T 5ESS, which is a TST switch. It has a capacity for 100,000 or more lines. They are able to accomplish this simpler architecture by using larger capacity time-slot interchangers (TSIs) and accordingly with higher bit rates in the space stage. A 5ESS TSI handles 512 time slots.[19] However, each TSI port has an incoming/outgoing time-slot rate of 256 time slots. Two ports are required (in one direction) to handle the 512 time slots: one for odd-numbered channels and one for even-numbered channels. Thus the bit rate at a TSI port is $256 \times 16 \times 8000 = 32.768$ Mbps. This odd-channel, even-channel arrangement carries through the entire switching fabric, with each port handling 256 time slots or 32.768 Mbps.

Another example of a widely implemented modern digital switch is the Northern Telecom DMS-100 with supernode/ENET. They modified the older DMS100 conventional switch, which had a TSTS-folded architecture. Like the 5ESS, they also moved into the 2048-time-slot domain in the ENET (extended network). But their time slot is 10 bits, and the ENET uses a 10-bit parallel format, so each line (i.e., there are 10 lines) has 2048×8000 or 16.384 Mbps.

6.12 DIGITAL NETWORK

6.12.1 Introduction

The North American public switched telecommunications network (PSTN) is 100% digital, with some possible holdouts in the local exchange area with small, independent telephone companies. The interexchange carrier (IXC) portion has been 100% digital for some years. The world network is expected to be all-digital by the first decade of the twenty-first century. That network is still basically hierarchical, and the structure changes slowly. There are possibly only two factors that change network structure:

1. Political; and
2. Technological.

In the United States, certainly divestiture of the Bell System/AT&T affected network structure with the formation of LECs (local exchange carriers) and IXCs. Outside North America, the movement toward privatization of government telecommunication monopolies in one way or another will affect network structure. As mentioned in Section 1.3, and to be discussed further in Chapter 8, there is a decided trend away from strict hierarchical structures, particularly in routing schemes; less so in topology.

Technology and its advances certainly may be equally or even more important than political causes. Satellite communications, we believe, brought about the move by CCITT away from any sort of international network hierarchy. International high-usage and direct routes became practical. We should not lose sight of the fact that every digital exchange has powerful computer power, permitting millisecond routing decisions for each call. This was greatly aided by the implementaton of CCITT Signaling System No. 7 (Chapter 13). Another evident factor certainly is fiber optic cable for a majority of trunk routes. It has also forced the use of geographic route diversity to improve survivability and availability. What will be the impact of the asynchronous transfer mode (ATM) (Chapter 18) on the evolving changes in network structure (albeit slowly)? The

[19]Remember that a 5ESS time slot has 16 bits (see Figure 6.23).

Internet certainly is forcing changes in data route capacity, right up to the subscriber. Privatization schemes now being implemented in many countries around the world will indeed have impact, as well, on network structure.

In the following section we discuss the digital network from the perspective of the overall PSTN. Certainly the information is valid for private networks as well, particularly if private networks are backed up by the local PSTN.

6.12.2 Technical Requirements of the Digital Network

6.12.2.1 Network Synchronization Rationale and Essentials. When a PCM bit stream is transmitted over a telecommunication link, there must be synchronization at three different levels: (1) bit, (2) time slot, and (3) frame. Bit synchronization refers to the need for the transmitter (coder) and receiver (decoder) to operate at the same bit rate. It also refers to the requirement that the receiver decision point be exactly at the mid-position of the incoming bit. Bit synchronization assures that the bits will not be misread by the receiver.

Obviously a digital receiver must also know where a time slot begins and ends. If we can synchronize a frame, time-slot synchronization can be assured. Frame synchronization assumes that bit synchronization has been achieved. We know where a frame begins (and ends) by some kind of marking device. With DS1 it is the framing bit. In some frames it appears as a 1 and in others it appears as a 0. If the 12-frame superframe is adopted, it has 12 framing bits, one in each of the 12 frames. This provides the 000111 framing pattern (Ref. 3). In the case of the 24-frame extended superframe, the repeating pattern is 001011, and the framing bit occurs only once in four frames.

E1, as we remember from Section 6.2, has a separate framing and synchronization channel, namely, channel 0. In this case the receiver looks in channel 0 for the framing sequence in bits 2 through 8 (bit 1 is reserved) of every other frame. The framing sequence is 0011011. Once the framing sequence is acquired, the receiver knows exactly where frame boundaries are. It is also time-slot aligned.

All digital switches have a master clock. Outgoing bit streams from a switch are slaved to the switch's master clock. Incoming bit streams to a switch derive timing from bit transitions of that incoming bit stream. It is mandatory that each and every switch in a digital network generate outgoing bit streams whose bit rate is extremely close to the nominal bit rate. To achieve this, network synchronization is necessary. Network synchronization can be accomplished by synchronizing all switch (node) master clocks so that transmissions from these nodes have the same average line bit rate. Buffer storage devices are judiciously placed at various transmission interfaces to absorb differences between the actual line bit rate and the average rate. Without this network-wide synchronization, *slips* will occur. Slips are a major impairment in digital networks. Slip performance requirements are discussed in Section 6.12.3.5. A properly synchronized network will not have slips (assuming negligible phase wander and jitter). In the next paragraph we explain the fundamental cause of slips.

As mentioned, timing of an outgoing bit stream is governed by the switch clock. Suppose a switch is receiving a bit stream from a distant source and expects this bit stream to have a transmission rate of $F(0)$ in Mbps. Of course, this switch has a buffer of finite storage capacity into which it is streaming these incoming bits. Let's further suppose that this incoming bit stream is arriving at a rate slightly greater than $F(0)$, yet the switch is draining the buffer at exactly $F(0)$. Obviously, at some time, sooner or later, that buffer must overflow. That overflow is a *slip*. Now consider the contrary condition: The incoming bit stream has a bit rate slightly less than $F(0)$. Now we will have an underflow condition. The buffer has been emptied and for a moment in time there are no further bits to be streamed out. This must be compensated for by the insertion of idle bits, false bits, or frame. However, it is more common just to repeat the previous frame. This is also a slip. We may remember the discussion of stuffing in Section 6.8.1 in the description higher-order multiplexers. Stuffing allows some variance of incoming bit rates without causing slips.

When a slip occurs at a switch port buffer, it can be controlled to occur at frame boundaries. This is much more desirable than to have an uncontrolled slip that can occur anywhere. Slips occur for two basic reasons:

1. Lack of frequency synchronization among clocks at various network nodes; and
2. Phase wander and jitter on the digital bit streams.

Thus, even if all the network nodes are operating in the synchronous mode and synchronized to the network master clock, slips can still occur due to transmission impairments. An example of environmental effects that can produce phase wander of bit streams is the daily ambient temperature variation affecting the electrical length of a digital transmission line.

Consider this example: A 1000-km coaxial cable carrying 300 Mbps (3×10^8 bps) will have about 1 million bits in transit at any given time, each bit occupying about one meter of the cable. A 0.01% increase in propagation velocity, as would be produced by a 1°F decrease in temperature, will result in 100 fewer bits in the cable; these bits must be absorbed to the switch's incoming elastic store buffer. This may end up causing an underflow problem forcing a controlled slip. Because it is underflow, the slip will be manifested by a frame repeat; usually the last frame just before the slip occurs is repeated.

In speech telephony, a slip only causes a click in the received speech. For the data user, the problem is far more serious. At least one data frame or packet will be corrupted.

Slips due to wander and jitter can be prevented by adequate buffering. Therefore adequate buffer size at the digital line interfaces and synchronization of the network node clocks are the basic means by which to achieve the network slip rate objective (Ref. 9).

6.12.2.2 Methods of Network Synchronization. There are a number of methods that can be employed to synchronize a digital network. Six such methods are shown graphically in Figure 6.24.

Figure 6.24*a* illustrates plesiochronous operation. In this case each switch clock is free-running (i.e., it is *not* synchronized to the network master clock.) Each network nodal switch has identical, high-stability clocks operating at the same nominal rate. When we say high stability, we mean a clock stability range from 1×10^{-11} to 5×10^{-13} per month. Such stabilities can only be achieved with atomic clocks, rubidium, or cesium. The accuracy and stability of each clock are such that there is almost complete coincidence in time-keeping. And the phase drift among many clocks is, in theory, avoided or the slip rate is acceptably low. This requires that all switching nodes, no matter how small, have such high-precision clocks. For commercial telecommunication networks, this is somewhat of a high cost burden.

Another synchronization scheme is mutual synchronization, which is illustrated in Figure 6.24*e* and 6.24*f*. Here all nodes in the network exchange frequency references, thereby establishing a common network clock frequency. Each node averages the incom-

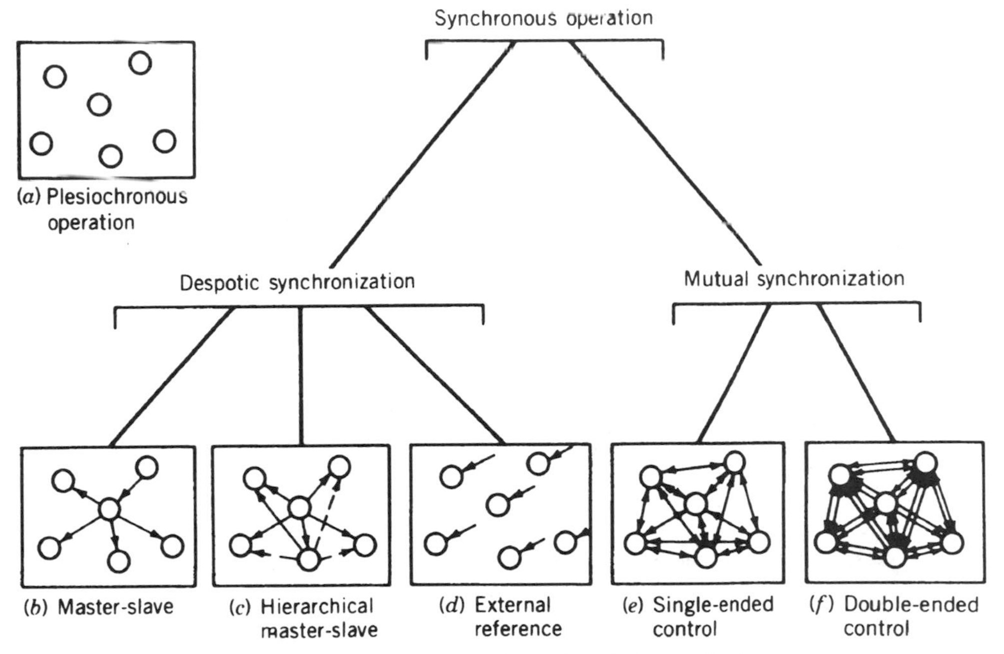

Figure 6.24 Digital network synchronization methods.

ing references and uses the result to correct its local transmitted clock. After an initialization period, the network aggregate clock converges to a single stable frequency.

It is important here to understand how we can "exchange frequency references." One method would be to have a separate synchronization channel connected to all nodes in the network. This is wasteful of facility assets. We can do just as well by sychronizing the switch clock from incoming bit streams carrying traffic, such as a DS1 or E1 bit stream. However this (these) incoming bit stream(s) must derive from a source (a switch), which has an equal or higher-level clock. One method of assigning clock levels based on clock stability is described later in this section. The synchronization information is carried in the transitions of the bit stream of interest. A phase-lock loop slaves the local clock to these transitions. Remember that a *transition* is a change of state in the bit stream, a change from a binary 1 to a binary 0, and vice versa.

A number of military systems as well as a growing number of civilian systems (e.g., Bell South in the United States; TelCel in Venezuela) use external synchronization, as illustrated in Figure 6.24*d*. Switch clocks use *disciplined oscillators* slaved to an external radio source. One of the most popular today is GPS (geographical positioning system), which disseminates universal coordinated time called UTC, an acronym deriving from the French. GPS is a multiple-satellite system such that there are always three or four satellites in view at once anywhere on the earth's surface. Its time-transfer capability is in the 10-ns to 100-ns range from UTC. In North American synchronization parlance, it provides timing at the stratum-1 level. The stratum levels are described in Section 6.12.2.2.1. We expect more and more digital networks to adopt the GPS external synchronization scheme. It adds notably to a network's survivability.

Other time-dissemination methods by radio are also available, such as satellite-based Transit and GOES, or terrestrially based Omega and Loran C, which has spotty worldwide coverage. HF radio time transfer is deprecated.

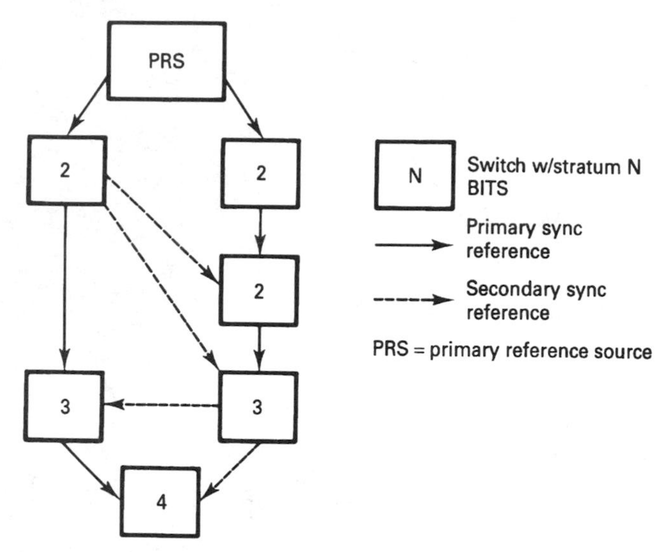

Figure 6.25 North American hierarchical network synchronization. (From Ref. 9, Figure 11-2.)

6.12.2.2.1 North American Synchronization Plan Stratum Levels. The North American network uses a hierarchical timing distribution system, as shown in Figure 6.24*c*. It is based on a four-level hierarchy and these levels are called *strata* (stratum in the singular). This hierarchical timing distribution system is illustrated in Figure 6.25. The timing requirements of each strata level are shown in Table 6.4. The parameters given in the table are defined as follows:

1. *Free-Run Accuracy.* This is the maximum fractional frequency offset that a clock may have when it has never had a reference or has been in holdover for an extended period, greater than several days or weeks.
2. *Holdover Stability.* This is the amount of frequency offset that a clock experiences after it has lost its synchronization reference. Holdover is specified for stratum 2. The stratum-3 holdover extends beyond one day and it breaks up the requirement into components for initial offset, drift, and temperature.
3. *Pull-in/Hold-in.* This is a clock's ability to achieve or maintain synchronization with a reference that may be off-frequency. A clock is required to have a pull-in/hold-in range at least as wide as its free-run accuracy. This ensures that a clock of a given stratum level can achieve and maintain synchronization with the clock of the same or higher stratum level.

Table 6.4 Stratum-Level Specifications

Stratum Level	Free-Run Accuracy	Holdover Stability	Pull-in/Hold-in
1	$\pm 10^{-11}$	N/A	N/A
2	$\pm 1.6 \times 10^{-8}$	$\pm 1 \times 10^{-10}$ per day	$\pm 1.6 \times 10^{-8}$
3E	$\pm 4.6 \times 10^{-6}$	$\pm 1 \times 10^{-8}$ day 1	4.6×10^{-6}
3	$\pm 4.6 \times 10^{-6}$	<255 slips during first day of holdover	4.6×10^{-6}
4	$\pm 32 \times 10^{-6}$	No holdover	32×10^{-6}

Source: Ref. 9, Table 3-1, p. 3-3.

Table 6.5 Expected Slip Performance in Holdover

Stratum Level	Slips in Day 1	Slips in Week 1
2	1 or less	2
3E	1 or less	13
3	17	266

Source: Ref. 9, Table 5-1, p. 5-2.

6.12.2.2.2 North American Holdover and Slip Performance. When a network clock loses its references, it enters holdover and drifts off frequency. The magnitude of this frequency drift determines the average slip rate experienced by equipment that depends on that clock timing source. Table 6.5 shows the number of slips expected after one day and one week of holdover given limited ambient temperature variations of $\pm 1°F$ in the switching center. The table shows the difference between stratum levels for performance during holdover. If maintenance actions are prompt when the unusual holdover occurs and we base a network on stratum-2 or -3E clocks, a virtually slip-free network can be expected (Ref. 9).

6.12.2.3 CCITT Synchronization Plans. CCITT Rec. G.811 (Ref. 10) deals with synchronization of international links. Plesiochronous operation is preferred (see Section 6.12.2). The recommendation states the problem at the outset:

> International digital links will be required to interconnect a variety of national and international networks. These networks may be of the following form:
>
> (a) a wholly synchronized network in which the timing is controlled by a single reference clock.
>
> (b) a set of synchronized subnetworks in which the timing of each is controlled by a reference clock but with plesiochronous operation between the subnetworks.
>
> (c) a wholly plesiochronous network (i.e., a network where the timing of each node is controlled by a separate reference clock).

Plesiochronous operation is the only type of synchronization that can be compatible with all three types listed. Such operation requires high-stability clocks. Thus Rec. G.811 states that all clocks at network nodes that terminate international links will have a long-term frequency departure of not greater than 1×10^{-11}. This is further described in what follows.

The theoretical long-term mean rate of occurrence of controlled frame or octet (time slot) slips under ideal conditions in any 64-kbps channel is consequently not greater than *1 in 70 days* per international digital link.

Any phase discontinuity due to the network clock or within the network node should result only in the lengthening or shortening of a time signal interval and should not cause a phase discontinuity in excess of one-eighth of a unit interval on the outgoing digital signal from the network node.

Rec. G.811 states that when plesiochronous and synchronous operation coexist within the international network, the nodes will be required to provide both types of operation. It is therefore important that the synchronization controls do not cause short-term frequency departure of clocks, which is unacceptable for plesiochronous operation.

6.12.3 Digital Network Performance Requirements

6.12.3.1 Blocking Probability. A blocking probability of B = 0.01 is the quality of service (QoS) objective. With judicious use of alternative routing, a blocking probability of 0.005 might be expected.

6.12.3.2 Error Performance from a Bellcore Perspective

Definitions

BER. The BER is the ratio of the number of bits in error to the total number of bits transmitted during a measurement period.

Errored Seconds (ES). An errored second is any 1-s interval containing at least one error.

Burst Errored Seconds. A burst errored second is any errored second containing at least 100 errors.

1. The BER at the interface levels DSX-1, DSX-1C, DSX-2, and DSX-3 shall be less than 2×10^{-10}, excluding all burst errored seconds in the measurement period.[20] During a burst errored second, neither the number of bit errors nor number of bits is counted. This requirement applies in a normal operating environment, and it shall be met by every channel in each protection switching section.
2. The frequency of burst errored seconds, other than those caused by protection switching induced by hard equipment failures, shall average no more than four per day at each of the interface levels DSX-1, DSX-1C, DSX-2, and DSX-3.[21] This requirement applies in a normal operating environment and must be met by every channel in each protection switching system.
3. For systems interfacing at the DS1 level, the long-term percentage of errored seconds (measured at the DS1 rate) shall not exceed 0.04%. This is equivalent to 99.96% error-free seconds (EFS). This requirement applies in a normal operating environment and is also an acceptance criterion. It is equivalent to no more than 10 errored seconds during a 7-h, one-way (loopback) test.
4. For systems interfacing at the DS3 level, the long-term percentage of errored seconds (measured at the DS3 rate) shall not exceed 0.4%. This is equivalent to 99.6% error-free seconds. This requirement applies in a normal operating environment and is also an acceptance criterion. It is equivalent to no more than 29 errored seconds during a 2-h, one-way (loopback) test (Ref. 11).

6.12.3.3 Error Performance from a CCITT Perspective. The CCITT cornerstone for error performance is Rec. G.821 (Ref. 12). Here error performance objectives are based on a 64-kbps circuit-switched connection used for voice traffic or as a "bearer circuit" for data traffic.

The CCITT error performance parameters are defined as follows (CCITT Rec. G.821): "The percentage of averaging periods each of time interval $T(0)$ during which the bit error rate (BER) exceeds a threshold value. The percentage is assessed over a much longer time interval $T(L)$." A suggested interval for $T(L)$ is 1 month.

It should be noted that total time $T(L)$ is broken down into two parts:

[20]DSX means digital system cross-connect.

[21]This is a long-term average over many days. Due to day-to-day variation, the number of burst errored seconds occurring on a particular day may be greater than the average.

Table 6.6 CCITT Error Performance Objectives for International ISDN Connections

Performance Classification	Objective[c]
a Degraded minutes[a,b]	Fewer than 10% of 1-min intervals to have a bit error ratio worse than 1×10^{-6}[d]
b Severely errored seconds[a]	Fewer than 0.2% of 1-s intervals to have a bit error ratio worse than 1×10^{-3}
c Errored seconds[a]	Fewer than 8% of 1-s intervals to have any errors (equivalent to 92% error-free seconds)

[a] The terms *degraded minutes*, *severely errored seconds*, and *errored seconds* are used as a convenient and concise performance objective "identifier." Their usage is not intended to imply the acceptability, or otherwise, of this level of performance.

[b] The 1-min intervals mentioned in the table and in the notes are derived by removing unavailable time and severely errored seconds from the total time and then consecutively grouping the remaining seconds into blocks of 60. The basic 1-s intervals are derived from a fixed time pattern.

[c] The time interval $T(L)$, over which the percentages are to be assessed, has not been specified since the period may depend on the application. A period of the order of any one month is suggested as a reference.

[d] For practical reasons, at 64 kbps, a minute containing four errors (equivalent to an error ratio of 1.04×10^{-6}) is not considered degraded. However, this does not imply relaxation of the error ratio objective of 1×10^{-6}.

Source: CCITT Rec. G.821 (Ref. 12).

1. Time that the connection is available; and
2. Time that the connection is unavailable.

The following BERs and intervals are used in CCITT Rec. G.821 in the statement of objectives (Ref. 12):

- A BER of less than 1×10^{-6} for $T(0) = 1$ min;
- A BER of less than 1×10^{-3} for $T(0) = 1$ s; and
- Zero errors for $T(0) = 1$ s.

Table 6.6 provides CCITT error performance objectives.

6.12.3.4 Jitter Jitter was discussed in Section 6.9.2, where we stated that it was a major digital transmission impairment. We also stated that jitter magnitude is a function of the number of regenerative repeaters there are in tandem. Guidelines on jitter objectives may be found in Ref. 15.

6.12.3.5 Slips

6.12.3.5.1 From a Bellcore Perspective. Slips, as a major digital network impairment, are explained in Section 6.12.2.1. When stratum-3 slip conditions are trouble-free, the nominal clock slip rate is 0. If there is trouble with the primary reference, a maximum of one slip on any trunk will result from a switched reference or any other rearrangement. If there is a loss of all references, the maximum slip rate is 255 slips the first day for any trunk. This occurs when the stratum-3 clocks drift a maximum of 0.37 parts per million from their reference frequency (Ref. 13).

6.12.3.5.2 From a CCITT Perspective. With plesiochronous operation, the number of slips on international links will be governed by the sizes of buffer stores and the

Table 6.7 Controlled Slip Performance on a 64-kbps International Connection Bearer Channel

Performance Category	Mean Slip Rate	Proportion of Time[a]
(a)[b]	≤5 slips in 24 h	>98.9%
(b)	>5 slips in 24 h and ≤30 slips in 1 h	<1.0%
(c)	>30 slips in 1 h	<0.1%

[a]Total time ≥1 year.

[b]The nominal slip performance due to plesiochronous operation alone is not expected to exceed 1 slip in 5.8 days.

Source: CCITT Rec. G.822 (Ref. 14).

accuracies and stabilities of the interconnecting national clocks.[22] The end-to-end slip performance should satisfy the service requirements for telephone and nontelephone services on a 64-kbps digital connection in an ISDN.

The slip rate objectives for an international end-to-end connection are specified with reference to the standard hypothetical reference connection (HRX)-which is 27,500 km in length.

The theoretical slip rate is one slip in 70 days per plesiochronous interexchange link assuming clocks with specified accuracies (see Section 6.12.2.2) and provided that the performance of the transmission and switching requirements remain within their design limits.

In the case where the international connection includes all of the 13 nodes identified in the HRX and those nodes are all operating together in a plesiochronous mode, the nominal slip performance of a connection could be 1 in 70/12 days (12 links in tandem) or 1 in 5.8 days. In practice, however, some nodes in such a connection would be part of the same synchronized network. Therefore, a better nominal slip performance can be expected (e.g., where the national networks at each end are synchronized). The nominal slip performance of the connection would be 1 in 70/4 or 1 in 17.5 days. Note that these calculations assume a maximum of four international links.

The performance objectives for the rate of *octet* slips on an international connection of 27,500 km in length of a corresponding bearer channel are given in Table 6.7 CCITT (Ref. 14) adds that further study is required to confirm that these values are compatible with other objectives such as error performance given in Section 6.12.3.3.

REVIEW EXERCISES

1. What is the major overriding advantage of binary digital transmission? Give at least two secondary advantages.

2. Name the three steps to develop a PCM signal from an analog signal.

3. Based on the Nyquist sampling theorem, what is the sampling rate for a nominal 4-kHz voice channel? for a 56-kHz radar product signal?

4. If I'm transmitting 8000 frames per second, what is the duration of one frame?

[22]CCITT is looking at the problem from an international switching center gateway. It will connect via digital trunks to many national networks, each with their own primary reference source (PRS).

5. Give a simple definition of *quantization distortion.*

6. Our system uses linear quantization. How can we reduce quantization distortion (noise)?

7. What is the negative downside to increasing quantization steps?

8. Companding used in PCM systems follows one of two laws. Identify each law—its name and where is it applied.

9. How many bits are there in a PCM word? How many different binary possibilities can be derived from a PCM word or codeword?

10. If a PCM code received starts with a 0, what do we know about the derived voltage sample?

11. Bits 2, 3, and 4 of the PCM code word identify the segment. How many total segments are there?

12. In DS1, of what use is the *framing bit*?

13. How is the identification of frame beginning carried out in E1?

14. How is signaling carried out in DS1? in E1?

15. How do we arrive at 1.544 Mbps for DS1?

16. With the AMI line code, how is the zero coded?

17. If, with bipolar transmission, under "normal" circumstances, the first 1 is a −1.0 V, what would the second "1" be?

18. Where does a wire-pair cable or light-wave regenerative repeater derive its timing from?

19. What is the purpose of *stuffing* on a higher-order PCM multiplex?

20. In the North American digital hierarchy, just from its nomenclature, what does an M34 multiplex do?

21. On a digital link, how would excessive loss affect signal quality?

22. Define *jitter* from the perspective of a PCM link.

23. The principal cause of systematic phase jitter is a function of________.

24. What value BER can we expect as a mean for the North American digital network?

25. What is the threshold BER for the digital network? Why is it set at that point?

26. Give at least three "economic" advantages of digital switching.

27. Define the function of a *time-slot interchanger*.

28. If we have a switch that is only a single time-slot interchanger for an E1 system, how many different subscribers could I be connected to?

29. What are the three functional blocks of a conventional time-slot interchanger (i.e., a time switch)?

30. How can the capacity of a time-slot interchanger be increased? Give two methods.

31. How can a switch manufacturer sell just one switch to cover both E1 and DS1 regimes?

32. What is the function of a *junctor* in a digital switch?

33. What are the two primary factors that can change a national digital network structure?

34. What causes a *slip* on a digital network?

35. What is the difference between *controlled slips* and *uncontrolled slips*?

36. Argue the efficacy of using an external timing source to synchronize network elements.

37. What is a *disciplined oscillator*?

38. What is *holdover stability*?

39. According to CCITT, if we can maintain long-term frequency departure to better than 1×10^{-11}, what kind of slip performance can we expect?

40. Define a *burst errored second.*

41. What kind of slip rate could we expect if all network timing references were lost?

REFERENCES

1. *IEEE Standard Dictionary of Electrical and Electronic Terms*, 6th ed., IEEE Std. 100-1996, IEEE, New York, 1996.
2. *Reference Data for Radio Engineering*, 5th ed., ITT, Howard W. Sams, Indianapolis, Indiana, 1968.
3. J. Bellamy, *Digital Telephony*, Wiley, New York, 1991.
4. D. R. Smith, *Digital Transmission Systems*, 2nd ed., Van Nostrand Reinhold, New York, 1993.
5. *Transmission Systems for Communications*, 5th ed., Bell Telephone Laboratories, Holmdel, NJ, 1982.
6. *Second Order Digital Multiplex Equipment Operating at 8448 kbps Using Positive/Zero/ Negative Justification*, CCITT Rec. G.745, Fascicle II.4, IXth Plenary Assembly, Melbourne, 1988.
7. *Functional Criteria for Digital Loop Carrier Systems*, Bellcore Technical Reference TR-NWT-000057, Issue 2, Bellcore, Piscataway, NJ, 1993.
8. J. C. McDonald, ed., *Fundamentals of Digital Switching*, 2nd ed., Plenum Press, New York, 1990.
9. *Digital Network Synchronization Plan*, Bellcore Generic Requirements, GR-435-CORE, Bellcore, Piscataway, NJ, 1994.
10. *Timing Requirements at the Outputs of Reference Clocks and Network Nodes Suitable for Plesiochronous Operation of International Digital Links*, CCITT Rec. G.811, Fascicle III.5, IXth Plenary Assembly, Melbourne, 1988.
11. *Transport Systems Generic Requirements (TSGR): Common Requirements*, Bellcore GR-499-CORE, Issue 1, Bellcore, Piscataway, NJ, Dec. 1995.
12. *Error Performance of an International Digital Connection Forming Part of an ISDN*, CCITT Rec. G.821, Fascicle III.5, IXth Plenary Assembly, Melbourne, 1988.

13. *BOC Notes on the LEC Networks—1994*, Special Report SR-TSV-002275, Issue 2, Bellcore, Piscataway, NJ, 1994.
14. *Controlled Slip Rate Objectives on an International Digital Connection*, CCITT Rec. G.822, Fascicle III.5, IXth Plenary Assembly, Melbourne, 1988.
15. R. L. Freeman, *Telecommunication Transmission Handbook*, 4th ed., Wiley, New York, 1988.
16. *Physical/Electrical Characteristics of Hierarchical Digital Interfaces*, CCITT Rec. G.703, ITU Geneva, 1991.
17. J. C. McDonald, ed., *Fundamentals of Digital Switching*, 2nd ed., Plenum Press, New York, 1990.

7

SIGNALING

7.1 WHAT IS THE PURPOSE OF SIGNALING?

The IEEE (Ref. 1) defines *signaling* as the "exchange of information specifically concerned with the establishment and control of connections and the transfer of user-to-user and management information in a telecommunication network."

Conventional signaling has evolved with the telephone network. Many of the techniques we deal with in this chapter are applicable to a telecommunication network which is principally involved with telephone calls. With telephony, signaling is broken down in three functional areas:

1. Supervisory;
2. Address; and
3. Call progress audible-visual.

Another signaling breakdown is:

- Subscriber signaling;
- Interswitch (interregister) signaling.

7.2 DEFINING THE FUNCTIONAL AREAS

7.2.1 Supervisory Signaling

Supervisory signaling provides information on line or circuit condition. It informs a switch whether a circuit (internal to the switch) or a trunk (external to the switch) is busy or idle; when a called party is off-hook or on-hook, and when a calling party is on-hook or off-hook.

Supervisory information (status) must be maintained end-to-end on a telephone call, whether voice, data, or facsimile is being transported. It is necessary to know when a calling subscriber lifts her/his telephone off-hook, thereby requesting service. It is equally important that we know when the called subscriber answers (i.e., lifts the telephone off-hook) because that is when we may start metering the call to establish charges. It is also important to know when the calling and called subscribers return their telephones to the on-hook condition. That is when charges stop, and the intervening trunks comprising the talk path as well as the switching points are then rendered idle for use by another pair of subscribers. During the period of occupancy of a speech path end-to-end, we must know that this particular path is busy (i.e., it is occupied) so no other call attempt can seize it.

7.2.2 Address Signaling

Address signaling directs and routes a telephone call to the called subscriber. It originates as dialed digits or activated push-buttons from a calling subscriber. The local switch accepts these digits and, by using the information contained in the digits, directs the call to the called subscriber. If more than one switch is involved in the call setup, signaling is required between switches (both address and supervisory). Address signaling between switches is called *interregister signaling*.

7.2.3 Call Progress—Audible-Visual

This type of signaling we categorize in the *forward direction* and in the *backward direction*. In the forward direction there is *alerting*. This provides some sort of audible-visual means of informing the called subscriber that there is a telephone call waiting. This is often done by ringing a telephone's bell. A buzzer, chime, or light may also be used for alerting.

The remainder of the techniques we will discuss are used in the backward direction. Among these are audible tones or voice announcements that will inform the calling subscriber of the following:

1. *Ringback.* This tells the calling subscriber that the distant telephone is ringing.
2. *Busyback.* This tells the calling subscriber that the called line is busy.
3. *ATB—All Trunks Busy.* There is congestion on the routing. Sometimes a recorded voice announcement is used here.
4. *Loud warble on telephone instrument—Timeout.* This occurs when a telephone instrument has been left off-hook unintentionally.

7.3 SIGNALING TECHNIQUES

7.3.1 Conveying Signaling Information

Signaling information can be conveyed by a number of means from a subscriber to the serving switch and between (among) switches. Signaling information can be transmitted by means such as:

- Duration of pulses (pulse duration bears a specific meaning);
- Combination of pulses;
- Frequency of signal;
- Combination of frequencies;
- Presence or absence of a signal;
- Binary code; and
- Direction and/or level of transmitted current (for dc systems).

7.3.2 Evolution of Signaling

Signaling and switching are inextricably tied together. Switching automated the network. But without signaling, switching systems could not function. Thus it would be better said that switching with signaling automated the network.

Conventional subscriber line signaling has not changed much over the years, with the exception of the push-button tones, which replaced the dial for address signaling. ISDN, being a full digital service to the subscriber, uses a unique digital signaling system called DSS-1 (Digital Subscriber Signaling No. 1).

In the 1930s and 1940s interregister and line signaling evolved into many types of signaling systems, which made international automatic working a virtual nightmare.[1] Nearly every international circuit required special signaling interfaces. The same was true, to a lesser extent, on the national level.

In this section we will cover several of the more common signaling techniques used on the analog network which operated with frequency division multiplex equipment (Section 4.5.2). Although these signaling systems are obsolete in light of the digital network, the concepts covered here will help in understanding how signaling works.

7.3.2.1 Supervisory/Line Signaling

7.3.2.1.1 Introduction. Line signaling on wire trunks was based essentially on the presence or absence of dc current. Such dc signals are incompatible with FDM equipment where the voice channel does not extend to 0 Hz. Remember, the analog voice channel occupies the band from 300 Hz to 3400 Hz. So the presence or absence of a dc current was converted to an ac tone for one of the states and no-tone for the other state. There were two ways to approach the problem. One was called *in-band signaling* and the other was called *out-of-band signaling*.[2]

7.3.2.1.2 In-Band Signaling. In-band signaling refers to signaling systems using an audio tone, or tones, inside the conventional voice channel to convey signaling information. There are two such systems we will discuss here: (1) one-frequency (SF or single frequency), and (2) two-frequency (2VF). These signaling systems used one or two tones in the 2000 Hz to 3000 Hz portion of the band, where less speech energy is concentrated.

Single-frequency (SF) signaling is used exclusively for supervision, often with its adjunct called *E&M signaling*, which we cover in Section 7.3.2.1.4. It is used with FDM equipment, and most commonly the tone frequency was 2600 Hz. Of course this would be in four-wire operation. Thus we would have a 2600-Hz tone in either/both directions. The direction of the tone is important, especially when working with its E&M signaling adjunct. A diagram showing the application of SF signaling on a four-wire trunk is shown in Figure 7.1.

Two-frequency (2VF) signaling can be used for both supervision (line signaling) and address signaling. Its application is with FDM equipment. Of course when discussing such types of line signaling (supervision), we know that the term *idle* refers to the on-hook condition, while *busy* refers to the off-hook condition. Thus, for such types of line signaling that are governed by audio tones of which SF and 2VF are typical, we have the conditions of "tone on when idle" and "tone on when busy." The discussion

[1] *Line signaling* is the supervisory signaling used among switches.

[2] Called *out-band* by CCITT and in nations outside of North America.

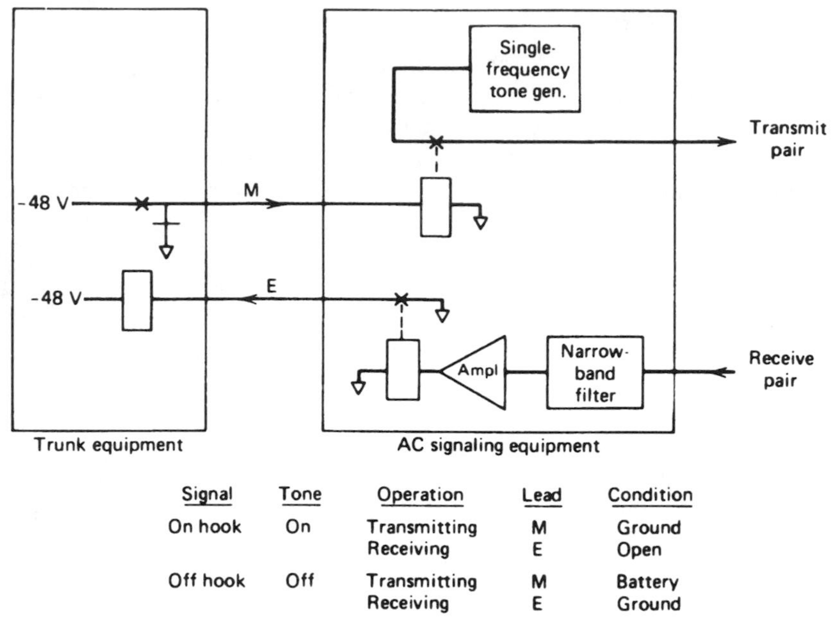

Signal	Tone	Operation	Lead	Condition
On hook	On	Transmitting	M	Ground
		Receiving	E	Open
Off hook	Off	Transmitting	M	Battery
		Receiving	E	Ground

Figure 7.1 Functional block diagram of an SF signaling circuit. *Note:* Wire pairs "receive" and "transmit" derive from the FDM multiplex equipment. Note also the E-lead and M-lead.

holds equally well for in-band and out-of-band signaling methods. However, for in-band signaling, supervision is by necessity tone-on idle; otherwise subscribers would have an annoying 2600-Hz tone on throughout the call.

A major problem with in-band signaling is the possibility of "talk-down," which refers to the premature activation or deactivation of supervisory equipment by an inadvertent sequence of voice tones through the normal use of the channel. Such tones could simulate the SF tone, forcing a channel dropout (i.e., the supervisory equipment would return the channel to the idle state). Chances of simulating a 2VF tone set are much less likely. To avoid the possibility of talk-down on SF circuits, a time-delay circuit or slot filters to by-pass signaling tones may be used. Such filters do offer some degradation to speech unless they are switched out during conversation. They must be switched out if the circuit is going to be used for data transmission (Ref. 2).

It becomes apparent why some administrations and telephone companies have turned to the use of 2VF supervision, or out-of-band signaling, for that matter. For example, a typical 2VF line signaling arrangement is the CCITT No. 5 code, where f_1 (one of the two VF frequencies) is 2400 Hz and f_2 is 2600 Hz. 2VF signaling is also used widely for address signaling (see Section 7.3.2.2 of this chapter; Ref. 3).

7.3.2.1.3 Out-of-Band Signaling. With out-of-band signaling, supervisory information is transmitted out of band (i.e., above 3400 Hz). In all cases it is a single-frequency system. Some out-of-band systems use "tone on when idle," indicating the on-hook condition, whereas others use "tone off." The advantage of out-of-band signaling is that either system, tone on or tone off, may be used when idle. Talk-down cannot occur because all supervisory information is passed out of band, away from the speech-information portion of the channel.

The preferred CCITT out-of-band frequency is 3825 Hz, whereas 3700 Hz is commonly used in the United States. It also must be kept in mind that out-of-band signaling

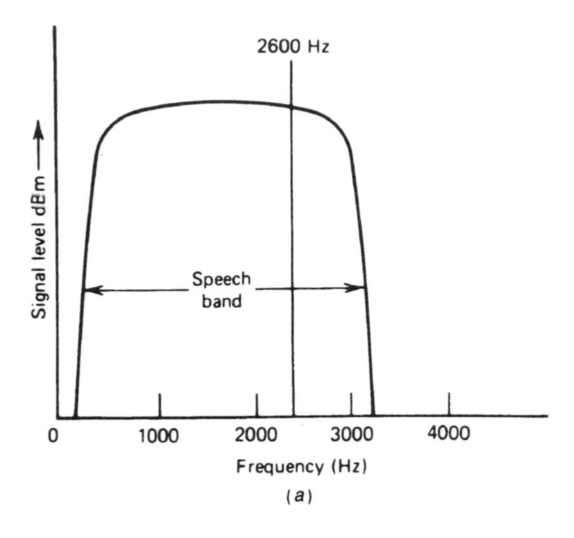

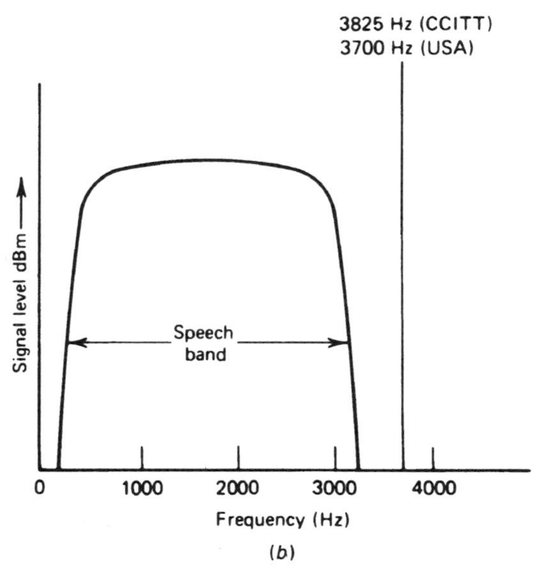

Figure 7.2 SF signaling (*a*) in-band and (*b*) out-of-band.

is used exclusively on carrier systems, not on wire trunks. On the wire side, inside an exchange, its application is E&M signaling. In other words, out-of-band signaling is one method of extending E&M signaling over a carrier system.

In the short run, out-of-band signaling is attractive in terms of both economy and design. One drawback is that when channel patching is required, signaling leads have to be patched as well. In the long run, the signaling equipment required may indeed make out-of-band signaling even more costly because of the extra supervisory signaling equipment and signaling lead extensions required at each end, and at each time that the carrier (FDM) equipment demodulates to voice. The major advantage of out-of-band signaling is that continuous supervision is provided, whether tone on or tone off, during the entire telephone conversation. In-band SF signaling and out-of-band signaling are illustrated in Figure 7.2. An example of out-of-band signaling is the regional signaling system R-2, prevalent in Europe and nations under European hegeonomy (see Table 7.1.)

Table 7.1 R-2 Line Signaling (3825 Hz)

	Direction	
Circuit State	Forward (Go)	Backward (Return)
Idle	Tone on	Tone on
Seized	Tone off	Tone on
Answered	Tone off	Tone off
Clear back	Tone off	Tone on
Release	Tone on	Tone on or off
Blocked	Tone on	Tone off

7.3.2.1.4 ***E&M Signaling.*** The most common form of trunk supervision in the analog network was E&M signaling. It derived from the SF or 2VF equipment, as shown in Figure 7.1. It only becomes true E&M signaling where the trunk interfaces with the switch (see Figure 7.3). E-lead and M-lead signaling systems are semantically derived from the historical designation of signaling leads on circuit drawings covering these systems. Historically, the E&M interface provides two leads between the switch and what we call the *trunk signaling equipment* (signaling interface). One lead is called the E-lead, which carries signals *to* the switching equipment. Such signal directions are shown in Figure 7.3, where we see that signals from switch A and switch B leave A on the M-lead and are delivered to B on the E-lead. Likewise, from B to A, supervisory information leaves B on the M-lead and is delivered to A on the E-lead.

For conventional E&M signaling (referring to electromechanical exchanges), the following supervisory conditions are valid:

DIRECTION		CONDITION AT *A*		CONDITION AT *B*	
Signal A to B	*Signal B to A*	*M-Lead*	*E-Lead*	*M-Lead*	*E-Lead*
On hook	On hook	Ground	Open	Ground	Open
Off hook	On hook	Battery	Open	Ground	Ground
On hook	Off hook	Ground	Ground	Battery	Open
Off hook	Off hook	Battery	Ground	Battery	Ground

Source: Ref. 8.

7.3.2.2 Address Signaling. Address signaling originates as dialed digits (or activated push buttons) from a calling subscriber, whose local switch accepts these digits and, using that information, directs the telephone call to the desired distant subscriber. If more than one switch is involved in the call setup, signaling is required between switches (both address and supervisory). Address signaling between switches in conventional systems is called *interregister signaling*.

The paragraphs that follow discuss various more popular standard ac signaling tech-

Figure 7.3 E&M signaling.

Table 7.2 North American R-1 Code[a]

Digit	Frequency Pair (Hz)	
1	700 + 900	
2	700 + 1100	
3	900 + 1100	
4	700 + 1300	
5	900 + 1300	
6	1100 + 1300	
7	700 + 1500	
8	900 + 1500	
9	1100 + 1500	
10 (0)	1300 + 1500	
Use	**Frequency Pair**	**Explanation**
KP	1100 + 1700	Preparatory for digits
ST	1500 + 1700	End-of-pulsing sequence
STP	900 + 1700	Used with TSPS (traffic service position system)
ST2P	1300 + 1100	Used with TSPS (traffic service position system)
ST3P	700 + 1700	Used with TSPS (traffic service position system)
Coin collect	700 + 1100	Coin control
Coin return	1100 + 1700	Coin control
Ring-back	700 + 1700	Coin control
Code 11	700 + 1700	Inward operator (CCITT No. 5)
Code 12	900 + 1700	Delay operator
KP1	1100 + 1700	Terminal call
KP2	1300 + 1700	Transit call

[a] Pulsing of digits is at the rate of about seven digits per second with an interdigital period of 68 ± 7 ms. For intercontinental dialing for CCITT No. 5 code compatibility, the R-1 rate is increased to 10 digits per second. The KP pulse duration is 100 ms.

Source: Ref. 4.

niques such as 2VF and MF tone. Although interregister signaling is stressed where appropriate, some supervisory techniques are also reviewed. Common-channel signaling is discussed in Chapter 13, where we describe the CCITT No. 7 signaling system.

7.3.2.2.1 Multifrequency Signaling. Multifrequency (MF) signaling has been in wide use around the world for interregister signaling. It is an in-band method using five or six tone frequencies, two tones at a time. It works well over metallic pair, FDM, and TDM systems. MF systems are robust and difficult to cheat. Three typical MF systems are reviewed in the following:

MULTIFREQUENCY SIGNALING IN NORTH AMERICA—THE R-1 SYSTEM. The MF signaling system principally employed in the United States and Canada is recognized by the CCITT as the R-1 code (where R stands for "regional"). It is a two-out-of-five frequency pulse system. Additional signals for control functions are provided by frequency combination using a sixth basic frequency. Table 7.2 shows the ten basic digits (0–9) and other command functions with their corresponding two-frequency combinations, as well as a brief explanation of "other applications." We will call this system a "spill forward" system. It is called this because few backward acknowledgment signals are required. This is in contraposition to the R-2 system, where every transmitted digit must be acknowledged.

Table 7.3 CCITT No. 5 Code Showing Variations with the R-1 Code[a]

Signal	Frequencies (Hz)	Remarks
KP1	1100 + 1700	Terminal traffic
KP2	1300 + 1700	Transit traffic
1	700 + 900	
2	700 + 1100	
3–0	Same as Table 7.2	
ST	1500 + 1700	
Code 11	700 + 1700	Code 11 operator
Code 12	900 + 1700	Code 12 operator

[a] Line signaling for CCITT No. 5 code is 2VF, with f_1 2400 Hz and f_2 2600 Hz. Line-signaling conditions are shown in Table 7.4.

Source: Ref. 5.

CCITT NO. 5 SIGNALING CODE. Interregister signaling with the CCITT No. 5 code is very similar to the North American R-1 code. Variations with the R-1 code are shown in Table 7.3. The CCITT No. 5 line signaling code is also shown in Table 7.4.

R-2 CODE. The R-2 code has been denominated by CCITT (CCITT Rec. Q.361) as a European regional signaling code. Taking full advantage of combinations of two-out-of-six tone frequencies, 15 frequency pair possibilities are available. This number is doubled in each direction by having meaning in groups I and II in the forward direction (i.e., toward the called subscriber) and groups A and B in the backward direction, as shown in Table 7.5.

Groups I and A are said to be of primary meaning, and groups II and B are said to be of secondary meaning. The change from primary to secondary meaning is commanded by the backward signal A-3 or A-5. Secondary meanings can be changed back to primary meanings only when the original change from primary to secondary was made by the use of the A-5 signal. Turning to Table 7.5, the 10 digits to be sent in the forward direction in the R-2 system are in group I and are index numbers 1 through 10 in the table. The index 15 signal (group A) indicates "congestion in an international exchange or at its output." This is a typical backward information signal giving circuit status information. Group B consists of nearly all "backward information" and, in particular, deals with subscriber status.

Table 7.4 CCITT No. 5 Line Signaling Code

Signal	Direction	Frequency	Sending Duration	Recognition Time (ms)
Seizing	→	f_1	Continuous	40 ± 10
Proceed to send	←	f_2	Continuous	40 ± 10
Busy flash	←	f_2	Continuous	125 ± 25
Acknowledgment	→	f_1	Continuous	125 ± 25
Answer	←	f_1	Continuous	125 ± 25
Acknowledgment	→	f_1	Continuous	125 ± 25
Clear back	←	f_2	Continuous	125 ± 25
Acknowledgment	→	f_1	Continuous	125 ± 25
Forward transfer	→	f_2	850 ± 200 ms	125 ± 25
Clear forward	→	$f_1 + f_2$	Continuous	125 ± 25
Release guard	←	$f_1 + f_2$	Continuous	125 ± 25

Source: Ref. 4.

Table 7.5 European R-2 Signaling System

	Frequencies (Hz)						
	1380	1500	1620	1740	1860	1980	Forward Direction I/II
Index No. for Groups I/II and A/B	1140	1020	900	780	660	540	Backward Direction A/B
1	×	×					
2	×		×				
3		×	×				
4	×			×			
5		×		×			
6			×	×			
7	×				×		
8		×			×		
9			×		×		
10				×	×		
11	×					×	
12		×				×	
13			×			×	
14				×		×	
15					×	×	

Source: Ref. 6.

The R-2 line-signaling system has two versions: the one used on analog networks is discussed here; the other, used on E-1 PCM networks, was briefly covered in Chapter 6. The analog version is an out-of-band tone-on-when idle system. Table 7.6 shows the line conditions in each direction, forward and backward. Note that the code takes advantage of a signal sequence that has six characteristic operating conditions. Let us consider several of these conditions.

Seized. The outgoing exchange (call-originating exchange) removes the tone in the forward direction. If seizure is immediately followed by release, removal of the tone must be maintained for at least 100 ms to ensure that it is recognized at the incoming end.

Answered. The incoming end removes the tone in the backward direction. When another link of the connection using tone-on-when-idle continuous signaling precedes the outgoing exchange, the "tone-off" condition must be established on the link as soon as it is recognized in this exchange.

Table 7.6 Line Conditions for the R-2 Code

Operating Condition of the Circuit	Signaling Conditions	
	Forward	Backward
1. Idle	Tone on	Tone on
2. Seized	Tone off	Tone on
3. Answered	Tone off	Tone off
4. Clear back	Tone off	Tone on
5. Release	Tone on	Tone on or off
6. Blocked	Tone on	Tone off

Source: Ref. 6.

Table 7.7 Audible Call Progress Tones Commonly Used in North America

Tone	Frequencies (Hz)	Cadence
Dial	350 + 440	Continuous
Busy (station)	480 + 620	0.5 s on, 0.5 s off
Busy (network congestion)	480 + 620	0.2 s on, 0.3 s off
Ring return	440 + 480	2 s on, 4 s off
Off-hook alert	Multifrequency howl	1 s on, 1 s off
Recording warning	1400	0.5 s on, 15 s off
Call waiting	440	0.3 s on, 9.7 s off

Source: Ref. 7.

Clear Back. The incoming end restores the tone in the backward direction. When another link of the connection using tone-on-when-idle continuous signaling precedes the outgoing exchange, the "tone-off" condition must be established on this link as soon as it is recognized in this exchange.

Clear Forward. The outgoing end restores the tone in the forward direction.

Blocked. At the outgoing exchange the circuit stays blocked as long as the tone remains off in the backward direction.

7.3.3 Subscriber Call Progress Tones and Push-Button Codes (North America)

Table 7.7 shows the audible call progress tones commonly used in North America as presented to a subscriber. Subscriber subsets are either dial or push button, and they will probably be all push button in the next ten years. A push button actuates two audio tones simultaneously, similar to the multifrequency systems described previously with interregister signaling. However, the tone library used by the subscriber is different than the tone library used with interregister signaling. Table 7.8 compares digital dialed, dial pulses (breaks), and multifrequency (MF) push-button tones.

7.4 COMPELLED SIGNALING

In many of the signaling systems discussed thus far, signal element duration is an important parameter. For instance, in a call setup an initiating exchange sends a 100-ms

Table 7.8 North American Push-Button Codes

Digit	Dial Pulse (Breaks)	Multifrequency Push-Button Tones
0	10	941,1336 Hz
1	1	697,1209 Hz
2	2	697,1336 Hz
3	3	697,1474 Hz
4	4	770,1209 Hz
5	5	770,1336 Hz
6	6	770,1477 Hz
7	7	852,1209 Hz
8	8	852,1336 Hz
9	9	852,1477 Hz

Source: Ref. 7.

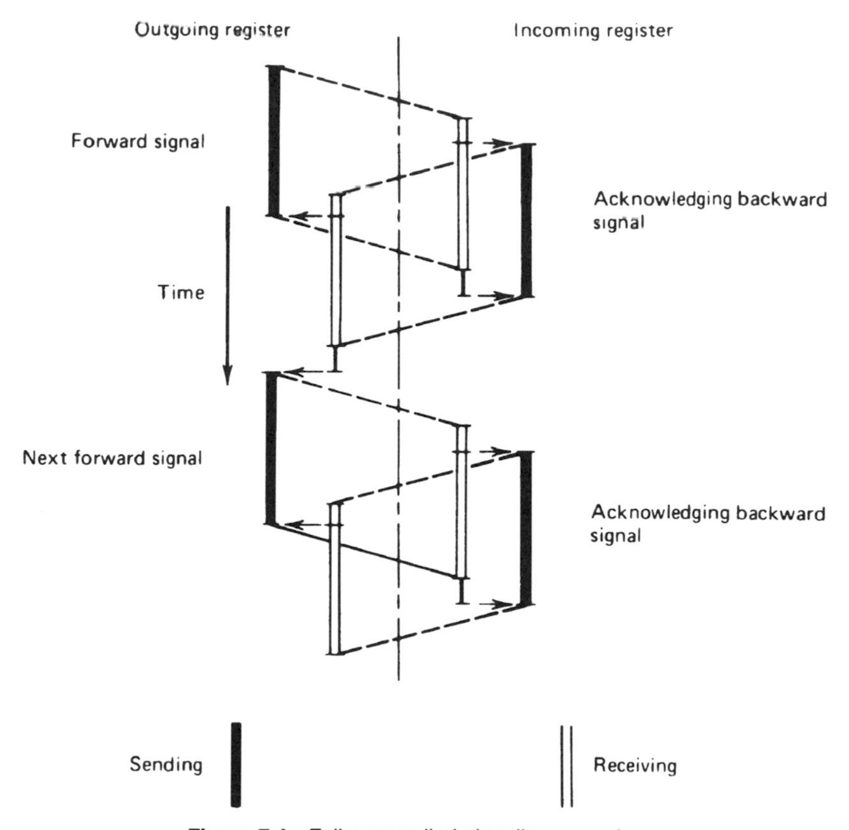

Figure 7.4 Fully compelled signaling procedure.

seizure signal. Once this signal is received at the distant end, the distant exchange sends a "proceed to send" signal back to the originating exchange; in the case of the R-1 system, this signal is 140 ms or more in duration. Then, on receipt of "proceed to send" the initiating exchange spills all digits forward. In the case of R-1, each digit is an MF pulse of 68-ms duration with 68 ms between each pulse. After the last address digit an ST (end-of-pulsing) signal is sent. In the case of R-1 the incoming (far-end) switch register knows the number of digits to expect. Consequently there is an explicit acknowledgment that the call setup has proceeded satisfactorily. Thus R-1 is a good example of noncompelled signaling.

A fully compelled signaling system is one in which each signal continues to be sent until an acknowledgment is received. Thus signal duration is not significant and bears no meaning. The R-2 and SOCOTEL are examples of fully compelled signaling systems.[3] Figure 7.4 illustrates a fully compelled signaling sequence. Note the small overlap of signals, causing the acknowledging (reverse) signal to start after a fixed time on receipt of the forward signal. This is because of the minimum time required for recognition of the incoming signal. After the initial forward signal, further forward signals are delayed for a short recognition time (see Figure 7.4). Recognition time is normally less than 80 ms.

Fully compelled signaling is advantageous in that signaling receivers do not have to measure duration of each signal, thus making signaling equipment simpler and more

[3]SOCOTEL is a European multifrequency signaling system used principally in France and Spain.

economical. Fully compelled signaling adapts automatically to the velocity of propagation, to long circuits, to short circuits, to metallic pairs, or to carrier and is designed to withstand short interruptions in the transmission path. The principal drawback of compelled signaling is its inherent lower speed, thus requiring more time for setup. Setup time over space-satellite circuits with compelled signaling is appreciable and may force the system engineer to seek a compromise signaling system.

There is also a partially compelled type of signaling, where signal duration is fixed in both forward and backward directions according to system specifications; or the forward signal is of indefinite duration and the backward signal is of fixed duration. The forward signal ceases once the backward signal has been received correctly. CCITT Signaling System No. 4 (not discussed in this text; see CCITT Recs. Q.120 to 130) is an example of a partially compelled signaling system.

7.5 CONCEPTS OF LINK-BY-LINK VERSUS END-TO-END SIGNALING

An important factor to be considered in switching system design that directly affects both signaling and customer satisfaction is postdialing delay. This is the amount of time it takes after the calling subscriber completes dialing until ring-back is received. Ring-back is a backward signal to the calling subscriber indicating that the dialed number is ringing. Postdialing delay must be made as short as possible.

Another important consideration is register occupancy time for call setup as the set-up proceeds from originating exchange to terminating exchange. Call-setup equipment, that equipment used to establish a speech path through a switch and to select the proper outgoing trunk, is expensive. By reducing register occupancy per call, we may be able to reduce the number of registers (and markers) per switch, thus saving money.

Link-by-link and end-to-end signaling each affect register occupancy and postdialing delay, each differently. Of course, we are considering calls involving one or more tandem exchanges in a call setup, because this situation usually occurs on long-distance or toll calls. Link-by-link signaling may be defined as a signaling system where *all* interregister address information must be transferred to the subsequent exchange in the call-setup routing. Once this information is received at this exchange, the preceding exchange control unit (register) releases. This same operation is carried on from the originating exchange through each tandem (transit) exchange to the terminating exchange of the call. The R-1 system is an example of link-by-link signaling.

End-to-end signaling abbreviates the process such that tandem (transit) exchanges receive only the minimum information necessary to route the call. For instance, the last four digits of a seven-digit telephone number need be exchanged only between the originating exchange (e.g., the calling subscriber's local exchange or the first toll exchange in the call setup) and the terminating exchange in the call setup. With this type of signaling, fewer digits are required to be sent (and acknowledged) for the overall call-setup sequence. Thus the signaling process may be carried out much more rapidly, decreasing postdialing delay. Intervening exchanges on the call route work much less, handling only the digits necessary to pass the call to the next exchange in the sequence.

The key to end-to-end signaling is the concept of "leading register." This is the register (control unit) in the originating exchange that controls the call routing until a speech path is setup to the terminating exchange before releasing to prepare for another call setup. For example, consider a call from subscriber X to subscriber Y:

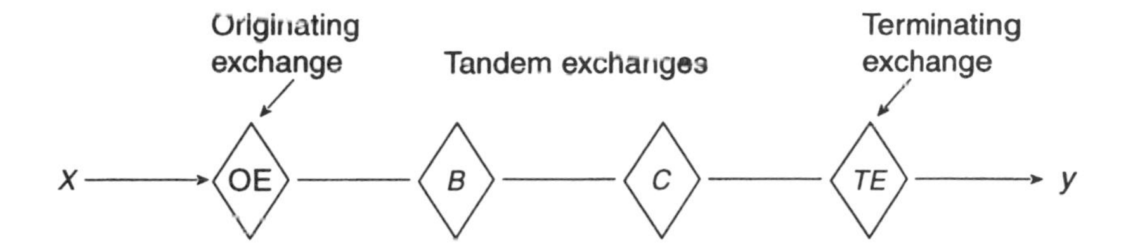

The telephone number of subscriber Y is 345–6789. The sequence of events is as follows using end-to-end signaling:

- A register at exchange OE receives and stores the dialed number 345–6789 from subscriber X.
- Exchange OE analyzes the number and then seizes a trunk (junction) to exchange B. It then receives a "proceed-to-send" signal indicating that the register at B is ready to receive routing information (digits).
- Exchange OE then sends digits 34, which are the minimum necessary to effect correct transit.
- Exchange B analyzes the digits 34 and then seizes a trunk to exchange C. Exchanges OE and C are now in direct contact and exchange B's register releases.
- Exchange OE receives the "proceed-to-send" signal from exchange C and then sends digits 45, those required to effect proper transit at C.
- Exchange C analyzes digits 45 and then seizes a trunk to exchange TE. Direct communication is then established between the leading register for this call at OE and the register at TE being used on this call setup. The register at C then releases.
- Exchange OE receives the "proceed-to-send" signal from exchange TE, to which it sends digits 5678, the subscriber number.
- Exchange TE selects the correct subscriber line and returns to A ring-back, line busy, out of order, or other information after which all registers are released.

Thus we see that a signaling path is opened between the leading register and the terminating exchange. To accomplish this, each exchange in the route must "know" its local routing arrangements and request from the leading register those digits it needs to route the call further along its proper course.

Again, the need for backward information becomes evident, and backward signaling capabilities must be nearly as rich as forward signaling capabilities when such a system is implemented.

R-1 is a system inherently requiring little backward information (interregister). The little information that is needed, such as "proceed to send," is sent via line signaling. The R-2 system has major backward information requirements, and backward information and even congestion and busy signals sent back by interregister signals (Ref. 5).

7.6 EFFECTS OF NUMBERING ON SIGNALING

Numbering, the assignment and use of telephone numbers, affects signaling as well as switching. It is the number or the translated number, as we found out in Section 1.3.2, that routes the call. There is "uniform" numbering and "nonuniform" numbering. How does each affect signaling? Uniform numbering can simplify a signaling system. Most uniform systems in the nontoll or local-area case are based on seven digits, although some are based on six. The last four digits identify the subscriber. The first three digits (or the first two in the case of a six-digit system) identify the exchange. Thus the local exchange or transit exchanges know when all digits are received. There are two advantages to this sort of scheme:

1. The switch can proceed with the call once all digits are received because it "knows" when the last digit (either the sixth or seventh) has been received.
2. "Knowing" the number of digits to expect provides inherent error control and makes "time out" simpler.[4]

For nonuniform numbering, particularly on direct distance dialing in the international service, switches require considerably more intelligence built in. It is the initial digit or digits that will tell how many digits are to follow, at least in theory.

However, in local or national systems with nonuniform numbering, the originating register has no way of knowing whether it has received the last digit, with the exception of receiving the maximum total used in the national system. With nonuniform numbering, an incompletely dialed call can cause a useless call setup across a network up to the terminating exchange, and the call setup is released only after time out has run its course. It is evident that with nonuniform numbering systems, national (and international) networks are better suited to signaling systems operating end to end with good features of backward information, such as the R-2 system (Ref. 5).

7.7 ASSOCIATED AND DISASSOCIATED CHANNEL SIGNALING

Here we introduce a new concept: disassociated channel signaling. Up to now we have only considered associated channel signaling. In other words, the signaling is carried right on its associated voice channel, whether in-band or out-of-band. Figure 7.5 illustrates two concepts: associated channel and separate channel signaling, but still associated. E-1 channel 16 is an example. It is indeed a separate channel, but associated with the 30-channel group of traffic channels. We will call this *quasi-associated channel signaling*.

Disassociated channel signaling is when signaling travels on a separate and distinct route than the traffic channels for which it serves. CCITT Signaling System No. 7 uses either this type of signaling or quasi-associated channel signaling. Figure 7.6 illustrates quasi-associated channel signaling, whereas Figure 7.7 shows fully disassociated channel signaling.

7.8 SIGNALING IN THE SUBSCRIBER LOOP

7.8.1 Background and Purpose

In Section 5.4 we described loop-start signaling, although we did not call it that. When a subscriber takes a telephone off-hook (out of its cradle), there is a switch closure at the subset (see the hook-switch in Figure 5.3), current flows in the loop alerting the serving

[4] "Time out" is the resetting of call-setup equipment and return of dial tone to subscriber as a result of incomplete signaling procedure, subset left off hook, and so forth.

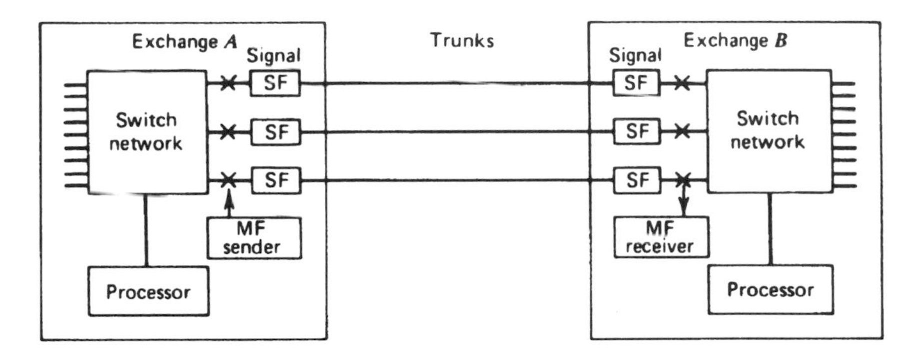

Associated Channel Signaling
(Conventional SF-MF)

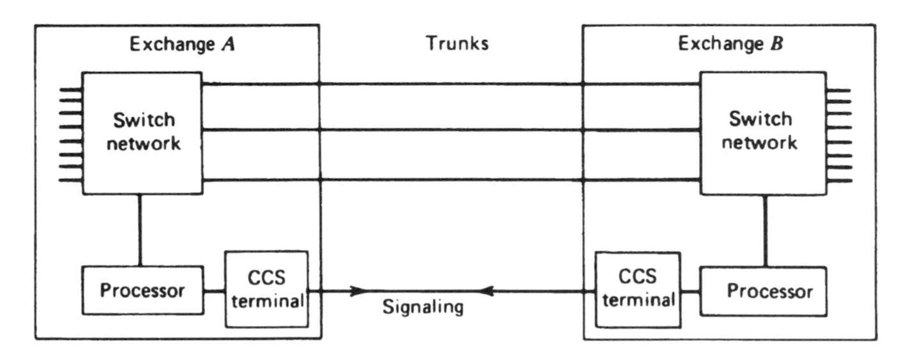

Separate Channel Signaling

Figure 7.5 Conventional analog associated channel signaling (*upper*) versus separate channel signaling (which we call quasi-associated channel signaling) (*lower*). *Note:* Signaling on upper drawing accompanies voice paths; signaling on the lower drawing is conveyed on a separate circuit (or time slot). CCS = common channel signaling such as CCITT Signaling System No. 7.

exchange that service is desired on that telephone. As a result, dial tone is returned to the subscriber. This is basic supervisory signaling on the subscriber loop.

A problem can arise from this form of signaling. It is called *glare*. Glare is the result of attempting to seize a particular subscriber loop from each direction. In this case it would be an outgoing call and an incoming call nearly simultaneously. There is a much greater probability of glare with a PABX than with an individual subscriber.

Ground-start signaling is the preferred signaling system when lines terminate in a switching system such as a PABX. It operates as follows: When a call is from the local serving switch to the PABX, the local switch immediately grounds the conductor tip to seize the line. With some several seconds delay, ringing voltage is applied to the line (where required). The PABX immediately detects the grounded tip conductor and will not allow an outgoing call from the PABX to use this circuit, thus avoiding glare.

In a similar fashion, if a call originates at the PABX and is outgoing to the local serving exchange, the PABX grounds the ring conductor to seize the line. The serving switch recognizes this condition and prevents other calls from attempting to terminate the circuit. The switch now grounds the tip conductor and returns dial tone after it connects a digit receiver. There can be a rare situation when double seizure occurs, causing glare. Usually one or the other end of the circuit is programmed to

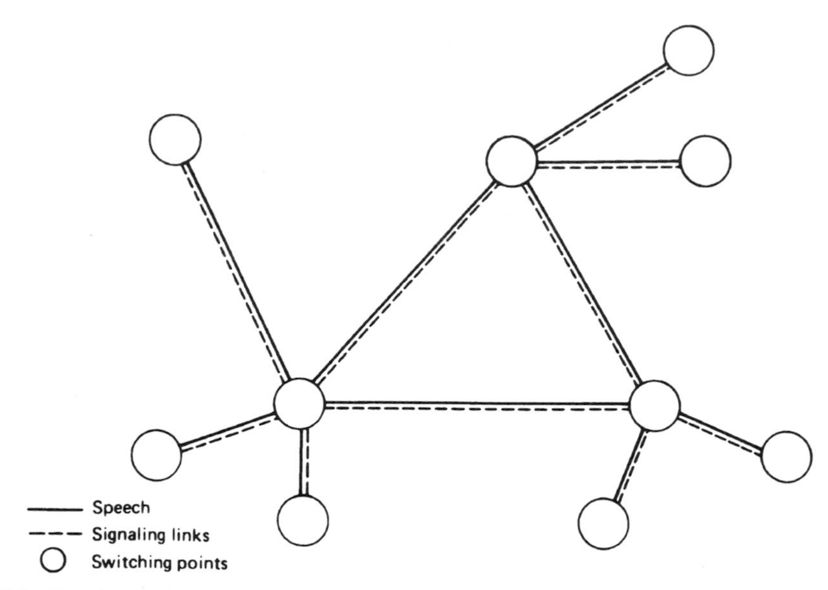

Figure 7.6 Quasi-associated channel signaling, typical of E-1 channel 16. As shown, the signaling travels on a separate channel but associated with its group of traffic channels for which it serves. If it were conventional analog signaling, it would be just one solid line, where the signaling is embedded with its associated traffic.

back down and allow the other call to proceed. A ground start interface is shown in Figure 7.8.

Terminology in signaling often refers back to manual switchboards or, specifically, to the plug used with these boards and its corresponding jack as illustrated in Figure 7.9. Thus we have tip (T), ring (R), and sleeve (S). Often only the tip and ring are used, and the sleeve is grounded and has no real electrical function.

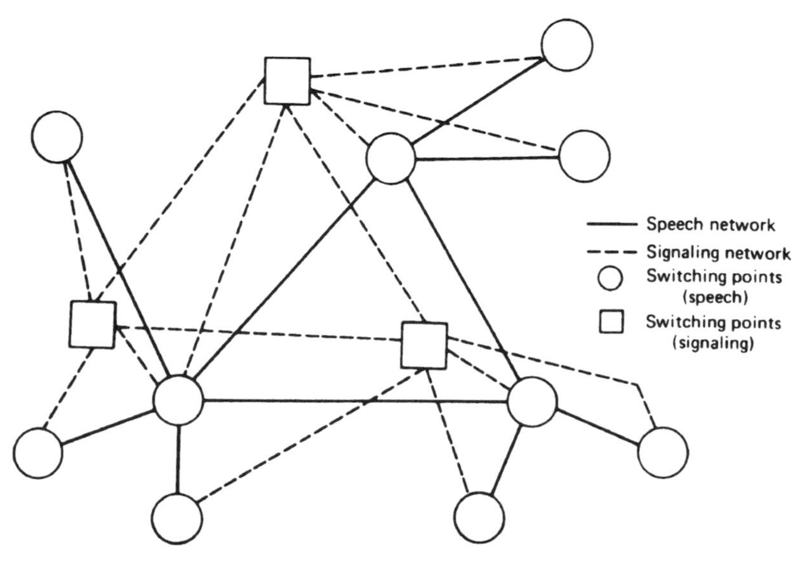

Figure 7.7 Fully disassociated channel signaling. This signaling may be used with CCITT Signaling System No. 7, described in Chapter 13.

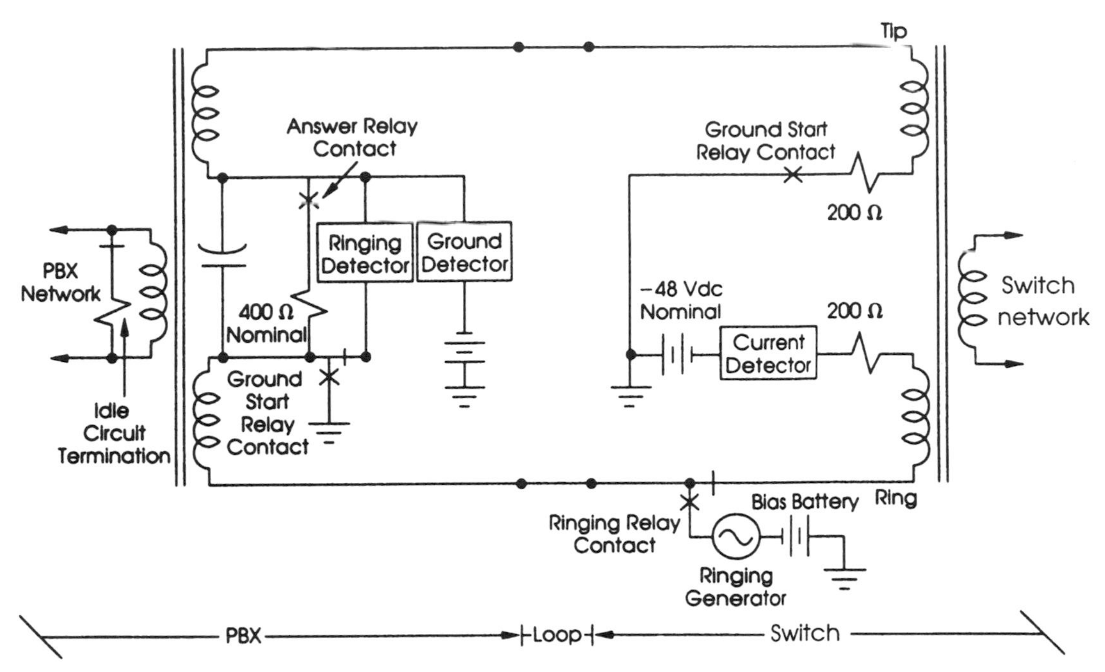

Figure 7.8 Ground-start interface block diagram. (From Figure 2-7 of Ref. 8, reprinted with permission.)

7.9 METALLIC TRUNK SIGNALING

7.9.1 Basic Loop Signaling

As mentioned earlier, many trunks serving the local area are metallic-pair trunks. They are actually loops much like the subscriber loop. Some still use dial pulses for address signaling along with some form of supervisory signaling.

Loop signaling is commonly used for supervision. As we would expect, it provides two signaling states: one when the circuit is opened and one when the circuit is closed. A third signaling state is obtained by reversing the direction or changing the magnitude of the current in the circuit. Combinations of (1) open/close, (2) polarity reversal, and (3) high/low current are used for distinguishing signals intended for one direction of signaling (e.g., dial-pulse signals) from those intended for the opposite direction (e.g., answer signals). We describe the most popular method of supervision on metallic pair trunks below, namely, reverse-battery signaling.

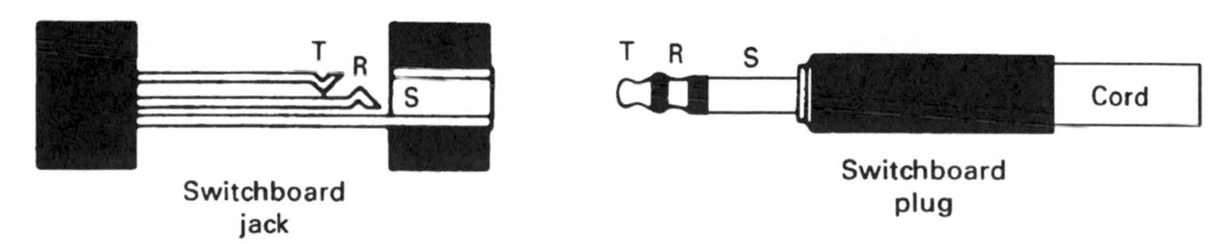

Figure 7.9 Switchboard plug with corresponding jack (R, S, and T are ring, sleeve, and tip, respectively).

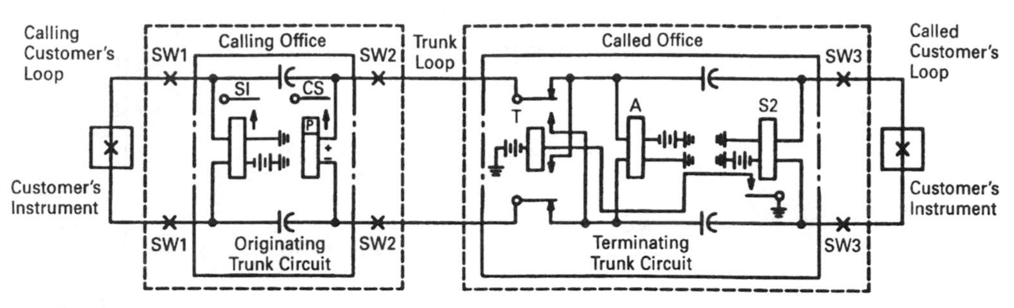

Figure 7.10 Reverse-battery signaling (From Figure 6-27 of Ref. 7, reprinted with permission.)

7.9.2 Reverse-Battery Signaling

Reverse-battery signaling employs basic methods (1) and (2) just mentioned, and takes its name from the fact that battery and ground are reversed on the tip and ring to change the signal toward the calling end from on-hook to off-hook. Figure 7.10 shows a typical application of reverse-battery signaling in a common-control path.

In the idle or on-hook condition, all relays are unoperated and the switch (SW) contacts are open. Upon seizure of the outgoing trunk by the calling switch (exchange) (trunk group selection based on the switch or exchange code dialed by the calling subscriber), the following occur:

- SW1 and SW2 contacts close, thereby closing loop to called office (exchange) and causing the A relay to operate.
- Operation of the A relay signals off-hook (connect) indication to the called switch (exchange).
- Upon completion of pulsing between swtiches, SW3 contacts close and the called subscriber is alerted. When the called subscriber answers, the S2 relay is operated.
- Operation of the S2 relay operates the T relay, which reverses the voltage polarity on the loop to the calling end.
- The voltage polarity causes the CS relay to operate, transmitting an off-hook (answer) signal to the calling end.

When the calling subscriber hangs up, disconnect timing starts (between 150 ms and 400 ms). After the timing is completed, SW1 and SW2 contacts are released in the calling switch. This opens the loop to the A relay in the called switch and releases the calling subscriber. The disconnect timing (150–400 ms) is started in the called switch as soon as the A relay releases. When the disconnect timing is completed,the following occur:

- If the called subscriber has returned to on hook, SW3 contacts release. The called subscriber is now free to place another call.
- If the called subscriber is still off-hook, disconnect timing is started in the called switch. On the completion of the timing interval, SW3 contacts open. The called subscriber is then returned to dial tone. If the circuit is seized again from the calling switch during the disconnect timing, the disconnect timing is terminated and the called subscriber is returned to dial tone. The new call will be completed without interference from the previous call.

When the called subscriber hangs up, the CS relay in the calling switch releases. Then the following occur:

- If the calling subscriber has also hung up, disconnection takes place as previously described.
- If the calling subscriber is still off-hook, disconnect timing is started. On the completion of the disconnect timing, SW1 and SW2 contacts are opened. This returns the calling subscriber to dial tone and releases the A relay in the called switch. The calling subscriber is free to place a new call at this time. After the disconnect timing, the SW3 contacts are released, which releases the called subscriber. The called subscriber can place a new call at this time.

REVIEW EXERCISES

1. Give the three generic signaling functions, and explain the purpose of each.
2. Differentiate between line signaling and interregister signaling.
3. There are seven ways to transmit signaling information, one is frequency. Name five others.
4. How does a switch know whether a particular talk path is busy or idle?
5. A most common form of line signaling is E&M signaling. Describe how it works in three sentences or less.
6. Compare in-band and out-of-band supervisory signaling regarding tone-on idle/busy, advantages, disadvantages.
7. What is the most common form of in-band signaling in North America?
8. What is the standard out-of-band signaling frequency in the United States?
9. Give the principal advantage of 2VF supervisory signaling over SF.
10. Compare CCITT No. 5/R1 signaling with R2 signaling.
11. Describe and compare end-to-end signaling with link-by-link signaling.
12. Describe at least four types of backward information.
13. Distinguish between associated channel signaling and separate channel signaling.
14. What is disassociated-channel signaling?
15. What is *glare*?

REFERENCES

1. *IEEE Standard Dictionary of Electrical and Electronic Terms*, 6th ed., IEEE Std 100-1996, IEEE, New York, 1996.
2. R. L. Freeman, *Telecommunication Transmission Handbook*, 4th ed., Wiley, New York, 1998.
3. *National Networks for the Automatic Service*, CCITT-ITU Geneva, 1968.

4. *Specifications of Signaling Systems 4 and 5*, CCITT Recommendations, Fascicle VI.4, IXth Plenary Assembly, Melbourne, 1988.
5. "Signaling" from Telecommunications Planning Documents, ITT Laboratories, Madrid, November, 1974.
6. *Specifications for Signaling Systems R1 and R2*, CCITT Recommendations, Fascicle VI.4, IXth Plenary Assembly, Melbourne, 1988.
7. *BOC Notes on the LEC Networks—1994*, Special Report SR-TSV-002275 Issue 2, Bellcore, Piscataway, NJ, 1994.
8. W. D. Reeve, *Subscriber Loop Signaling and Transmission Handbook—Analog*, IEEE Press, New York, 1992.

8

LOCAL AND LONG-DISTANCE NETWORKS

8.1 CHAPTER OBJECTIVE

This chapter concentrates on network design of the PSTN, how it is structured, and why. Routing techniques have a strong influence on how a network is structured. Thus we also discuss routing and, in particular, dynamic routing. The third topic deals with transmission, namely, assigning losses in the network to eliminate any possibility of singing and to keep echo inside some tolerable limits.

8.2 MAKEUP OF THE PSTN

As discussed in Section 1.3, the PSTN consists of a group of local networks connected by a long-distance network. In countries where competition is permitted, there may be two or many long-distance networks. Some of these may cover the nation, whereas others are regional long-distance networks.

The heart of a local network is the *subscriber plant*. This consists of customer premise equipment (CPE), a copper wire distribution network made up of subscriber loops which connect to a local serving switch via the main distribution frame (MDF). The concept of the subscriber plant feeding the local network, which, in turn, feeds the long-distance network to a distant local network and associated subscriber plant, is illustrated in Figure 8.1.

8.2.1 Evolving Local Network

Over 65% of the investment in a PSTN is in the local network. More cost-effective and efficient means are now being implemented to connect a subscriber to the local serving exchange. In Section 4.3.8. concentrators and remote switching were introduced. The digital subscriber line (DSL) concept was covered in Section 6.12.

With ISDN/BRI we bring digital service directly to the subscriber premise. If the two B-channels are combined (see Section 12.4), 128 kbps service can be provided. Users of the Internet desire downstream bit rates in excess of this value, up to 1.544 Mbps or better.[1] The shorter we make a conventional subscriber loop, the greater the bit rate it can support.

[1]*Downstream* in the direction local serving switch to the subscriber.

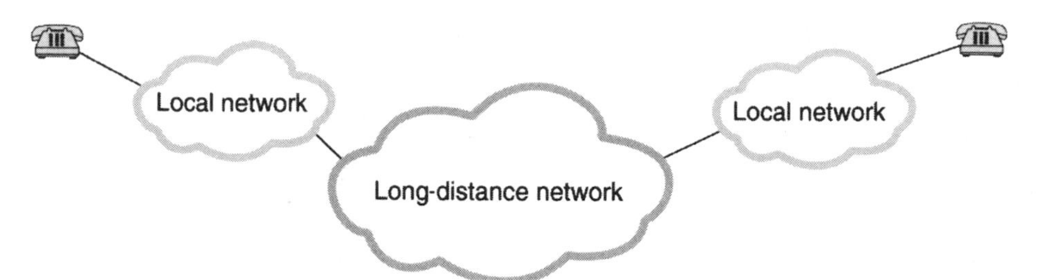

Figure 8.1 General structure of a PSTN.

We will incorporate a new, generic device in the subscriber plant. It is called a *remote subscriber unit* (RSU). It distributes service to the subscriber plant customer. The basic service it provides is plain old telephone service (POTS), using a typical subscriber line interface card (SLIC). The RSU has a large group of optional capabilities, listed below:

- It may or may not have a local switching capability.
- In most cases the RSU will carry out a concentrator function.
- It may be a fiber-to-wire interface point for a hybrid fiber-wire pair system.
- It may be a node for wireless local loop employing point-to-multipoint radio.
- It may provide an add-drop multiplex (ADM) capability on a synchronous optical network of 16.32, or 48 km (SONET) or SDH self-healing ring (SHR).

Several of these RSU concepts are illustrated in Figure 8.2. Figure 8.2*a* shows the conventional subscriber distribution plant and identifies its various functional parts. Figure 8.2*b* illustrates a digital subscriber line (DSL) feeding an RSU. The DSL might consist of one or several DS1 or E-1 bit streams. In this case the DSL would be on two wire pair, one for downstream and one for upstream. Figure 8.2*c* is the same as Figure 8.2*b*, but in this case the DSL is carried on two fibers plus a spare in a fiber optic cable configuration. Figure 8.2*d* illustrates the use of a fiber optic or wire-pair bus to feed several RSUs. Figure 8.2*e* shows a simple self-healing ring architecture employing either SONET or SDH (see Chapter 17).

In sparsely populated rural areas, point-to-multipoint full-duplex radio systems are particularly applicable. Such systems would operate in the 2 GHz or 4 GHz band with 10-, 20-, or 30-mile links. Access to the system would probably be time division multiple access. However, in densely populated urban areas, the 23-, 25-, or 38-GHz frequency bands should be considered to keep link length well under 3.5 miles (5.5 km). Because of expected high interference levels from nearby users, code division multiple access should be considered for this application. The TDMA and CDMA access schemes are described in Chapter 16.

8.2.2 What Affects Local Network Design?

There are a number of important factors that will influence the design of a local network. Among these factors are:

- Subscriber density and density distribution;
- Breakdown between residential and business subscribers;

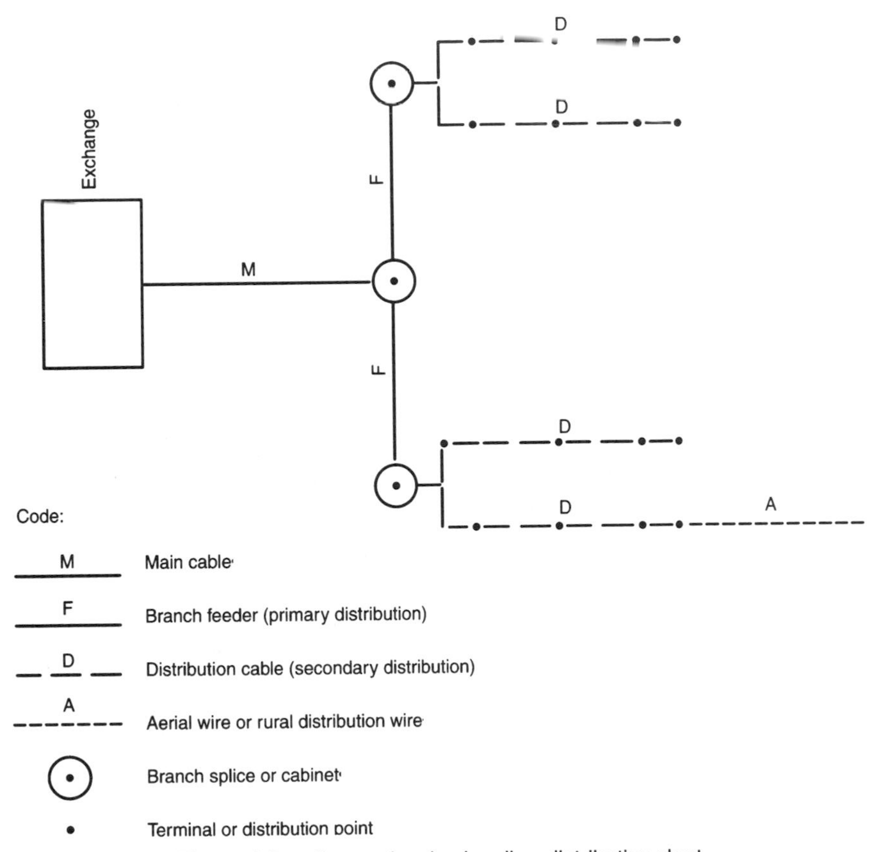

Figure 8.2*a* Conventional subscriber distribution plant.

- Further breakdown among types business subscribers as a function of white-collar and blue-collar workers. For example, a large insurance company will have much greater calling activity than a steel mill with the same number of workers; and
- Cultural factors.

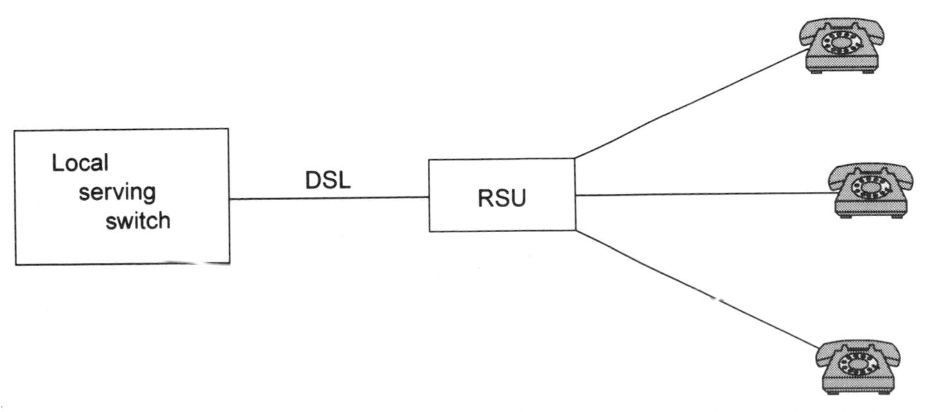

Figure 8.2*b* A digital subscriber line carrying a DS1 or E-1 configuration feeds an RSU. The RSU provides the necessary functions for subscriber loop interface.

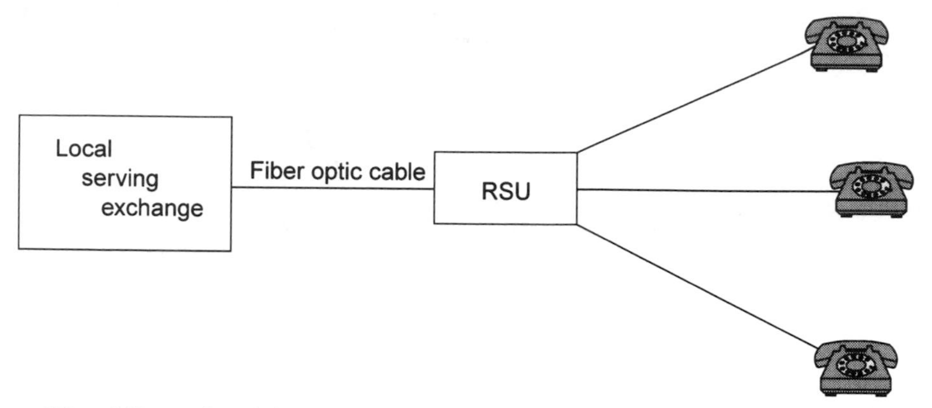

Figure 8.2*c* A fiber-optic pair feeds an RSU. This configuration is particularly useful for nests of subscribers beyond 12,000 feet (3700 m) from the local exchange.

It follows that in regions of high subscriber density typical of urban areas, there would be many exchanges, each with comparatively short subscriber loops. Such a network would trend toward mesh connectivities. The opposite would be in rural regions where we would expect few exchanges, and a trend toward a greater use of tandem working. Long route design would be the rule rather than the exception. Suburban bedroom communities present still a different problem, where we would expect calling trends toward urban, industrial centers. Many calls in this case will be served by just one exchange. Judicious use of tandem working would be advisable.

Cultrual factors can give the network designer insight into expected calling habits. The affluence of a region may well be a factor. However, consider countries where affluence goes hand-in-hand with volunteerism, an organizational spirit and a socialization factor. Certain cultures encourage strong family ties, where one can expect high intrafamily calling activity.

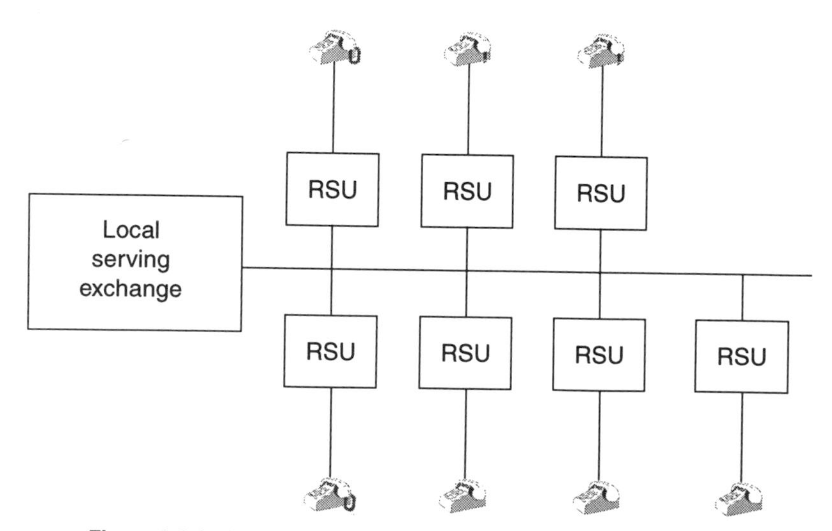

Figure 8.2*d* A fiber optic or wire-pair cable bus feeds several RSUs.

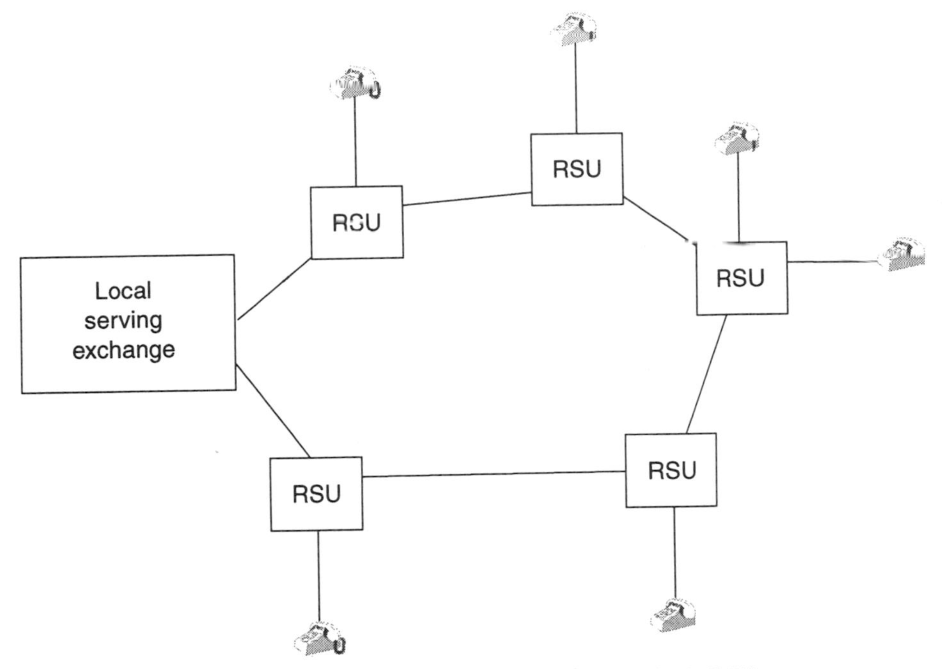

Figure 8.2e A fiber optic cable self-healing ring feeds RSUs.

8.3 DESIGN OF LONG-DISTANCE NETWORKS

8.3.1 Introduction

A long-distance network connects an aggregate of local networks. Any subscriber in the nation should be able to reach any other subscriber. Likewise, any subscriber in the nation should be able to connect with any other subscriber in the world. So we have need not only of local and long-distance (toll) exchanges, but international exchanges as well.

8.3.2 Three Design Steps

The design of a long-distance network basically involves three considerations:

1. Routing scheme given inlet and outlet points and their traffic intensities;
2. Switching scheme and associated signaling; and
3. Transmission plan.

In the design each design step will interact with the other two. In addition, the system designer must specify the type of traffic, lost-call criterion or grade of service, a survivability criterion, forecast growth, and quality of service (QoS). The trade-off of these factors with *economy* is probably the most vital part of initial planning and downstream system design.

Consider transcontinental communications in the United States. Service is now available for people in New York to talk to people in San Francisco. From the history of this service, we have some idea of how many people wish to talk, how often, and for how long. These factors are embodied in traffic intensity and calling rate. There are also other cities on the West Coast to be served and other cities on the East Coast. In addition, there are existing traffic nodes at intermediate points such as Chicago and St. Louis. An obvious approach would be to concentrate all traffic into one transcontinental route with drops and inserts at intermediate points.

Again, we must point out that switching enhances the transmission facilities. From an economic point of view, it would be desirable to make transmission facilities (carrier, radio, and cable systems) adaptive to traffic load. These facilities taken alone are inflexible. The property of adaptivity, even when the transmission potential for it has been predesigned through redundancy, cannot be exercised, except through the mechanism of switching in some form. It is switching that makes transmission adaptive.

The following requirements for switching ameliorate the weaknesses of transmission systems: concentrate light, discretely offered traffic from a multiplicity of sources and thus enhance the utilization factor of transmission trunks; select and make connections to a statistically described distribution of destinations per source; and restore connections interrupted by internal or external disturbances, thus improving reliabilities (and survivability) from the levels on the order of 90% to 99% to levels on the order of 99% to 99.9% or better. Switching cannot carry out this task alone. Constraints have to be iterated or fed back to the transmission systems, even to the local area. The transmission system must not excessively degrade the signal to be transported; it must meet a reliability constraint expressed in MTBF (mean time between failures) and availability and must have an alternative route scheme in case of facility loss, whether switching node or trunk route. This latter may be termed *survivability* and is only partially related to overflow (e.g., alternative routing).

The single transcontinental main traffic route in the United States suggested earlier has the drawback of being highly vulnerable. Its level of survivability is poor. At least one other route would be required. Then why not route that one south to pick up drops and inserts? Reducing the concentration in the one route would result in a savings. Capital, of course, would be required for the second route. We could examine third and fourth routes to improve reliability–survivability and reduce long feeders for concentration at the expense of less centralization. In fact, with overflow, one to the other, dimensioning can be reduced without reduction of overall grade of service.

8.3.3 Link Limitation

From a network design perspective a connectivity consists of one or more links in tandem.[2] We define a *link* as the transmission facilities connecting two adjacent switches. CCITT in Rec. E.171 (Ref. 1) states that for an international connection there shall be no more than 12 links in tandem. This is apportioned as follows:

- 4 links in the calling party's country;
- 4 links in the called party's country; and
- 4 international links.

This concept is illustated in Figure 8.3.

The PSTN network designer should comply with this CCITT criterion, in that for a national connection, there should be no more than four links in tandem. The reason

[2]It should be noted that there are connectivities with "no links in trandem." This is an own-exchange connectivity, where the calling and called subscriber terminate their subscriber loops in the same exchange.

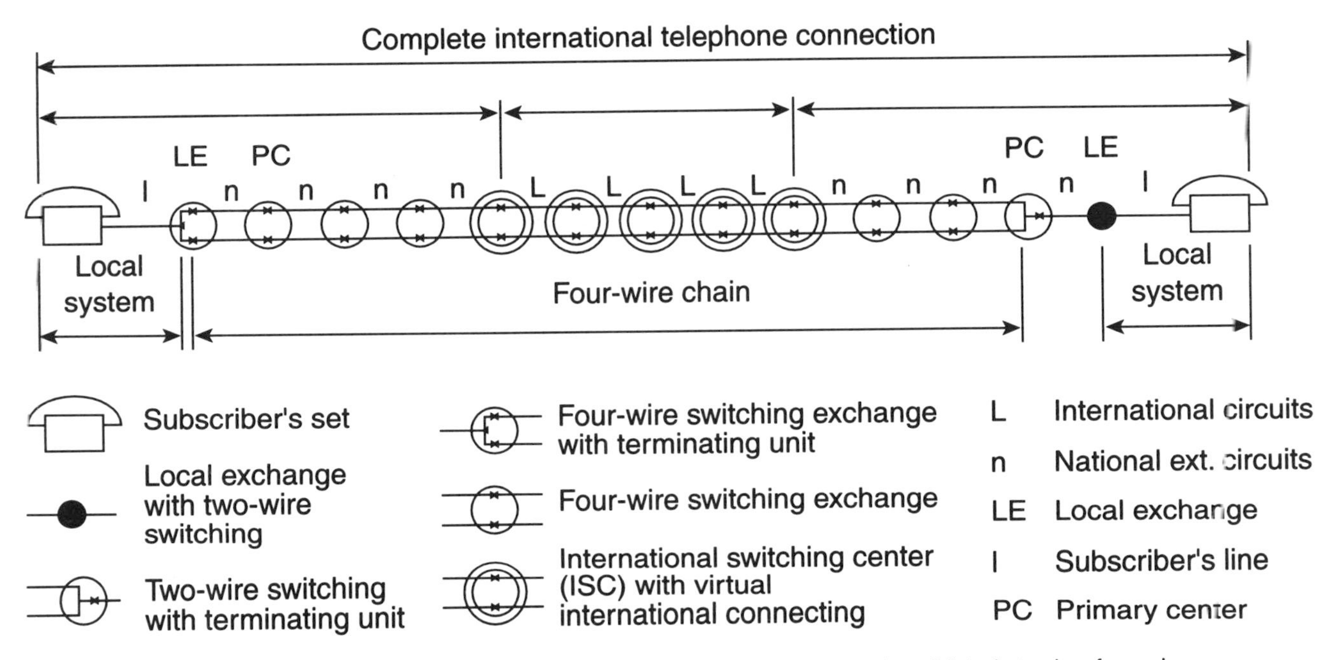

Figure 8.3 An international connection to illustrate the maximum number of links in tandem for such a connection. (From Figure 6/G.101 of Ref. 2.)

CCITT/ITU-T set this limit was to ensure transmission QoS. As we add links in tandem, transmission quality deteriorates. Delay increases and we include here processing delay because of the processing involved with a call passing through each switch. End-to-end bit error rate deteriorates and jitter and wander accumulate. Transcontinental calls in North America generally need no more than three links in tandem, except during periods of heavy congestion when a fourth link may be required for an alternate route.

8.3.4 Numbering Plan Areas

The geographical territory covered by the long-distance network will be broken up into numbering plan areas (NPAs). In North America, each NPA is assigned a three-digit area code. In other parts of the world, two- and even one-digit area codes are used. NPA size and shape are driven more by numbering capacity and future numbering requirements. Numbering plan administrators are encouraged to design an NPA such that it coincides with political and/or administrative boundaries. For example, in the United States, an NPA should not cross a state boundary; in Canada, it should not cross a provincial boundary. NPAs are also important for establishing a rates and tariffs scheme.[3]

We know a priori that each NPA will have *at least* one long-distance exchange. It may be assigned more. This long-distance exchange may or may not colocate with the POP (point of presence).[4] We now have made the first steps in determining exchange location. In other countries this exchange may be known as a *toll-connecting exchange*.

8.3.5 Exchange Location

We have shown that the design of the long-distance network is closely related to the layout of numbering plan areas or simply numbering areas. These exchanges are ordinarily placed near a large city. The number of long-distance exchanges in a numbering area is dependent on exchange size and certain aspects of survivability. This is the idea of "not having all one's eggs in one basket." There may be other reasons to have a second or even a third exchange in a numbering area (NPA in the United States). Not only does it improve survivability aspects of the network, but it also may lead the designer to place a second exchange near another distant large city.

Depending on long-distance calling rates and holding times, and if we assume 0.004 erlangs per line during the busy hour, a 4000-line long-distance exchange could serve some 900,000 subscribers. The exchange capacity should be dimensioned to the forecast long-distance traffic load 10 years after installation. If the system goes through a 15% expansion in long-distance traffic volume per year, it will grow to over four times its present size in 10 years. Exchange location in the long-distance network is not very sensitive to traffic.

8.3.6 Hierarchy

Hierarchy is another essential aspect in long-distance (toll) network design. One important criterion is establishing the number of hierarchical levels in a national network. The United States has a two-level hierarchy: the local exchange carrier (or LATA [local

[3]This deals with how much a telephone company charges for a telephone call.

[4]*POP*, remember, is where the local exchange carrier interfaces with long-distance carriers. This whole concept of the POP is peculiar to the United States and occurred when the Bell System was divested. In other parts of the world it may be called a *toll-connecting exchange*.

access and transport area]) and the interexchange carrier network. Our concern here is the interexchange carrier network, which is synonymous with the long-distance network. So the question remains: how many hierarchical levels in the long-distance or toll network?

There will be "trandem" exchanges in the network, which we will call *transit exchanges*. These switches may or may not be assigned a higher hierarchical level. Let us assume that we will have at least a two-level hierarchy.

Factors that may lead to more than two levels are:

- Geographical size;
- Telephone density, usually per 100 inhabitants;[5]
- Long-distance traffic trends; and
- Political factors (such as Bell System divestiture in the United States, privatization in other countries).

The trend toward greater use of direct HU (high-usage) routes tends to keep the number of hierarchical levels low (e.g., at two levels). The employment of *dynamic routing* can have a similar effect.

We now deal with *fan-out*. A higher-level exchange, in the hierarchical sense, fans out to the next lower level. This level, in turn, fans out to still lower levels in the hierarchy. It can be shown that 6- and 8-fan-outs are economic and efficient.

Look at this example. The highest level, one exchange, fans out to six exchanges in the next level. This level, in turn fans out to eight exchanges. Thus there is connectivity to 48 exchanges (8×6), and if the six exchanges in the higher level also serve as third-level exchanges, then we have the capability of $48 + 6$, or 54 toll exchanges.

Suppose that instead of one exchange in the highest level, there were four interconnected in mesh for survivability and improved service. This would multiply the number of long-distance exchanges served to $48 \times 4 = 192$, and if we use the 56 value it would be $56 \times 4 = 224$ total exchanges. In large countries we deal with numbers like this. If we assign a long-distance exchange in each NPA, and assume all spare NPA capacity is used, there would be 792 NPAs in the United States, each with a toll exchange. Allow for a three- or four-level hierarchy and the importance of fan-out becomes evident.

Figure 8.4 shows one-quarter of a three-level hierarchy network, where the top level is mesh connected with four transit exchanges.

The fan-out concept assumes a pure hierarchy without high-usage routes. HU routes tend to defeat the fan-out concept and are really mandatory to reduce the number of links in tandem to a minimum.

8.3.7 Network Design Procedures

A national territory consists of a large group of contiguous local areas, each with a toll/toll-connecting exchange. There will also be at least one international switching center (ISC). In larger, more populous countries there may be two or more such ISCs. Some may call these switching centers *gateways*. They need not necessarily be near a coastline. Chicago is an example in North America. So we now have established three bases to work from:

[5]The term *telephone density* should not mislead the reader. Realize that some "telephone lines" terminate in a modem in a computer or server, in a facsimile machine, and so on.

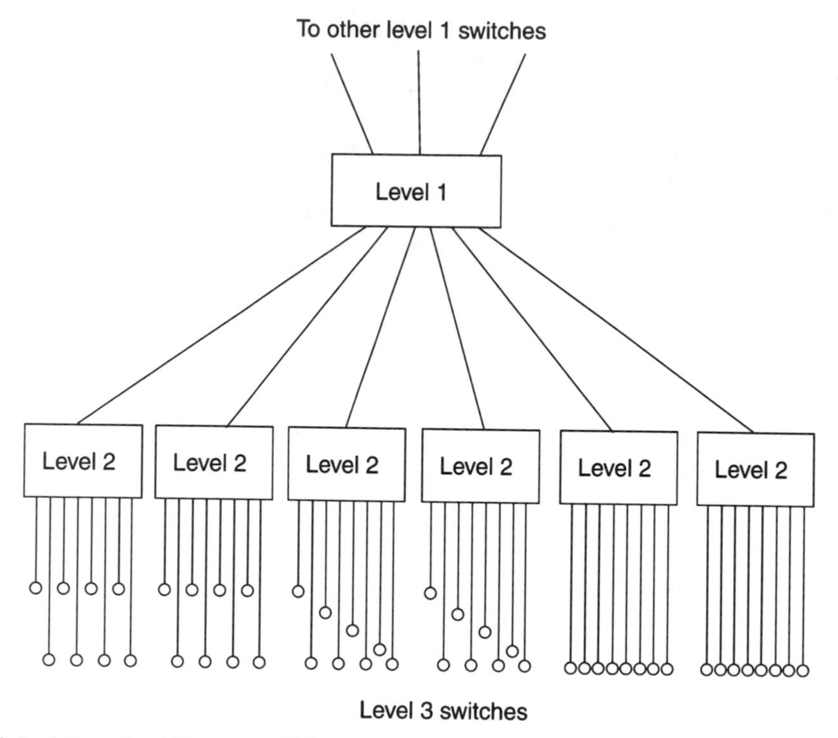

Figure 8.4 A three-level hierarchy with initial fan-out of six and subsequent fan-out of eight. The highest level consists of four transit exchanges, but only one is shown.

1. There are existing local areas, each with a long-distance exchange.
2. There is one or more ISCs placed at the top of the network hierarchy.
3. There will be no more than four links in tandem on any connection to reach an ISC.

As mentioned previously, Point 1 may be redefined as a long-distance network consisting of a grouping of local areas probably coinciding with a numbering (plan) area. This is illustrated in a very simplified manner in Figure 8.5, where T, in CCITT terminology, is a higher-level center, a "Level 1" or "Level 2 center." Center T, of course, is a long-distance transit exchange with a fan-out of four; these are four local exchanges (A,

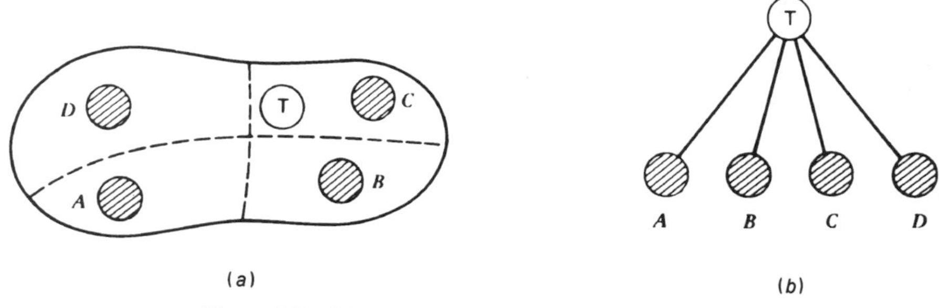

Figure 8.5 (*a*) Areas and (*b*) exchange relationships.

Table 8.1 Traffic Matrix Example—Long-Distance Service (in erlangs)

From Exchange	To Exchange 1	2	3	4	5	6	7	8	9	10
1		57	39	73	23	60	17	21	23	5
2	62		19	30	18	26	25	2	9	6
3	42	18		28	17	31	19	8	10	12
4	70	31	23		6	7	5	8	4	3
5	25	19	32	5		22	19	31	13	50
6	62	23	19	8	20		30	27	19	27
7	21	30	17	40	16	32		15	16	17
8	21	5	12	3	25	19	17		18	29
9	25	10	9	1	16	22	18	19		19
10	7	8	7	2	47	25	13	30	17	

B, C, and D) connect to T.[6] The entire national geographic area is made up of such small segments as shown in Figure 8.5, and each may be represented by a single exchange T, which has some higher level or rank.

The next step is to examine traffic flows to and from (originating and terminating) each T. This information is organized and tabulated on a *traffic matrix*. A simplified example is illustrated in Table 8.1. Care must be taken in the preparation and subsequent use of such a table. The convention used here is that values (in erlangs or ccs) are read *from* the exchange in the left-hand column *to* the exchange in the top row. For example, traffic from exchange 1 to exchange 5 is 23 erlangs, and traffic from exchange 5 to exchange 1 is 25 erlangs. It is often useful to set up a companion matrix of distances between exchange pairs. The matrix (Table 8.1) immediately offers candidates for HU routes. Nonetheless, this step is carried out after a basic hierarchical structure is established.

We recommend that a hierarchical structure be established at the outset, being fully aware that the structure may be modified or even done away with entirely in the future as dynamic routing disciplines are incorporated (see Section 8.4). At the top of a country's hierarchy is (are) the international switching center(s). The next level down, as a minimum, would be the long-distance network, then down to a local network consisting of local serving exchanges and tandem exchanges. The long-distance network itself, as a minimum, might be divided into a two-layer hierarchy.

Suppose, for example, that a country had four major population centers and could be divided into four areas around each center. Each of the four major population centers would have a Level 1 switching center assigned. One of these four would be the ISC. Each Level 1 center would have one or several Level 2 or secondary centers homing on it.[7] Level 3 or tertiary centers home on a Level 2 switching center. This procedure is illustrated in Figure 8.6. Its hierarchical representation is illustrated in Figure 8.7 setting out the final route. One of the Level 1 switching centers is assigned as the ISC. We define a final route as a route from which no traffic can overflow to an alternative route. It is a route that connects an exchange immediately above or below it in the network hierarchy and there is also a connection of the two exchanges at the top hierarchical level of the network. Final routes are said to make up the "backbone" of a network. Calls that are offered to the backbone but cannot be completed are lost calls.[8]

[6]Of course, in the United States, T would be the POP (point of presence).

[7]*Homing on* meaning subsidiary to in a hierarchical sense. It "reports to."

[8]*Completed* calls are those where a full connectivity is carried out indicated by both calling and called subscriber in the off-hook condition.

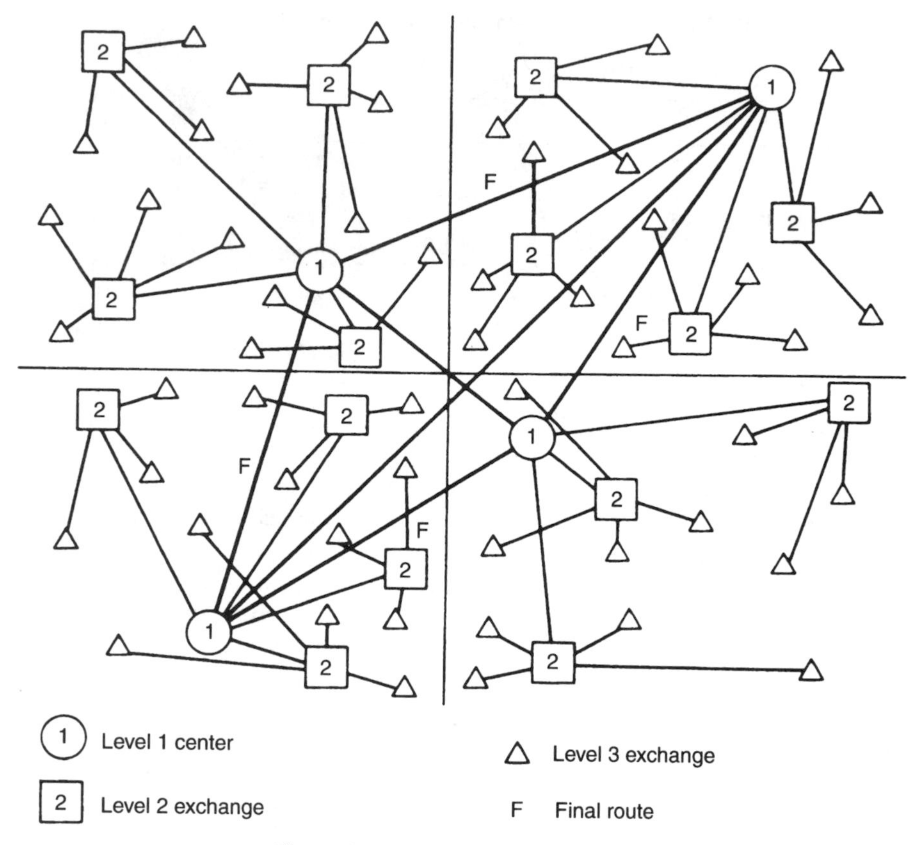

Figure 8.6 A sample network design.

A *high-usage* (HU) route is defined as any route that is not a final route; it may connect exchanges at a level of the network hierarchy *other than* the top level, such as between 1_1 and 1_2 in Figure 8.7. It may also be a route between exchanges on different hierarchical levels when the lower-level exchange (higher level number) does home on a higher level. A *direct route* is a special type of HU route connecting exchanges in the local area. Figure 8.8 shows a hierarchical network with alternative routing. Note that it employs CCITT nomenclature.

Before final dimensioning can be carried out of network switches and trunks, a grade

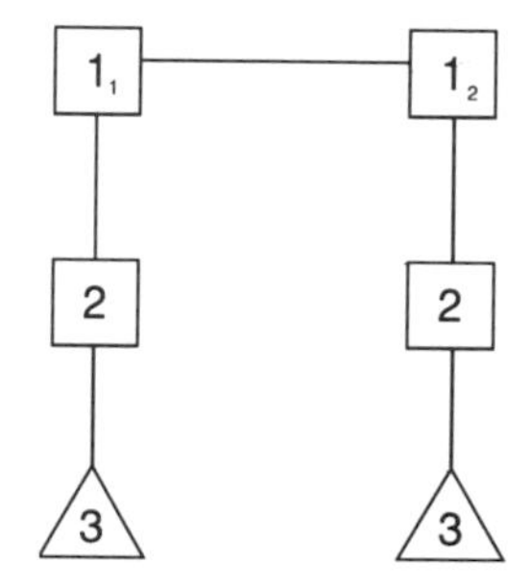

Figure 8.7 Hierarchical representation showing final routes.

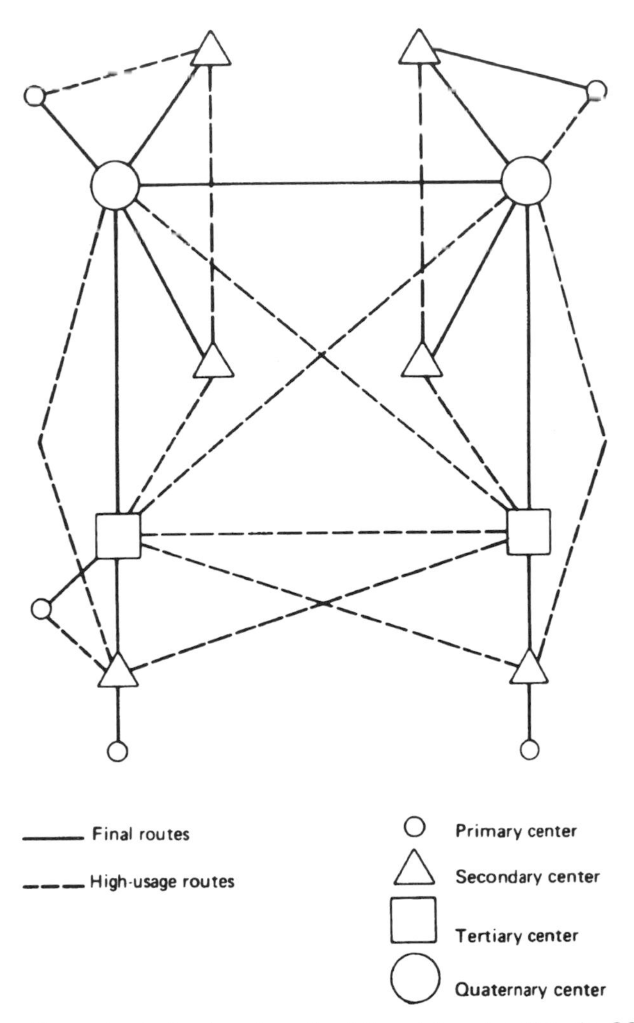

Figure 8.8 A hierarchical network showing alternative (alternate) routing. Note the CCITT nomenclature.

of service criterion must be established.[9] If we were to establish a grade of service as p = 0.01 per link on a final route, and there were four links in tandem, then the grade of service end-to-end would be 4 × 0.01 or 0.04. In other words, for calls traversing this final route, one in 25 would meet congestion during the busy hour. The use of HU connections reduces tandem operation and tends to improve overall grade of service.

The next step in the network design is to lay out HU routes. This is done with the aid of a traffic matrix. A typical traffic matrix is shown in Table 8.1. Some guidelines may be found in Section 4.2.4. Remember that larger trunk groups are more efficient. As a starting point (Section 4.2.4) for those traffic relations where the busy hour traffic intensity was >20 erlangs, establish a HU route; for those relations <20 erlangs, the normal hierarchical routing should remain in placc.

National network design as described herein lends itself well to computer-based design techniques. The traffic intensity values used in traffic matrices, such as Table 8.1, should be taken from a 10-year forecast.

[9]*Grade of service* is the probability of meeting congestion (blockage) during the busy hour (BH).

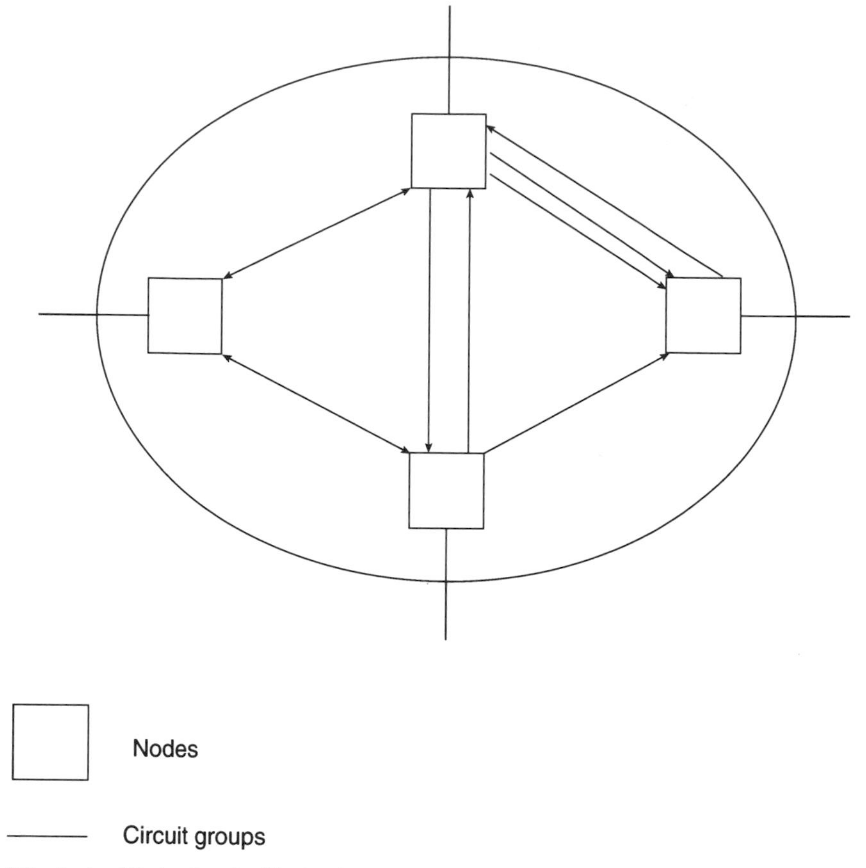

Figure 8.9 A simplified network with circuit groups connecting pairs of nodes with one-way and both-way working.

8.4 TRAFFIC ROUTING IN A NATIONAL NETWORK

8.4.1 New Routing Techniques

8.4.1.1 Objective of Routing. The objective of routing is to establish a successful connection between any two exchanges in the network. The function of traffic routing is the selection of a particular circuit group, for a given call attempt or traffic stream, at an exchange in the network. The choice of a circuit group may be affected by information on the availability of downstream elements of the network on a quasi-real-time basis.

8.4.1.2 Network Topology. A network comprises a number of nodes (i.e., switching centers) interconnected by circuit groups. There may be several direct circuit groups between a pair of nodes and these may be one-way or both-way (two-way). A simplified illustration of this idea is shown in Figure 8.9.

Remember that a direct route consists of one or more circuit groups connecting adjacent nodes. We define an indirect route as a series of circuit groups connecting two nodes providing end-to-end connection via other nodes.

An ISC is a node in a national network, which in all probability will have some sort of hierarchial structure as previously discussed. An ISC is also a node on the international network that has no hierarchical structure. It consists entirely of HU direct routes.

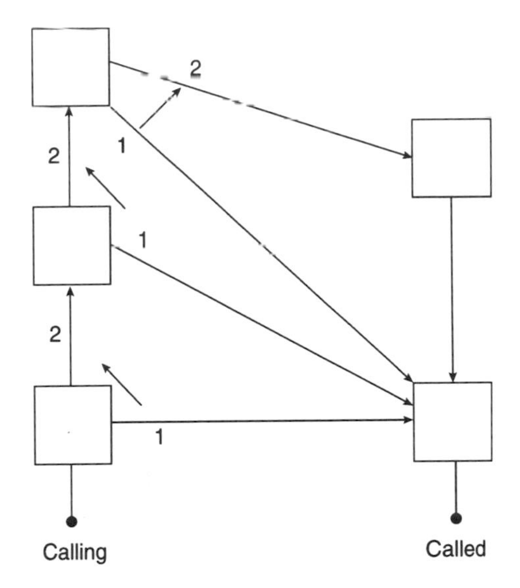

Note — All nodes are of equal status.

Figure 8.10 Hierarchical routing in a nonhierarchical network of exchanges.

8.4.2 Logic of Routing

8.4.2.1 Routing Structure. Conceptually, hierarchical routing need not be directly related to a concept of a hierarchy of switching centers, as just described. A routing structure is hierarchical if, for all traffic streams, all calls offered to a given route, at a specific node, overflow to the same set of routes irrespective of the routes already tested.[10] The routes in the set will always be tested in the same sequence, although some routes may not be available for certain types of calls. The last choice route is final (i.e., the final route), in the sense that no traffic streams using this route may overflow further.

A routing structure is nonhierarchical if it violates the previously mentioned definition (e.g., mutual overflow between circuit groups originating at the same exchange). An example of hierarchical routing in a nonhierarchical network of exchanges is shown in Figure 8.10.

8.4.2.2 Routing Scheme. A routing scheme defines how a set of routes is made available for calls between pairs of nodes. The *scheme* may be *fixed* or *dynamic*. For a fixed scheme the set of routes in the routing pattern is always the same. In the case of a dynamic scheme, the set of routes in the pattern varies.

8.4.2.2.1 Fixed Routing Scheme. Here routing patterns in the network are fixed, in that changes to the route choice for a given type of call attempt require manual intervention. If there is a change it represents a "permanent change" to the routing scheme. Such changes may be the introduction of new routes.

[10]*Tested* means that at least one free circuit is available to make a connectivity. This "testing" is part and parcel of CCITT Signaling System No. 7, which is discussed in Chapter 13.

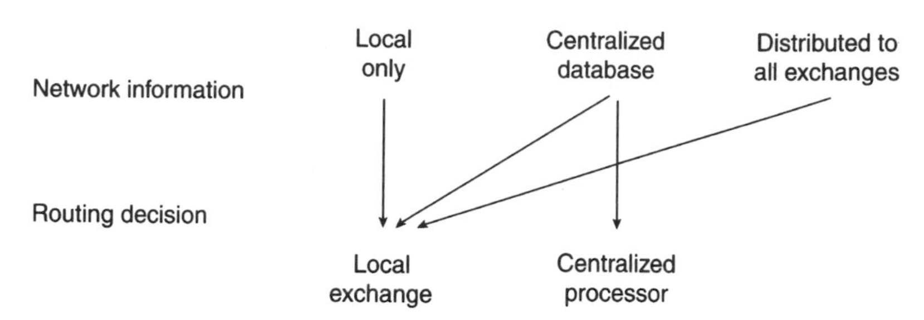

Figure 8.11 Adaptive or state-dependent routing, network information (status) versus routing decisions.

8.4.2.2.2 Dynamic Routing Scheme. Routing schemes may also incorporate frequent automatic variations. Such changes may be time-dependent, state-dependent, and/or event-dependent. The updating of routing patterns make take place periodically, aperiodically, predetermined, depending on the state of the network, or depending on whether calls succeed or fail in the setup of a route.

In time-dependent routing, routing patterns are altered at fixed times during the day or week to allow for changing traffic demands. It is important to note that these changes are preplanned and are implemented consistently over a long time period.

In state-dependent routing, routing patterns are varied automatically according to the state of the network. These are called *adaptive routing schemes*. To support such a routing scheme, information is collected about the status of the network. For example, each toll exchange may compile records of successful calls or outgoing trunk occupancies. This information may then be distributed through the network to other exchanges or passed to a centralized database. Based on this network status information, routing decisions are made either in each exchange or at a central processor serving all exchanges. The concept is shown in Figure 8.11.

In event-dependent routing, patterns are updated locally on the basis of whether calls succeed or fail on a given route choice. Each exchange has a list of choices, and the updating favors those choices that succeed and discourages those that suffer congestion.

8.4.2.3 Route Selection. Route selection is the action taken to actually select a definite route for a specific call. The selection may be *sequential* or *nonsequential*. In the case of sequential selection, routes in a set are always tested in sequence and the first available route is selected. For the nonsequential case, the routes are tested in no specific order.

The decision to select a route can be based on the state of the outgoing circuit group or the states of series of circuit groups in the route. In either case, it can also be based on the incoming path of entry, class of service, or type of call to be routed.

8.4.3 Call-Control Procedures

Call-control procedures define the entire set of interactive signals necessary to establish, maintain, and release a connection between exchanges. Two such call-control procedures are *progressive call control* and *originating call control*.

8.4.3.1 Progressive Call Control. This type of call control uses link-by-link signaling (see Chapter 7) to pass supervisory controls sequentially from one exchange to the next. Progressive call control can be either irreversible or reversible. In the irreversible case, call control is always passed downstream toward the destination exchange. Call control is reversible when it can be passed backwards (maximum of one node) using automatic rerouting or crankback actions.

8.4.3.2 Originating Call Control. In this case the originating exchange maintains control of the call set-up until a connection between the originating and terminating exchanges has been completed.

8.4.4 Applications

8.4.4.1 Automatic Alternative Routing. One type of progressive (irreversible) routing is automatic alternative routing (AAR). When an exchange has the option of using more than one route to the next exchange, an alternative routing scheme can be employed. The two principal types of AAR that are available are:

1. When there is a choice of direct-circuit groups between two exchanges; and
2. When there is a choice of direct and indirect routes between the two exchanges.

Alternative routing takes place when all appropriate circuits in a group are busy. Several circuit groups then may be tested sequentially. The test order is fixed or time-dependent.

8.4.4.2 Automatic Rerouting (Crankback). Automatic rerouting (ARR) is a routing facility enabling connection of call attempts encountering congestion during the initial call setup phase. Thus, if a signal indicating congestion is received from Exchange B, subsequent to the seizure of an outgoing trunk from Exchange A, the call may be rerouted at A. This concept is shown in Figure 8.12. ARR performance can be improved through the use of different signals to indicate congestion—S1 and S2 (see Figure 8.12).

- S1 indicates that congestion has occurred on outgoing trunks from exchange B.
- S2 indicates that congestion has occurred further downstream—for example, on outgoing trunks from D.

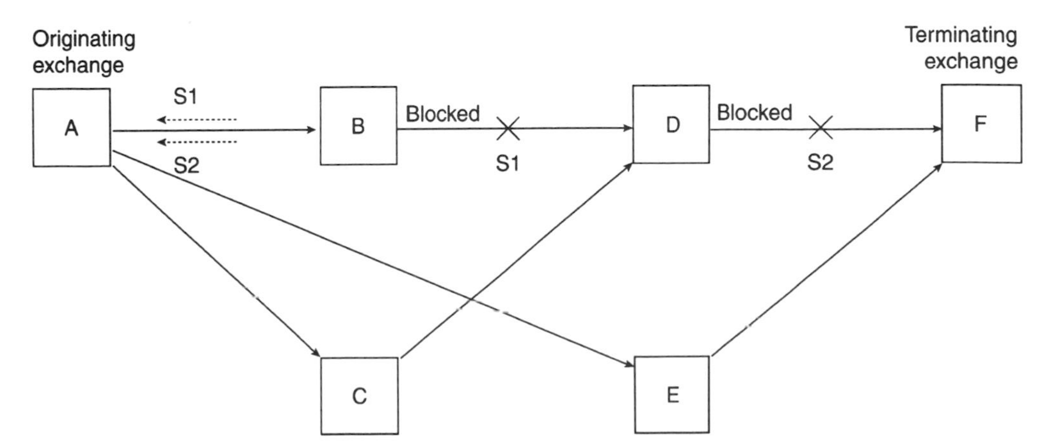

Figure 8.12 The automatic rerouting (ARR) or crankback concept. *Note:* Blocking from B to D activates signal S1 to A. Blocking from D to F activates signal S2 to A. (From Figure 4/E.170 of Ref. 3.)

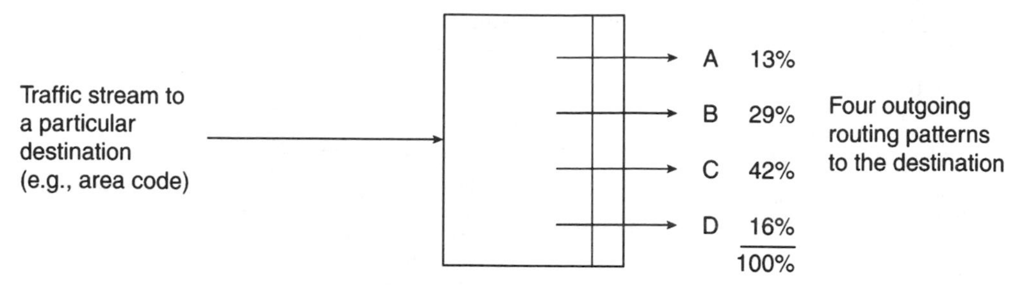

Figure 8.13 An example of preplanned distribution of load sharing. *Note:* Each outgoing routing pattern (A, B, C, D) may include alternative routing options.

The action to be taken at exchange A upon receiving S1 or S2 may be either to block the call or to reroute it.

In the example illustrated in Figure 8.12, a call from A to D is routed via C because the circuit group B-D is congested (S1-indicator) and a call from A to F is routed via E because circuit group D-F is congested (S2 indicator).

One positive consequence of this alternative is to increase the signaling load and number of call set-up operations resulting from the use of these signals. If such an increase is unacceptable, it may be advisable to restrict the number of reroutings or limit the signaling capability to fewer exchanges. Of course, care must be taken to avoid circular routings ("ring-around-the-rosy"), which return the call to the point at which blocking previously occurred during call setup.

8.4.4.3 Load Sharing. All routing schemes should result in the sharing of traffic load between network elements. Routing schemes can, however, be developed to ensure that call attempts are offered to route choices according to a preplanned distribution. Figure 8.13 illustrates this application to load sharing, which can be made available as a software function of SPC exchanges. The system works by distributing the call attempts to a particular destination in a fixed ratio between the specified routing patterns.

8.4.4.4 Dynamic Routing. Let us look at an example of state-dependent routing. A centralized routing processor is employed to select optimum routing patterns on the basis of actual occupancy level of circuit groups and exchanges in the network which are monitored on a periodic basis (e.g., every 10 s). Figure 8.14 illustrates this concept.

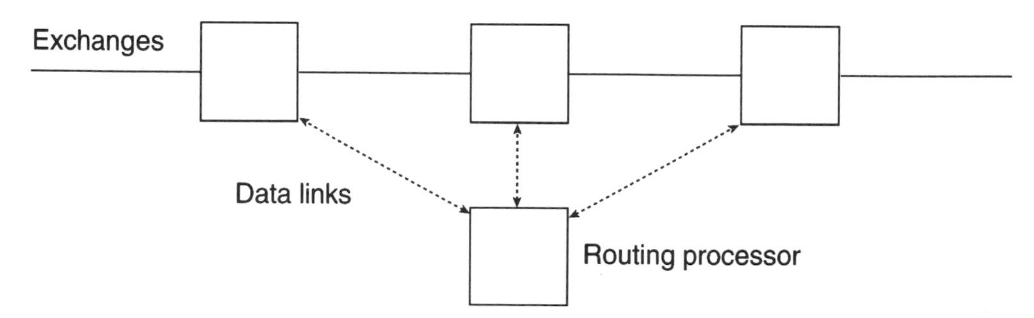

Figure 8.14 State-dependent routing example with centralized processor.

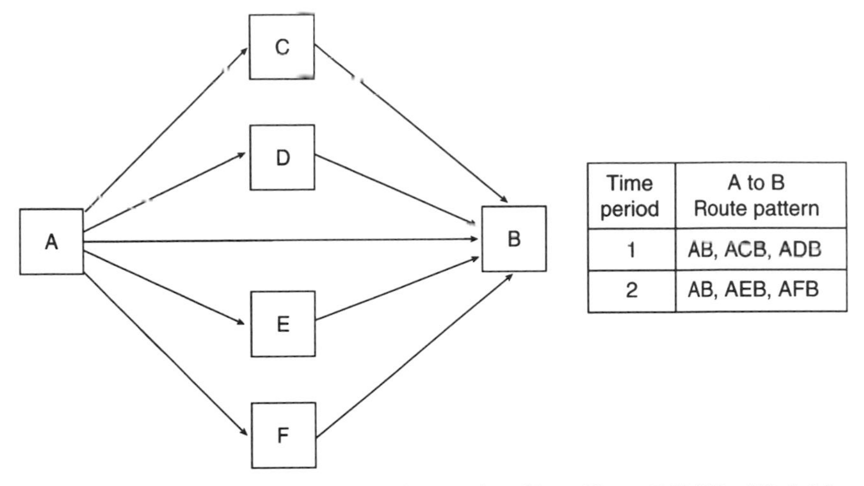

Time period	A to B Route pattern
1	AB, ACB, ADB
2	AB, AEB, AFB

Figure 8.15 Example of time-dependent routing. (From Figure 7/E.170 of Ref. 3.)

In addition, qualitative traffic parameters may also be taken into consideration in the determination of the optimal routing pattern.

This routing technique inherently incorporates fundamental principles of network management in determining routing patterns. These principles include:

- Avoiding occupied circuit groups;
- Not using overloaded exchanges for transit; and
- In overload circumstances, restriction of routing direct connections.

Now let's examine an example of time-dependent routing. For each originating and terminating exchange pair, a particular route pattern is planned depending on the time of day and day of the week. This is illustrated in Figure 8.15. A weekday, for example, can be divided into different time periods, with each time period resulting in different route patterns being defined to route traffic streams between the same pair of exchanges.

This type of routing takes advantage of idle circuit capacity in other possible routes between originating and terminating exchanges which may exist due to noncoincident busy hours.[11] Crankback may be utilized to identify downstream blocking on the second link of each two-link alternative path.

The following is an example of event-dependent routing. In a fully connected (mesh) network, calls between each originating and terminating exchange pair try the direct route with a two-link alternative path selected dynamically. While calls are successfully routed on a two-link path, that alternative is retained. Otherwise, a new two-link alternative path is selected. This updating, for example, could be random or weighted by the success of previous calls. This type of routing scheme routes traffic away from congested links by retaining routing choices where calls are successful. It is simple, adapts quickly to changing traffic patterns, and requires only local information. Such a scheme is illustrated in Figure 8.16 (Refs. 3, 4).

[11]Noncoincident busy hours: in large countries with two or more time zones.

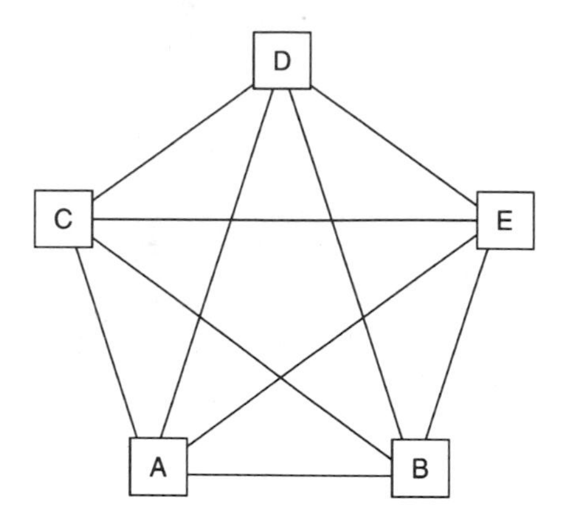

	A to B route pattern	
Choice	Current	After call failure
1	AB	AB
2	AEB	ACB

Figure 8.16 Event-dependent routing in a mesh network.

8.5 TRANSMISSION FACTORS IN LONG-DISTANCE TELEPHONY

8.5.1 Introduction

Long-distance analog communication systems require some method to overcome losses. As a wire-pair telephone circuit is extended, there is some point where loss accumulates such as to attenuate signals to such a degree that the far-end subscriber is dissatisfied. The subscriber cannot hear the near-end talker sufficiently well. Extending the wire connections still further, the signal level can drop below the noise level. For a good received signal level, a 40-dB signal-to-noise ratio is desirable (see Sections 3.2.1 and 3.2.2.4). To overcome the loss, amplifiers are installed on many wire-pair trunks. Early North American transcontinental circuits were on open-wire lines using amplifiers quite widely spaced. However, as BH demand increased to thousand of circuits, the limited capacity of such an approach was not cost effective.

System designers turned to wideband radio and coaxial cable systems where each bearer or pipe carried hundreds or thousands of simultaneous telephone conversations.[12] Carrier (frequency division) multiplex techniques made this possible (see Section 4.5). Frequency division multiplex (FDM) requires separation of transmit and receive voice paths. In other words, the circuit must convert from two-wire to four-wire transmission. Figure 8.17 is a simplified block diagram of a telephone circuit with transformation from two-wire to four-wire operation at one end and conversion back to two-wire operation at the other end. This concept was introduced in Section 4.4.

[12]On a pair of coaxial cables, a pair of fiber optic light guides, or a pair of radio-frequency carriers, one coming and one going.

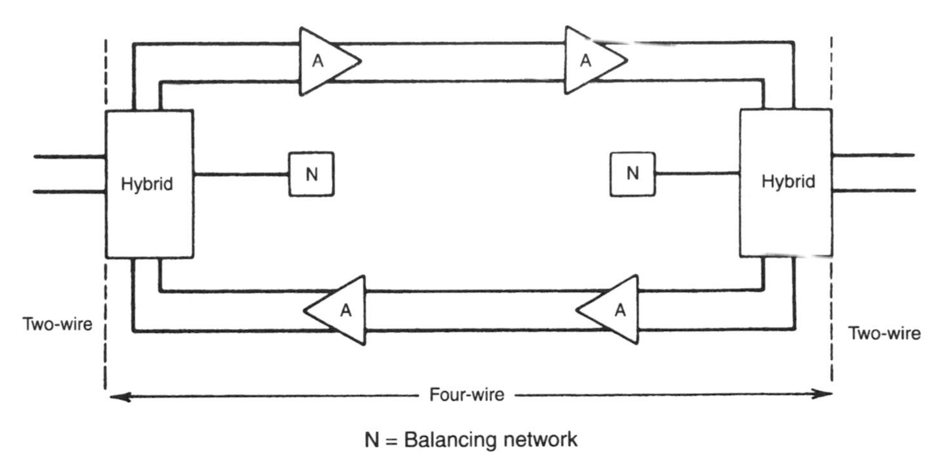

Figure 8.17 Simplified schematic of two-wire/four-wire operation.

The two factors that must be considered that greatly affect transmission design in the long-distance network are *echo* and *singing*.

8.5.2 Echo

As the name implies, echo in telephone systems is the return of a talker's voice. To be an impairment, the returned voice must suffer some noticeable delay. Thus we can say that echo is a reflection of the voice. Analogously, it may be considered as that part of the voice energy that bounces off obstacles in a telephone connection. These obstacles are impedance irregularities, more properly called *impedance mismatches*. Echo is a major annoyance to the telephone user. It affects the talker more than the listener. Two factors determine the degree of annoyance of echo: its loudness and the length of its delay.

8.5.3 Singing

Singing is the result of sustained oscillation due to positive feedback in telephone amplifiers or amplifying circuits. The feedback is the result of excessive receive signal feeding back through the hybrid to the transmit side, which is then amplified setting up oscillations. Circuits that sing are unusuable and promptly overload multichannel carrier (FDM) equipment.

Singing may be regarded as echo that is completely out of control. This can occur at the frequency at which the circuit is resonant. Under such conditions the circuit losses at the singing frequency are so low that oscillation will continue, even after cessation of its original impulse.

8.5.4 Causes of Echo and Singing

Echo and singing can generally be attributed to the impedance mismatch between the balancing network of a hybrid and its two-wire connection associated with the subscriber loop. It is at this point that we can expect the most likelihood of impedance mismatch which may set up an echo path. To understand the cause of echo, one of two possible conditions may be expected in the local area network:

1. There is a two-wire (analog) switch between the two-wire/four-wire conversion point and the subscriber plant. Thus, a hybrid may look into any of (say) 10,000 deifferent subscriber loops. Some of these loops are short, other are of medium length, and still others are long. Some are in excellent condition, and some are in dreadful condition. Thus the possibility of mismatch at a hybrid can be quite high under these circumstances.
2. In the more modern network configuration, subscriber loops may terminate in an analog concentrator before two-wire/four-wire conversion in a PCM channel bank. The concentration ratio may be anywhere from 2 : 1 to 10 : 1. For example, in the 10 : 1 case a hybrid may connect to any one of a group of ten subscriber loops. Of course, this is much better than selecting any one of a population of thousands of subscriber loops as in condition 1, above.

Turning back to the hybrid, we can keep excellent impedance matches on the four-wire side; it is the two-wire side that is troublesome. So our concern is the match (balance) between the two-wire subscriber loop and the balancing network (N in Figure 8.17). If we have a hybrid term set assigned to each subscriber loop, the telephone company (administration) could individually balance each loop, greatly improving impedance match. Such activity has high labor content. Secondly, in most situations there is a concentrator with from 4 : 1 to 10 : 1 concentration ratios (e.g., AT&T 5ESS).

With either condition 1 or condition 2 we can expect a fairly wide range of impedances of two-wire subscriber loops. Thus, a compromise balancing network is employed to cover this fairly wide range of two-wire impedances.

Impedance match can be quantified by *return loss*. The higher the return loss, the better the impedance match. Of course we are referring to the match between the balancing network (N) and the two-wire line (L) (see Figure 8.17).

$$\text{Return Loss}_{\text{dB}} = 20 \log_{10}(Z_N + Z_L)/(Z_N - Z_L). \quad (8.1)$$

If the balancing network (N) perfectly matches the impedance of the two-wire line (L), then $Z_N = Z_L$, and the return loss would be infinite.[13]

We use the term *balance return loss* (Ref. 5) and classify it as two types:

1. Balance return loss from the point of view of echo.[14] This is the return loss across the band of frequencies from 300 to 3400 Hz.[15]
2. Balance return loss from the point of view of stability.[16] This is the return loss between 0 and 4000 Hz.

"Stability" refers to the fact that loss in a four-wire circuit may depart from its nominal value for a number of reasons:

- Variation of line losses and amplifier gains with time and temperature;
- Gain at other frequencies being different from that measured at the test frequency. (This test frequency may be 800, 1000, or 1020 Hz.)

[13]Remember, for any number divided by zero, the result is infinity.
[14]Called *echo return loss* (ERL) in North America, but with a slightly different definition.
[15]Recognize this as the CCITT definition of the standard analog voice channel.
[16]From the point of view of stability—for this discussion, it may be called from the point of view of singing.

- Errors in making measurements and lining up circuits.

The band of frequencies most important in terms of echo for the voice channel is that from 300 Hz to 3400 Hz. A good value for echo return loss for toll telephone plant is 11 dB, with values on some connections dropping to as low as 6 dB. For further information, the reader should consult CCITT Recs. G.122 and G.131 (Refs. 5, 6).

Echo and singing may be controlled by:

- Improved return loss at the term set (hybrid);
- Adding loss on the four-wire side (or on the two-wire side); and
- Reducing the gain of the individual four-wire amplifiers.

The annoyance of echo to a subscriber is also a function of its delay. Delay is a function of the velocity of propagation of the intervening transmission facility. A telephone signal requires considerably more time to traverse 100 km of a voice-pair cable facility, particularly if it has inductive loading, than it requires to traverse 100 km of radio facility (as low as 22,000 km/s for a loaded cable facility and 240,000 km/s for a carrier facility). Delay is measured in one-way or round-trip propagation time measured in milliseconds. The CCITT recommends that if the mean round-trip propagation time exceeds 50 ms for a particular circuit, an echo suppressor or echo canceler should be used. Practice in North America uses 45 ms as a dividing line. In other words, where echo delay is less than that stated previously here, echo can be controlled by adding loss.

An echo suppressor is an electronic device inserted in a four-wire circuit that effectively blocks passage of reflected signal energy. The device is voice operated with a sufficiently fast reaction time to "reverse" the direction of transmission, depending on which subscriber is talking at the moment. The block of reflected energy is carried out by simply inserting a high loss in the return four-wire path. Figure 8.18 shows the echo path on a four-wire circuit. An echo canceller generates an echo-canceling signal.[17]

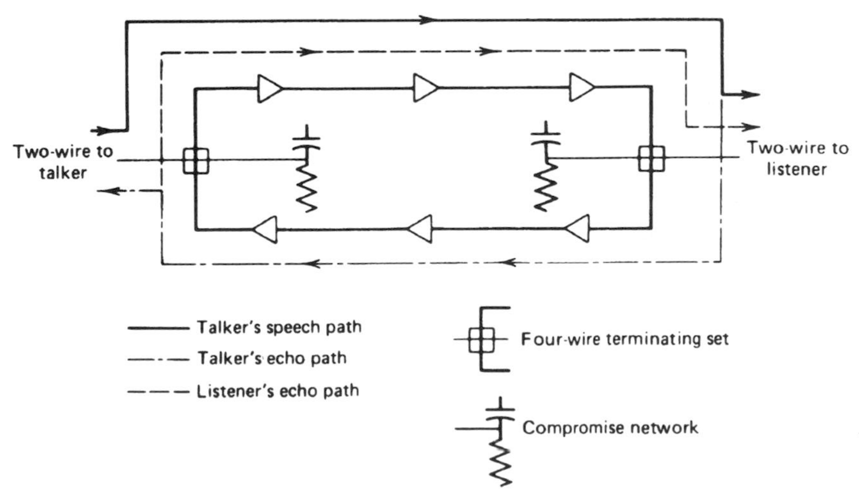

Figure 8.18 Echo paths in a four-wire circuit.

[17]*Echo canceller*, as defined by CCITT, is a voice-operated device placed in the four-wire portion of a circuit and used for reducing near-end echo present on the send path by subtracting an estimation of that echo from the near-end echo (Ref. 7).

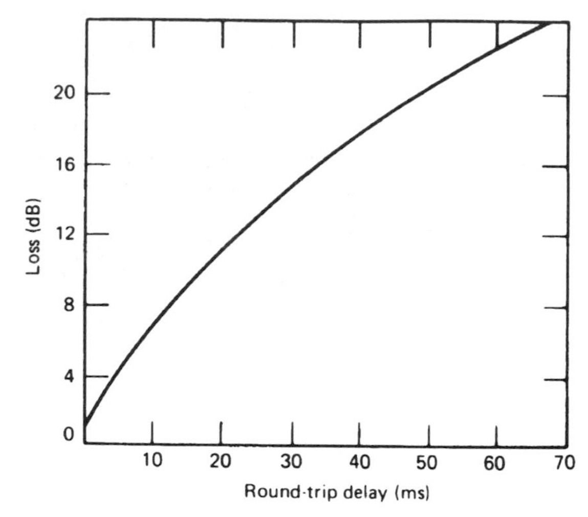

Figure 8.19 Talker echo tolerance for average telephone users.

8.5.5 Transmission Design to Control Echo and Singing

As stated previously, echo is an annoyance to the subscriber. Figure 8.19 relates echo path delay to echo path loss. The curve in Figure 8.19 traces a group of points at which the average subscriber will tolerate echo as a function of its delay. Remember that the longer the return signal is delayed, the more annoying it is to the telephone talker (i.e., the more the signal has to be attenuated). For example, if the echo delay on a particular circuit is 20 ms, an 11-dB loss must be inserted to make the echo tolerable to the talker. Be careful here. The reader should note that the 11 dB designed into the circuit to control echo will increase the end-to-end loudness loss (see Section 3.2.2.4) an equal amount, which is quite undesirable. The effect of loss design on loudness ratings and the trade-offs available are discussed in the paragraphs that follow.

If singing is to be controlled, all four-wire paths must have some amount of loss. Once they go into a gain condition, and we refer here to overall circuit gain, positive feedback will result and the amplifiers will begin to oscillate or "sing." For an analog network, North American practice called for a minimum of 4-dB loss on all four-wire circuits to ensure against singing. CCITT recommends 10 dB for minimum loss on the national network. (Ref. 5, p. 3).

The modern digital network with its A/D (analog-to-digital) circuits in PCM channel banks provides signal isolation, analog-to-digital, and digital-to-analog. As a result, the entire loss scenario has changed. This new loss plan for digital networks is described in Section 8.5.7.

8.5.6 Introduction to Transmission-Loss Engineering

One major aspect of transmission system design for a telephone network is to establish a transmission-loss plan. Such a plan, when implemented, is formulated to accomplish three goals:

1. Control singing (stability);
2. Keep echo levels within limits tolerable to the subscriber; and
3. Provide an acceptable overall loudness rating to the subscriber.

From preceding discussions we have much of the basic background necessary to develop a transmission-loss plan. We know the following:

- A certain minimum loss must be maintained in four-wire circuits to ensure against singing.
- Up to a certain limit of round-trip delay, echo may be controlled by adding loss (i.e., inserting attenuators, sometimes called *pads*).
- It is desirable to limit these losses as much as possible, to improve the loudness rating of a connection.

National transmission plans vary considerably. Obviously the length of a circuit is important, as well as the velocity of propagation of the transmission media involved.

Velocity of propagation. A signal takes a finite amount of time to traverse from point A to point B over a specific transmission medium. In free space, radio signals travel at 3×10^8 m/sec or 186,000 mi/sec; fiber optic light guide, about 2×10^8 m/s or about 125,000 mi/sec; on heavily loaded wire-pair cable, about 0.22×10^8 m/sec or 14,000 mi/sec; and 19-gauge nonloaded wire-pair cable, about 0.8×10^8 m/sec or 50,000 mi/sec. So we see that the velocity of propagation is very dependent on the types of transmission media being employed to carry a signal.

Distances covered by network connectivities are in hundreds or thousands of miles (or kilometers). It is thus of interest to convert velocities of propagation to miles or kilometers per millisecond. Let's use a typical value for carrier (multiplex) systems of 105,000 miles/sec or 105 miles per millisecond (169 km/ms).

First let's consider a country of small geographic area such as Belgium, which could have a very simple transmission-loss plan. Assume that the 4-dB minimum loss for singing is inserted in all four-wire circuits. Based on Figure 8.19, a 4-dB loss will allow up to 4 ms of round-trip delay. By simple arithmetic, we see that a 4-dB loss on all four-wire circuits will make echo tolerable for all circuits extending 210 mi (338 km) (i.e., 2×105). This could be an application of a fixed-loss type transmission plan. In the case of small countries or telephone companies covering a rather small geographic expanse, the minimum loss to control singing controls echo as well for the entire system.

Let us try another example. Assume that all four-wire connections have a 7-dB loss. Figure 8.20 indicates that 7 dB permits an 11-ms round-trip delay. Again assume that the velocity of propagation is 105,00 mi/sec. Remember that we are dealing with *round-trip* delay. The talker's voices reach the far-end hybrid and some of the signal is reflected back to the talker. This means that the signal traverses the system twice, as shown in Figure 8.20. Thus 7 dB of loss for the given velocity of propagation allows about 578 mi (925 km) of extension or, for all intents and purposes, the distance between subscribers, and will satisfy the loss requirements with a country of maximum extension of 578 mi (925 km).

It is interesting to note that the talker's signal is attenuated only 7 dB toward the distant-end listener; but the reflected signal is not only attenuated the initial 7 dB, but attenuated by 7 dB still again, on its return trip.

It has become evident by now that we cannot continue increasing losses indefinitely to compensate for echo on longer circuits. Most telephone companies and administrations have set a 45- or 50-ms round-trip delay criterion, which sets a top figure above which echo suppressors are to be used. One major goal of the transmission-loss plan is to improve overall loudness rating or to apportion more loss to the subscriber plant so that subscriber loops can be longer or to allow the use of less copper (i.e., smaller-diameter

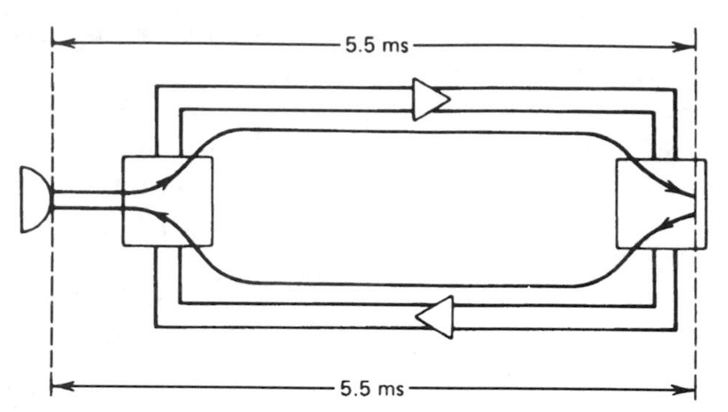

Figure 8.20 Example of echo round-trip delay (5.5 + 5.5 = 11 ms round-trip delay).

conductors). The question arises as to what measures can be taken to reduce losses and still keep echo within tolerable limits. One obvious target is to improve return losses at the hybrids. If all hybrid return losses are improved, the echo tolerance curve shifts; this is because improved return losses reduce the intensity of the echo returned to the talker. Thus the talker is less annoyed by the echo effect.

One way of improving return loss is to make all two-wire lines out of the hybrid look alike—that is, have the same impedance. The switch at the other end of the hybrid (i.e., on the two-wire side) connects two-wire loops of varying length, thus causing the resulting impedances to vary greatly. One approach is to extend four-wire transmission to the local office such that each hybrid can be better balanced. This is being carried out with success in Japan. The U.S. Department of Defense has its Autovon (automatic voice network), in which every subscriber line is operated on a four-wire basis. Two-wire subscribers connect through the system via PABXs (private automatic branch exchanges).

As networks evolve to all-digital, four-wire transmission is carried directly through the local serving switch such that subscriber loops terminate through a hybrid directly to a PCM channel bank. Hybrid return losses could now be notably improved by adjusting the balancing network for its dedicated subscriber loop.

8.5.7 Loss Plan for Digital Networks (United States)

For digital connections terminated in analog access lines, the required loss values are dependent on the connection architecture:

- For interLATA or interconnecting network connections, the requirement is 6 dB.
- For intraLATA connections involving different LECs (local exchange carriers), 6 dB is the preferred value, although 3 dB may apply to connections not involving a tandem switch.
- For intraLATA connections involving the same LEC, the guidelines are 0–6 dB (typically 0 dB, 3 dB, or 6 dB).

The choice of network loss value depends on performance considerations, administrative simplicity, and current network design (Ref. 4).

Loss can be inserted in a digital bit stream by using a digital signal processor involving a look-up table. By doing this, the bit sequence integrity is broken for each digital 8-bit time slot. Some of these time slots may be carrying data bit sequences. For this reason we cannot break up this bit integrity. To avoid this intermediate digital processing (which destroys bit integrity), loss is inserted on all-digital connections on the receiving end only, where the digital-to-analog conversion occurs (i.e., after the signal has been returned to its analog equivalent). Devices such as echo cancelers, which utilize digital signal processing, need to have the capability of being disabled when necessary, to preserve bit integrity.

In Section 8.5.6 round-trip delay was brought about solely by propagation delay. In digital networks there is a small incremental delay due to digital switching and digital multiplexing. This is due to buffer storage delay, more than anything else.

REVIEW EXERCISES

1. Name at least three factors that affect local network design.

2. What are the three basic underlying considerations in the design of a long-distance (toll) network?

3. What is the fallacy of providing just one high-capacity trunk group across the United States to serve all major population centers by means of tributaries off the main trunk?

4. How can the utilization factor on trunks be improved?

5. For long-distance (toll) switching centers, what is the principal factor in the placement of such exchanges (differing from local exchange placement substantially)?

6. How are the highest levels of a national hierarchical network connected, and why is this approach used?

7. On a long-distance toll connection, why must the number of links in tandem be limited?

8. What type of routing is used on the majority of international connections?

9. Name two principal factors used in deciding how many and where long-distance (toll) exchanges will be located in a given geographic area.

10. Discuss the impacts of fan-outs on the number of hierarchical levels.

11. Name the three principal bases required at the outset for the design of a long-distance network.

11. Once the hierarchical levels have been established and all node locations identified, what is assembled next?

12. Define a final route.

13. A grade of service no greater than ____% per link is recommended on a final route.

14. There are two generic types of routing schemes. What are they?

15. Name three different types of dynamic routing and explain each in one sentence.

16. What is *crankback*?

17. Give an example of state-dependent routing.

18. What is the principal cause of echo in the telephone network?

19. What causes singing in the telephone network?

20. Differentiate balance return loss from the point of view of stability (singing) from echo return loss.

21. How can we control echo? (Two answers required).

22. The stability of a telephone connection depends on three factors. Give two of these factors.

23. Based on the new loss plan for North America for the digital network, how much loss is inserted for interLATA connections?

REFERENCES

1. *International Telephone Routing Plan*, CCITT Rec. E.171, Vol. II, Fascicle II.2, IXth Plenary Assembly, Melbourne, 1988.
2. *The Transmission Plan*, ITU-T Rec. G.101, ITU Geneva, 1994.
3. *Traffic Routing*, CCITT Rec. E.170, ITU Geneva, 1992.
4. *BOC Notes on the LEC Networks—1994*, Special Report SR-TSV-002275, Issue 2, Bellcore, Piscataway, NJ, 1994.
5. *Influence of National Systems on Stability and Talker Echo in International Connections*, ITU-T Rec. G.122, ITU Geneva, 1993.
6. *Control of Talker Echo*, ITU-T Rec. G.131, ITU Geneva, 1996.
7. *Terms and Definitions*, CCITT Rec. B.13, Fascicle I.3, IXth Plenary Assembly, Melbourne, 1988.

9

CONCEPTS IN TRANSMISSION TRANSPORT

9.1 CHAPTER OBJECTIVE

A telecommunication network consists of customer premise equipment (CPE), switching nodes, and transmission links, as illustrated in Figure 9.1. Chapter 5 dealt with one important type of CPE, namely, the telephone subset. The chapter also covered wire-pair connectivity from the telephone subset to the local serving switch over a subscriber loop. Basic concepts of switching were reviewed in Chapter 4, and Chapter 6 covered digital switching. In this chapter we introduce the essential aspects for the design of long-distance links.

There are four different ways by which we can convey signals from one switching node to another:

1. Radio;
2. Fiber optics;
3. Coaxial cable; and
4. Wire pair.

Emphasis will be on radio and fiber optics. The use of coaxial cable for this application is deprecated. However, it was widely employed from about 1960 to 1985 including some very-high-capacity systems. One such system (L5) crossed the United States from coast-to-coast with a capacity in excess of 100,000 simultaneous full-duplex voice channels in FDM configurations (see Section 4.5.2). Fiber optic cable has replaced the greater portion of these coaxial cable systems. There is one exception. Coaxial cable is still widely employed in cable television configurations (see Chapter 15). Wire pair remains the workhorse in the subscriber plant.

At the outset, we can assume that these transmission links are digital and will be based on the PCM configurations covered in Chapter 6, namely, either of the DS1 (T1) or E1 families of formats. However, the more advanced, higher-capacity digital formats such as synchronous optical network (SONET) or synchronous digital hierarchy (SDH) (see Chapter 17) are now being widely deployed on many if not most new fiber optic systems, and with lower capacity configurations on certain radio systems.

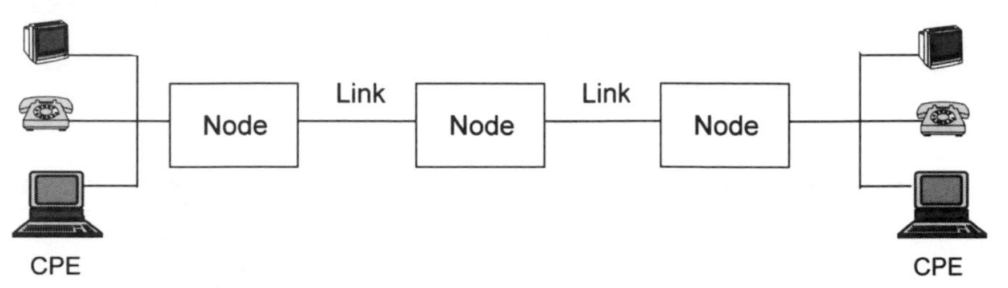

Figure 9.1 A telecommunication network consists of customer premise equipment (CPE), switching nodes, and interconnecting transmission links.

9.2 RADIO SYSTEMS

9.2.1 Scope

The sizes, capacities, ranges, and operational frequency bands for radio systems vary greatly. Our discussion will be limited to comparatively high-capacity systems. Only two system types meet the necessary broadband requirements of the long-distance network. These are line-of-sight (LOS) microwave and satellite communications. Satellite communications is really nothing more than an extension of LOS microwave.

9.2.2 Introduction to Radio Transmission

Wire, cable, and fiber are well-behaved transmission media, and they display little variability in performance. The radio medium, on the other hand, displays notable variability in performance. The radio-frequency spectrum is shared with others and requires licensing. Metallic and fiber media need not be shared and do not require licensing (but often require right-of-way).

A major factor in the selection process is information bandwidth. Fiber optics seems to have nearly an infinite bandwidth. Radio systems have very limited information bandwidths. It is for this reason that radio-frequency bands 2 GHz and above are used for PSTN and private network applications. In fact, the U.S. Federal Communications Commission (FCC) requires that users in the 2-GHz band must have systems supporting 96 digital voice channels where bandwidths are still modest. In the 4- and 6-GHz bands, available bandwidths are 500 MHz, allocated in 20- and 30-MHz segments for each radio-frequency carrier.

One might ask, why use radio in the first place if it has so many drawbacks? Often, it turns out to be less expensive compared with fiber optic cable. But there are other factors such as:

- No requirement for right-of-way;
- Less vulnerable to vandalism;
- Not susceptible to "accidental" cutting of the link;
- Often more suited to crossing rough terrain;
- Often more practical in heavily urbanized areas; and
- As a backup to fiber-optic cable links.

Fiber-optic cable systems provide strong competition with LOS microwave, but LOS microwave does have a place and a good market.

Satellite communications is an extension of LOS microwave. It is also feeling the "pinch" of competition from fiber-optic systems. It has two drawbacks. First, of course, is limited information bandwidth. The second is excessive delay when the popular geostationary satellite systems are utilized. It also shares frequency bands with LOS microwave.

One application showing explosive growth is very small aperture terminal (VSAT) systems. It is very specialized and has great promise for certain enterprise networks, and there are literally thousands of these networks now in operation.

Another application that is becoming widely deployed is large families of low earth orbit (LEO) satellites such as Motorola's Iridium, which provide worldwide cellular/PCS coverage. Because of its low altitude orbit (about 785 km above earth's surface), the notorious delay problem typical of GEO (geostationary satellite) is nearly eliminated.

9.2.3 Line-of-Sight Microwave

9.2.3.1 Introduction. Line-of-sight (LOS) microwave provides a comparative broadband connectivity over a single link or a series of links in tandem. We must be careful on the use of language here. First a link, in the sense we use it, connects one radio terminal to another or to a repeater site. The term *link* was used in Figure 9.1 in the "network" sense. Figure 9.2 illustrates the meaning of a "link" in LOS microwave. Care must also be taken with the use of the expression *line-of-sight.* Because we can "see" a distant LOS microwave antenna does not mean that we're in compliance with line-of-sight clearance requirements.

We can take advantage of this "line-of-sight" phenomenon at frequencies from about 150 MHz well into the millimeter-wave region.[1] Links can be up to 30 miles long depending on terrain topology. I have engineered some links well over 100 miles long. In fact, links with geostationary satellites can be over 23,000 miles long.

On conventional LOS microwave links, the length of a link is a function of antenna height. The higher the antenna, the further the reach. Let us suppose *smooth earth.* This means an earth surface with no mountains, ridges, buildings, or sloping ground whatsoever. We could consider an over-water path as a smooth earth path. Some paths on the North American prairie approach smooth earth. In the case of smooth earth, the LOS distance from an antenna is limited by the horizon. Given an LOS microwave

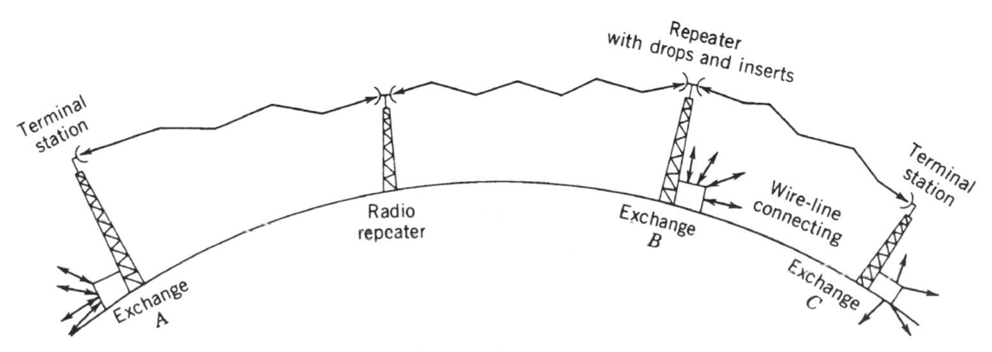

Figure 9.2 A sketch of an LOS microwave radio relay system.

[1] *Millimeter-wave region* is where the wavelength of an equivalent frequency is less than 1 cm.

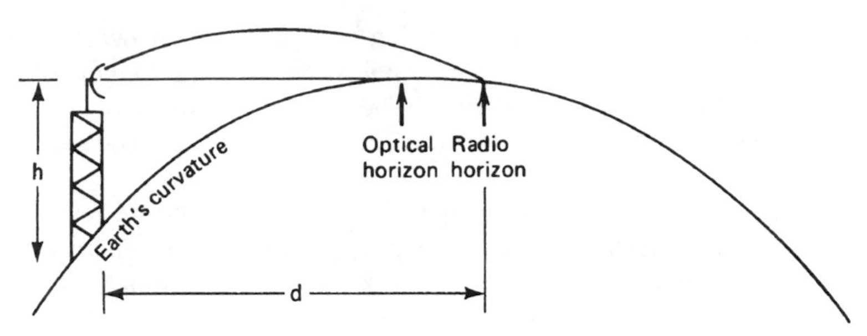

Figure 9.3 Radio and optical horizon (smooth earth).

antenna of h_{ft} or h'_{m} above ground surface, the distance d_{mi} or d_{km} to the horizon just where the ray beam from the transmitting antenna will graze the rounded surface of the horizon can be calculated using one of the formulas given as follows:

To optical horizon ($k = 1$):

$$d = \sqrt{\frac{3h}{2}} \tag{9.1a}$$

and to the radio horizon ($k = 4/3$):

$$d = \sqrt{2h} \tag{9.1b}$$

$$d' = 2.9(2h')^{1/2}, \tag{9.1c}$$

where k expresses the bending characteristic of the path.

These formulas should only be used for rough estimates of distance to the horizon under smooth earth conditions. As we will find out later, the horizon clearance must be something greater than (n feet or meters) of grazing. The difference between formulas (9.1a) and (9.1b) and (9.1c) is that formula (9.1a) is "true" line-of-sight and expresses the optical distance. Here the radio ray beam follows a straight line. Under most circumstances the microwave ray beam is bent toward the earth because of characteristics of the atmosphere. This is expressed in formulas (9.1b) and (9.1c), and assumes the most common bending characteristic. Figure 9.3 is a model that may be used for formulas (9.1). It also shows the difference between the *optical distance* to the horizon and the *radio distance* to the horizon.

The design of an LOS microwave link involves five basic steps:

1. Setting performance requirements;
2. Selecting site and preparing a *path profile* to determine antenna tower heights;
3. Carrying out a path analysis, often called a *link budget*. (Here is where we dimension equipment to meet the performance requirements set in step 1);
4. Physically running a path/site survey; and
5. Installing equipment and testing the system prior to cutting it over to carry traffic.

In the following subsections we review the first four steps.

9.2.3.2 Setting Performance Requirements. As we remember from Chapter 6, the performance of a digital system is expressed in a bit error rate (bit error ratio) (BER). In our case here, it will be expressed as a BER with a given time distribution. A *time distribution* tells us that a certain BER value is valid for a certain percentage of time, percentage of a year, or percentage of a month.

Often a microwave link is part of an extensive system of multiple links in tandem. Thus we must first set system requirements based on the output of the far-end receiver of the several or many links in tandem. If the system were transmitting in an analog format, typically FDM using frequency modulation (FM), the requirement would be given for noise in the derived voice channel; if it were video, a signal-to-noise ratio specification would be provided. In the case we emphasize here, of course, it will be BER on the far-end receiver digital bit stream.

The requirements should be based on existing standards. If the link (or system) were to be designed as part of the North American PSTN, we would use a Bellcore standard. (Ref. 1).[2] In this case the BER at the digital interface level shall be less than 2×10^{-10}, excluding burst error seconds. Another source is CCIR/ITU-R. For example, CCIR Rec. 594-3 (Ref. 2) states that the BER should not exceed 1×10^{-6} during more than 0.4% of any month and 1×10^{-3} during more than 0.054% of any month. We will recall that the bottom threshold for bit error performance in the PSTN is 1×10^{-3} to support supervisory signaling, even though 8-bit PCM (Chapter 6) is intelligible down to a BER of 1×10^{-2}.

A common time distribution is 99.99% of a month to be in conformance with ITU-R/CCIR recommendations (e.g., 0.054% of a month). This time distribution translates directly into *time availability*, which is the percentage of time a link meets its performance criteria.

9.2.3.3 Site Selection and Preparation of a Path Profile

9.2.3.3.1 Site Selection. In this step we will select operational sites where we will install and operate radio equipment. After site selection, we will prepare a path profile of each link to determine the heights of radio towers to achieve "line of sight." Sites are selected using large topographical maps. If we are dealing with a long system crossing a distance of hundreds of miles or kilometers, we should minimize the number of sites involved. There will be two terminal sites, where the system begins and ends. Along the way, repeater sites will be required. At some repeater sites, we may have need to drop and insert traffic. Other sites will just be repeaters. This concept is illustrated in Figure 9.4. The figure shows the drops and inserts (also called *add-drops*) of traffic at telephone exchanges. These drop and insert points may just as well be buildings or other facilities in a private/corporate network. There must be considerable iteration between site selection and path profile preparation to optimize the route.

In essence, the sites selected for drops and inserts will be points of traffic concentration. There are several trade-offs to be considered:

1. Bringing traffic in by wire or cable rather than adding additional drop and insert (add-drop) capabilities at relay point, which provides additional traffic concentration;
2. Siting based on propagation advantages (or constraints) only, versus colocation with exchange (or corporate facility) (saving money for land and buildings); and

[2]Bellcore is Bell Communications Research, Piscataway, NJ.

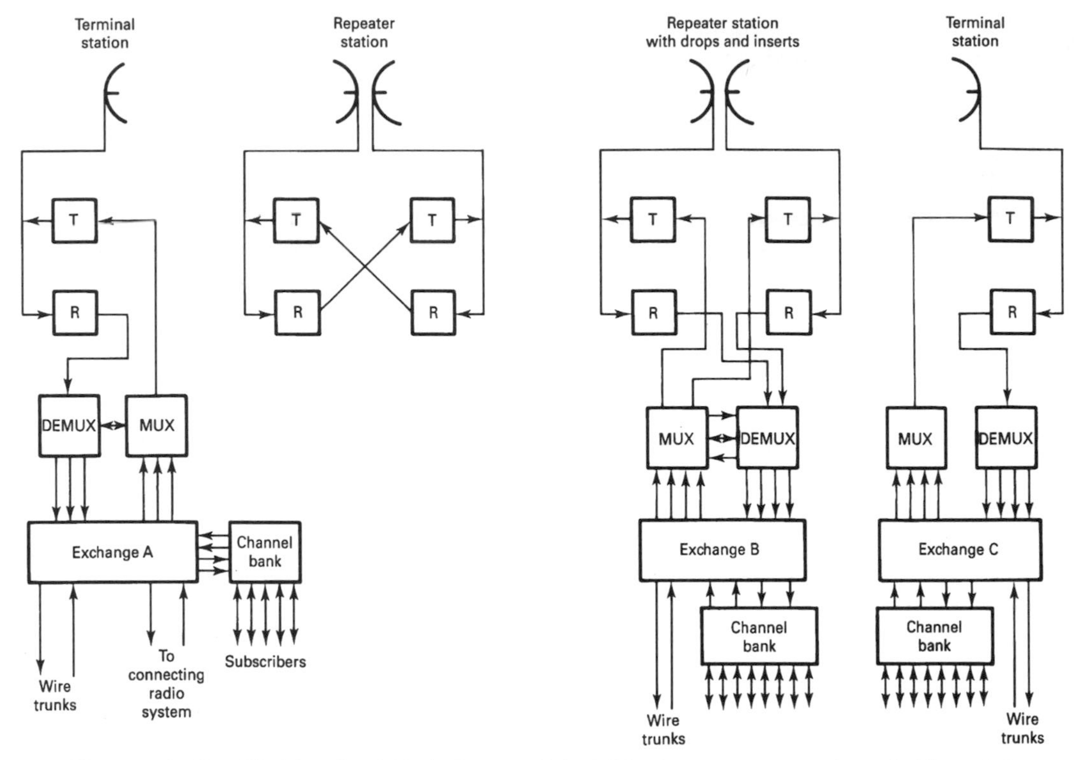

Figure 9.4 Simplified functional block diagram of the LOS microwave system shown in Figure 9.2.

3. Choosing a method of feeding (feeders): by light-route radio, fiber optic cable, or wire-pair cable.[3]

9.2.3.3.2 Calculation of Tower Heights. LOS microwave antennas are mounted on towers. Formula (9.1) allowed us to calculate a rough estimate of tower height. Towers and their installation are one of the largest cost factors in the installation of LOS microwave systems. Thus we recommend that actual tower heights do not exceed 300 ft (92 m). Of course, the objective is to keep the tower height as low as possible and still maintain effective communication. The towers must be just high enough to surmount obstacles in the path. *High enough* must be carefully defined. What sort of obstacles might we encounter in the path? To name some: terrain such as mountains, ridges, hills, and earth curvature—which is highest at midpath—and buildings, towers, grain elevators, and so on. The path designer should consider using natural terrain such as hilltops for terminal/relay sites. The designer should also consider leasing space on the top of tall buildings or on TV broadcast towers. In the following paragraphs we review a manual method of plotting a path profile.

From a path profile we can derive tower heights. Path profiles may be prepared by a PC with a suitable program and the requisite topological data for the region stored on a disk. Our recommendation is to use ordinary rectangular graph paper such as "millimeter" paper or with gradations down to 1/16 inch or better. "B-size" is suggested. There are seven steps required to prepare a path profile:

[3]Here the word *feeders* refers to feeding a mainline trunk radio systems. Feeders may also be called *spurs*.

1. Obtain good topo(logical) maps of the region, at least 1 : 62, 500 and identify the two sites involved, one of which we arbitrarily call a "transmit" site and the other a "receive" site.
2. Draw a straight line with a long straightedge connecting the two sites identified.
3. Follow along down the line identifying obstacles and their height. Put this information on a table, labeling the obstacles "A," "B," etc.
4. Calculate earth curvature (or earth bulge) (EC). This is maximum at midpath. On the same table in the next column write the EC value for each obstacle.
5. Calculate the Fresnel zone clearance for each obstacle. The actual value here will be 0.6 of the first Fresnel zone.
6. Add a value of additional height for vegetation such as trees; add a growth factor as well (10 ft or 3 m if actual values are unavailable).
7. Draw a straight line from left to right connecting the two highest obstacle locations on the profile. Do the same from right to left. Where this line intersects the vertical extension of the transmit site and the vertical extension of the receive site defines tower heights.

In step 4, the calculation of EC, remember that the earth is a "sphere." Our path is a tiny arc on that sphere's surface. Also in this calculation we must account for the radio ray path bending. To do this we use a tool called *K-factor*. When the K-factor is greater than 1, the ray beam bends toward the earth, as illustrated in Figure 9.3. When the K-factor is less than 1, the ray beam bends away from the earth.

The EC value (h) is the amount we will add to the obstacle height in feet or meters to account for that curvature or bulge. The following two formulas apply:

$$h_{\text{ft}} = 0.667 d_1 d_2 / K \ (d \text{ in miles}) \tag{9.2a}$$

$$h_{\text{m}} = 0.078 d_1 d_2 / K \ (d \text{ in km}) \tag{9.2b}$$

where d_1 is the distance from the "transmit" site to the obstacle in question and d_2 is the distance from that obstacle to the receive site.

Table 9.1 is a guide for selecting the K-factor value. For a more accurate calculation of the K-factor, consult Ref. 3. Remember that the value obtained from Eq. (9.2) is to be added to the obstacle height.

In step 5, calculation of the Fresnel zone clearance, 0.6 of the value calculated is added to the obstacle height in addition to earth curvature. It accounts for the expanding properties of a ray beam as it passes over an obstacle. Use the following formulas to calculate Fresnel zone (radius) clearance:

$$R_{\text{ft}} = 72.1 \sqrt{\frac{d_1 d_2}{FD}}, \tag{9.3a}$$

where F is the frequency in gigahertz, d_1 is the distance from transmit antenna to obstacle (statute miles), d_2 is the distance from path obstacle to receive antenna (statute miles), and $D = d_1 + d_2$. For metric units:

Table 9.1 ***K*-Factor Guide**[a]

	Propagation Conditions				
	Perfect	Ideal	Average	Difficult	Bad
Weather	Standard atmosphere	No surface layers or fog	Substandard, light log	Surface layers, ground fog	Fog moisture over water
Typical	Temperate zone, no fog, no ducting, good atmospheric mix day and night	Dry, mountainous, no fog	Flat, temperate, some fog	Coastal	Coastal, water, tropical
K factor	1.33	1–1.33	0.66–1.0	0.66–0.5	0.5–0.4

[a] For 99.9% to 99.99% time availability.

$$R_{\mathrm{m}} = 17.3 \sqrt{\frac{d_1 d_2}{FD}}, \tag{9.3b}$$

where F is the frequency (the microwave transmitter operating frequency) in GHz, and d_1, d_2, and D are in now kilometers with R in meters.

The three basic increment factors that must be added to obstacle heights are now available—earth curvature (earth bulge), Fresnel zone clearance, and trees and growth (T&G). These are marked on the path profile chart. On the chart a straight line is drawn from right to left just clearing the obstacle points as corrected for the three factors. Another similar line is drawn from left to right. A sample profile is shown in Figure 9.5. The profile now gives us two choices, the first based on the right-to-left line and the second based on the left-to-right line. However, keep in mind that some balance is desirable so that at one end we do not have a very tall tower and at the other a small, stubby tower. Nevertheless, an imbalance may be desirable when a reflection point exists at an inconvenient spot along the path so we can steer the reflection point off the reflecting medium such as smooth desert or body of water.

9.2.3.4 Path Analysis or Link Budget

9.2.3.4.1 Introduction. A path analysis or link budget is carried out to dimension the link. What is meant here is to establish operating parameters such as transmitter power output, parabolic antenna aperture (diameter), and receiver noise figure, among others. The link is assumed to be digitally based on one of the formats discussed in Chapter 6 or possibly some of the lower bit rate formats covered in Chapter 17. The type of modulation, desired BER, and modulation rate (i.e., the number of transitions per second) are also important parameters.

Table 9.2 shows basic LOS microwave equipment/system parameters in two columns. The first we call "normal" and would be the most economic; the second column is titled "special," giving improved performance parameters, but at an increased price.

Diversity reception is another option to be considered. It entails greater expense. The options in Table 9.2 and diversity reception will be addressed further on in our discussion.

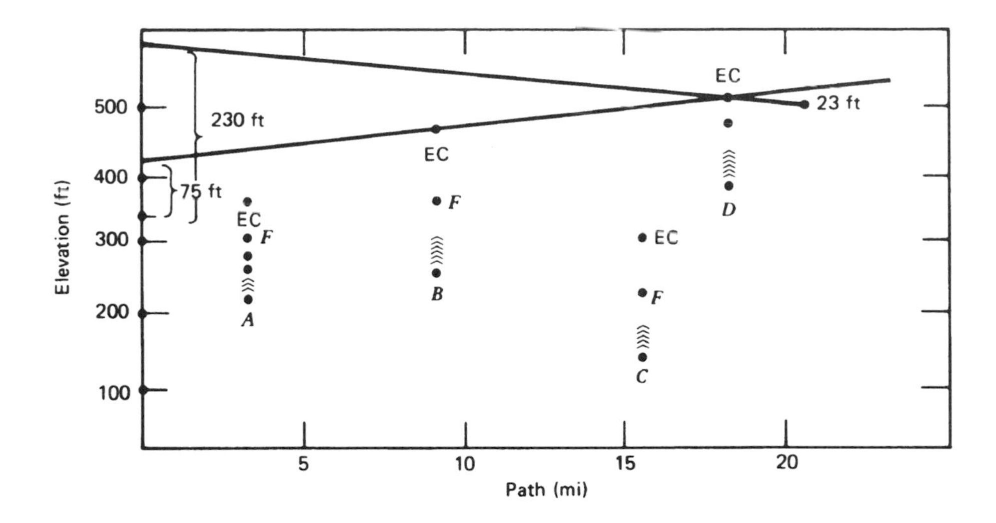

	Obstacle	d_1	d_2	Basic Height (ft)	F (Fresnel) (ft)	E.C. (ft)	T and G (ft)	Adjusted Total Height (ft)
Tree conditions: 40 + 10 ft growth (T and G)	A	3.5	19.0	220	30	49	50	349
Frequency band: 6 GHz	B	10	12.5	270	41	91.7	50	452.7
Midpath Fresnel (0.6)	C	17	5.5	160	36	68.6	50	314.6
= 42 ft	D	20	2.5	390	25.2	36.6	50	501.8

Figure 9.5 Practice path profile. (The x-axis is in miles, the y-axis is in feet; assume that $K = 0.9$; EC = earth curvature and F is the dimension of the first Fresnel zone.)

9.2.3.4.2 Approach. We can directly relate the desired performance to the receive signal level (RSL) at the first active stage of the far-end receiver and that receiver's noise characteristics. To explain: The RSL is the level or power of the received signal in dBW or dBm, as measured at the input of the receiver's mixer, or, if the receiver has an LNA (low-noise amplifier) at the LNA's input. This is illustrated in the block diagram of a typical LOS microwave receiver shown in Figure 9.6.

Table 9.2 Digital LOS Microwave Basic Equipment Parameters

Parameter	Normal	Special	Comments
Transmitter power	1 W	10 W	500 mW common above 10 GHz
Receiver noise figure	4–8 dB	1–2.5 dB	Use of low-noise amplifier
Antenna	Parabolic, 2–12 ft diameter	Same	Antennas over 12 ft not recommended
Modulation	64–128 QAM	Up 512-QAM, or QPR, or QAM/trellis	Based on bandwidth bit rate constraints, or bandwidth desired in case of SONET/SDH

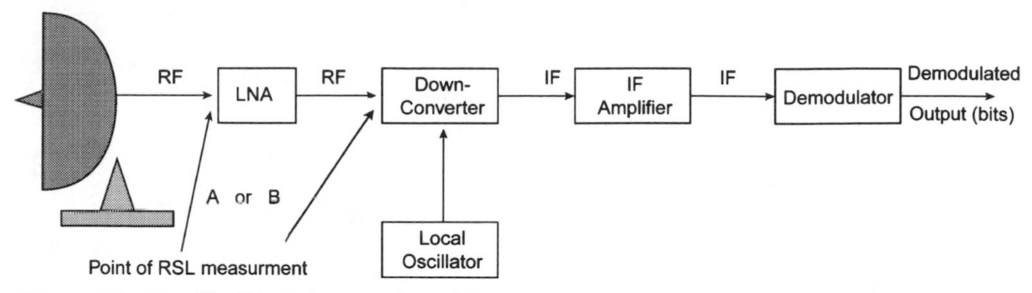

Figure 9.6 Simplified block diagram of an LOS microwave receiver. RF = radio frequency; IF = intermediate frequency; LNA = low noise amplifier. Point A is used to measure RSL when an LNA is employed, which is optional. Otherwise, the measurement point is point B, at the input of the downconverter.

In Figure 9.6 the incoming signal from the antenna (RF) is amplified by the LNA and then fed to the downconverter, which translates the signal to the intermediate frequency (IF), often 70 MHz. The IF is amplified and then inputs the demodulator. The demodulator output is the serial bit stream, replicating the input serial bit stream at the far-end transmitter.

The next step in the path analysis (link budget) is to calculate the free-space loss between the transmit antenna and the receive antenna. This is a function of distance and frequency (i.e., the microwave transmitter operational frequency). We then calculate the EIRP (effective isotropically radiated power) at the transmit antenna. The EIRP (in dBm or dBW, Appendix C) is the sum of the transmitter power output, minus the transmission line losses plus the antenna gain, all in decibel units. The units of power must be consistent, either in dBm or dBW. If the transmitter power is in dBW, the EIRP will be in dBW and the distant-end RSL must also be in dBW.

We then algebraically add the EIRP to the free-space loss in dB (often called *path loss*), and the result is the isotropic receive level (IRL).[4] When we add the receive antenna gain to the IRL and subtract the receive transmission line losses, we get the RSL. This relationship of path losses and gains is illustrated in Figure 9.7.

Path loss. For operating frequencies up to about 10 GHz, path loss is synonymous with *free-space loss*. This represents the steady decrease of power flow as the wave expands out in space in three dimensions. The formula for free-space loss is:

$$L_{\text{dB}} = 96.6 + 20\log_{10} F_{\text{GHz}} + 20\log_{10} D \tag{9.4a}$$

where L is the free-space loss between isotropic antennas, F is measured in GHz, and D is in statute miles. In the metric system:

$$L_{\text{dB}} = 92.4 + 20\log F_{\text{GHz}} + 20\log D_{\text{km}}, \tag{9.4b}$$

where D is in km.

Calculation of EIRP. Effective isotropically radiated power is calculated by adding decibel units: transmitter power (in dBm or dBW), the transmission line losses (in dB;

[4]An isotropic antenna is an antenna that is uniformly omnidirectional and thus, by definition, it has a 0 dB gain. It is a hypothetical reference antenna. The isotropic receive level (IRL) is the power level we would expect to achieve at that point using an isotropic antenna.

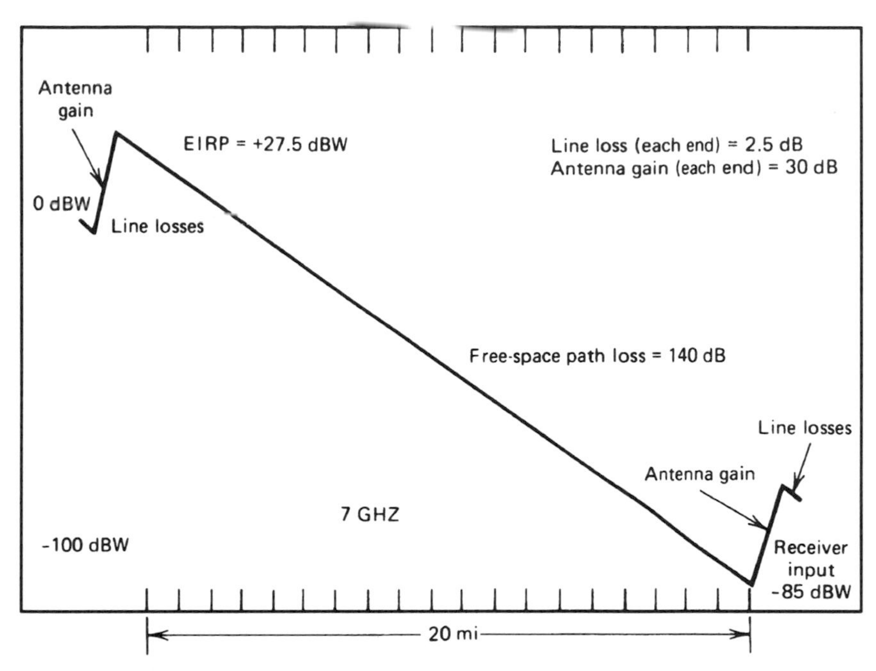

Figure 9.7 LOS microwave link gains and losses (simplified). Transmitter power output is 1 W or 0 dBW.

a negative value because it is a loss), and the antenna gain in dBi.[5]

$$EIRP_{dBW} = \text{trans. output power}_{dBW} - \text{trans. line losses}_{dB} + \text{ant. gain}_{dB}. \tag{9.5}$$

Figure 9.8 shows this concept graphically.

Example. If a microwave transmitter has 1-W (0-dBW) power output, the waveguide loss is 3 dB, and the antenna gain is 34 dBi, what is the EIRP in dBW?

$$\begin{aligned} EIRP &= 0\ \text{dBW} - 3\ \text{dB} + 34\ \text{dBi} \\ &= +31\ \text{dBW}. \end{aligned}$$

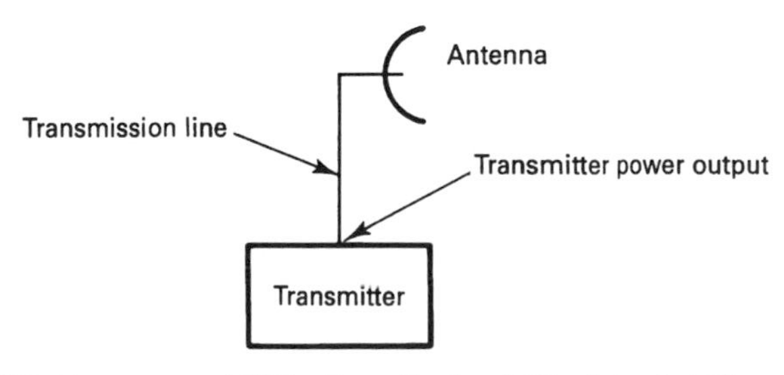

EIRP = Trans. output (dBW) – Trans. line loss (dB) + Ant. gain (dB)

Figure 9.8 Elements in the calculation of EIRP.

[5]dBi = decibels referenced to an isotropic (antenna).

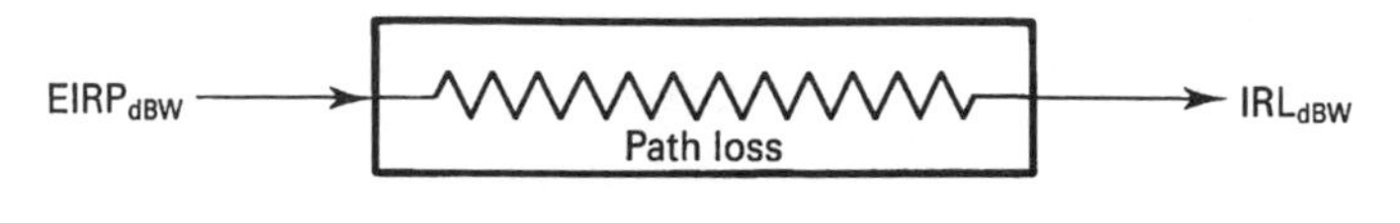

$IRL_{dBW} = EIRP_{dBW} - \text{Path loss}_{dB}$
below 10-GHz Path loss = Free-space loss (FSL)

Figure 9.9 Calculation of isotropic receive level (IRL).

Calculation of isotropic receive level (IRL). The IRL is the RF power level impinging on the receive antenna. It would be the power we would measure at the base of an isotropic receive antenna.

$$\text{IRL}_{\text{dBW}} = \text{EIRP}_{\text{dBW}} - \text{path loss}_{\text{dB}}. \tag{9.6}$$

This calculation is shown graphically in Figure 9.9.

Calculation of receive signal level (RSL). The RSL is the power level at the input port of the first active stage in the receiver. The power level is conventionally measured in dBm or dBW.

$$\text{RSL}_{\text{dBW}} = \text{IRL}_{\text{dBW}} + \text{rec. ant. gain (dB)} - \text{rec. trans. line losses (dB)}. \tag{9.7}$$

(*Note:* Power levels can be in dBm as well, but we must be consistent.)

Example. Suppose the IRL was −121 dBW, the receive antenna gain was 31 dB, and the line losses were 5.6 dB. What would the RSL be?

$$\begin{aligned} \text{RSL} &= -121 \text{ dBW} + 31 \text{ dB} - 5.6 \text{ dB} \\ &= -95.6 \text{ dBW}. \end{aligned}$$

Calculation of receiver noise level. The thermal noise level of a receiver is a function of the receiver noise figure and its bandwidth. For analog radio systems, receiver thermal noise level is calculated using the bandwidth of the intermediate frequency (IF). For digital systems, the noise level of interest is in only 1 Hz of bandwidth using the notation N_0, the noise level in a 1-Hz bandwidth.

The noise that a device self-generates is given by its noise figure (dB) or a noise temperature value. Any device, even passive devices, above absolute zero generates thermal noise. We know the thermal noise power level in a 1-Hz bandwidth of a perfect receiver operating at absolute zero. It is:

$$P_{\text{n}} = -228.6 \text{ dBW/Hz}, \tag{9.8}$$

where P_{n} is the noise power level. Many will recognize this as Boltzmann's constant expressed in dBW.

We can calculate the thermal noise level of a perfect receiver operating at room temperature using the following formula:

$$\begin{aligned} P_{\text{n}} &= -228.6 \text{ dBW/Hz} + 10 \log 290 \text{ (K)} \\ &= -204 \text{ dBW/Hz}. \end{aligned} \tag{9.9}$$

The value, 290 K (kelvins), is room temperature, or about 17°C or 68°F.

Noise figure simply tells us how much noise has been added to a signal while passing through a device in question. Noise figure (dB) is the difference in signal-to-noise ratio between the input to the device and the output of that same device.

We can convert noise figure to noise temperature in kelvins with the following formula:

$$\mathrm{NF_{dB}} = 10\ \log(1 + T_e/290), \tag{9.10}$$

where T_e is the effective noise temperature of a device. Suppose the noise figure of a device is 3 dB. What is the noise temperature?

$$\begin{aligned} 3\mathrm{dB} &= 10\ \log(1 + T_e/290) \\ 0.3 &= \log(1 + T_e/290) \\ 1.995 &= 1 + T_e/290. \end{aligned}$$

We round 1.995 to 2; thus

$$\begin{aligned} 2 - 1 &= T_e/290 \\ T_e &= 290\mathrm{K}. \end{aligned}$$

The thermal noise power level of a device operating at room temperature is:

$$P_n = -204\ \mathrm{dBW/Hz} + \mathrm{NF_{dB}} + 10\ \log\ \mathrm{BW_{Hz}}, \tag{9.11}$$

where BW is the bandwidth of the device in Hz.

Example. A microwave receiver has a noise figure of 8 dB and its bandwidth is 10 MHz. What is the thermal noise level (sometimes called the thermal noise threshold)?

$$\begin{aligned} P_n &= -204\ \mathrm{dBW/Hz} + 8\ \mathrm{dB} + 10\ \log(10 \times 10^6) \\ &= -204\ \mathrm{dBW/Hz} + 8\ \mathrm{dB} + 70\ \mathrm{dB} \\ &= -126\ \mathrm{dBW}. \end{aligned}$$

Calculation of E_b/N_0 in digital radio systems. In Section 3.2.1 signal-to-noise ratio (S/N) was introduced. S/N is widely used in analog transmission systems as one measure of signal quality. In digital systems the basic measure of transmission quality is BER. With digital radio links, we will introduce and employ the ratio E_b/N_0 as a measure of signal quality. Given a certain modulation type, we can derive BER from an E_b/N_0 curve.

In words, E_b/N_0 means *energy per bit per noise spectral density ratio.* N_0 is simply the thermal noise in 1 Hz of bandwidth or:

$$N_0 = -204\ \mathrm{dBW/Hz} + \mathrm{NF_{dB}}. \tag{9.12}$$

NF, as used above, is the noise figure of the receiver in question. The noise figure tells us the amount of thermal noise a device injects into a radio system.

Example. Suppose a receiver has a noise figure (NF) of 2.1 dB. What is its thermal noise level in 1 Hz of bandwidth? In other words, what is N_0?

$$
\begin{aligned}
N_0 &= -204 \text{ dBW} + 2.1 \text{ dB} \\
&= -201.9 \text{ dBW/Hz.}
\end{aligned}
$$

E_b is the signal energy per bit. We apply this to the receive signal level (RSL). The RSL represents the total power (in dBm or dBW) entering the receiver front end, during, let's say, 1-sec duration (energy). We want the power carried by just 1 bit. For example, if the RSL were 1 W, and the signal was at 1000 bps, the energy per bit would be 1/1000 or 1 mW per bit. However, it will be more convenient here to use logarithms and decibel values (which are logarithmic). Then we define E_b as:

$$E_b = \text{RSL}_{\text{dBm or dBW}} - 10 \log(\text{bit rate}). \tag{9.13}$$

Here's an example using typical values. The RSL into a certain receiver was −89 dBW and bit rate was 2.048 Mbps. What is the value of E_b?

$$
\begin{aligned}
E_b &= -89 \text{ dBW} - 10 \log(2.048 \times 10^6) \\
&= -89 \text{ dBW} - 63.11 \text{ dB} \\
&= -152.11 \text{ dBW.}
\end{aligned}
$$

We can now develop a formula for E_b/N_0:

$$E_b/N_0 = \text{RSL}_{\text{dBW}} - 10 \log(\text{bit rate}) - (-204 \text{ dBW} + \text{NF}_{\text{dB}}). \tag{9.14}$$

Simplifying we obtain

$$E_b/N_0 = \text{RSL}_{\text{dBW}} - 10 \log(\text{bit rate}) + 204 \text{ dBW} - \text{NF}_{\text{dB}}. \tag{9.15}$$

Some notes on E_b/N_0 and its use. E_b/N_0, for a given BER, will be different for different types of modulation (e.g., FSK, PSK, QAM, etc.). When working with E_b, we divide RSL by the bit rate, not the symbol rate nor the baud rate. There is a theoretical E_b/N_0 and a practical E_b/N_0. The practical is always a greater value than the theoretical, greater by the *modulation implementation loss* in dB, which compensates for system imperfections.

Figure 9.10 is an example of where BER is related to E_b/N_0. There are two curves in the figure. The first from the left is for BPSK/QPSK (binary phase shift keying/quadrature phase shift keying), and the second is for 8-ary PSK (an eight-level PSK modulation scheme). The values are for coherent detection. Coherent detection means that the receiver has a built-in phase reference as a basis to make its binary or higher-level decisions.

9.2.3.5 Digital Modulation of LOS Microwave Radios. Digital systems, typically standard PCM, as discussed in Chapter 6, are notoriously wasteful of bandwidth compared with their analog counterparts.[6] For example, the analog voice channel is

[6]A cogent example is FDM using frequency modulation; another is single sideband modulation.

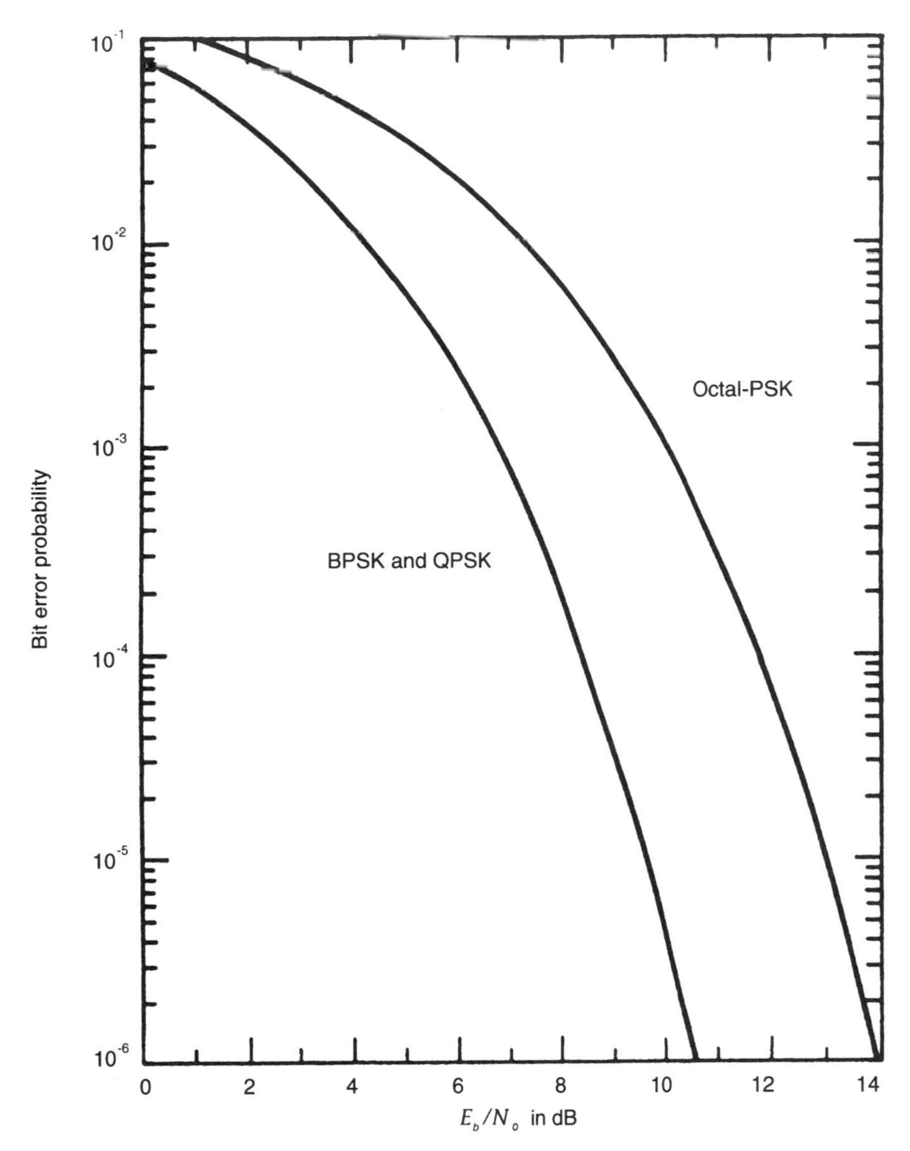

Figure 9.10 Bit error probability (BER) versus E_b/N_0 performance for BPSK/QPSK and 8-ary PSK (octal PSK).

nominally of 4 kHz bandwidth, whereas the digital voice channel requires a 64-kHz bandwidth, assuming 1 bit per Hz occupancy. This is a 16-to-1 difference in required bandwidth. Thus national regulatory authorities such as the FCC require that digital systems be bandwidth conservative. One means that is used to achieve bandwidth conservation is *bit packing*. This means packing more bits into 1 Hz of bandwidth. Another driving factor for bit packing is the need to transmit such higher bit rate formats such as SONET and SDH (Chapter 17). Some radio systems can transmit as much as 622 Mbps using advanced bit packing techniques.

How does bit packing work? In the binary domain we can estimate bandwidth to approximately equate to 1 bit/Hz. For example, if we were transmitting at 1.544 Mbps, following this premise, we'd need 1.544 MHz of bandwidth. Suppose now that we turn

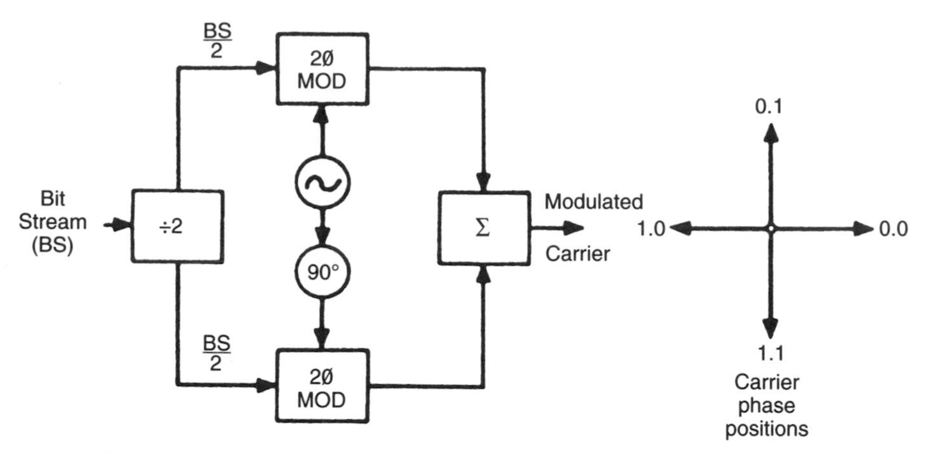

Figure 9.11 Conceptual block diagram of a QPSK modulator. Note that these are two BPSK modulators working jointly, and one modulator is 90 degrees out of phase with the other. The first bit in the bit stream is directed to the upper modulator, the second to the lower, the third to the upper, and so on.

to higher levels of modulation. Quadrature phase shift (QPSK) keying is one example. In this case we achieve a theoretical packing of 2 bits/Hz. Again, if we are transmitting 1.544 Mbps, with QPSK we would need 1.544 MHz/2 or 0.772 MHz. QPSK is one of a family of modulation schemes that are based on phase-shift keying (PSK). With binary PSK (BPSK) we might assign a binary 1 to the 0° position (i.e., no phase retardation) and a binary 0 to the 180° phase retardation point. For QPSK, the phase-circle is broken up into 90° segments, rather than 180° segments as we did with binary PSK. In this case, for every transition we transmit 2 bits at a time. Figure 9.11 is a functional block diagram of a QPSK modulator. It really only consists of two BPSK modulators where one is out of phase with the other by 90°.

Eight-ary PSK modulation is not uncommon. In this case the phase circle is broken up into 45° phase segments. Now for every transition, 3 bits at a time are transmitted. The bit packing in this case as 3 bits per Hz theoretical.

Now add two amplitude levels to this making a hybrid waveform covering both amplitude modulation as well as phase modulation. This family of waveforms is called *quadrature amplitude modulation* (QAM). For example, 16-QAM has 16 different state possibilities: eight are derived for 8-ary PSK and two are derived from the two amplitude levels. We call this 16-QAM, where for each state transition, 4 bits are transmitted at once. The bit packing in this case is theoretically 4 bits/Hz. Certain digital LOS microwave system used 256-QAM and 512-QAM, theoretically achieving 8 bits/Hz and 9 bits/Hz of bit packing. The difference between theoretical bit packing and the practical deals with filter design. For QAM-type waveforms, depending on design, practical bit packing may vary, from 1.25 to 1.5, the *baud-rate-bandwidth*. The extra bandwidth required provides a filter with spectral space to roll-off. In other words, a filter's skirts are not perfectly vertical. Figure 9.12 is a space diagram for 16-QAM. The binary values for each of the 16 states are illustrated.

Suppose we are using a 48-Mbps bit stream to input to our transmitter which was using 16-QAM modulation. Its baud rate, which measures transitions per second, would be 48/4 megabauds/sec. If we allowed 1 baud/Hz, then 12 MHz bandwidth would be required. If we used a roll-off factor of 1.5, then the practical bandwidth required would be 18 MHz. Carry this one step further to 64-QAM. Here the theoretical bit packing is 6 bits/Hz and for the 48-Mbps bit stream, a 12-MHz bandwidth would be required (practical).

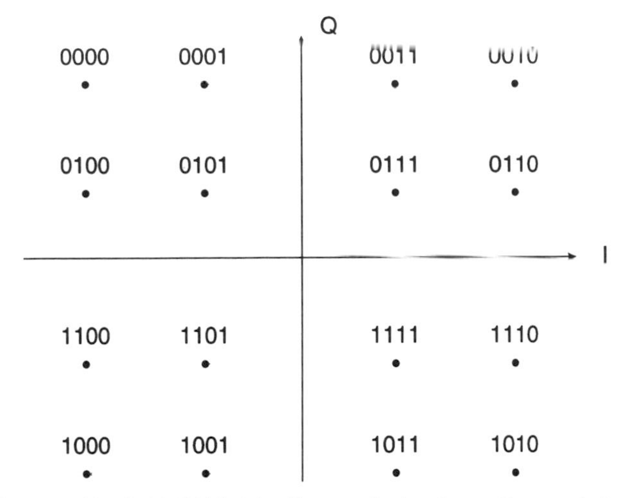

Figure 9.12 A 16-QAM state diagram. I = in-phase; Q = quadrature.

There are no free lunches. As M increases (e.g., $M = 64$), for a given error rate, E_b/N_0 increases. Figure 9.13 illustrates a family of E_b/N_0 curves for various M-QAM modulation schemes plotted against BER.

In summary, to meet these bandwidth requirements, digital LOS microwave will commonly use some form of QAM, and as a minimum at the 64-QAM level, or 128-QAM, 256-QAM, or 512-QAM. The theoretical bit packing capabilities are 6 bits/Hz, 7 bits/Hz, 8 bits/Hz, and 9 bits/Hz, respectively. Figure 9.13 compares BER performance versus E_b/N_0 for various QAM schemes.

9.2.3.6 Parabolic Dish Antenna Gain. At a given frequency the gain of a parabolic antenna is a function of its effective area and may be expressed by the formula:

$$G = 10 \log_{10}(4\pi A\eta/\lambda^2), \tag{9.16}$$

where G is the gain in decibels relative to an isotropic antenna, A is the area of antenna aperture, η is the aperture efficiency, and λ is the wavelength at the operating frequency. Commercially available parabolic antennas with a conventional horn feed at their focus usually display a 55% efficiency or somewhat better. With such an efficiency, gain (G, in dB) is then

$$G_{\mathrm{dB}} = 20 \log_{10} D + 20 \log_{10} F + 7.5, \tag{9.17a}$$

where F is the frequency in GHz, and D is the diameter of the parabolic reflector in feet. In metric units we have:

$$G_{\mathrm{dB}} = 20 \log_{10} D + 20 \log_{10} F + 17.8, \tag{9.17b}$$

where D is in meters and F is in GHz.

9.2.3.7 Running a Path/Site Survey. This can turn out to be the most important step in the design of an LOS microwave link (or hop). We have found through expe-

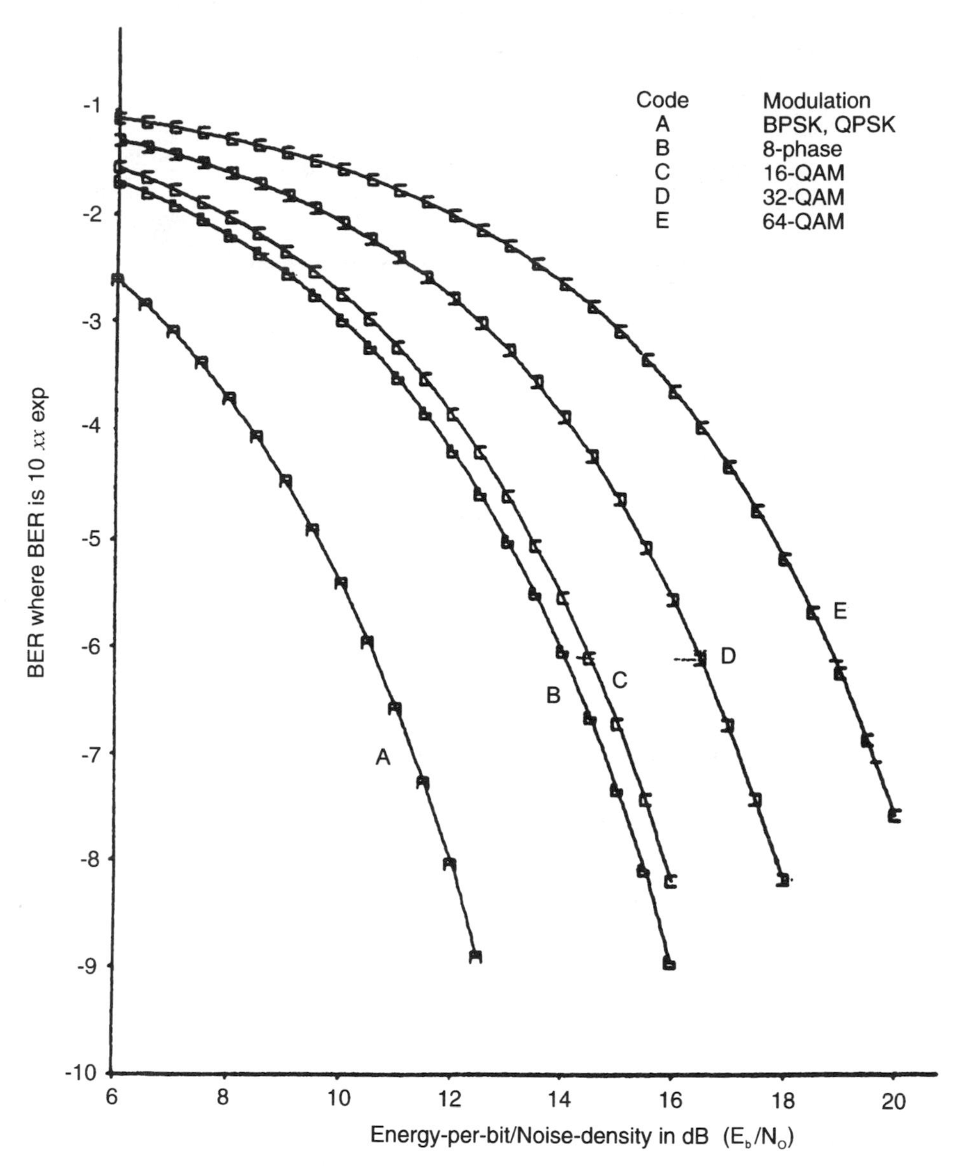

Figure 9.13 BER performance for several modulation types.

rience that mountains move (i.e., map error), buildings grow, grain elevators appear where none were before, east of Madrid a whole high-rise community goes up, and so forth.

Another point from experience: If someone says "line-of-sight" conditions exist on a certain path, *don't believe it*! Line of sight must be precisely defined. We reiterate that for each obstacle in the LOS microwave path, earth curvature with proper K-factor must be added to obstacle height, 0.6 of the first Fresnel zone must be added on top of that, and then 50 ft for trees and 10 ft more for growth must be added if in a vegetated area (to avoid foliage-loss penalties).[7]

[7]Often it is advisable to add 10 ft (or 3 m) of safety factor on top of the 0.6 first Fresnel zone clearance to avoid any diffraction-loss penalties.

Much of the survey is to verify findings and conclusions of the path profile. Of course, each site must be visited to determine the location of the radio equipment shelter, the location of the tower, whether site improvement is required, the nearest prime power lines, and site access, among other items to be investigated.

Site/path survey personnel must personally inspect the sites in question, walking/driving the path or flying the path in a helicopter, or a combination thereof. The use of GPS receivers are helpful to verify geographical positions along the path, including altitudes.[8]

9.2.4 Fades, Fading, and Fade Margins

In Section 9.2.3.4.2 we showed how path loss can be calculated. This was a fixed loss that can be simulated in the laboratory with an attenuator. On very short radio paths below about 10 GHz, the signal level impinging on the distant-end receiving antenna, assuming full LOS conditions, can be calculated to less than 1 dB. If the transmitter continues to give the same output, the RSL will remain uniformly the same over long periods of time, for years. As the path is extended, the measured RSL will vary around a median. The signal level may remain at that median for minutes or hours, and then suddenly drop and then return to the median again. In other periods and/or on other links, this level variation can be continuous for periods of time. Drops in level can be as much as 30 dB or more. This phenomenon is called *fading*. The system and link design must take fading into account when sizing or dimensioning the system/link.

As the RSL drops in level, so does the E_b/N_0. As the E_b/N_0 decreases, there is a deterioration in error performance; the BER degrades. Fades vary in depth, duration, and frequency (i.e., number of fade events per unit of time). We cannot eliminate the fades, but we can mitigate their effects. The primary tool we have is to overbuild each link by increasing the margin.

Link margin is the number of dB we have as surplus in the link design. We could design an LOS microwave link so we just achieve the RSL at the distant receiver to satisfy the E_b/N_0 (and BER) requirements using free-space loss as the only factor in link attenuation (besides transmission line loss). Unfortunately we will only meet our specified requirements about 50% of the time. So we must add margin to compensate for the fading.

We have to determine what percentage of the time the link meets BER performance requirements. We call this *time availability*.[9] If a link meets its performance requirements 99% of the time, then it does not meet performance requirements 1% of the time. We call this latter factor *unavailability*.

To improve time availability, we must increase the link margin, often called the *fade margin*. How many additional dB are necessary? There are several approaches to the calculation of a required fade margin. One of the simplest and most straightforward approaches is to assume that the fading follows a Rayleigh distribution, often considered worst-case fading. If we base our premise on a Rayleigh distribution, then the following fade margins can be used:

[8]GPS stands for *geographical positioning system*, a satellite navigation system that is extremely accurate.

[9]Other texts call this "reliability." The use of this term should be deprecated because it is ambiguous and confusing. In our opinion, reliability should relate to equipment failure rate, not propagation performance.

TIME AVAILABILITY (%)	REQUIRED FADE MARGIN (dB)
90	8
99	18
99.9	28
99.99	38
99.999	48

More often than not, LOS microwave systems consist of multiple hops. Here our primary interest is the time availability at the far-end receiver in the system after the signal has progressed across all of the hops. From this time availability value we will want to assign an availability value for each hop or link.

Suppose a system has nine hops and the system time availability specified is 99.95%, and we want to calculate the time availability per hop or link. The first step is to calculate the system time unavailability. This is simply $1.0000 - 0.9995 = 0.0005$. We now divide this value by 9 (i.e., there are nine hops or links):

$$0.0005/9 = 0.0000555.$$

Now we convert this value to time availability:

$$\begin{aligned}\text{Per-hop time availability} &= 1.0000000 - 0.0000555\\ &= 0.9999445 \text{ or } 99.99445\%.\end{aligned}$$

We recommend that for digital links, an additional 2 dB of fade margin be added to the Rayleigh values to compensate for path anomalies which could not be handled by automatic equalizers.

The most common cause of fading is called *multipath*. Refer to Figure 9.14. As the term implies, signal energy follows multiple paths from the transmit antenna to the receive antenna. Two additional paths, besides the main ray beam, are shown in Figure 9.14. Most of the time the delayed signal energy (from the reflected/refracted paths) will be out of phase with the principal ray beam. These out-of-phase conditions are what cause fading. On digital links we also have to concern ourselves about dispersion. This is delayed signal energy caused by the multipath conditions. Of course, the delayed

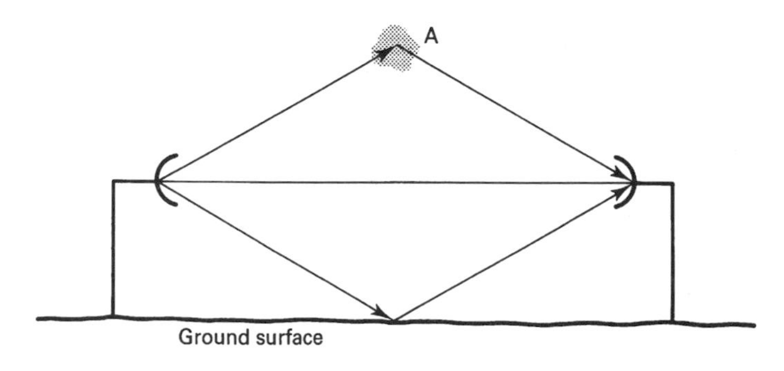

A = layers of different refractive index

Figure 9.14 Multipath is the most common cause of fading.

energy arrives after the main ray beam pulse energy, spilling into the subsequent bit position, greatly increasing the probability that the bit position two will be in error. This is called *intersymbol interference* (ISI).

To achieve the fade margin we must overbuild the link, which will increase the RSL above that of the RSL if we just designed the link to meet the objectives of the unfaded conditions. We must add to the link margin that number of decibels indicated by the Raleigh fading table on the facing page, plus the recommended additional 2 dB.

Probably the most economic way to overbuild a link is to increase the antenna aperture. Every time we double the aperture (i.e., in this case, doubling the diameter of the parabolic dish), we increase the antenna gain 6 dB (see Eqs. (9.17a) and (9.17b)). We recommend the apertures for LOS microwave antennas not exceed 12 ft (3.7 m). Not only does the cost of the antenna get notably greater as aperture increases over 8 ft (2.5 m), but the equivalent sail area of the dish starts to have an impact on system design. Wind pressure on large dishes increases tower twist and sway, resulting in movement out of the capture area of the ray beam at the receive antenna. This forces us to stiffen the tower, which could dramatically increase system cost. Also, as antenna aperture increases, gain increases and beamwidth decreases.

Other measures we can take to overbuild a link are:

- Insert a low-noise amplifier (LNA) in front of the receiver-mixer. Improvement: 6–12 dB.
- Use an HPA (high-power amplifier). Usually a traveling-wave tube (TWT) amplifier; 10 W output. Improvement: 10 dB.
- Implement FEC (forward error correction). Improvement: 1–5 dB. Involves adding a printed circuit board at each end. It will affect link bandwidth (See Ref. 3 for description of FEC.)
- Implement some form of diversity. Space diversity is preferable in many countries. Can be a fairly expensive measure. Improvement: 5–20 dB or more. Diversity is described in Section 9.2.5.

It should be appreciated that fading varies with path length, frequency, climate, and terrain. The rougher the terrain, the more reflections are broken up. Flat terrain, and especially paths over water, tends to increase the incidence of fading. For example, in dry, windy, mountainous areas the multipath fading phenomenon may be nonexistent. In hot, humid coastal regions a very high incidence of fading may be expected.

9.2.5 Diversity and Hot-Standby

Diversity reception means the simultaneous reception of the same radio signal over two or more paths. Each "path" is handled by a separate receiver chain and then combined by predetection or postdetection combiners in the radio equipment so that effects of fading are mitigated. The separate diversity paths can be based on space, frequency, and/or time diversity. The simplest and preferred form of diversity for LOS microwave is space diversity. Such a configuration is illustrated in Figure 9.15.

The two diversity paths in space diversity are derived at the receiver end from two separate receivers with a combined output. Each receiver is connected to its own antenna, separated vertically on the same tower. The separation distance should be at least 70 wavelengths and preferably 100 wavelengths. In theory, fading will not occur on both paths simultaneously.

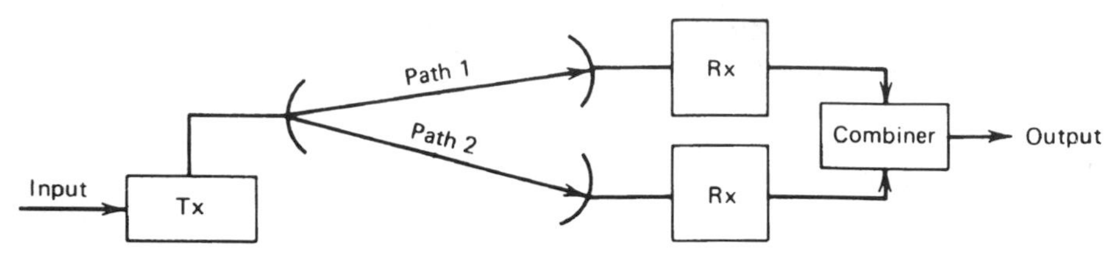

Figure 9.15 A space diversity configuration. The vertical distance between the upper and lower antennas is of key importance.

Frequency diversity is more complex and more costly than space diversity. It has advantages as well as disadvantages. Frequency diversity requires two transmitters at the near end of the link. The transmitters are modulated simultaneously by the same signal but transmit on different frequencies. Frequency separation must be at least 2%, but 5% is preferable. Figure 9.16 is an example of a frequency-diversity configuration. The two diversity paths are derived in the frequency domain. When a fade occurs one on frequency, it will probably not occur on the other frequency. The more one frequency is separated from the other, the less chance there is that fades occur simultaneously on each path.

Frequency diversity is more expensive, but there is greater assurance of path reliability. It provides full and simple equipment redundancy and has the great operational advantage of two complete end-to-end electrical paths. In this case, failure of one transmitter or one receiver will not interrupt service, and a transmitter and/or a receiver can be taken out of service for maintenance. The primary disadvantage of frequency diversity is that it doubles the amount of frequency spectrum required in this day and age when spectrum is at a premium. In many cases it is prohibited by national licensing authorities. For example, the FCC does not permit frequency diversity for industrial users. It also should be appreciated that it will be difficult to get the desired frequency spacing.

The full equipment redundancy aspect is very attractive to the system designer. Another approach to achieve diversity improvement in propagation plus reliability improvement by fully redundant equipment is to resort to the "hot-standby" technique. On the receive end of the path, a space-diversity configuration is used. On the transmit end a second transmitter is installed, as in Figure 9.16, but the second transmitter is on *hot standby*. This means that the second transmitter is on but its signal is not radiated by the antenna. On a one-for-one basis, the second transmitter is on the same frequency as the first transmitter. On the failure of transmitter 1, transmitter 2 is switched in automatically, usually with no dropout of service at all.

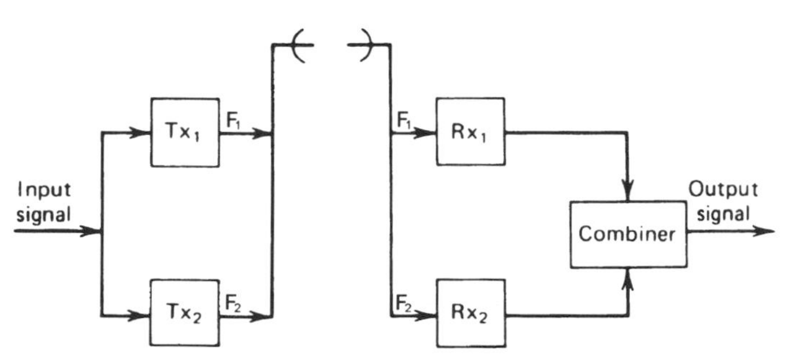

Figure 9.16 A frequency-diversity configuration.

Table 9.3 LOS Microwave Frequency Bands

2110–2130 MHz	18,920–19,160 MHz
3700–4200 MHz	19,260–19,700 MHz
5925–6425 MHz	21,200–23,600 MHz
6525–6875 MHz	27,500–29,500 MHz
10,700–11,700 MHz	31,000–31,300 MHz
17,700–18,820 MHz	38,600–40,000 MHz

9.2.6 Frequency Planning and Frequency Assignment

9.2.6.1 Introduction. To derive optimum performance from an LOS microwave system, the design engineer must set out a frequency-usage plan that may or may not have to be approved by the national regulatory organization. The problem has many aspects. First, the useful RF spectrum is limited from above dc (0 Hz) to about 150 GHz. The upper limit is technology-restricted. To some extent it is also propagation-restricted. The frequency ranges of interest for this discussion cover the bands listed in Table 9.3. The frequencies above 10 GHz could also be called rainfall-restricted, because at about 10 GHz is where excess attenuation due to rainfall can become an important design factor.

Then there is the problem of frequency congestion. Around urban and built-up areas, frequency assignments below 10 GHz are difficult to obtain from national regulatory authorities. If we plan properly for excess rainfall attenuation, nearly equal performance is available at those higher frequencies.

9.2.6.2 Radio-Frequency Interference (RFI): There are three facets to RFI in this context: (1) own microwave can interfere with other LOS microwave and satellite communication earth stations nearby, (2) nearby LOS microwave and satellite communication facilities can interfere with own microwave, and (3) own microwave can interfere with itself. To avoid self-interference (3), it is advisable to use frequency plans of CCIR (ITU-R organization) as set forth in the RF Series (Fixed Service). Advantage is taken of proper frequency separation, transmit and receive, and polarization isolation. CCIR also provides methods for interference analysis (coordination contour), also in the RF series. Another alternative is specialist companies, which provide a service of electromagnetic compatibility analysis.

The IEEE defines *electromagnetic compatibility* (EMC) as "The requirements for electromagnetic emission and susceptibility dictated by the physical environment and regulatory governing bodies in whose jurisdiction a piece of equipment is operated." We'll call electromagnetic emission (EMI) RFI, meaning the level of RF interference caused by a certain piece of equipment such as a microwave terminal. *Susceptibility* deals with how well a piece of equipment can operate in an RFI environment. EMC can be a real headache for a microwave engineer.

9.3 SATELLITE COMMUNICATIONS

9.3.1 Introduction

Satellite communications is an extension of LOS microwave technology covered in Section 9.2. The satellite must be within line-of-sight of each participating earth terminal. We are more concerned about noise in satellite communication links than we were with LOS microwave. In most cases, received signals will be of a much lower level. On satellite systems operating below 10 GHz, very little link margin is required; there is

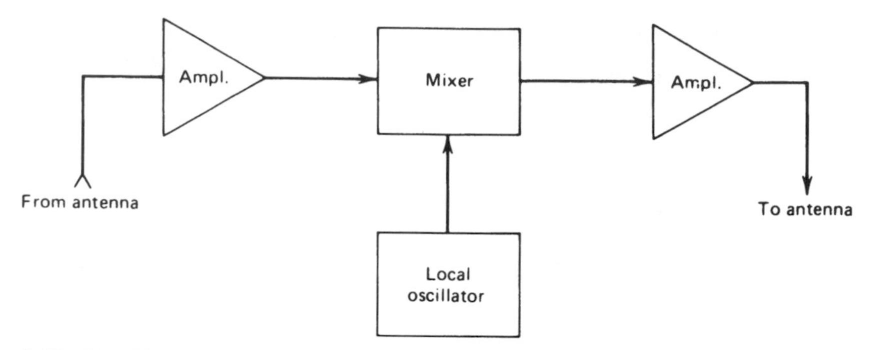

Figure 9.17 Simplified functional block diagram of a transponder of a typical communication satellite.

essentially no fading, as experienced with LOS microwave. The discussion here only deals with geostationary orbit (GEO) communication satellites.

Satellite communications presents another method of extending the digital network (Chapter 6). These digital trunks may be used as any other digital trunks for telephony, data, the Internet, facsimile, and video. However, fiber optics has become a strong competitor of satellite communications. Only very small aperture terminal (VSAT) systems are showing any real growth in the GEO arena. A new type of communication satellite is being fielded. This is the LEO class of satellites, which we discuss in Chapter 16.

Three quarters of the satellite transponders over North America are used to provide entertainment services such as direct broadcast television, cable system headend feeds, and for private broadcaster connectivity.

9.3.2 Satellite

Most of the commercial communication satellites that are currently employed are RF repeaters. A typical RF repeater used in a communication satellite is illustrated in Figure 9.17. The tendency is to call these types of satellite *bent pipe* as opposed to *processing satellites*. A processing satellite, as a minimum, demodulates and regenerates the received digital signal. It may also decode and recode (FEC) a digital bit stream.[10] It also may have some bulk switching capability, switching to crosslinks connecting to other satellites. Theoretically, three GEO satellites placed correctly in equatorial orbit could provide connectivity from one earth station to any other located anywhere on the surface of the earth (see Figure 9.18). However, high-latitude service is marginal and nil north of 80°N and south of 80°S.

9.3.3 Three Basic Technical Problems

As the reader can appreciate, satellite communication is nothing more than LOS microwave using one (or two) satellites located at great distances from the terminal earth stations, as illustrated in Figure 9.18.[11] Because of the distance involved, consider the slant range from the earth station to the satellite to be the same as the satellite altitude above the equator. This would be true if the antenna were pointing at zenith (0° ele-

[10] *FEC*, forward error correction. (See Ref. 3, Chapter 4.)

[11] For voice communications, connectivity is limited to only one GEO satellite link because of the delay involved.

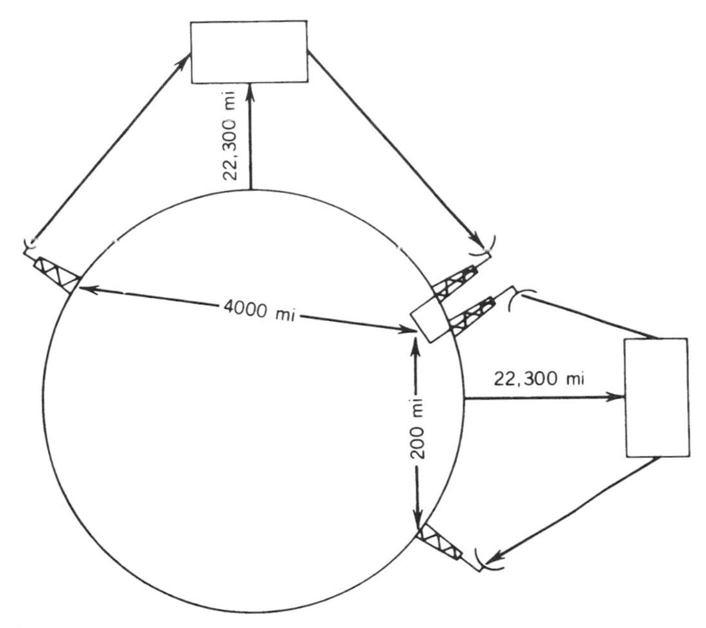

Figure 9.18 Distances involved in satellite communications. One is looking down at or up at the equator (i.e., the circle).

vation angle) to the satellite. Distance increases as the pointing angle to the satellite decreases (elevation angle).

We thus are dealing with very long distances. The time required to traverse these distances—namely, earth station to satellite to another earth station—is on the order of 250 ms. Round-trip delay will be 2 × 250 or 500 ms. These propagation times are much greater than those encountered on conventional terrestrial systems. So one major problem is propagation time and resulting echo on telephone circuits. It influences certain data circuits in delay to reply for block or packet transmission systems and requires careful selection of telephone signaling systems, or call-setup time may become excessive.

Naturally, there are far greater losses. For LOS microwave we encounter free-space losses possibly as high as 145 dB. In the case of a satellite with a range of 22,300 mi operating on 4.2 GHz, the free-space loss is 196 dB and at 6 GHz, 199 dB. At 14 GHz the loss is about 207 dB. This presents no insurmountable problem from earth to satellite, where comparatively high-power transmitters and very-high-gain antennas may be used. On the contrary, from satellite to earth the link is power-limited for two reasons: (1) in bands shared with terrestrial services such as the popular 4-GHz band to ensure noninterference with those services, and (2) in the satellite itself, which can derive power only from solar cells. It takes a great number of solar cells to produce the RF power necessary; thus the downlink, from satellite to earth, is critical, and received signal levels will be much lower than on comparative radiolinks, as low as −150 dBW. A third problem is crowding. The equatorial orbit is filling with geostationary satellites. Radio-frequency interference from one satellite system to another is increasing. This is particularly true for systems employing smaller antennas at earth stations with their inherent wider beamwidths. It all boils down to a frequency congestion of emitters.

It should be noted that by the year 2000, we can expect to see several low earth-orbit (LEO) satellite systems in operation. These satellites typically orbit some 500 km above the earth.

9.3.4 Frequency Bands: Desirable and Available

The most desirable frequency bands for commercial satellite communication are in the spectrum 1000–10,000 MHz. These bands are:

3700–4200 MHz (satellite-to-earth or downlink);
5925–6425 MHz (earth-to-satellite or uplink);
7250–7750 MHz (downlink);[12] and
7900–8400 MHz (uplink).[12]

These bands are preferred by design engineers for the following primary reasons:

- Less atmospheric absorption than higher frequencies;
- Rainfall loss not a concern;
- Less noise, both galactic and man-made;
- Well-developed technology; and
- Less free-space loss compared with the higher frequencies.

There are two factors contraindicating application of these bands and pushing for the use of higher frequencies:

1. The bands are shared with terrestrial services.
2. There is orbital crowding (discussed earlier).

Higher-frequency bands for commercial satellite service are:

10.95–11.2 GHz (downlink);
11.45–12.2 GHz (downlink);
14.0–14.5 GHz (uplink);
17.7–20.2 GHz (downlink); and
27.5–30.0 GHz (uplink).

Above 10 GHz rainfall attenuation and scattering and other moisture and gaseous absorption must be taken into account. The satellite link must meet a BER of 1×10^{-6} at least 99.9% of the time. One solution is a space-diversity scheme where we can be fairly well assured that one of the two antenna installations will not be seriously affected by the heavy rainfall cell affecting the other installation. Antenna separations of 4–10 km are being employed. Another advantage with the higher frequencies is that requirements for downlink interference are less; thus satellites may radiate more power. This is often carried out on the satellite using spot-beam antennas rather than general-coverage antennas.

9.3.5 Multiple Access to a Communication Satellite

Multiple access is defined as the ability of a number of earth stations to interconnect their respective communication links through a common satellite. Satellite access is classified (1) by assignment, whether quasi-permanent or temporary, namely, (a) pre-

[12]These two bands are intended mainly for military application.

Table 9.4 INTELSAT VI, VII, and VIII Regular FDM/FM Carriers, FDMA Voice-Channel Capacity versus Bandwidth Assignments (Partial Listing)

Carrier capacity (number of voice channels)	24.0	60.0	96.0	132.0	252.0	432.0	792.0
Top baseband frequency (kHz)	108.0	252.0	408.0	552.0	1052.0	1796.0	3284.0
Allocated satellite bandwidth (MHz)	2.5	2.5	5.0	10.0	10.0	15.0	36.0
Occupied bandwidth (MHz)	2.00	2.25	4.5	7.5	8.5	12.4	32.4

Source: Ref. 4.

assigned multiple access or (b) demand-assigned multiple access (DAMA); and (2) according to whether the assignment is in the frequency domain or the time domain, namely, (a) frequency-division multiple access (FDMA) or (b) time-division multiple access (TDMA). On comparatively heavy routes (≥10 erlangs), preassigned multiple access may become economical. Other factors, of course, must be considered, such as whether the earth station is "INTELSAT" standard as well as the space-segment charge that is levied for use of the satellite. In telephone terminology, "preassigned" means dedicated circuits. DAMA is useful for low-traffic multipoint routes where it becomes interesting from an economic standpoint. Also, an earth station may resort to DAMA as a remedy to overflow for its FDMA circuits.

9.3.5.1 Frequency Division Multiple Access (FDMA). Historically, FDMA has the highest usage and application of the various access techniques considered here. The several RF bands available (from Section 9.3.4) each have a 500-MHz bandwidth. A satellite contains a number of transponders, each of which covers a frequency segment of the 500-MHz bandwidth. One method of segmenting the 500 MHz is by utilizing 12 transponders, each with a 36-MHz bandwidth. Sophisticated satellites, such as INTELSAT VIII, segment the 500-MHz bandwidth available with transponders with 36, 72, and 77 MHz in the 6/4 GHz frequency band pair and 72, 77, and 112 MHz in the 14/12 GHz frequency band pair.[13]

With FDMA operation, each earth station is assigned a segment or a portion of a frequency segment. For a nominal 36-MHz transponder, 14 earth stations may access in an FDMA format, each with 24 voice channels (two groups) in a standard CCITT modulation plan (analog) (see Section 4.5.2.3). The INTELSAT VIII assignments for a 36-MHz transponder are shown in Table 9.4, where it can be seen that when larger channel groups are used, fewer earth stations can access the same transponder.

9.3.5.2 Time Division Multiple Access (TDMA). Time-division multiple access operates in the time domain and may only be used for digital network connectivity. Use of the satellite transponder is on a time-sharing basis. Individual time slots are assigned to earth stations in a sequential order. Each earth station has full and exclusive use of the transponder bandwidth during its time-assigned segment. Depending on the bandwidth of the transponder, bit rates of 10–100 Mbps (megabits per second) are used.

With TDMA operation, earth stations use digital modulation and transmit with bursts

[13] 6/4 GHz frequency band pair, meaning 5925–6425 MHz uplink and 3700–4200 MHz downlink.

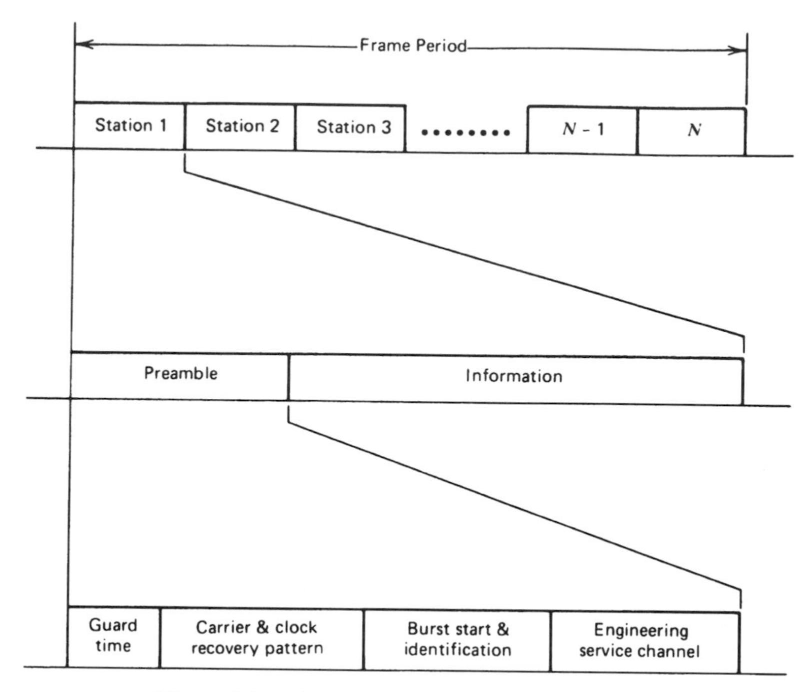

Figure 9.19 Example of TDMA burst frame format.

of information. The duration of a burst lasts for the time period of the slot assigned. Timing synchronization is a major problem.

A frame, in digital format, may be defined as a repeating cycle of events. It occurs in a time period containing a single digital burst from each accessing earth station and the guard periods or guard times between each burst. A sample frame is shown in Figure 9.19 for earth stations 1, 2, and 3 to earth station N. Typical frame periods are 750 μsec for INTELSAT and 250 μsec for the Canadian Telesat.

The reader should appreciate that timing is crucial to effective TDMA operation. The greater N becomes (i.e., the more stations operating in the frame period), the more clock timing affects the system. The secret lies in the *carrier* and *clock timing recovery pattern*, as shown in Figure 9.19. One way to ensure that all stations synchronize to a master clock is to place a synchronization burst as the first element in the frame format. INTELSAT's TDMA does just this. The burst carries 44 bits, starting with 30 bits carrier and bit timing recovery, 10 bits for the *unique word*, and 4 bits for the station identification code in its header.

Why use TDMA in the first place? It lies in a major detraction of FDMA. Satellites use traveling-wave tubes (TWTs) in their transmitter final amplifiers. A TWT has the undesirable property of nonlinearity in its input–output characteristics. When there is more than one carrier accessing the transponder simultaneously, high levels of intermodulation (IM) products are produced, thus increasing noise and crosstalk. When a transponder is operated at full power output, such noise can be excessive and intolerable. Thus input must be backed off (i.e., level reduced) by $\geq$3 dB. This, of course, reduces the EIRP and results in reduced efficiency and reduced information capacity. Consequently, each earth station's uplink power must be carefully coordinated to ensure proper loading of the satellite. The complexity of the problem increases when a large number of earth stations access a transponder, each with varying traffic loads.

On the other hand, TDMA allows the transponder's TWT to operate at full power because only one earth-station carrier is providing input to the satellite transponder at any one instant.

To summarize, consider the following advantages and disadvantages of FDMA and TDMA. The major advantages of FDMA are as follows:

- No network timing is required; and
- Channel assignment is simple and straightforward.

The major disadvantages of FDMA are as follows:

- Uplink power levels must be closely coordinated to obtain efficient use of transponder RF output power.
- Intermodulation difficulties require power back-off as the number of RF carriers increases with inherent loss of efficiency.

The major advantages of TDMA are as follows:

- There is no power sharing and IM product problems do not occur.
- The system is flexible with respect to user differences in uplink EIRP and data rates.
- Accesses can be reconfigured for traffic load in almost real time.

The major disadvantages of TDMA are as follows:

- Accurate network timing is required.
- There is some loss of throughput due to guard times and preambles.
- Large buffer storage may be required if frame lengths are long.

9.3.5.3 Demand-Assigned Multiple Access (DAMA). The DAMA access method has a pool of single voice channel available for assignment to an earth station on demand. When a call has been completed on the channel, the channel is returned to the idle pool for reassignment. The DAMA system is a subset of FDMA, where each voice channel is assigned its own frequency slot, from 30 kHz to 45 kHz wide.

The DAMA access method is useful at earth stations that have traffic relations of only several erlangs. DAMA may also be used for overflow traffic. It operates something like a telephone switch with its pool of available circuits. A call is directed to an earth station, where through telephone number analysis, that call will be routed on a DAMA circuit. With centralized DAMA control, the earth station requests a DAMA channel from the master station. The channel is assigned and connectivity is effected. When the call terminates (i.e., there is an on-hook condition), the circuit is taken down and the DAMA channel is returned to the pool of available channels. Typical DAMA systems have something under 500 voice channels available in the pool. They will occupy one 36-MHz satellite transponder.

9.3.6 Earth Station Link Engineering

9.3.6.1 Introduction. Up to this point we have discussed basic satellite communication topics such as access and coverage. This section covers satellite link engineering with emphasis on the earth station, the approach used to introduce the reader to essential path engineering. It expands on the basic principles introduced in Section 9.2, dealing with LOS microwave. As we saw in Section 9.3.2, an earth station is a distant RF repeater. By international agreement the satellite transponder's EIRP is limited because nearly all bands are shared by terrestrial services, principally LOS microwave. This is one reason we call satellite communication *downlink limited.*

9.3.6.2 Satellite Communications Receiving System Figure of Merit, G/T.

The figure of merit of a satellite communications receiving system, G/T, has been introduced into the technology to describe the capability of an earth station or a satellite to *receive* a signal. It is also a convenient tool in the link budget analysis.[14] A link budget is used by the system engineer to size components of earth stations and satellites, such as RF output power, antenna gain and directivity, and receiver front-end characteristics.

G/T can be written as a mathematical identity:

$$G/T = G_{\mathrm{dB}} - 10\log T_{\mathrm{sys}}, \tag{9.18}$$

where G is the net antenna gain up to an arbitrary reference point or reference plane in the downlink receive chain (for an earth station). Conventionally, in commercial practice the reference plane is taken at the input of the low-noise amplifier (LNA). Thus G is simply the gross gain of the antenna minus all losses up to the LNA. These losses include feed loss, waveguide loss, bandpass filter loss, and, where applicable, directional coupler loss, waveguide switch insertion loss, radome loss, and transition losses.

T_{sys} is the effective noise temperature of the receiving system and

$$T_{\mathrm{sys}} = T_{\mathrm{ant}} + T_{\mathrm{recvr}}. \tag{9.19}$$

T_{ant} or the antenna noise temperature includes all noise-generating components up to the reference plane. The reference plane is a dividing line between the antenna noise component and the actual receiver noise component (T_{recvr}). The antenna noise sources include sky noise (T_{sky}) plus the thermal noise generated by ohmic losses created by all devices inserted into the system. T_{recvr} or the actual receiver noise temperature, which has equivalence to the receiver noise figure. A typical earth station receiving system is illustrated in Figure 9.20 for a 12-GHz downlink. Earth stations generally have minimum elevation angles. At 4 GHz the minimum elevation angle is 5°; at 12 GHz, 10°. The elevation angle is that angle measured from the horizon (0°) to the antenna main beam when pointed at the satellite. As the elevation angle decreases below these values, noise (sky noise) increases radically.

Take note that we are working with noise temperatures here. *Noise temperature* is another way of expressing thermal noise levels of a radio system, subsystem, or component. In Section 9.2 we used noise figure for this function. Noise figure can be related to noise temperature by the following formula:

$$\mathrm{NF}_{\mathrm{dB}} = 10\log(1 + T_{\mathrm{e}}/290) \tag{9.20}$$

where T_{e} is the effective noise temperature measured in kelvins. Note that the kelvin temperature scale is based on absolute zero.

[14]We called this *path analysis* in LOS microwave terminology.

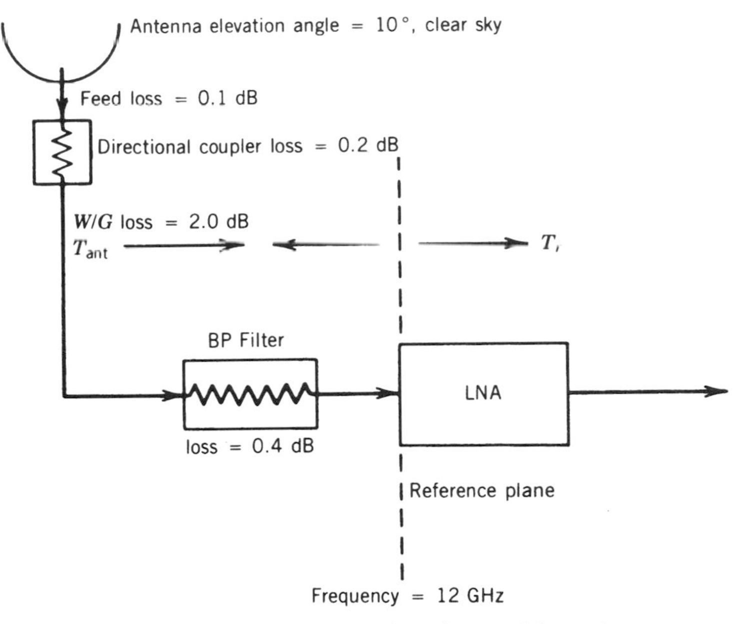

Figure 9.20 Model for an earth station receiving system.

Example. If the noise figure of a device is 1.2 dB, what is its equivalent noise temperature?

$$1.2\ \text{dB} = 10\log(1 + T_e/290)$$
$$0.2 = \log(1 + T_e/290).$$

Remember, 0.2 is the logarithm of the number. What is the number? Take the antilog of 0.2 by using the 10^x function on your calculator. This turns out to be 1.58.

$$1.58 = 1 + T_e/290$$
$$T_e = 168.2\ \text{K}.$$

Antenna noise (T_{ant}) is calculated by the following formula:

$$T_{ant} = \frac{(l_a - 1)290 + T_{sky}}{l_a}, \tag{9.21}$$

where l_a is the numeric equivalent of the sum of the ohmic losses up to the reference plane and is calculated by

$$l_a = \log_{10}^{-1} \frac{L_a}{10}, \tag{9.22}$$

where L_a is the sum of the losses in decibels.

Sky noise varies directly with frequency and inversely with elevation angle. Some typical sky noise values are given in Table 9.5.

Table 9.5 Sky Noise Values for Several Frequencies and Elevation Angles

Frequency (GHz)	Elevation Angle (°)	Sky Noise (K)
4.0	5	28
4.0	10	16
7.5	5	33
7.5	10	18
11.7	10	23
11.7	15	18
20.0	10	118
20.0	15	100
20.0	20	80

Example. An earth station operating at 12 GHz with a 10° elevation angle has a 47-dB gain and a 2.5-dB loss from the antenna feed to the input of the LNA. The sky noise is 25 K developing an antenna noise temperature of 240 K. The noise figure of the LNA is 1.5 dB. Calculate the G/T.

Convert the 2.5 dB noise figure value to its equivalent noise temperature. Use formula (9.20). For this sample problem $T_e = T_{recvr}$.

$$\begin{aligned} 1.5 \text{ dB} &= 10 \log(1 + T_e/290) \\ &= 119.6 \text{ K} = T_{recvr} \\ T_{sys} &= T_{ant} + T_{recvr} \\ &= 240 + 119.6 \\ &= 359.6 \text{ K}. \end{aligned}$$

Now we can calculate the G/T. Derive the net antenna gain (up to the reference plane—at the input of the LNA).

$$\begin{aligned} G_{net} &= 47 - 2.5 \\ &= 44.3 \text{ dB} \\ G/T &= 44.3 \text{ dB} - 10 \log T_{sys} \\ &= 44.3 - 10 \log 359.6 \\ &= +18.74 \text{ dB/K} \qquad \text{or just } +18.74 \text{ dB/K} \end{aligned}$$

For earth stations operating below 10 GHz, it is advisable to have a link margin of 4 dB to compensate for propagation anomalies and deterioration of components due to aging.

9.3.6.3 Typical Downlink Power Budget. A link budget is a tabular method of calculating space communication system parameters. The approach is very similar to that used for LOS microwave links (see Section 9.2.3.4). We start with the EIRP of the satellite for the downlink or the EIRP of the earth station for the uplink. The bottom line is C/N_0 and the link margin, all calculated with dB notation. C/N_0 is the carrier-to-noise ratio in 1 Hz of bandwidth at the input of the LNA. (*Note:* RSL, or receive signal level, and C are synonymous.) Expressed as an equation:

$$\frac{C}{N_0} = \text{EIRP} - \text{FSL}_{\text{dB}} - (\text{other losses}) + G/T_{\text{dB/K}} \quad k, \tag{9.23}$$

where FSL is the free-space loss to the satellite for the frequency of interest and k is Boltzmann's constant expressed in dBW. Remember in Eq. (9.9) we used Boltzmann's constant, which gives the thermal noise level at the output of a "perfect" receiver operating at absolute zero in 1 Hz of bandwidth (or N_0).[15] Its value is −228.6 dBW/Hz. "Other losses" may include:

- Polarization loss (0.5 dB);
- Pointing losses, terminal and satellite (0.5 dB each);
- Off-contour loss (depends on satellite antenna characteristics);
- Gaseous absorption loss (varies with frequency, altitude, and elevation angle); and
- Excess attenuation due to rainfall (for systems operating above 10 GHz).

The loss values in parentheses are conservative estimates and should be used only if no definitive information is available.

The off-contour loss refers to spacecraft antennas that provide a spot or zone beam with a footprint on a specific geographical coverage area. There are usually two contours, one for G/T (uplink) and the other for EIRP (downlink). Remember that these contours are looking from the satellite down to the earth's surface. Naturally, an off-contour loss would be invoked only for earth stations located outside of the contour line. This must be distinguished from satellite pointing loss, which is a loss value to take into account that satellite pointing is not perfect. The contour lines are drawn as if the satellite pointing were "perfect."

Gaseous absorption loss (or atmospheric absorption) varies with frequency, elevation angle, and altitude of the earth station. As one would expect, the higher the altitude, the less dense the air and thus the less loss. Gaseous absorption losses vary with frequency and inversely with elevation angle. Often, for systems operating below 10 GHz, such losses are neglected. Reference 3 suggests a 1-dB loss at 7.25 GHz for elevation angles under 10° and for 4 GHz, 0.5 dB below 8° elevation angle.

Example of a Link Budget. Assume the following: a 4-GHz downlink, 5° elevation angle, EIRP is +30 dBW; satellite range is 25,573 statute miles (sm), and the terminal G/T is +20.0 dB/K. Calculate the downlink C/N_0.

First calculate the free-space loss. Use Eq. (9.4):

$$\begin{aligned} L_{\text{dB}} &= 96.6 + 20 \log F_{\text{GHz}} + 20 \log D_{\text{sm}} \\ &= 96.6 + 20 \log 4.0 + 20 \log 25{,}573 \\ &= 96.6 + 12.04 + 88.16 \\ &= 196.8 \text{ dB}. \end{aligned}$$

[15]Remember that geostationary satellite range varies with elevation angle and is minimum at zenith.

EXAMPLE LINK BUDGET: DOWNLINK

EIRP of satellite	+30 dBW
Free-space loss	−196.8 dB
Satellite pointing loss	−0.5 dB
Off-contour loss	0.0 dB
Excess attenuation rainfall	0.0 dB
Gaseous absorption loss	−0.5 dB
Polarization loss	−0.5 dB
Terminal pointing loss	−0.5 dB
Isotropic receive level	−168.8 dBW
Terminal G/T	+20.0 dB/K
Sum	−148.8 dBW
Boltzmann's constant (dBW)	−(−228.6 dBW)
C/N_0	79.8 dB

On repeatered satellite systems, sometimes called "bent-pipe satellite systems" (those that we are dealing with here), the link budget is carried out only as far as C/N_0, as we did above. It is calculated for the uplink and for the downlink separately. We then calculate an equivalent C/N_0 for the system (i.e., uplink and downlink combined). Use the following formula to carry out this calculation:

$$\left(\frac{C}{N_0}\right)_{(s)} = \frac{1}{1/(C/N_0)_{(u)} + 1/(C/N_0)_{(d)}}. \tag{9.24}$$

Example. Suppose that an uplink has a C/N_0 of 82.2 dB and its companion downlink has a C/N_0 of 79.8 dB. Calculate the C/N_0 for the system, $(C/N_0)_s$. First calculate the equivalent numeric value (NV) for each C/N_0 value:

$$\begin{aligned}
\text{NV}(1) &= \log^{-1}(79.8/10) = 95.5 \times 10^6 \\
\text{NV}(2) &= \log^{-1}(82.2/10) = 166 \times 10^6 \\
C/N_0 &= 1/[(10^{-6}/95.5) + (10^{-6}/166)] \\
&= 1/(0.016 \times 10^{-6}) = 62.5 \times 10^6 = 77.96 \text{ dB}.
\end{aligned}$$

This is the carrier-to-noise ratio in 1 Hz of bandwidth. To derive C/N for a particular RF bandwidth, use the following formula:

$$C/N = C/N_0 - 10 \log BW_{\text{Hz}}.$$

Suppose the example system had a 1.2-MHz bandwidth with the C/N_0 of 77.96 dB. What is the C/N?

$$\begin{aligned}
C/N &= 77.96 \text{ dB} - 10 \log(1.2 \times 10^6) \\
&= 77.96 - 60.79 \\
&= 17.17 \text{ dB}.
\end{aligned}$$

9.3.6.4 Uplink Considerations. A typical specification for INTELSAT states that the EIRP per voice channel must be +61 dBW (example); thus, to determine the EIRP for a specific number of voice channels to be transmitted on a carrier, we take the required output per voice channel in dBW (the +61 dBW in this case) and add logarithmically 10 log N, where N is the number of voice channels to be transmitted.

For example, consider the case for an uplink transmitting 60 voice channels:

$$+61\ \text{dBW} + 10 \log 60 = 61 + 17.78 = +78.78\ \text{dBW}.$$

If the nominal 50-ft (15-m) antenna has a gain of 57 dB (at 6 GHz) and losses typically of 3 dB, the transmitter output power, P_t, required is

$$\text{EIRP}_{\text{dBW}} = P_t + G_{\text{ant}} - \text{line losses}_{\text{dB}},$$

where P_t is the output power of the transmitter (in dB/W) and G_{ant} is the antenna gain (in dB) (uplink). Then in the example we have

$$\begin{aligned} +78.78\ \text{dBW} &= P_t + 53 - 3 \\ P_t &= +24.78\ \text{dBW} \\ &= 300.1\ \text{W}. \end{aligned}$$

9.3.7 Digital Communication by Satellite

There are three methods to handle digital communication by satellite: (1) TDMA, (2) FDMA, and (3) over a VSAT network. TDMA was covered in Section 9.3.5.2 and VSATs will be discussed in Section 9.3.8. Digital access by FDMA is handled in a similar fashion as with an analog FDM/FM configuration (Section 9.3.5.1). Several users may share a common transponder and the same backoff rules hold; in fact they are even more important when using a digital format because the IM products generated in the satellite TWT high-power amplifier (HPA) can notably degrade error performance. In the link budget, once we calculate C/N_0 (Eq. (9.24)), we convert to E_b/N_0 with the following formula:

$$E_b/N_0 = C/N_0 - 10 \log(\text{bit rate}). \tag{9.25}$$

The E_b/N_0 value can now be applied to the typical curves found in Figure 9.10 to derive the BER.

As mentioned previously, satellite communication is downlink limited because downlink EIRP is strictly restricted. Still we want to receive sufficient power to meet error performance objectives. One way to achieve this goal is to use forward error correction on the links where the lower E_b/N_0 ratios will still meet error objectives. Thus INTELSAT requires coding on their digital accesses. Some typical INTELSAT digital link parameters are given in Table 9.6. These parameters are for the intermediate data rate (IDR) digital carrier system. All IDR carriers are required to use at least $R = 3/4$ where R is the code rate.[16] A detailed description of various FEC channel coding schemes is provided in Ref. 3.

[16] R = (information bit rate)/(coded symbol rate). When $R = 3/4$ and the information rate is 1.544 Mbps, the coded symbol rate is 4/3 that value or 2,058,666 symbols a second. FEC coding simply adds redundant bits in a systematic manner such that errors may be corrected by the distant-end decoder.

Table 9.6 QPSK Characteristics and Transmission Parameters for IDR Carriers

Parameter	Requirement
1. Information rate (IR)	64 kbit/s to 44.736 Mbit/s
2. Overhead data rate for carriers with IR ≥ 1.544 Mbit/s	96 kbit/s
3. Forward error correction encoding	Rate 3/4 convolutional encoding/Viterbi decoding
4. Energy dispersal (scrambling)	As per ITU-R5.524-4
5. Modulation	Four-phase coherent PSK
6. Ambiguity resolution	Combination of differential encoding (180°) and FEC (90°)
7. Clock recovery	Clock timing must be recovered from the received data stream
8. Minimum carrier bandwidth (allocated)	0.7 R Hz of [0.933 (IR + Overhead)]
9. Noise bandwidth (and occupied bandwidth)	0.6 R Hz or [0.8 (IR + Overhead)]
10. E_b/N_0 at BER (Rate 3/4 FEC)	10^{-3} 10^{-7} 10^{-8}
a. Modems back-to-back	5.3 dB 8.3 dB 8.8 dB
b. Through satellite channel	5.7 dB 8.7 dB 9.2 dB
11. C/T at nominal operating point	$-219.9 + 10 \log_{10}$ (IR + OH), dBW/K
12. C/N in noise bandwidth at nominal operating point (BER $\leq 10^{-7}$)	9.7 dB
13. Nominal bit error rate at operating point	1×10^{-7}
14. C/T at threshold (BER = 1×10^{-3})	$-222.9 + 10 \log_{10}$ (IR + OH), dBW/K
15. C/N in noise bandwidth at threshold (BER = 1×10^{-3})	6.7 dB
16. Threshold bit error rate	1×10^{-3}

Notes: IR is the information rate in bits per second. R is the transmission rate in bits per second and equals (IR + OH) times 4/3 for carriers employing Rate 3/4 FEC. The allocated bandwidth will be equal to 0.7 times the transmission rate, rounded up to the next highest odd integer multiple of 22.5 kHz increment (for information rates less than or equal to 10 Mbit/s) or 125 kHz increment (for information rates greater than 10 Mbit/s). Rate 3/4 FEC is mandatory for all IDR carriers. OH = overhead.

Source: IESS-308, Rev. 7, Ref. 5. Courtesy of INTELSAT.

The occupied satellite bandwidth unit for IDR carriers is approximately equal to 0.6 times the transmission rate. The transmission rate is defined as the coded symbol rate. To provide guardbands between adjacent carriers on the same transponder, the nominal satellite bandwidth unit is 0.7 times the transmission rate.

9.3.8 Very Small Aperture Terminal (VSAT) Networks

9.3.8.1 Rationale. VSATs are defined by their antenna aperture (diameter of the parabolic dish), which can vary from 0.5 m (1.6 ft) to 2.5 m (8.125 ft). Such apertures are considerably smaller than conventional earth stations. A VSAT network consists of one comparatively large hub earth terminal and remote VSAT terminals. Some networks in the United States have more than 5000 outlying VSAT terminals (a large drugstore chain). Many such networks exist.

There are three underlying reasons for the use of VSAT networks:

1. An economic alternative to establish a data network, particularly if traffic flow is to/from a central facility, usually a corporate headquarters to/from outlying remotes;
2. To by-pass telephone companies with a completely private network; and
3. To provide quality telecommunication connectivity where other means are sub-standard or nonexistent.

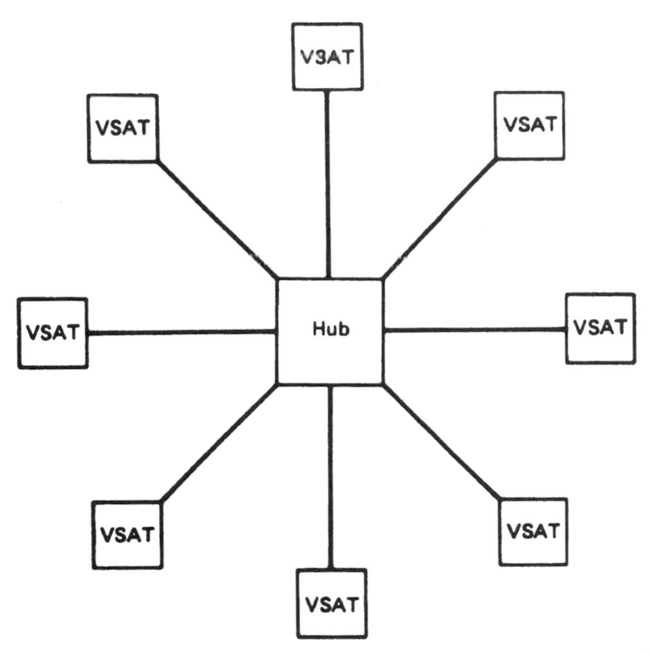

Figure 9.21 Typical VSAT network topology. Note the star network configuration. The outlying VSAT terminals can number in the thousands.

Regarding reason 3, the author is aware of one emerging nation where 124 bank branches had no electrical communication whatsoever with the headquarters institution in the capital city.

9.3.8.2 Characteristics of Typical VSAT Networks. On conventional VSAT networks, the hub is designed to compensate for the VSAT handicap (i.e., its small size). For example, a hub antenna aperture is 5 m to 11 m (16 ft to 50 ft) (Ref. 12). High-power amplifiers (HPAs) run from 100 W to 600 W output power. Low-noise amplifiers (LNAs), typically at 12 GHz, display (a) noise figures from 0.5 dB to 1.0 dB and (b) low-noise downconverters in the range of 1.5-dB noise figure. Hub G/T values range from +29 dB/K to +34 dB/K.

VSAT terminals have transmitter output powers ranging from 1 W to 50 W, depending on service characteristics. Receiver noise performance using a low-noise downconverter is about 1.5 dB; otherwise 1 dB with an LNA. G/T values for 12.5-GHz downlinks are between +14 dB/K and +22 dB/K, depending greatly on antenna aperture. The idea is to make a VSAT terminal as inexpensive as possible. Figure 9.21 illustrates the conventional hub/VSAT concept of a star network. The hub is at the center.

9.3.8.3 Access Techniques. *Inbound* refers to traffic from VSAT(s) to hub, and *outbound* refers to traffic from hub to VSAT(s). The outbound link is commonly a time-division multiplex (TDM) serial bit stream, often 56 kbps, and some high-capacity systems reach 1.544 Mbps or 2.048 Mbps. The inbound links can take on any one of a number of flavors, typically 9600 bps.

More frequently VSAT systems support interactive data transactions, which are very short in duration. Thus, we can expect bursty operation from a remote VSAT terminal. One application is to deliver, in near real time, point-of-sale (POS) information, forwarding it to headquarters where the VSAT hub is located. Efficiency of bandwidth use is not a primary motivating factor in system design. Thus for the interactive VSAT data network environment, low delay, simplicity of implementation, and robust operation are generally of greater importance than the bandwidth efficiency achieved.

Message access on any shared system can be of three types: (1) fixed assigned, (2) contention (random access), or (3) reservation (controlled access). There are hybrid schemes between contention and reservation.

In the fixed assigned multiple access, VSAT protocols are SCPC/FDMA, CDMA (a spread spectrum technique), and TDMA.[17] All three are comparatively inefficient in the bursty environment with hundreds of thousands of potential users.

9.4 FIBER OPTIC COMMUNICATION LINKS

9.4.1 Applications

Fiber optics as a transmission medium has a comparatively unlimited bandwidth. It has excellent attenuation properties, as low as 0.25 dB/km. A major advantage fiber has when compared with coaxial cable is that no equalization is necessary. Also, repeater separation is on the order of 10–100 times that of coaxial cable for equal transmission bandwidths. Other advantages are:

- Electromagnetic immunity;
- Ground loop elimination;
- Security;
- Small size and lightweight;
- Expansion capabilities requiring change out of electronics only, in most cases; and
- No licensing required.

Fiber has analog transmission application, particularly for video/TV. However, for this discussion we will be considering only digital applications, principally as a PCM highway or "bearer."

Fiber optic transmission is used for links under 1 ft in length all the way up to and including transoceanic undersea cable. In fact, all transoceanic cables currently being installed and planned for the future are based on fiber optics.

Fiber optic technology was developed by physicists and, following the convention of optics, wavelength rather than frequency is used to denote the position of light emission in the electromagnetic spectrum. The fiber optics of today uses three wavelength bands: (1) around 800 nm (nanometers), (2) 1300 nm, and (3) 1600 nm or near-visible infrared. This is illustrated in Figure 9.22.

This section includes an overview of how fiber optic links work, including the more common types of fiber employed, a discussion of light sources (transmitters), light detectors (receivers), optical amplifiers, and how lengths of fiber optic cable are joined. There is also a brief discussion of wavelength-division multiplexing (WDM).

[17]SCPC stands for single channel per carrier.

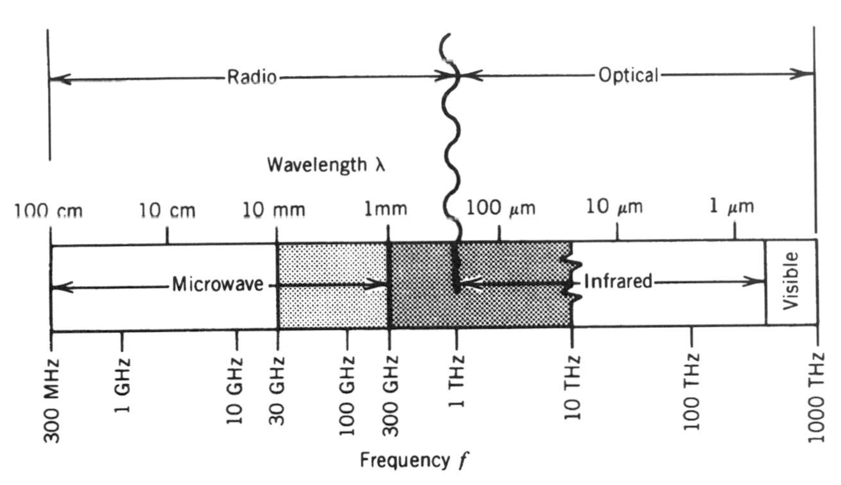

Figure 9.22 Frequency spectrum above 300 MHz. The usable wavelengths are just above and below 1 μm (1×10^{-6} meters).

9.4.2 Introduction to Optical Fiber as a Transmission Medium

Optical fiber consists of a core and a cladding, as illustrated in Figure 9.23. At present the most efficient core material is silica (SiO_2). The *cladding* is a dielectric material that surrounds the core of an optical fiber.

The practical propagation of light through an optical fiber may best be explained using ray theory and Snell's law. Simply stated, we can say that when light passes from a medium of higher refractive index (n_1) into a medium of lower refractive index (n_2), the refractive ray is bent away from the normal. For instance, a ray traveling in water and passing into an air region is bent away from the normal to the interface between the two regions. As the angle of incidence becomes more oblique, the refracted ray is bent more until finally the refracted energy emerges at an angle of 90° with respect to the normal and just grazes the surface. Figure 9.24 shows various incident angles of light entering a fiber. Figure 9.24*b* illustrates what is called the *critical angle*, where the refracted ray just grazes the surface. Figure 9.24*c* is an example of total internal reflection. This occurs when the angle of incidence exceeds the critical angle. A glass fiber, for the effective transmission of light, requires total internal reflection.

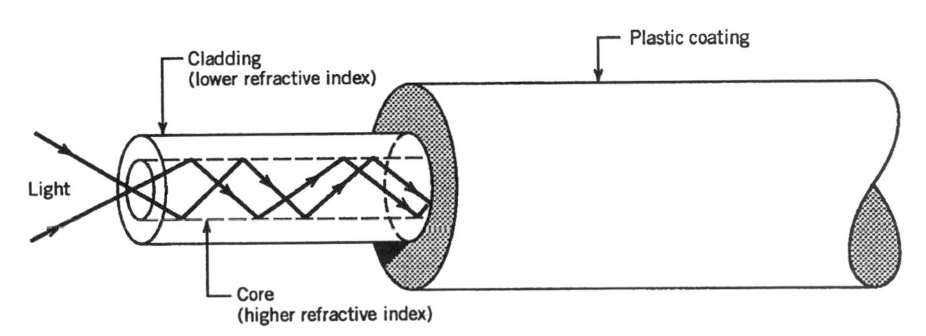

Figure 9.23 Structure of optical fiber consisting of a central core and a peripheral transparent cladding surrounded by protective packaging.

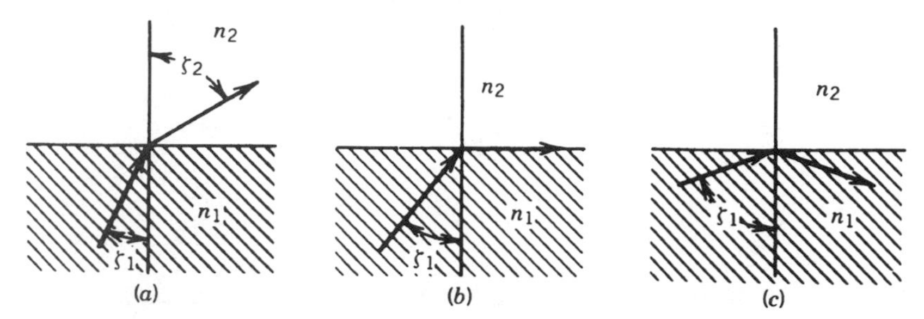

Figure 9.24 Ray paths for several angles of incidence ($n_1 > n_2$).

Figure 9.25 illustrates a model of a fiber optic link. Besides the supporting electrical circuitry, it shows the three basic elements in an optical fiber transmission system: (1) the optical source, (2) the fiber link, and (3) optical detector. Regarding the fiber optic link itself, there are two basic impairments that limit the length of such a link without resorting to repeaters or that can limit the distance between repeaters. These impairments are loss (attenuation), usually expressed in decibels per kilometer, and dispersion, usually expressed as bandwidth per unit length, such as megahertz per kilometer. A particular fiber optic link may be *power-limited* or *dispersion-limited.*

Dispersion, manifesting itself in intersymbol interference at the receive end, can be brought about by several factors. There is material dispersion, modal dispersion, and chromatic dispersion. Material dispersion can manifest itself when the emission spectral line is very broad, such as with a light-emitting diode (LED) optical source. Certain frequencies inside the emission line travel faster than others, causing some transmitted energy from a pulse to arrive later than other energy. This causes intersymbol interference. Modal dispersion occurs when several different modes are launched. Some have more reflections inside the fiber than other modes, thus, again, causing some energy from the higher-order modes to be delayed, compared with lower-order modes.

Let us examine the effect of dispersion on a train of pulses arriving at a light detector. Essentially, the light is "on" for a binary 1 and "off" for a binary 0. As shown in Figure 9.26, the delayed energy from bit position 1 falls into bit position 2 (and possibly 3,

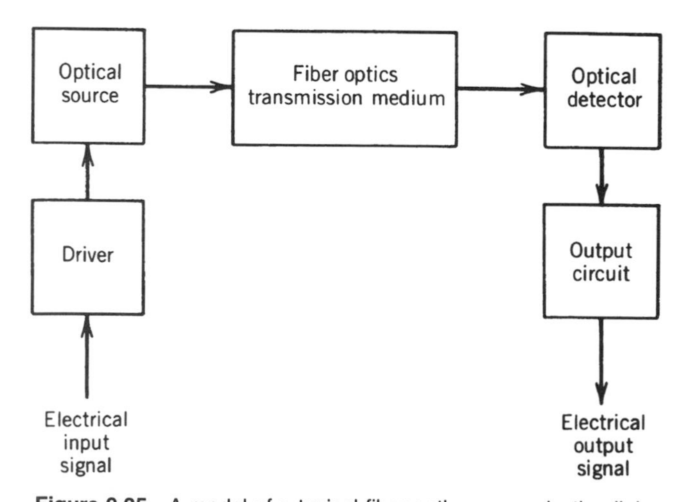

Figure 9.25 A model of a typical fiber optic communication link.

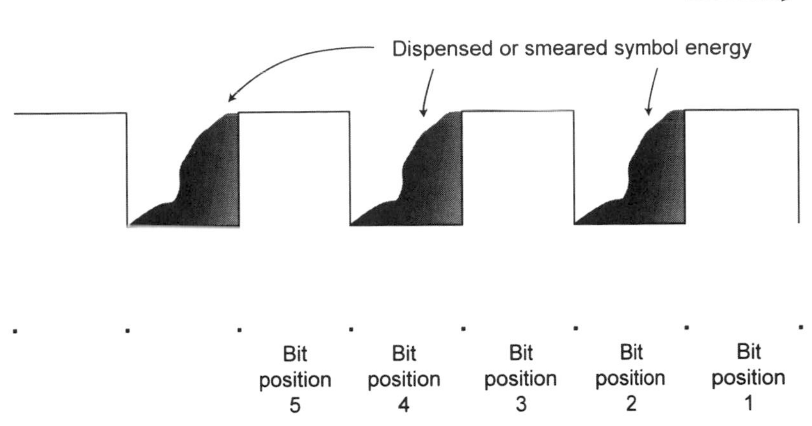

Figure 9.26 A simplified sketch of delayed symbol (bit) energy of bit 1 spilling into bit position 2. Alternating 1s and 0s are shown. It should be noted that as the bit rate increases, the bit duration (period) decreases, exacerbating the situation.

4, etc.) confusing the decision circuit. Likewise, delayed energy from bit position 2 falls into bit position 3 (possibly 4, 5, etc.), and so on. This is aptly called *intersymbol interference* (ISI), which was previously introduced.

One way we can limit the number of modes propagated down a fiber is to make the fiber diameter very small. This is called *monomode fiber*, whereas the larger fibers are called multimode fibers. For higher bit rate (e.g., > 622 Mbps), long-distance fiber optic links, the use of monomode fiber is mandatory. This, coupled with the employment of the longer wavelengths (e.g., 1330 nm and 1550 nm), allows us to successfully transmit bit rates greater than 622 Mbps, and with certain care the new 10-Gbps rate can be accommodated.

9.4.3 Types of Optical Fiber

There are three categories of optical fiber, as distinguished by their modal and physical properties:

1. Step index (multimode);
2. Graded index (multimode); and
3. Single mode (also called monomode).

Step-index fiber is characterized by an abrupt change in refractive index, and graded index is characterized by a continuous and smooth change in refractive index (i.e., from n_1 to n_2). Figure 9.27 shows the fiber construction and refractive index profile for step-index fiber (Fig. 9.27*a*) and graded-index fiber (Fig. 9.27*b*). Both step-index and graded-index light transmission are characterized as *multimode* because more than one mode propagates. (Two modes are shown in the figure.) Graded-index fiber has a superior bandwidth-distance produce compared to that of step-index fiber. In other words, it can transport a higher bit rate further than step index. It is also more expensive. We can eliminate this cause of dispersion if we use single mode fiber. Figure 9.28 shows a typical five-fiber cable for direct burial.

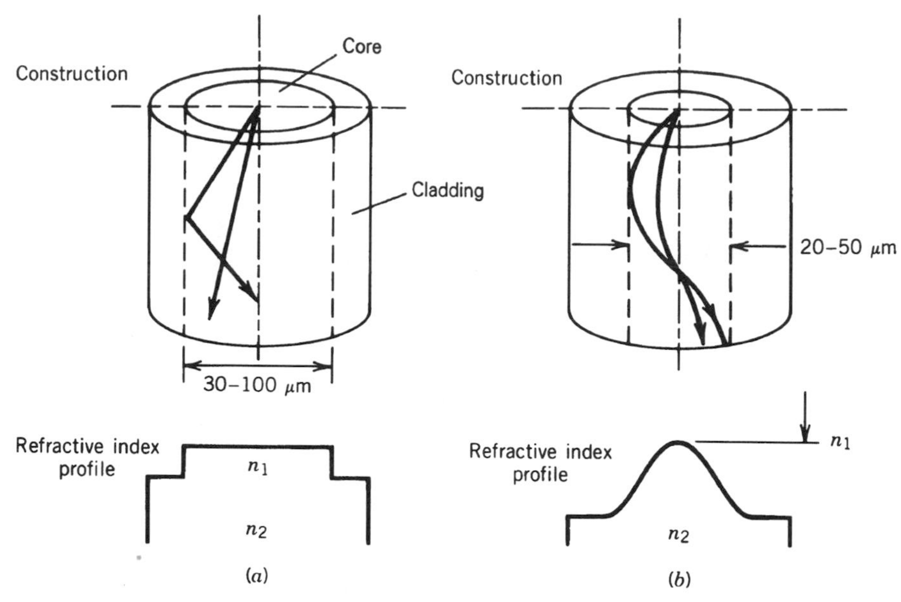

Figure 9.27 Construction and refractive index properties for (*a*) step-index fiber and (*b*) graded-index fiber.

9.4.4 Splices and Connectors

Optical fiber cable is commonly available in 1-km sections; it is also available in longer sections, in some types up to 10 km or more. In any case there must be some way of connecting the fiber to the source and to the detector as well as connecting the reels of cable together, whether in 1 km or more lengths, as required. There are two methods

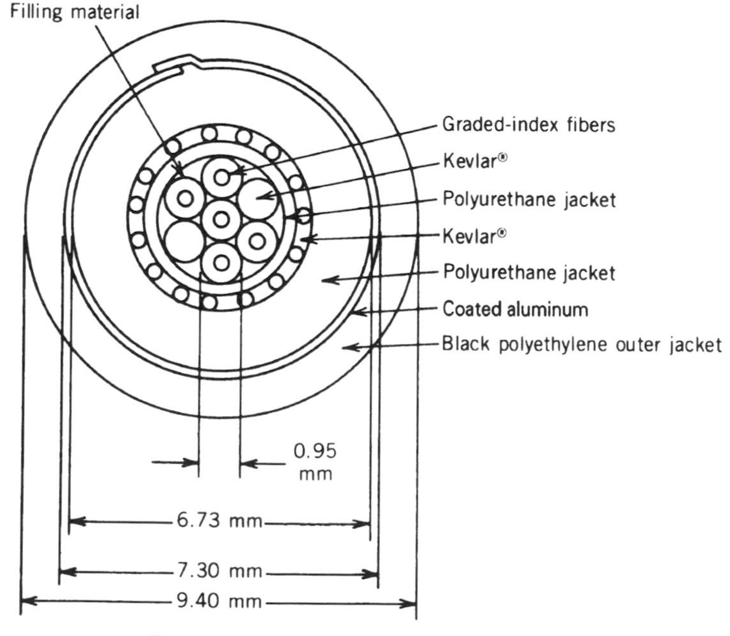

Figure 9.28 Direct burial optical fiber cable.

of connection, namely, splicing or using connectors. The objective in either case is to transfer as much light as possible through the coupling. A good splice couples more light than the best connectors. A good splice can have an insertion loss as low as 0.09 dB, whereas the best connector loss can be as low as 0.3 dB. An optical fiber splice requires highly accurate alignment and an excellent end finish to the fibers. There are three causes of loss at a splice:

1. Lateral displacement of fiber axes;
2. Fiber end separation; and
3. Angular misalignment.

There are two types of splice now available, the mechanical splice and the fusion splice. With a mechanical splice an optical matching substance is used to reduce splicing losses. The matching substance must have a refractive index close to the index of the fiber core. A cement with similar properties is also used, serving the dual purpose of refractive index matching and fiber bonding. The fusion splice, also called a *hot splice*, is where the fibers are fused together. The fibers to be spliced are butted together and heated with a flame or electric arc until softening and fusion occur.

Splices require special splicing equipment and trained technicians. Thus it can be seen that splices are generally hard to handle in a field environment such as a cable manhole. Connectors are much more amenable to field connecting. However, connectors are lossier and can be expensive. Repeated mating of a connector may also be a problem, particularly if dirt or dust deposits occur in the area where the fiber mating takes place.

However, it should be pointed out that splicing equipment is becoming more economic, more foolproof, and more user-friendly. Technician training is also becoming less of a burden.

Connectors are nearly universally used at the source and at the detector to connect the main fiber to these units. This makes easier change-out of the detector and source when they fail or have degraded operation.

9.4.5 Light Sources

A light source, perhaps more properly called a *photon source*, has the fundamental function in a fiber optic communication system to efficiently convert electrical energy (current) into optical energy (light) in a manner that permits the light output to be effectively launched into the optical fiber. The light signal so generated must also accurately track the input electrical signal so that noise and distortion are minimized.

The two most widely used light sources for fiber optic communication systems are the light-emitting diode (LED) and the semiconductor laser, sometimes called a *laser diode* (LD). LEDs and LDs are fabricated from the same basic semiconductor compounds and have similar heterojunction structures. They do differ in the way they emit light and in their performance characteristics.

An LED is a forward-biased *p–n* junction that emits light through spontaneous emission, a phenomenon referred to as *electroluminescence*. LDs emit light through stimulated emission. LEDs are less efficient than LDs but are considerably more economical. They also have a longer operational life. The emitted light of an LED is incoherent with a relatively wide spectral line width (from 30 nm to 60 nm) and a relatively large angular spread, about 100°. On the other hand, a semiconductor laser emits a comparatively narrow line width (from <2 nm to 4 nm). Figure 9.29*a* shows the spectral line for an

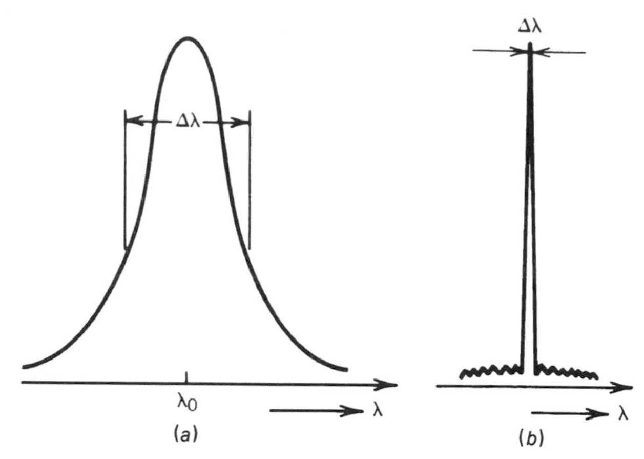

Figure 9.29 Spectral distribution (line width) of the emission from (*a*) an LED and (*b*) a semiconductor laser (LD), where λ is the optical wavelength and Δλ, where is the spectral or line width.

LED, and Figure 9.29*b* shows the spectral line for a semiconductor laser (i.e., a laser diode or LD).

What is a *spectral line*? Many of us imagine that if we were to view a radio carrier (without modulation) on an oscilloscope, it would be a vertical line that appeared to be of infinitely narrow width. This thinking tends to be carried into the world of light in a fiber optic light-guide. In neither case is this exactly true. The emission line or light carrier has a finite width, as does a radio carrier. The IEEE (Ref. 6) defines *spectral width, full-width half maximum* as "The absolute difference between the wavelengths at which the spectral radiant intensity is 50% of the maximum."

With present technology the LED is capable of launching about 100 μW (−10 dBm) or less of optical power into the core of a fiber with a numerical aperture of 0.2 or better. A semiconductor laser with the same input power can couple up to 7 mW (+8.5 dBm) into the same cable. The coupling efficiency of an LED is on the order of 2%, whereas the coupling efficiency of an LD (semiconductor laser) is better than 50%.

Methods of coupling a source into an optical fiber vary, as do coupling efficiencies. To avoid ambiguous specifications on source output powers, such powers should be stated at the *pigtail.* A pigtail is a short piece of optical fiber coupled to the source at the factory and, as such, is an integral part of the source. Of course, the pigtail should be the same type of fiber as that specified for the link.

LED lifetimes are in the order of 100,000 hr mean time between failures (MTBF) with up to a million hours reported in the literature. Many manufacturers guarantee a semiconductor laser for 20,000 hr or more. About 150,000 hours can be expected from semiconductor lasers after stressing and culling of unstable units. Such semiconductors are used in the latest TAT and PTAT series of undersea cables connecting North America and Europe.

Current fiber optic communication systems operate in the nominal wavelength regions of 820 nm, 1330 nm, and 1550 nm. Figure 9.30 is a plot of attenuation per unit length versus wavelength. Based on this curve we can expect about 3 dB per kilometer at 820 nm, 0.50 dB per kilometer at 1330 nm, and at 1550 nm some 0.25 dB per kilometer of attenuation. Also take note that at about 1300 nm is a region of zero dispersion. For added expense, fiber is available with the dispersion minimum shifted to the 1550-nm band, where attenuation per unit length is minimum. There is mature technology at all three wavelengths.

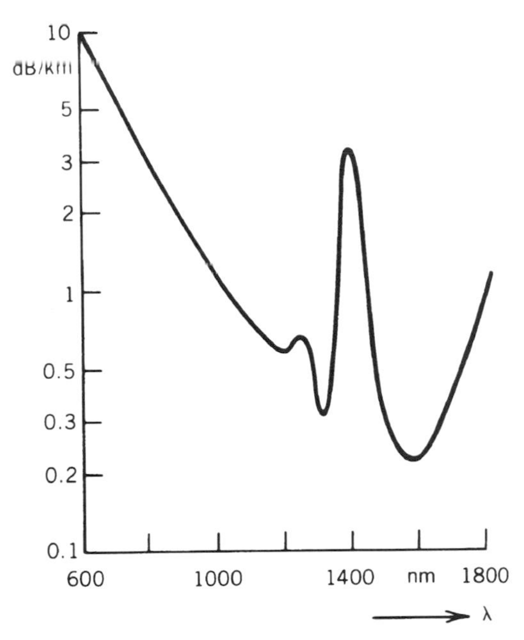

Figure 9.30 Attenuation per unit length versus wavelength of glass fiber (Ref. 7).

9.4.6 Light Detectors

The most commonly used detectors (receivers) for fiber optic communication systems are photodiodes, either PIN or APD. The terminology *PIN* derives from the semiconductor construction of the device where an intrinsic (I) material is used between the *p–n* junction of the diode.

Another type of detector is avalanche photodiode (APD), which is a gain device displaying gains on the order of 15 dB to 20 dB. The PIN diode is not a gain device. Table 9.7 summarizes various detector sensitivities with a "standard" BER of 1×10^{-9} for some common bit rates.[18]

Of the two types of photodiodes discussed here, the PIN detector is more economical and requires less complex circuitry that does its APD counterpart. The PIN diode has peak sensitivity from about 800 nm to 900 nm for silicon devices.

The overall response time for the PIN diode is good for about 90% of the transient but sluggish for the remaining 10%, which is a "tail." The power response of the tail portion of a pulse may limit the net bit rate on digital systems.

The PIN detector does not display gain, whereas the APD does. The response time of the APD is far better than that of the PIN diode, but the APD displays certain temperature instabilities where responsivity can change significantly with temperature. Compensation for temperature is usually required in APD detectors and is often accomplished by a feedback control of bias voltage. It should be noted that bias voltages for APDs are much higher than for PIN diodes, and some APDs require bias voltages as high as 200 V. Both the temperature problem and the high-voltage bias supply complicate repeater design.

[18]Some entities such as SPRINT (US) specify 1×10^{-12}.

Table 9.7 Summary of Receiver Diode Sensitivities, Average Received Optical Power (in dBm, BER = 1×10^{-9})

Bit Rate (Mbps)	InGaAs PIN		InGaAs SAM/SAGM[a] APD		Ge APD		InGaAs Photoconductor	
	1.3 μm	1.55 μm	1.3 μm	1.55 μ	1.3 μm	1.55 μm	1.3 μm	1.55 μm
34	−52.5				−46	−55.8		
45	−49.9		−51.7		−51.9			
100					−40.5			
140	−46				−45.2	−49.3		
274	−43		−45	−38.7		−36		
320	−43.5							
420			−43	−41.5				
450			−42.5		−39.5	−40.5		
565					−33			
650	−36							
1000			−38	−37.5	−28		−34.4	
1200	−33.2	−36.5						
1800			−31.3		−30.1			
2000				−36.6		−31		−28.8
4000				−32.6				

[a]SAGM = separated absorption, grating, and mulitplication regions. SAM = separated absorption and multiplication regions.

Source: Ref. 8.

9.4.7 Optical Fiber Amplifiers

Optical amplifiers amplify incident light through stimulated emission, the same mechanism as used with lasers. These amplifiers are the same as lasers without feedback. Optical gain is achieved when the amplifier is pumped either electrically or optically to realize population inversion.

There are semiconductor laser amplifiers, Raman amplifiers, Brillouin amplifiers, and erbium-doped fiber amplifiers (EDFAs). Certainly the EDFAs show the widest acceptance. One reason is that they operate near the 1.55-μm wavelength region, where fiber loss is at a minimum. Reference 11 states that it is possible to achieve high amplifier gains in the range of 30 dB to 40 dB with only a few milliwatts of pump power when EDFAs are pumped by using 0.980-μm or 1.480-μm semiconductor lasers. Figure 9.31 is a block diagram of a low-noise EDFA.

In Figure 9.31, optical pumping is provided by fiber-pigtailed semiconductor lasers with typically 100 mW of power. Low-loss wavelength division multiplexers efficiently combine pump and signal powers and can also be used to provide a pump by-pass around the internal isolator. The EDFA has an input stage that is codirectionally pumped and an output stage that is counterdirectionally pumped. Such multistage EDFA designs have simultaneously achieved a low noise figure of 3.1 dB and a high gain of 54 dB (Refs. 10 and 11).

The loops of fiber should be noted in Figure 9.31. These are lengths of fiber with a dopant. The length of erbium-doped fiber required for a particular amplifier application depends on the available pump power, doping concentration, the design topology, and gain and noise requirements.

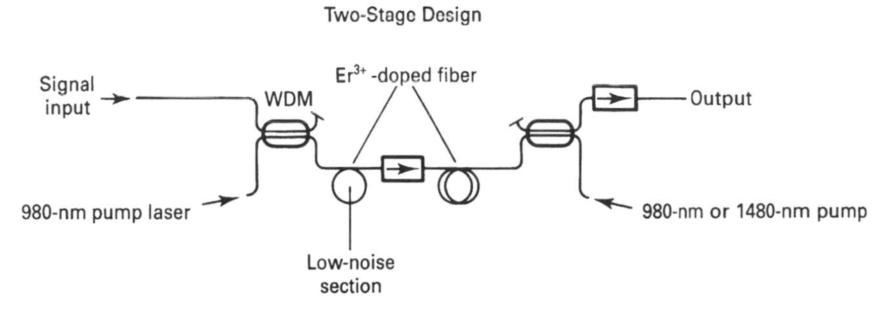

Figure 9.31 An EDFA block diagram (From Ref. 9. Courtesy of Hewlett-Packard.)

EDFAs are often installed directly after a semiconductor laser source (transmitter) and/or directly before the PIN or APD receiver at the distant end. Figure 9.32 illustrates this concept. With the implementation of EDFAs, the length of a fiber-optic link can be extended without repeaters and additional 100 km to 250 km, or it can extend the distance between repeaters a similar amount.

9.4.8 Wavelength Division Multiplexing

Wavelength division multiplexing (WDM) is just another name for frequency division multiplexing (FDM). In this section we will hold with the conventional term, WDM. The WDM concept is illustrated in Figure 9.33. WDM can multiply the transmission capacity of an optical fiber many-fold. For example, if we have a single fiber carrying 2.4 Gbps, and it is converted for WDM operation with 10 WDM channels, where each of these channels carry 2.4 Gbps, the aggregate capacity of this single fiber is now 2.4 $\times$ 10, or 24 Gbps.

The 1550-nm band is the most attractive for WDM applications because the aggregate wavelengths of a WDM signal can be amplified with its total assemblage together by a single EDFA. This is a great advantage. On the other hand, if we wish to regenerate the digital derivate of a WDM signal in a repeater, the aggregate must be broken down into its components, as shown in Figure 9.34.

The cogent question is: how many individual wavelength signals can we multiplex on a single fiber? It depends on the multiplexing/demultiplexing approach employed, the number of EDFAs that are in tandem, the available bandwidth, channel separation (in nm), and the impairments peculiar to light systems and the WDM techniques used.

As mentioned, the number of WDM channels that can be accommodated depends on the bandwidth available and the channel spacing among other parameters and characteristics. For example, in the 1330-nm band with about 80 nm bandwidth avail-

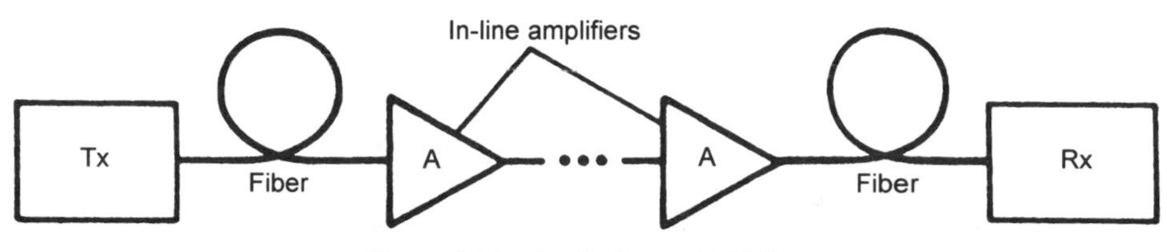

Figure 9.32 Applications of EDFAs.

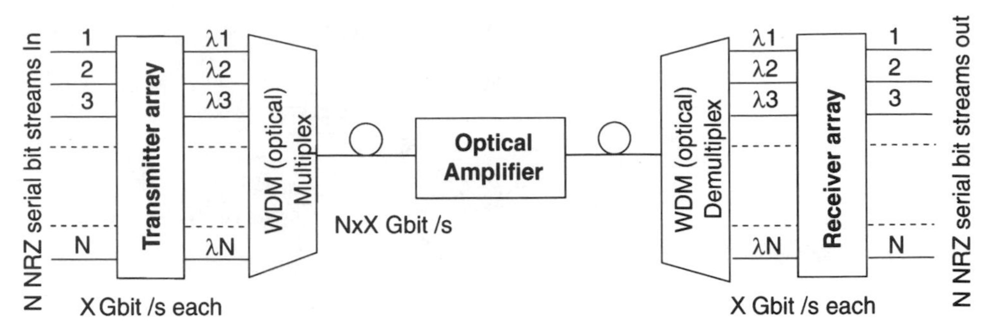

Figure 9.33 The WDM concept.

able, and in the 1550-nm band, with about 100 nm bandwidth available, which includes the operational bandwidth and guardbands, the capacity on a single fiber would be 40 WDM and 50 WDM channels, respectively, on each fiber.[19] Then each fiber could transport an aggregate of 100 Gbps and 120 Gbps, respectively, for each band. At present (1998), 16-channel systems are available, off-the-shelf.

9.4.9 Fiber Optic Link Design

The design of a fiber optic communication link involves several steps. Certainly the first consideration is to determine the feasibility of such a transmission system for a desired application. There are two aspects of this decision: (1) economic and (2) technical. Can we get equal or better performance for less money using some other transmission medium such as wire pair, coaxial cable, LOS microwave, and so on?

Fiber optic communication links have wide application. Analog applications for cable television (CATV) trunks are showing particularly rapid growth. Fiber is also used for low-level signal transmission in radio systems, such as for long runs of IF, and even for RF. However, in this text we stress digital applications, some of which are listed below:

- On-premises data bus;
- LANs (e.g., fiber distributed data interface);

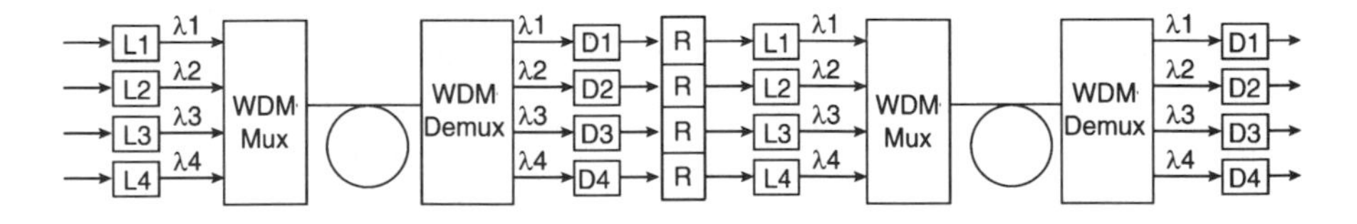

L = laser diode   R = Regenerator
D = PIN detector

Figure 9.34 The use of regenerative repeaters on a WDM link.

[19]Guardbands are empty spaces to provide isolation between adjacent channels. This helps minimize interference from one channel to the next. Of course, that bandwidth allocated to guardbands is nonrevenue-bearing, and thus must be minimized.

- High-level PCM or CVSD configurations;[20] SONET and SDH;
- Radar data links;
- Conventional data links where bit/symbol rates exceed 19.2 kbps; and
- Digital video including cable television.

It seems that the present trend of cost erosion will continue for fiber cable and components. Fiber optic repeaters are considerably more expensive than their PCM metallic counterparts. The powering of the repeaters can be more involved, particularly if power is to be taken from the cable itself. This means that the cable must have a metallic element to supply power to downstream repeaters, thereby losing a fiber optic advantage. Metal in the cable, particularly for supplying power, can be a conductive path for ground loops. Another approach is to supply power locally to repeaters with a floating battery backup.

A key advantage to fiber over metallic cable is *fewer repeaters* per unit length. In Chapter 6 we showed that repeaters in tandem are the principal cause of jitter, a major impairment to a digital system such as PCM. Reducing the number of repeaters reduces jitter accordingly. In fact, fiber optic systems require a small fraction of the number of repeaters, compared with PCM for the same unit length, either on wire-pair or coaxial cable.

9.4.9.1 Design Approach. The first step in the design of a fiber optic communication system is to establish the basic system parameters. Among these we would wish to know:

- Signal to be transmitted: digital or analog, video/CATV; bit rate and format;
- System length, fiber portion end-to-end;
- Growth requirements (additional circuits, increase of bit rates);
- Availability/survivability requirements; and
- Tolerable signal impairment level, stated as signal-to-noise ratio on BER at the output of the terminal-end detector.

The link BER should be established based on the end-to-end BER. In the past we had used 1×10^{-9} as a link BER. Bellcore now requires 2×10^{-10} end-to-end. One link value might be 1×10^{-12}.

However, there is still another saving factor or two. If we have operation at the zero-dispersion wavelength, about 1300 nm, dispersion may really not be a concern until about the 1-Gbps rate. Of course, at 1300 nm we have lost the use of the really low loss band at around 1550 nm. There is an answer to that, too. Use a fiber where the minimum dispersion window has been shifted to the 1550-nm region. Such fiber, of course, is more costly, but the cost may be worth it. It is another trade-off.

The designer must select the most economic alternatives among the following factors:

- Fiber parameters: single mode or multimode; if multimode, step index or graded index; number of fibers, cable makeup, strength;
- Transmission wavelength: 820 nm, 1330 nm, or 1550 nm;
- Source type: LED or semiconductor laser; there are subsets to each source type;

[20]CVSD stands for continuous variable slope delta (modulation), a form of digital modulation where the coding is 1 bit at a time. It is very popular with the armed forces.

- Detector type: PIN or APD;
- Use of EDFA (amplifiers);
- Repeaters, if required, and how they will be powered; and
- Modulation: probably intensity modulation (IM), but the electrical waveform entering the source is important; possibly consider Manchester coding.

9.4.9.2 Loss Design. As a first step, assume that the system is power limited. This means that are principal concern is loss. Probably a large number of systems being installed today can stay in the power-limited regime if monomode fiber is used with semiconductor lasers (i.e., LDs or laser diodes). When designing systems for bit rates in excess of 600 Mbps to 1000 Mbps, consider using semiconductor lasers with very narrow line widths (see Figure 9.29) and dispersion shifted monomode fiber. Remember that "dispersion shifting" moves the zero dispersion window from 1300 nm to 1550 nm.

For a system operating at these high bit rates, even with the attribute of monomode fiber, chromatic dispersion can become a problem, particularly at the desirable 1550-nm band. Chromatic dispersion is really a form of material dispersion described earlier. It is the sum of two effects: "material dispersion" and waveguide dispersion. As one would expect, with material dispersion, different wavelengths travel at different velocities of propagation. This is true even with the narrow line width of semiconductor lasers. *Waveguide dispersion* is a result of light waves traveling through single-mode fibers that extend into the cladding. Its effect is more pronounced at the longer wavelengths because there is more penetration of the cladding and the "effective" refractive index is reduced. This causes another wavelength dependence on the velocity of light through the fiber, and therefore another form of dispersion. Thus the use of semiconductor lasers with very narrow line widths (e.g., <0.5 nm) helps mitigate chromatic dispersion (Ref. 11).

Link margin is another factor for trade-off. We set this dB value aside in reserve for the following contingencies:

- Cable reel loss variability;
- Future added splices (due to cable repair) and their insertion loss; and
- Component degradation over the life of the system. This is particularly pronounced for LED output.

CCITT recommends 3 dB for link margin; others (Ref. 9) recommend 6 dB. Ideally, for system reliability, a large margin is desirable. To optimize system first cost, as low a value as possible would be desired.

The system designer develops a power budget, similar in many respects to the path analysis or link budget of LOS microwave and satellite communication link design. However, there is little variability in a fiber optic link budget; for example, there is no fading.

For a first-cut design, there are two source types, LED and semiconductor laser. Expect a power output of an LED in the range of −10 dBm; and for the semiconductor laser budget 0 dBm, although up to nearly +10 dBm is possible. There are two types of detectors, PIN and APD. For long links with high bit rates, the APD may become the choice. We would expect that the longer wavelengths would be used, but 820/850-nm links are still being installed. We must not forget reliability in our equation for choices. For lower bit rates and shorter links, we would give LEDs a hard look. They are cheaper and are much more reliable (MTBF).

Example Link Budget Exercise. The desired bit rate is 140 Mbps. What will be the maximum distance achievable without the use of repeaters? The detector is a PIN type. Turn to Table 9.6 and determine that the threshold (dBm) for a BER of 1×10^{-9} is −46 dBm at 1.3 μm. One EDFA is used with a gain of 40 dB. This now becomes the starting point for the link budget.

The light source is a laser diode with −0.3 dBm output. The receiver threshold is −46 dBm, leaving 45.7 dB in the power budget. Add to this value the EDFA gain of 40 dB, bringing the power budget up to 85.7 dB. We allocate this value as follows:

- Fiber at 0.25 dB/km;
- Two connectors at 0.5 dB each or a total of 1.0 dB;[21]
- Fusion splices every kilometer; allows 0.25 dB per splice; and
- A margin of 4 dB.

If we subtract the 1 dB for the connectors and the 4-dB margin from the 85.7 dB, we are left with 80.7 dB. Add the splice loss and the kilometer fiber loss for 1-km reels, the result is 0.5 dB. Divide this value into 80.7 dB, and the maximum length is 160 km between a terminal and first repeater or between repeaters. Of course, there is one less splice than 1-km lengths, plus the 0.45 dB left over from the 80.7 dB. This will be additional margin. Setting up these calculations in tabular form gives the following:

ITEM	LOSS
Connector loss @ 0.5 dB/conn, 2 connectors	1.0 dB
Margin	4.0 dB
Splice losses @ 0.25 dB/splice, 160 splices	40.0 dB
Fiber loss, 161-km fiber @ 0.25 dB/km	40.25 dB
Total	85.25 dB
Additional margin	0.45 dB

For an analysis of dispersion and system bandwidth, consult Ref. 7.

9.5 COAXIAL CABLE TRANSMISSION SYSTEMS

9.5.1 Introduction

The employment of coaxial cableused inthe telecommunication plant is now practically obsolete, with the following exceptions:

- The last mile or last 100 feet in the cable television (CATV) plant; and
- As an RF transmission for short distances.

It is being replaced in the enterprise network with high-quality twisted pair and fiber optic connectivities. Certainly in the long-distance network, the fiber optic solution is far superior in nearly every respect. In the next section we provide a brief review of coaxial cable systems.

[21]These connectors are used at the output pigtail of the source and at the input pigtail to the detector. Connectors are used for rapid and easy disconnect/connect because, at times during the life of these active devices, they must be changed out for new ones, having either failed or reached the end of their useful life.

9.5.2 Description

A coaxial cable is simply a transmission line consisting of an unbalanced pair made up of an inner conductor surrounded by a grounded outer conductor, which is held in a concentric configuration by a dielectric.[22] The dielectric can be of many different types such as solid "poly" (polyethylene or polyvinyl chloride), foam, Spiralfil, air, or gas. In the case of an air/gas dielectric, the center conductor is kept in place by spacers or disks.

Historically, coaxial cable systems carried large FDM configurations, over 10,000 voice channels per "tube." CATV (community antenna television or cable television) systems use single cables for transmitted bandwidths in excess of 750 MHz. Coaxial cable systems competed with analog LOS microwave and often were favored because of reduced noise accumulation.

9.5.3 Cable Characteristics

When employed in the long-distance telecommunication plant, standard coaxial cable sizes are as follows:

DIMENSION (in)	DIMENSION (mm)
0.047/0.174	1.2/4.4 (small diameter)
0.104/0.375	2.6/9.5

The fractions express the outside diameter of the inner conductor over the inside diameter of the outer conductor. For instance, for the large bore cable, the outside diameter of the inner conductor is 0.104 in. and the inside diameter of the outer conductor is 0.375 in. This is illustrated in Figure 9.35. As can be seen from Eq. (9.27) in Figure 9.35, the ratio of the diameters of the inner and outer conductors has an important bearing on attenuation (loss). If we can achieve a ratio of $b/a = 3.6$, a minimum attenuation per unit length results.

For air dielectric cable pair, $\epsilon = 1.0$
Outside diameter of inner conductor, $= 2a$
Inside diameter of outer conductor, $= 2b$
Attenuation constant (dB)/mi

$$\alpha = 2.12 \times 10^{-5} \frac{\sqrt{f}\left(\frac{1}{a} + \frac{1}{b}\right)}{\log b/a} \tag{9.26}$$

where a = radius of inner conductor and b = radius of outer conductor.

Chararacteristic impedances (Ω)

$$Z = \left(\frac{138}{\sqrt{\epsilon}}\right) \log \frac{b}{a} = 138 \log \frac{b}{a} \text{ in air} \tag{9.27}$$

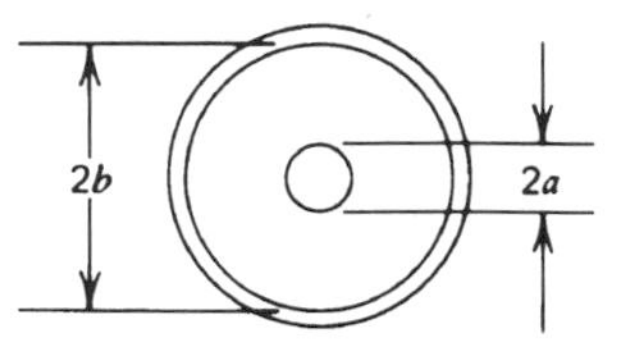

Figure 9.35 The basic electrical characteristics of coaxial cable.

[22] *Dielectric* means an insulator.

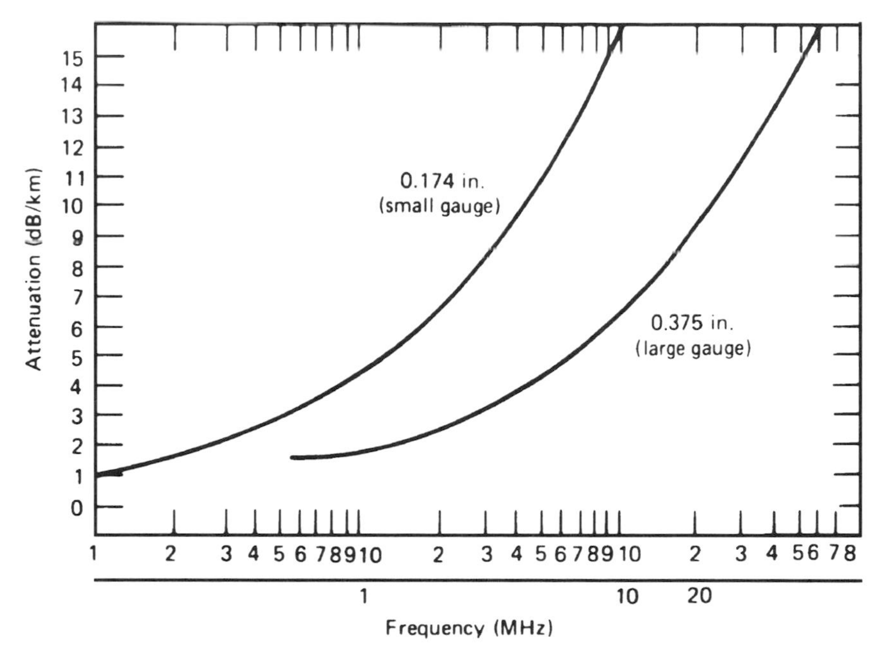

Figure 9.36 Attenuation-frequency response per kilometer of coaxial cable.

The characteristic impedance of coaxial cable is $Z_0 = 138 \log (b/a)$ for an air dielectric. If $b/a = 3.6$, then $Z_0 = 77\ \Omega$. Using dielectric other than air reduces the characteristic impedance. If we use the disks previously mentioned to support the center conductor, the impedance lowers to 75 Ω.

Figure 9.36 illustrates the attenuation-frequency characteristics of the coaxial cable discussed in the text. Attenuation increases rapidly as frequency is increased. It is a function of the square root of frequency, as shown in Figure 9.35. The telecommunication system designer is basically interested in how much bandwidth there is available to transmit a signal. For instance, the 0.375-in. cable has an attenuation of about 5.8 dB/mi at 2.5 MHz and the 0.174-in. cable, 12.8 dB/mi. At 5 MHz the 0.174-in. cable has about 19 dB/mi and the 0.375-in. cable, 10 dB/mi. Attenuation is specified for the highest frequency of interest.

Equalization (i.e., the use of equalizers) will tend to flatten the response curves in the figure at the expense of some added loss per unit length. *Equalization* is defined by the IEEE (Ref. 6) as "A technique used to modify the frequency response of an amplifier or network to compensate for variations in the frequency response across the network bandwidth." The ideal result is a flat overall response. The CATV plant makes wide use of such equalizers.

9.6 TRANSMISSION MEDIA SUMMARY

Table 9.8 presents a summary of the performance characteristics of the five basic transmission media.

Table 9.8 Characteristics of Transmission Media

Item	Wire Pair	LOS Microwave	Satellite Communications	Fiber Optics	Coaxial Cable
Bandwidth	2 MHz to 400 MHz	500/2500 MHz	500/2500 MHz	120 GHz per band	Up to 1 GHz
Commone bit rates	1.544/2.048 Mbps	155 Mpbs	2.048 Mbps	2.4/10 Gbps	100 Mbps
Achievable bit rates	100 Mbps	622 Mpbs	155 Mbps	200 Gbps	1 Gbps
Limitations	Length limited	By statute	By statute; delay	Severing cable	Serving cable
Applications	LANs, TelCo outside plant	Long-distance/short-distance links, TelCo and CATV, private networks	VSAT networks, long-distance links, video transport	For every broadband terrestrial application	CATV last mile/last 100 feet; RF transport short distances; otherwise limited

Notes: Wire pair is distance limited. The shorter the pair length, the higher the bit rate. Also, balance and accumulating capacitance with length affect bit rate.

LOS microwave is limited by statute, meaning by the ITU Radio Regulations and the national regulatory authority.

Satellite communication faces the same legal limitations. Geostationary orbit (GEO) satellites have long delays, which could affect interactive data systems. Only one GEO satellite relay allowed for a voice connectivity.

The limits of fiber optics are still being explored. All terrestrial buried and aerial cable systems are vulnerable to severing by natural disaster or by man.

Coaxial cable is limited by amplitude-frequency response characteristics. In nearly every instance, fiber optic cable connectivity is preferred.

REVIEW EXERCISES

1. For very-high-capacity transmission systems (e.g., >20,000 equivalent voice channel), what transmission medium should be selected?

2. What are the advantages of using the RF bands from 2 GHz to 10 GHz for trunk telephony/data? Name at least two.

3. Discuss the problem of delay in speech telephone circuits traversing a geostationary satellite. Will there be any problem with data and signaling circuits?

4. Give four of the five basic procedure steps in designing an LOS microwave link.

5. Where is earth curvature maximum on an LOS microwave path?

6. In a path profile, what are the three basic increment factors that are added to obstacle height?

7. When a K factor is 4/3, does the radio ray beam bend away or toward the earth?

8. Name at least four parameters that we will derive from a path analysis to design an LOS microwave link.

9. Calculate the free space loss of a radio link operating at 4100 MHz and 21 statute miles long.

10. What is the EIRP in dBW out of an LOS microwave antenna if the transmit power is 2 W, the transmission line losses are 3.5 dB, and the antenna gain is 36 dB?

11. A receiving system operating at room temperature has a 5 dB noise figure, and its bandwidth is 2000 kHz. What is its thermal noise threshold?

12. A receiver operating at room temperature in a digital LOS microwave link displays a noise figure of 2 dB. What is its N_0?

13. What theoretical bit packing can we achieve with QPSK? With 8-ary PSK?

14. What efficiency can we expect from an LOS microwave parabolic antenna bought off-the-shelf?

15. What is the cause of the most common form of fading encountered on an LOS microwave link?

16. What theoretical bit packing can be achieved from 64-QAM?

17. Why is *line-of-sight* something more than line-of-sight? Explain.

18. Based on the Rayleigh fade margin criterion, what fade margin will we need for a 99.975% time availability? For 99.9%?

19. There are two kinds of diversity that can be used with conventional LOS microwave. What are they? It is nearly impossible for one type to be licensed in the United States. Which one and why?

20. An LOS microwave link cannot meet performance requirements. What measures can we take to remedy this situation? List in ascending order of cost.

21. Satellite communications is just an extension of LOS microwave. Thus a satellite earth station must be within ____ of the satellite.

22. There are two basic generic methods of satellite access. What are they? List a third method which is a subset of one of them.

23. What are two major advantages of satellite TDMA?

24. Define G/T mathematically.

25. Why is the geostationary satellite downlink limited? Give two reasons.

26. Receiving system noise temperature has two components. What are they?

27. Name three applications for VSAT networks.

28. What sets a VSAT aside from conventional geostationary (fixed) satellite earth stations?

29. What sort of digital format might we expect on an outbound link for VSAT operation?

30. What is the great, overriding advantage of fiber optic communication links?

31. What are the two basic impairments that limit the length of a fiber optic link?

32. How does dispersion manifest itself on a digital bit stream?

33. Name the three basic components of a fiber optic link (in the light domain).

34. There are three wavelength bands currently in use on optical fiber networks. Identify the bands and give data on loss per unit distance.

35. What does a glass fiber consist of (as used for telecommunications)?

36. What are the two generic types of optical fiber?

37. Identify the two basic light sources. Compare.

38. What is a *pigtail*?

39. What are the two generic types of optical detectors? Give some idea of gain that can be achieved by each.

40. Where do we place fiber amplifiers (in most situations)?

REFERENCES

1. *Transport Systems Generic Requirements (TSGR): Common Requirements*, Bellcore GR-499-CORE, Issue 1, Bellcore, Piscataway, NJ, Dec. 1995.
2. *Allowable Bit Error Ratios at the Output of a Hypothetical Reference Digital Path for Radio-Relay Systems Which May Form Part of an Integrated Services Digital Network*, CCIR Rec. 594-3, 1994 F Series Volume, Part 1, ITU Geneva, 1994.
3. R. L. Freeman, *Radio System Design for Telecommunications*, 2nd ed., Wiley, New York, 1997.
4. *Performance Characteristics for Frequency Division Multiplex/Frequency Modulation (FDM/FM) Telephony Carriers*, INTELSAT IESS 301 (Rev. 3), INTELSAT, Washington, DC, May 1994.
5. *Performance Characteristics for Intermediate Data Rate (IDR) Digital Carriers*, INTELSAT IESS 308 with Rev. 7 and 7A, INTELSAT, Washington, DC, Aug. 1994.

6. *IEEE Standard Dictionary of Electrical and Electronic Terms*, 6th ed., IEEE Std. 100-1996, IEEE, New York, 1996.
7. R. L. Freeman, *Telecommunication Transmission Handbook*, 4th ed., Wiley, New York, 1998.
8. *Telecommunication Transmission Engineering*, 3rd ed., Vol. 2, Bellcore, Piscataway, NJ, 1991.
9. 1993 Lightwave Symposium, Hewlett-Packard, Burlington, MA, Mar. 23, 1993.
10. S. Shimada and H. Ishio, *Optical Amplifiers and Their Applications*, Wiley, Chichester (UK), 1992.
11. G. P. Agrawal, *Fiber-Optic Communication Systems*, Wiley, New York, 1992.
12. J. Everett, *VSATs: Very Small Aperture Terminals*, IEE/Peter Peregrinus, Stevenage, Herts, UK, 1992.

10

DATA COMMUNICATIONS

10.1 CHAPTER OBJECTIVE

Data communications is the fastest growing technology in the telecommunications arena. In the PSTN it is expected to equal or surpass voice communications in the next ten years. The widespread availability of the PC not only spurred data communications forward, but it also added a completely new direction, distributed processing. No longer are we tied to the mainframe computer; it has taken on almost a secondary role in the major scheme of things. Another major impetus in this direction is, of course, the Internet.

The IEEE (Ref. 1) defines *data communications* (data transmission) as "The movement of encoded information by means of communication techniques." The objective of this chapter is to introduce the reader to the technology of the movement of encoded information. Encoded information includes alpha-numeric data, which may broadly encompass messages that have direct meaning to the human user. It also includes the movement of strictly binary sequences that have meaning to a machine, but no direct meaning to a human being.

Data communications evolved from automatic telegraphy, which was so prevalent from the 1920s through the 1960s. We start the chapter with information coding or how can we express our alphabet and numeric symbols electrically without ambiguity. Data network performance is then covered with a review of the familiar BER. We then move on to the organization of data for transmission and introduce protocols including electrical and logical interfaces. Enterprise networks covering LAN and WAN technology, frame relay and ISDN are treated in Chapters 11 and 12.[1] The asynchronous transfer mode (ATM) is covered in Chapter 18. The principal objective of this chapter is to stress concepts, and to leave specific details to other texts.

10.2 THE BIT: A REVIEW

The *bit* is often called the most elemental unit of information. The IEEE (Ref. 1) defines it as a contraction of *binary digit*, a unit of information represented by either a 1 or a 0. These are the same bits that were introduced in Section 2.4.3 and later applied in Chapter 6, and to a lesser extent in Chapter 7. In Chapter 6, Digital Networks, the primary purpose of those bits was to signal the distant end the voltage level of an analog channel

[1]ISDN stands for Integrated Services Digital Networks.

Table 10.1 Equivalent Binary Designations: Summary of Equivalence

Symbol 1	Symbol 0
Mark or marking	Space or spacing
Current on	Current off
Negative voltage	Positive voltage
Hole (in paper tape)	No hole (in paper tape)
Condition *Z*	Condition *A*
Tone on (amplitude modulation)	Tone off
Low frequency (frequency shift keying)	High frequency
Inversion of phase	No phase inversion (differential phase shift keying)
Reference phase	Opposite to reference phase

Source: (Ref. 2).

at some moment in time. Here we will be assembling bit groupings that will represent letters of the alphabet, numerical digits 0 through 9, punctuation, graphic symbols, or just operational bit sequences that are necessary to make the data network operate with little or no ostensible outward meaning to us.

From old-time telegraphy the terminology has migrated to data communications. A *mark* is a binary 1 and a *space* is a binary 0. A space or 0 is represented by a positive-going voltage, and a mark or 1 is represented by a negative-going voltage. (Now *I* am getting confused. When I was growing up in the industry, a 1 or mark was a positive-going voltage, and so forth.)

10.3 REMOVING AMBIGUITY: BINARY CONVENTION

To remove ambiguity of the various ways we can express a 1 and a 0, CCITT in Rec. V.1 (Ref. 2) states clearly how to represent a 1 and a 0. This is summarized in Table 10.1, with several additions from other sources. Table 10.1 defines the *sense* of transmission so that the mark and space, the 1 and 0, respectively, will not be inverted. Inversion can take place by just changing the voltage polarity. We call it reversing the sense. Some data engineers often refer to such a table as a "table of mark-space convention."

10.4 CODING

Written information must be coded before it can be transmitted over a data network. One bit carries very little information. There are only those two possibilities: the 1 and the 0. It serves good use for supervisory signaling where a telephone line could only be in one of two states. It is either idle or busy. As a minimum we would like to transmit every letter of the alphabet and the 10 basic decimal digits plus some control characters, such as a space and hard/soft return, and some punctuation.

Suppose we join two bits together for transmission. This generates four possible bit sequences:[2]

00 01 10 11,

[2]To a certain extent this is a review of the argument presented in Section 6.2.3.

or four pieces of information, and each can be assigned a meaning such as 1, 2, 3, 4, or A, B, C, D. Suppose three bits are transmitted in sequence. Now there are eight possibilities:

$$
\begin{array}{cccc}
000 & 001 & 010 & 011 \\
100 & 101 & 110 & 111.
\end{array}
$$

We could continue this argument to sequences of four bits and it will turn out that there are now 16 different possibilities. It becomes evident that for a binary code, the number of distinct characters available is equal to two raised to a power equal to the number of elements (bits) per character. For instance, the last example was based on a four-element code giving 16 possibilities or information characters, that is, $2^4 = 16$.

The classic example is the ASCII code, which has seven information bits per character. Therefore the number of different characters available is $2^7 = 128$. The American Standard Code for Information Interchange (ASCII) is nearly universally used worldwide. Figure 10.1 illustrates ASCII. It will be noted in the figure that there are more than 30 special bit sequences such as SOH, NAK, EOT, and so on. These are/were used for data circuit control. For a full explanation of these symbols, refer to Ref. 3.

Another yet richer code was developed by IBM. It is the EBCDIC (extended binary coded decimal interchange code) code, which uses eight information bits per character. Therefore is has $2^8 = 256$ character possibilities. This code is illustrated in Figure 10.2. It should be noted that a number of the character positions are unassigned.

Bits $b_7 b_6 b_5$ →					0 0 0	0 0 1	0 1 0	0 1 1	1 0 0	1 0 1	1 1 0	1 1 1
b_4 ↓	b_3 ↓	b_2 ↓	b_1 ↓	Column → / Row ↓	0	1	2	3	4	5	6	7
0	0	0	0	0	NUL	DLE	SP	0	@	P	`	p
0	0	0	1	1	SOH	DC1	!	1	A	Q	a	q
0	0	1	0	2	STX	DC2	"	2	B	R	b	r
0	0	1	1	3	ETX	DC3	#	3	C	S	c	s
0	1	0	0	4	EOT	DC4	$	4	D	T	d	t
0	1	0	1	5	ENQ	NAK	%	5	E	U	e	u
0	1	1	0	6	ACK	SYN	&	6	F	V	f	v
0	1	1	1	7	BEL	ETB	'	7	G	W	g	w
1	0	0	0	8	BS	CAN	(	8	H	X	h	x
1	0	0	1	9	HT	EM	)	9	I	Y	i	y
1	0	1	0	10	LF	SUB	*	:	J	Z	j	z
1	0	1	1	11	VT	ESC	+	;	K	[	k	{
1	1	0	0	12	FF	FS	,	<	L	\	l	¦
1	1	0	1	13	CR	GS	—	=	M	]	m	}
1	1	1	0	14	SO	RS	.	>	N	^	n	~
1	1	1	1	15	SI	US	/	?	O	_	o	DEL

Figure 10.1 American Standard Code for Information Interchange (ASCII). (From MiL-STD-188C. Updated [26].)

B			4	0	0	0	0	0	0	0	0	1	1	1	1	1	1	1	1
	I		3	0	0	0	0	1	1	1	1	0	0	0	0	1	1	1	1
		T	2	0	0	1	1	0	0	1	1	0	0	1	1	0	0	1	1
		S	1	0	1	0	1	0	1	0	1	0	1	0	1	0	1	0	1
8	7	6	5																
0	0	0	0	NUL				PF	HT	LC	DEL								
0	0	0	1					RES	NL	BS	IL								
0	0	1	0					BYP	LF	EOB	PRE			SM					
0	0	1	1					PN	RS	UC	EOT								
0	1	0	0	SP										¢	.	<	(	+	|
0	1	0	1	&										!	$	*	)	;	¬
0	1	1	0	-	/									^	,	%	—	>	?
0	1	1	1											:	#	@	'	=	"
1	0	0	0		a	b	c	d	e	f	g	h	i						
1	0	0	1		j	k	l	m	n	o	p	q	r						
1	0	1	0			s	t	u	v	w	x	y	z						
1	0	1	1																
1	1	0	0		A	B	C	D	E	F	G	H	I						
1	1	0	1		J	K	L	M	N	O	P	Q	R						
1	1	1	0			S	T	U	V	W	X	Y	Z						
1	1	1	1	0	1	2	3	4	5	6	7	8	9						¤

PF – Punch Off
HT – Horiz. Tab
LC – Lower Case
DEL – Delete
SP – Space
UC – Upper Case

RES – Restore
NL – New Line
BS – Backspace
IL – Idle
PN – Punch On
EOT – End of Transmission

BYP – Bypass
LF – Line Feed
EOB – End of Block
PRE – Prefix
RS – Reader Stop
SM – Start Message

Figure 10.2 The extended binary-coded decimal interchange code (EBCDIC).

10.5 ERRORS IN DATA TRANSMISSION

10.5.1 Introduction

In data transmission one of the most important design goals is to minimize error rate. Error rate may be defined as the ratio of the number of bits incorrectly received to the total number of bits transmitted or to a familiar number such as 1000, 1,000,000, etc. CCITT (Ref. 6) holds with a design objective of better than one error in one million (bits transmitted). This is expressed as 1×10^{-6}. Many circuits in industrialized nations provide error performance one or two orders of magnitude better than this.

One method for minimizing the error rate would be to provide a "perfect" transmission channel, one that will introduce no errors in the transmitted information at the output of the receiver. However, that perfect channel can never be achieved. Besides improvement of the channel transmission parameters themselves, error rate can be reduced by forms of a systematic redundancy. In old-time Morse code, words on a bad circuit were often sent twice; this is redundancy in its simplest form. Of course, it took twice as long to send a message; this is not very economical if the number of useful words per minute received is compared to channel occupancy.

This illustrates the trade-off between redundancy and channel efficiency. Redundancy can be increased such that the error rate could approach zero. Meanwhile, the information transfer across the channel would also approach zero. Thus unsystematic redundancy is wasteful and merely lowers the rate of useful communication. On the other hand, maximum efficiency could be obtained in a digital transmission system if all redundancy and other code elements, such as "start" and "stop" elements, parity bits, and other "overhead" bits, were removed from the transmitted bit stream. In other words, the channel would be 100% efficient if all bits transmitted were information bits. Obviously, there is a trade-off of cost and benefits somewhere between maximum efficiency on a data circuit and systematically added redundancy.

10.5.2 Nature of Errors

In binary transmission an error is a bit that is incorrectly received. For instance, suppose a 1 is transmitted in a particular bit location and at the receiver the bit in that same location is interpreted as a 0. Bit errors occur either as single random errors or as bursts of errors.

Random errors occur when the signal-to-noise ratio deteriorates. This assumes, of course, that the noise is thermal noise. In this case noise peaks, at certain moments of time, are of sufficient level as to confuse the receiver's decision, whether a 1 or a 0.

Burst errors are commonly caused by fading on radio circuits. Impulse noise can also cause error bursts. Impulse noise can derive from lightning, car ignitions, electrical machinery, and certain electronic power supplies, to name a few sources.

10.5.3 Error Detection and Error Correction

Error detection just identifies that a bit (or bits) has been received in error. Error correction corrects errors at a far-end receiver. Both require a certain amount of redundancy to carry out the respective function. Redundancy, in this context, means those added bits or symbols that carry out no other function than as an aid in the error-detection or error-correction process.

One of the earliest methods of error detection was the *parity check*. With the 7-bit ASCII code, a bit was added for parity, making it an 8-bit code. This is character parity. It is also referred to as *vertical redundancy checking* (VRC).

We speak of *even parity* and *odd parity*. One system or the other may be used. Either system is based on the number of marks or 1s in a 7-bit character, and the eighth bit is appended accordingly, either a 0 or a 1. Let us assume even parity and we transmit the ASCII bit sequence 1010010. There are three 1s, an odd number. Thus a 1 is appended as the eighth bit to make it an even number.

Suppose we use odd parity and transmit the same character. There is an odd number of 1s (marks), so we append a 0 to leave the total number of 1s an odd number. With odd parity, try 1000111. If you added a 1 as the eighth bit, you'd be correct.

Character parity has the weakness that a lot of errors can go undetected. Suppose two bits are changed in various combinations and locations. Suppose a 10 became a 01; a 0 became a 1, and a 1 became a 0; and two 1s became two 0s. All would get by the system undetected.

To strengthen this type of parity checking, the *longitudinal redundancy check* (LRC) was included as well as the VRC. This is a summing of the 1s in a vertical column of all characters, including the 1s or 0s in each eighth bit location. The sum is now appended at the end of a message frame or packet. Earlier this bit sequence representing the sum was called the block check count (BCC). Today it may consist of two or four 8-bit sequences and we call it the FCS (frame check sequence), or sometimes the CRC (cyclic redundancy check). At the distant-end receiver, the same addition is carried out and if the sum agrees with the value received, the block is accepted as error-free. If not, it then contains at least one bit error, and a request is sent to the transmit end to retransmit the block (or frame).

Even with the addition of LRC, errors can slip through. In fact, no error-detection system is completely foolproof. There is another method, though, that has superior error detection properties. This is the CRC. It comes in a number of varieties.

10.5.3.1 *Cyclic Redundancy Check (CRC).* In very simple terms the CRC error detection technique works as follows: A data block or frame is placed in memory. We can call the frame a k-bit sequence and it can be represented by a polynomial which is called $G(x)$. Various modulo-2 arithmetic operations are carried out on $G(x)$ and the result is divided by a known generator polynomial called $P(x)$.[3] This results in a quotient $Q(x)$ and a remainder $R(x)$. The remainder is appended to the frame as an FCS, and the total frame with FCS is transmitted to the distant-end receiver where the frame is stored, then divided by the same generating polynomial $P(x)$. The calculated remainder is compared to the received remainder (i.e., the FCS). If the values are the same, the frame is error free. If they are not, there is at least one bit in error in the frame.

For many WAN applications the FCS is 16 bits long; on LANs it is often 32 bits long. Generally speaking, the greater the number of bits, the more powerful the CRC is for catching errors.

The following are two common generating polynomials:

1. ANSI CRC-16: $X^{16} + X^{15} + X^2 + 1$

2. CRC-CCITT: $X^{16} + X^{12} + X^5 + 1$

producing a 16-bit FCS.

CRC-16 provides error detection of error bursts up to 16 bits in length. Additionally, 99.955% of error bursts greater than 16 bits can be detected (Ref. 4).

10.5.3.2 *Forward-Acting Error Correction (FEC).* Forward-acting error correction (FEC) uses certain binary codes that are designed to be self-correcting for errors introduced by the intervening transmission media. In this form of error correction the receiving station has the ability to reconstitute messages containing errors.

The codes used in FEC can be divided into two broad classes: (1) block codes and (2) convolutional codes. In block codes information bits are taken k at a time, and c parity bits are added, checking combinations of the k information bits. A block consists of $n = k + c$ digits. When used for the transmission of data, block codes may be systematic. A systematic code is one in which the information bits occupy the first k positions in a block and are followed by the $(n - k)$ check digits.

A convolution(al) code is another form of coding used for error correction. As the word "convolution" implies, this is one code wrapped around or convoluted on another. It is the convolution of an input-data stream and the response function of an encoder. The encoder is usually made up of shift registers. Modulo-2 adders are used to form check digits, each of which is a binary function of a particular subset of the information digits in the shift register.

[3]Modulo-2 arithmetic is the same as binary arithmetic but without carries or borrows.

10.5.3.3 Error Correction with a Feedback Channel. Two-way or feedback error correction is used widely today on data circuits. Such a form of error correction is called ARQ. The letter sequence ARQ derives from the old Morse and telegraph signal, "automatic repeat request." There are three varieties of ARQ:

1. Stop-and-wait ARQ;
2. Selective or continuous ARQ; and
3. Go-back-*n* ARQ.

Stop-and-wait ARQ is simple to implement and may be the most economic in the short run. It works on a frame-by-frame basis. A frame is generated; it goes through CRC processing and an FCS is appended. It is transmitted to the distant end, where the frame runs through CRC processing. If no errors are found, an acknowledgment signal (ACK) is sent to the transmitter, which now proceeds to send the next frame—and so forth. If a bit error is found, a negative acknowledgment (NACK) is sent to the transmitter, which then proceeds to repeat that frame. It is the waiting time of the transmitter as it waits for either acknowledgment or negative acknowledgment signals. Many point to this wait time as wasted time. It could be costly on high-speed circuits. However, the control software is simple and the storage requirements are minimal (i.e., only one frame).

Selective ARQ, sometimes called *continuous ARQ*, eliminates the waiting. The transmit side pours out a continuous stream of contiguous frames. The receive side stores and CRC processes as before, but it is processing a continuous stream of frames. When a frame is found in error, it informs the transmit side on the return channel. The transmit side then picks that frame out of storage and places it in the transmission queue. Several points become obvious to the reader. First, there must be some way to identify frames. Second, there must be a better way to acknowledge or "negative-acknowledge." The two problems are combined and solved by the use of send sequence numbers and receive sequence numbers. The header of a frame has bit positions for a send sequence number and a receive sequence number. The send sequence number is inserted by the transmit side, whereas the receive sequence number is inserted by the receive side. The receive sequence numbers forwarded back to the transmit side are the send sequence of frame numbers acknowledged by the receive side. Of course, the receive side has to insert the corrected frame in its proper sequence before passing the data message to the end user.

Continuous or selective ARQ is more costly in the short run, compared with stop-and-wait ARQ. It requires more complex software and notably more storage on both sides of the link. However, there are no gaps in transmission and no time is wasted waiting for the ACK or NACK.

Go-back-n ARQ is a compromise. In this case, the receiver does not have to insert the corrected frame in its proper sequence, thus less storage is required. It works this way: When a frame is received in error, the receiver informs the transmitter to "go-back-*n*," *n* being the number of frames back to where the errored frame was. The transmitter then repeats all *n* frames, from the errored frame forward. Meanwhile, the receiver has thrown out all frames from the errored frame forward. It replaces this group with the new set of *n* frames it received, all in proper order.

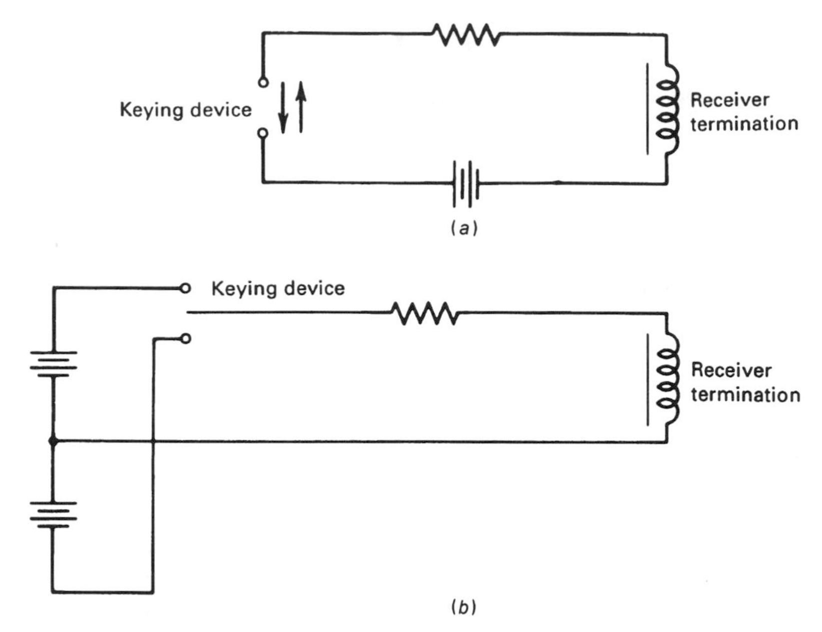

Figure 10.3 Simplified diagram illustrating a dc loop with (*a*) neutral keying and (*b*) polar keying.

10.6 dc NATURE OF DATA TRANSMISSION

10.6.1 dc Loops

Binary data are transmitted on a dc loop. More correctly, the binary data end instrument delivers to the line and receives from the line one or several dc loops. In its most basic form a dc loop consists of a switch, a dc voltage, and a termination. A pair of wires interconnects the switch and termination. The voltage source in data work is called the *battery*, although the device is usually electronic, deriving the dc voltage from an ac power line source. The battery is placed in the line to provide voltage(s) consistent with the type of transmission desired. A simplified drawing of a dc loop is shown in Figure 10.3*a*.

10.6.2 Neutral and Polar dc Transmission Systems

Older telegraph and data systems operated in the *neutral* mode. Nearly all present data transmission systems operate in some form of *polar* mode. The words "neutral" and "polar" describe the manner in which battery is applied to the dc loop. On a "neutral" loop, following the convention of Table 10.1, battery is applied during spacing (0) conditions and is switched off during marking (1). Current therefore flows in the loop when a space is sent and the loop is closed. Marking is indicated on the loop by a condition of no current. Thus we have two conditions for binary transmission, an open loop (no current flowing) and a closed loop (current flowing). Keep in mind that we could reverse this, namely, change the convention and assign marking to a condition of current flowing or closed loop, and spacing to a condition of no current or an open loop.[4] As mentioned, this is called "changing the sense." Either way, a neutral loop is a dc loop circuit where one binary condition is represented by the presence of voltage

[4]In fact, this was the older convention, prior to about 1960.

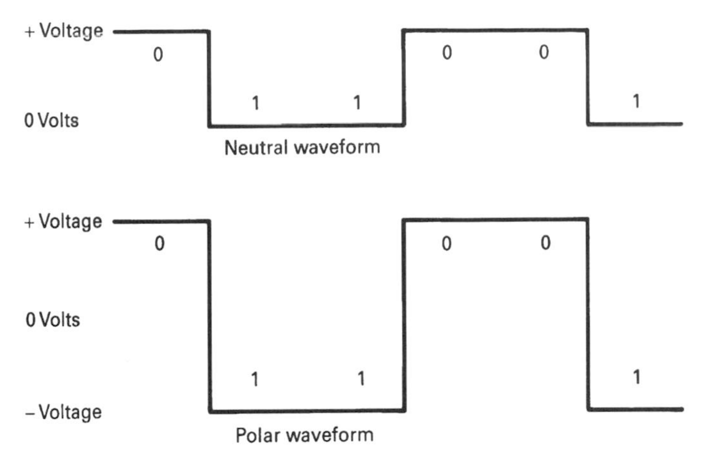

Figure 10.4 Neutral and polar waveforms.

and the flow of current, and the other condition is represented by the absence of voltage and current. Figure 10.3*a* illustrates a neutral loop.

Polar transmission approaches the problem differently. Two battery sources are provided, one "negative" and the other "positive." Following the convention in Table 10.1, during a condition of spacing (binary 0), a positive battery (i.e., a positive voltage) is applied to the loop, and a negative battery is applied during the marking (binary 1) condition. In a polar loop current is always flowing. For a mark or binary "1" it flows in one direction, and for a space or binary "0" it flows in the opposite direction. Figure 10.3*b* shows a simplified polar loop. Notice that the switch used to selected the voltage is called a *keying device*. Figure 10.4 illustrates the two electrical waveforms.

10.7 BINARY TRANSMISSION AND THE CONCEPT OF TIME

10.7.1 Introduction

As emphasized in Chapter 6, time and timing are most important factors in digital transmission. For this discussion consider a binary end instrument (e.g., a PC) sending out in series a continuous run of marks and spaces. Those readers who have some familiarity with the Morse code will recall that the spaces between dots and dashes told the operator where letters ended and where words ended. The sending device or transmitter delivers a continuous series of characters to the line, each consisting of five, six, seven, eight, or nine elements (bits) per character. A receiving device starts its print cycle when the transmitter starts sending and, if perfectly in step with the transmitter, can be expected to provide good printed copy and few, if any, errors at the receiving end.

It is obvious that when signals are generated by one machine and received by another, the speed of the receiving machine must be the same or very close to that of the transmitting machine. When the receiver is a motor-driven device, timing stability and accuracy are dependent on the accuracy and stability of the speed of rotation of the motors used. Most simple data-telegraph receivers sample at the presumed center of the signal element. It follows, therefore, that whenever a receiving device accumulates timing error of more than 50% of the period of one bit, it will print in error.

The need for some sort of synchronization is illustrated in Figure 10.5. A five-unit

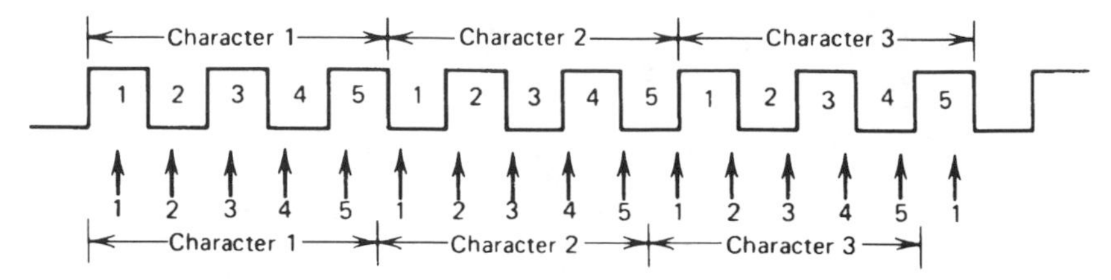

Figure 10.5 Five-unit synchronous bit stream with timing error.

code is employed, and it shows three characters transmitted sequentially.[5] The vertical arrows are receiver sampling points, which are points in time. Receiving timing begins when the first pulse is received. If there is a 5% timing difference between the transmitter and receiver, the first sampling at the receiver will be 5% away from the center of the transmitted pulse. At the end of the tenth pulse or signal element, the receiver may sample in error. Here we mean that timing error accumulates at 5% per received signal element and when there is a 50% accumulated error, the sampling will now be done at an incorrect bit position. The eleventh signal element will indeed be sampled in error, and all subsequent elements will be errors. If the timing error between transmitting machine and receiving machine is 2%, the cumulative error in timing would cause the receiving device to receive all characters in error after the 25th element (bit).

10.7.2 Asynchronous and Synchronous Transmission

In the earlier days of printing telegraphy, "start–stop" transmission, or asynchronous operation, was developed to overcome the problem of synchronism. Here timing starts at the beginning of a character and stops at the end. Two signal elements are added to each character to signal the receiving device that a character has begun and ended.

For example, consider the seven-element ASCII code (see Figure 10.1) configured for start–stop operation with a stop element which is of 2 bits duration. This is illustrated in Figure 10.6. In front of a character an element called a *start space* is inserted, and a *stop mark* is inserted at the end of a character. In the figure the first character is the ASCII letter upper case *U*(1010101). Here the receiving device knows (a priori) that it starts its timing 1 element (in this case a bit) after the mark-to-space transition—it counts out 8 unit intervals (bits) and looks for the stop-mark to end its counting. This is a transition from space-to-mark. The stop-mark, in this case, is two unit intervals long. It is followed by the mark-to-space transition of the next start space, whence it starts counting unit intervals up to 8. So as not to get confused, the first seven information bits are the ASCII bits, and the eighth bit is a parity bit. Even parity is the convention here.

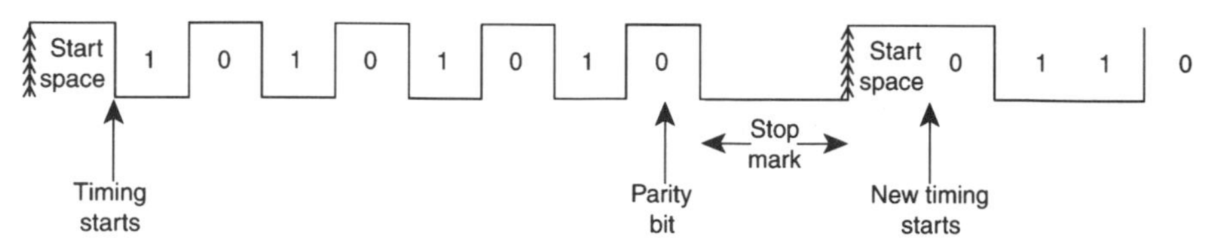

Figure 10.6 An 8-unit start-stop bit stream with a 2-unit stop element.

[5]A 5-bit code. The unit and bit are synonymous in this text. A code element carries out a function. It may be one or more bits in duration.

In such an operation timing error only accumulates inside a character. Suppose the receiving device is 5% slower or faster than the transmitting device. Only 40% (5×8) timing error accumulates, well inside of the 50% maximum. Remember that the sampling takes place at mid-bit position of 50% of its timing interval.

Minimum lengths of stop elements vary, depending on the convention used on a particular network. In the commercial world, the stop element can be one or two unit intervals (bits) duration. With some older systems, the stop element was 1.42 unit intervals duration. In military operation the stop element is 1.5 unit intervals long. The start space is always 1 unit interval duration.

As we are aware, a primary objective in the design of data systems is to minimize errors in a received bit stream or to minimize the error rate. Two of the prime causes of errors are noise and improper timing relationships. With start–stop systems a character begins with a mark-to-space transition at the beginning of the start-space. Then 1.5-unit intervals later, the timing causes the receiving device to sample the first information element, which is simply a 1 or 0 decision. The receiver continues to sample at one-bit intervals until the stop mark is received. It knows a priori where (when) it should occur. In start–stop systems the last information bit is the most susceptible to cumulative timing errors.

Synchronous data transmission is another story altogether. It is much more efficient because it does not have start and stop elements, which are really overhead. Synchronous data transmission consists of a continuous serial stream of information elements or bits, as illustrated in Figure 10.5. With start–stop systems, timing error could only accumulate inside a character, the eight bits of the example given previously. This is not so for synchronous systems. Timing error can accumulate for the entire length of a frame or over many frames.

With start–stop systems, the receiving device knows when a character starts by the mark-to-space transition at the start-space. In a synchronous transmission system, some marker must be provided to tell the receiver when a frame starts. This "marker" is the *unique field*. Every data frame starts with a unique field. A generic data frame is shown in Figure 10.7. View the frame from left to right. The first field is the unique field or flag, and it generally consists of the binary sequence 01111110. A frame always starts with this field and ends with the same field. If one frame follows another contiguously, the unique field ending frame No. 1 is the unique field starting frame No. 2, and so forth. Once a frame knows where it starts, it will know a priori where the following fields

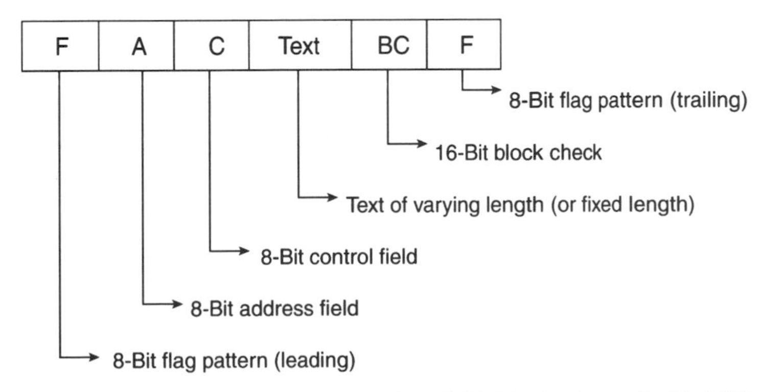

Figure 10.7 A generic data frame. Flag pattern = unique field. *Text* is often called "info" field or information field.

begin and end by simple bit/octet counting. However, with some data link protocols, the information field is of variable length. In this case, the information field length will appear as a subfield in the control field.

Synchronous data transmission systems not only require frame alignment, but also must be bit-aligned. This was an easy matter on start–stop systems because we could let the receive clock run freely inside the 5 bits or 8 bits of a character. There is no such freedom with synchronous systems. Suppose we assume a free-running receive clock. However, if there was a timing error of 1% between the transmit and receive clocks, not more than 100 bits could be transmitted until the synchronous receiving device would be off in timing by the duration of 1 bit from the transmitter, and all bits received thereafter would be in error. Even if the timing accuracy of one relative to the other were improved to 0.05%, the correct timing relationship between transmitter and receiver would exist for only the first 2000 bits transmitted. It follows, therefore, that no timing error whatsoever can be permitted to accumulate since anything but absolute accuracy in timing would cause eventual malfunctioning. In practice, the receiver is provided with an accurate clock that is corrected by small adjustments based on the transitions of the received bit stream, as explained in Section 10.7.3.

10.7.3 Timing

All currently used data-transmission systems are synchronized in phase and symbol rate in some manner. Start–stop synchronization has already been discussed. All fully synchronous transmission systems have timing generators or clocks to maintain stability. The transmitting device and its companion receiver at the far end of the circuit must maintain a timing system. In normal practice, the transmitter is the master clock of the system. The receiver also has a clock that in every case is corrected by some means to its transmitter's master clock equivalent at the far end.

Another important timing factor is the time it takes a signal to travel from the transmitter to the receiver. This is called *propagation time*. With velocities of propagation as low as 20,000 mi/s, consider a circuit 200 mi in length. The propagation time would then be 200/20,000 s or 10 ms. Ten milliseconds is the time duration of 1 bit at a data rate of 100 bps; thus the receiver in this case must delay its clock by 10 ms to be in step with its incoming signal. Temperature and other variations in the medium may also affect this delay, as well as variations in the transmitter master clock.

There are basically three methods of overcoming these problems. One is to provide a separate synchronizing circuit to slave the receiver to the transmitter's master clock. However, this wastes bandwidth by expending a voice channel or subcarrier just for timing. A second method, which was quite widely used until twenty years ago, was to add a special synchronizing pulse for groupings of information pulses, usually for each character. This method was similar to start–stop synchronization, and lost its appeal largely because of the wasted information capacity for synchronizing. The most prevalent system in use today is one that uses transition timing, where the receiving device is automatically adjusted to the signaling rate of the transmitter by sampling the transitions of the incoming pulses. This type of timing offers many advantages, particularly automatic compensation for variations in propagation time. With this type of synchronization the receiver determines the average repetition rate and phase of the incoming signal transition and adjusts its own clock accordingly by means of a phase-locked loop.

In digital transmission the concept of a transition is very important. The transition is what really carries the information. In binary systems the space-to-mark and mark-to-space transitions (or lack of transitions) placed in a time reference contain the information. In sophisticated systems, decision circuits regenerate and retime the pulses on the occurrence of a transition. Unlike decision circuits, timing circuits that reshape a pulse when a transition takes place must have a memory in case a long series of marks or spaces is received. Although such periods have no transitions, they carry meaningful information. Likewise, the memory must maintain timing for reasonable periods in case of circuit outage. Note that synchronism pertains to both frequency and phase and that the usual error in high-stability systems is a phase error (i.e., the leading edges of the received pulses are slightly advanced or retarded from the equivalent clock pulses of the receiving device). Once synchronized, high-stability systems need only a small amount of correction in timing (phase). Modem internal timing systems may have a long-term stability of 1×10^{-8} or better at both the transmitter and receiver. At 2400 bps, before a significant timing error can build up, the accumulated time difference between transmitter and receiver must exceed approximately 2×10^{-4} s. Whenever the circuit of a synchronized transmitter and receiver is shut down, their clocks must differ by at least 2×10^{-4} s before significant errors take place once the clocks start back up again. This means that the leading edge of the receiver–clock equivalent timing pulse is 2×10^{-4} in advance or retarded from the leading edge of the pulse received from the distant end. Often an idling signal is sent on synchronous data circuits during periods of no traffic to maintain the timing. Some high-stability systems need resynchronization only once a day.

Note that thus far in our discussion we have considered dedicated data circuits only. With switched (dial-up) synchronous circuits, the following problems exist:

- No two master clocks are in perfect phase synchronization.
- The propagation time on any two paths may not be the same.

Thus such circuits will need a time interval for synchronization for each call set-up before traffic can be passed.

To summarize, synchronous data systems use high-stability clocks, and the clock at the receiving device is undergoing constant but minuscule corrections to maintain an in-step condition with the received pulse train from the distant transmitter, which is accomplished by responding to mark-to-space and space-to-mark transitions. The important considerations of digital network timing were also discussed in Chapter 6.

10.7.4 Bits, Bauds, and Symbols

There is much confusion among professionals in the telecommunication industry over terminology, especially in differentiating, bits, bauds, and symbols. The bit, a binary digit, has been defined previously.

The baud is a unit of transmission rate or modulation rate. It is a measure of transitions per second. A transition is a change of state. In *binary* systems, bauds and bits per second (bps) are synonymous. In higher-level systems, typically m-ary systems, bits and bauds have different meanings. For example, we will be talking about a type of modulation called QPSK. In this case, every transition carries two bits. Thus the modulation rate in bauds is half the bit rate.

The industry often uses symbols per second and bauds interchangeably. It would be preferable, in our opinion, to use "symbols" for the output of a coder or other conditioning device. For the case of a channel coder (or encoder), bits go in and symbols come out. There are more symbols per second in the output than bits per second in the input. They differ by the coding rate. For example, a 1/2 rate coder (used in FEC) may have 4800 bps at the input and then would have 9600 symbols per second at the output.

10.7.4.1 Period of a Bit, Symbol, or Baud. The period of a bit is the time duration of a bit pulse. When we use NRZ (nonreturn-to-zero) coding (discussed in Section 10.7.5), the period of a bit, baud, or symbol is simply 1/(bit rate), 1/(baud rate) or 1/(symbol rate). For example, if we were transmitting 9600 bits per second, what is the period of a bit? It is 1/9600 = 104.16 μsec. For 2400 baud = 1/2400 = 416.6 μsec; for 33.6 kbps = 1/33,600 = 0.0297 μsec, or 29.7 ns.

10.7.5 Digital Data Waveforms

Digital symbols may be represented in many different ways by electrical signals to facilitate data transmission. All these methods for representing (or coding) digital symbols assign electrical parameter values to the digital symbols. In binary coding, of course, these digital symbols are restricted to two states, space (0) and mark (1). The electrical parameters used to code digital signals are levels (or amplitudes), transitions between different levels, phases (normally 0° and 180° for binary coding), pulse duration, and frequencies or a combination of these parameters. There is a variety of coding techniques for different areas of application, and no particular technique has been found to be optimum for all applications, considering such factors as implementing the coding technique in hardware, type of transmission technique employed, decoding methods at the data sink or receiver, and timing and synchronization requirements.

In this section we discuss several basic concepts of *electrical* coding of binary signals. In the discussion reference is made to Figure 10.8, which graphically illustrates several line coding techniques. Figure 10.8*a* shows what is still called by many today *neutral transmission* (also see Figure 10.4). This was the principal method of transmitting telegraph signals until about 1960. In many parts of the world, neutral transmission is still widely employed. First, the waveform is a nonreturn-to-zero (NRZ) format in its most elementary form. *Nonreturn-to-zero* simply means that if a string of 1s (marks) is transmitted, the signal remains in the mark state with no transitions. Likewise, if a string of 0s is transmitted, there is no transition and the signal remains in the 0 state until a 1 is transmitted. A "0" (zero) infers the 0 voltage line as shown in the figure. Figure 10.8*c* is the conventional NRZ waveform most encountered in the data industry.

Figures 10.8*b* and 10.8*d* show the typical *return-to-zero* (RZ) waveform, where, when a continuous string of marks (1s) or spaces (0s) is transmitted, the signal level (amplitude) returns to the zero voltage condition at each element or bit. Obviously RZ transmission is much richer in transitions that NRZ.

In Section 10.6.2 we discussed neutral and polar transmission systems. Figure 10.8*a* shows a typical neutral waveform where the two state conditions are 0 V for the mark or binary 1 condition and some positive voltage for the space or binary 0 condition. On the other hand, in polar transmission, as illustrated in Figures 10.8*c* and 10.8*d*, a positive voltage represents a space and a negative voltage a mark. With NRZ transmission, the pulse width is the same as the duration of a unit interval or bit. Not so with RZ transmission, where the pulse width is less than the duration of a unit interval. This is because we have to allow time for the pulse to return to the zero (voltage) condition.

Bi-phase-L or Manchester coding (Figure 10.8*e*) is a code format that is being used ever more widely on digital systems such as on certain local area networks (LANs) (see Chapter 11). Here binary information is carried in the transition. By convention, a logic 0 is defined as a positive-going transition and a logic 1 as a negative-going transition.

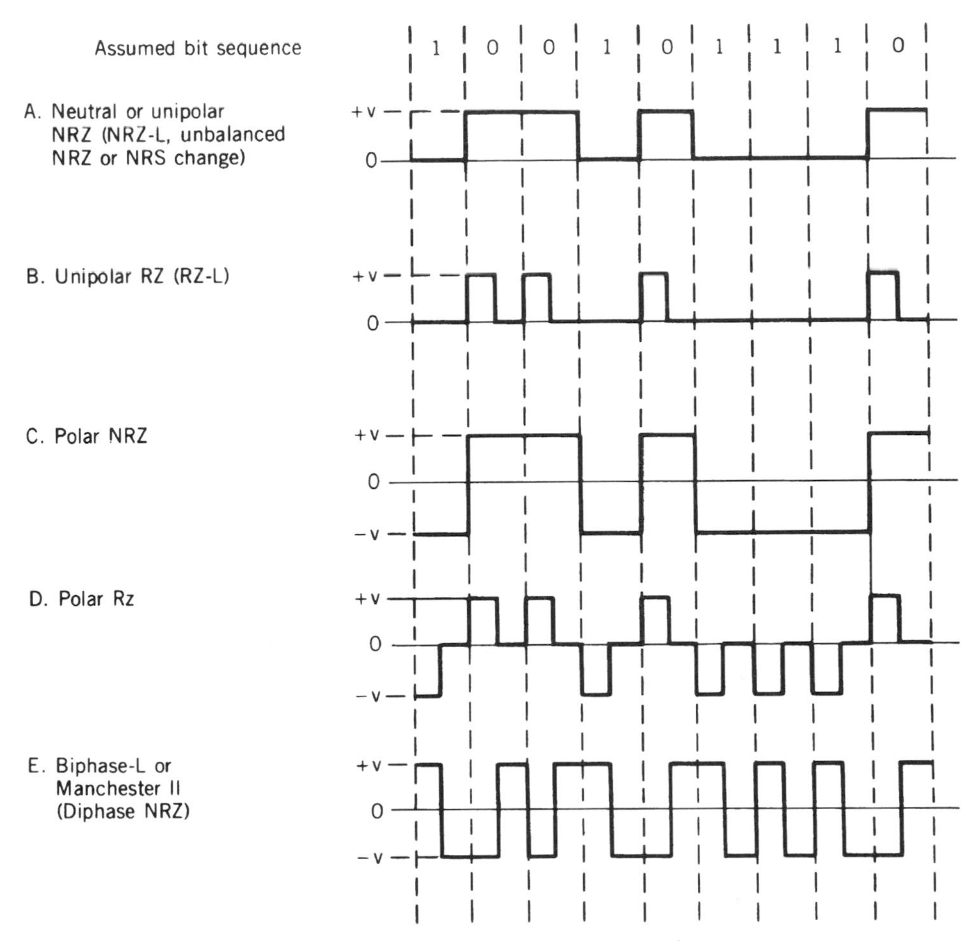

Figure 10.8 Digital data transmission waveforms.

Note that Manchester coding has a signal transition in the middle of each unit interval (bit). Manchester coding is a form of phase coding.

10.8 DATA INTERFACE: THE PHYSICAL LAYER

When we wish to transmit data over a conventional analog network, the electrical representation of the data signal is essentially direct current (i.e., has a frequency of 0 Hz). As such it is incompatible with that network that accepts information channels in the band 300–3400 Hz.[6] A data modem is a device that brings about this compatibility. It translates the electrical data signal into a modulated frequency tone in the range of 300–3400 Hz. In most cases with modern modems that tone is 1800 Hz. Also, more often than not, the digital network (as described in Chapter 6) has extensions that are analog and require the same type(s) of modem. If the digital network is extended to a user's premise, a digital conditioning device (CSU/DSU) is required for bit rate and waveform compatibility.

For this discussion of data interface, we will call the modem or digital conditioning

[6]This is the conventional CCITT voice channel.

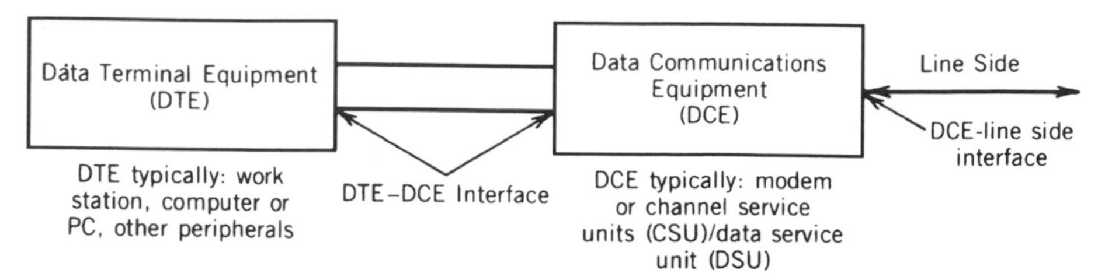

Figure 10.9 Data circuit interfaces, physical layer.

device *data communication equipment* (DCE). This equipment has two interfaces, one on each side, as shown in Figure 10.9. The first, which is discussed in this section, is on the user side, which is called *data terminating equipment* (DTE), and the applicable interface is the DTE–DCE interface. The second interface is on the line side, which is covered in Section 10.9. It should be noted that the DTE-DCE interface is well defined.

The most well-known DTE–DCE standard was developed by the (U.S.) Electronics Industries Association (EIA) and is called EIA-232E (Ref. 5). It is essentially equivalent to international standards covered by CCITT Recs. V.24 and V.28, and ISO IS2110.

EIA-232E and most of the other standards discussed here are applicable to the DTE–DCE interface employing serial binary data interchange. It defines signal characteristics, mechanical interface characteristics, and functional descriptions of the interchange circuits. EIA-232E is applicable for data transmission rates up to 20,000 bps and for synchronous/asynchronous serial binary data systems.

Section 2.1.3 is quoted from EIA-232E (Ref. 5). It is crucial to the understanding of signal state convention and level:

> For data interchange circuits, the signal shall be considered in the marking condition when the voltage (V_1) on the interchange circuit, measured at the interface point, is more negative than minus three volts with respect to circuit AB (signal ground). The signal shall be considered in the spacing condition when the voltage V_1 is more positive than plus three volts with respect to circuit AB. ... The region between plus three volts and minus three volts is defined as the transition region. The signal state is not uniquely defined when the voltage (V_1) is in this transition region.

During the transmission of data, the marking condition is used to denote the binary state ONE and the spacing condition is used to denote the binary state ZERO.

Besides EIA-232 there are many other interface standards issued by EIA, CCITT, U.S. federal standards, U.S. military standards, and ISO. Each defines the DTE–DCE interface. Several of the more current standards are briefly described in the following.

EIA-530 (Ref. 7) is a comparatively recent standard developed by the EIA. It provides for all data rates below 2.1 Mbps and it is intended for all applications requiring a balanced electrical interface.[7] It can also be used for unbalanced operation.

Let us digress for a moment. An unbalanced electrical interface is where one of the signal leads is grounded; for a balanced electrical interface, no ground is used.

EIA-530 applies for both synchronous and nonsynchronous (i.e., start–stop) operation. It uses a standard 25-pin connector; alternatively it can use a 26-pin connector. A list of interchange circuits showing circuit mnemonic, circuit name, circuit direction (meaning toward DCE or toward DTE), and circuit type is presented in Table 10.2.

[7]EIA-530 is more properly called ANSI/EIA/TIA-530-A.

Table 10.2 EIA-530 Interchange Circuits

Circuit Mnemonic	CCITT Number	Circuit Name	Circuit Direction	Circuit Type
AB	102	Signal common		Common
AC	102B	Signal common		
BA	103	Transmitted data	To DCE	Data
BB	104	Received data	From DCE	
CA	105	Request to send	To DCE	
CB	106	Clear to send	From DCE	
CF	109	Received line signal detector	From DCE	
CJ	133	Ready for receiving	To DCE	
CE	125	Ring indicator	From DCE	Control
CC	107	DTE ready	From DCE	
CD	108/1, /2	DTE ready	To DCE	
DA	113	Transmit signal element timing (DTE source)	To DCE	
DB	114	Transmit signal element timing (DCE source)	From DCE	Timing
DD	115	Receiver signal element timing (DCE source)	From DCE	
LL	141	Local loopback	To DCE	
RL	140	Remote loopback	To DCE	
TM	142	Test mode	From DCE	

Source: (Ref. 7).

10.9 DIGITAL TRANSMISSION ON AN ANALOG CHANNEL

10.9.1 Introduction

Two fundamental approaches to the practical problem of data transmission are (1) to design and construct a complete, new network expressly for the purpose of data transmission, and (2) to adapt the many existing telephone facilities for data transmission. The following paragraphs deal with the latter approach.

Analog transmission facilities designed to handle voice traffic have characteristics that hinder the transmission of dc binary digits or bit streams. To permit the transmission of data over voice facilities (i.e., the telephone network), it is necessary to convert the dc data into a signal within the voice-frequency range. The equipment that performs the necessary conversion to the signal is generally called a *modem*, an acronym for *mo*dulator–*dem*odulator.

10.9.2 Modulation–Demodulation Schemes

A modem modulates and demodulates a carrier signal with digital data signals. In Section 2.4.2 the three basic types of modulation were introduced. These are listed as follows, along with their corresponding digital terminology:

MODULATION TYPE	CORRESPONDING DIGITAL TERMINOLOGY
Amplitude modulation (AM)	Amplitude shift keying (ASK)
Frequency modulation (FM)	Frequency shift keying (FSK)
Phase modulation (PM)	Phase shift keying (PSK)

In amplitude shift keying, the two binary conditions are represented by a tone on and a tone off. ASK, used alone, is not really a viable alternative in today's digital network. It is employed in many advanced hybrid modulation schemes.

Frequency shift keying is still used for lower data rate circuits ($\leq$ 1200 bps). The two binary states are each represented by a different frequency. At the demodulator the frequency is detected by a tuned circuit. So, of course, there are two tuned circuits, one for each frequency. Binary 0 is represented by the higher frequency and binary 1 by the lower frequency.

Phase shift keying is widely used with data modems, both alone and with hybrid modulation schemes. The most elementary form is binary PSK. In this case there are just two phases, which are separated one from the other by 180° phase difference. The demodulator is based on a synchronous detector using a reference signal of known phase. This known signal operates at the same frequency of the incoming signal carrier and is arranged to be in phase with one of the binary signals. In the relative-phase system a binary 1 is represented by sending a signal burst of the same phase as that of the previous signal burst sent. A binary 0 is represented by a signal burst of a phase opposite to that of the previous signal transmitted. The signals are demodulated at the receiver by integrating and storing each signal burst of 1-bit period for comparison in phase with the next signal burst.

10.9.3 Critical Impairments to the Transmission of Data

The effect of various telephone circuit parameters on the capability of a circuit to transmit data is a most important consideration. The following discussion is intended to familiarize the reader with problems most likely to be encountered in the transmission of data over analog or mixed analog/digital circuits. We make certain generalizations in some cases, which can be used to facilitate planning the implementation of data systems.

10.9.3.1 Phase Distortion. Phase distortion "constitutes the most limiting impairment to data transmission, particularly over telephone voice channels" (Ref. 8). When specifying phase distortion, the terms *envelope delay distortion* (EDD) and *group delay* are often used. The IEEE (Ref. 1) states that envelope delay is often defined the same as group delay; that is, the rate of change, with angular frequency, of the phase shift between two points in a network.[8]

Let us try to put this another way that will be easier to understand. We are dealing with the voice channel which will be transported in an analog network or in a mixed analog/digital network. It is a band-limited system. Phase distortion arises from the fact that not all frequency components of the input signal will propagate to the receiving end in exactly the same elapsed time. This is true when the signal passes through filters or their equivalent. Particularly troublesome are loaded cable circuits and FDM carrier circuits. Even PCM channel banks have filters.

Figure 10.10 shows a typical frequency-delay response curve in milliseconds of a voice channel due to FDM equipment only. For the voice channel (or any symmetrical passband, for that matter), delay increases toward band edge and is minimum about band center (around 1800 Hz). As the bit rate is increased on the channel, there is more and more opportunity for the delayed energy from bit 1 to spill into the principal energy component of bit 2. When it does, we call it ISI (intersymbol interference).

Phase or delay distortion is the major limitation of modulation rate on the voice

[8]*Angular frequency* is $2\pi \times$ frequency in Hz. It is sometimes called the *radian frequency*.

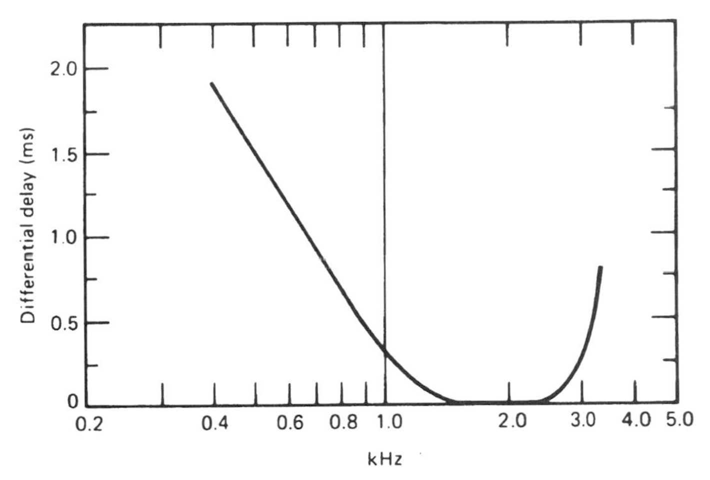

Figure 10.10 Typical differential delay across a voice channel, FDM equipment back-to-back. "Back-to-back" means that the transmit and receive portions of the equipment are placed on a test bench and appropriately interconnected.

channel. The shorter the pulse width (the width or period of 1 bit in binary systems), the more critical will be the EDD parameters. As we will discuss in Section 10.9.5, it is desirable to keep delay distortion in the band of interest below the period of 1 baud.

10.9.3.2 Attenuation Distortion (Frequency Response). Another parameter that seriously affects the transmission of data and can place definite limits on the modulation rate is attenuation distortion. Ideally, all frequencies across the passband of a channel of interest should undergo the same loss or attenuation. For example, let a −10-dBm signal enter a channel at any given frequency between 300 Hz and 3400 Hz. If the channel has 13 dB of flat attenuation, we would expect an output at the distant end of −23 dBm at any and all frequencies in the band. This type of channel is ideal but unrealistic in a real working system.

In Rec. G.132 (Ref. 9), the CCITT recommends no more than 9 dB of attenuation distortion relative to 800 Hz between 400 Hz and 3000 Hz. This figure, 9 dB, describes the maximum variation that may be expected from the reference level at 800 Hz. This variation of amplitude response is often called *attenuation distortion.* A conditioned channel, such as a Bell System C-4 channel, will maintain a response of −2 dB to +3 dB from 500 Hz to 3000 Hz and −2 dB to +6 dB from 300 Hz to 3200 Hz.

Considering tandem operation, the deterioration of amplitude response is arithmetically cumulative when sections are added. This is particularly true at band edge in view of channel unit transformers and filters that account for the upper and lower cutoff characteristics. Figure 10.11 illustrates a typical example of attenuation distortion (amplitude response) across carrier equipment back-to-back. Attenuation distortion, phase distortion, and noise were introduced in Section 3.3.

10.9.3.3 Noise. Another important consideration in the transmission of data is noise. All extraneous elements appearing at the voice channel output that were not due to the input signal are considered to be noise. For convenience, noise is broken down into four categories: (1) thermal, (2) crosstalk, (3) intermodulation, and (4) impulse (Ref. 27). Thermal noise, often called "resistance noise," "white noise," or "Johnson noise,"

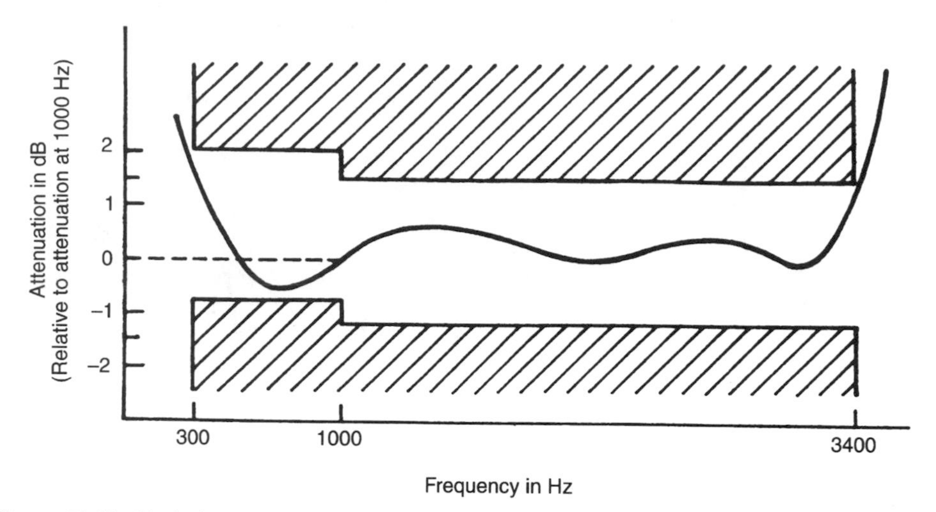

Figure 10.11 Typical attenuation distortion across a voice channel, carrier equipment back-to-back.

is of a Gaussian nature or completely random. Any system or circuit operating at a temperature above absolute zero inherently will display thermal noise. The noise is caused by the random noise of discrete electrons in the conduction path. Crosstalk is a form of noise caused by unwanted coupling from one signal path into another. It may be caused by direct inductive or capacitive coupling between conductors or between radio antennas. Intermodulation noise is another form of unwanted coupling, usually caused by signals mixing in nonlinear elements of a system. Carrier and radio systems are highly susceptible to intermodulation noise, particularly when overloaded. Impulse noise can be a primary source of errors in the transmission of data over telephone networks. It is sporadic and may occur in bursts or discrete impulses called "hits." Some types of impulse noise are natural, such as that from lightning. However, man-made impulse noise is ever increasing, such as that from automobile ignition systems and power lines. Impulse noise may be of a high level in conventional analog telephone switching centers as a result of dialing, supervision, and switching impulses that may be induced or otherwise coupled into the data-transmission channel. The worst offender in the switching area is the step-by-step exchange, some of which are still around. Impulse noise in digital exchanges is almost nonexistent.

For our discussion of data transmission, two types of noise are considered, (1) random (or Gaussian) noise and (2) impulse noise. Random noise measured with a typical transmission measuring set appears to have a relatively constant value. However, the instantaneous value of the noise fluctuates over a wide range of amplitude levels. If the instantaneous noise voltage is of the same magnitude as the received signal, the receiving detection equipment may yield an improper interpretation of the received signal and an error or errors will occur. Thus we need some way of predicting the behavior of data transmission in the presence of noise. Random noise or white noise has a Gaussian distribution and is considered representative of the noise encountered on a communications channel, which, for this discussion is the voice channel.[9] From the probability distribution curve of Gaussian noise shown in Figure 10.12, we can make some statistical

[9]The definition of the term *Gaussian distribution* is beyond the scope of this book. It essentially means to us that this type of noise is well characterized and we will use the results of this characterization such as found in Figure 10.12.

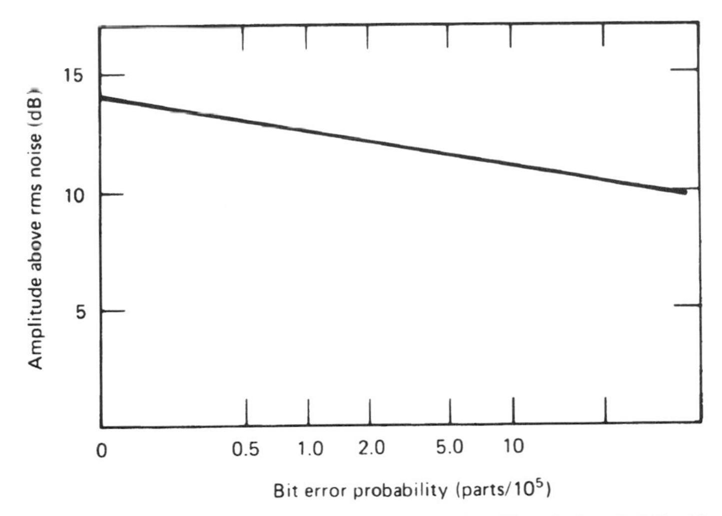

Figure 10.12 Probability of bit error in Gaussian (random) noise. Nyquist bandwidth, binary polar transmission. The reader should note that this is for a baseband signal.

predictions. It may be noted from this curve that the probability of occurrence of noise peaks that have amplitudes 12.5 dB above the rms level is 1 in 10^5 or about 14 dB for 1 in 10^6.[10,11] Hence, if we wish to ensure an error rate of 10^{-5} in a particular system using binary polar modulation, the rms noise should be at least 12.5 dB below the signal level (Ref. 8, p. 114). This simple analysis is only valid for the type of modulation used (i.e., binary polar baseband modulation), assuming that no other factors are degrading the operation of the system and that a cosine-shaped receiving filter is used. If we were to interject distortion such as EDD into the system, we could translate the degradation into an equivalent signal-to-noise ratio improvement necessary to restore the desired error rate. For example, if the delay distortion were the equivalent of one pulse width, the signal-to-noise ratio improvement required for the same error rate would be about 5 dB, or the required signal-to-noise ratio would now be 17.5 dB.

Unlike random noise, which is measured by its rms value when we measure level, impulse noise is measured by the number of "hits" or "spikes" per interval of time above a certain threshold. In other words, it is a measurement of the recurrence rate of noise peaks over a specified level. The word "rate" should not mislead the reader. The recurrence is not uniform per unit time, as the word "rate" may indicate, but we can consider a sampling and convert it to an average.

Remember that random noise has a Gaussian distribution and will produce noise peaks at 12.5 dB above the rms value 0.001% of the time on a data bit stream for an equivalent error rate of 1×10^{-5}. The 12.5 dB above rms random noise floor should establish the impulse noise threshold for measurement purposes. We should assume that a well-designed data-transmission system traversing the telephone network, the signal-to-noise ratio of the data signal will be well in excess of 12.5 dB. Thus impulse noise may well be the major contributor to the degradation of the error rate.

[10]The abbreviation *rms* stands for root-mean square, the square root of the average (mean) of the square(s) of the values (of level in this case).

[11]These dB values are for a "Nyquist bandwidth" accommodating 2 bits per Hz; for a bit-rate bandwidth, 1 bit/Hz, the values would be 9.5 dB and 12.5 dB, respectively.

10.9.4 Channel Capacity

A leased or switched voice channel represents a financial investment. Therefore one goal of the system engineer is to derive as much benefit as possible from the money invested. For the case of digital transmission, this is done by maximizing the information transfer across the system. This section discusses how much information in bits can be transmitted, relating information to bandwidth, signal-to-noise ratio, and error rate. These matters are discussed empirically in Section 10.9.5.

First, looking at very basic information theory, Shannon stated in his classic paper (Ref. 10) that if input information rate to a band-limited channel is less than C (bps), a code exists for which the error rate approaches zero as the message length becomes infinite. Conversely, if the input rate exceeds C, the error rate cannot be reduced below some finite positive number.

The usual voice channel is approximated by a Gaussian band-limited channel (GBLC) with additive Gaussian noise.[12] For such a channel, consider a signal wave of mean power of S watts applied at the input of an ideal low-pass filter that has a bandwidth of W (Hz) and contains an internal source of mean Gaussian noise with a mean power of N watts uniformly distributed over the passband. The capacity in bits per second is given by

$$C = W \log_2 \left(1 + \frac{\mathrm{S}}{\mathrm{N}} \right).$$

Applying Shannon's "capacity" formula to an ordinary voice channel (GBLC) of bandwidth (W) 3000 Hz and a signal-to-noise S/N ratio of 1023 (about 30 dB), the capacity of the channel is 30,000 bps. (Remember that bits per second and bauds are interchangeable in binary systems.) Neither S/N nor W is an unreasonable value. Seldom, however, can we achieve a modulation rate greater than 3000 bauds. The big question in advanced design is how to increase the data rate and keep the error rate reasonable.

One important item not accounted for in Shannon's formula is intersymbol interference. A major problem of a pulse in a band-limited channel is that the pulse tends not to die out immediately, and a subsequent pulse is interfered with by "tails" from the preceding pulse, as illustrated in Figure 10.13.

10.9.5 Modem Selection Considerations

The critical parameters that affect data transmission have been discussed; these are amplitude-frequency response (sometimes called "amplitude distortion"), envelope delay distortion, and noise. Now we relate these parameters to the design of data

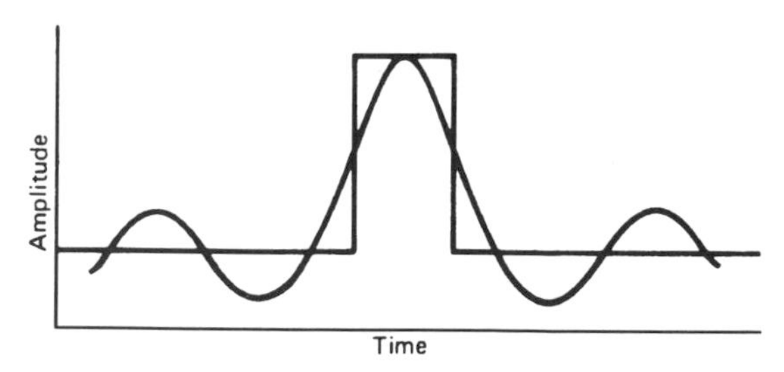

Figure 10.13 Pulse response through a Gaussian band-limited channel (GBLC). Gaussian refers to a channel limited by random noise.

[12]Thermal noise.

modems to establish some general limits or "boundaries" for equipment of this type. The discussion that follows purposely avoids HF radio considerations.

As stated earlier in the discussion of envelope delay distortion, it is desirable to keep the transmitted pulse (bit) length equal to or greater than the residual differential EDD. Since about 1.0 ms is assumed to be reasonable residual delay after equalization (conditioning), the pulse length should be no less than approximately 1 ms. This corresponds to a modulation rate of 1000 pulses per second (binary). In the interest of standardization [CCITT Rec. V.22 (Ref. 11)], this figure is modified to 1200 bps.

The next consideration is the *usable* bandwidth required for the transmission of 1200 bps. The figure is approximately 1800 Hz using modulation such as FSK (frequency shift keying) or PSK (phase shift keying). Since delay distortion of a typical voice channel is at its minimum between 1700 Hz and 1900 Hz, the required band, when centered around these points, extends from 800 Hz to 2600 Hz or 1000 Hz to 2800 Hz. Remember, the delay should not exceed duration of a bit (baud). At 1200 bps, the duration of a bit is 0.833 milliseconds. Turning to Figure 10.10, the delay distortion requirement is met over the range of 800 Hz to 2800 Hz.

Bandwidth limits modulation rate. However, the modulation rate in bauds and the data rate in bits per second need not necessarily be the same. This is a very important concept. The "baud rate" is the measure of transitions per second. We then must turn to bit packing much as we did in Section 9.3. There we briefly discussed PSK and especially quadrature phase shift keying (QPSK). Let us review it here.

The tone frequency for this example is 1800 Hz. The phase of that tone can be retarded. For binary phase shift keying, we assign binary 0 to 0° (no phase retardation) and binary 1 to 180° (retardation), as illustrated in Figure 10.14.

Suppose we divide the circle in Figure 10.14 into quadrants with the unretarded phase at 0° as before (unretarded), then with 90°, 180°, and 270° of retardation. Let us assign two binary digits (bits) to each of the four phase possibilities. These could be as follows:

PHASE CHANGE (DEGREES)	EQUIVALENT BINARY NUMBER
0	00
90	01
180	10
270	11

we could call it a 4-ary scheme. More commonly it is called *quadrature phase shift keying* (QPSK). This scheme is illustrated in Figure 10.15.

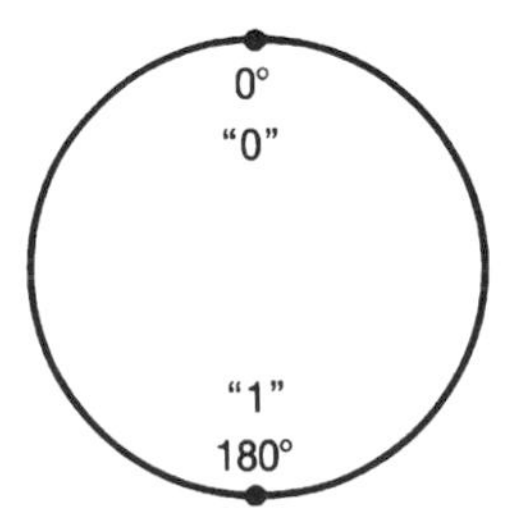

Figure 10.14 A spatial representation of a data tone with BPSK modulation. Commonly a circle is used when we discuss phase modulation representing the phase retardation relationship up to 360°.

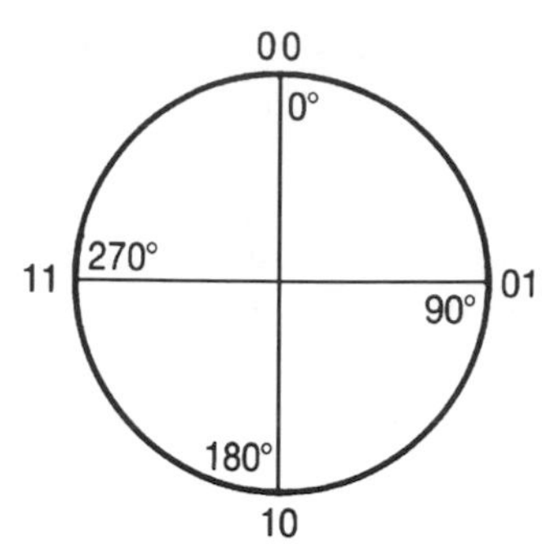

Figure 10.15 A spatial representation of a data tone with QPSK modulation.

Nearly all data modems operating at 2400 bps use QPSK modulation. The phases do not have to start at 0°. The only requirement is that each contiguous phase be separated from its neighbor by 90° in the case of QPSK.

Of course we can carry the idea still further by breaking the circle up into 45° phase segments. There are now eight phase segments, each segment representing a distinct grouping of three bits. This is shown in Figure 10.16. It is called *8-ary PSK* (i.e., $M = 8$).

Several important points arise here which must be made clear. First, this is how we can achieve a data-transmission system where the data rate exceeds the bandwidth. We stated that the data rate on a voice channel could not exceed 3100 bps, assigning 1 bit per hertz of bandwidth. We mentioned in passing that this held for the binary condition only. Theoretically, for QPSK, we can achieve 2 bits per hertz of bandwidth, but practically it is more like 1.2 bits per hertz. With 8-ary PSK the theoretical bit packing is 3 bits per hertz.

Bandwidth is a function of transitions per second. Transitions per second are commonly referred to as *bauds*. Only in the binary domain are bauds and bits per second synonymous. With QPSK operation, the baud rate is half the bit rate. For example, with a 2400-bps modem, the baud rate is 1200 bauds. The bandwidth required for a certain data transmission system is dependent on the baud rate or the number of transitions per second. For example, for BPSK a transition carries just one bit of information; for QPSK a transition carries two bits of information, and for 8-ary PSK a transition carries three bits of information. The CCITT 4800-bps modem uses 8-ary PSK and its modulation rate is then 1600 baud.

Higher data rate modems utilize a hybrid modulation scheme, combining phase mod-

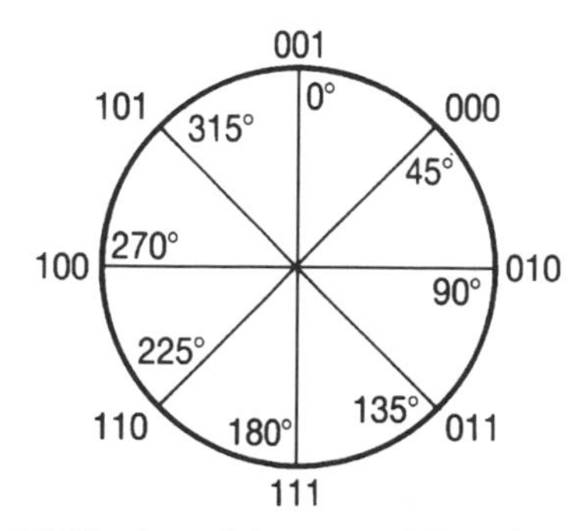

Figure 10.16 A spatial representation of 8-ary PSK.

Table 10.3 Phase Encoding for the CCITT Rec. V.29 Modem

Q_2	Q_3	Q_4	Phase Change (°)
0	0	1	0
0	0	0	45
0	1	0	90
0	1	1	135
1	1	1	180
1	1	0	225
1	0	0	270
1	0	1	315

ulation and amplitude modulation. One typical modem in this category is based on CCITT Rec. V.29 (Ref. 12) and operates at 9600 bps and its baud rate is 2400 baud. Its theoretical bit packing capability is 4 bits per Hz of bandwidth. Each transition carries 4 bits of information. The first three bits derive from 8-ary PSK and the fourth bit derives from using two equivalent amplitude levels. Table 10.3 gives the phase encoding for the V.29 modem and Table 10.4 shows its amplitude–phase relationships. The four bits are denominated Q_1, Q_2, Q_3, and Q_4. Figure 10.17 shows the signal space diagram of the V.29 modem. There are 16 points in the diagram representing the 16 quadbit possibilities.

Other modems of the CCITT V.-series are designed to operate at 14,400 bps, 28,800 bps, and 33.6 kbps over the standard voice channel. A proprietary modem of North American manufacture can operate up to 56 kbps on the standard voice channel. In all cases at these data rates the waveforms become extremely complex.

10.9.6 Equalization

Of the critical circuit parameters mentioned in Section 10.9.3, two that have severely deleterious effects on data transmission can be reduced to tolerable limits by *equalization*. These two are amplitude–frequency response (amplitude distortion) and EDD (delay distortion).

The most common method of performing equalization is the use of several networks in tandem. Such networks tend to flatten response and, in the case of amplitude response, add attenuation increasingly toward channel center and less toward its edges. The overall effect is one of making the amplitude response flatter. The delay equalizer operates in a similar manner. Delay increases toward channel edges parabolically from the center. To compensate, delay is added in the center much like an inverted parabola, with less and less delay added as the band edge is approached. Thus the delay response is

Table 10.4 Amplitude–Phase Relationships for the CCITT Rec. V.29 Modem

Absolute Phase (°)	Q_1	Relative Signal Element Amplitude
0, 90, 180, 270	0	3
	1	5
45, 135, 225, 315	0	$\sqrt{2}$
	1	$3\sqrt{2}$

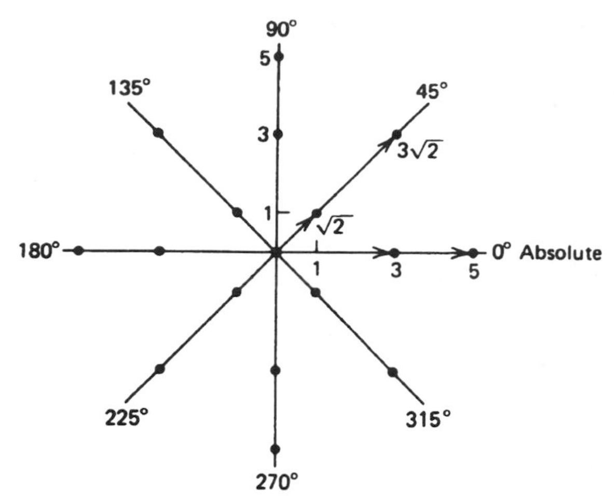

Figure 10.17 Signal space diagram for the CCITT Rec. V.29 modem when operating at 9600 bps (Ref. 12).

flattened at some small cost to absolute delay, which has no effect in most data systems. However, care must be taken with the effect of a delay equalizer on an amplitude equalizer and, conversely, of an amplitude equalizer on the delay equalizer. Their design and adjustment must be such that the flattening of the channel for one parameter does not entirely distort the channel for the other.

Automatic equalization for both amplitude and delay are effective, particularly for switched data systems. Such devices are self-adaptive and require a short adaptation period after switching, on the order of <1 s (Ref. 13). This can be carried out during synchronization. Not only is the modem clock being "averaged" for the new circuit on transmission of a synchronous idle signal, but the self-adaptive equalizers adjust for optimum equalization as well. The major drawback of adaptive equalizers is cost.

Figure 10.18 shows typical envelope delay response of a voice channel along with the opposite response of a delay equalizer to flatten the envelope delay characteristics of the voice channel.

10.9.7 Data Transmission on the Digital Network

Many data users only have analog access to the digital network. In other words, their connectivity to the network is via a subscriber loop to the local serving exchange. It is commonly at this point where the analog channel enters a PCM channel bank and the signal is converted to the standard digital signal. This class of data users will utilize conventional data modems, as described in Section 10.9.5.

Other data users will be within some reasonable distance (some hundreds of feet) from a digital network terminal. This may be a PABX so equipped, providing access to a DS0 or E0 line (Chapter 6), where the line rate is 64 kbps, the standard digital voice channel. In some instances in North America the digital line will only provide 56 kbps.

The digital network transmission rates are incompatible with standard transmission rates in the data environment. Standard data rates based on CCITT Recs. V.5 and V.6 (Refs. 15 and 16) as well as EIA-269 (Ref. 17). These data rates are 600, 1200, 2400, 4800, 9600, 14,400, 19,200, 28,800, and 33,600 bps. They are not evenly divisible into 64,000 bps, the standard digital voice channel. Two methods are described in the fol-

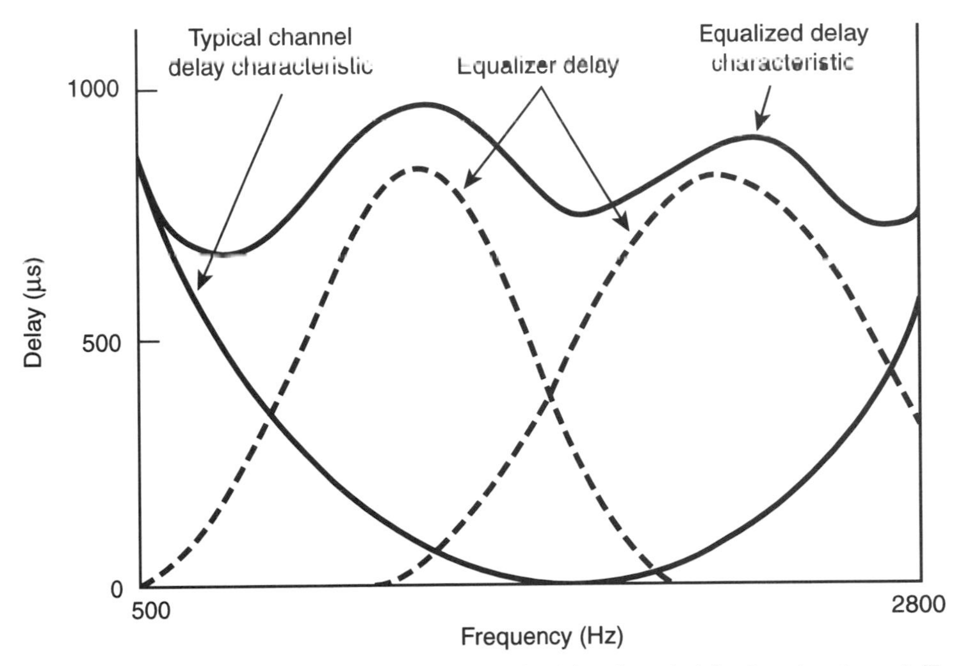

Figure 10.18 A delay (phase) equalizer tends to flatten the delay characteristic of a voice channel. (From Ref. 14. Copyright BTL. Reprinted with permission.)

lowing to interface these standard data rates with the 56/64 small kbps digital channel. The first is AT&T's Digital Data System (DDS) and the second is based on CCITT Rec. V.110 (Ref. 18).

10.9.7.1 AT&T Digital Data System (DDS). The AT&T digital data system (DDS) provides duplex point-to-point and multipoint private line digital data transmission at a number of synchronous data rates. This system is based on the standard 1.544-Mbps DS1 PCM line rate, where individual bit streams have data rates that are submultiples of that line rate (i.e., based on 64 kbps). However, pulse slots are reversed for identification in the demultiplexing of individual user bit streams as well as for certain status and control signals and to ensure that sufficient line pulses are transmitted for receive clock recovery and pulse regeneration. The maximum data rate available to a subscriber to the system is 56 kbps, some 87.5% of the 64-kbps theoretical maximum.

The 1.544-Mbps line signal as applied to DDS service consists of 24 sequential 8-bit words (i.e., channel time slots) plus one additional framing bit. This entire sequence is repeated 8000 times per second. Note that again we have $(192 + 1)8000 = 1.544$ Mbps, where the value 192 is 8×24 (see Chapter 6). Thus the line rate of a DDS facility is compatible with the DS1 (T1) PCM line rate and offers the advantage of allowing a mix of voice (PCM) and data where the full dedication of a DS1 facility to data transmission would be inefficient in most cases.

AT&T calls the basic 8-bit word a *byte*. One bit of each 8-bit word is reserved for network control and for stuffing to meet nominal line bit rate requirements. This control bit is called a *C*-bit. With the *C*-bit removed we see where the standard channel bit rate is derived, namely, 56 kbps or 8000×7. Four subrate or submultiple data rates are also available: 2.4, 4.8, 9.6, and 19.2 kbps. However, when these rates are implemented, an additional bit must be robbed from the basic byte to establish flag patterns to route

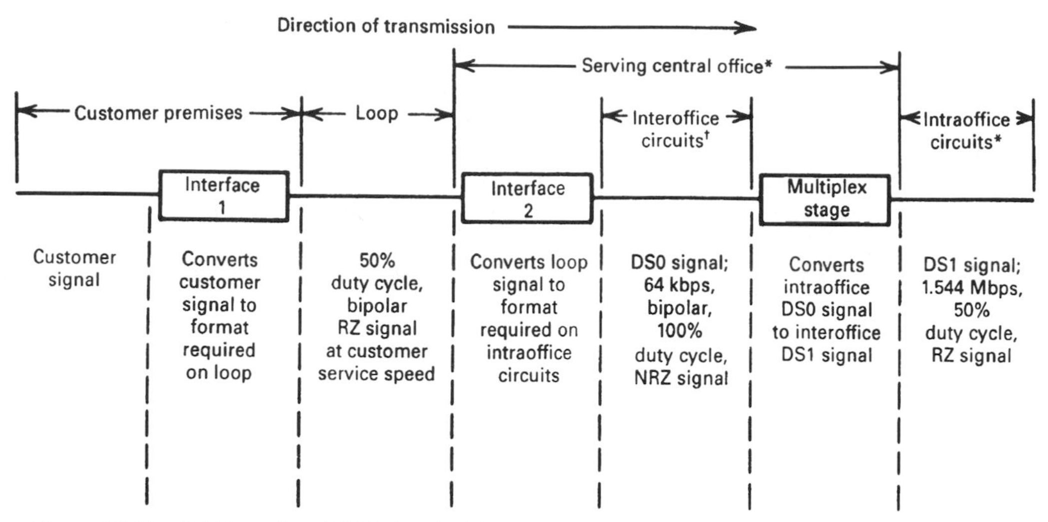

Figure 10.19 Subhierarchy of DDS signals. *Note:* Inverse processing must be provided for the opposite direction of transmission. Four-wire transmission is used throughout. (From Ref. 19. Copyright American Telephone and Telegraph Company.)

each substrate channel to its proper demultiplexer port. This allows only 48 kbps out of the original 64 kbps for the transmission of user data. The 48 kbps composite total may be divided into five 9.6-kbps channels, ten 4.8-kbps or twenty 2.4-kbps channels, or two 19.2-kbps channels plus a 9.6-kbps channel. The subhierarchy of DDS signals is illustrated in Figure 10.19.

10.9.7.2 Transmitting Data on the Digital Network Based on CCITT Rec. V.110. This CCITT recommendation covers data rate adaption for standard rates up through 19.2 kbps by means of a two-stage process. It also includes adaptation for 48 kbps and 56 kbps to the 64 kbps E0 or clear DS0 channel.[13]

In the two-step case, the first conversion is to take the incoming data rate and convert it to an appropriate intermediate rate expressed by $2^k \times 8$ kbps, where $k = 0, 1$, or 2. The second conversion takes the intermediate rate and converts it to 64 kbps.

Simple division of 64,000 bps by standard data rates shows that as a minimum, a lot of bit stuffing would be required. These would essentially be wasted bits. CCITT makes use of these bits to provide framing overhead, status information, and control information. A frame (step 1 conversion) consists of 10 octets ($10 \times 8 = 80$ bits). Six octets carry user data, the first octet is all 0s for frame alignment, and bit 1 of the remaining 9 octets is set to 1. There are 15 overhead bits. Nine different frames accommodate the various data rates, and each frame has 80 bits (10 octets). The recommendation also covers conversion of start–stop data rates including 50, 75, 110, 150, 300 bps, and the standard rates up through 19.2 kbps (Ref. 18).

10.10 WHAT ARE DATA PROTOCOLS?

To get the most out of a network, certain operational rules, procedures, and interfaces have to be established. A data network is a big investment, and our desire is to get

[13]*Clear DS0 channel* is a DS0 channel with the signaling bits disabled.

the best return on that investment. One way to optimize return on a data network is by the selection of operational protocols. We must argue that there are multiple tradeoffs involved, all interacting one with another. Among these are data needs to be satisfied, network topology and architecture, selection of transmission media, switching and network management hardware and software, and operational protocols. In this section we will focus on the protocols.

In the IEEE dictionary (Ref. 1), one definition of a protocol is: "a set of rules that govern functional units to achieve communication." We would add interfaces to the definition to make it more all-encompassing. In this section we will trace some of the evolution of protocols up through the International Standards Organization (ISO), OSI and its seven layers. Emphasis will be placed on the first three layers because they are more directly involved in communication.

In this section we will familiarize the reader with basic protocol functions. This is followed by a discussion of the Open System Interconnection (OSI), which has facilitated a large family of protocols. A brief discussion of HDLC (high-level data-link control) is provided. This particular protocol was selected because it spawned so many other link layer protocols. Some specific higher layer protocols are described in Chapter 11.

10.10.1 Basic Protocol Functions

There are a number of basic protocol functions. Typical among these are:

- Segmentation and reassembly (SAR);
- Encapsulation;
- Connection control;
- Ordered delivery;
- Flow control; and
- Error control.

A short description of each follows.

Segmentation and reassembly. Segmentation refers to breaking up the data message or file into blocks, packets, or frames with some bounded size. Which term we use depends on the semantics of the system. There is a new data segment called a *cell*, used in asynchronous transfer mode (ATM) and other digital systems. *Reassembly* is the reverse of segmentation, because it involves putting the blocks, frames, or packets back into their original order. The device that carries out segmentation and reassembly in a packet network is called a PAD (packet assembler–disassembler).

Encapsulation. Encapsulation is the adding of header and control information in front of the text or info field and parity information, which is generally carried behind the text or info fields.

Connection control. There are three stages of connection control:

1. Connection establishment;
2. Data transfer; and
3. Connection termination.

Some of the more sophisticated protocols also provide connection interrupt and recovery capabilities to cope with errors and other sorts of interruptions.

Ordered delivery. Packets, frames, or blocks are often assigned sequence numbers to ensure ordered delivery of the data at the destination. In a large network with many nodes and possible routes to a destination, especially when operated in a packet mode, the packets can arrive at the destination out of order. With a unique segment (packet) numbering plan using a simple numbering sequence, it is a rather simple task for a long data file to be reassembled at the destination it its original order.

Flow control. Flow control refers to the management of the data flow from source to destination such that buffer memories do not overflow, but maintain full capacity of all facility components involved in the data transfer. Flow control must operate at several peer layers of protocols, as will be discussed later.

Error control. Error control is a technique that permits recovery of lost or errored packets (frames, blocks). There are four possible functions involved in error control:

1. Numbering of packets (frames, blocks) (e.g., missing packet);
2. Incomplete octets (a bit sequence does not carry the proper number of bits, in this case 8 bits);
3. Error detection. Usually includes error correction. (See Section 10.5.3); and
4. Acknowledgment of one, several, or a predetermined string of packets (blocks, frames). Acknowledgment may be carried out by returning to the source (or node) the send sequence number as a receive sequence number.

10.10.2 Open Systems Interconnection (OSI)

10.10.2.1 Rationale and Overview of OSI

Data communication systems can be very diverse and complex. These systems involve elaborate software that must run on equipment having ever-increasing processing requirements. Under these conditions, it is desirable to ensure maximum independence among the various software and hardware elements of a system for two reasons:

- To facilitate intercommunication among disparate elements; and
- To eliminate the "ripple effect" when there is a modification to one software element that may affect all elements.

The ISO set about to make this data intercommunication problem more manageable. It developed its famous OSI reference model (Ref. 20). Instead of trying to solve the global dilemma, it decomposed the problem into more manageable parts. This provided standard-setting agencies with an architecture that defines communication tasks. The OSI model provides the basis for connecting open systems for distributed applications processing. The term *open* denotes the ability of any two systems conforming to the reference model and associated standards to interconnect. OSI thus provides a common groundwork for the development of families of standards permitting data assets to communicate.

ISO broke data communications down into seven areas or layers, arranged vertically starting at the bottom with layer 1, the input/output ports of a data device. The OSI reference model is shown in Figure 10.20. It takes at least two to communicate. Thus we consider the model in twos, one entity to the left in the figure and one peer entity to the right. ISO and the ITU-T organization use the term *peers*. Peers are corresponding entities on either side of Figure 10.20. A peer on one side of the system (system A) communicates with its peer on the other side (system B) by means of a common proto-

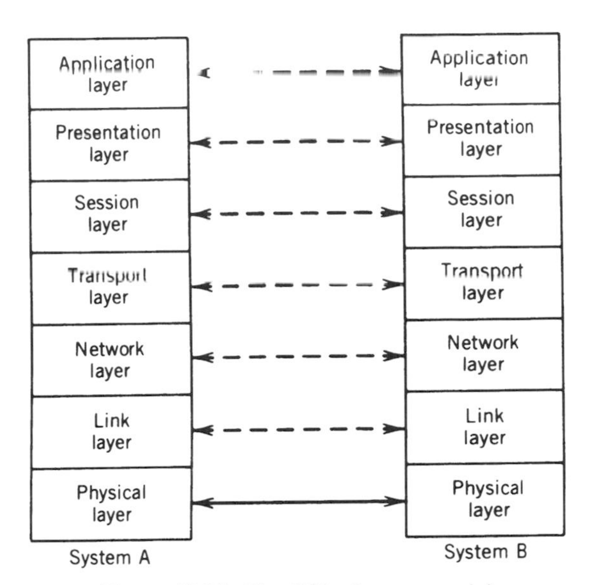

Figure 10.20 The OSI reference model.

col. For example, the transport layer of system A communicates with its peer transport layer at system B. It is important to note that there is no direct communication between peer layers except at the physical layer (layer 1). That is, above the physical layer, each protocol entity sends data down to the next lower layer, and so on to the physical layer, then across and up to its peer on the other side. Even the physical layer may not be directly connected to its peer on the other side of the "connection" such as in packet communications. This we call *connectionless service* when no physical connection is set up.[14] However, peer layers must share a common protocol in order to communicate.

There are seven OSI layers, as shown in Figure 10.20. Any layer may be referred to as an N-layer. Within a particular system there are one or more active entities in each layer. An example of an entity is a process in a multiprocessing system. It could simply be a subroutine. Each entity communicates with entities above it and below it across an interface. The interface is at a service access point (SAP).

The data that pass between entities are a bit grouping called a *protocol data unit* (PDU). Data units are passed downward from a peer entity to the next OSI layer, called the $(N-1)$ layer. The lower layer calls the PDU a *service data unit* (SDU). The $(N-1)$ layer adds control information, transforming the SDU into one or more PDUs. However, the identity of the SDU is preserved to the corresponding layer at the other end of the connection. This concept is illustrated in Figure 10.21.

OSI has considerable overhead. By overhead, we mean bit sequences that are used for logical interfaces or just simply to make the system work. Overhead does not carry revenue-bearing traffic. Overhead has a direct bearing on system efficiency: as overhead increases, system efficiency decreases.

OSI layering is widely accepted in the world of data communications even with its considerable overhead. Encapsulation is a good example. Encapsulation is used on all OSI layers above layer 1, as shown in Figure 10.22.

[14]*Connectionless service* is a type of delivery service that treats each packet, datagram, or frame as a separate entity containing the source and destination address. An analogy in everyday life is the postal service. We put a letter in the mail and we have no idea how it is routed to its destination. The address on the letter serves to route the letter.

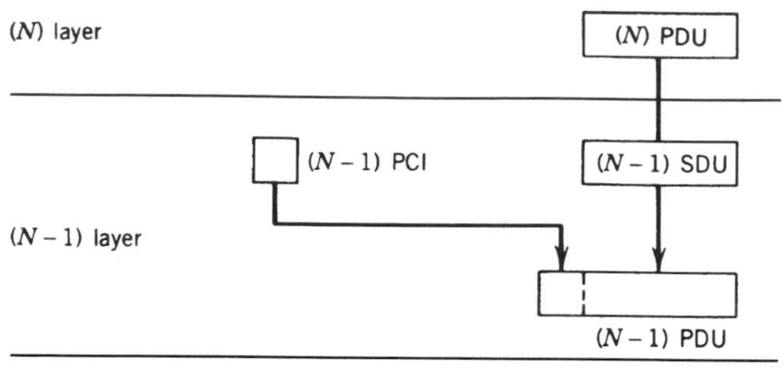

Figure 10.21 An illustration of mapping between data units on adjacent layers.

10.10.2.2 Functions of the First Four OSI Layers. Only the first four OSI layers are described in the following paragraphs. Layers 5, 6, and 7 are more in the realm of software design.

Physical layer. The physical layer is layer 1, the lowest OSI layer. It provides the physical connectivity between two data terminals who wish to communicate. The services it provides to the data link layer (layer 2) are those required to connect, maintain the connection, and disconnect the physical circuits that form the physical connectivity. The physical layer represents the traditional interface between the DCE and DTE, described in Section 10.8.

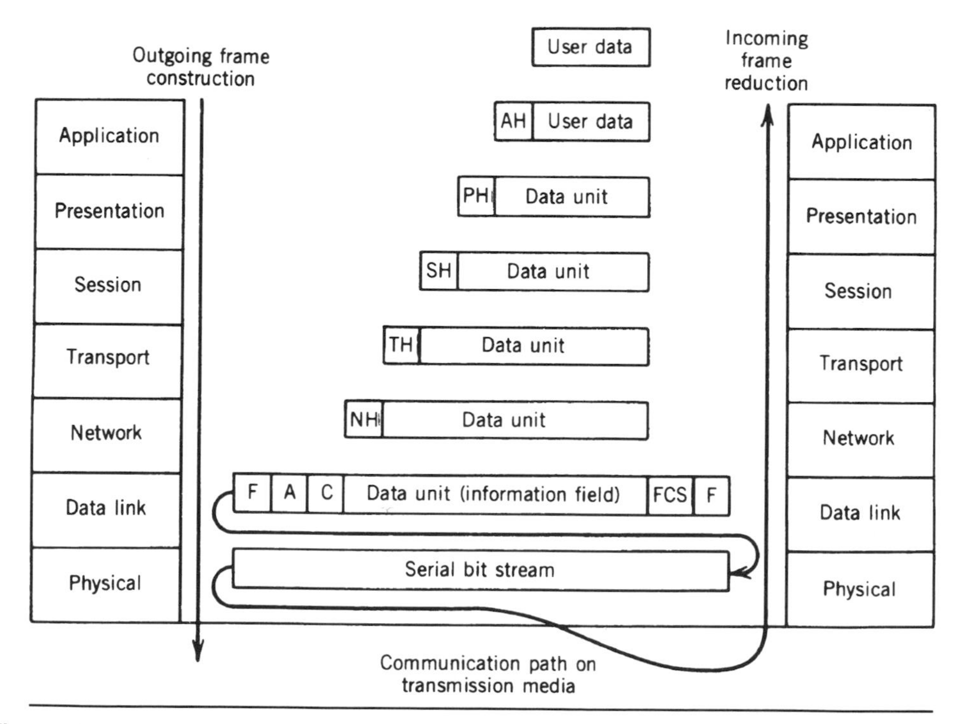

Figure 10.22 Buildup and breakdown of a data message based on the OSI model. OSI encapsulates at every layer except layer 1.

The physical layer has four important characteristics:

1. Mechanical;
2. Electrical;
3. Functional; and
4. Procedural.

The mechanical aspects include the actual cabling and connectors necessary to connect the communications equipment to the media. Electrical characteristics cover voltage and impedance, balanced and unbalanced. Functional characteristics include connector pin assignments at the interface and the precise meaning and interpretation of the various interface signals and data set controls. Procedures cover sequencing rules that govern the control functions necessary to provide higher-layer services such as establishing a connectivity across a switched network.

Data-link layer. The data-link layer provides services for reliable interchange of data across a data link established by the physical layer. Link-layer protocols manage the establishment, maintenance, and release of data-link connections. These protocols control the flow of data and supervise error recovery. A most important function of this layer is recovery from abnormal conditions. The data-link layer services the network layer or logical link control (LLC; in the case of LANs) and inserts a data unit into the INFO portion of the data frame or block. A generic data frame generated by the link layer is illustrated in Figure 10.7.

Several of the more common data-link layer protocols are: CCITT LAPB, LAPD; IBM SDLC; and ANSI ADCCP (also the U.S. government standard).

Network layer. The network layer moves data through the network. At relay and switching nodes along the traffic route, layering concatenates. In other words, the higher layers (above layer 3) are not required and are utilized only at user end-points.

The concept of relay open system is shown in Figure 10.23. At the relay switching point, only the first three layers of OSI are required.

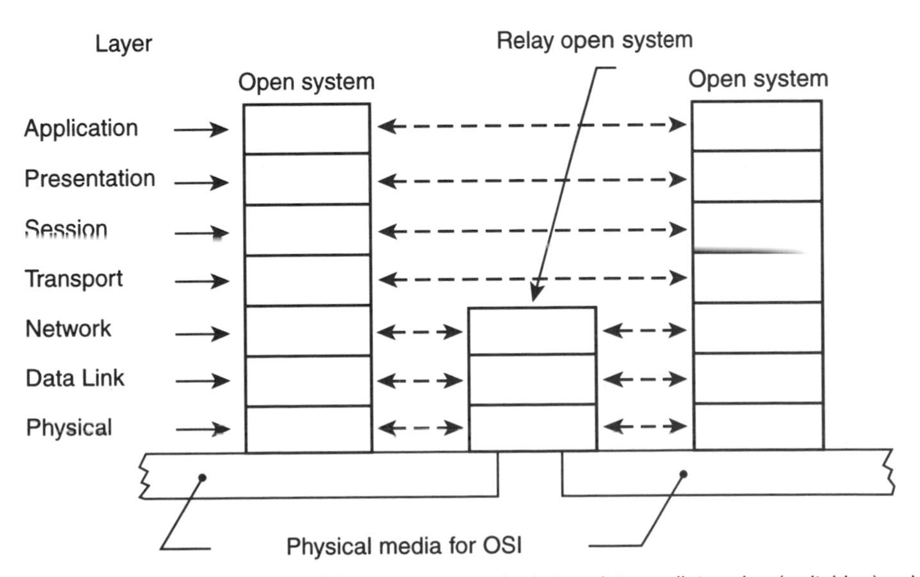

Figure 10.23 Only the first three OSI layers are required at an intermediate relay (switching) point.

The network layer carries out the functions of switching and routing, sequencing, logical channel control, flow control, and error-recovery functions. We note the duplication of error recovery in the data-link layer. However, in the network layer error recovery is network-wide, whereas on the data-link layer error recovery is concerned only with the data link involved.

The network layer also provides and manages logical channel connections between points in a network such as virtual circuits across the public switched network (PSN). It will be appreciated that the network layer concerns itself with the network switching and routing function. On simpler data connectivities, where a large network is not involved, the network layer is not required and can be eliminated. Typical of such connectivities are point-to-point circuits, multipoint circuits, and LANs. A packet-switched network is a typical example where the network layer is required.

The best-known layer 3 standard is CCITT Rec. X.25. (Ref. 22).

Transport layer. The transport layer (layer 4) is the highest layer of the services associated with the provider of communication services. One can say that layers 1–4 are the responsibility of the communication system engineer. Layers 5, 6, and 7 are the responsibility of the data end-user. However, we believe that the telecommunication system engineer should have a working knowledge of all seven layers.

The transport layer has the ultimate responsibility for providing a reliable end-to-end data-delivery service for higher-layer users. It is defined as an end-system function, located in the equipment using network service or services. In this way its operations are independent of the characteristics of all the networks that are involved. Services that a transport layer provides are as follows:

- *Connection Management.* This includes establishing and terminating connections between transport users. It identifies each connection and negotiates values of all needed parameters.
- *Data Transfer.* This involves the reliable delivery of transparent data between the users. All data are delivered in sequence with no duplication or missing parts.
- *Flow Control.* This is provided on a connection basis to ensure that data are not delivered at a rate faster than the user's resources can accommodate.

The TCP (transmission control protocol) was the first working version of a transport protocol and was created by DARPA for DARPANET.[15] All the features in TCP have been adopted in the ISO version. TCP is often combined with the Internet protocol (IP) and is referred to as TCP/IP.

10.10.3 High-Level Data-Link Control: A Typical Link-Layer Protocol

High-level data-link control (HDLC) was developed by the ISO. It has spawned many related or nearly identical protocols. Among these are ANSI ADCCP, CCITT LAPB and LAPD, IEEE Logical Link Control (LLC), and IBM SDLC. (Refs. 20–25).

The following HDLC definitions include stations, configurations, and three modes of operation.

[15]DARPA stands for Defense Advanced Research Projects Agency, under the U.S. Department of Defense.

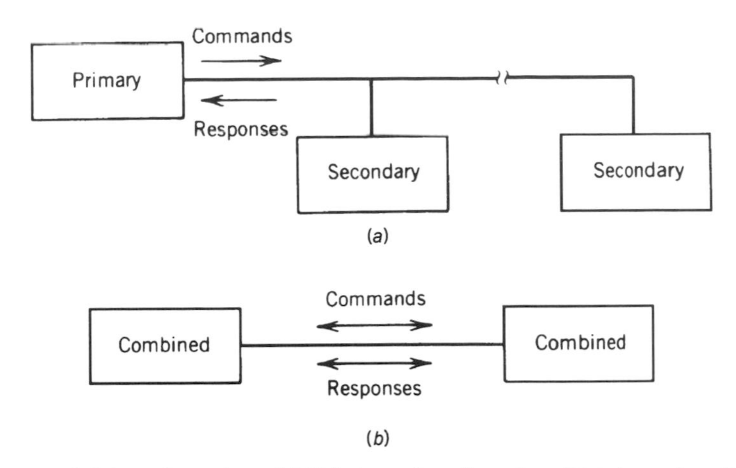

Figure 10.24 HDLC link configurations. (*a*) Unbalanced configuration; (*b*) balanced configuration.

Primary Station. A logical primary station is an entity that has primary link control responsibility. It assumes responsibility for organization of data flow and for link level error recovery. Frames issued by the primary station are called *commands.*

Secondary Station. A logical secondary station operates under control of a primary station. It has no direct responsibility for control of the link, but instead responds to primary station control. Frames issued by a secondary station are called *responses.*

Combined Station. A combined station combines the features of primary and secondary stations. It may issue both commands and responses.

Unbalanced Configuration. An unbalanced configuration consists of a primary station and one or more secondary stations. It supports full-duplex and half-duplex operation, point-to-point, and multipoint circuits. An unbalanced configuration is illustrated in Figure 10.24*a*.

Balanced Configuration. A balanced configuration consists of two combined stations in which each station has equal and complementary responsibility of the data link. A balanced configuration, shown in Figure 10.24*b*, operates only in the point-to-point mode and supports full-duplex operation.

Modes of Operation. With *normal response mode* (NRM) a primary station initiates data transfer to a secondary station. A secondary station transmits data only in response to a poll from the primary station. This mode of operation applies to an unbalanced configuration. With *asynchronous response mode* (ARM) a secondary station may initiate transmission without receiving a poll from a primary station. It is useful on a circuit where there is only one active secondary station. The overhead of continuous polling is thus eliminated. *Asynchronous balanced mode* (ABM) is a balanced mode that provides symmetric data transfer capability between combined stations. Each station operates as if it were a primary station, can initiate data transfer, and is responsible for error recovery. One application of this mode is hub polling, where a secondary station needs to initiate transmission.

10.10.3.1 HDLC Frame. Figure 10.25 shows the HDLC frame format. Note the similarity to the generic data-link frame illustrated in Figure 10.8. Moving from left to right in the figure, we have the flag field (F), which delimits the frame at both ends with the unique pattern 01111110. This unique field or flag was described in Section 10.7.2.

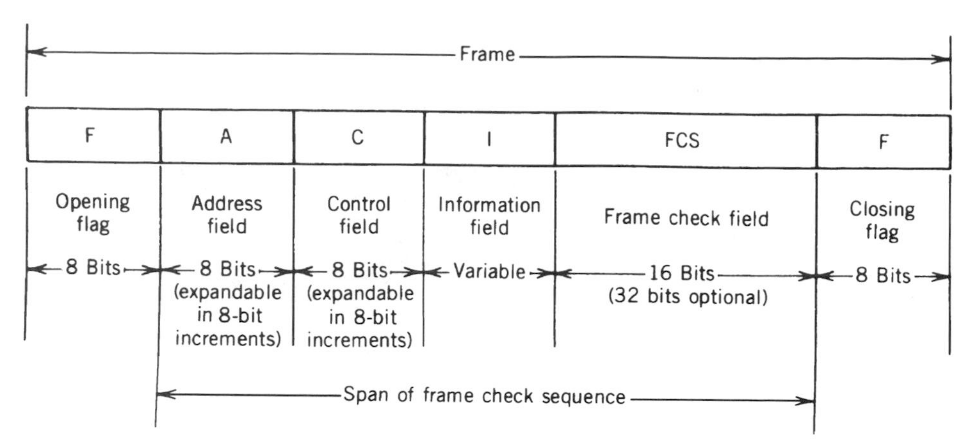

Figure 10.25 The HDLC frame format (Ref. 25).

The *address field* (A) immediately follows the opening flag of a frame and precedes the control field (C). Each station in the network normally has an individual address and a group address. A group address identifies a family of stations. It is used when data messages must be accepted from or destined to more than one user. Normally the address is 8 bits long, providing 256 bit combinations or addresses ($2^8 = 256$). In HDLC (and ADCCP) the address field can be extended in increments of 8 bits. When this is implemented, the least significant bit is used as an extension indicator. When that bit is 0, the following octet is an extension of the address field. The address field is terminated when the least significant bit of an octet is 1. Thus we can see that the address field can be extended indefinitely.

The *control field* (C) immediately follows the address field (A) and precedes the information field (I). The control field conveys commands, responses, and sequence numbers to control the data link. The basic control field is 8 bits long and uses modulo 8 sequence numbering. There are three types of control field: (1) I frame (information frame), (2) S frame (supervisory frame), and (3) U frame (unnumbered frame). The three control field formats are illustrated in Figure 10.26.

Consider the basic 8-bit format shown in Figure 10.26. The information flows from left to right. If the frame shown in Figure 10.25 has a 0 as the first bit in the control field, the frame is an I frame (see Figure 10.26*a*). If the bit is a 1, the frame is an S or a U frame, as illustrated in Figure 10.26*b* and 10.26*c*. If the first bit is followed by a 0, it is an S frame, and if the bit again is a 1 followed by a 1, it is a U frame. These bits are called *format identifiers*.

Turning now to the information (I) frame (Figure 10.26*a*), its purpose is to carry user data. Bits 2, 3, and 4 of the control field in this case carry the *send sequence number* N(S) of the transmitted messages (i.e., I frames). N(S) is the frame sequence number of the next frame to be transmitted and N(R) is the sequence number of the frame to be received.

Each frame carries a poll/final (P/F) bit. It is bit 5 in each of the three different types of control fields shown in Figure 10.26. This bit serves a function in both command and response frames. In a command frame it is referred to as a poll (P) bit; in a response frame as a final (F) bit. In both cases the bit is sent as a 1.

The P bit is used to solicit a response or sequence of responses from a secondary or balanced station. On a data link only one frame with a P bit set to 1 can be outstanding at any given time. Before a primary or balanced station can issue another frame with a P

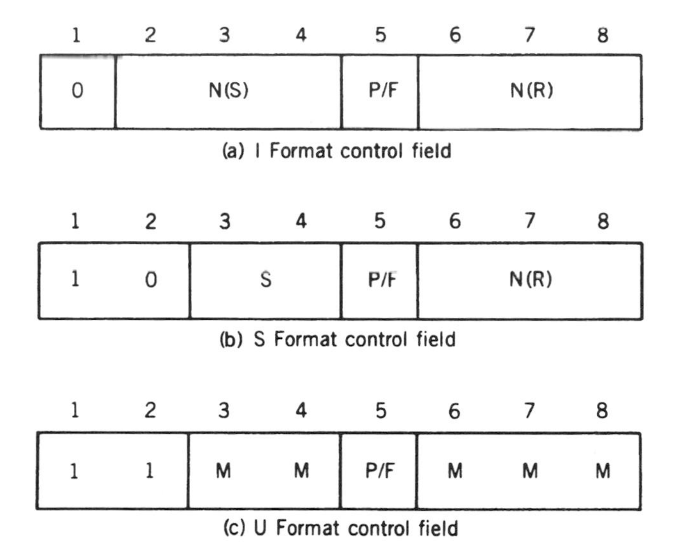

Figure 10.26 The three control field formats of HDLC.

bit set to 1, it must receive a response frame from a secondary or balanced station with the F bit set to 1. In the NRM mode, the P bit is set to 1 in command frames to solicit response frames from the secondary station. In this mode of operation the secondary station may not transmit until it receives a command frame with the P bit set to 1.

Of course, the F bit is used to acknowledge an incoming P bit. A station may not send a final frame without prior receipt of a poll frame. As can be seen, P and F bits are exchanged on a one-for-one basis. Thus only one P bit can be outstanding at a time. As a result the N(R) count of a frame containing a P or F bit set to 1 can be used to detect sequence errors. This capability is called *check pointing*. It can be used not only to detect sequence errors but to indicate the frame sequence number to begin retransmission when required.

Supervisory frames, shown in Figure 10.26*b*, are used for flow and error control. Both go-back-n and continuous (selective) ARQ can be accommodated. There are four types of supervisory frames:

1. Receive ready (RR): 1000 P/F N(R);
2. Receive not ready (RNR): 1001 P/F N(R);
3. Reject (Rej): 1010 P/F N(R); and
4. Selective reject (SRej): 1011 P/F N(R).

The RR frame is used by a station to indicate that it is ready to receive information and acknowledge frames up to and including N(R) − 1. Also, a primary station may use the RR frame as a command with the poll (P) bit set to 1.

The RNR frame tells a transmitting station that it is not ready to receive additional incoming I frames. It does acknowledge receipt of frames up to and including sequence number N(R) − 1. I frames with sequence number N(R) and subsequent frames, if any, are not acknowledged. The Rej frame is used with go-back-n ARQ to request retransmission of I frames with frame sequence number N(R), and N(R) − 1 frames and below are acknowledged.

Unnumbered frames are used for a variety of control functions. They do not carry sequence numbers, as the name indicates, and do not alter the flow or sequencing of I frames. Unnumbered frames can be grouped into the following four categories:

1. Mode-setting commands and responses;
2. Information transfer commands and responses;
3. Recovery commands and responses; and
4. Miscellaneous commands and responses.

The information field follows the control field (Figure 10.25) and precedes the frame check sequence (FCS) field. The I field is present only in information (I) frames and in some unnumbered (U) frames. The I field may contain any number of bits in any code, related to character structure or not. Its length is not specified in the standard (ISO 3309, Ref. 25). Specific system implementations, however, usually place an upper limit on I field size. Some versions require that the I field contain an integral number of octets.

Frame check sequence (FCS). Each frame includes an FCS field. This field immediately follows the I field, or the C field if there is no I field, and precedes the closing flag (F). The FCS field detects errors due to transmission. The FCS field contains 16 bits, which are the result of a mathematical computation on the digital value of all bits excluding the inserted zeros (zero insertion) in the frame and including the address, control, and information fields.

REVIEW EXERCISES

1. What is the basic element of information in a binary system? How much information does it contain?

2. How does one extend the information content of that basic information element (question 1), for example, to construct a binary code that represents, as a minimum, our alphabet?

3. There are many ways we can express the binary 1 and the binary 0. Give at least four. How do we remove ambiguity about meaning (reversing the sense)?

4. How many distinct characters or symbols can be represented by a 4-unit binary code? a 7-unit binary code? an 8-unit binary code? (*Hint:* Consider a unit as a bit for this argument.)

5. How many information elements (bits) are there is an ASCII character?

6. Give two causes of burst errors.

7. Give the two generic methods of *correcting* errors on a datalink.

8. Name the three different types of ARQ and define each.

9. Describe the difference between neutral and polar transmission.

10. On a start–stop circuit, where does a receiver start counting information bits?

11. There are three major causes of error on a data link. Name two of them.

12. On start–stop transmission, the mark-to-space transition on the start element tells the receiver when to start counting bits. How does a synchronous data system know when to start counting bits in a frame or packet?

13. How does a synchronous data receiver keep in synchronization with an incoming bit stream?

14. A serial synchronous NRZ bit stream has a data rate of 19.2 kbps. What is the period of one bit?

15. What is notably richer in transitions per unit time: RZ or NRZ coding?

16. The CCITT V.29 modem operates on the standard analog voice channel at 9600 bps. How can it do this on a channel with a 3100-Hz bandwidth?

17. Name the three basic impairments for data transmission.

18. Name the four types of noise. Indicate the two that a data circuit is sensitive to and explain.

19. Phase distortion, in general, has little effect on speech transmission. What can we say about it for data transmission?

20. Shannon's formula for capacity (bps) for a particular bandwidth was based only on one parameter. What was/is it?

21. We usually equalize two voice channel impairments. What are they and how does the equalization work?

22. Why do higher-speed modems use a center frequency around 1800 Hz?

23. Why is the PSTN digital network not compatible with data bit streams?

REFERENCES

1. *IEEE Standard Dictionary of Electrical and Electronic Terms*, 6th ed, IEEE Std. 100-1996, IEEE, New York, 1996.
2. *Equivalence between Binary Notation Symbols and the Significant Conditions of a Two-Condition Code*, CCITT Rec. V.1, Fascicle VIII.1, IXth Plenary Assembly, Melbourne 1988.
3. R. L. Freeman, *Practical Data Communications*, Wiley, New York, 1995.
4. R. L. Freeman, *Telecommunication Transmission Handbook*, 4th ed., Wiley, New York, 1998.
5. *Interface between Data Terminal Equipment and Data Circuit-Terminating Equipment Employing Serial Binary Data Interchange*, EIA/TIA-232E, Electronics Industries Assoc., Washington, DC, July 1991.
6. *Error Performance on an International Digital Connection Forming Part of an Integrated Services Digital Network*, CCITT Rec. G.821, Fascicle III.5, IXth Plenary Assembly, Melbourne, 1988.
7. *High-Speed 25-Position Interface for Data Terminal Equipment and Data Circuit Terminating Equipment Including Alternative 26-Position Connector*, EIA/TIA-530-A, Electronic Industries Assoc., Washington, DC, June 1992.
8. W. R. Bennett and J. R. Davey, *Data Transmission*, McGraw-Hill, New York, 1965.
9. *Attenuation Distortion*, CCITT Rec. G132, Fascicle III.1, IXth Plenary, Melbourne, 1988.
10. C. E. Shannon, "A Mathematical Theory of Communications." *BSTJ* **27**, 1948.
11. *1200 bps Duplex Modem Standardized for Use in the General-Switched Telephone Network*

and on Point-to-Point 2-Wire Leased Telephone Type Circuits, CCITT Rec. V.22, Fascicle VIII.1, IXth Plenary Assembly, Melbourne, 1988.

12. *9600 bps Modem Standardized for Use on Point-to-Point 4-Wire Leased Telephone Type Circuits*, CCITT Rec. V.29, Fascicle VIII.1, IXth Plenary Assembly, Melbourne, 1988.
13. K. Pahlavan and J. L. Holsinger, "Voice Band Communication Modems: An Historical Review, 1919–1988," *IEEE Communications Magazine*, **261**(1), 1988.
14. *Transmission Systems for Communications*, 5th ed., Bell Telephone Laboratories, Holmdel, NJ, 1982.
15. *Standardization of Data Signaling Rates for Synchronous Data Transmission in the General Switched Telephone Network*, CCITT Rec. V.5, Fascicle VIII.1, IXth Plenary Assembly, Melbourne, 1988.
16. *Standardization of Data Signaling Rates for Synchronous Data Transmission on Leased Telephone-Type Circuits*, CCITT Rec. V.6, IXth Plenary Assembly, Melbourne, 1988.
17. *Synchronous Signaling Rates for Data Transmission*, EIA-269A, Electronic Industries Assoc., Washington, DC, May 1968.
18. *Support of Data Terminal Equipments with V-Series Type Interfaces by an Integrated Services Digital Network*, CCITT Rec. V.110, ITU Geneva, 1992.
19. *Digital Data System Data Service Unit Interface Specification*, Bell System Reference 41450, AT&T, New York, 1981.
20. *Information Processing Systems Open Systems Interconnection—Basic Reference Model*, ISO 7498, Geneva, 1984.
21. *Advanced Data Communications Control Procedures*, X.3.66, ANSI, New York, 1979.
22. *Interface between Data Terminal Equipment (DTE) and Data Circuit-Terminating Equipment (DCE) for Terminals Operating in the Packet Mode and Connected to Public Data Networks by a Dedicated Circuit*, ITU-T Rec. X.25, ITU Geneva, March 1993.
23. *ISDN User Network Interface—Data Link Layer*, CCITT Rec. Q.921, Fascicle VI.10, IXth Plenary Assembly, Melbourne, 1988.
24. *Information Processing Systems—Local Area Networks, Part 2, Logical Link Control*, IEEE Std. 802.3, 1994 ed., IEEE, New York, 1994.
25. *High-Level Data Link Control Procedures—Frame Structure*, ISO 3309, International Standards Organization, Geneva, 1979.
26. *"Military Communication System" Technical Standard*, MiL-STD-188C, U.S. Dept. of Defense, Washington, DC 1966.
27. *Reference Data for Engineers: Radio, Electronics, Computers and Communications*, 8th ed, SAM Publishing, Carmel, IN, 1993.

11

ENTERPRISE NETWORKS I: LOCAL AREA NETWORKS

11.1 WHAT DO ENTERPRISE NETWORKS DO?

An *enterprise network* consists of an interconnected group of telecommunication facilities that are confined to a singular entity. There can be broad, midrange, and narrow interpretations of the definition. The "singular entity" may be government or an industrial enterprise. FTS2000, the large, GSA (General Services Administration) network, which was chartered to serve all U.S. government organizations, is an example of a broad interpretation. This network provides multiple services: voice, data, and image. It is the U.S. government's own PSTN. A narrow interpretation might be Corrugated Box Works' local area network. Or Walgreens' large VSAT network. This latter example is included because the network serves a singular purpose, for logistics and cash flow.

This chapter will be based on the narrower interpretation and will be confined to the various data networks that may be employed in government and industry. We first present an overview of local area networks (LANs), followed by a discussion of wide area networks (WANs).

11.2 LOCAL AREA NETWORKS (LANs)

The IEEE (Ref. 1) defines a local area network as "A communication network to interconnect a variety of devices (e.g., personal computers, workstations, printers, file storage devices) that can transmit data over a limited area, typically within a facility."

The geographical extension or "local area" may extend from less than 100 ft (<30 m) to over 6 mi (>10 km). More commonly we can expect a LAN to extend over a floor in a building and, in some cases, over a portion of a floor. Other LANs may cover a college or industrial campus.

The transmission media encompass wire-pair, coaxial cable, fiber-optic cable, and radio. The implementation of coaxial cable systems is slowing in favor of high-quality twisted wire pair. Wireless LANs are just coming on line. Data rates vary from 1 Mbps to 1000 Mbps. LAN data rates, the number of devices connected to a LAN, the spacing of those devices and the network extension, all depend on:

- The transmission medium employed;
- The transmission technique (i.e., baseband or broadband);

- Network access protocol; and
- The incorporation of such devices as repeaters, bridges, routers and switching hubs.

The preponderance of LANs operate without error correction with BERs specified in the range of 1×10^{-8} to 1×10^{-12} or better.

The most common application of a LAN is to interconnect data terminals and other processing resources, where all the devices reside in a single building or complex of buildings, and usually these resources have a common owner. A LAN permits effective cost sharing of high-value data-processing equipment, such as mass storage media, mainframe computers or minicomputers, and high-speed printers. Resource sharing is probably equally as important where a LAN serves as the access vehicle for an *intranet.* Resource sharing in this context means that each LAN user has access to all other users' files and other data resources.

The interconnection of LANs in the local area (e.g., the floor of a building) with a high-speed backbone (e.g., between building floors) is very prevalent. LANs may connect through wide area networks (WANs) to other distant LANs. This is frame relay's principal application. The interface to the WAN may be via a smart bridge or a router.

There are two generic transmission techniques utilized by LANs: baseband and broadband. Baseband transmission can be defined as the direct application of the baseband signal to the transmission medium. Broadband transmission, in this context, is where the baseband signal from the data device is translated in frequency to a particular frequency slot in the RF spectrum. Broadband transmission requires a modem to carry out the translation. Baseband transmission may require some sort of signal-conditioning device. With broadband LAN transmission we usually think of simultaneous multiple RF carriers that are separated in the frequency domain. Present broadband technology comes from the cable television (CATV) industry.

The discussion in this section will essentially cover baseband LANs. An important aspect is that only one user at a time may access the LAN.

11.3 LAN TOPOLOGIES

There are three types of basic LAN topology: bus, ring, and star.[1] These are illustrated in Figure 11.1 along with the *tree network*, (Figure 11.1*d*), which is a simple derivative of the conventional bus topology.

A bus is a length of transmission medium which users tap into, as shown in Figure 11.1*a*. Originally the medium was coaxial cable. Today coaxial cable is being phased out in favor of unshielded twisted pair (UTP) or fiber-optic cable.

A ring is simply a bus that is folded back onto itself. A ring topology is illustrated in Figure 11.1*b*. User traffic flows in one direction around the ring. In another approach (FDDI), which is discussed later in Section 11.6.4, a second ring is added where the traffic flow is in the opposite direction. Such a dual counterrotating ring concept improves reliability in case of a failed station or a cut in the ring.

A star network is illustrated in Figure 11.1*c*. At the center of the star is a switching device. This could be a switching hub. Users can be paired, two at a time, three at a time, or all at a time, segmented into temporary families of users, depending on the configuration of the switch at that moment in time. Such a concept lends itself particu-

[1] *Topology* means the logical and/or physical arrangement of stations on a network. In other words, topology tells how these assets are connected together.

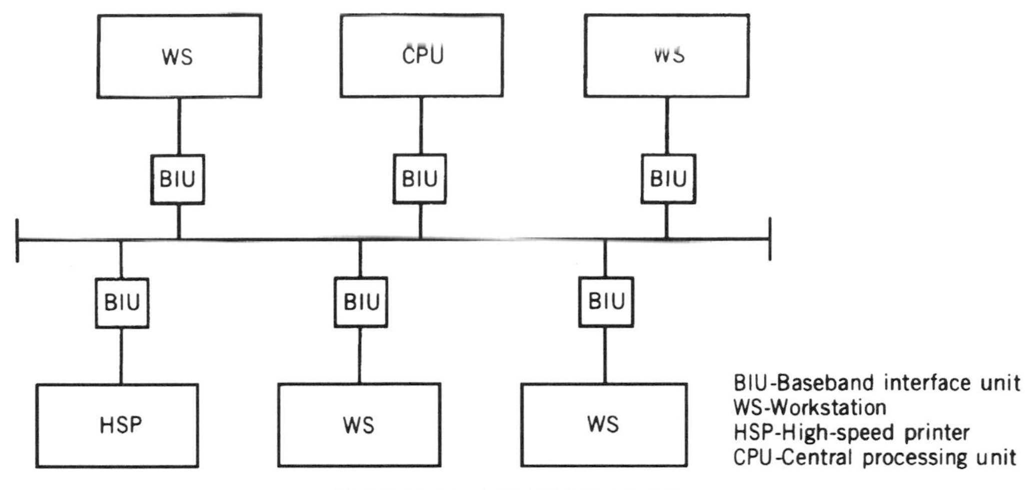

Figure 11.1*a* A typical bus network.

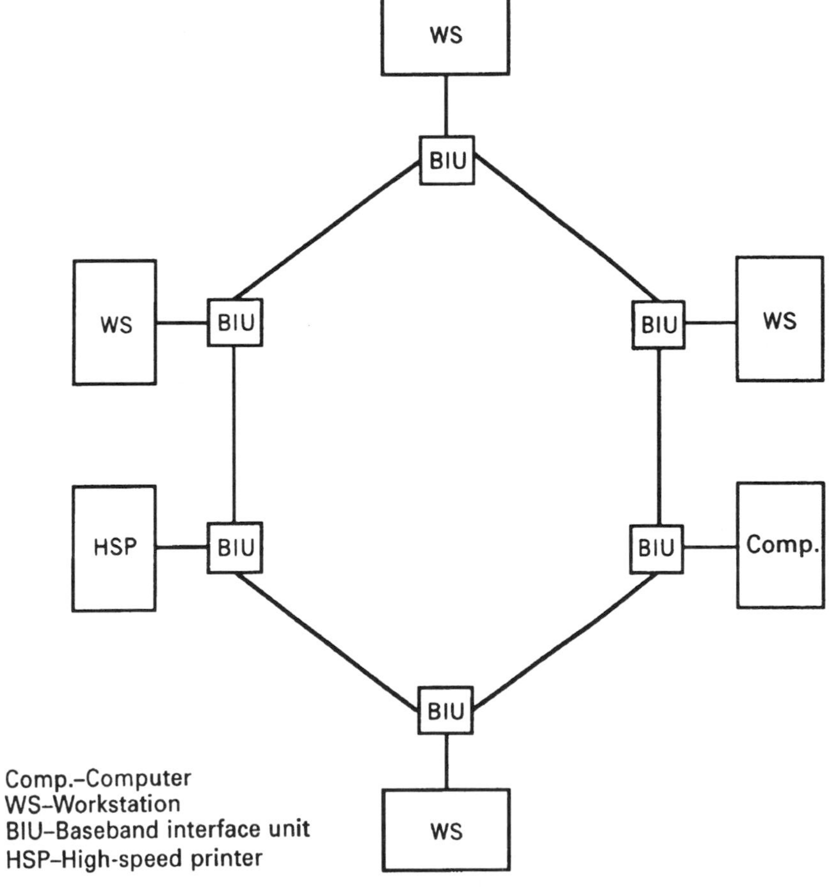

Figure 11.1*b* A typical ring network.

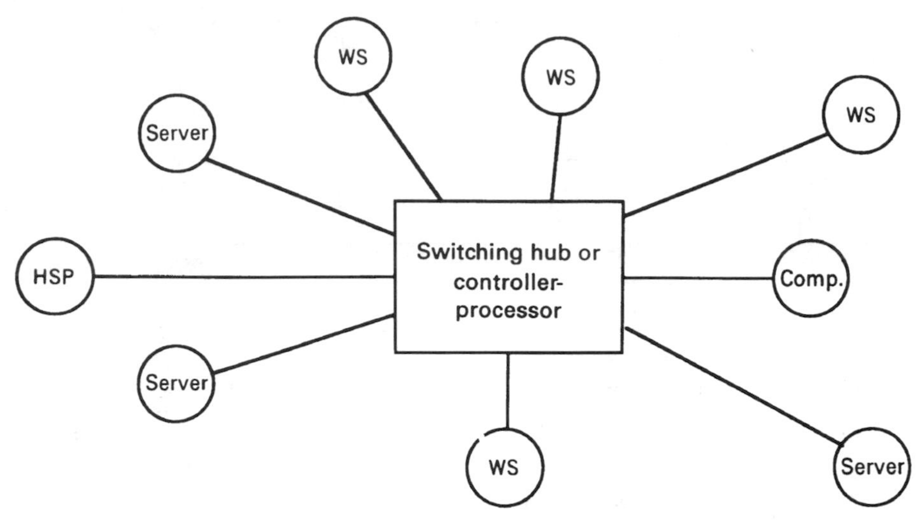

Figure 11.1c A star network.

larly well to ATM (asynchronous transfer mode). Each user is connected to the switch on a point-to-point basis.

11.4 BASEBAND LAN TRANSMISSION CONSIDERATIONS

A baseband LAN is a point-to-point or point-to-multipoint network. Two transmission problems arise as a result. The first deals with signal level and signal-to-noise (S/N) ratio, and the second deals with standing waves. Each access on a common medium

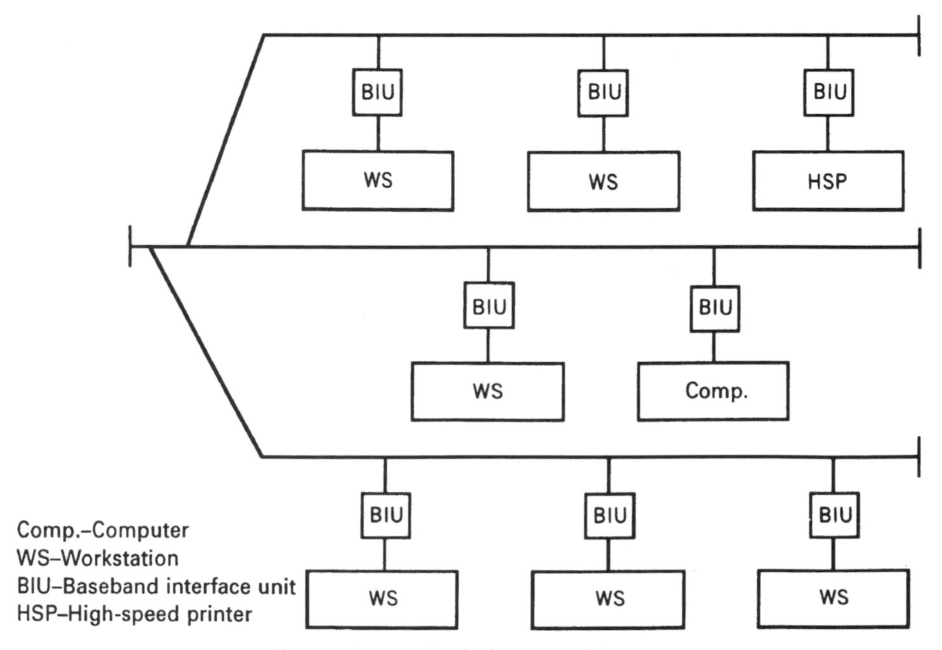

Figure 11.1d A typical tree configuration.

must have sufficient signal level and S/N such that copied signals have a BER in the range of 1×10^{-8} to 1×10^{-12}. If the medium is fairly long in extension and there are many accesses, the signal level must be high for a transmitting access to reach its most distant destination. The medium is lossy, particularly at the higher bit rates, and each access tap has an insertion loss. This leads to very high signal levels. These may be rich in harmonics and spurious emissions, degrading bit error rate. On the other hand, with insufficient level, the S/N ratio degrades, which will also degrade error performance. A good level balance must be achieved for all users. Every multipoint connectivity must be examined. The number of multipoint connectivities can be expressed by $n(n-1)$, where n is the number of accesses. If, on a particular LAN, 100 accesses are planned, there are 9900 possible connectivities to be analyzed to carry out signal level balance. One way to simplify the job is to segment the network, placing a regenerative repeater (or bridge) at each boundary. This reduces the signal balance job to realizable proportions and ensures that a clean signal of proper level is available at each access tap. For baseband LANs, 50-Ω coaxial cable is favored over the more common 75-Ω cable. The lower-impedance cable is less prone to signal reflections from access taps and provides better protection against low-frequency interference.

11.5 OVERVIEW OF ANSI/IEEE LAN PROTOCOLS

11.5.1 Introduction

Many of the widely used LAN protocols have been developed in North America through the offices of the Institute of Electrical and Electronic Engineers (IEEE). The American National Standards Institute (ANSI) has subsequently accepted and incorporated these standards, and they now bear the ANSI imprimatur.

The IEEE develops LAN standards in the IEEE 802 family of committees. Of interest in this chapter are the following IEEE committees, each with published standards, which continue to evolve:

- 802.1 High-Level Interface;
- 802.2 Logical Link Control (LCC);
- 802.3 CSMA/CD Networks;
- 802.4 Token Bus Networks;
- 802.5 Token Ring Networks;
- 802.6 Metropolitan Area Networks (MANs); and
- 802.7 Broadband Technical Advisory Group.

The Fiber Distributed Data Interface (FDDI) standard, which is addressed in Section 11.6.4, was developed and is maintained by ANSI.

11.5.2 How LAN Protocols Relate to OSI

LAN protocols utilize only OSI layers 1 and 2, the physical and data-link layers, respectively. The data-link layer is split into two sublayers: logical link control (LLC) and medium access control (MAC). These relationships are shown in Figure 11.2. The principal functions of OSI layer 3, namely, switching, relaying, and network end-to-end control, are not necessary in this simple, closed network. Remaining layer 3 functions

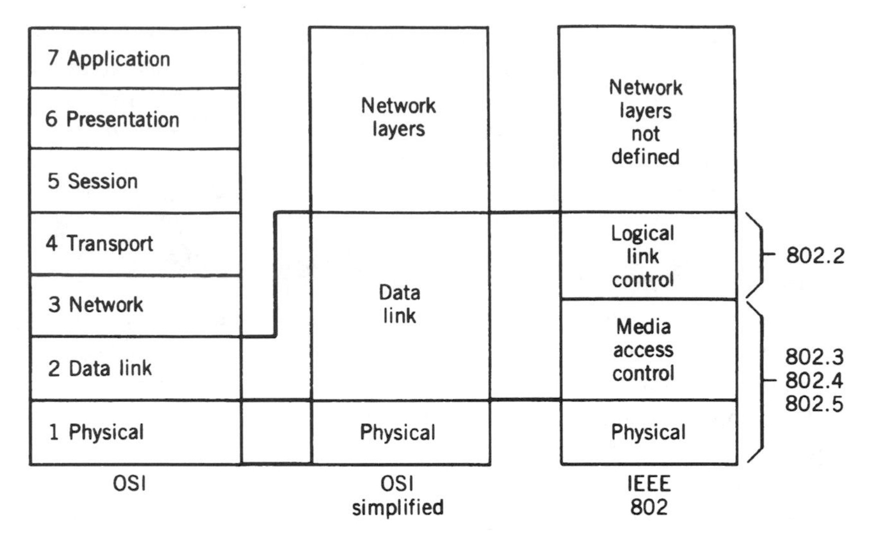

Figure 11.2 LAN 802 architecture related to OSI.

that are necessary are incorporated in layer 2. The two layer-2 sublayers (LLC and MAC) carry out four functions:

1. Provide one or more service access points (SAPs). A SAP is a logical interface between two adjacent layers.
2. Before transmission, assemble data into a frame with address and error-detection fields.
3. On reception, disassemble the frame and perform address recognition and error detection.
4. Manage communications over the link.

The first function and those related to it are performed by the LLC sublayer. The last three functions are handled by the MAC sublayer.

In the following subsections we will describe four common IEEE and ANSI standardized protocols. Logical link control (LLC) is common to all four. They differ in the medium access control (MAC) protocol.

A station on a LAN may have multiple users; oftentimes these are just processes, such as processes on a host computer. These processes may wish to pass traffic to another LAN station that may have more than one "user" in residence. We will find that LLC produces a protocol data unit (PDU) with its own source and destination address. The source address, in this case, is the address of the originating user. The destination address is the address of a user in residence at a LAN station. Such a user is connected through a service access point (SAP) at the upper boundary of the LLC layer. The resulting LLC PDU is then embedded in the information field of a MAC frame. This is shown in Figure 11.3. The MAC frame also has source and destination addresses. These direct traffic to a particular LAN station or stations.

11.5.3 Logical Link Control (LLC)

The LLC provides services to the upper layers at a LAN station. It provides two forms of services for its users:

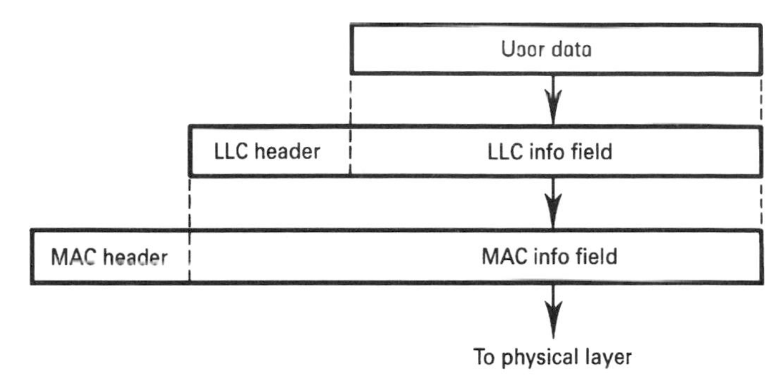

Figure 11.3 A user passes traffic to an LLC where encapsulation takes place forming an LLC PDU. The LLC PDU is embedded in a MAC frame info(rmation) field. The resulting MAC frame is passed to the physical layer, which transmits the traffic on the LAN.

1. Unacknowledged connectionless service; and
2. Connection mode services.

Some brief comments are required to clarify the functions and limitations of each service. With unacknowledged connectionless service a single service access initiates the transmission of a data unit to the LLC, the service provider. From the viewpoint of the LLC, previous and subsequent data units are unrelated to the present unit. There is no guarantee by the service provider of the delivery of the data unit to its intended user, nor is the sender informed if the delivery attempt fails. Furthermore, there is no guarantee of ordered delivery. This type of service supports point-to-point, multipoint, and broadcast modes of operation.

As we might imagine with connection mode service, a logical connection is established between two LLC users. During the data-transfer phase of the connection, the service provider at each end of the connection keeps track of the data units transmitted and received. The LLC guarantees that all data will be delivered and that the delivery to the intended user will be ordered (e.g., in the sequence as presented to the source LLC for transmission). When there is a failure to deliver, it is reported to the sender.

IEEE (Ref. 2) defines the LLC as that part of a data station that supports the LLC functions of one or more logical links. The LLC generates command PDUs and response PDUs for transmission and interprets received command PDUs and response PDUs. Specific responsibilities assigned to the LLC include:

- Initiation of control signal interchange;
- Interpretation of received command PDUs and generation of appropriate response PDUs;
- Organization of data flow; and
- Actions regarding error-control and error-recovery functions in the LLC sublayer.

As shown in Figure 11.3, the LLC accepts higher level user data and encapsulates it, forming an LLC PDU. The resulting LLC frame is embedded into the MAC user field for transmission.

The LLC is another derivative of HDLC, which was discussed in Section 10.10.3. It is based on the balanced mode of that link-layer protocol with similar formats and functions. This is particularly true when operating in the connection mode.

DSAP Address	SSAP Address	Control	Information
8 bits	8 bits	8 or 16 bits	M * 8 bits

DSAP Address = destination service access point address field
SSAP Address = source service access point address field
Control = control field (16 bits for formats that include sequence numbering, and 8 bits for formats that do not)
Information = information field
* = multiplication
M = an integer value equal to or greater than 0. (Upper bound of M is a function of the medium access control methodology used.)

Figure 11.4 LLC PDU frame format.

11.5.3.1 LLC PDU Structure. As shown in Figure 11.3, the LLC appends a header forming the LLC PDU. The PDU frame format is illustrated in Figure 11.4. The header consists of address and control information; the information field contains the user data. The control field is identical with HDLC control field, illustrated in Figure 10.29 and described in Section 10.10.3.1. However, the LLC control field is two octets long and there is no provision to extend it to 3 or 4 octets in length as there is in HDLC.

As mentioned previously, the LLC destination address is the user address at an SAP inside the destination LAN station. It is called the *destination service access point* (DSAP). The SSAP is the *source service address point* and it indicates the message originator inside a particular LAN station. Each has fields of 8 bits, as shown in Figure 11.5.

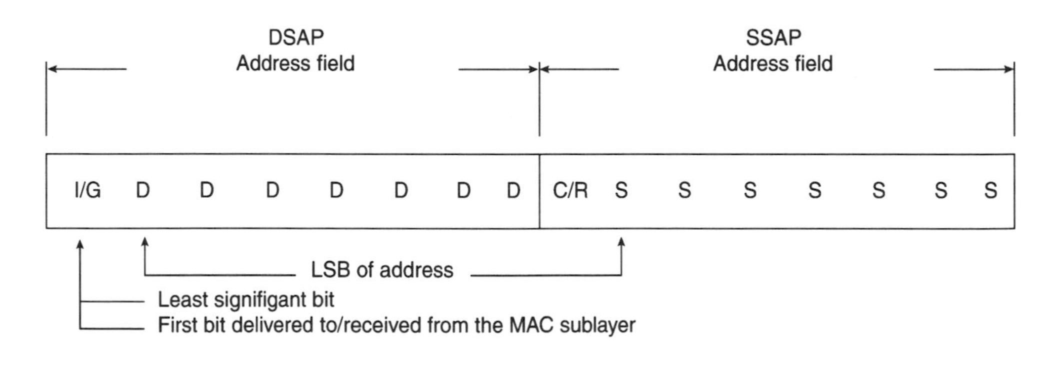

I/G = 0	Individual DSAP
I/G = 1	Group DSAP
C/R = 0	Command
C/R = 1	Response
XODDDDDD	DSAP Address
XOSSSSSS	SSAP Address
X1DDDDDD	Reserved for ISO definition
X1SSSSSS	Reserved for ISO definition

Figure 11.5 DSAP and SSAP address field formats (Ref. 2).

LLC PDU Control field bits

	1	2	3	4	5	6 7 8	9	10-16
Information transfer command/response (I-format PDU)	0	N(S)					P/F	N(R)
Supervisory command/response (S-format PDUs)	1	0	S	S	X	X X X	P/F	N(R)
Unnumbered command/response (U-format PDUs)	1	1	M	M	P/F	M M M		

N(S) = Transmitter send sequence number (Bit 2 = low-order bit)
N(R) = Transmitter receive sequence number (Bit 10 = low-order bit)
S = Supervisory function bit
M = Modifier function bit
X = Reserved and set to zero
P/F = Poll bit—command LLC PDU transmissions
Final bit—response LLC PDU transmissions
(1 = Poll/final)

Figure 11.6 LLC PDU control field formats (Ref. 2).

However, only the last seven of these bits are used for the actual address. The first bit in the DSAP indicates whether the address is an individual address or a group address (i.e., addressed to more than one SAP). The first bit in the SSAP is the C/R bit, which indicates whether a frame is a command frame or a response frame. The control field is briefly described in Section 11.5.3.2.

11.5.3.2 LLC Control Field and Its Operation. The LLC control field is illustrated in Figure 11.6. It is 16 bits long for formats that include sequence numbering and 8 bits long for formats that do not. The three formats described for the control field are used to perform numbered information transfer (I-frame), unnumbered control (S-frames), and unnumbered information transfer (U-frames) functions. These functions are described in Section 10.10.3.

11.6 LAN ACCESS PROTOCOLS

11.6.1 Introduction

In this context a protocol includes a means of permitting all users to access a LAN fairly and equitably. Access can be random or controlled. The random access schemes to be discussed include CSMA (carrier sense multiple access) and CSMA/CD, where CD stands for collision detection. The controlled access schemes that are described are token bus and token ring. It should be kept in mind that users accessing the network are unpredictable, and the transmission capacity of the LAN should be allocated in a dynamic fashion in response to those needs.

11.6.2 CSMA and CSMA/CD Access Techniques

Carrier sense multiple access (CSMA) is a LAN access technique that some simplistically call "listen before transmit." This "listen before transmit" idea gives insight into the control mechanism. If user 2 is transmitting, user 1 and all others hear that the medium is occupied and refrain from using it. In actuality, when an access with traffic senses that the medium is busy, it backs off for a period of time and tries again. How does one control that period of time? There are three methods of control commonly used. These methods are called *persistence algorithms* and are outlined briefly below:

- *Nonpersistent.* The accessing station backs off a random period of time and then reattempts access.
- *1-Persistent.* The station continues to sense the medium until it is idle and then proceeds to send its traffic.
- *p-Persistent.* The accessing station continues to sense the medium until it is idle, then transmits with some preassigned probability p. Otherwise it backs off a fixed amount of time, then transmits with a probability p or continues to back off with a probability of $(1 - p)$.

The algorithm selected depends on the desired efficiency of the medium usage and the complexity of the algorithm and resulting impact on firmware and software. With the nonpersistent algorithm, collisions are effectively avoided because the two stations attempting to access the medium will back off, most probably with different time intervals. The result is wasted idle time following each transmission. The 1-persistent algorithm is more efficient by allowing one station to transmit immediately after another transmission. However, if more than two stations are competing for access, collision is virtually assured.

The p-persistent algorithm lies between the other two and is a compromise, attempting to minimize collisions and idle time.

It should also be noted that with CSMA, after a station transmits a message, it must wait for an acknowledgment from the destination. Here we must take into account the round-trip delay (2 × propagation time) and the fact that the acknowledging station must also contend for medium access. Another important point is that collisions can occur only when more than one user begins transmitting within the period of propagation time. Thus CSMA is an effective access protocol for packet transmission systems where the packet transmission time is much longer than the propagation time.

The inefficiency of CSMA arises from the fact that collisions are not detected until the transmissions from the two offenders have been completed. With CSMA/CD, which has collision detection, a collision can be recognized early in the transmission period and the transmissions can be aborted. As a result, channel time is saved and overall available channel utilization capability is increased.

CSMA/CD is sometimes called "listen while transmitting." It must be remembered that collisions can occur at any period during channel occupancy, and this includes the total propagation time from source to destination. Even at multimegabit data rates, propagation time is not instantaneous; it remains constant for a particular medium, no matter what the bit rate is. With CSMA the entire channel is wasted. With CSMA/CD one offending station stops transmitting as soon as it detects the second offending station's signal. It can do this because all accesses listen *while* transmitting.

11.6.2.1 CSMA/CD Description. Carrier sense multiple access with collision detection (CSMA/CD) is defined by ISO/IEC 8802-3 international standard and by the reference ANSI/IEEE Std 802.3 (Refs. 3 and 4). Some refer to this as *Ethernet*, which was initiated by Xerox Corporation, Digital Equipment Corporation, and Intel. The version

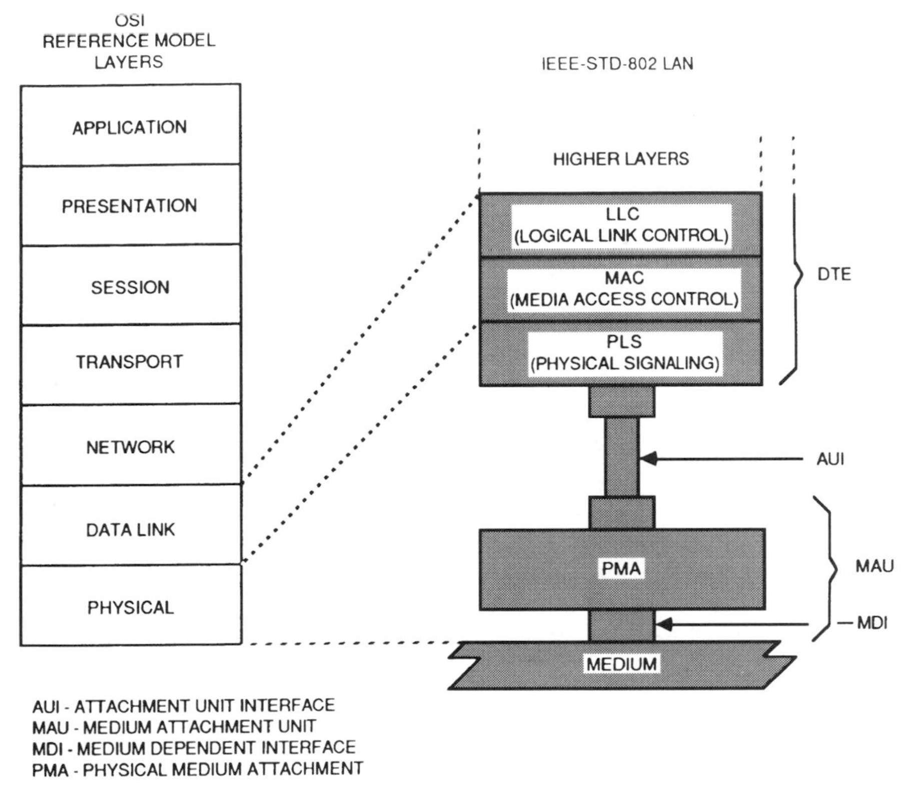

Figure 11.7 CSMA/CD LAN relationship to the OSI model as well as the functional blocks required. (From IEEE 802.3 [Ref. 3]; Courtesy of the IEEE, New York.)

discussed here is based on IEEE 802.3, which closely resembles Ethernet with changes in frame structure and an expanded set of physical layer options. Figure 11.7 relates CSMA/CD protocol layers to the conventional OSI reference model described in the previous chapter. The figure also identifies acronyms that we use in this description. The bit rates generally encompassed are 10 Mbps and 20 Mbps. We will also briefly cover an IEEE 802.3 subset standard for 100-Mbps operation. The model used in this present discussion covers the 10-Mbps data rate. There is also a 100-Mbps CSMA/CD option, which is described in Section 11.6.2.2. There will also be a 1000-Mbps variance of this popular MAC protocol.

The medium described here is coaxial cable. However, there is a trend to use twisted wire pair. The LAN station connects to the cable by means of a medium access unit (MAU). This connects through an attachment unit interface (AUI) to the data terminal equipment (DTE). As illustrated in Figure 11.7, the DTE consists of the physical signaling sublayer (PLS), the medium access control (MAC), and the logical link control (LLC). The PLS is responsible for transferring bits between the MAC and the cable. It uses differential Manchester coding for the data transfer. With such coding the binary 0 has a transition from high to low at midcell, while the binary 1 has the opposite transition.

CSMA/CD LAN systems (or Ethernet) are probably the most widely used type of LANs worldwide. We would say that this is due to their relatively low cost to implement and maintain and to their simplicity. The down side is that their efficiency starts to drop off radically as the number of users increases, as well as increased user activity. Thus the frequency of collisions and backoffs increases to a point that throughput can drop to zero. Some users argue that efficiency starts to drop off at around 30% capacity, while others argue that that point is nearer 50%. We will discuss ways to mitigate this problem in our coverage of bridges.

11.6.2.1.1 System Operation

Transmission without contention. A MAC frame is generated from data from the LLC sublayer. This frame is handed to the transmit media access management component of the MAC sublayer for transmission. To avoid contention with other traffic on the medium, the transmit medium access management monitors the carrier sense signal provided by the physical layer signaling (PLS) component. When the medium is clear, frame transmission is initiated through the PLS interface. When the transmission has been completed without contention (a collision event), the MAC sublayer informs the LLC and awaits the next request for frame transmission.

Reception without contention. At each receiving station, the arrival of a frame is first detected by the PLS, which responds by synchronizing with the incoming preamble and by turning on the carrier sense signal. The PLS passes the received bits up to the MAC sublayer where the leading bits are discarded, up to and including the end of the preamble and start frame delimiter (SFD). In this period the receive media access management component of the MAC sublayer has detected the carrier sense and is waiting for the incoming bits to be delivered. As long as the carrier sense is on, the receive media access management collects bits from the PLS. Once the carrier sense signal has been removed, the frame is truncated at an octet boundary, if required, and then passed to the receive data decapsulation for processing.

It is in receive data decapsulation where the destination address is checked to determine if this frame is destined for this particular LAN station. If it is, the destination address (DA) and source address (SA) and the LLC data unit are passed to the LLC sublayer. It also passes along the appropriate status code indicating that reception is complete or reception too long. It also checks for invalid MAC frames by inspecting the frame check sequence (FCS) to detect any damage to the frame en route, as well as by checking for proper octet-boundary alignment of the end of frame.

Collision handling. A collision is caused by multiple stations attempting to transmit at the same time, in spite of their attempts to avoid this by deferring. A given station can experience a collision during the initial part of its transmission (the collision window) before its transmitted signal has had time to propagate to all stations on the CSMA/CD medium. Once the collision window has passed, a transmitting station is said to have acquired the medium. Once all stations have noticed that there is a signal on the medium (by way of carrier sense), they defer to it by not transmitting, avoiding any chance of subsequent collision. The time to acquire the medium is thus based on the round-trip propagation time of the physical layer whose elements include the PLS, the physical medium attachment (PMA), and the physical medium itself.

In the event of collision, the transmitting station's physical layer notices a marked increase in standing waves on the medium[2] and turns on the collision detect (CD) signal. The collision-handling process now starts. First, the transmit media access management enforces the collision by transmitting a bit sequence called jam. This "jam," specified as 32 bits long, ensures that all stations involved in the collision are aware that a collision has occurred. After the jam has been sent, the transmit media access

[2]Standing waves are called "interference" in the ISO/IEC reference standard. Also, the signal swings because two signals are reinforcing and then nulling out each other.

PREAMBLE (7 OCTETS)	SFD (1 OCTET)	DESTINATION ADDRESS (2 OR 6 OCTETS)	SOURCE ADDRESS (2 OR 6 OCTETS)	LENGTH (2 OCTETS)	LLC DATA	PAD	FCS (4 OCTETS)

LENGTH - GIVES NUMBER OF OCTETS IN DATA FIELD

LLC - LOGICAL LINK CONTROL (OSI LAYER 3 AND ABOVE)

FCS - FRAME CHECK SEQUENCE - A 32-BIT CRC

PAD - ADDS OCTETS TO ACHIEVE MINIMUM FRAME LENGTH WHERE NECESSARY

SFD - START FRAME DELIMITER

Figure 11.8 MAC frame format (Ref. 3).

management component terminates the transmission and schedules another transmission attempt after a randomly selected time interval. Retransmission is attempted again in the face of repeated collisions. If, on this second attempt, another collision occurs, the transmit media access management attempts to reduce the medium's load by backing off, meaning it voluntarily delays its own retransmissions to reduce the load on the medium. This is accomplished by expanding the interval from which the random retransmission time is selected on each successive transmission attempt. Eventually, either the transmission succeeds or the attempt is abandoned on the assumption that the medium has failed or has become overloaded.

The MAC frame is shown in Figure 11.8. There are eight fields in the frame: preamble, SFD, the addresses of the frame's destination(s) and source, a length field to indicate the length of the following field containing the LLC data, a field that contains padding (PAD) if required,[3] and the FCS for error detection. All eight fields are of fixed size except the LLC data and PAD fields, which may contain any integer number of octets (bytes) between the minimum and maximum values determined by a specific implementation.

The minimum and maximum frame size limits refer to that portion of the frame from the destination address field through the frame check sequence field, inclusive. The default maximum frame size is 1518 octets; the minimum size is 64 octets.

The preamble field is 7 octets in length and is used so that the receive PLS can synchronize to the transmitted symbol stream. The SFD is the binary sequence 10101011. It follows the preamble and delimits the start of frame.

There are two address fields: the source address and the destination address. The address field length is an implementation decision. It may be 16 or 48 bits long. In either field length, the first bit specifies whether the address is an individual address (bit set to 0) or group address (bit set to 1). In the 48-bit address field, the second bit specifies whether the address is globally administered (bit set to 0) or locally administered (bit set to 1). For broadcast address, the bit is set to 1.

The length field is 2 octets long and indicates the number of LLC data octets in the data field. If the value is less than the minimum required for proper operation of the protocol, a PAD field (sequence of octets) is appended at the end of the data field and prior to the FCS field.[4] The length field is transmitted and received with the high-order octet first.

The data (LLC data) field contains a sequence of octets that is fully transparent in that any arbitrary sequence of octet values may appear in the data field up to the maximum number specified by the implementation of this standard that is used. The maximum

[3]Padding means the adding of dummy octets (bytes) to meet minimum frame length requirements.

[4]Minimum frame length is 64 octets.

size of the data field supplied by the LLC is determined by the maximum frame size and address size parameters of a particular implementation.

The FCS field contains four octets (32 bits) CRC value. This value is computed as a function of the contents of the source address, destination address, length, LLC data, and pad—that is, all fields except the preamble, SFD, and FCS. The encoding is defined by the following generating polynomial:

$$G(x) = x^{32} + x^{26} + x^{23} + x^{22} + x^{16} + x^{12} + x^{11} + x^{10} + x^{8} + x^{7} + x^{5} + x^{4} + x^{2} + x + 1.$$

An invalid MAC frame meets at least one of the following conditions:

1. The frame length is inconsistent with the length field.
2. It is not an integral number of octets in length.
3. The bits of the received frame (exclusive of the FCS itself) do not generate a CRC value identical to the one received. An invalid MAC frame is not passed to the LLC.

The minimum frame size is 512 bits for the 10-Mbps data rate (Ref. 3). This requires a data field of either 46 or 54 octets, depending on the size of the address field used. The minimum frame size is based on the *slot time*, which for the 10-Mbps data rate is 512 bit times. Slot time is the major parameter controlling the dynamics of collision handling and it is:

- An upper bound on the acquisition time of the medium;
- An upper bound on the length of a frame fragment generated by a collision;
- The scheduling quantum for retransmission.

To fulfill all three functions, the slot time must be larger than the sum of the physical round-trip propagation time and the MAC sublayer jam time. The propagation time for a 500-m segment of 50-Ω coaxial cable is 2165 ns, assuming that the velocity of propagation of this medium is $0.77 \times 300 \times 10^6$ m/s (Ref. 3).

11.6.2.1.2 Transmission Requirements

System model. Propagation time is critical for the CSMA/CD access method. The major contributor to propagation time is the coaxial cable and its length. The characteristic impedance of the coaxial cable is 50 Ω ± 2 Ω. The attenuation of a 500-m (1640-ft) segment of the cable should not exceed 8.5 dB (17 db/km) measured with a 10-MHz sine wave. The velocity of propagation is $0.77c$.[5] The referenced maximum propagation times were derived from the physical configuration model described here. The maximum configuration is as follows:

1. A trunk coaxial cable, terminated in its characteristic impedance at each end, constitutes a coax segment. A coax segment may contain a maximum of 500 m of coaxial cable and a maximum of 100 MAUs. The propagation velocity of the coaxial cable is assumed to be $0.77c$ minimum (c = 300,000 km/s). The maximum end-to-end propagation delay for a coax segment is 2165 ns.

[5]Where c = velocity of light in a vacuum.

2. A point-to-point link constitutes a link segment. A link segment may contain a maximum end-to-end propagation delay of 2570 ns and shall terminate in a repeater set at each end. It is not permitted to connect stations to a link segment.
3. Repeater sets are required for segment interconnection. Repeater sets occupy MAU positions on coax segments and count toward the maximum number of MAUs on a coax segment. Repeater sets may be located in any MAU position on a coax segment but shall only be located at the ends of a link segment.
4. The maximum length, between driver and receivers, of an AUI cable is 50 m. The propagation velocity of the AUI cable is assumed to be 0.65*c* minimum. The maximum allowable end-to-end delay for the AUI cable is 257 ns.
5. The maximum transmission path permitted between any two stations is five segments, four repeater sets (including optional AUIs), two MAUs, and two AUIs. Of the five segments, a maximum of three may be coax segments; the remainder are link segments.

The maximum transmission path consists of 5 segments, 4 repeater sets (with AUIs), 2 MAUs, and 2 AUIs, as shown in Figure 11.9. If there are two link segments on the transmission path, there may be a maximum of three coaxial cable segments on that path. If there are no link segments on a transmission path, there may be a maximum of three coaxial cable segments on that path given current repeater technology. Figure 11.10 shows a large system with maximum length transmission paths. It also shows the application of link segments versus coaxial cable segments. The bitter ends of coaxial cable segments are terminated with the coaxial cable characteristic impedance. The coaxial cable segments are marked at 2.5-m intervals. MAUs should only be attached at these 2.5-m interval points. This assures nonalignment at fractional wavelength boundaries.

11.6.2.2 CSMA/CD at 100 Mbps

11.6.2.2.1 Introduction. The demand for greater information capacity and higher transmission data rates has brought about the development of a 100-Mbps CSMA/CD option for the enterprise network. There are several versions. The version presented here is based on IEEE Std. 802.3u-1995. The MAC and LLC are identical to those discussed in Section 11.6.2.1. To make the MAC and LLC compatible with 100-Mbps operation a new interface is provided between the PHY layer and the MAC sublayer. It consists of the *reconciliation sublayer* (RS) and the *medium independent interface* (MII). The PHY layer has been reconfigured for 100-Mbps operation on several optional transmission media configurations. The relationship among the LLC, MAC, reconciliation, MII, and PHY components is illustrated in Figure 11.10. The generic designation

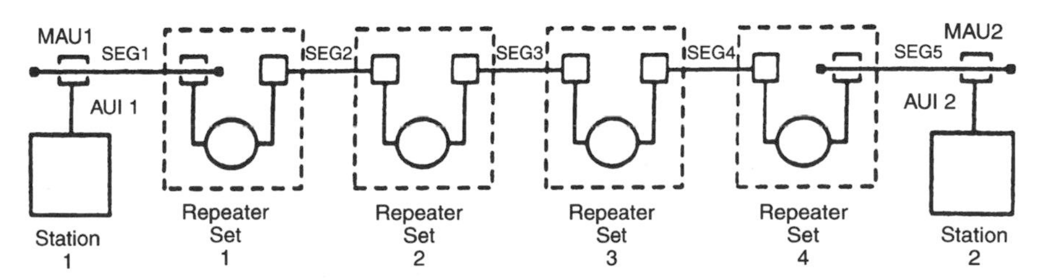

Figure 11.9 Maximum transmission path.

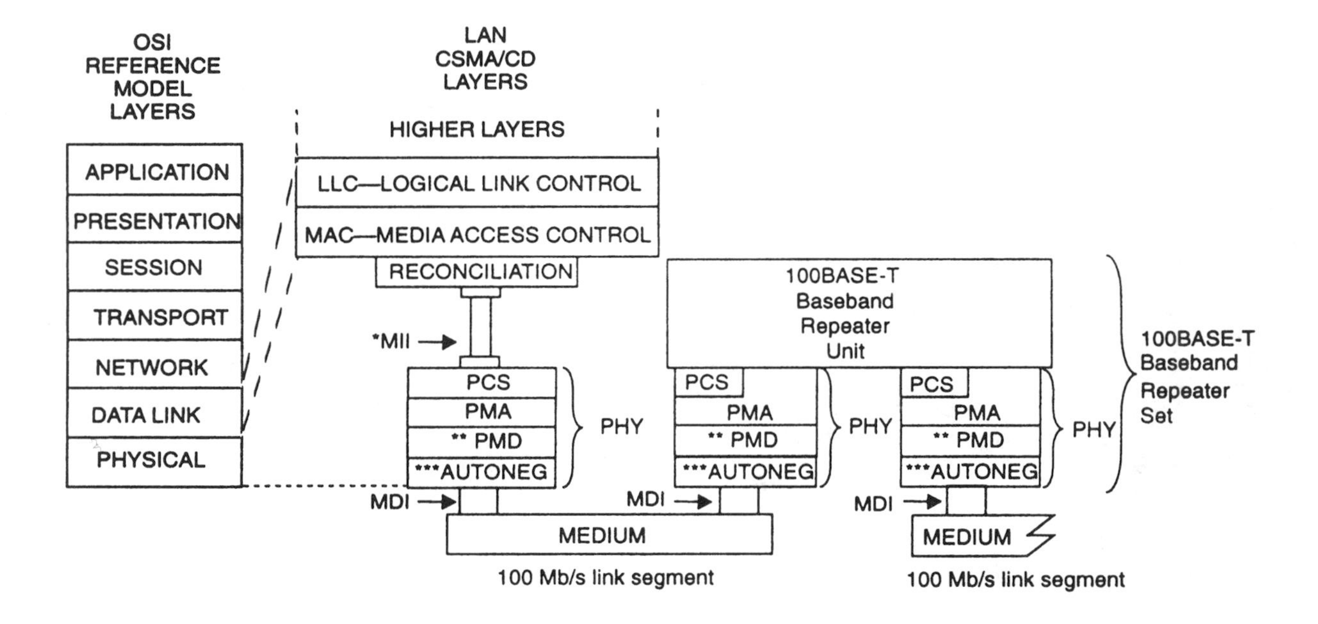

MDI = MEDIUM DEPENDENT INTERFACE
MII = MEDIA INDEPENDENT INTERFACE

PCS = PHYSICAL CODING SUBLAYER
PMA = PHYSICAL MEDIUM ATTACHMENT
PHY = PHYSICAL LAYER DEVICE
PMD = PHYSICAL MEDIUM DEPENDENT

* MII is optional for 10 Mb/s DTEs and for 100 Mb/s systems and is not specified for 1 Mb/s systems.

** PMD is specified for 100BASE-X only; 100BASE-T4 does not use this layer.
Use of MII between PCS and Baseband Repeater Unit is optional.

*** AUTONEG is optional.

Figure 11.10 Architectural positioning of 100BASE-T. (From Ref. 4, Figure 21-1, reprinted with permission of the IEEE.)

for 100-Mbps CSMA/CD is 100BASE-T (Ref. 4), which is available in several configurations:

- 100BASE-T4 uses four pair of Category 3, 4, or 5 UTP (unshielded twisted pair) balanced cable.
- 100BASE-TX uses two pair of Category 5 UTP balanced cable or 150-Ω STP (shielded twisted pair) balanced cable.
- 100BASE-FX uses two multimode fibers employing FDDI physical layer.

100BASE-T extends the bit rate of 10BASE-T to 100 Mbps. Of course, the bit rates are faster, the bit period is shorter and the frame transmission times are reduced, and the cable delay budgets are smaller, all in proportion to the change in bit rate.

11.6.2.2.2 Reconciliation Sublayer (RS) and Media Independent Interface (MII). The purpose of this interface is to provide a simple and easy way to implement interconnection between the MAC sublayer described in Section 11.6.2.1 and the PHY (physical layer entity), and between the PHY and station-management entities. The interface has the following characteristics:

- Supports both 10-Mbps and 100-Mbps operation;
- Data and delimiters are synchronous to clock reference;
- Provides independent four-bit-wide transmit and receive paths[6]
- Provides simple management interface;
- Capable of driving a limited length of shielded cable.

The major concepts of the MII and RS are:

- Each direction of data transfer is serviced with seven (making a total of 14) signals including: data (a 4-bit bundle), delimiter, error, and clocks.[7]
- Two media status symbols are used: one indicates the presence of carrier and the other indicates the occurrence of a collision.
- A management interface, composed of two signals, provides access to management parameters and services.
- The RS maps the signal set provided at the MII to the PLS service definition.

11.6.2.2.3 Physical Coding Sublayer (PCS), Physical Medium Attachment (PMA) Sublayer, and Baseband Medium Type 100BASE-T4. The objectives of 100BASE-T4 are:

- To support CSMA/CD MAC (see Section 11.6.2.1);
- To support 100BASE-T MII, repeater and optional autonegotiation;
- To provide a 100-Mbps data rate at the MII;

[6]"4-bit wide" is called a *nibble*, which is the unit of data exchange on the MII. Of course an octet consists of two consecutive nibbles.

[7]*Delimiter*: a bit, character, or set of characters used to denote the beginning or end of a group of related bits, characters, words, or statements. Synonym: separator.

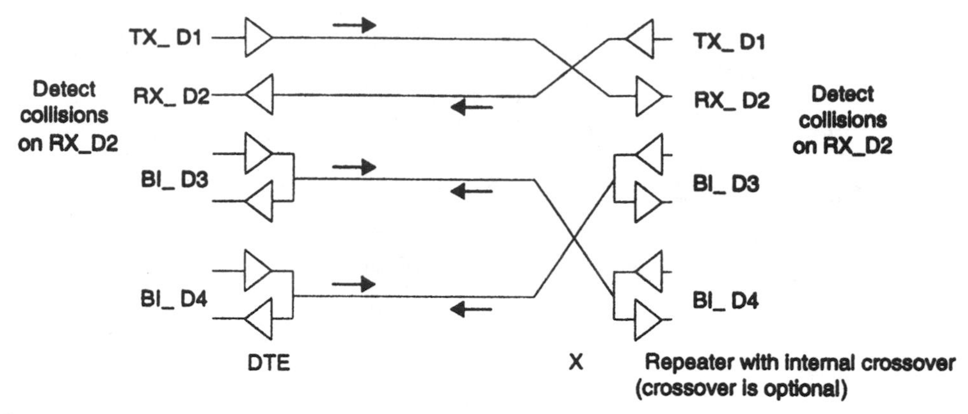

Figure 11.11 Use of wire pairs. Tx = transmit, Rx = receive, BI = bidirectional. (From Ref. 4, Figure 23-3, reprinted with permission from the IEEE.)

- To provide for operation over UTP Category 3, 4, or 5 cable, installed as horizontal runs at distances up to 100 m (328 feet);
- To allow for a nominal network extension of 200 m (656 ft), including
  - UTP links of 100 m; and
  - Two repeater networks of approximately 200-m span.
- To provide a communication channel with a mean ternary symbol error rate,[8] at the PMA service interface, of less than 1×10^{-8}.

The PCS transmit function accepts data nibbles from the MII, and encodes these nibbles in a 8B6T coding (described in the following) and passes the resulting ternary symbols to the PMA. In the reverse direction, the PMA conveys received ternary symbols to the PCS receive function. The PCS receive function decodes them into octets, then passes the octets one nibble at a time to the MII. The PCS also contains a PCS carrier sense function, a PCS error sense function, a PCS collision presence function, and a management interface.

The physical level communication between PHY entities (LAN station to LAN station) takes place over four twisted pairs. Figure 11.11 shows how the 4-pairs are employed.

The 100BASE-T transmission algorithm leaves one pair open for detecting carrier from the far-end (see Figure 11.11). Leaving one pair open for carrier detection greatly simplifies media access control. All collision-detection functions are accomplished using only the unidirectional pairs TX_D1 and RX_D2, a manner similar to 10BASE-T. This collision detection strategy leaves three pair in each direction free for data transmission, which uses an 8B6T block code, illustrated in Figure 11.12.

The 8B6T coding maps data octets into ternary symbols. Each octet is mapped into a pattern of six ternary symbols, called a 6T code group. The 6T code groups are fanned out to three independent serial channels. The effective data rate carried on each pair is one third of 100 Mbps, which is 33.333 ... Mbps. The ternary symbol transmission rate on each pair is 6/8 times 33.333, or precisely 25.000 megasymbols per second.

[8]A *ternary symbol* can have one of three states (in voltage): +, 0, −. It is a means of baseband transmission.

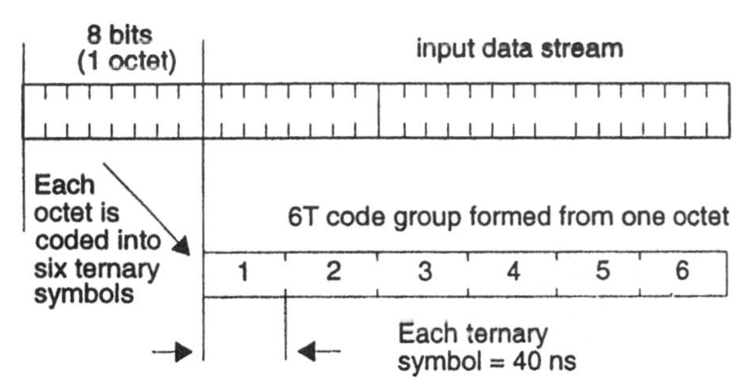

Figure 11.12 8B6T coding. (From Ref. 4, Figure 23-4, reprinted with permission from the IEEE.)

11.6.3 Token Ring

A typical token-passing ring is shown in Figure 11.13. The token ring operation, as specified in IEEE Std 802.5 (Ref. 5), has the capability of 4 Mbps or 16 Mbps data rate. A ring is formed by physically folding the medium back on itself. Each LAN station regenerates and repeats each bit and serves as a means of attaching one or more data terminals (e.g., workstations, PCs, servers) to the ring for the purpose of communicating with other devices on the network. As a traffic frame passes around the ring, all stations, in turn, copy the traffic. Only those stations included in the destination address field pass that traffic on to the appropriate users that are attached to the station. The traffic frame continues onward back to the originator, who then removes the traffic from the ring. The *pass-back* to the originator acts as a form of acknowledgment that the traffic had at least passed by the destination(s).

With token ring a reservation scheme is used to accommodate priority traffic. Also, one station acts as a ring monitor to ensure correct network operation. A monitor devolvement scheme to other stations is provided in case a monitor fails or drops off the ring (i.e., shuts down). Any station on the ring can become inactive (i.e., close down), and a physical by-pass is provided for this purpose.

A station gains the right to transmit frames onto the medium when it detects a token passing on the medium. Any station with traffic to transmit, on detection of the appropriate token, may capture the token by modifying it to a start-of-frame sequence and append the proper fields to transmit the first frame. At the completion of its information transfer and after appropriate checking for proper operation, the station initiates

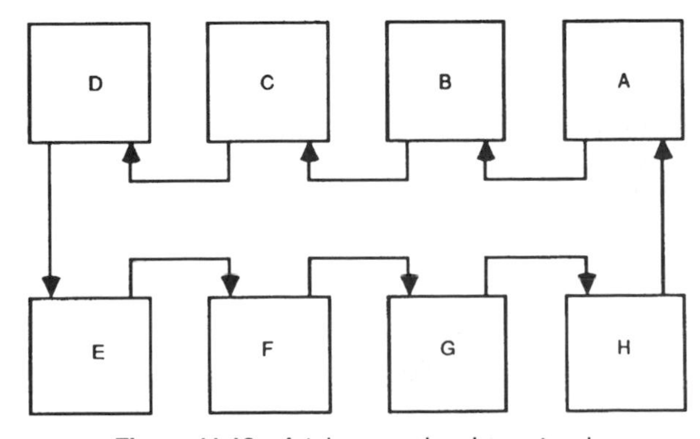

Figure 11.13 A token-passing ring network.

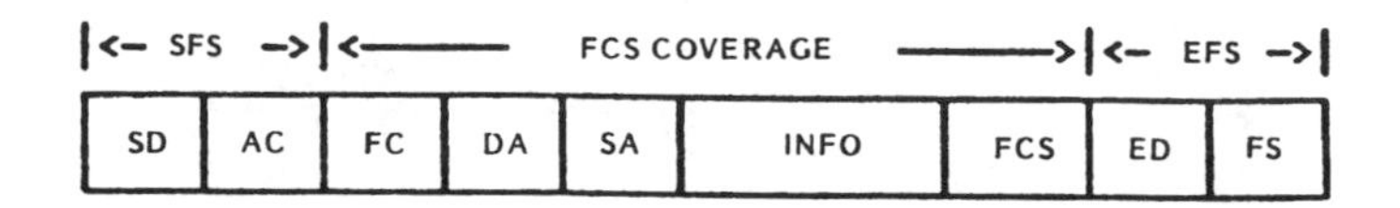

WHERE: SFS - START-OF-FRAME SEQUENCE
SD - STARTING DELIMITER (1 OCTET)
AC - ACCESS CONTROL (1 OCTET)
FC - FRAME CONTROL (1 OCTET)
DA - DESTINATION ADDRESS (2 OR 6 OCTETS)
SA - SOURCE ADDRESS (2 OR 6 OCTETS)
INFO - INFORMATION (0 OR MORE OCTETS)*
FCS - FRAME CHECK SEQUENCE (4 OCTETS)
EFS - END-OF-FRAME SEQUENCE
ED - ENDING DELIMITER (1 OCTET)
FS - FRAME STATUS (1 OCTET)

* ALTHOUGH THERE IS NO MAXIMUM LENGTH FOR THE INFORMATION FIELD, THE TIME REQUIRED TO TRANSMIT A FRAME MAY BE NO GREATER THAN THE TOKEN HOLDING PERIOD ESTABLISHED FOR THE STATION.

NOTE: TOKEN CONSISTS OF SD, AC AND ED FIELDS

Figure 11.14 Frame format for token ring.

a new token, which provides other stations with the opportunity to gain access to the ring. Each station has a token-holding timer that controls the maximum period of time a station may occupy the medium before passing the token on.

Figure 11.14 shows the frame format for token ring, and Figure 11.15 illustrates the token format. In these figures, the left-most bit is transmitted first. The frame format in Figure 11.14 is used for transmitting both MAC and LLC messages to destination station(s). A frame may or may not carry an INFO field.

The starting delimiter (SD) consists of the symbol sequence JK0JK000, where J and K are nondata symbols. Both frames and tokens start with the SD sequence.

The access control (AC) is 1 octet long and contains 8 bits that are formatted PPPTMRRR. The first three bits, PPP, are the priority bits. These are used to indicate the priority of a token and therefore which stations are allowed to use the token. In a system designed for multiple priority, there are eight levels of priority available, where the lowest priority is PPP = 000 and the highest is PPP = 111. The AC field contains the token bit T and the monitor bit M. If T = 0, then the frame is a token as shown in Figure 11.15. The T bit is a 1 in all other frames. The M bit is set by the monitor as part of the procedures for recovering from malfunctions. The bit is transmitted as a 0 in all frames and tokens. The active monitor inspects and modifies this bit. All other stations repeat this bit as received. The three R bits are reservation bits. These bits allow stations with high-priority PDUs to request that the next token issued be at the requested priority.

The next field in Figure 11.14 is the frame control (FC) field. It is one octet long and defines the type of frame and certain MAC and information frame functions. The first two bits in the FC are designated FF bits, and the last six bits are called ZZZZZZ bits. If FF is 00, the frame contains a MAC PDU, and if FF is 01, the frame contains an LLC PDU.

Figure 11.15 Token format for token ring.

Each frame (not token) contains a destination address (DA) field and a source address (SA) field. Depending on the token ring LAN system, the address fields may be 16 bits or 48 bits in length. In either case, the first bit indicates whether the frame is directed to an individual station (0) or to a group of stations (1).

The INFO field carries zero, one, or more octets of user data intended for the MAC, NMT (network management), or LLC. Although there is no maximum length specified for the INFO field, the time required to transmit a frame may be no greater than the token-holding period that has been established for the station. For LLC frames, the format of the information field is not specified in the referenced standard (Ref. 5). However, in order to promote interworking among stations, all stations should be capable of receiving frames whose information field is up to and including 133 octets in length.

The frame check sequence is a 32-bit sequence based on the standard CRC generator polynomial. The error-detection process utilizing a CRC was described in Section 10.5.3.1 using a 16-bit generator polynomial. The 32-bit polynomial was introduced in Section 11.6.2.1. As the order of the polynomial increases, the error-detection capability increases markedly.

The end delimiter (ED) is one octet in length and is transmitted as the sequence JK1JK1IE. The transmitting station transmits the delimiter as shown. Receiving stations consider the ED valid if the first six symbols JK1JK1 are received correctly. The I is the intermediate frame bit and is used to indicate whether a frame transmitted is a singular frame or whether it is a multiple frame transmission. The I bit is set at 0 for the singular frame case. The E bit is the error-detected bit. The E bit is transmitted as 0 by the station that originates the token, abort sequence, or frame. All stations on the ring check tokens and frames for errors such as FCS errors and nondata symbols. The E bit of tokens and frames that are repeated is set to 1 when a frame with an error is detected; otherwise the E bit is repeated as received.

The last field in the frame is the frame status (FS) field. It consists of one octet of the sequence ACrrACrr. The r bits are reserved for future standardization and are transmitted as 0s, and their value is ignored by the receiver. The A bit is the address-recognized bit, and the C bit is the frame-copied bit. These two bits are transmitted as 0 by the frame originator. The A bit is changed to 1 if another station recognizes the destination address as its own or relevant group address. If it copies the frame into its buffer, it then sets the C bit to 1. When the frame reaches the originator again, it may differentiate among three conditions:

1. Station nonexistent or nonactive on the ring;
2. Station exists but frame was not copied; or
3. Frame copied.

Fill is used when a token holder is transmitting preceding or following frames, tokens, or abort sequences to avoid what would otherwise be an inactive or indeterminate transmitter state. Fill can be either 1s or 0s or any combination thereof and can be any number of bits in length within the constraints of the token-holding timer.

IEEE Std 802.5 describes a true baseband-transmitting waveform using differential Manchester coding. It is characterized by the transmission of two line signal elements per symbol. An example of this coding is shown in Figure 11.16. The figure shows only the data symbols 1 and 0 where a signal element of one polarity is transmitted for one-half the duration of the symbol (bit) to be transmitted, followed by the contiguous transmission of a signal element of the opposite polarity for the remainder

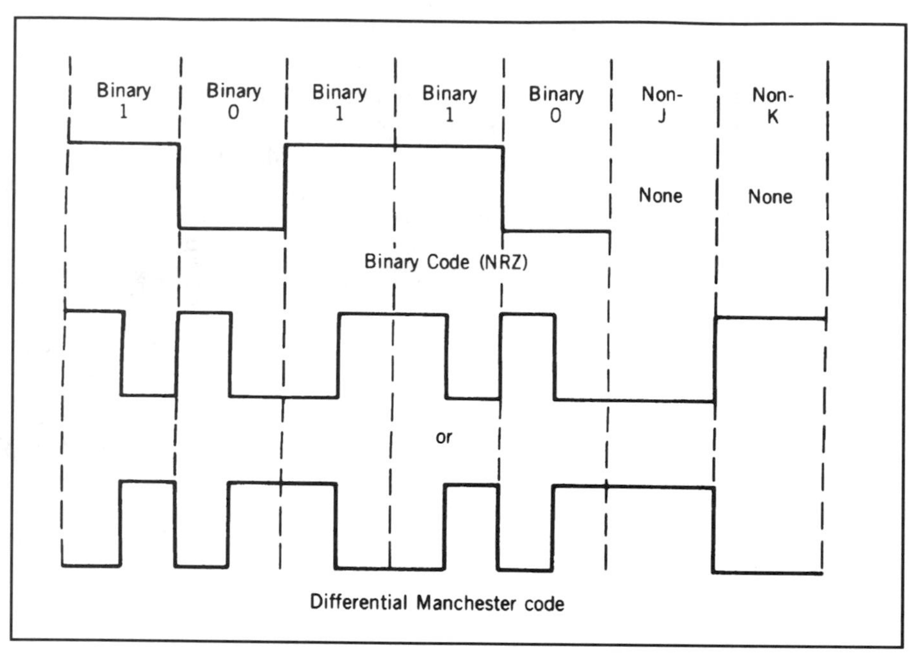

Figure 11.16 Differential Manchester coding format for symbols 1, 0, nondata J, and nondata K. (From IEEE Std. 802.5, [Ref. 5], reprinted with permission.)

of the symbol duration.[9] The following advantages accrue for using this type of coding:

- The transmitted signal has no dc component and can be inductively or capacitively coupled.
- The forced midsymbol transition provides inherent timing information on the channel.

All stations on the LAN ring are slaved to the active monitor station. They extract timing from the received data by means of a phase-locked loop. *Latency* is the time, expressed in number of bits transmitted, for a signal element to proceed around the entire ring. In order for the token to circulate continuously around the ring when all stations are in the repeat mode, the ring must have a latency of at least the number of bits in the token sequence—that is, 24. Since the latency of the ring varies from one system to another and no a priori knowledge is available, a delay of at least 24 bits should be provided by the active monitor.

A LAN station provides an output with an error rate of less than or equal to 1×10^{-9} when the signal-to-noise ratio at the output of the equalizer, specified in paragraph 7.5.2 of the reference document (Ref. 5), is 22 dB.

11.6.4 Fiber Distributed Data Interface (FDDI)

11.6.4.1 Overview. A fiber distributed data interface (FDDI) network consists of a set of nodes (e.g., LAN stations) connected by an optical transmission medium (or other

[9] *Polarity* refers to whether a signal is a positive-going or negative-going voltage or simply a positive voltage or a negative voltage.

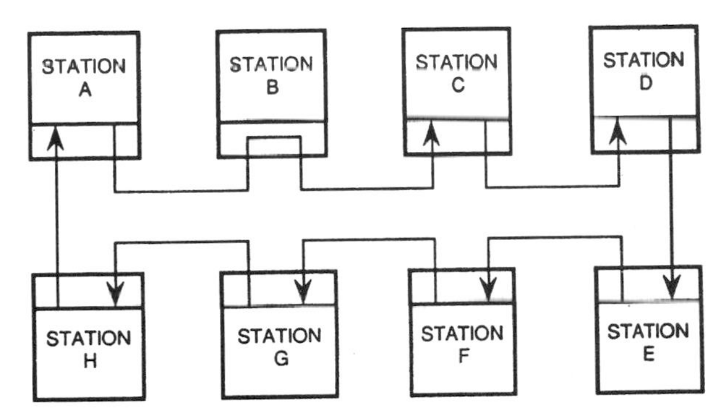

Figure 11.17 FDDI token ring: example of logical configuration.

medium) into one or more logical rings. A logical ring consists of a set of stations connected as an alternating series of nodes and transmission medium to form a closed loop. This is shown in Figure 11.17. Information is transmitted as a stream of suitably encoded symbols from one active node to the next. Each active node regenerates and repeats each symbol and serves as a means for attaching one or more devices to the ring for the purpose of communicating with other devices on the ring.

FDDI provides equivalent bandwidth to support a peak data rate of 100 Mbps and a sustained data transfer rate of at least 80 Mbps. With equivalent 4B/5B coding, FDDI line transmission rate is 125 Mbaud (peak). It provides connectivity for many nodes over distances of many kilometers in extent. Certain default parameter values for FDDI (such as timer settings) are calculated on the basis of up to 1000 transmission links or up to 200 km total fiber path length (typically corresponding to 500 nodes and 100 km of dual-fiber cable, respectively). However, the FDDI protocols can support much larger networks by increasing these parameter values.

Two kinds of data service can be provided in a logical ring: packet service and circuit service. With packet service, a given station that holds the token transmits information on the ring as a series of data packets, where each packet circulates from one station to the next. The stations that are addressed copy the packets as they pass. Finally, the station that transmitted the packets effectively removes them from the ring.

In the case of circuit service, some of the logical ring bandwidth is allocated to independent channels. Two or more stations can simultaneously communicate via each channel. The structure of the information stream within each channel is determined by the stations sharing the channel.

Conventional FDDI provides packet service via a token ring. A station gains the right to transmit its information on to the medium when it detects a token passing on the medium. The token is a control signal comprising a unique symbol sequence that circulates on the medium following each series of transmitted packets. Any active station, upon detection of a token, may capture the token by removing it from the ring. The station may then transmit one or more data packets. After transmitting its packets, the station issues a new token, which provides other stations the opportunity to gain access to the ring.

Each station has a token-holding timer (or equivalent means) incorporated, which limits the length of time a station may occupy the medium before passing the token onwards.

FDDI provides multiple levels of priority for independent and dynamic assignment depending on the relative class of service required. The classes of service may be synchronous, which may be typically used for applications such as real-time packet voice; asynchronous, which is typically used for interactive applications; or immediate, which is used for extraordinary applications such as ring recovery. The allocation of ring bandwidth occurs by mutual agreement among users of the ring.

Error-detection and recovery mechanisms are provided to restore ring operation in the event that transmission errors or medium transients (e.g., those resulting from station insertion or deletion) cause the access method to deviate the normal operation. Detection and recovery of these cases utilize a recovery function that is distributed among the stations attached to the ring.

One of the more common topologies of FDDI is the counterrotating ring. Here two classes of station may be defined: dual (attachment) and single (attachment). FDDI trunk rings may be composed only of dual-attachment stations which have two PMD (physical layer medium dependent) entities (and associated PHY entities) to accommodate the dual ring. Concentrators provide additional PMD entities beyond those required for their own attachment to the FDDI network, for the attachment of single-attachment stations which have only one PMD and thus cannot directly attach to the FDDI trunk ring. A dual-attachment station, or one-half of it, may be substituted for a single-attachment station in attaching to a concentrator. The FDDI network consists of all attached stations.

The example illustrated in Figure 11.18 shows the concept of multiple physical connections used to create logical rings. As shown in the figure, the logical sequence of MAC connections is stations, 1, 3, 5, 8, 9, 10, and 11. Station 2, 3, 4, and 6 form an FDDI trunk ring. Stations 1, 5, 7, 10, and 11 are attached to the ring by lobes branching out from stations that form it. Stations 8 and 9, in turn, are attached by lobes branching out from station 7. Stations 2, 4, 6, and 7 are concentrators, serving as the means of attaching MAC entities and station functionality. The concentrator examples of Figure 11.18 do not show any MACs, although their presence is implied by the designation of these concentrators as stations.

Symbol set. A symbol is the smallest signaling element used by the FDDI MAC. Symbols can be used to convey three types of information:

1. Line states such as halt (H), quiet (Q), and idle (I);
2. Control sequences such as the starting delimiter (SD), ending delimiter (ED) symbol (T), initial SD symbol (J), and final symbol (K); and
3. Data quartets, each representing a group of four ordered data bits.

Peer MAC entities on the ring communicate via a set of fixed-length symbols. These symbols are passed across the MAC-to-PHY interface via defined primitives. The MAC generates PDUs as matched pairs of symbols in accordance with the referenced FDDI standards.

11.6.4.2 FDDI Protocol Data Units. Two types of PDUs are used by the MAC: tokens and frames. In the figures that follow, formats are depicted of PDUs in the order of transmission on the medium, with the leftmost symbol transmitted first. Figure 11.19 illustrates the token format. The token is the means by which the right to transmit MAC

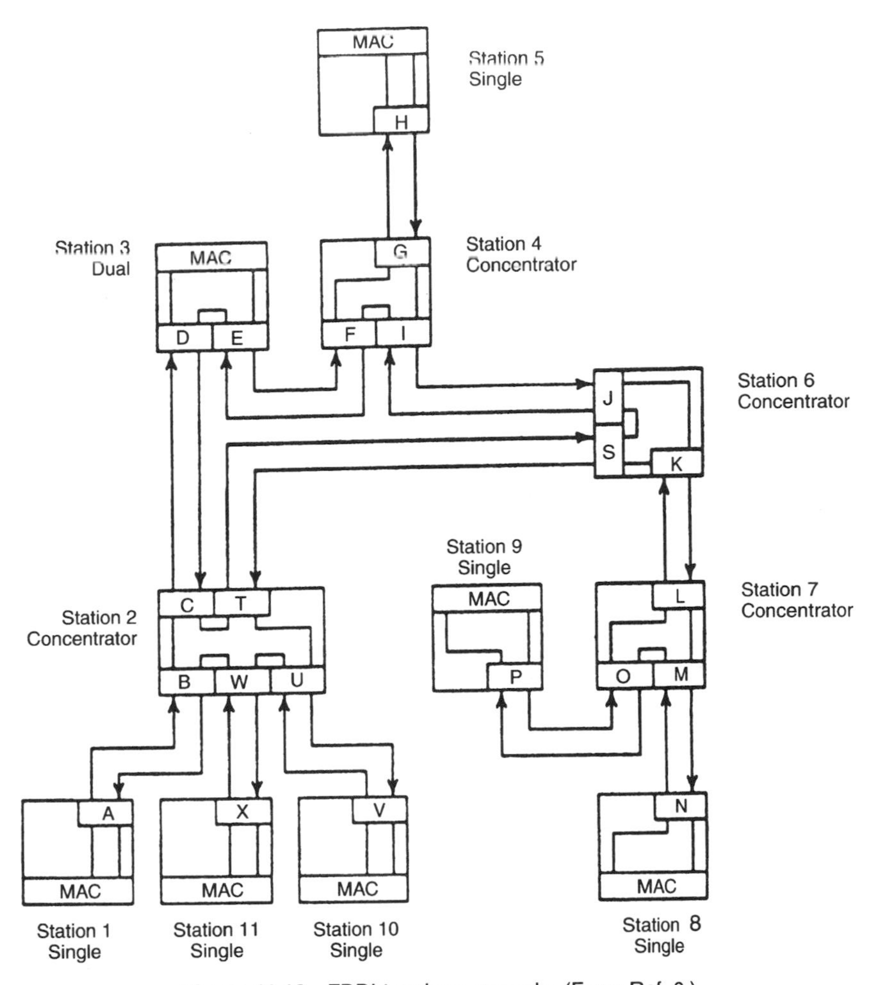

Figure 11.18 FDDI topology example. (From Ref. 6.)

SDUs (service data units) (as opposed to the normal process of repeating) is passed from one MAC to another.

FDDI frame. Figure 11.20 shows the FDDI frame format. The frame format is used for transmitting both MAC recovery information and MAC SDUs between peer MAC entities. A frame may or may not have an information field.

Frame length. The maximum frame length is 9000 symbols or 4500 octets.

Destination and source addrsses. The approach is very similar to token ring. Addresses may be either 16 bits or 48 bits in length.

Frame check sequence. The FDDI FCS is identical to the FCS used in CSMA/CD and token ring.

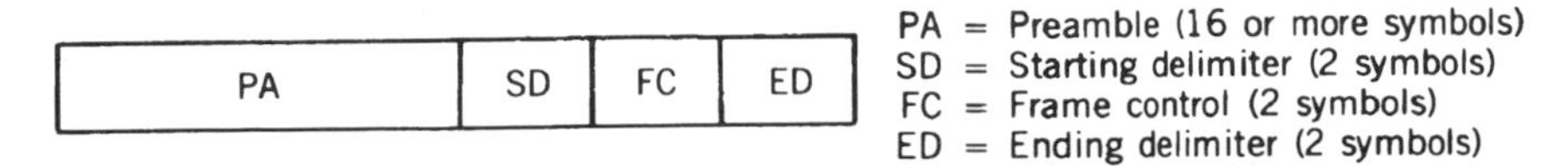

Figure 11.19 The FDDI token format.

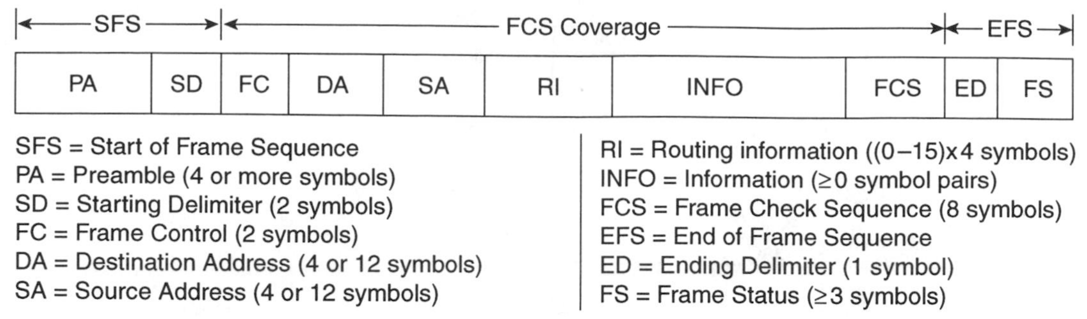

Figure 11.20 The FDDI frame format (Ref. 7).

11.6.4.3 FDDI Timers. Each MAC maintains three timers to regulate the operation of the ring. These timers are: (1) token holding timer (THT), (2) valid transmission timer (TVT), and (3) token rotation timer (TRT).

11.6.4.4 FDDI Operation. Access to the physical medium (i.e., the ring) is controlled by passing a token around the ring. The token gives the downstream MAC (receiving relative to the MAC passing the token) the opportunity to transmit a frame or sequence of frames. The time a MAC may hold a token is limited by the token-holding timer. If a MAC wants to transmit, it strips the token from the ring before the frame control field of the token is repeated. After the token is completely received, the MAC begins transmitting its eligible queued frames. After transmission, the MAC issues a new token for use by a downstream MAC.

MACs that are not transmitting repeat the incoming symbol stream. While repeating the incoming symbol stream, the MAC determines whether frames are intended for this MAC. This is done by matching the DA to its own address or a relevant group address. If a match occurs, the frame is processed by the MAC or sent to SMT (station management) or LLC.

Frame transmission. Upon request for SDU transmission, the MAC constructs the PDU or frame from the SDU by placing the SDU in the INFO field of the frame. The SDU remains queued by the requesting entity awaiting the receipt of a token that may be used to transmit it. Upon reception and capture of an appropriate token, the MAC begins transmitting its queued frame(s) in accordance with the rules of token holding. During transmission, the FCS for each frame is generated and appended to the end of the PDU. After transmission of the frame(s) is completed, the MAC immediately transmits a new token.

Frame stripping. Each transmitting station is responsible for stripping from the ring the frames that it originated. A MAC strips each frame that it transmits beginning not later than the seventh symbol after the end of the SA field. Normally, this is accomplished by stripping the remainder of each frame whose source address matches the MAC's address from the ring and replacing it with idle symbols.

The process of stripping leaves remnants of frames, consisting at most of PA, SD, FC, DA, and SA and six symbols after the SA field, followed by idle symbols. These remnants exist because the decision to strip a frame is normally based upon recognition of the MAC's address in the SA field, which cannot occur until after the initial part of the frame has already been repeated. These remnants are not recognized as frames because they lack an ending delimiter (ED). The limit of remnant length also prevents remnants from satisfying the minimum frame length criteria. To the level of accuracy required for statistical purposes, they can be distinguished from errored or lost frames because they are always followed by the idle symbol. Remnants are removed from the ring when they encounter a transmitting MAC. Remnants may also be removed by the smoothing function of PHY.

Ring scheduling. Transmission of normal PDUs (i.e., PDUs formed from SDUs) on the ring is controlled by a timed token-rotation protocol. This protocol supports two major classes of service:

1. Synchronous—guaranteed bandwidth and response time; and
2. Asynchronous—dynamic bandwidth sharing.

The synchronous class of service is used for those applications whose bandwidth and response limits are predictable in advance, permitting them to be preallocated (via the SMT). The asynchronous class of service is used for those applications whose bandwidth requirements are less critical (e.g., bursty or potentially unlimited) or whose response time requirements are less critical. Asynchronous bandwidth in FDDI is instantaneously allocated from the pool of remaining bandwidth that is unallocated, unused, or both. Section 11.6.4 is based on ANSI X3.231-1994 (Ref. 7) and ANSI X3.239-1994 (Ref. 8).

11.7 LAN INTERWORKING VIA SPANNING DEVICES

11.7.1 Repeaters

A repeater is nothing more than a regenerative repeater (see Section 6.6). It extends a LAN. It does not provide any kind of segmentation of a LAN, except the physical regeneration of the signal. Multiple LANs with common protocols can be interconnected with repeaters, in effect making just one large segment. A network using repeaters must avoid multiple paths, as any kind of loop would cause data to circulate indefinitely and could ultimately make the network crash. The multiple path concept is shown in Figure 11.21.

The following example shows how a loop can be formed. Suppose two repeaters connect CSMA/CD LAN segments as shown in Figure 11.21. Station #1 initiates an interchange with station #3, both on the same segment (upper in the figure). As data packets or frames are transmitted on the upper segment, each repeater will transmit them unnecessarily to the lower segment. Each repeater will receive the repeated packet on the lower segment and retransmit it once again on the upper segment. As one can see, any traffic introduced into this network will circulate indefinitely around the loop created by the two repeaters. On larger networks the effects can be devastating, although perhaps less apparent (Ref. 9).

11.7.2 LAN Bridges

Whereas repeaters have no intelligence, bridges do. Bridges can connect two LANs, at the data-link or MAC protocol level. There are several varieties of bridges, depending on the intelligence incorporated.

There is the *transparent bridge* that builds a list of nodes the bridge sees transmitting on either side. It isolates traffic and will not forward traffic that it knows is destined to another station on the same side of the bridge as the sending station. The bridge is able to isolate traffic according to the MAC source and destination address(es) of each individual data frame. MAC-level broadcasts, however, are propagated through

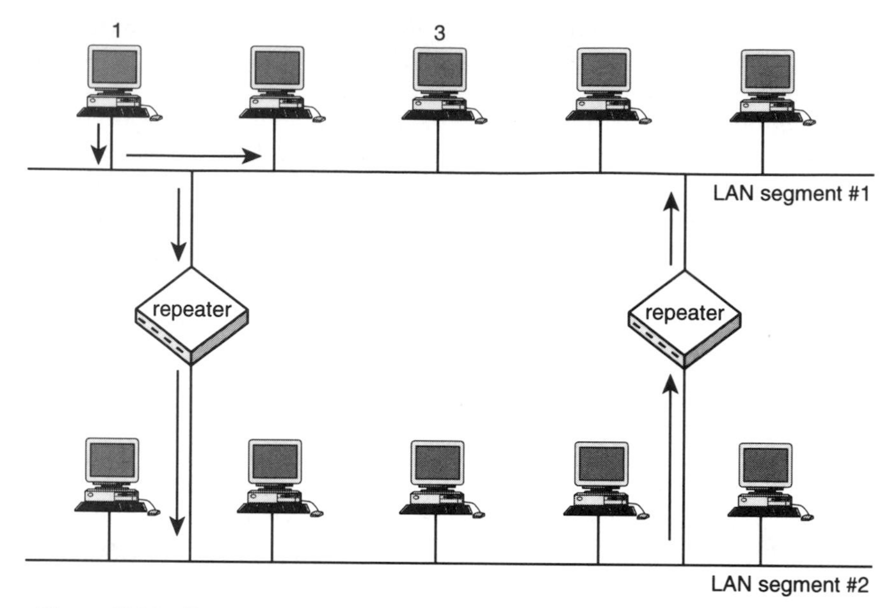

Figure 11.21 Repeaters in multiple paths. (Courtesy of Hewlett-Packard Co., Ref. 9.)

the network by the bridges. A bridge can be used for segmenting and extending LAN coverage. Thus it lowers traffic volume for each segment. A transparent bridge does not modify any part of a message that it forwards.

The second bridge is the *translation bridge*. It is used to connect two dissimilar LANS, such as a token ring to CSMA/CD. In order to do this it must modify the MAC-level header and FCS of each frame it fowards in order to make it compatible with the receiving LAN segment. The MAC addresses and the rest of the data frame are unchanged. Translation bridges are far less common than transparent bridges.

The third type of bridge, as shown in Figure 11.22, is the *encapsulation bridge*. It is also used to connect LANs of dissimilar protocols. But rather than translate the MAC header and FCS fields, it simply appends a second MAC layer protocol around the original frame for transport over the intermediate LAN with a different protocol. There is the destination bridge which strips off this additional layer and extracts the original frame for delivery to the destination network segment.

The fourth type of bridge is a *source routing* bridge. It is commonly used in token ring networks. With source routing bridges, each frame carries within it a route identifier (RI) field, which specifies the path which that frame is to take through the network. This concept is illustrated in Figure 11.22 (bottom).

Up to this point we have been discussing local bridges. A local bridge spans LANs in the same geographical location. A remote bridge spans LANs in different geographic locations. In this case, an intervening WAN (wide area network) is required. The remote bridge consists of two separate devices that are connected by a WAN, affording transport of data frames between the two. This concept is shown in Figure 11.23.

As illustrated in Figure 11.23, the LAN data packet/frame is encapsulated by the remote bridge adding the appropriate WAN header and trailer. The WAN transports the data packet/frame to the distant-end remote bridge, which strips the WAN header and trailer, and delivers the data packet/frame to the far-end LAN. Remote bridges typically use proprietary protocols such that, in most cases, remote bridges from different vendors do not interoperate.

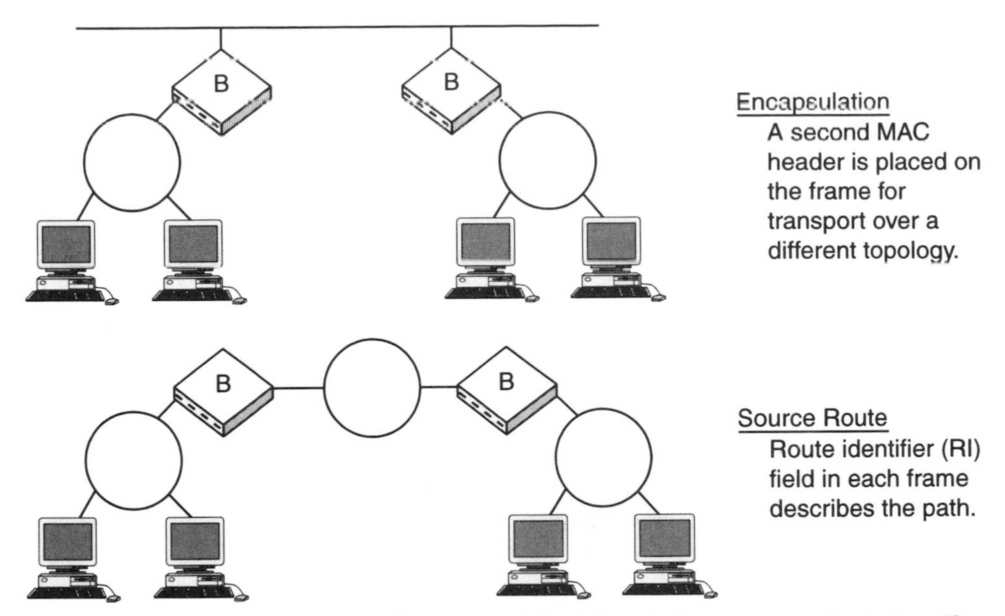

Figure 11.22 The concept of bridging. *Top*: encapsulation bridge; *bottom*: source routing bridge. (Courtesy of Hewlett-Packard Co., Ref. 9.)

Bridges are good devices to segment LANS, particularly CSMA/CD LANs. Segmenting breaks up a LAN into user families. It is expected that there is a high community of interest among members of a family, but a low community of interest among different families. There will be large traffic volumes intrasegment and low traffic volumes intersegment. It should be pointed out that routers are more efficient at segmenting than bridges (Ref. 9).

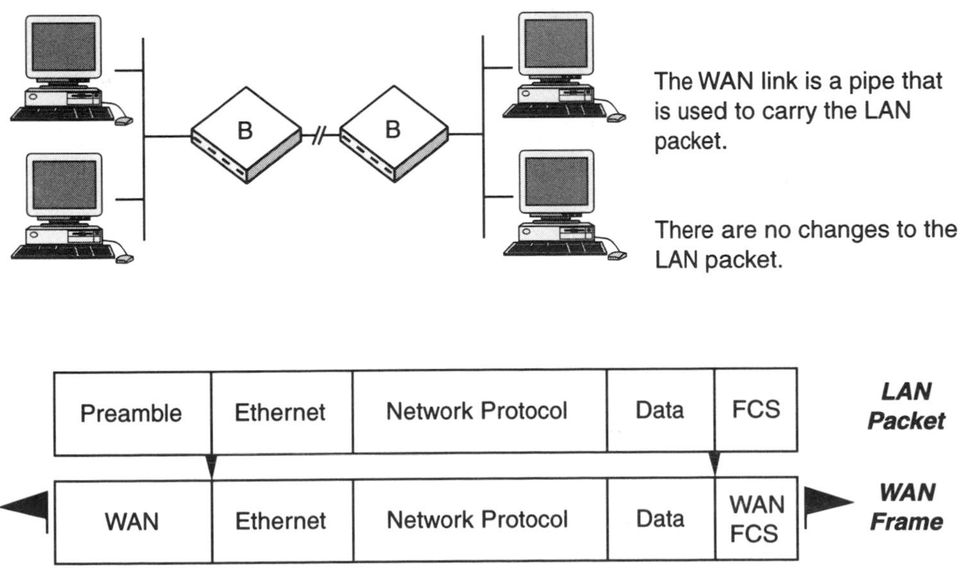

Figure 11.23 The concept of remote bridging. The LAN frame/packet is encapsulated in a WAN frame. (Courtesy of Hewlett-Packard Co., Ref. 9.)

A major limitation of bridges is the inability to balance traffic across two or more redundant routes in a network. The existence of multiple paths in a bridged network can prove to be a bad problem. In such a case, we are again faced with the endless route situation as we were with repeaters. One way to avoid the problem is to use the *spanning tree algorithm*. This algorithm is implemented by having bridges communicate with each other to establish a subset of the actual network topology that is loop-free (often called a *tree*). The idea, of course, is to eliminate duplicate paths connecting one LAN to another, or one segment to another. If there is only one path from one LAN to another, there can be no loop formed (Ref. 9).

11.7.3 Routers

Routers carry more intelligence than bridges. Like a bridge, a router forwards data packets/frames. Routers make forwarding decisions based on the destination network layer address. Whereas a bridge worked on the data-link layer, a router operates at the network-layer level. Routers commonly connect disparate LANs such as CSMA/CD to token ring and FDDI to CSMA/CD.

Routers are addressable nodes in a network. They carry their own MAC address(es) as well as a network address for each protocol handled. Because routers are addressable, a station desiring the facility of a router must direct its packets/frames to the router in question so that the traffic can be forwarded to the appropriate network. As one would expect, networking software at each station is more complex with a network using routers than one using bridges.

Routers handle only traffic addressed to them. They make decisions about forwarding data packets/frames based on one or several criteria. The decisions may be based on the cost of the link, the number of hops on each path, and the time-to-live.

Routers change packets/frames that pass through them such as MAC source and destination address; they may also modify the network protocol header of each frame (typically decrementing the time-to-live in the case of IP and other protocol fields).

Because routers have more intelligence than bridges, routers will typically have better network management agents installed. This enables them to be remotely configured, to be programmed to pass or not to pass data for security purposes, and to be monitored for performance, particularly error performance. Due to the additional processing performed at routers, they tend to be slower than bridges. Reference 9 suggests that some protocols do not lend themselves to routing, such as IBM's SNA and NetBios, among others.

11.7.4 Hubs and Switching Hubs

A hub is a multiport device that allows centralization. A hub is usually mounted in a wiring closet or other central location. Signal leads are brought in from workstations/PCs and other data devices, one for each hub port. Physical rings or buses are formed by internally configuring the hub ports. A typical hub may have 8 or 16 ports. Suppose we wished to incorporate 24 devices on our LAN using the hub. We can stack two hubs, one on top of the other (stackables), using one of the hub ports on each interconnection. In this case we would have a hub with a 30-port capacity ($2 \times 16 - 2$).[10] Hubs may also have a certain amount of intelligence, such as the incorporation of a network management capability. Also, each hub can include a repeater.

[10] Two 16-port hubs are used for a total of 32 ports. However, two ports are required to connect one hub to the other. This leaves just 30 ports for equipment connections.

There are also hubs with higher levels of intelligence. These are typically modular, multiprotocol, multimedia, multichannel, fault-tolerant, manageable devices where one can concentrate all the LAN connections into a wiring closet or data center. Since these types of hubs are modular (i.e., they have various numbers of slots to install LAN interface boards), they can support CSMA/CD, token ring, FDDI, or ATM simultaneously as well as various transmission media such as twisted pair, fiber cable, and others.[11]

Switching hubs are high-speed interconnecting devices with still more intelligence than the garden-variety hub or the intelligent hub. They typically interconnect entire LAN segments and nodes. Full LAN data rate is provided at each port of a switching hub. They are commonly used on CSMA/CD LANs, providing a node with the entire 10-Mbps data rate. Because of a hub's low latency, high data rates and throughputs are achieved.

With a switching hub, nodes are interconnected within the hub itself using its high-speed backplane. As a result, the only place the entire aggregate LAN traffic appears is on that backplane. Traffic between ports on a single card does not even appear on the backplane (Refs. 9–11).

REVIEW EXERCISES

1. Contrast a LAN with a WAN.
2. What are the two basic underlying transmission techniques for a LAN?
3. Name the three basic LAN topologies. Identify a fourth that is a subset of one of the three.
4. What range of BER can we expect on LANs? How are such good BERs achieved?
5. If a LAN has 50 accesses, how many connectivities must be theoretically analyzed for sufficient S/N?
6. What two basic transmission problems must a designer face with a baseband LAN?
7. LLC derives from what familiar link-layer protocol?
8. Relate the IEEE 802 LAN standard model with the ISO seven-layer OSI model.
9. What is a LAN access protocol?
10. Name at least three responsibilities of the LLC.
11. What are the two services the LLC provides its users?
12. How are collisions detected with CSMA/CD?
13. What is the purpose of the *jam signal* on CSMA/CD?
14. What is the function of a frame check sequence (FCS)?
15. Give three reasons why a MAC frame may be invalid?
16. How does a LAN know that a traffic frame is destined to it?
17. What is the bit rate of FDDI? What is its baud rate?

[11]ATM is covered in Chapter 18.

18. What is *frame stripping*?

19. How are collisions avoided using a token-passing scheme?

20. What is the function of a LAN repeater?

21. What are the four types of bridges covered in the text?

22. On what OSI layer do routers operate?

REFERENCES

1. *IEEE Standard Dictionary of Electrical and Electronic Terms*, 6th ed., IEEE Std. 100-1996, IEEE, New York, 1996.
2. *Information Technology—Part 2, Logical Link Control*, ANSI/IEEE 802.2, 1994 ed., with Amd. 3, IEEE, New York, 1994.
3. *Information Technology—Part 3: Carrier Sense Multiple Access with Collision Detection (CSMA/CD) Access Method*, ANSI/IEEE Std. 802.3, 1996 ed., IEEE, New York, 1996.
4. *Supplement to Carrier Sense Multiple Access with Collision Detection (CSMA/CD) Access Method—100 Mbps Operation, Type 100BaseT*, IEEE Std. 802.3u-1995, IEEE, New York, 1995.
5. *Information Technology—Part 5: Token Ring Access Method and Physical Layer Specification*, ANSI/IEEE 802.5, 1995 ed., IEEE, New York, 1995.
6. *Fiber Distributed Data Interface (FDDI)—Token Ring Physical Layer Medium Dependent*, ANSI X3.166-1990, ANSI, New York, 1990.
7. *Fiber Distributed Data Interface (FDDI) Physical Layer Protocol (PHY-2)*, ANSI X3.231-1994, ANSI, New York, 1994.
8. *Fiber Distributed Data Interface (FDDI): Token Ring Media Access Control—2, MAC-3*, ANSI X.239-1994, ANSI, New York, 1994.
9. Internetworking Troubleshooting Seminar Presentation, Hewlett-Packard Co., Tempe, AZ, Jan. 1995.
10. ChipCom promotional material, ChipCom, Southboro, MA, 1995.
11. *High-Speed Networking—Options and Implications*, ChipCom, Southboro, MA, 1995.

12

ENTERPRISE NETWORKS II: WIDE AREA NETWORKS

12.1 WIDE AREA NETWORK DEPLOYMENT

Wide area networks (WANs) provide data connectivity over much greater expanses than their local area network counterparts. Data rates on WANs are lower. One reason is that in many cases WANs are transported over the PSTN voice channels, either analog or digital.

In this chapter we will cover four types of WANs. These are:

1. X.25 packet communications;
2. TCP/IP protocol family overview;
3. Integrated Services Digital Networks (ISDN); and
4. Frame relay.

Three of these four are actually families of protocols involved with a particular WAN. The fourth allows some small insight regarding transmission requirements. All four utilize the resources of the PSTN one way or another.

12.1.1 Introductory Comments

Whereas the conventional LAN discussed in Chapter 11 provides data communication capabilities among a comparatively small and closed user group covering a very limited geographical area, a WAN has the potential of not only covering the entire world, even outer space, but an extremely large and diverse user group (e.g., the Internet). With these facts in mind, what are the really key essentials that we must understand that will make such a system provide us the capabilities we would expect? Let's brainstorm and prepare a short list of requirements of a data network to communicate *data messages*:

1. Data messages should have a high probability of reaching recipient(s) intact and comparatively error-free.
2. There may be an issue of urgency. Here we mean how soon after transmission a data message will reach its recipient.
3. The recipient(s) must be prepared to receive the message and "understand" its contents.

12.1.1.1 Data Message Must Reach the Recipient(s). How is a data message routed such that the indicated recipient(s) receive the message? With conventional telephony, signaling carries out this function. It sets up a circuit (Chapter 7), maintains the connectivity throughout the call, and then takes the circuit down when one or both subscribers go on-hook. With data communications these same functions must be carried out. One form of data connectivity is called *connection oriented*, where indeed a circuit is set up, traffic is passed, and then the circuit is taken down. There is another form of data communications, where a frame is launched by the originator, and that frame must find its own way to the destination, similar to what happens using the postal service. A letter is dropped in an outgoing box, and it finds its way (with the help of the postal service) to the destination. We, the originators, are completely unaware of the letter's routing; we do not really care. In the data world, this is called *connectionless service*. In the case of the postal service, the address on the envelope routed the letter.

For the data message case, it is the header that carries the routing information. In many situations, a data message may be made up of a number of frames. Each frame carries a field set aside for the recipient address. More often than not it is called the *destination address*. In many cases, a frame also carries the originator's address. In either case, these are 8-bit fields, often extendable in increment of 8 bits.

Three key questions come to mind: (1) What part of a frame's header can be easily identified as the originator and destination address fields? (2) What is the addressing capacity of an address field? Rephrased: how many distinct addresses can be accommodated in an 8-bit field? (3) Once a router, smart bridge, or data switch recognizes the boundaries of a destination address(es), how does it know how to route the frame?

1. What part of a frame's header can be easily identified as the originator and destination address fields? A family of protocols governs the operation of a particular data network. Our interest here is in the network-layer and data-link layer protocols. These protocols carefully lay out the data-link frame and the network-layer frame (see Section 10.10). For example, in Section 10.7.2 we introduced synchronous transmission and the generic data-link layer frame (Figure 10.7). One thing a digital processor can do and do very well is count bits, and groups of 8 bits, which we call an octet (others call it a byte). A specific network utilizes a particular data-link layer protocol. Thus a processor knows a priori where field boundaries are, because it is designed to meet the requirements of a particular protocol. With some data-link protocols, however, there may be a variable length info field. It is obvious that if this is the case, the processor must be informed of the length of a particular info field. When this is so, info field length information is often found in a subfield inside the control field (see Figures 10.7 and 10.25). So to answer question 1, the digital processor knows a priori exactly which 8 (or 16) bits comprise the destination address field by simply counting down from the unique start-of-frame octet. It knows a priori because the router processor conforms to a particular protocol.

2. What is the addressing capacity of an address field? We will let each address consist of a distinct 8- or 16-bit binary sequence. For an 8-bit binary group, how many distinct 8-bit sequences are there? Remember, on developing code capacity (Section 10.4), that we can have 2^n sequences where n is the number of bits in a particular sequence. In our sample case it is 8 bits. Thus, its addressing capacity is 2^8 or 256 distinct addresses. If extended addressing is employed, namely, where we use two octets for a destination address, often the least significant bit of the first octet is reserved to tell the receiver processor to expect the next octet also to be dedicated to destination

address. Now we have only 7 bits in the first octet left for addressing and all 8 bits in the second octet. Adding these two together, we get 15 bits for addressing. Thus, there are 2^{15} or 32,768 distinct addresses, quite a respectable number.

3. How does a router or switching data node know how to route to a particular address? Simply by consulting a *lookup table*. Now this lookup table may have fixed routing entries, or entries that can be updated manually or routing entries that are updated dynamically. Here we must recognize three possible conditions:

1. A node/router is added or dropped from the network;
2. New routing patterns are established; other routing patterns may be discontinued; and
3. Congestion, route/node degradation and/or failure may occur.

Read Section 12.3 on the TCP/IP protocol family because IP has some very interesting means and methods to update routing/lookup tables as well as finding routes to "unknown" destination addresses. Error detection and correction are discussed in Section 10.5.

12.1.1.2 There May Be an Issue of Urgency. There are several "families" of data messages that have limited or no real issue regarding urgency. For example, long accounting files, including payroll, may only require 24-hour or 48-hour delivery times. In this case, why not use the postal service, Federal Express, or UPS? On the other hand, credit card verification has high urgency requirements. A customer is waiting (probably impatiently) to have her/his credit verified during the process of buying some item. Most *transaction* data messages are highly urgent.

One reason for implementing frame relay is its low *latency*. Let us call latency the time it takes to complete a data message transaction. There are four causes for an increase in transaction time:

1. Propagation delay;
2. The number of message exchanges required to complete a transaction (e.g., handshakes, circuit set-up, ARQ exchanges, and so on; X.25 is rich in such transactions);
3. Processing time and processing requirements (e.g., every effort has been made to reduce processing requirements in frame relay; there are virtually no message exchanges with frame relay); and
4. Secondarily, the quality of a circuit (if the circuit is noisy, many ARQ exchanges will occur increasing latency dramatically).

12.1.1.3 Recipient Must Be Prepared to Receive and "Understand" a Data Message. This is simply a question of compatibility. The data message receiver must be compatible with the far-end transmitter and intermediate nodes. We cannot have one transmitting the ASCII code and the companion receiver only able to receive EBCDIC. Such compatibility *may* extend through all seven OSI layers. For example, frame relay is based *only* on OSI layers 1 and 2. It is the responsibility of the frame relay user to provide the necessary compatibility of the upper OSI layers.

Now that the readers understand the three major points discussed above, we proceed with the following sections dealing with the most well-known generic WAN protocols.

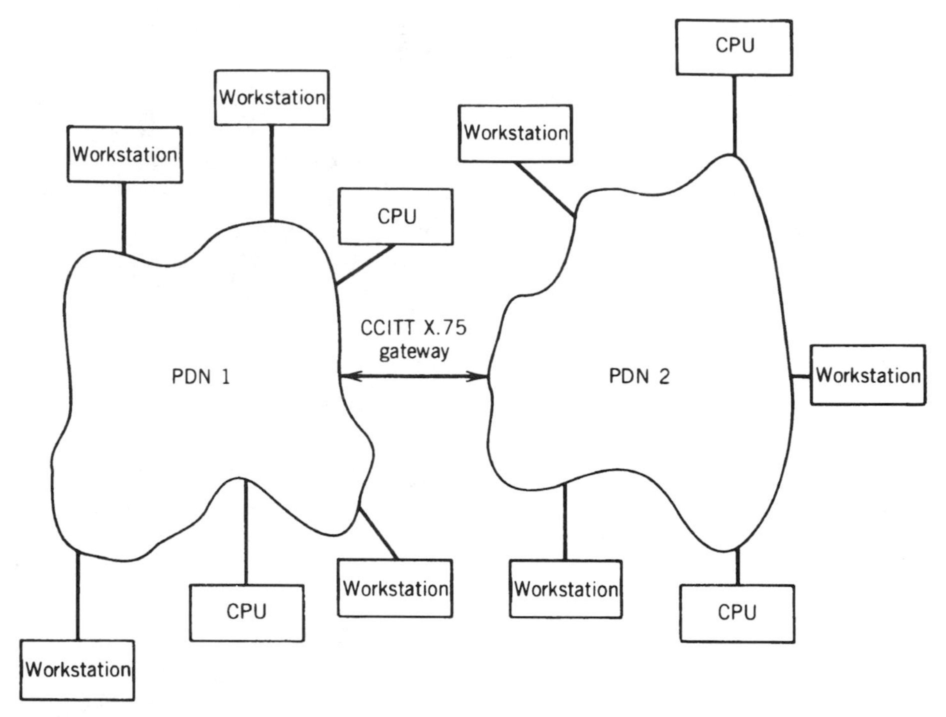

Figure 12.1 X.25 packet communications operates with the public switched data network (PDN). CPU = central processing unit or mainframe computer.

12.2 PACKET DATA COMMUNICATIONS BASED ON CCITT REC. X.25

The concept of a packet-switched network is based on the idea that the network switching nodes will have multiple choices for routing of data packets. If a particular route becomes congested or has degraded operation, a node can send a packet on another route, and if that route becomes congested, possibly a third route will be available to forward the packet to its destination.

At a data source, a file is segmented into comparatively short data packets, each of the same length and each with its own header and trailer. As we mentioned, these packets may take diverse routes through various nodes to their destination. The destination node is responsible for data message reassembly in its proper order.

12.2.1 Introduction to CCITT Rec. X.25

The stated purpose of CCITT Rec. X.25 (Ref. 1) is to define an interface between the DTE and the DCE at the first three OSI layers. Ideally the DCE resides at the local data switching exchange and the DTE is located on customer premises. In addition X.25 defines the procedures necessary for accessing a packet-switched public data network (PDN). Figure 12.1 shows the X.25 concept of accessing the PDN. An example of the PDN is ISDN described in Section 12.4.

Data terminals defined by X.25 operate in a synchronous full-duplex mode with data rates of 2400, 4800, 9600, 14,400, 28,800, and 33,600 bps; 48, 64, 128, 192, 256, 384, 512, 1024, 1536, and 1920 kbps.

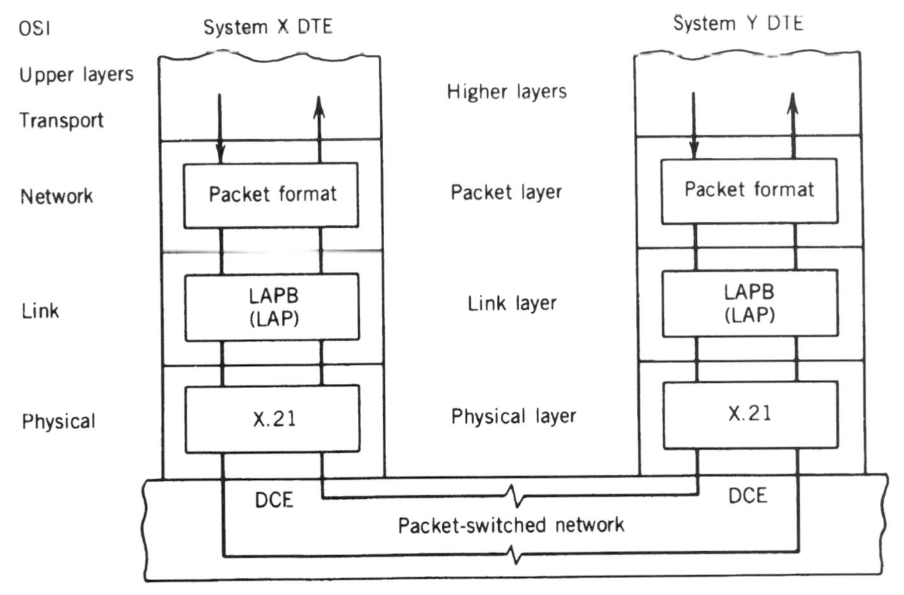

Figure 12.2 X.25 relationship with the OSI reference model.

12.2.2 X.25 Architecture and Its Relationship to OSI

X.25 spans the lowest three layers of the OSI reference model, as illustrated in Figure 12.2. It can be seen in the figure that X.25 is compatible with OSI up to the network layer. In this context there are differences at the network/transport layer boundary. CCITT leans toward the view that the network and transport layer services are identical and that these are provided by X.25 virtual circuits.

12.2.2.1 User Terminal Relationship to the PDN. CCITT Rec. X.25 calls the user terminal the DTE, and the DCE resides at the related PDN node. The entire recommendation deals with this DTE–DCE interface, not just the physical layer interface. For instance, a node (DCE) may connect to a related user (DTE) with one digital link, which is covered by SLP (single-link procedure) or several links covered by MLP (multilink procedure). Multiple links from a node to a DTE are usually multiplexed on one transmission facility.

The user (DTE) to user (DTE) connectivity through the PDN based on OSI is shown in Figure 12.3. A three-node connection is illustrated in this example. Of course, OSI layers 1–3 are Rec. X.25 specific. Note that the protocol peers for these lower three layers are located in the PDN nodes and not in the distant DTE. Further, the Rec. X.25 protocol operates only at the interface between the DTE and its related PDN node and does not govern internodal network procedures.

12.2.2.2 Three Layers of X.25

12.2.2.2.1 ***Physical Layer.*** The physical layer is layer 1, where the requirements are defined for the functional, mechanical, procedural, and electrical interfaces between the DTE and DCE. CCITT Rec. X.21 or X.21 bis is the applicable standard for this interface. X.21 bis is similar to EIA-232 (See Section 10.8). CCITT Rec. X.21 specifies a 15-pin DTE–DCE interface connector. The electrical characteristics for this interface are the

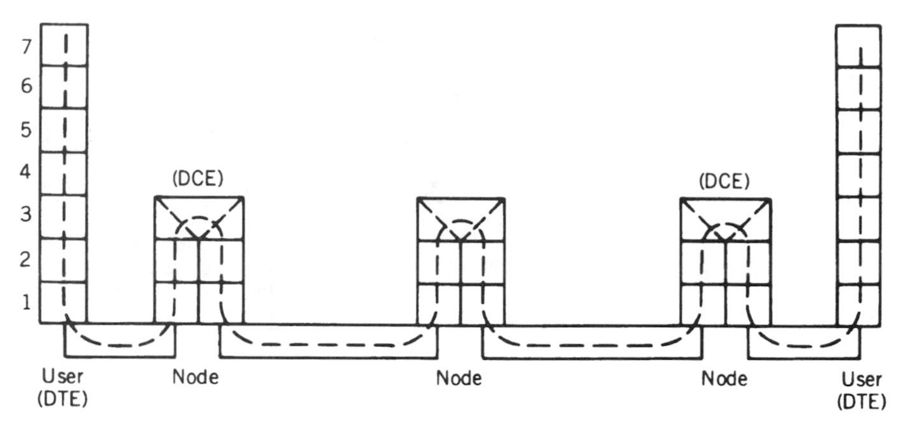

Figure 12.3 X.25 user (DTE) connects through the PDN to a distant user (DTE).

same as CCITT Recs. V.10 and V.11, depending on whether electrically balanced or unbalanced operation is desired. These two recommendations have some similarity to the electrical portion of EIA-232.

12.2.2.2.2 CCITT X.25 Link Layer. The link layer of X.25 uses the LAPB protocol. LAPB is fully compatible with HDLC link-layer access protocol (see Section 10.10.3). The information field in the LAPB frame carries the user data, in this case the layer 3 packet.

LAPB provides several options for link operation. These include two versions of the control field: standard and extended. It also supports multilink procedures. MLP allows a group of links to be managed as a single transmission facility. It carries out the function of resequencing packets in the proper order at the desired destination. When MLP is implemented, an MLP control field of two octets in length is inserted as the first 16 bits of the information field. This field contains a multilink sequence number and four control bits (see Figure 12.4).

12.2.2.2.3 Datagrams, Virtual Circuits, and Logical Connections. There are three approaches used with X.25 operation to manage the transfer and routing of packet streams: datagrams, virtual connections (VCs), and permanent virtual connections (PVCs). Datagram service uses optimal routing on a packet-by-packet basis, usually over diverse routes. In the virtual circuit approach, there are two operational "modes": virtual connection and permanent virtual connection. These two are analogous to a dial-up telephone connection and a leased line connection, respectively. With the virtual connection a logical connection is established before any packets are sent. The packet originator sends a call request to its serving node, which sets up a route in advance to the desired destination. All packets of a particular message traverse this route, and each packet of the message contains a virtual circuit identifier (logical channel number) and the packet data. At any one time each station can have more than one virtual circuit to any other station and can have virtual circuits to more than one station. With virtual circuits routing decisions are made in advance. With the datagram approach ad hoc decisions are made for each packet at each node. There is no call-setup phase with datagrams as there is with virtual connections. Virtual connections are advantageous for high community-of-interest connectivities, datagram service for low community-of-interest relations.

Datagram service is more reliable because traffic can be alternately routed around

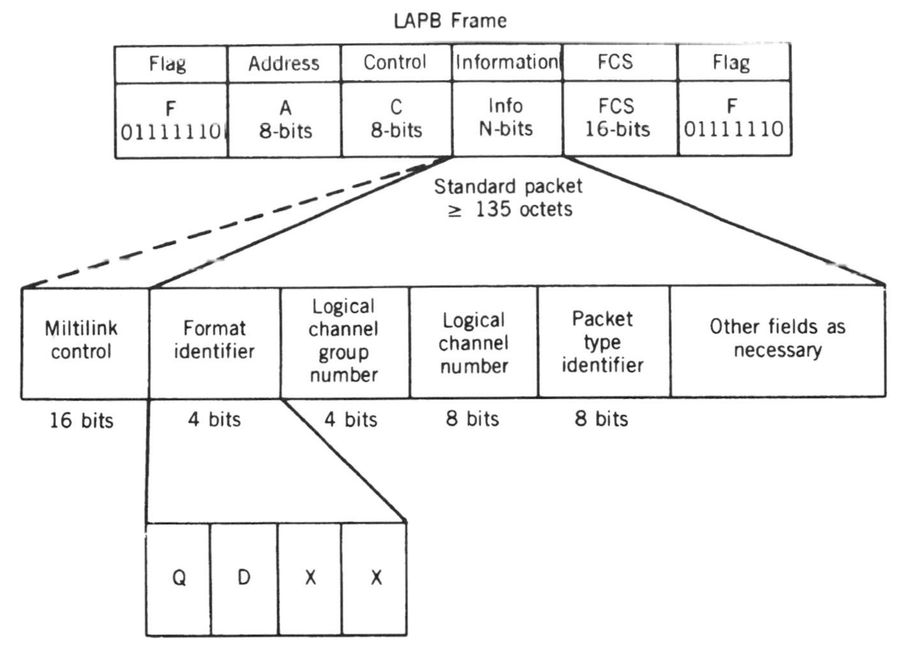

Figure 12.4 Basic X.25 frame structure. Note how the X.25 packet is embedded in the LAPB frame. For the extended LAPB structure, the control field will have 16 bits. There is a basic packet structure and an extended packet structure. For the basic data packet, XX = 01. For an extended data packet, XX = 10. Q = qualifier bit; D is the delivery confirmation bit.

network congestion points. Virtual circuits are fixed-routed for a particular call. Call-setup time at each node is eliminated on a packet basis as with the virtual connection technique. X.25 also allows the possibility of setting up permanent virtual connections and is network assigned. This latter alternative is economically viable only for very-high-traffic relations; otherwise these permanently assigned logical channels will have long dormant periods.

12.2.2.3 X.25 Frame Structure: Layer 3, the Packet Layer. The basic data-link layer (LAPB) frame structure is given in Figure 12.4. Its similarity to the HDLC frame structure is apparent. For the X.25 case the packet is embedded in the LAPB information field, as mentioned. When applicable, the other part of the information field contains the MLP, which is appended in front of the X.25 packet and is the first subfield in the information (I) field. In fact, the MLP is not part of the actual packet and is governed by the layer 2 LAPB protocol.[1]

12.2.2.3.1 Structure Common to All Packets. Table 12.1 shows 17 packet types involved in X.25. Every packet transferred across the X.25 DTE–DCE interface consists of at least three octets.[2] These three octets contain a general format identifier, a logical channel identifier, and a packet type identifier. Other fields are appended as required, as illustrated in Figure 12.4.

[1]*LAPB* is often called "link access protocol B-channel." B-channel is ISDN nomenclature for a 64-kbps channel designated to carry revenue-bearing traffic.

[2]An octet is an 8-bit sequence. There is a growing tendency to use octet rather than byte when describing such a sequence. It removes ambiguity on the definition of byte.

Table 12.1 Packet Type Identifier

Packet Type		Octet 3							
		Bit							
From DCE to DTE	From DTE to DCE	8	7	6	5	4	3	2	1
	Call Setup and Cleaning								
Incoming call	Call request	0	0	0	0	1	0	1	1
Call connected	Call accepted	0	0	0	0	1	1	1	1
Clear indication	Clear request	0	0	0	1	0	0	1	1
DCE clear confirmation	DTE clear confirmation	0	0	0	1	0	1	1	1
	Data and Interrupt								
DCE data	DTE data	×	×	×	×	×	×	×	0
DCE interrupt	DTE interrupt	0	0	1	0	0	0	1	1
DCE interrupt confirmation	DTE interrupt confirmation	0	0	1	0	0	1	1	1
	Flow Control and Reset								
DCE RR (modulo 8)	DTE RR (modulo 8)	×	×	×	0	0	0	0	1
DCE RR (modulo 128)[a]	DTE RR (modulo 128)[a]	0	0	0	0	0	0	0	1
DCE RNR (modulo 8)	DTE RNR (modulo 8)	×	×	×	0	0	1	0	1
DCE RNR (modulo 128)[a]	DTE RNR (modulo 128)[a]	0	0	0	0	0	1	0	1
	DTE REJ (modulo 8)[a]	×	×	×	0	1	0	0	1
	DTE REJ (modulo 128)[a]	0	0	0	0	1	0	0	1
Reset indication	Reset request	0	0	0	1	1	0	1	1
DCE reset confirmation	DTE reset confirmation	0	0	0	1	1	1	1	1
	Restart								
Restart indication	Restart request	1	1	1	1	1	0	1	1
DCE restart confirmation	DTE restart confirmation	1	1	1	1	1	1	1	1
	Diagnostic								
Diagnostic[a]		1	1	1	1	0	0	0	1
	Registration[a]								
	Registration request	1	1	1	1	0	0	1	1
Registration confirmation		1	1	1	1	0	1	1	1

[a] Not necessarily available on every network.

Note: A bit that is indicated as × may be set to either 0 or 1.

Source: ITU-T Rec. X.25, Table 5-2/X.25, p. 52 (Ref. 1).

Now consider the general format identifier field in Figure 12.4. This is a 4-bit sequence. Bit 8, counting from right to left in Figure 12.4, is the qualifier bit (Q) found only in data packets. In the call-setup and clearing packets it is the A-bit, and in all other packets it is set to 0. The D-bit, when set to 1, specifies end-to-end delivery confirmation. This confirmation is provided through the packet receive number [P(R)]. When set to 01, the XX bits indicate a basic packet and when set to 10, an extended packet is indicated. The extension number involves sequence number lengths. (i.e., a longer sequence number can be accommodated).

Logical channel assignment is shown in Figure 12.5. The logical channel group and logical channel number subfields identify logical channels with the capability of identifying up to 4096 channels (2^{12}). This permits the DTE to establish up to 4095 simul-

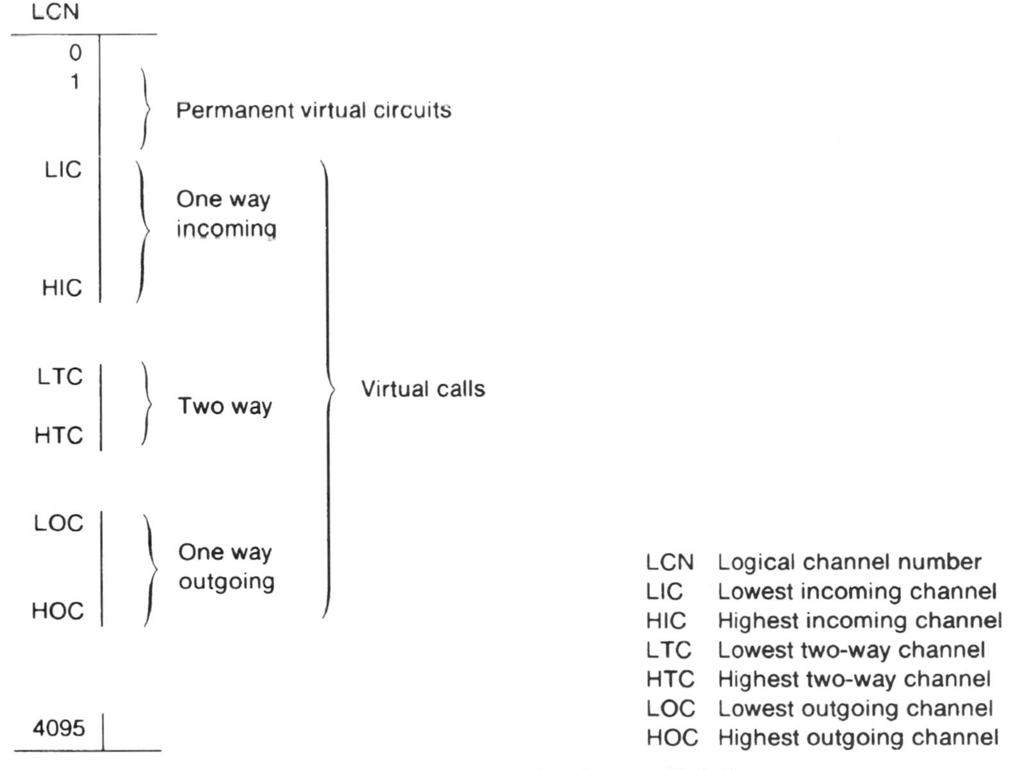

Figure 12.5 Logical channel assignment (Ref. 1).

taneous virtual circuits through its DCE to other DTEs. As mentioned, this is usually done by multiplexing these circuits over a single transmission facility.

Permanent virtual circuits (PVCs) have permanently assigned logical channels, whereas those for virtual calls are assigned channels only for the duration of a call, as shown in Figure 12.5. Channel 0 is reserved for restart and diagnostic functions. To avoid collisions, the DCE starts assigning logical channels at the lowest number end and the DTE from the highest number end. There are one-way and both-way (two-way) circuits. The both-way circuits are reserved for overflow to avoid chances of double seizures (i.e., both ends seize the same circuit).

Octet 5 in Figure 12.4 is the packet type identifier subfield. The packet type and its corresponding coding are shown in Table 12.1. We can see in the table that packet types are identified in associated pairs carrying the same packet identifier (bit sequence). A packet from the calling terminal (DTE) to the network (DCE) is identified by one name. The associated packet delivered by the network to the called terminal (DTE) is referred to by another associated name.

12.2.2.4 Two Typical Packets

Call request and incoming call packet. This type of packet sets up the call for the virtual circuit. The format of the call request and incoming packet is illustrated in Figure 12.6. Octets 1–3 have been described in the previous subsection. Octet 4 consists of the address length field indicators for the called and calling DTE addresses. Each address length indicator is binary coded, and bit 1–5 is the lower-order bit of the indicator.

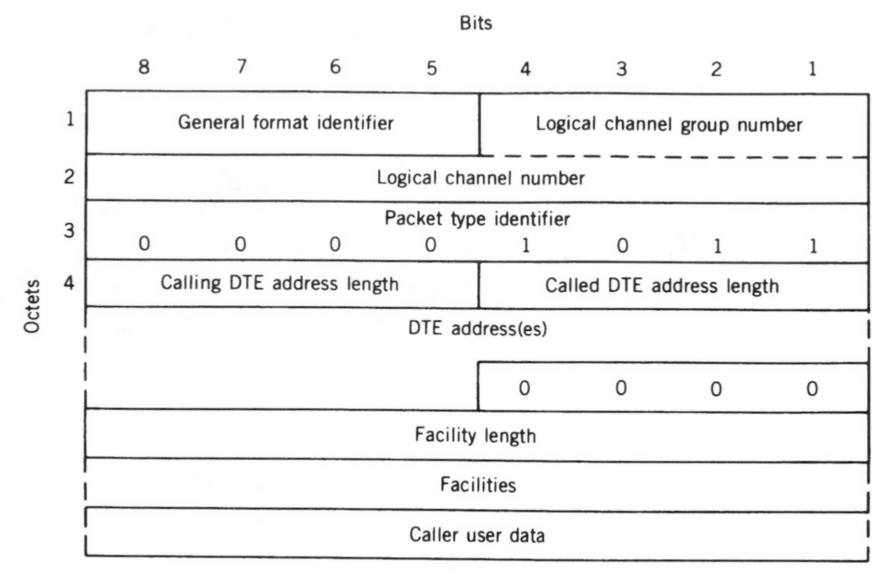

Figure 12.6 Call request and incoming call packet format. The general format identifier is coded 0X01 (basic) and 0X10 (extended format). (From ITU-T Rec. X.25, Figure 5-3/X,25, p. 58, ITU-T Organization, Helsinki, March 1993, [Ref. 1].)

Octet 5 and the following octets consist of the called DTE address, when present, and then the calling DTE address, when present.

The facilities length field (one octet) indicates the length of the facilities field that follows. The facility field is present only when the DTE is using an optional user facility requiring some indication in the call request and incoming call packets. The field must contain an integral number of octets with a maximum length of 109 octets.

Optional user facilities are listed in CCITT Rec. X.2. There are 45 listed. Several examples are listed here to give some idea of what is meant by *facilities* in CCITT Rec. X.25:

- Nonstandard default window;
- Flow control parameter negotiation;
- Throughput class negotiation;
- Incoming calls barred;
- Outgoing calls barred;
- Closed user group (CUG);
- Reverse charging acceptance; and
- Fast select.

DTE and DCE data packet. Of the 17 packet types listed in Table 12.1, only one truly carries user information, the DTE and DCE data packet. Figure 12.7 illustrates the packet format.

Octets 1 and 2 have been described. Bits 6, 7, and 8 of octet 3 or bits 2–8 of octet 4, when extended, are used for indicating the packet received sequence number $P(R)$. It is binary coded and bit 6, or bit 2 when extended, is the low-order bit.

In Figure 12.7, M, which is bit 5 in octet 3 or bit 1 in octet 4 when extended, is used for more data (M bit). It is coded 0 for "no more data" and 1 for "more data to follow."

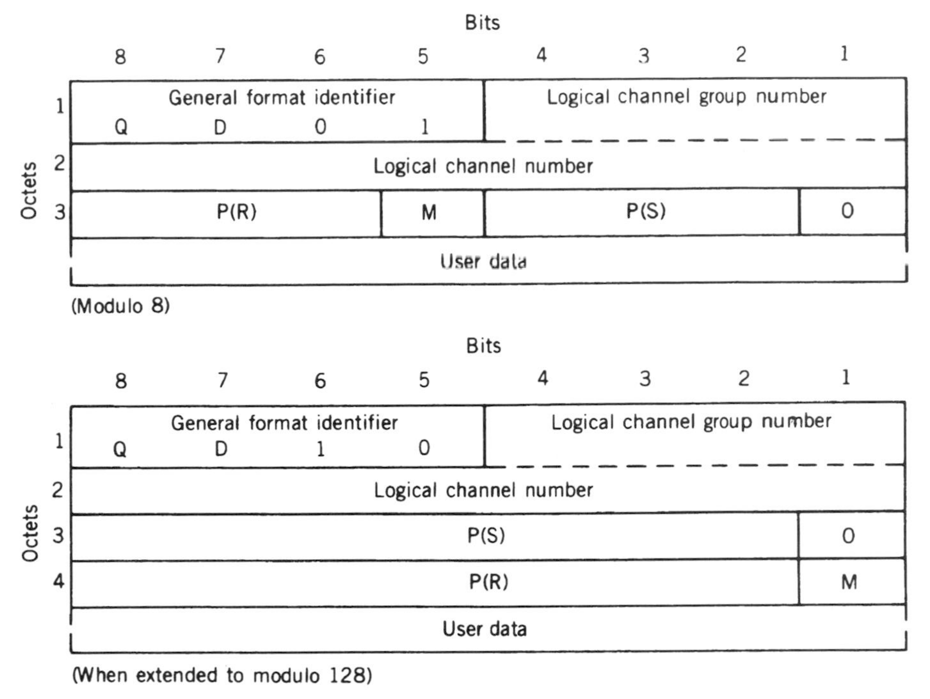

Figure 12.7 DTE and DCE packet format. (From ITU-T Rec. X.25, Figure 5-7/X.25, p. 64, ITU-T Organization, Helsinki, March 1993, [Ref. 1]. D = Delivery confirmation bit; M = More databit; Q = Qualifier bit.)

Bits 2, 3, and 4 of octet 3, or bits 2–8 of octet 3 when extended, are used for indicating the packet send sequence number $P(S)$. Bits following octet 3, or octet 4 when extended, contain the user data.

The standard maximum user data field length is 128 octets. CCITT Rec. X.25 (para. 4.3.2) states: "In addition, other maximum user data field lengths may be offered by Administrations from the following list: 16, 32, 64, 256, 512, 1024, 2048, and 4096 octets . . . . Negotiation of maximum user data field lengths on a per call basis may be made with the flow parameter negotiation facility."

12.2.3 Tracing the Life of a Virtual Call

A call is initiated by a DTE by the transfer to the network of a call request packet. It identifies the logical channel number selected by the originating DTE, the address of the called DTE (destination), and optional facility information, and can contain up to 16 octets of user information. The facility and user fields are optional at the discretion of the source DTE. The receipt by the network of the call request packet initiates the call-setup sequence. This same call request packet is delivered to the destination DTE as an incoming call packet. The destination DTE in return sends a "call accepted" packet, and the source DTE receives a "call confirmation" packet. This completes the call setup phase, and the data transfer can begin.

Data packets (Figure 12.7) carry the user data to be transferred. There may be one or a sequence of packets transferred during a virtual call. It is the M bit that tells the destination that the next packet is a logical continuation of the previous packet(s). Sequence numbers verify correct packet order and are the packet acknowledgment tools.

The last phase in the life of a virtual call is the clearing (takedown). Either the DTE or the network can clear a virtual call. The "clear" applies only to the logical channel that was used for that call. Three different packet types are involved in the call-clearing phase. The clear request packet is issued by the DTE initiating the clear. The remote DTE receives it as a clear indication packet. Both the DTE and DCE then issue clear confirmation packets to acknowledge receipt of the clear packets.

Flow control is called out by the following packet types: receiver ready (RR), receive not ready (RNR), and reject (Rej). Each of these packets is normally three octets long, or four octets long using the extended version. Sequence numbering also assists flow control (Ref. 1).

12.3 TCP/IP AND RELATED PROTOCOLS

12.3.1 Background and Scope

The transmission control protocol/Internet protocol (TCP/IP) family was developed for the ARPANET (Advanced Research Projects Agency Network). ARPANET was one of the first large advanced packet-switched networks. It was initially designed and operated to interconnect the very large university and industrial defense community to share research resources. It dates back to 1968 and was well into existence before ISO and CCITT took interest in layered protocols.

The TCP/IP suite of protocols (Refs. 2, 3) has wide acceptance today, especially in the commercial and industrial community worldwide. These protocols are used on both LANs and WANs. They are particularly attractive for their internetworking capabilities. The IP (Ref. 2) competes with CCITT Rec. X.75 protocol (Ref. 4), but is notably more versatile and has much wider application.

The architectural model of the IP (Ref. 2) uses terminology that differs from the OSI reference model.[3] Figure 12.8 shows the relationship between TCP/IP and related DoD

OSI	TCP/IP and Related Protocols		
Application	File transfer	Electronic mail	Terminal emulation
Presentation	File transfer protocol (FTP)	Simple mail transfer protocol (SMTP)	Telnet protocol
Session			
Transport	Transmission control protocol (TCP)		User datagram protocol (UDP)
Network	Address resolution protocol (ARP)	Internet protocol (IP)	Internet control message protocol (ICMP)
Data link	Network interface cards CSMA/CD (Ethernet), Token Ring, ARCNET, StarLan		
Physical	Transmission media Wire pair, fiber optics, coaxial cable, radio		

Figure 12.8 How TCP/IP and associated protocols relate to OSI.

[3]IP predates OSI.

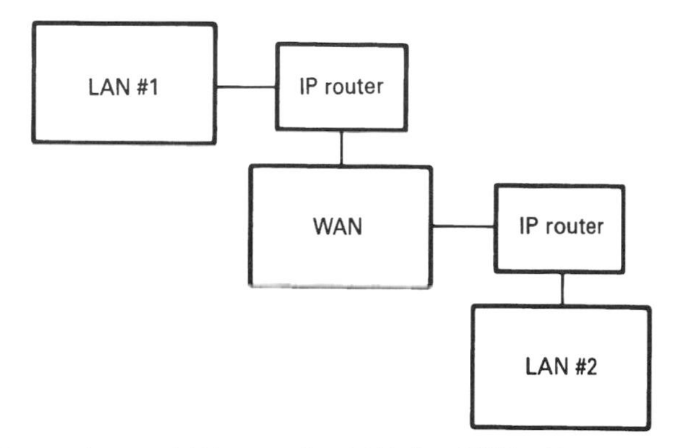

Figure 12.9 Connecting one LAN to another LAN via a WAN with routers equipped with IP.

(Department of Defense) protocols and the OSI reference model. Tracing data traffic from an originating host, which runs an application program, to another host in another network, is shown in Figure 12.9. This may be a LAN-to-WAN-to-LAN connectivity, as shown in the figure. It may also be just a LAN-to-WAN or it may be a WAN-to-WAN connectivity. The host would enter its own network by means of a network access protocol such as HDLC or an IEEE 802 series protocol (Chapter 11).

A LAN connects via a router (or gateway) to another network. Typically a router (or gateway) is loaded with three protocols. Two of these protocols connect to each of the attached networks (e.g., LAN and WAN), and the third protocol is the IP, which provides the network-to-network interface.

Hosts typically are equipped with four protocols. To communicate with routers or gateways, a network access protocol and Internet protocol are required. A transport layer protocol assures reliable communication between hosts because end-to-end capability is not provided in either the network access or Internet protocols. Hosts also must have application protocols such as e-mail or file transfer protocols (FTPs).

12.3.2 TCP/IP and Data-Link Layers

TCP/IP is transparent to the type of data-link layer involved, and it is also transparent whether it is operating in a LAN or WAN domain or among them. However, there is document support for Ethernet, IEEE 802 series, ARCNET LANs, and X.25 for WANs (Refs. 2, 5).

Figure 12.10 shows how upper OSI layers are encapsulated with TCP and IP header information and then incorporated into a data-link layer frame.

For the case of IEEE 802 series LAN protocols, advantage is taken of the LLC common to all 802 LAN protocols. The LLC extended header contains the SNAP (subnetwork access protocol) such that we have three octets of the LLC header and five octers in the SNAP. The LLC header has its fields fixed as follows (LLC is discussed in Chapter 11):

DSAP = 10101010;
SSAP = 10101010;
Control = 00000011.

The five octets in the SNAP have three assigned for protocol ID or organizational

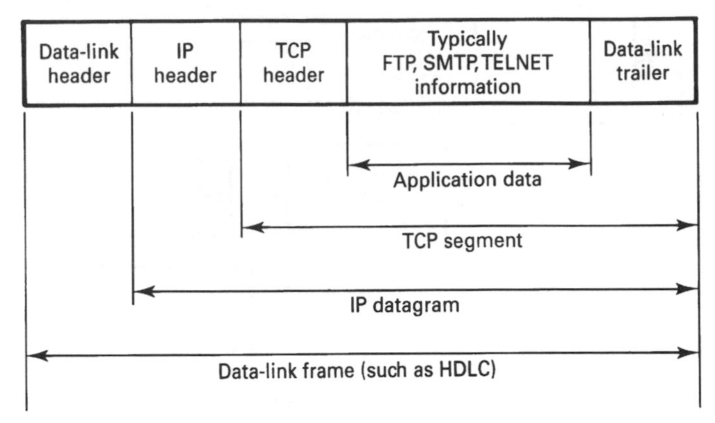

Figure 12.10 The incorporation of upper-layer PDUs into a data-link layer frame showing the relationship with TCP and IP.

code and two octets for "EtherType." EtherType assignments are shown in Table 12.2. EtherType refers to the general class of LANs based on CSMA/CD (see Chapter 11 for a discussion of CSMA/CD).

Figure 12.11 shows the OSI relationships with TCP/IP working with the IEEE 802 LAN protocol group. Figure 12.12 illustrates an IEEE 802 frame incorporating TCP, IP, and LLC.

Often addressing formats are incompatible from one protocol in one network to

Table 12.2 Ether-Type Assignments

Ethernet Decimal	Hex	Description
512	0200	XEROX PUP
513	0201	PUP address translation
1536	0600	XEROX NS IDP
2048	0800	DOD Internet protocol (IP)
2049	0801	X.75 Internet
2050	0802	NBS Internet
2051	0803	ECMA Internet
2052	0804	Chaosnet
2053	0805	X.25 level 3
2054	0806	Address resolution protocol (ARP)
2055	0807	XNS compatibility
4096	1000	Berkeley trailer
21000	5208	BBN Simnet
24577	6001	DEC MOP dump/load
24578	6002	DEC MOP remote control
24579	6003	DEC DECnet phase IV
24580	6004	DEC LAT
24582	6005	DEC
24583	6006	DEC
32773	8005	HP probe
32784	8010	Excelan
32821	8035	Reverse ARP
32824	8038	DEC LANBridge
32823	8098	Appletalk

Source: Ref. 6.

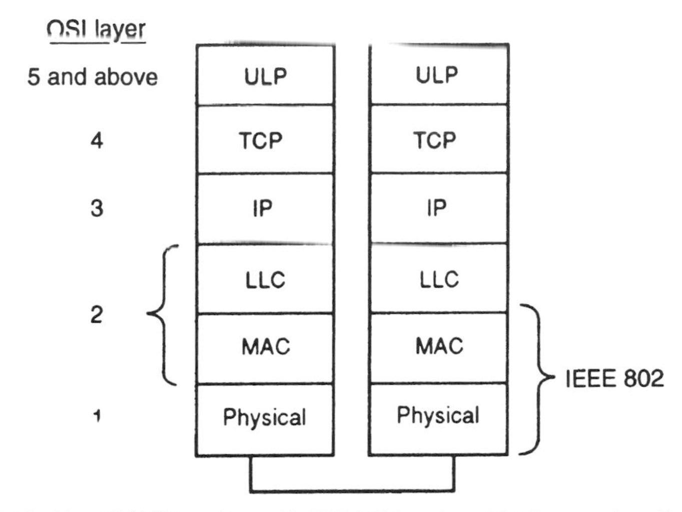

Figure 12.11 How TCP/IP, working with IEEE 802 series of LAN protocols, relates to OSI.

another protocol in another network. A good example is the mapping of the 32-bit Internet address into a 48-bit IEEE 802 address. This problem is resolved with ARP (address resolution protocol). Another interface problem is limited IP datagram length of 576 octets, where with the 802 series, frames have considerably longer length limits (Ref. 7).

12.3.3 IP Routing Algorithm

In OSI the network layer functions include routing and switching of a datagram through the telecommunications subnetwork. The IP provides this essential function. It forwards the datagram based upon the network address contained within the IP header. Each data-

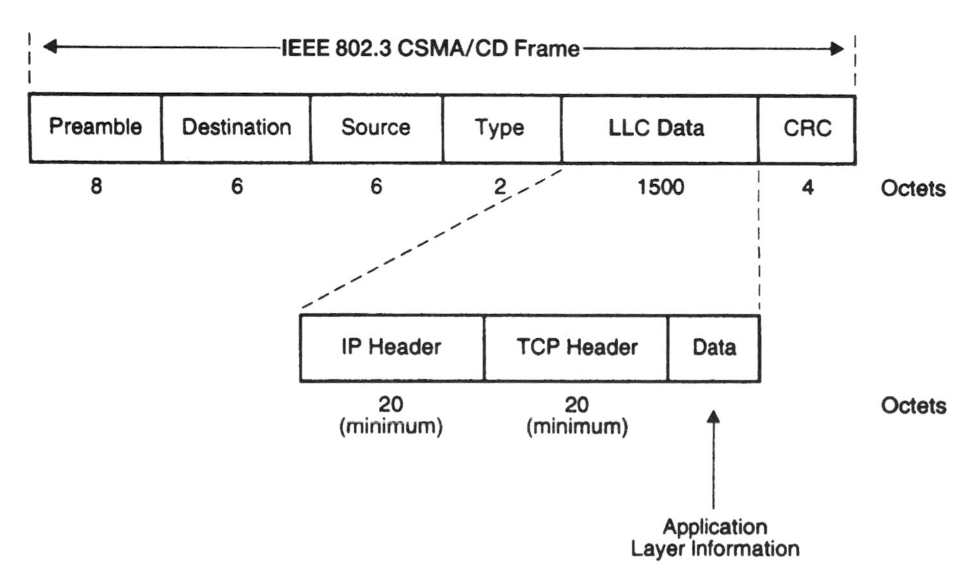

Figure 12.12 A typical IEEE 802 frame showing LLC and TCP/IP functions.

gram is independent and has no relationship with other datagrams. There is no guaranteed delivery of the datagrams from the standpoint of the Internet protocol. However, the next higher layer, the TCP layer, provides for the reliability that the IP lacks. It also carries out segmentation and reassembly function of a datagram to match frame sizes of data-link layer protocols.

Addresses determine routing and, at the far end, equipment (hardware). Actual routing derives from the IP address, and equipment addresses derive from the data-link layer header (typically the 48-bit Ethernet address) (Ref. 5).

User data from upper-layer protocols is passed to the IP layer. The IP layer examines the network address (IP address) for a particular datagram and determines if the destination node is on its own local area network or some other network. If it is on the same network, the datagram is forwarded directly to the destination host. If it is on some other network, it is forwarded to the local IP router (gateway). The router, in turn, examines the IP address and forwards the datagram as appropriate. Routing is based on a lookup table residing in each router or gateway.

12.3.3.1 IP Routing Details. A gateway (router) needs only the network ID portion of the address to perform its routing function. Each router or gateway has a routing table, which consists of destination network addresses and specified next-hop gateway.

Three types of routing are performed by the routing table:

1. Direct routing to locally attached devices;
2. Routing to networks that are reached via one or more gateways; and
3. Default routing to destination network in case the first types of routing are unsuccessful.

Suppose a datagram (or datagrams) is (are) directed to a host which is not in the routing table resident in a particular gateway. Likewise, there is a possibility that the network address for that host is also unknown. These problems may be resolved with the *address resolution protocol* (ARP) (Ref. 7).

First the ARP searches a mapping table which relates IP addresses with corresponding physical addresses. If the address is found, it returns the correct address to the requester. If it cannot be found, the ARP broadcasts a *request* containing the IP target address in question. If a device recognizes the address, it will reply to the request where it will update its ARP *cache* with that information. The ARP cache contains the mapping tables maintained by the ARP module.

There is also a *reverse address resolution protocol* (RARP) (Ref. 8). It works in a fashion similar to that of the ARP, but in reverse order. RARP provides an IP address to a device when the device returns its own hardware address. This is particularly useful when certain devices are booted and only know their own hardware address.

Routing with IP involves a term called *hop*. A hop is defined as a link connecting adjacent nodes (gateways) in a connectivity involving IP. A *hop count* indicates how many gateways (nodes) must be traversed between source and destination.

One part of an IP routing algorithm can be *source routing*. Here an upper-layer protocol (ULP) determines how an IP datagram is to be routed. One option is that the ULP passes a listing of Internet addresses to the IP layer. In this case information is provided on the intermediate nodes required for transit of a datagram in question to its final destination.

Each gateway makes its routing decision based on a resident routing list or routing table. If a destination resides in another network, a routing decision is required by the IP gateway to implement a route to that other network. In many cases, multiple hops are involved and each gateway must carry out routing decisions based on its own routing table.

A routing table can be static or dynamic. The table contains IP addressing information for each reachable network and closest gateway for the network, and it is based on the concept of shortest routing, thus routing through the closest gateway.

Involved in IP shortest routing is the *distance metric*, which is a value expressing minimum number of hops between a gateway and a datagram's destination. An IP gateway tries to match the destination network address contained in the header of a datagram with a network address entry contained in its routing table. If no match is found, the gateway discards the datagram and sends an ICMP (see next subsection) message back to the datagram source.

12.3.3.2 Internet Control Message Protocol (ICMP). ICMP is used as an adjunct to IP when there is an error in datagram processing. ICMP uses the basic support of IP as if it were a higher-level protocol; however, ICMP is actually an integral part of IP and is implemented by every IP module.

ICMP messages are sent in several situations: for example, when a datagram cannot reach its destination, when a gateway (router) does not have the buffering capacity to forward a datagram, and when the gateway (router) can direct the host to send traffic on a shorter route.

ICMP messages typically report errors in the processing of datagrams. To avoid the possibility of infinite regress of messages about messages, and so on, no ICMP messages are sent about ICMP messages. There are eight distinct ICMP messages:

1. Destination unreachable message;
2. Time exceeded message;
3. Parameter problem message;
4. Source quench message;
5. Redirect message;
6. Echo or echo reply message;
7. Timestamp or timestamp reply message; and
8. Information request or information reply message.

12.3.3.3 IP Summary. The IP provides connectionless service, meaning that there is no call-setup phase prior to the exchange of traffic. There are no flow control or error control capabilities incorporated in IP. These are left to the next higher layer, the transmission control protocol (TCP). The IP is transparent to subnetworks connecting at lower layers; thus different types of networks can attach to an IP gateway or router. To compensate for these deficiencies in IP, the TCP (transmission control protocol) was developed as an upper layer to IP. It should be noted that TCP/IP can be found in both the LAN and WAN environments.

12.3.4 Transmission Control Protocol (TCP)

12.3.4.1 TCP Defined. TCP (Refs. 3, 9) was designed to provide reliable communication between pairs of processes in logically distinct hosts on networks and sets of

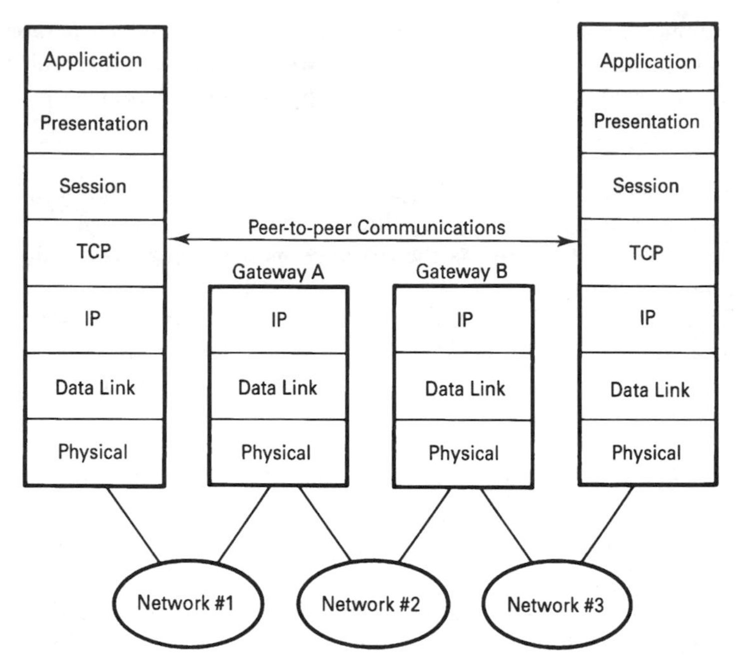

Figure 12.13 Protocol layers showing the relationship of TCP with other layered protocols.

interconnected networks.[4] TCP operates successfully in an environment where the loss, damage, duplication, or misorder of data, and network congestion can occur. This robustness in spite of unreliable communications media makes TCP well suited to support commercial, military, and government applications. TCP appears at the transport layer of the protocol hierarchy. Here TCP provides connection-oriented data transfer that is reliable, ordered, full duplex, and flow controlled. TCP is designed to support a wide range of upper-layer protocols (ULPs). The ULP can channel continuous streams of data through TCP for delivery to peer ULPs. The TCP breaks the streams into portions that are encapsulated together with appropriate addressing and control information to form a segment—the unit of exchange between TCPs. In turn, the TCP passes the segments to the network layer for transmission through the communication system to the peer TCP.

As shown in Figure 12.13, the layer below the TCP in the protocol hierarchy is commonly the IP layer. The IP layer provides a way for the TCP to send and receive variable-length segments of information enclosed in Internet datagram "envelopes." The Internet datagram provides a means for addressing source and destination TCPs in different networks. The IP also deals with fragmentation or reassembly of TCP segments required to achieve transport and delivery through the multiple networks and interconnecting gateways (routers). The IP also carries information on the precedence, security classification, and compartmentation of the TCP segments, so this information can be communicated end-to-end across multiple networks.

12.3.4.2 TCP Mechanisms. TCP builds it services on top of the network layer's potentially unreliable services with mechanisms such as error detection, positive ack-

[4]The IEEE (Ref. 10) defines a *host* as "A device to which other devices (peripherals) are connected and that generally controls those devices" and a *host computer* as "A computer, attached to a network, providing primary services such as computation, data base access or special programs or programming languages."

nowledgments, sequence numbers and flow control. These mechanisms require certain addressing and control information to be initialized and maintained during data transfer. This collection of information is called a *TCP connection*. The following paragraphs describe the purpose and operation of the major TCP mechanisms.

Par mechanism. TCP uses a positive acknowledgment with retransmission (PAR) mechanism to recover from the loss of a segment by the lower layers. The strategy with PAR is for a sending TCP to retransmit a segment at timed intervals until a positive acknowledgment is returned. The choice of retransmission interval affects efficiency. An interval that is too long reduces data throughput while one that is too short floods the transmission media with superfluous segments. In TCP, the timeout is expected to be dynamically adjusted to approximate the segment round-trip time plus a factor for internal processing; otherwise performance degradation may occur. TCP uses a simple checksum to detect segments damaged in transit. Such segments are discarded without being acknowledged. Hence, damaged segments are treated identically to lost segments and are compensated for by the PAR mechanism. TCP assigns sequence numbers to identify each octet of the data stream. These enable a receiving TCP to detect duplicate and out-of-order segments. Sequence numbers are also used to extend the PAR mechanism by allowing a single acknowledgment to cover many segments worth of data. Thus, a sending TCP can still send new data, although previous data have not been acknowledged.

Flow control mechanism. TCP's flow control mechanism enables a receiving TCP to govern the amount of data dispatched by a sending TCP. The mechanism is based on a *window*, which defines a contiguous interval of acceptable sequence-numbered data. As data are accepted, TCP slides the window upward in the sequence number space. This window is carried in every segment, enabling peer TCPs to maintain up-to-date window information.

Multiplexing mechanism. TCP employs a multiplexing mechanism to allow multiple ULPs within a single host and multiple processes in a ULP to use TCP simultaneously. This mechanism associates identifiers, called *ports*, to ULP processes accessing TCP services. A ULP connection is uniquely identified with a *socket*, the concatenation of a port and an Internet address. Each connection is uniquely named with a socket pair. This naming scheme allows a single ULP to support connections to multiple remote ULPs. ULPs which provide popular resources are assigned permanent sockets, called *well-known sockets.*

12.3.4.3 ULP Synchronization. When two ULPs (upper-layer protocols) wish to communicate (see Figure 12.11), they instruct their TCPs to initialize and synchronize the mechanism information on each to open the connection. However, the potentially unreliable network layer (i.e., the IP layer) can complicate the process of synchronization. Delayed or duplicate segments from previous connection attempts might be mistaken for new ones. A handshake procedure with clock-based sequence numbers is used in connection opening to reduce the possibility of such false connections. In the simplest handshake, the TCP pair synchronizes sequence numbers by exchanging three segments, thus the name *three-way handshake.*

12.3.4.4 ULP Modes. A ULP can open a connection in one of two modes: passive or active. With a passive open, a ULP instructs its TCP to be *receptive* to connections with other ULPs. With an active open, a ULP instructs its TCP to actually initiate a three-way handshake to connect to another ULP. Usually an active open is targeted to a passive open. This active/passive model supports server-oriented applications where a permanent resource, such as a database-management process, can always be accessed by remote users. However, the three-way handshake also coordinates two simultaneous active opens to open a connection. Over an open connection, the ULP pair can exchange a continuous stream of data in both directions. Normally, TCP groups the data into TCP segments for transmission at its own convenience. However, a ULP can exercise a *push* service to force TCP to package and send data passed up to that point without waiting for additional data. This mechanism is intended to prevent possible deadlock situations where a ULP waits for data internally buffered by TCP. For example, an interactive editor might wait forever for a single input line from a terminal. A push will force data through the TCPs to the awaiting process. A TCP also provides the means for a sending ULP to indicate to a receiving ULP that "urgent" data appear in the upcoming data stream. This urgent mechanism can support, for example, interrupts or breaks. When a data exchange is complete, the connection can be closed by either ULP to free TCP resources for other connections. Connection closing can happen in two ways. The first, called a *graceful close*, is based on the three-way handshake procedure to complete data exchange and coordinate closure between the TCPs. The second, called an *abort*, does not allow coordination and may result in the loss of unacknowledged data. [*Note:* There is a certain military flavor in the TCP/IP protocol family. There is good reason; its development (around 1975) was supported by the U.S. Department of Defense for ARPANET. Since then, this protocol family has become extremely popular in the commercial world (e.g., Internet), both for LAN and WAN operations.]

12.4 INTEGRATED SERVICES DIGITAL NETWORKS (ISDN)

12.4.1 Background and Objectives

The original concept of ISDN dates back to the early 1970s. Its design, in the context of the period, was built around the copper distribution plant (subscriber loop and local trunk plant). The designers saw and understood that by the early 1980s there would be a digital network in place controlled by CCITT Signaling System No. 7 (Chapter 13). It was revolutionary for its time by bringing 64-kbps digital channels right into the home and office. With the ISDN design, the 64-kbps digital channel handles:

- Voice telephony (digital);
- Digital data, both packet switched and circuit switched;
- Telex/teletext;
- Facsimile (e.g., CCITT Group 4); and
- Conference television (56, 64, or 128 kbps).

The goal of ISDN is to provide an integrated facility to incorporate each of the services just listed on a common 64-kbps channel. This section provides an overview of that integration.

In North America, one gets the distinct feeling the certain technologies [e.g., ATM (Chapter 18), frame relay (Section 12.5) and gigabit enterprise networks (Chapter 11), and EHF wireless local loops)] have leap-frogged ISDN. Yet ISDN keeps seeming to have second, third, ... renaissances in North America and has had a fairly high penetration in Europe. It seems that the ISDN market is very price/cost driven.

ISDN, in itself, is really only an interface specification at customer premises and at the first serving digital exchange. This digital exchange is part of a larger digital network that must have CCITT Signaling System No. 7 implemented and operational for ISDN to work end-to-end successfully (Ref. 11).

12.4.2 ISDN Structures

12.4.2.1 ISDN User Channels. Here we look from the user into the network. We consider two user classes: residential and commercial. The following are the standard bit rates for user access links:

- B-channel: 64 kbps;
- D-channel: 16 kbps or 64 kbps, and
- H-channels (discussed in the following).

The B-channel is the basic user channel. It is transparent to bit sequences. It serves all the traffic types listed in Section 12.4.1.

In one configuration, called the *basic rate*, the D-channel has a 16-kbps data rate; in another, called the *primary rate*, it is 64 kbps. Its primary use is for signaling. The 16-kbps version, besides signaling, may serve as transport for low-speed data applications, particularly those using X.25 packet data (see Section 12.2).

There are a number of H-channels:

- H_0 channel: 384 kbps;
- 1536 kbps (H_{11}) and 1920 kbps (H_{12}).

The H-channel is intended to carry a variety of user information streams. A distinguishing characteristic is that an H-channel does not carry signaling information for circuit switching by the ISDN. User information streams may be carried on a dedicated, alternate (within one call or separate calls), or simultaneous basis, consistent with the H-channel bit-rates. The following are examples of user information streams:

- Fast facsimile;
- Video, such as video conferencing;
- High-speed data;
- High-quality audio or sound program channel;
- Information streams, each at rates lower than the respective H-channel bit rate (e.g., 64-kbps voice), which have been rate-adapted or multiplexed together; and
- Packet-switched information.

12.4.2.2 Basic and Primary User Interfaces. The *basic* rate interface structure is composed of two B-channels and a D-channel referred to as "2B + D." The D-channel at this interface is 16 kbps. The B-channels may be used independently (i.e., two different simultaneous connections). Industry and much of the literature call the basic rate interface the *BRI*.

Appendix I to CCITT Rec. I.412 (Ref. 11) states that alternatively the basic access may be just one B-channel and a D-channel, or just a D-channel.

The *primary* rate interface (PRI) structures are composed of *n* B-channels and one

D-channel, where the D-channel in this case is 64 kbps. There are two primary data rates:

1. 1.544 Mbps = 23B + D (from the North American T1 configuration);
2. 2.048 Mbps = 30B + D (from the European E1 configuration).

For the user–network access arrangement containing multiple interfaces, it is possible for the D-channel in one structure not only to serve the signaling requirements of its own structure but also to serve another primary rate structure without an activated D-channel. When a D-channel is not activated, the designated time slot may or may not be used to provide an additional B-channel, depending on the situation, such as 24B with 1.544 Mbps.

The primary rate interface H_0-channel structures are composed of H_0 channels with or without a D-channel. When present in the same interface structure the bit rate of the D-channel is 64 kbps.

At the 1544-kbps primary rate interface, the H_0-channel structures are $4H_0$ and $3H_0$ + D. When the D-channel is not provided, signaling for the H_0-channels is provided by the D-channel in another interface.

At the 2048-kbps primary rate interface, the H_0 structure is $5H_0$ + D. In the case of a user–network access arrangement containing multiple interfaces, it is possible for the D-channel in one structure to carry the signaling for H_0-channels in another primary rate interface without a D-channel in use.

The 1536-kbps H_{11}-channel structure is composed of one 1536-kbps H_{11}-channel. Signaling for the H_{11}-channel, if required, is carried on the D-channel of another interface structure within the same user-network access arrangement.

The 1920-kbps H_{12} structure is composed of one 1920-kbps H_{12}-channel and a D-channel. The bit rate of the D-channel is 64 kbps. Signaling for the H_{12}-channel, if required, is carried in this D-channel or the D-channel of another interface structure within the same user–network access arrangement.

12.4.3 User Access and Interface Structures

12.4.3.1 General. Figure 12.14 shows generic ISDN user connectivity to the network. We can select either the basic or primary rate service (e.g., 2B+D, basic rate; 23B+D or 30B+D, primary rates) to connect to the ISDN network. The objectives of any digital interface design, and specifically of ISDN access and interface are as follows:

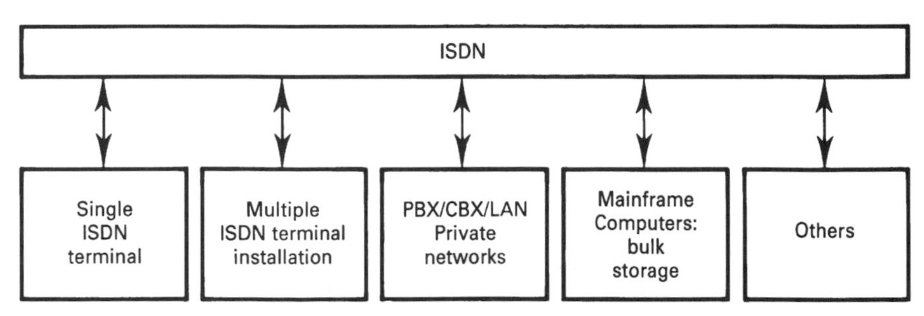

Figure 12.14 ISDN generic users.

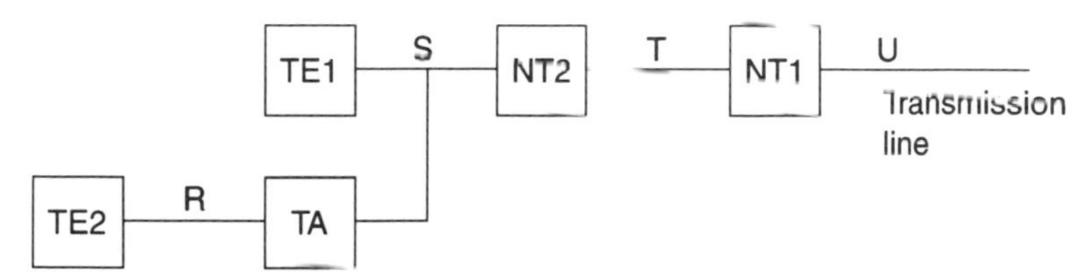

Figure 12.15 ISDN reference model.

- Electrical and mechanical specification;
- Channel structure and access capabilities;
- User–network protocols;
- Maintenance and operation;
- Performance; and
- Services.

ISDN specifications as set out by the ITU-T I Recommendations and relevant Bellcore/ANSI specifications cover these six items.

Figure 12.15 shows the conventional ISDN reference model. It delineates interface points for the user. In the figure NT1, or network termination 1, provides the physical layer interface; it is essentially equivalent to OSI layer 1. Functions of the physical layer include:

- Transmission facility termination;
- Layer 1 maintenance functions and performance monitoring;
- Timing;
- Power transfer;
- Layer 1 multiplexing; and
- Interface termination, including multidrop termination employing layer 1 contention resolution.

Network termination 2 (NT2) can be broadly associated with OSI layers 1, 2, and 3. Among the examples of equipment that provide NT2 functions are user controllers, servers, LANs, and PABXs. Among the NT2 functions are:

- Layers 1, 2, and 3 protocol processing;
- Multiplexing (layers 2 and 3);
- Switching;
- Concentration;
- Interface termination and other layer 1 functions; and
- Maintenance functions.

A distinction must be drawn here between North American and European practice. Historically, telecommunication administrations in Europe have been, in general, national monopolies that are government controlled. In North America (i.e., United States and Canada) they are private enterprises, often very competitive. In Europe, NT1 is considered as part of the digital network and belongs to the telecommunications administration. The customer ISDN equipment ISDN equipment starts at the T interface. In North America, both NT1 and NT2 belong to the ISDN user, and the U interface defines the network entry point.

It should be noted that there is an overall trend outside of North America to privatize telecommunications such as has happened in the United Kingdom and is scheduled to take place in other countries such as Germany, Mexico, and Venezuela.

TE1 in Figure 12.15 is the terminal equipment, which has an interface that complies with the ISDN terminal–network interface specifications at the S interface. We will call this equipment *ISDN compatible*. TE1 covers functions broadly belonging to OSI layer 1 and higher OSI layers. Among the TE1 equipment are digital telephones, computer work stations, and other devices in the user end-equipment category that is ISDN compatible.

TE2 in Figure 12.15 refers to equipment that does *not* meet the ISDN terminal–network interface at point S. TE2 adapts the equipment to meet that ISDN terminal–network interface. This process is assisted by the TA, the terminal adapter.

Reference points T, S, and R are used to identify the interface available at those points. T and S are identical electrically and mechanically, and from the point of view of protocol. Point R relates to the TA interface or, in essence, it is the interface of that nonstandard (i.e., non-ISDN) device. The U-interface is peculiar to the North American version of ISDN.

We will return to user–network interfaces once the stage is set for ISDN protocols looking into the network from the user.

12.4.4 ISDN Protocols and Protocol Issues

When fully implemented, ISDN will provide both circuit and packet switching. For the circuit-switching case, now fairly broadly installed in North America, the B-channel is fully transparent to the network, permitting the user to utilize any protocol or bit sequence so long as there is end-to-end agreement on the protocol utilized.[5] Of course, the protocol itself should be transparent to bit sequences.

It is the D-channel that carries the circuit-switching control function for its related B-channels. Whether it is the 16-kbps D-channel associated with BRI or the 64-kbps D-channel associated with PRI, it is that channel which transports the signaling information from the user's ISDN terminal from NT to the first serving telephone exchange of the telephone company or administration. Here the D-channel signaling information is converted over to CCITT No. 7 signaling data employing ISUP (ISDN User Part) of SS No. 7. Thus it is the D-channel's responsibility for call establishment (setup), supervision, termination (takedown), and all other functions dealing with network access and signaling control.

The B-channel in the case of circuit switching is serviced by NT1 or NT2 using OSI layer 1 functions only. The D-channel carries out OSI layers 1, 2, and 3 functions such that the B-channel protocol established by a family of ISDN end-users will generally make layer 3 null in the B-channel where the networking function is carried out by the associated D-channel.

With packet switching two possibilities emerge. The first basically relies on the B-channel to carry out OSI layers 1, 2, and 3 functions at separate packet-switching facilities (PSFs). The D-channel is used to set up the connection to the local switching exchange at each end of the connection. This type of packet-switched offering provides 64-kbps service. The second method utilizes the D-channel exclusively for

[5]"Broadly installed" means that most of the public carriers can offer the service to their customers.

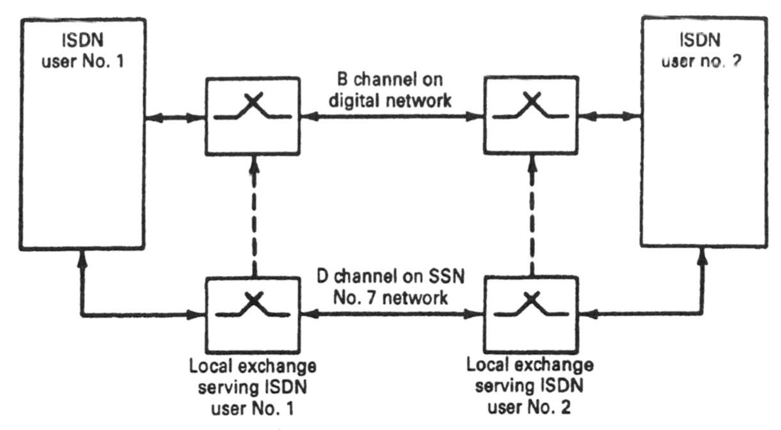

Figure 12.16 Simplified concept of ISDN circuit switching.

lower data rate packet-switched service, where the local interface can act as a CCITT (ITU-T) X.25 data communication equipment (DCE) device.

Figure 12.16 is a simplified conceptual diagram of ISDN circuit switching. It shows the B-channel riding on the public digital network and the D-channel, which is used for signaling. Of course the D-channel is a separate channel. It is converted to a CCITT No. 7 signaling structure and in this may traverse several signal transfer points (STPs; see Chapter 13) and may be quasiassociated or fully disassociated from its companion B-channel(s). Figure 12.17 is a more detailed diagram of the same ISDN circuit-switching concept. The reader should note the following in the figure: (1) Only users at each end of the connection have a peer-to-peer relationship available for all seven OSI layers of the B-channel. As a call is routed through the system, there is only layer 1 (physical layer) interaction at each switching node along the call route. (2) The D-channel requires

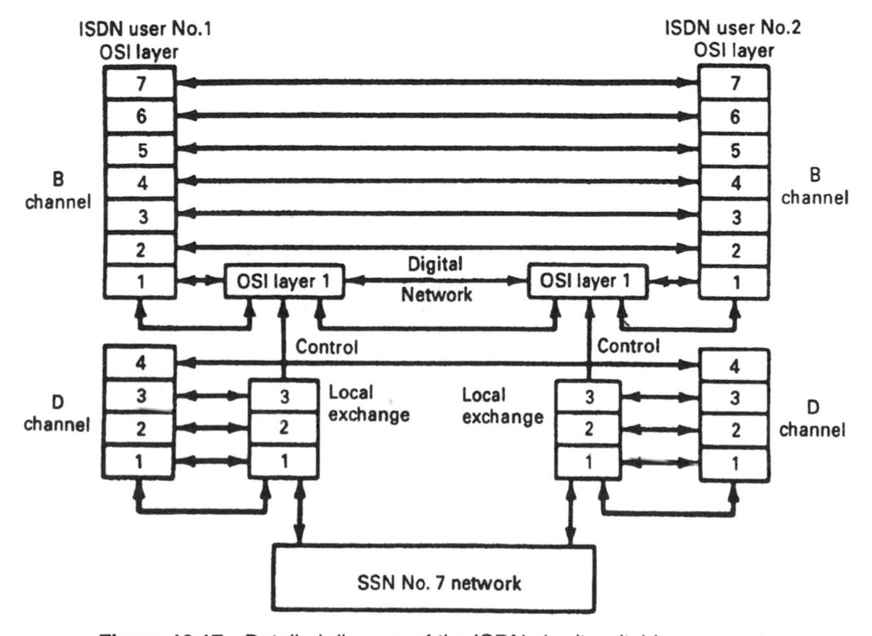

Figure 12.17 Detailed diagram of the ISDN circuit-switching concept.

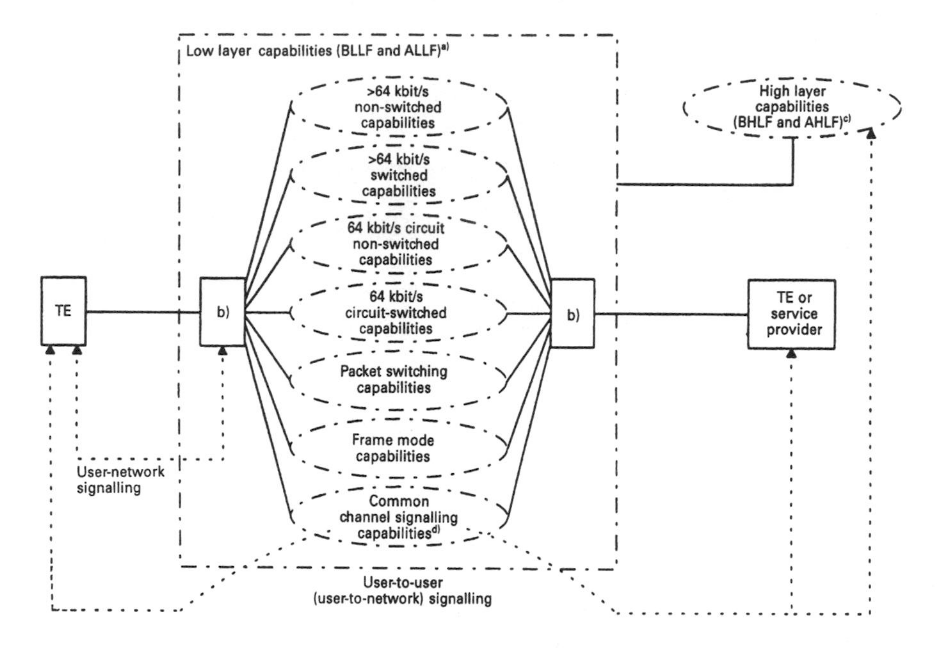

TE Terminal equipment
BLLF Basic low layer functions
ALLF Additional low layer functions
BHLF Basic high layer functions
AHLF Additional high layer functions

a) In certain national situations, ALLF may also be implemented outside the ISDN, in special nodes or in certain categories of terminals.
b) The ISDN local functional capabilities correspond to functions provided by a local exchange and possibly including other equipment, such as electronic cross connect equipment, muldexes, etc.
c) These functions may either be implemented within ISDN or be provided by separate networks. Possible applications for basic high layer functions and for additional high layer functions are contained in Recommendation 1.210.
d) For signalling between international ISDNs, CCITT Signalling System No. 7 shall be used.

Figure 12.18 Basic architectural model of ISDN. (From ITU-T Rec. I.324, Figure 1/I.324, p. 3, [Ref. 12].)

the first three OSI layers for call setup to the local switching center at each end of the circuit. (3) The D-channel signaling data are turned over to CCITT Signaling System No. 7 (SS No. 7) at the near- and far-end local switching centers. (4) SS No. 7 also utilizes the first three OSI layers for circuit establishment, which requires the transfer of control information. In SS No. 7 terminology, this is called the *message transfer part*. There is a fourth layer called the *user part* in SS No. 7. There are three user parts: telephone user part, data user part, and ISDN user part, depending on whether the associated B-channel is in telephone, data, or ISDN service for the user.

12.4.5 ISDN Networks

In this context, ISDN networking is seen as a group of access attributes connecting an ISDN user at either end to the local serving exchange (i.e., the local digital switch). This is illustrated in Figure 12.18, the basic architectural model of ISDN. It is here that the ISDN circuits meet the national transit network (CCITT terminology). This is the public switched digital telephone network. That network, whether DS1-DS4-based or E1-E5-based, provides two necessary attributes for ISDN compatibility:[6]

[6]We should add "as well as SONET-based or SDH-based."

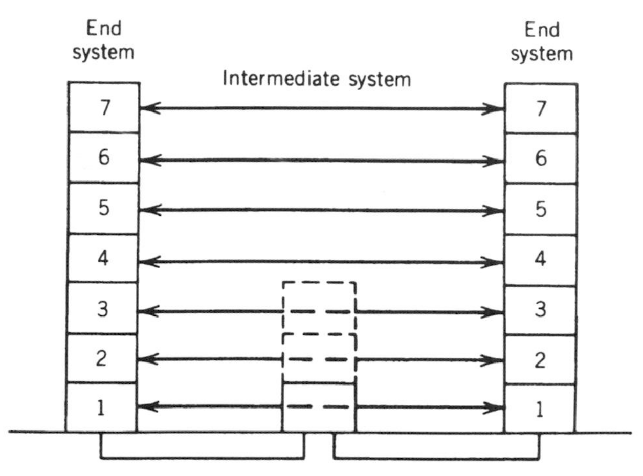

Figure 12.19 A generic communication context showing ISDN's relationship with the seven-layer OSI reference model. Note that the end-system protocol blocks may reside in the subscriber's TE or network exchanges or other ISDN equipment.

1. 64-kbps channelization (*Note:* In North America this may be 56 kbps);
2. Separate channel signaling based on CCITT Signaling System No. 7 (Chapter 13).

Connections from the ISDN user at the local connecting exchange interface include:

- Basic service (BRI) 2B + D = 192 kbps (CCITT specified)
  = 160 kbps (North American/Bellcore specified).
  Both rates include overhead bits.
- Primary rate service (PRI) 23B + D/30B + D = 1.544/2.048 Mbps.

Note: For E-1 service, it is assumed that the user will provide the synchronization channel, Channel 0. It is not the responsibility of the ISDN service provider.

12.4.6 ISDN Protocol Structures

12.4.6.1 ISDN and OSI. Figure 12.19 shows the ISDN relationship with OSI. (OSI was discussed in Chapter 10.) As is seen in the figure, ISDN concerns itself with only the first three OSI layers. OSI layers 4 through 7 are peer-to-peer connections and are the end-user's responsibility. Remember that the B-channel deals with OSI layer 1 exclusively. There is one exception. That is when the B-channel is used for packet data service. In this case it will deal with the first three OSI layers.

Of course, with the BRI service, the D-channel is another exception. The D-channel interfaces with CCITT Signaling System No. 7 at the first serving exchange. These (BRI) D-channels handle three types of information: (1) signaling (s), (2) interactive data (p), and (3) telemetry (t).

The layering of the D-channel has followed the intent of the OSI reference model. The handling of the p and t data can be adapted to the OSI model; the s data, by its very

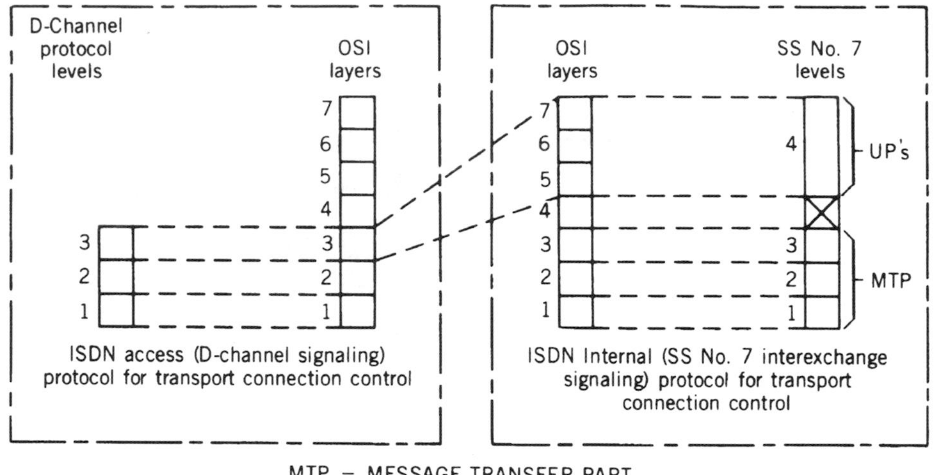

Figure 12.20 Correspondence among the ISDN D-channel, CCITT Signaling System No. 7, and the OSI mode. (©1985 IEEE Computer Soc. Press, Washington, DC [Ref. 13].)

nature, cannot.[7] Figure 12.20 shows the correspondence between D-channel signaling protocols, SS No. 7 levels, and the OSI seven-layer model.

12.4.6.2 Layer 1 Interface, Basic Rate. The S/T interface of the reference model, Figure 12.15 (or layer 1, physical interface), requires a balanced metallic transmission medium (i.e., copper pair) in each direction of transmission (four-wire) capable of supporting 192 kbps. Again, this is the NT interface of the ISDN reference model.

Layer 1 provides the following services to layer 2 for ISDN operation:

- The transmission capability by means of appropriately encoded bit streams of the B- and D-channels and also any timing and synchronization functions that may be required.
- The signaling capability and the necessary procedure to enable customer terminals and/or network terminating equipment to be deactivated when required and reactivated when required.
- The signaling capability and necessary procedures to allow terminals to gain access to the common resource of the D-channel in an orderly fashion while meeting the performance requirements of the D-channel signaling system.
- The signaling capability and procedures and necessary functions at layer 1 to enable maintenance functions to be performed.
- An indication to higher layers of the status of layer 1.

12.4.6.2.1 Interface Functions. The S and T functions for the BRI consist of three bit streams that are time-division multiplexed: two 64-kbps B-channels and one 16-kbps D-channel for an aggregate bit rate of 192 kbps. Of this 192 kbps, the 2B + D configura-

[7]For CCITT Signaling System No. 7, like any signaling system, the primary quality-of-service measure is "post dial delay." This is principally the delay in call setup. To reduce the delay time as much as possible, it is incumbent upon system engineers to reduce processing time as much as possible. Thus SS 7 truncates OSI to four layers, because each additional layer implies more processing time.

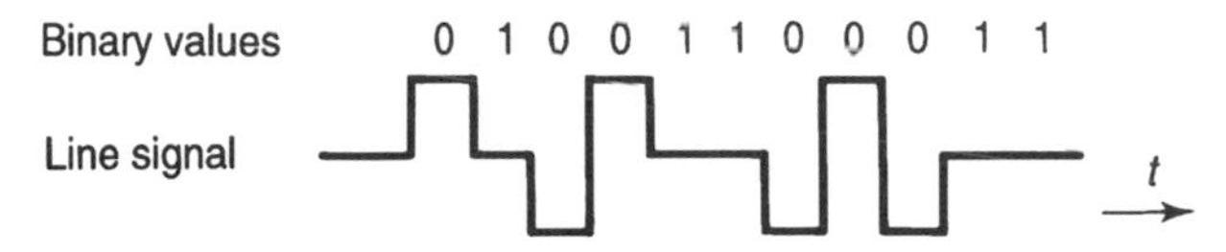

Figure 12.21 Pseudoternary line code, example of application.

tion accounts for only 144 kbps. The remaining 48 kbps are overhead bits, whose functions are briefly reviewed as follows.

The functions covered at the interface include bit timing at 192 kbps to enable the TE and NT to recover information from the aggregate bit stream. This octet timing provides 8-kHz octet timing for the NT and TE to recover the time-division multiplexed channels (i.e., 2B + D multiplexed). Other functions include D-channel access control, power feeding, deactivation, and activation.

Interchange circuits are required, of which there is one in either direction of transmission (i.e., to and from the NT); they are used to transfer digital signals across the interface. All of the functions described previously, except for power feeding, are carried out by means of a digitally multiplexed signal. In both directions of transmission the bits are grouped into frames of 48 bits each. However, the frame structures are different in each direction of transmission. A dc balancing bit is periodically inserted to move the signal energy away from 0 Hz.

12.4.6.2.2 Line Code. For both directions of transmission, a pseudoternary coding is used with 100% pulse width, as illustrated in Figure 12.21. Coding is performed such that a binary 1 is represented by no line signal, whereas a binary 0 is represented by a positive or negative pulse. The first binary signal following the framing balance bit is the same polarity as the balance bit. Subsequent binary 0s alternate in polarity. A balance bit is a 0 if the number of 0s following the previous balance bit is odd. A balance bit is a binary 1 if the number of 0s following the previous balance bit is even. As mentioned, balance bits tend to limit the build-up of a dc component on the line.

12.4.6.2.3 Timing Considerations. The NT derives its timing from the network clock. A TE synchronizes its bit, octet, and frame timing from the NT, which has derived its timing from the ISDN bit stream being received from the network. The NT uses this derived timing to synchronize its transmitter clock.

12.4.6.2.4 BRI Differences in the United States. The Bellcore/ANSI ISDN standards differ considerably from their CCITT counterparts in the I and several Q recommendations. The various PSTN administrations (telephone companies) in the United States are at variance with most other countries of the world. Bellcore stated its intention at the outset to produce equipment that was cost effective and marketable and that would easily interface with existing North American telephone plant.

One point of variance, of course, is where the telephone company responsibilities end and customer responsibilities begin. This is called the U-interface, which is peculiar to U.S. ISDN operation (see Figure 12.15). For example, U.S. ISDN calls for a two-wire customer interface; CCITT uses a four-wire connectivity. Rather than a pseudoternary line waveform, the United States uses 2B1Q, a four-level waveform. The line bit rate is 160 kbps rather than the CCITT recommended 192 kbps. The 2B + D frame overhead differs significantly from its CCITT counterpart. Bellcore uses the generic term DSL for digital subscriber line, and the ISDN is one of a large class of digital subscriber lines.

The effective 160 kbps signal is divided up into 12 kbps for synchronization words, 144 kbps for 2B + D customer data, and 4 kbps of DSL overhead.

The synchronization technique used is based on transmission of nine (quarternary) symbols every 1.5 ms, followed by 216 bits of 2B + D data and 6 bits of overhead. The synchronization word provides a robust method of conveying line timing and establishes a 1.5-ms DSL "basic frame" for multiplexing subrate signals. Every eighth synchronization word is inverted (i.e., the 1s become 0s and the 0s become 1s) to provide a boundary for a 12-ms superframe composed of eight basic frames. This 12-ms interval defines an appropriate block of customer data for performance monitoring and permits a more efficient suballocation of the overhead bits among various operational functions.

2B + D Customer data bit pattern. There are 216 2B + D bits placed in each 1.5-ms basic frame, for a customer data rate of 144 kbps. The bit pattern (before conversion to quaternary form and after reconversion to binary form) for the 2B + D data is

$$B_1B_1B_1B_1B_1B_1B_1B_1B_2B_2B_2B_2B_2B_2B_2B_2DD,$$

where B_1 and B_2 are bits from the B_1- and B_2-channels and D is a bit from the D-channel. This 18-bit pattern is repeated 12 times per DSL basic frame.

DSL Line code—2B1Q. The average power of a 2B1Q transmitted signal is between +13 dBm and +14 dBm over a frequency band from 0 Hz and 80 kHz, with the nominal peak of the largest pulse being 2.5 V. The maximum signal power loss at 40 kHz is about 42 dB. As mentioned earlier, the bit rate is 160 kbps and the modulation rate is 80 kbaud.

2B1Q Waveform. It is convenient to express the 2B1Q waveform as +3, +1, −1, −3 because this indicates symmetry about zero, equal spacing between states, and convenient integer magnitudes. The block synchronization word (SW) contains nine quaternary elements repeated every 1.5 ms:

$$+3, +3, -3, -3, -3, +3, -3, +3, +3.$$

As we are aware, North American BRI connects to an ISDN user on a two-wire basis, which operates full-duplex. It can do this without interference between transmit and receive sides by the use of local echo suppressors at the U-interface.

12.4.7 Primary Rate Interfaces

The primary rate interfaces cover two standard bit rates: 1.544 Mbps in North America and 2.048 Mbps when in areas of the world that utilize E1.

12.4.7.1 Interface at 1.544 Mbps

12.4.7.1.1 Bit Rate and Synchronization—Network Connection Characteristics. The network delivers (except as noted in the following) a signal synchronized from a clock having a minimum accuracy of 1×10^{-11} (stratum 1; see Section 6.12 for strata definitions). When synchronization by a stratum 1 clock has been interrupted, the signal delivered by the network to the interface should have a minimum accuracy of 4.6×10^{-6} (stratum 3).

12.4.7.1.2 DS1 Interface. The ISDN interface for the 1.544 Mbps rate is the standard line interface described in Chapter 6. Note that time slot 24 is assigned to the D-channel when the D-channel is present (thus 23B + D).

A channel occupies an integer number of time slots and in the same time slot positions in every frame. A B-channel may be assigned any time slot in the frame, an H_0-channel may be assigned any six slots in a frame in numerical order (not necessarily consecutive), and an H_{11}-channel may be assigned slots 1 to 24. The assignments may vary on a call-by-call basis.

12.4.7.2 Interface at 2.048 Mbps

12.4.7.2.1 Frame Structure. This is the standard E1 frame structure described in Chapter 6. Channel 16, in compliance with the E1 standard (CCITT Rec. G.703) carries the signaling information. Channel 0, the synchronization channel, is the responsibility of the user. Thus primary service is 30B + D (i.e., not 30B + D + "S," where S means synchronization).

12.4.7.2.2 Timing Considerations. The NT derives its timing from the network clock. The TE synchronizes its timing (bit, octet, and frame) from the signal received from the NT and synchronizes its transmitted signal accordingly. In an unsynchronized condition—that is, when the access that normally provides network timing is unavailable—the frequency deviation of the free-running clock shall not exceed ±50 ppm.

12.4.8 Overview of Layer 2, ISDN D-Channel, LAPD Protocol

The link access procedure (LAP) for the D-channel (LAPD) is used to convey information between layer 3 entities across the ISDN user-network interface (UNI) using the D-channel.[8]

A *service access point* (SAP) is a point at which the data-link layer provides services to its next higher OSI layer or layer 3.[9] Associated with each data-link layer is one or more data-link connection endpoints (see Figure 12.22). A data-link connection endpoint is identified by a data-link connection endpoint identifier, as seen from layer 3, and a data-link connection identifier (DLCI), as seen from the data-link layer.

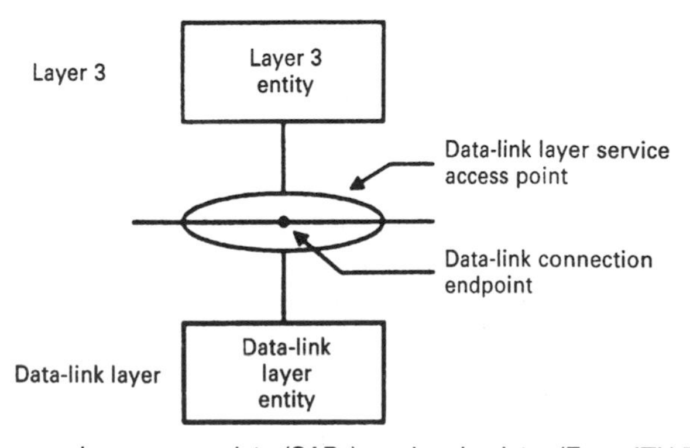

Figure 12.22 Entities, service access points (SAPs), and end-points. (From ITU-T Rec. Q.920, Figure 2/Q.920, [Ref. 14].)

[8]The discussion here only covers aspects of BRI service, namely, the 16-kbps signaling channel. It does not cover the PRI service.
[9]Remember that the *data-link layer* is synonymous with OSI layer 2.

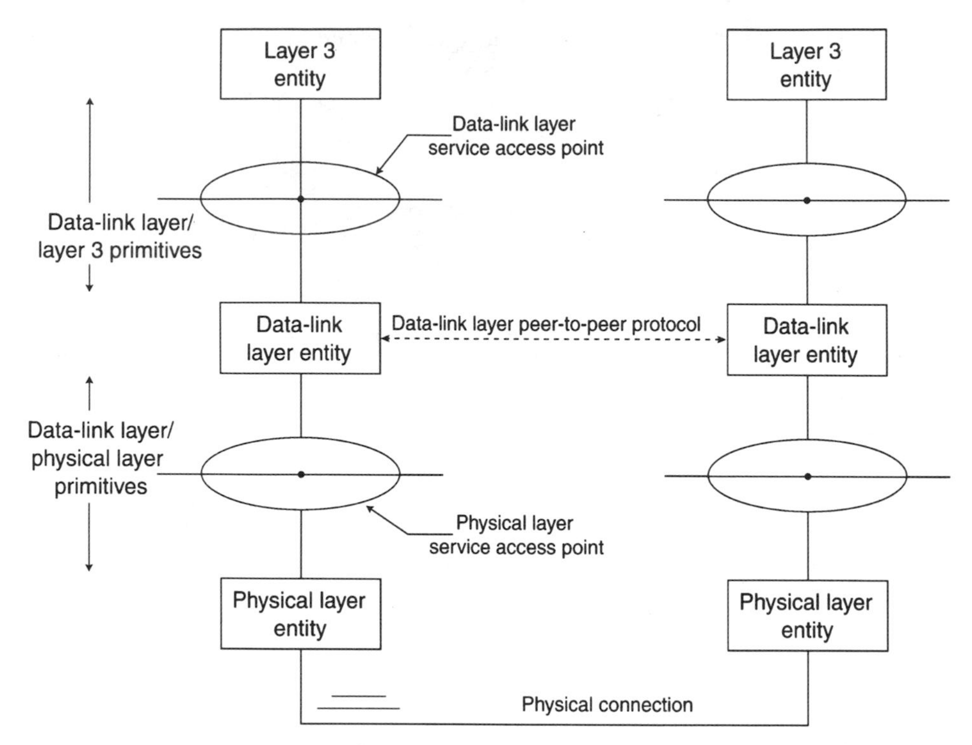

Figure 12.23 Data-link layer reference model. (From ITU-T Rec. Q.920, Figure 5/Q.920, p. 5, [Ref. 14].)

Cooperation between data-link layer entities is governed by a specific protocol to the applicable layer. In order for information to be exchanged between two or more layer 3 entities, an association must be established between layer 3 entities in the data-link layer using a data-link layer protocol.

Figure 12.23 shows the data-link layer reference model. All data-link layer messages are transmitted in frames delimited by flags, where a flag is a unique binary sequence pattern (i.e., 01111110). The governing data-link layer protocol is LAPD (link access protocol D-channel). The frame structure is described in ITU-T Q.921 (Ref. 15).[10]

The LAPD includes functions for:

1. The provision of one or more data-link connections on a D-channel; discrimination between the data-link connections is by means of a data-link connection identifier (DLCI) contained in each frame;
2. Frame delimiting, alignment, and transparency, allowing recognition of a sequence of bits transmitted over a D-channel as a frame;
3. Sequence control, which maintains the sequential order of frames across a data-link connection;
4. Detection of transmission, format, and operational errors on a data link;
5. Recovery from detected transmission, format, and operational errors, and notification to the management entity of unrecovered errors; and
6. Flow control.

[10]LAPD is a direct descendent of HDLC (see Section 10.10.3). LAPD modified is at the very heart of frame relay described in Section 12.5.

There is unacknowledged and acknowledged operation. With unacknowledged operation, information is transmitted in unnumbered information (UI) frames. At the data-link layer the UI frames are unacknowledged. Transmission and format errors may be detected, but no recovery mechanism is defined. Flow control mechanisms are also not defined. With acknowledged operation, layer 3 information is transmitted in frames that are acknowledged at the data-link layer. Error-recovery procedures based on retransmission of unacknowledged frames are specified. For errors that cannot be corrected by the data-link layer, a report to the management entity is made. Flow-control procedures are also defined.

Unacknowledged operation is applicable for point-to-point and broadcast information transfer. However, acknowledged operation is applicable only for point-to-point information transfer.

There are two forms of acknowledged information that are defined:

1. Single-frame operation; and
2. Multiframe operation.

12.4.8.1 Layer 2 Frame Structure for Peer-to-Peer Communications. There are two frame formats used for layer 2 frames:

1. Format A, for frames where there is no information field; and
2. Format B, for frames containing an information field.

These two frame formats are illustrated in Figure 12.24. The following discussion briefly describes the frame content (sequences and fields) for the LAPD layer 2 frames.

Format A

Bits 8 7 6 5 4 3 2 1	Octet
Flag 0 1 1 1 1 1 1 0	Octet 1
Address (high-order octet)	2
Address (low-order octet)	3
Control [a)]	4
Control [a)]	
FCS (first octet)	N - 2
FCS (second octet)	N - 1
Flag 0 1 1 1 1 1 1 0	N

Format B

Bits 8 7 6 5 4 3 2 1	Octet
Flag 0 1 1 1 1 1 1 0	Octet 1
Address (high-order octet)	2
Address (low-order octet)	3
Control [a)]	
Control [a)]	4
Information	⋮
FCS (first octet)	N - 2
FCS (second octet)	N - 1
Flag 0 1 1 1 1 1 1 0	N

[a)] Unacknowleded operation – one octet
Multiple-frame operation – two octets for frames with sequence numbers;
– one octet for frames without sequence numbers.

Figure 12.24 Frame formats for LAPD frames. (From ITU-T Rec. Q.921, Figure 1/Q.921, p. 20, [Ref. 15].)

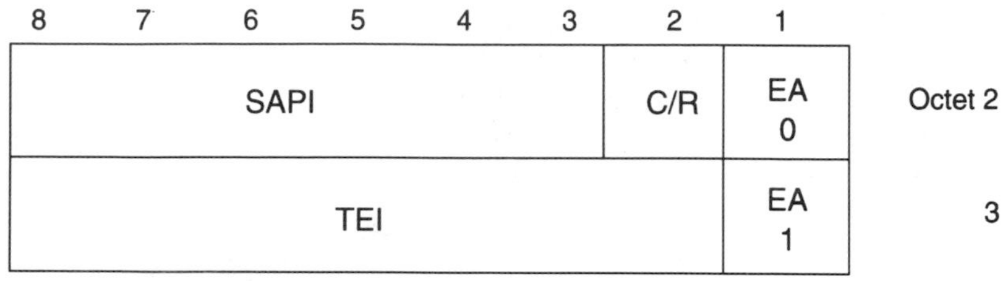

8	7	6	5	4	3	2	1	
SAPI						C/R	EA 0	Octet 2
TEI							EA 1	3

EA = Address field extension bit
C/R = Command/response field bit
SAPI = Service access point identifier
TEI = Terminal endpoint identifier

Figure 12.25 LAPD address field format. (From CCITT Rec. Q.921, Figure 5/Q.921, p. 23, [Ref. 15].)

Flag sequence. Identical to HDLC described in Section 10.10.3, it is the binary sequence 01111110. The flag opens and closes individual frames. For LAPD frames sent in sequence, the closing flag of one frame is the opening flag of the next frame.

Address field. As shown in Figure 12.25, the address field consists of two octets and identifies the intended receiver of a command frame and the transmitter of a response frame.

Control field. The control field consists of one or two octets. It identifies the type of frame, either command or response. It contains sequence numbers where applicable. Three types of control field formats are specified:

1. Numbered information transfer (I format);
2. Supervisory functions (S format); and
3. Unnumbered information transfers and control functions (U format).

Information field. The information field of a frame, when present, follows the control field and precedes the frame check sequence (FCS). The information field contains an integer number of octets:

- For a SAP supporting signaling, the default value is 128 octets.
- For SAPs supporting packet information, the default value is 260 octets.

Frame check sequence (FCS) field. Identical to HDLC, Section 10.10.3, it is 16 bits long and is based on the generating polynomial:

$$X^{16} + X^{12} + X^{5} + 1.$$

Transparency, mentioned previously, ensures that a flag or abort sequence is not initiated within a frame. On the transmit site the data-link layer examines the frame content between the opening and closing flag sequences and inserts a 0 bit after all sequences with five contiguous 1 bits (including the last five bits of the FCS). On the receive side the data-link layer examines the frame contents between the opening and closing flag sequences and discards any 0 bit that directly follows five contiguous 1 bits.

Address field format. The address field is illustrated in Figure 12.25. It contains address field extension bits (EA), command/response indication bit (C/R), a data-link layer service access point identifier (SAPI) subfield, and a terminal end point identifier (TEI) subfield.

Address field extension bit (EA). The address field range is extended by reserving the first transmitted bit of the address field to indicate the final octet of the address field. The presence of a 1 in the first bit position of an address field octet signals that it is the final octet of the address field. The double octet address field for LAPD operation has bit 1 of the first octet set to a 0 and bit 1 of the second octet set to a 1.

Command/response field bit (C/R). The C/R bit identifies a frame as either a command or a response. The user side sends commands with the C/R bit set to 0, and it sends responses with the C/R bit set to 1. The network side does the opposite; that is, commands are sent with the C/R bit set to 1, and responses are sent with the C/R bit set to 0.

In keeping with HDLC rules, commands use the address of the peer data-link entity while responses use the address of their own data-link layer entity.[11] In accordance with these rules, both peer entities on a point-to-point data-link connection use the same data-link connection identifier (DLCI) composed of an SAPI and TE1.

12.4.9 Overview of Layer 3

The layer 3 protocol, of course, deals with the D-channel and its signaling capabilities. It provides the means to establish, maintain, and terminate network connections across an ISDN between communicating application entities. A more detailed description of the layer 3 protocol may be found in ITU-T Rec. Q.931 (Ref. 16). Layer 3 utilizes functions and services provided by its data-link layer, as described in Section 12.4.8 under LAPD functions.

Layer 3 performs two basic categories of functions and services in the establishment of network connections. The first category directly controls the connection establishment. The second category includes those functions relating to the transport of messages in addition to the functions provided by the data-link layer. Among these additional functions are the provision of rerouting of signaling messages on an alternative D-channel (where provided) in the event of D-channel failure. Other possible functions include multiplexing and message segmenting and blocking. The D-channel layer 3 protocol is designed to carry out establishment and control of circuit-switched and packet-switched connections. Also, services involving the use of connections of different types, according to user specifications, may be provided through "multimedia" call control procedures. Functions performed by layer 3 include:

1. The processing of primitives for communicating with the data-link layer;
2. Generation and interpretation of layer 3 messages for peer-level communications;
3. Administration of timers and logical entities (e.g., call references) used in call-control procedures;
4. Administration of access resources, including B-channels and packet-layer logical channels (e.g., ITU-T X.25); and
5. Checking to ensure that services provided are consistent with user requirements, such as compatibility, address, and service indicators.

The following functions may also be performed by layer 3:

[11] LAPD, as we know, is a derivative of HDLC.

1. *Routing and Relaying.* Network connections exist either between users and ISDN exchanges or between users. Network connections may involve intermediate systems that provide relays to other interconnecting subnetworks and that facilitate interworking with other networks. Routing functions determine an appropriate route between layer 3 addressees.
2. *Network Connection.* This function includes mechanisms for providing network connections making use of data-link connections provided by the data-link layer.
3. *Conveying User Information.* This function may be carried out with or without the establishment of a circuit-switched connection.
4. *Network Connection Multiplexing.* Layer 3 provides multiplexing of call control information for multiple calls onto a single data-link connection.
5. *Segmentation and Reassembly (SAR).* Layer 3 may segment and reassemble layer 3 messages to facilitate their transfer across user–network interface.
6. *Error Detection.* Error-detection functions are used to detect procedural errors in the layer 3 protocol. Error detection in layer 3 uses, among other information, error notification from the data-link layer.
7. *Error Recovery.* This includes mechanisms for recovering from detected errors.
8. *Sequencing.* This includes mechanisms for providing sequenced delivery of layer 3 information over a given network connection when requested. Under normal conditions, layer 3 ensures the delivery of information in the sequence it is submitted by the user.
9. *Congestion Control and User Data Flow Control.* Layer 3 may indicate rejection or unsuccessful indication for connection establish requests to control congestion within a network. Typical is the congestion control message to indicate the establishment or termination of flow control on the transmission of user information messages.
10. *Restart.* This function is used to return channels and interfaces to an idle condition to recover from certain abnormal conditions.

12.4.10 ISDN Packet Mode Review

12.4.10.1 Introduction. Two main services for packet-switched data transmission are defined for packet-mode terminals connected to the ISDN:

Case A: Access to a PSPDN (PSPDN services) (PSPDN = packet-switched public data network); and

Case B: Use of an ISDN virtual circuit service (Refs. 11, 18).

12.4.10.2 Case A: Configuration When Accessing PSPDN Services. This configuration is shown in Figure 12.26 and refers to Case A, which implies a transparent handling of packet calls through an ISDN. Only access via the B-channels is possible. In this context, the only support that an ISDN gives to packet calls is a physical 64-kbps circuit-mode semipermanent or demand transparent network connection type between appropriate PSPDN port and the X.25 DTE + TA or + TE1 at the customer premises.

In the case of semipermanent access, the X.25 DTE + TA or TE1 is connected to the corresponding ISDN port at the PSPDN (AU [access unit]). The TA, when present, performs only the necessary physical channel rate adaption between the user at the R reference point and the 64-kbps B-channel rate. D-channel layer 3 messages are not used in this case.

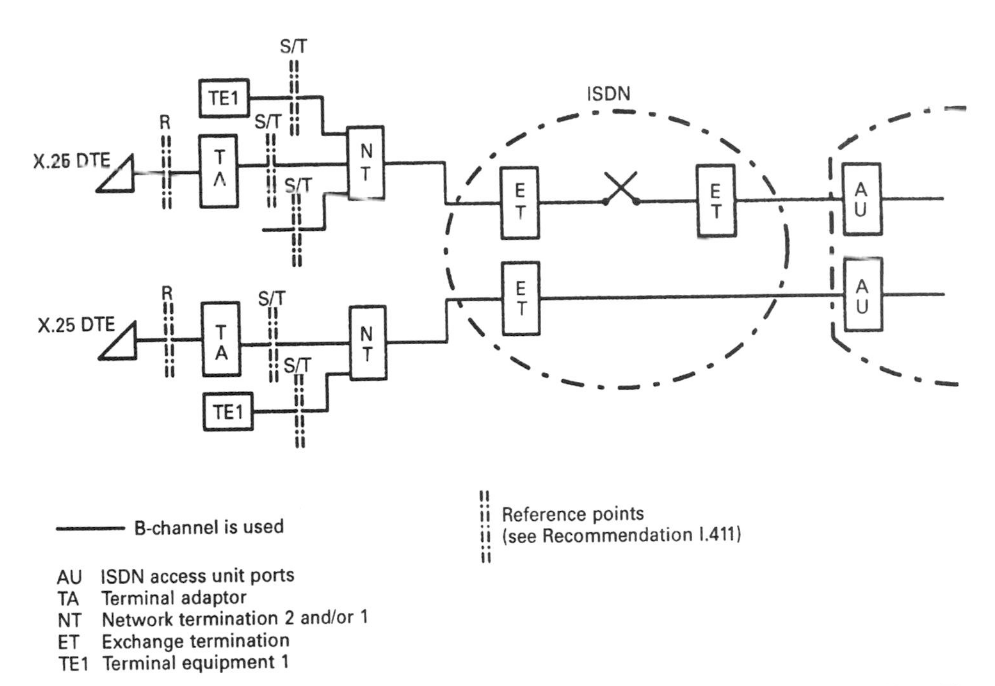

Figure 12.26 Case A, configuration when accessing PSPDN services. (From ITU-T Rec. X.31, Figure 2-1/X.31, p. 3, [Ref. 18].)

In the case of demand access to the PSPDN, which is shown in the upper portion in Figure 12.26, the X.25 DTE + TA or TE1 is connected to an ISDN port at the PSPDN (AU). The AU is also able to set up 64-kbps physical channels through the ISDN.

In this type of connection, an originating call will be set up over the B-channel toward the PSPDN port using the ISDN signaling procedure prior to starting X.25 layer 2 and layer 3 functions. This is done by using either hot-line (e.g., direct call) or complete selection methods. Moreover, the TA, when present, performs user rate adaption to 64 kbps. Depending on the data rate adaption technique employed, a complementary function may be needed at the AU of the PSPDN.

In the complete selection case, two separate numbers are used for outgoing access to the PSPDN:

1. The ISDN number of the access port of the PSPDN, given in the D-channel layer 3 setup message (Q.931); and
2. The address of the called DTE indicated in the X.25 call request packet.

The corresponding service requested in the D-channel layer 3 setup message is ISDN circuit-mode bearer services.

For calls originated by the PSPDN, the same considerations apply. In fact, with reference to Figure 12.26, the ISDN port of the PSPDN includes both rate adaption (if required) and path setting-up functions. When needed, DTE identification may be provided to the PSPDN by using call establishment signaling protocols in D-Channel layer 3 (ITU-T Rec. Q.931). Furthermore, DCE identification may be provided to the DTE, when needed, by using the same protocols.

For the demand access case, X.25 layer 2 and layer 3 operation in the B-channel, as well as service definitions, are found in ITU-T Rec. X.32. Some PSPDNs may operate

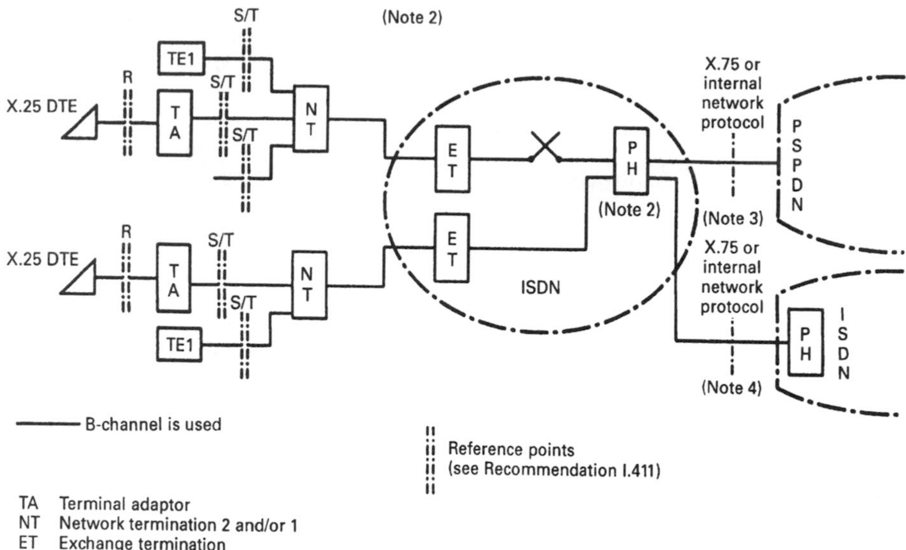

NOTES

1 This figure is only an example of many possible configurations and is included as an aid to the text describing the various interface functions.
2 In some implementations, the PH functions logically belonging to the ISDN may reside physically in a node of the PSPDN. The service provided is still the ISDN virtual circuit service.
3 See Recommendation X.325.
4 See Recommendation X.320.

Figure 12.27 Case B, configuration for the ISDN virtual circuit services (access via a B-channel). (From ITU-T Rec. X.31, Figure 2-2/X.31, p. 5, [Ref. 18].)

the additional DTE identification procedures defined in Rec. X.32 (Ref. 17) to supplement the ISDN-provided information in Case A.

12.4.10.3 Case B: Configuration for the ISDN Virtual Service. This configuration refers to the case where a packet handling (PH) function is provided within the ISDN. The configuration shown in Figure 12.27 relates to the case of X.25 link and packet procedures conveyed through the B-channel. In this case, the packet is routed, within the ISDN, to a PH function, where the complete processing of the X.25 call can be carried out.

There is still another configuration where X.25 packet procedures are conveyed through the D-channel. In this case a number of DTEs can operate simultaneously through a D-channel by using connection identifier discrimination at ISDN layer 2. The accessed port of the PH is still able to support X.25 layer 3 procedures.

It should be pointed out that the procedures for accessing the PSDTS (packet-switched data transmission services) through the ISDN user–network interface over a B- or D-channel are independent of where the service provider chooses to locate PH functions such as:

- In a remote exchange or packet-switching module in an ISDN; or
- In the local exchange.

However, the procedures for packet access through the B-channel or the D-channel are different.

In both cases of B- and D-channel accesses, in the service of Case B, the address of the called DTE is contained in the X.25 call request packet. The establishment of the physical connection from the TA/TE1 to the packet handling function is done on the basis of the requested bearer service (ISDN virtual circuit service), and therefore the user does not provide any addressing information in the layer procedures (D-channel, Q.931).

12.5 SPEEDING UP THE NETWORK: FRAME RELAY

12.5.1 Rationale and Background

It would seem from the terminology and the popular press that somehow we were making bits travel faster down the pipe; that by some means we had broken the velocity barrier by dramatically increasing the velocity of propagation. Of course, this is patently not true.

If bandwidth permits, bit rate can be increased.[12] That certainly will speed things up. One way to get around the bandwidth crunch is to work around the analog voice channel; use some other means. ISDN was a step in that direction; use coaxial cable and fiber optics as the cable television people do.

Probably the greatest impetus to speed up the network came from LAN users who wished to extend LAN connectivity to distant destinations. Ostensibly this traffic, as described in Chapter 11, has local transmission rates from 4 Mbps to 100 Mbps, with gigabit rates in the near future. X.25 WAN connectivity was one possible answer. Its packet circuits are slow and tedious, but very robust, even on circuits with poor error performance. There must be a better way with our digital network and its sterling error performance.

What slows down X.25 service? X.25 commonly operates at 64 kbps, and is even feasible at T1/E1 transmission rates. What slows things down is X.25's intensive processing at every node and continual message exchange as to the progress of packets, from node to originator and from node to destination. See Table 12.3 for a comparison between X.25 and frame relay.

On many X.25 connectivities, multiple nodes are involved, slowing service still further. One clue lies in the fact that X.25 was designed for circuits with poor transmission performance. This is manifested in degraded error rates, on the order of 1×10^{-4}. Meanwhile, the underlying digital networks in North America display end-to-end error performance better than 1×10^{-7}. Sprint calls for digital-link performance of 1×10^{-12}. Such excellent error performance begs the questions of removing the responsibility of error recovery from the service provider. If errors statistically occur in about 1 in 10 million bits, there is a strong argument for removing error recovery, at least in the data-link layer. In fact, with frame relay the following salient points emerge:

- There is no process for error recovery by the frame relay service provider.
- The service provider does not guarantee delivery nor are there any sort of acknowledgments provided.

[12]"If bandwidth permits:" We cannot increase bandwidth as the popular press insinuates. We *can* make better use of bandwidth. With data transmission, particularly as the bit rate increases, other bandwidth constraints, besides amplitude response, may become the limiting factor on bit rate. Typically this may be group delay (envelope delay distortion), bandwidth coherence.

Table 12.3 Functional Comparison of X.25 and Frame Relay

Function	X.25 in ISDN (X.31)	Frame Relay
Flag recognition/generation	×	×
Transparency	×	×
FCS checking/generation	×	×
Recognize invalid frames	×	×
Discard incorrect frames	×	×
Address translation	×	×
Fill inter-frame time	×	×
Manage V(S) state variable	×	
Manage V(R) state variable	×	
Buffer packets awaiting acknowledgment	×	
Manager timer T-1	×	
Acknowledge received I-frames	×	
Check received N(S) against V(R)	×	
Generation of rejection message	×	
Respond to poll/final bit	×	
Keep track of number of retransmissions	×	
Act upon reception of rejection message	×	
Respond to receiver not ready (RNR)	×	
Respond to receiver ready (RR)	×	
Multiplexing of logical channels	×	
Management of D bit	×	
Management of M bit	×	
Management of W bit	×	
Management of P(S) packets sent	×	
Management of P(R) packets received	×	
Detection of out-of-sequence packets	×	
Management of network layer RR	×	
Management of network layer RNR	×	

- It only uses the first two OSI layers (physical and data-link layer), thus removing layer 3 and its intensive processing requirements.
- Frame overhead is kept to a minimum, to minimize processing time and to increase useful throughput.
- There is no control field, and no sequence numbering.
- Frames are discarded without notifying originator for such reasons as congestion and having encountered an error.
- It operates on a statmultiplex concept.

In sum, the service that the network provides can be speeded up by increasing data rate, eliminating error-recovery procedures, and reducing processing time. One source states that a frame relay frame takes some 20 ms to reach the distant end (statistically), where an X.25 packet of similar size takes in excess of 200 ms on terrestrial circuits inside CONUS (CONUS stands for contiguous United States).

Another advantage of frame relay over a conventional static TDM connection is that it uses virtual connections. Data traffic is often bursty and normally would require much larger bandwidths to support the short data messages and much of the time that bandwidth would remain idle. Virtual connections of frame relay only use the required bandwidth for the period of the burst or usage. This is one reason why frame relay is used so widely to interconnect LANs over a wide area network (WAN). Figure 12.28 shows a tpyical frame relay network.

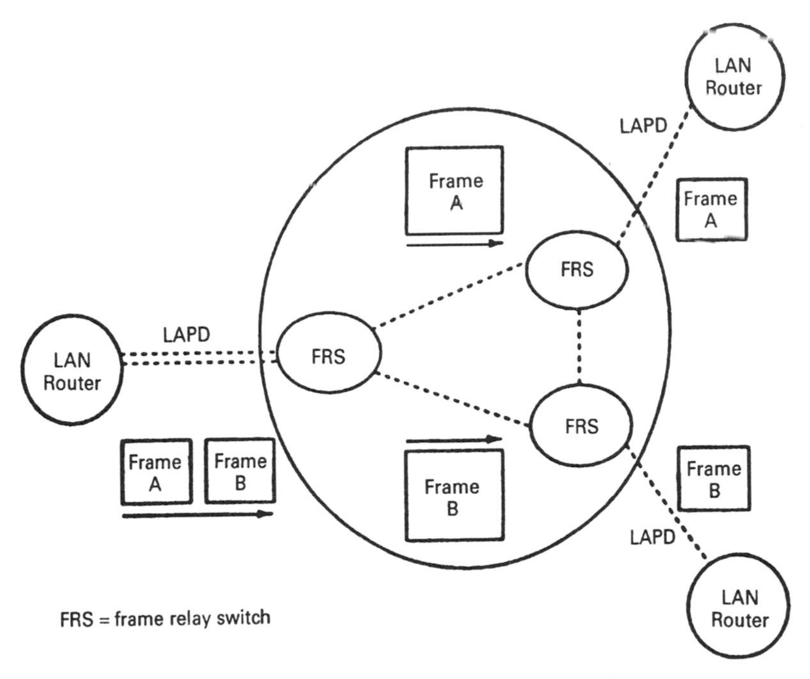

Figure 12.28 A typical frame relay network.

12.5.2 Genesis of Frame Relay

Frame relay extends only through the data-link layer (i.e., OSI layer 2). It has derived its data-link layer protocol from the ISDN D-channel LAPD.[13] We discussed LAPD in Section 12.4.8. Frame relay's importance has taken on such a magnitude (it was developed in North America) that the ITU-T organization formulated Rec. I.122, *Framework for Frame Mode Bearer Services* (Ref. 19) and I.233, *Frame Mode Bearer Services* (Ref. 20). Even the term LAPD, although modified in many cases for frame relay application, continues to be used.

Frame relay has become an ANSI initiative. There is also the Frame Relay Forum, consisting of manufacturers of frame relay equipment, that many feel is leading this imaginative initiative. So when we discuss frame relay, we must consider what specifications a certain system is designed around:

- ANSI, based on ANSI specifications and their publication dates;
- Frame Relay Forum with publication dates; and
- ITU-T organization and its most current recommendations.

There are also equivalent ANSI specifications directly derived from ITU-T recommendations such as ANSI T1.617-1991 (Ref. 21). We will see the term *core aspects* of ISDN LAPD or *DL-CORE*. This refers to a reduced subset of LAPD found in Annex A of ITU-T Rec. Q.922 (Ref. 22). The basic body of Q.922 presents CCITT/ITU-T specification for frame relay. This derivative is called LAPF rather than LAPD. The material found in ANSI T1.618-1991 (Ref. 23) is identical for all intents and purposes with Annex A of Q.922.

[13]LAPD = link access protocol D-channel.

To properly describe frame relay from our perspective, we will briefly give an overview of the ANSI T1.618-1991 (Ref. 23) and T1.606-1990 (Ref. 24). This will be followed by some fairly well identified variants.

12.5.3 Introduction to Frame Relay Operation

Frame relay may be considered a cost-effective outgrowth of ISDN, meeting high data rate (e.g., 2 Mbps) and low delay data communications requirements. Frame relay encapsulates data files. These may be considered "packets," although they are called frames. Thus frame relay is compared to CCITT Rec. X.25 packet service. Frame relay was designed for current transmission capabilities of the network with its relatively wider bandwidths[14] and excellent error performance (e.g., BER better than 1×10^{-7}).

The incisive reader will note the use of the term *bandwidth*. It is used synonymously with bit rate. If we were to admit at first approximation 1 bit per hertz of bandwidth, such use is acceptable. We are mapping frame relay bits into bearer channel bits probably on a one-for-one basis. The bearer channel may be a DS0/E0 64-kbps channel, a 56-kbps channel of a DS1 configuration, or multiple DS0/E0 channels in increments of 64 kbps up to 1.544/2.048 Mbps. We may also map the frame relay bits into a SONET or SDH configuration (Chapter 17). The final bearer channel may require more or less bandwidth than that indicated by the bit rate. This is particularly true for such bearer channels riding on radio systems and, to a lesser extent, on a fiber optic medium or other transmission media. The reader should be aware of certain carelessness of language used in industry publications.

Frame relay works well in the data rate range from 56 kbps up to 1.544/2.048 Mbps. It is being considered for the 45-Mbps DS3 rate for still additional *speed*.

ITU-T's use of the ISDN D-channel for frame relay favors X.25-like switched virtual circuits (SVCs). However, ANSI recognized that the principal application of frame relay was interconnection of LANs, and not to replace X.25. Because of the high data rate of LANs (megabit range), dedicated connections are favored. ANSI thus focused on permanent virtual connections (PVCs). With PVCs, routes are provisioned at the time of frame relay contract. This notably simplified the signaling protocol. Also, ANSI frame relay does not support voice or video.

As mentioned, the ANSI frame relay derives from ISDN LAPD core functions. The core functions of the LAPD protocol that are used in frame relay (as defined here) are as follows:

- Frame delimiting, alignment, and transparency provided by the use of HDLC flags and zero bit insertion/extraction;[15]
- Frame multiplexing/demultiplexing using the address field;[16]
- Inspection of the frame to ensure that it consists of an integer number of octets prior to zero bit insertion or following zero bit extraction;
- Inspection of the frame to ensure that it is not too long or too short;
- Detection of (but *not* recovery from) transmission errors; and
- Congestion control functions.

[14]We would rather use the term *greater bit rate capacity*.

[15]Zero bit insertion is a technique used to assure that the unique beginning flag of a frame is not imitated inside the frame that allows full transparency.

[16]Where the DLCI indicates a particular channel or channel group in the multiplex aggregate for PVC operation.

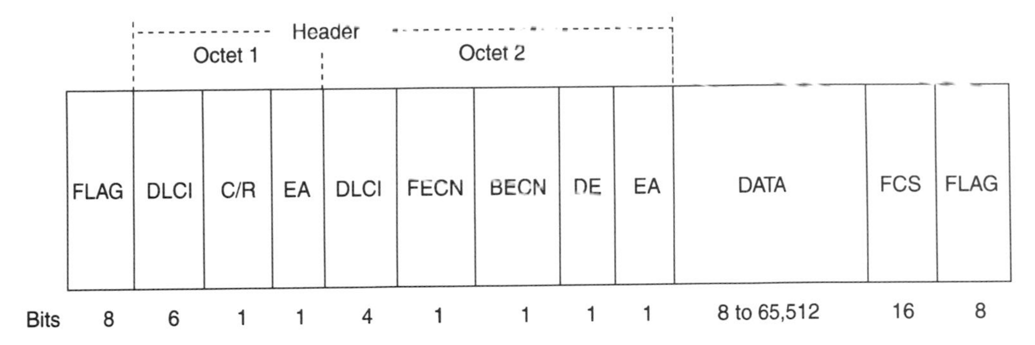

Figure 12.29 Frame relay ANSI frame format with a two-octet address. DLCI—data link connection identifier, C/R—command response indicator, EA–address field extension bit, FECN/BECN—see text, DE—discard eligibility.

In other words, ANSI has selected certain features from the LAPD structure/protocol, rejected others, and added some new features. For instance, the control field was removed, but certain control functions have been incorporated as single bits in the address field. These are the C/R bit (command/response), DE (discard eligibility), FECN bit (forward explicit congestion notification), and BECN bit (backward explicit congestion notification).

12.5.4 Frame Structure

User traffic passed to a FRAD (frame relay access device) is segmented into frames with a maximum length information field or with a default length of 262 octets. The minimum information field length is one octet.

Figure 12.29 illustrates the frame relay frame structure. As mentioned before in Section 12.4.8 (LAPD), it uses HDLC flags (01111110) as opening and closing flags. A closing flag may also serve as the opening flag of the next frame.

Address field. This consists of two octets, but may be extended to three or four octets. However, there is no control field as there is in HDLC, LAPB, and ISDN LAPD.

In its most reduced version, there are just 10 bits allocated to the address field in two octets (the remainder of the bits serve as control functions) supporting up to 1024 logical connections.

It should be noted that the number of addressable logical connections is multiplied because they can be reused at each nodal (switch) interface. That is, an address in the form of a data-link connection identifier (DLCI) has meaning only on one trunk between adjacent nodes. The switch (node) that receives a frame is free to change the DLCI before sending the frame onwards over the next link. Thus, the limit of 1024 DLCIs applies to the link, not the network.

Information field. This follows the address field and precedes the frame check sequence (FCS). The maximum size of the information field is an implementation parameter, and the default maximum is 262 octets. ANSI chose this default maximum to be compatible with LAPD on the ISDN D-channel, which has a two-octet control field and a 260-octet maximum information field. All other maximum values are negotiated between users and networks and between networks. The minimum information field size is one octet. The field must contain an integer number of octets; partial octets are not allowed. A maximum of 1600 octets is encouraged for applications such as LAN interconnects to minimize the need for segmentation and reassembly by user equipment.

Transparency. As with HDLC, X.25 (LAPB), and LAPD, the transmitting data-link layer must examine the frame content between opening and closing flags and inserts a 0 bit after all sequences of five contiguous 1s (including the last five bits of the FCS) to ensure a flag or an abort sequence is not simulated within the frame. At the other side of the link, a receiving data-link layer must examine the frame contents between the opening and closing flags and must discard any 0 bit that directly follows five contiguous 1s.

Frame check sequence (FCS). This is based on the generator polynomial $X^{16} + X^{12} + X^5 + 1$. The CRC processing includes the content of the frame existing between, but not including, the final bit of the opening flag and the first bit of the FCS, excluding the bits inserted for transparency. The FCS, of course, is a 16-bit sequence. If there are no transmission errors (detected), the FCS at the receiver will have the sequence 00011101 00001111.

Invalid frames. An invalid frame is a frame that:

- Is not properly bounded by two flags (e.g., a frame abort);
- Has fewer than three octets between the address field and the closing flag;
- Does not consist of an integral number of octets prior to zero bit insertion or following zero bit extraction;
- Contains a frame check sequence error;
- Contains a single octet address field; and
- Contains a data-link connection identifier (DLCI) that is not supported by the receiver.

Invalid frames are discarded without notification to the sender, with no further action.

12.5.4.1 Address Field Variables

12.5.4.1.1 Address Field Extension Bit (EA). The address field range is extended by reserving the first transmitted bit of the address field octets to indicate the final octet of the address field. If there is a 0 in this bit position, it indicates that another octet of the address field follows this one. If there is a 1 in the first bit position, it indicates that this octet is the final octet of the address field. As an example, for a two-octet address field, bit one of the first octet is set to 0 and bit one of the second octet is set to 1.

It should be understood that a two-octet address field is specified by ANSI. It is a user's option whether a three- or four-octet field is desired.

12.5.4.1.2 Command/Response Bit (C/R). The C/R bit is not used by the ANSI protocol, and the bit is conveyed transparently.

12.5.4.1.3 Forward Explicit Congestion Notification (FECN) Bit. This bit may be set by a congested network to notify the user that congestion avoidance procedures should be initiated, where applicable, for traffic in the direction of the frame carrying the FECN indication. This bit is set to 1 to indicate to the receiving end-system that the frames it receives have encountered congested resources. The bit may be used to adjust the rate of destination-controlled transmitters. While setting this bit by the network or user is optional, no network shall ever clear this bit (i.e., set to 0). Networks that do not provide FECN shall pass this bit unchanged.

12.5.4.1.4 Backward Explicit Congestion Notification (BECN). This bit may be set by a congested network to notify the user that congestion avoidance procedures should be initiated, where applicable, for traffic in the opposite direction of the frame carrying the BECN indicator. This bit is set to 1 to indicate to the receiving end-system that the frames it transmits may encounter congested resources. The bit may be used to adjust the rate of source-controlled transmitters. While setting this bit by the network or user is optional according to the ANSI specification, no network shall ever clear (i.e., set to 0) this bit. Networks that do not provide BECN shall pass this bit unchanged.

12.5.4.1.5 Discard Eligibility Indicator (DE) Bit. This bit, if used, is set to 1 to indicate a request that a frame should be discarded in preference to other frames in a congestion situation. Setting this bit by the network or user is optional. No network shall ever clear (i.e., set to 0) this bit. Networks that do not provide DE capability shall pass this bit unchanged. Networks are not constrained to only discard frames with DE equal to 1 in the presence of congestion.

12.5.4.1.6 Data-Link Connection Identifier (DLCI). This is used to identify the logical connection, multiplexed within the physical channel, with which a frame is associated. All frames carried within a particular physical channel and having the same DLCI value are associated with the same logical connection. The DLCI is an unstructured field. For two-octet addresses, bit 5 of the second octet is the least significant bit. For three- and four-octet addresses, bit 3 of the last octet is the least significant bit. In all cases, bit 8 of the first octet is the most significant bit. The structure of the DLCI field may be established by the network at the user–network interface subject to bilateral agreements.

12.5.5 Traffic and Billing on a Frame Relay Network.

Figure 12.30*a* shows a typical traffic profile on a conventional PSTN, whereas Figure 12.30*b* illustrates a typical profile of bursty traffic over a frame relay network. Such a traffic profile is also typical of a LAN. Of course the primary employment of frame relay is to interconnect LANs at a distance.

With conventional leased data circuits we have to pay for the bit rate capacity whether it is used or not. On the other hand, with frame relay, we only pay for the "time" used. Billing can be handled in one of three ways:

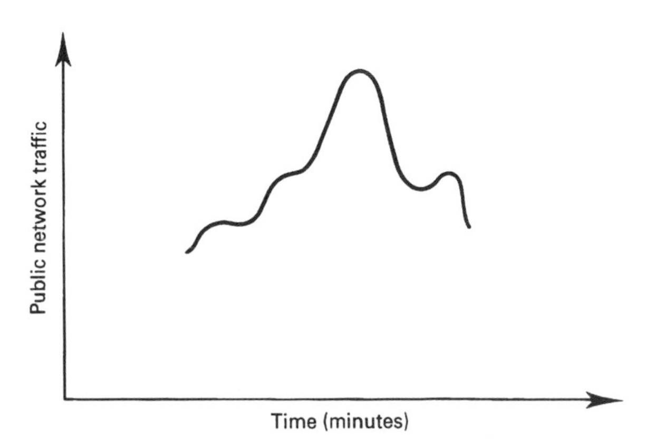

Figure 12.30*a* A typical traffic profile of a public switched telephone network. (Courtesy of Hewlett-Packard Co., Ref. 25.)

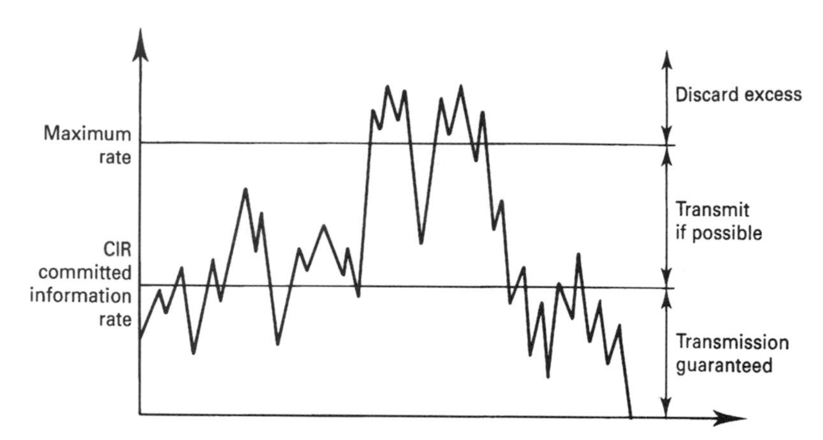

Figure 12.30*b* Typical bursty traffic of a frame relay circuit. Note the traffic levels indicated. (Courtesy of Hewlett-Packard Co., Ref. 25.)

1. CIR (committed information rate) is a data rate subscribed to by a user. This rate may be exceeded for short bursts during peak period(s) as shown in Figure 13.30*b*.
2. We can just pay a flat rate.
3. We can pay per packet (i.e., frame).

Now turn to Figure 12.30*b*. Note that on the right-hand side of the figure there is the guaranteed transmission bit rate equivalent to the CIR. Depending on the traffic load and congestion, during short periods the user may exceed the CIR. However, there is a point where the network cannot sustain further increases in traffic without severe congestion resulting. Traffic above such levels is arbitrarily discarded by the network without informing the originator.

12.5.6 Congestion Control: A Discussion

Congestion in the user plane occurs when traffic arriving at a resource exceeds the network's capacity. It can also occur for other reasons such as equipment failure. Network congestion affects the throughput, delay, and frame loss experienced by the end-user.

End-users should reduce their offered load in the face of network congestion. Reduction of offered load by an end-user may well result in an increase in the effective throughput available to the end-user during congestion.

Congestion avoidance procedures, including optional explicit congestion notification, are used at the onset of congestion to minimize its negative effects on the network and its users. Explicit notification is a procedure used for congestion avoidance and is part of the data-transfer phase. Users should react to explicit congestion notification (i.e., optional but highly desirable). Users who are not able to act on explicit congestion notification shall have the capability to receive and ignore explicit notification generated by the networks.

Congestion recovery and the associated implicit congestion indication due to frame discard are used to prevent network collapse in the face of severe congestion. Implicit congestion detection involves certain events available to the protocols operating above the core function to detect frame loss (e.g., receipt of a REJECT frame, timer recovery). Upon detection of congestion, the user reduces the offered load to the network. Use of such reduction by users is optional.

12.5.6.1 Network Response to Congestion. Explicit congestion signals are sent in both the forward direction (toward the frame destination) and in the backwards direction (toward the frame source or originator). Forward explicit congestion notification (FECN) is provided by using the FECN bit (see Figure 12.29) in the address field. Backward explicit congestion notification (BECN) is provided by one of two methods. When timely reverse traffic is available, the BECN bit in the appropriate address field may be used. Otherwise, a consolidated link layer management message may be generated by the network. The consolidated link layer management (CLLM) message travels on the network as though it were a conventional frame relay frame. The generation and transport of CLLM by the network are optional. All networks transport the FECN and BECN bits without resetting.

12.5.6.2 User Response to Congestion. Reaction by the end-user to the receipt of explicit congestion notification is rate-based. Annex A to ANSI T1.618-1991 (Ref. 23) describes user reaction to FECN and BECN.

12.5.6.2.1 End-User Equipment Employing Destination-Controlled Transmitters. End-user reaction to implicit congestion detection or explicit congestion notification (FECN indications), when supported, is based on the values of FECN indications that are received over a period of time. The method is consistent with commonly used destination-controlled protocol suites, such as OSI class 4 transport protocol operated over the OSI connectionless service.

12.5.6.2.2 End-User Equipment Employing Source-Controlled Transmitters. End-user reaction to implicit congestion notification (BECN indication), when supported, is immediate when a BECN indication or a CLLM is received. This method is consistent with implementation as a function of data-link layer elements of procedure commonly used in source-controlled protocols such as CCITT Rec. Q.922 elements of procedure.

12.5.6.3 Consolidated Link Layer Management (CLLM) Message. The CLLM utilizes a special type of frame which has been appropriated from the HDLC protocol. This is the XID frame, commonly called an *exchange identification* frame. In HDLC it was used just as the name implied, for exchange identification. In frame relay, however, it may be used for network management as an alternative to congestion control. CLLM messages originate at network nodes, near the frame relay interface, usually housed in a router or otherwise incorporated with operational equipment.

As mentioned, BECN/FECN bits in frames must pass congested nodes in the forward or backward direction. Suppose that, for a given user, no frames pass in either direction, and that the user therefore has no knowledge of network congestion because at that moment the user is not transmitting or receiving frames. Frame relay standards do not permit a network to generate frames with the DLCI of the congested circuit. CLLM covers this contingency. It has DLCI = 1023 reserved.

The use of CLLM is optional. If it is used, it may or may not operate in conjunction with BECN/FECN. The CLLM frame format has one octet for the cause of congestion such as excessive traffic, equipment or facility failure, preemption, or maintenance action.

This same octet indicates whether the cause is expected to be short or long term. Short term is on the order of seconds or minutes, and anything greater is long term. There is also a bit sequence in this octet indicating an unknown cause of congestion and whether short or long term.

CLLM octets 19 and above give the DLCI values that identify logical links that have encountered congestion. This field must accommodate DLCI length such as two-octet, three-octet, and four-octet DLCI fields.

12.5.6.4 Action at a Congested Node. When a node is congested, it has several alternatives it may use to mitigate or eliminate the problem. It may set the FECN and BECN bits to a binary 1 in the address field and/or use the CLLM message. Of course, the purpose of explicit congestion notification is:

- To inform the "edge" node at the network ingress of congestion so that edge node can take the appropriate action to reduce the congestion; or
- To notify the source that the negotiated throughput has been exceeded; or
- To do both.

One of the strengths of the CLLM is that it contains a list of DLCIs that correspond to the congested frame relay bearer connections. These DLCIs indicate not only the sources currently active causing the congestion, but also those sources that are not active. The reason for the latter is to prevent those sources that are not active from becoming active and thus causing still further congestion. It may be necessary to send more than one CLLM message if all the affected DLCIs cannot fit into a single frame.

12.5.7 Quality of Service Parameters

The quality that frame-relaying service provides is characterized by the values of the following parameters. ANSI adds in Ref. 24 that the specific list of negotiable parameters is for further study.

1. Throughput;
2. Transit delay;
3. Information integrity;
4. Residual error rate;
5. Delivered error(ed) frames;
6. Delivered duplicated frames;
7. Delivered out-of-sequence frames;
8. Lost frames;
9. Misdelivered frames;
10. Switched virtual call establishment delay;
11. Switched virtual call clearing delay;
12. Switched virtual call establishment failure;
13. Premature disconnect; and
14. Switched virtual call clearing failure.

12.5.7.1 Network Responsibilities. Frame relay frames are routed through the network on the basis of an attached label (i.e., the DLCI value of the frame). This label is a logical identifier with local significance. In the virtual call case, the value of the logical identifier and other associated parameters such as layer 1 channel delay, and so on, may be requested and negotiated during call setup. Depending on the value of the parameters, the network may accept or reject the call. In the case of the permanent virtual circuit, the logical identifier and other associated parameters are defined by means of administrative procedures (e.g., at the time of subscription).

The user–network interface structure allows for the establishment of multiple virtual calls or permanent virtual circuits, or both, to many destinations over a single access channel. Specifically, for each connection, the bearer service:

1. Provides bidirectional transfer of frames;
2. Preserves their order as given at one user–network interface if and when they are delivered at the other end. (*Note:* No sequence numbers are kept by the network. Networks are implemented in such a way that frame order is preserved);
3. Detects transmission, format, and operational errors such as frames with an unknown label;
4. Transports the user data contents of a frame transparently; only the frame's address and FCS fields may be modified by network nodes; and
5. Does not acknowledge frames.

At the user–network interface, the FRAD (frame relay access device), as a minimum, has the following responsibilities:

1. Frame delimiting, alignment, and transparency provided by the use of HDLC flags and zero bit insertion;
2. Virtual circuit multiplexing/demultiplexing using the address field of the frame;
3. Inspection of the frame to ensure that it consists of an integer number of octets prior to zero bit insertion or following zero bit extraction;
4. Inspection of the frame to ensure it is not too short or too long; and
5. Detection of transmission, format, and operational errors.

A frame received by a frame handler may be discarded if the frame:

1. Does not consist of an integer number of octets prior to zero bit insertion or following zero bit extraction;
2. Is too long or too short; and
3. Has an FCS that is in error.

The network will discard a frame if it:

1. Has a DLCI value that is unknown; or
2. Cannot be routed further due to internal network conditions. A frame can be discarded for other reasons, such as exceeding negotiated throughput.

Section 12.5 is based on ANSI standards T1.618-1991, T1.606-1990, and T1.606a-1992 (Refs. 23, 24, 26).

REVIEW EXERCISES

1. X.25 deals primarily with which OSI layer? and LAPB?

2. How does a processor know where a frame's field boundaries are?

3. Where is an X.25 DTE located? its DCE located?

4. What are the three approaches used with X.25 to manage the transfer and routing of data packets?
5. What is the purpose of a *call request and incoming call packet*?
6. Name at least two types of flow control packets used with X.25.
7. Even though TCP/IP predates OSI, in what OSI layers would we expect to find TCP and IP?
8. What is the purpose of the ARP (address resolution protocol)?
9. What is the primary function of IP?
10. What is the function of a router in an IP network?
11. What are the three types of routing carried out by an IP routing table?
12. How is ICMP used as an adjunct to IP?
13. What is the term used in TCP/IP parlance for segmentation?
14. What important mechanisms does TCP offer IP with its potentially unreliable services?
15. What is the purpose of the *three-way handshake*?
16. Name at least four communication services that ISDN will support.
17. What is the purpose of the D-channel with ISDN?
18. What are the three basic variants of the H-channel?
19. Using your imagination and what you have previously learned, relate the two higher bit rate H-channels to the digital network.
20. How does an ISDN user derive its timing?
21. How many B-channels can carry traffic in PRI service in North America?
22. What is the function of the balancing bit in the BRI configuration?
23. Is North American ISDN BRI 2-wire or 4-wire?
24. What are the three types of LAPD control field formats?
25. Name and describe at least three functions carried out by ISDN layer 3.
26. What BRI line code is used with CCITT ISDN? with North American ISDN?
27. What is the first field in an LAPD frame?
28. Compare frame relay and X.25 for "speed" of operation.
29. What does a frame relay network do with errored frames?
30. Frame relay operation derives from which predecessor?
31. There are six possible causes for declaring a frame relay frame invalid. Name four of them.
32. Discuss the use of CLLM as an alternative for congestion control.
33. Name some actions a congested node can take to alleviate the problems.

34. If no sequence numbers are used with frame relay, how are frames kept in order?

35. How does an end-user know that a frame or frames have been lost or discarded?

REFERENCES

1. *Interface between Data Terminal Equipment (DTE) and Data Circuit-Terminating Equipment (DCE) for Terminals Operating in the Packet Mode and Connected to the Public Data Networks by Dedicated Circuit*, ITU-T Rec. X.25, Helsinki, 1993.
2. *Internet Protocol*, RFC 791, DDN Network Information Center, SRI International, Menlo Park, CA, 1981.
3. *Transmission Control Protocol*, RFC 793, DDN Network Information Center, SRI International, Menlo Park, CA, 1981.
4. *Packet-Switched Signaling System between Public Networks Providing Data Transmission Services*, ITU-T Rec. X.75, Helsinki, 1993.
5. *Internet Protocol Transition Workbook*, SRI International, Menlo Park, CA, 1982.
6. *Assigned Numbers*, RFC 1060, DDN Network Information Center, SRI International, Menlo Park, CA, 1990.
7. *An Ethernet Address Resolution Protocol*, RFC 826, DDN Network Information Center, SRI International, Menlo Park, CA, 1984.
8. *A Reverse Address Resolution Protocol*, RFC 903, DDN Network Information Center, SRI International, Menlo Park, CA, 1984.
9. *Military Standard, Transmission Control Protocol*, MIL-STD-1778, U.S. Dept. of Defense, Washington, DC, 1983.
10. *IEEE Standard Dictionary of Electrical and Electronic Terms*, 6th ed., IEEE Std. 100-1996, IEEE, New York, 1996.
11. ISDN *User-Network Interfaces—Interface Structure and Access Capabilities*, CCITT Rec. I.412, Fascicle III.8, IXth Plenary Assembly, Melbourne, 1988.
12. *ISDN Network Architecture*, CCITT Rec. I.324, ITU Geneva, 1991.
13. W. Stallings, ed., *Integrated Services Digital Networks (ISDN)*, IEEE Computer Society Press, Washington, DC, 1985.
14. *Digital Subscriber Signaling System No. 1 (DSS1): ISDN User-Network Interface, Data Link Layer—General Aspects*, ITU-T Rec. Q.920, ITU, Geneva, 1993.
15. *ISDN User-Network Interface—Data Link Layer Specification*, ITU-T Rec. Q.921, ITU, Geneva, 1993.
16. *ISDN User-Network Interface: Layer 3—for Basic Call Control*, ITU-T Rec. Q.931, ITU, Geneva, 1993.
17. *Interface between Data Terminal Equipment (DTE) and Data Circuit-Terminating Equipment (DCE) for Terminals Operating in the Packet Mode and Accessing a Public-Switched Telephone Network or a Circuit-Switched Public Data Network*, CCITT Rec. X.32, Fascicle VIII.2, IXth Plenary Assembly, Melbourne, 1988.
18. *Support of Packet Mode Terminal Equipment by an ISDN*, ITU-T Rec. X.31, ITU, Geneva, 1993.
19. *Framework for Frame Mode Bearer Services*, ITU-T Rec. I.122, ITU, Geneva, 1993.
20. *Frame Mode Bearer Services*, CCITT Rec. I.233, Geneva, 1992.
21. *ISDN Signaling Specification for Frame Relay Bearer Service for Digital Subscriber Signaling System No. 1 (DDS1)*, ANSI T1.617-1991, ANSI, New York, 1991.
22. *ISDN Data Link Layer Service for Frame Mode Bearer Services*, CCITT Rec. Q.922, ITU, Geneva, 1992.

23. *Integrated Services Digital Network (ISDN)—Core Aspects of Frame Protocol for Use with Frame Relay Bearer Service*, ANSI T1.618-1991, ANSI, New York, 1991.
24. *ISDN—Architectural Framework and Service Description for Frame Relay Bearer Service*, ANSI T1.606-1990, ANSI, New York, 1990.
25. Frame Relay and SMDS seminar, Hewlett-Packard Co., Burlington, MA, Oct. 1993.
26. *Integrated Services Digital Network (ISDN)—Architectural Framework and Service Description for Frame-Relay Bearer Service (Congestion Management and Frame Size)*, ANSI T1.606a-1992, ANSI, New York, 1992.

13

CCITT SIGNALING SYSTEM NO. 7

13.1 INTRODUCTION

CCITT Signaling System No. 7 (SS No. 7) was developed to meet the stringent signaling requirements of the all-digital network based on the 64-kbps channel. It operates in quite a different manner from the signaling discussed in Chapter 7. Nevertheless, it must provide for supervision of circuits and address signaling, and carry call progress signals and alerting notification to be eventually passed to the called subscriber. These requirements certainly look familiar and are no different than the ones discussed in Chapter 7. The difference is in how these requirements are met. CCITT No. 7 is a data network entirely dedicated to interswitch signaling.[1]

Simply put, CCITT SS No. 7 is described as an international standardized common-channel signaling system that:

- Is optimized for operation with digital networks where switches use stored-program control (SPC), such as the DMS-100 series switches and the 5ESS, among others, which were discussed in Section 6.11;
- Can meet present and future requirements of information transfer for interprocessor transactions with digital communications networks for call control, remote control, network database access and management, and maintenance signaling;
- Provides a reliable means of information transfer in correct sequence without loss or duplication (Ref. 1).

CCITT SS No. 7, in the years since 1980, has become known as the *signaling system for ISDN*. This it is. Without the infrastructure of SS No. 7 embedded in the digital network, there will be no ISDN with ubiquitous access. One important point is to be made. CCITT SS No. 7, in itself, is the choice for signaling in the digital PSTN without ISDN. It can and does stand on its own in this capacity.

As mentioned, SS No. 7 is a data communication system designed for only one purpose: signaling. It is *not* a general-purpose system. We then must look at CCITT SS No. 7 as (1) a specialized data network and (2) a signaling system (Ref. 2).

[1]This would be called *interoffice signaling* in North America.

13.2 OVERVIEW OF SS NO. 7 ARCHITECTURE

The SS No. 7 network model consists of network nodes, termed *signaling points* (SPs), interconnected by point-to-point signaling links, with all the links between two SPs called a *link set*. When the model is applied to a physical network, most commonly there is a one-to-one correspondence between physical nodes and logical entities. But when there is a need (e.g., a physical gateway node needs to be a member of more than one network), a physical network node may be logically divided into more than one SP, or a logical SP may be distributed over more than one physical node. These artifices require careful administration to ensure that management procedures within the protocol work correctly.

Messages between two SPs may be routed over a link set directly connecting the two points. This is referred to as the *associated mode* of signaling. Messages may also be routed via one or more intermediate SPs that relay messages at the network layer. This is called *nonassociated mode* of signaling. SS No. 7 supports only a special case of this routing, called *quasiassociated mode*, in which routing is static except for relatively infrequent changes in response to events such as link failures or addition of new SPs. SS No. 7 does not include sufficient procedures to maintain in-sequence delivery of information if routing were to change completely on a packet-by-packet basis.

The function of relaying messages at the network layer is called the *signaling transfer point* (STP) function.[2] Although this practice results in some confusion, the logical and physical network nodes at which this function is performed are frequently called STPs, even though they may provide other functions as well. An important part in designing an SS No. 7 network is including sufficient equipment redundancy and physical-route diversity so that the stringent availability objectives of the system are met. The design is largely a matter of locating signaling links and SPs with the STP function, so that performance objectives can be met for the projected traffic loads at minimum cost.

Figure 13.1 is an SS No. 7 network structure model. The STP function is concentrated in a relatively small number of nodes that are essentially dedicated to that function. The STPs are paired or *mated*, and pairs of STPs are interconnected with a quad configuration, as shown in the figure. We could also say that the four STPs are connected in mesh. This has proved to be an extremely reliable and survivable backbone network. Other nodes, such as switching centers and service control points (SCPs), are typically homed on one of the mated pairs of STPs, with one or more links to each of the mates, depending on traffic volumes (Ref. 3).

13.3 SS NO. 7: RELATIONSHIP TO OSI

SS No. 7 relates to OSI (Section 10.10.2) up to a certain point. During the development of SS No. 7, one group believed that there should be complete compatibility with all seven OSI layers. However, the majority of the CCITT working group responsible for the concept and design of SS No. 7 was concerned with delay, whether for the data, telephone, or ISDN user of the digital PSTN. Recall from Chapter 7 that *post-dial delay* is probably the most important measure of performance of a signaling system. To minimize delay, the seven layers of OSI were truncated at layer 4. In fact, CCITT Rec. Q.709 specifies no more than 2.2 seconds of postdial delay for 95% of calls. To accomplish

[2]Bellcore (Ref. 6) reports that "purists restrict this further to MTP relaying." (MTP stands for message transfer part.)

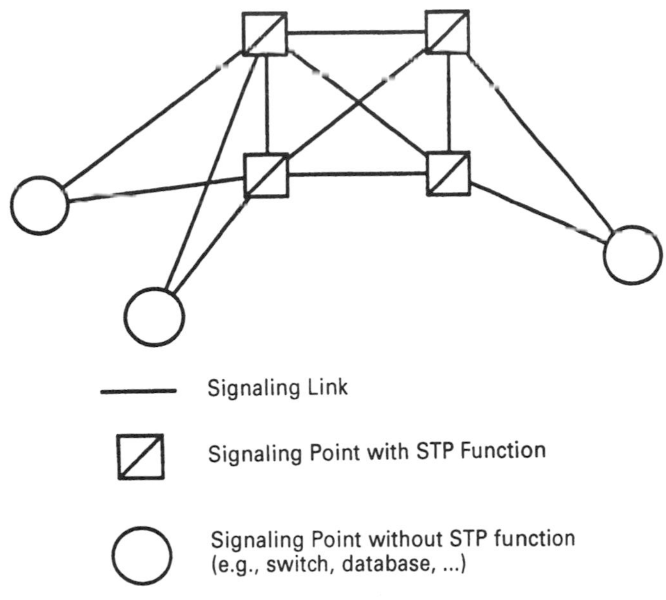

Figure 13.1 Signaling System No. 7 network structure model.

this, a limit is placed on the number of relay points, called STPs, that can be traversed by a signaling message and by the inherent design of SS No. 7 as a four-layer system. Figure 13.2 relates SS No. 7 protocol layers to the OSI reference model. Remember that reducing the number of OSI layers reduces processing, and thus processing time. As a result postdial delay is also reduced.

We should note that SS No. 7 layer 3 signaling network functions include signaling message-handling functions and network management functions. Figure 13.3 shows the general structure of the SS No. 7 signaling system.

There seem to have been various efforts to force-fit SS No. 7 into OSI layer 4 upwards. These efforts have resulted in the sublayering of layer 4 into user parts and the SCCP (signaling connection control part).

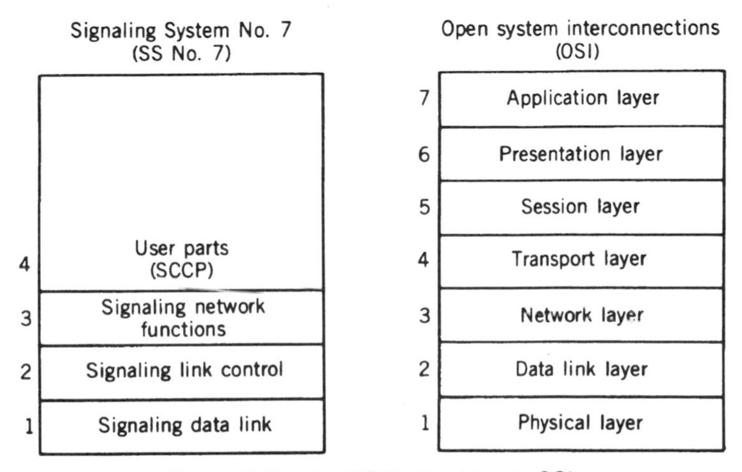

Figure 13.2 How SS No. 7 relates to OSI.

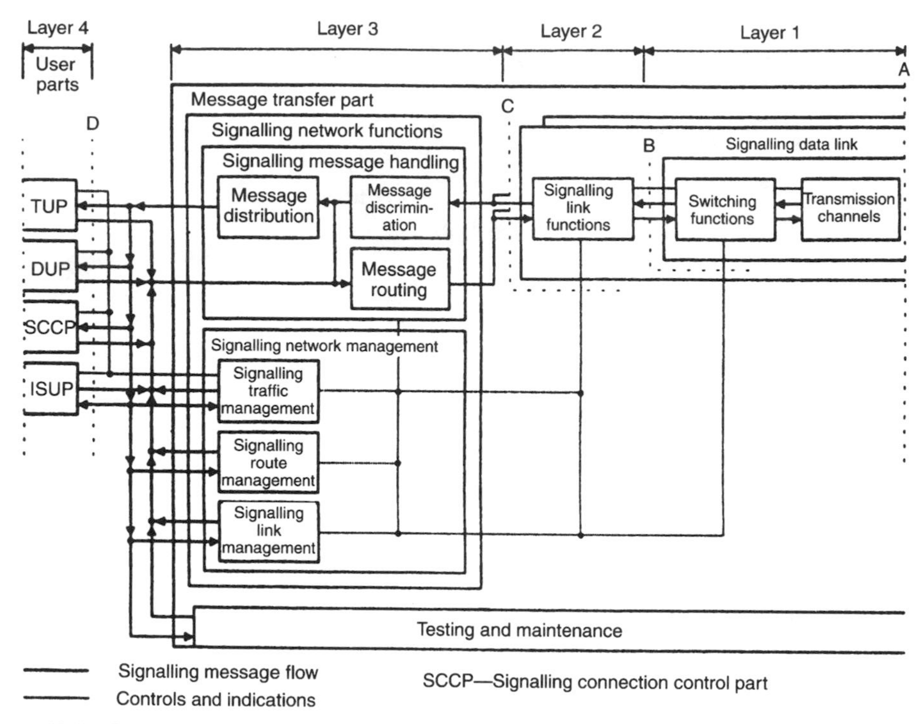

Figure 13.3 General structure of signaling functions. (From ITU-T Rec. Q.701, Figure 6/Q.701, p. 8 [Ref. 4].)

In Section 13.4 we briefly describe the basic functions of the four SS No. 7 layers, which are covered in more detail in Sections 13.5 through 13.7.

13.4 SIGNALING SYSTEM STRUCTURE

Figure 13.3, which illustrates the basic structure of SS No. 7, shows two parts to the system: the message transfer part (MTP) and the user parts. There are three user parts: (1) telephone user part (TUP), (2) data user part (DUP), and (3) the ISDN user part (ISUP). Figures 13.2 and 13.3 show OSI layers 1, 2, and 3, which make up the MTP. The following paragraphs describe the functions of each of these layers from a system viewpoint.

Layer 1 defines the physical, electrical, and functional characteristics of the signaling data link and the means to access it. In the digital network environment the 64-kbps digital path is the normal basic connectivity. The signaling link may be accessed by means of a switching function that provides the capability of automatic reconfiguration of signaling links.

Layer 2 carries out the signaling link function. It defines the functions and procedures for the transfer of signaling messages over one individual signaling data link. A signaling message is transferred over the signaling link in variable-length signal units. A signal unit consists of transfer control information in addition to the information content of the signaling message. The signaling link functions include:

- Delimitation of a signal unit by means of flags;
- Flag imitation prevention by bit stuffing;

- Error detection by means of check bits included in each signal unit;
- Error control by retransmission and signal unit sequence control by means of explicit sequence numbers in each signal unit and explicit continuous acknowledgments; and
- Signaling link failure detection by means of signal unit error monitoring, and signaling link recovery by means of special procedures.

Layer 3, signaling network functions, in principle, defines such transport functions and procedures that are common to and independent of individual signaling links. There are two categories of functions in layer 3:

1. *Signaling Message-Handling Functions.* During message transfer, these functions direct the message to the proper signaling link or user part.
2. *Signaling Network Management Functions.* These control real-time routing, control, and network reconfiguration, if required.

Layer 4 is the user part. Each user part defines the functions and procedures peculiar to the particular user, whether telephone, data, or ISDN user part.

The *signal message* is defined by CCITT Rec. Q.701 as an assembly of information, defined at layer 3 or 4, pertaining to a call, management transaction, and so on, which is then transferred as an entity by the message transfer function. Each message contains "service information," including a service indicator identifying the source user part and possibly whether the message relates to international or national application of the user part.

The *signaling information* portion of the message contains user information, such as data or call control signals, management and maintenance information, and type and format of message. It also includes a "label." The label enables the message to be routed by layer 3 through the signaling network to its destination and directs the message to the desired user part or circuit.

On the signaling link such signaling information is contained in the *message signal units* (MSUs), which also include transfer control functions related to layer 2 functions on the link.

There are a number of terms used in SS No. 7 literature that should be understood before we proceed further:

Signaling Points. Nodes in the network that utilize common-channel signaling;

Signaling Relation (similar to traffic relation). Any two signaling points for which the possibility of communication between their corresponding user parts exist are said to have a signaling relation;

Signaling Links. Signaling links convey signaling messages between two signaling points;

Originating and Destination Points. The originating and destination points are the locations of the source user part function and location of the receiving user part function, respectively;

Signaling Transfer Point (STP). An STP is a point where a message received on one signaling link is transferred to another link;

Message Label. Each message contains a label. In the standard label, the portion that is used for routing is called the *routing label.* The routing label includes:

- Destination and originating points of the message;
- A code used for load sharing, which may be the least significant part of a label component that identifies a user transaction at layer 4.

The standard label assumes that each signaling point in a signaling network is assigned an identification code, according to a code plan established for the purpose of labeling.

Message Routing. Message routing is the process of selecting the signaling link to be used for each signaling message. Message routing is based on analysis of the routing label of the message in combination with predetermined routing data at a particular signaling point.

Message Distribution. Message distribution is the process that determines to which user part a message is to be delivered. The choice is made by analysis of the service indicator.

Message Discrimination. Message discrimination is the process that determines, on receipt of a message at a signaling point, whether or not the point is the destination point of that message. This decision is based on analysis of the destination code of the routing label in the message. If the signaling point is the destination, the message is delivered to the message destination function. If not, the message is delivered to the routing function for further transfer on a signaling link.

13.4.1 Signaling Network Management

Three signaling network-management functional blocks are shown in Figure 13.3. These are signaling traffic management, signaling link management, and signaling route management.

13.4.1.1 Signaling Traffic Management. The signaling traffic management functions are:

1. To control message routing. This includes modification of message routing to preserve, when required, accessibility of all destination points concerned or to restore normal routing;
2. In conjunction with modifications of message routing, to control the resulting transfer of signaling traffic in a manner that avoids irregularities in message flow; and
3. Flow control.

Control of message routing is based on analysis of predetermined information about all allowed potential routing possibilities in combination with information, supplied by the signaling link management and signaling route management functions, about the status of the signaling network (i.e., current availability of signaling links and routes).

Changes in the status of the signaling network typically result in modification of current message routing and thus in the transfer of certain portions of the signaling traffic from one link to another. The transfer of signaling traffic is performed in accordance with specific procedures. These procedures are *changeover*, *changeback*, *forced rerouting*, and *controlled rerouting*. The procedures are designed to avoid, as far as circumstances permit, such irregularities in message transfer as loss, missequencing, or multiple delivery of messages.

The changeover and changeback procedures involve communication with other signaling point(s). For example, in the case of changeover from a failing signaling link, the two ends of the failing link exchange information (via an alternative path) that normally enables retrieval of messages that otherwise would have been lost on the failing link.

A signaling network has to have a signaling traffic capacity that is higher than the normal traffic offered. However, in overload conditions (e.g., due to network failures or extremely high traffic peaks) the signaling traffic management function takes flow control actions to minimize the problem. An example is the provision of an indication to the local user functions concerned that the MTP is unable to transport messages to a particular destination in the case of total breakdown of all signaling routes to that destination point. If such a situation occurs at an STP, a corresponding indication is given to the signaling route management function for further dissemination to other signaling points in the network.

13.4.1.2 Signaling Link Management. Signaling link management controls the locally connected signaling link sets. In the event of changes in the availability of a local link set, it initiates and controls actions with the objective of restoring the normal availability of that link set.

The signaling link management interacts with the signaling link function at level 2 by receipt of indications of the status of signaling links. It also initiates actions, also at level 2, such as initial alignment of an out-of-service link.

The signaling system can be applied in the method of provision of signaling links. Consider that a signaling link probably will consist of a terminal device and data link. It is also possible to employ an arrangement in which any switched connection to the far end may be used in combination with any local signaling terminal device. Here the signaling link management initiates and controls reconfigurations of terminal devices and signaling data links to the extent such reconfigurations are automatic. This implies some sort of switching function at layer 1.

13.4.1.3 Signaling Route Management. Signaling route management only relates to the quasiassociated mode of signaling (see Section 7.7). It transfers information about changes in availability of signaling routes in the signaling network to enable remote signaling points to take appropriate signaling traffic actions. For example, a signaling transfer point may send message indicating inaccessibility of a particular signaling point via that signal transfer point, thus enabling other signaling points to stop routing messages to an inoperative route.

13.5 SIGNALING DATA-LINK LAYER (LAYER 1)

A signaling data link is a bidirectional transmission path for signaling, comprising two data channels operating together in opposite directions at the same data rate. It constitutes the lowest layer (layer 1) in the SS No. 7 functionality hierarchy.

A digital signaling data link is made up of digital transmission channels and digital switches or their terminating equipment, providing an interface to SS No. 7 signaling terminals. The digital transmission channels may be derived from a digital multiplex signal at 1.544, 2.048, or 8.448 Mbps having a frame structure as defined in CCITT Rec. G.704 (see Chapter 6) or from digital multiplex bit streams having a frame structure specified for data circuits in CCITT Recs. X.50, X.51, X.50 bits, and X.51 bit.

The operational signaling data link is exclusively dedicated to the use of SS No. 7

signaling between two signaling points. No other information may be carried by the same channels together with the signaling information.

Equipment such as echo suppressors, digital pads, or A/μ-law converters attached to the transmission link must be disabled in order to ensure full-duplex operation and bit count integrity of the transmitted data stream. In this situation, 64-kbps digital signaling channels are used which are switchable as semipermanent channels in the exchange.

The standard bit rate on a digital bearer is 64 kbps. The minimum signaling bit rate for telephone call control applications is 4.8 kbps. For other applications such as network management, bit rates lower than 4.8 kbps may also be used.

The following is applicable for a digital signaling data link derived from a 2.048-Mbps digital path (i.e., E1). At the input/output interface, the digital multiplex equipment or digital switch block will comply with CCITT Recs. G.703 for electrical characteristics and G.704 for the functional characteristics—in particular, the frame structure. The signaling bit rate is 64 kbps. The standard time slot for signaling is time slot 16. When time slot 16 is not available, any time slot available for 64-kbps user transmission rate may be used. No bit inversion is performed.

For a signaling data link derived from an 8.448-Mbps (E2) digital link, the following applies: At the multiplex input/output interface, there should be compliance with CCITT Recs. G.703 for electrical characteristics and G.704 for functional characteristics—in particular, the frame structure. The signaling bit rate is 64 kbps. The standard time slots for use of a signaling data link are time slots 67–70 in descending order of priority. When these time slots are not available, any channel time slot available for 64-kbps user transmission rate may be used. No bit inversion is performed (Ref. 5).

For North American applications of SS No. 7, *BOC Notes on the LEC Networks—1994* (Ref. 6) states that data rates from 4.8 kbps to 64 kbps may be used.

13.6 SIGNALING LINK LAYER (LAYER 2)

This section deals with the transfer of signaling messages over one signaling link directly connecting two signaling points. Signaling messages delivered by upper hierarchical layers are transferred over the signaling link in variable-length signal units. The signal units include transfer control information for proper operation of the signaling link in addition to the signaling information. The signaling link (layer 2) functions include:

- Signaling unit delimitation;
- Signal unit alignment;
- Error detection;
- Error correction;
- Initial alignment;
- Signal link error monitoring; and
- Flow control.

All of these functions are coordinated by the link state control, as shown in Figure 13.4.

13.6.1 Signal Unit Delimitation and Alignment

The beginning and end of a signal unit are indicated by a unique 8-bit pattern, called the *flag*. Measures are taken to ensure that the pattern cannot be imitated elsewhere in

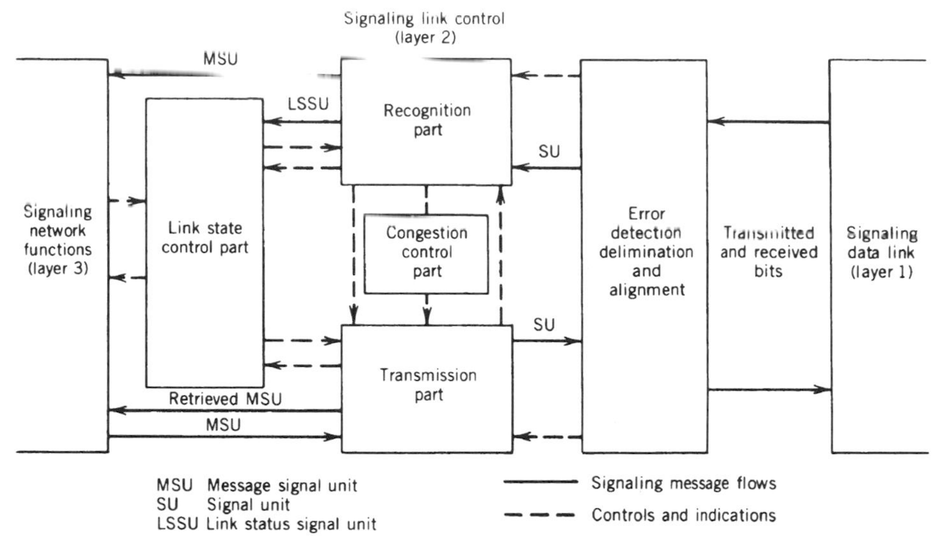

Figure 13.4 Interactions of functional specification blocks for signaling link control. *Note:* The MSUs, LSSUs, and SUs do not include error-control information. (From ITU-T Rec. Q.703, Figure 1/Q.703, p. 2 [Ref. 7].)

the unit. Loss of alignment occurs when a bit pattern disallowed by the delimitation procedure (i.e., more than six consecutive 1s) is received, or when a certain maximum length of signal unit is exceeded. Loss of alignment will cause a change in the mode of operation of the signal unit error rate monitor.

13.6.2 Error Detection

The error detection function is performed by means of the 16 check bits provided at the end of each signal unit. The check bits are generated by the transmitting signaling link terminal by operating on the preceding bits of the signal unit following a specified algorithm. At the receiving signaling link terminal, the received check bits are operated by using specified rules which correspond to that algorithm. If consistency is not found between the received check bits and the preceding bits of the signal unit according to the algorithm, then the presence of errors is indicated and the signal unit is discarded.

13.6.3 Error Correction

Two forms of error correction are provided: the *basic method* and the *preventive cyclic retransmission method.* The basic method applies to (a) signaling links using nonintercontinental terrestrial transmission means and (b) intercontinental signaling links where one-way propagation is less than 15 ms.

The preventive cycle retransmission method applies to (a) intercontinental signaling links where the one-way delay is equal to or greater than 15 ms and (b) signaling links established via satellite.

In cases where one signaling link with an intercontinental link set is established via satellite, the preventive cycle retransmission method is used for all signaling links of that set.

The basic method is a noncompelled, positive/negative acknowledgment, retransmission error correction system. A signal unit that has been transmitted is retained at the transmitting signaling link terminal until a positive acknowledgment for that signal unit is received. If a negative acknowledgment is received, then the transmission of new signal units is interrupted and those signal units which have been transmitted but not yet positively acknowledged (starting with that indicated by the negative acknowledgment) will be transmitted once, in the order in which they were first transmitted.

The preventive cyclic retransmission method is a noncompelled, positive acknowledgment, cyclic retransmission forward error correction system. A signal unit which has been transmitted is retained at the transmitting signaling unit terminal until a positive acknowledgment for that signaling unit is received. During the period when there are no new signal units to be transmitted, all signal units which have not been positively acknowledged are retransmitted cyclically.

The forced retransmission procedure is defined to ensure that forward error correction occurs in adverse conditions (e.g., degraded BER and/or high-traffic loading). When a predetermined number of retained, unacknowledged signal units exist, the transmission of new signal units is retransmitted cyclically until the number of acknowledged signal units is reduced.

13.6.4 Flow Control

Flow control is initiated when congestion is detected at the receiving end of the signaling link. The congested receiving end of the link notifies the remote transmitting end of the condition by means of an appropriate link status signal and it withholds acknowledgments of all incoming message signal units. When congestion abates, acknowledgments of all incoming signal units are resumed. When congestion exists, the remote transmitting end is periodically notified of this condition. The remote transmitting end will indicate that the link has failed if the congestion continues too long.

13.6.5 Basic Signal Unit Format

Signaling and other information originating from a user part is transferred over the signaling link by means of signal units. There are three types of signal units used in SS No. 7:

1. Message signal unit (MSU);
2. Link status signal unit (LSSU); and
3. Fill-in signal unit (FISU).

These units are differentiated by means of the *length indicator*. MSUs are retransmitted in case of error; LSSUs and FISUs are not. The MSU carries signaling information; the LSSU provides link status information; and the FISU is used during the link idle state—it fills in.

The signaling information field is variable in length and carries the signaling information generated by the user part. All other fields are fixed in length. Figure 13.5 illustrates the basic formats of the three types of signal units. As shown in the figure, the message transfer control information encompasses eight fixed-length fields in the signal unit that contains information required for error control and message alignment. These eight fields are described in the following. In Figure 13.5 we start from right to left, which is the direction of transmission.

The opening *flag* indicates the start of a signal unit. The opening flag of one sig-

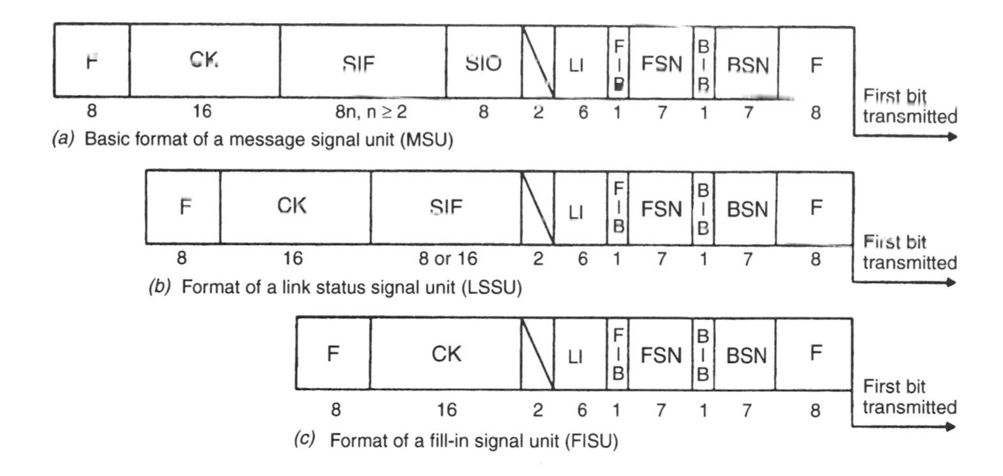

BIB Backward indicator bit
BSN Backward sequence number
CK Check bits
F Flag
FIB Forward indicator bit
FSN Forward sequence number
LI Length indicator
n Number of octets in the SIF
SF Status field
SIF Signalling information field
SIO Service information octet

Figure 13.5 Signal unit formats. (From ITU-T Rec. Q.703, Figure 3/Q.703, p. 5 [Ref. 7].)

nal unit is normally the closing flag of the previous signal unit. The flag bit pattern is 01111110. The *forward sequence number* (FSN) is the sequence number of the signal unit in which it is carried. The *backward sequence number* (BSN) is the sequence number of a signal unit being acknowledged. The value of the FSN is obtained by incrementing (modulo 128) the last assigned value by 1. The FSN value uniquely identifies a message signal unit until its delivery is accepted without errors and in correct sequence by the receiving terminal. The FSN of a signal unit other than an MSU assumes the value of the FSN of the last transmitted MSU. The maximum capacity of sequence numbers is 127 message units before reset (modulo 128) (i.e., a 7-bit binary sequence, $2^7 = 128 - 1$).

Positive acknowledgment is accomplished when a receiving terminal acknowledges the acceptance of one or more MSUs by assigning an FSN value of the latest accepted MSU to the BSN of the next signal unit sent in the opposite direction. The BSNs of subsequent signal units retain this value until a further MSU is acknowledged, which will cause a change in the BSN sent. The acknowledgment to an accepted MSU also represents an acknowledgment to all, if any, previously accepted, though not yet acknowledged, MSUs.

Negative acknowledgment is accomplished by inverting the backward indicator bit (BIB) value of the signal unit transmitted. The BIB value is maintained in subsequently sent signal units until a new negative acknowledgment is to be sent. The BSN assumes the value of the FSN of the last accepted signal unit.

As we can now discern, the forward indicator bit (FIB) and the backward indicator bit together with the FSN and BSN are used in the basic error-control method to perform signal unit sequence control and acknowledgment functions.

Table 13.1 Three-Bit Link Status Indications

Bits			Status	
C	B	A	Indication	Meaning
0	0	0	0	Out of alignment
0	0	1	N	Normal alignment
0	1	0	E	Emergency alignment
0	1	1	OS	Out of service
1	0	0	PO	Processor outage
1	0	1	B	Busy

Source: From para. 11.1.3, ITU-T Rec.Q.703, Ref. 7.

The *length indicator* (LI) is used to indicate the number of octets following the length indicator octet and preceding the check bits and is a binary number in the range of 0–63. The length indicator differentiates between three types of signal units as follows:

Length indicator = 0	Fill-in signal unit
Length indicator = 1 or 2	Link status signal unit
Length indicator ≥ 2	Message signal unit

The *service information octet* (ISO) is divided into a *service indicator* and a *subservice field.* The service indicator is used to associate signaling information for a particular user part and is present only in MSUs. Each is 4 bits long. For example, a service indicator with a value 0100 relates to the telephone user part, and 0101 relates to the ISDN user part. The subservice field portion of the SIO contains two network indicator bits and two spare bits. The network indicator discriminates between international and national signaling messages. It can also be used to discriminate between two national signaling networks, each having a different routing label structure. This is accomplished when the network indicator is set to 10 or 11.

The *signaling information field* (SIF) consists of an integral number of octets greater than or equal to 2 and less than or equal to 62. In national signaling networks it may consist of up to 272 octets. Of these 272 octets, information blocks of up to 256 octets in length may be accommodated, accompanied by a label and other possible housekeeping information that may, for example, be used by layer 4 to link such information blocks together.

The *link status signal unit* (LSSU) provides link status information between signaling points. The status field can be made up of one or two octets. CCITT Rec. Q.703 indicates application of the one-octet field in which the first three bits (from right to left) are used (bits A, B, and C) and the remaining five bits are spare. The values of the first 3 bits are given in Table 13.1.

13.7 SIGNALING NETWORK FUNCTIONS AND MESSAGES (LAYER 3)

13.7.1 Introduction

In this section we describe the functions and procedures relating to the transfer of messages between signaling points (i.e., signaling network nodes). These nodes are connected by signaling links involving layers 1 and 2 described in Sections 13.5 and 13.6. Another important function of layer 3 is to inform the appropriate entities of a fault and, as a consequence, carry out a rerouting of messages through the network. The signaling network functions are broken down into two basic categories:

1. Signaling message handling; and
2. Signaling network management (see Section 13.4.1 for description).

13.7.2 Signaling Message-Handling Functions

The signaling message-handling function ensures that a signaling message originated by a particular user part at an originating signaling point is delivered to the same user part at the destination point as indicated by the sending user part. Depending on the particular circumstances, the delivery may be made through a signaling link directly interconnecting the originating and destination points or via one or more intermediate signaling transfer points (STPs).

The signaling message-handling functions are based on the label contained in the messages which explicitly identifies the destination and origination points. The label part used for signaling message handling by the MTP is called the *routing label.* As shown in Figure 13.3 (upper-left portion), the signaling message-handling is divided up into the following:

- The *message routing* function, used at each signaling point to determine the outgoing signaling link on which a message is to be sent toward its destination point;
- The *message discrimination* function, used at a signaling point to determine whether or not a received message is destined to that point itself. When the signaling point has the transfer capability, and a message is not destined for it, that message is transferred to the message routing function; and
- The *message distribution* function, used at each signaling point to deliver the received messages (destined to the point itself) to the appropriate user part.

13.7.2.1 Routing Label. The label contained in a signaling message and used by the relevant user part to identify a particular task to which the message refers (e.g., a telephone circuit) is also used by the message transfer part to route the message towards its destination point. The part of the message that is used for routing is called the *routing label*, and it contains the information necessary to deliver the message to its destination point. Normally the routing label is common to all services and applications in a given signaling network, national or international. (However, if this is not the case, the particular routing label of a message is determined by means of the service indicator.) The standard routing label should be used in the international signaling network and is applicable in national applications. The standard routing label is 32 bits long and is placed at the beginning of the signaling information field (SIO). Its structure is illustrated in Figure 13.6.

The *destination point code* (DPC) indicates the destination of the message. The *originating point code* (OPC) indicates the originating point of the message. The coding of these codes is pure binary. Within each fold, the least significant bit occupies the first position and is transmitted first.

A unique numbering scheme for the coding of the fields is used for the signaling points of the international network irrespective of the user parts connected to each signaling point. The *signaling link selection* (SLS) field is used, where appropriate, in performing load sharing. This field exists in all types of messages and always in the same

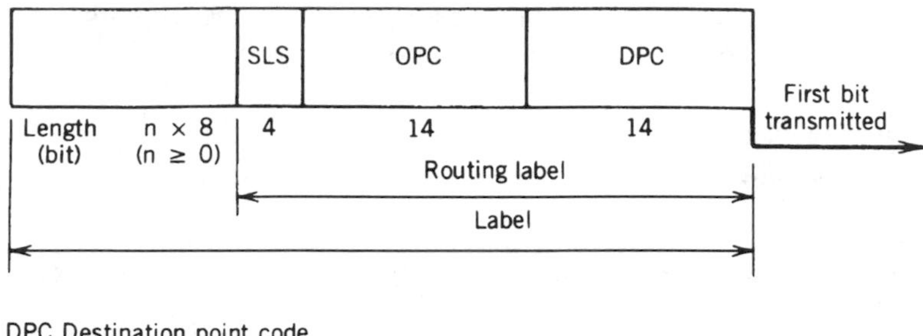

DPC Destination point code
OPC Originating print code
SLS Signaling link selection

Figure 13.6 Routing label structure. (Based on Figure 3/Q.704, p. 5, CCITT Rec. Q.704 [Ref. 8].)

position. The only exception to this rule is some message transfer part layer 3 messages (e.g., changeover order) for which the message routing function in the signaling point of origin of the message is not dependent on the field. In this particular case the field does not exist as such, but is replaced by other information (e.g., in the case of the changeover order, the identity of the faulty link).

In the case of circuit-related messages of the TUP, the field contains the least significant bits of the circuit identification code [or the bearer identification code in the case of the data user part (DUP)], and these bits are not repeated elsewhere. In the case of all other user parts, the SLS is an independent field. In these cases it follows that the signaling link selection of messages generated by any user part will be used in the load-sharing mechanism. As a consequence, in the case of the user parts which are not specified (e.g., transfer of charging information) but for which there is a requirement to maintain order of transmission of messages, the field is coded with the same value for all messages belonging to the same transaction, sent in a given direction.

In the case of message transfer part layer 3 messages, the signaling link selection field exactly corresponds to the signaling link code (SLC) which indicates the signaling link between destination point and originating point to which the message refers.

13.8 SIGNALING NETWORK STRUCTURE

13.8.1 Introduction

In this section several aspects in the design of signaling networks are treated. These networks may be national or international. The national and international networks are considered to be structurally independent and, although a particular signaling point (SP) may belong to both networks, SPs are allocated *signaling point codes* according to the rules of each network.

Signaling links are basic components in a signaling network connecting signaling points. The signaling links encompass layer 2 functions that provide for message error control. In addition, provision for maintaining the correct message sequence is provided.

Signaling links connect signaling points at which signaling network functions such as message routing are provided at layer 3 and at which the user functions may be provided at layer 4 if it is also an originating or destination point. An SP that *only* transfers messages from one signaling link to another at level 3 serves as a signaling transfer point (STP). The signaling links, STPs, and signaling (originating or destination) points may be combined in many different ways to form a *signaling network.*

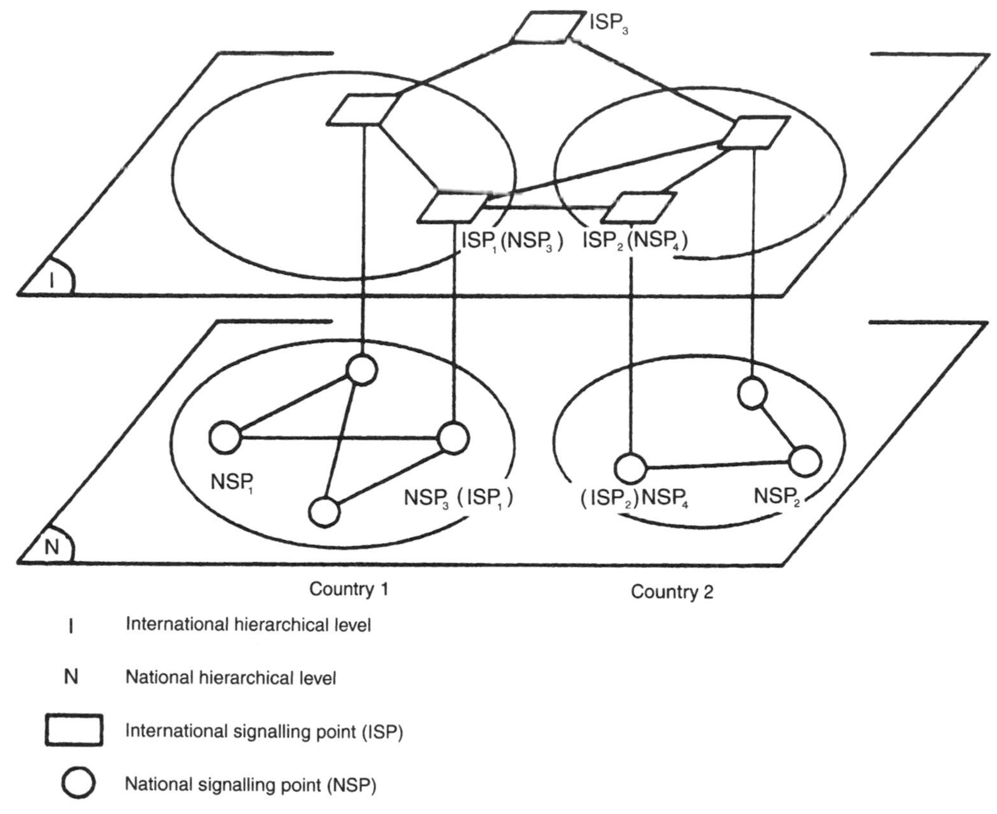

Figure 13.7 International and national signaling networks. (From ITU-T Rec. Q.705, Figure 1/Q.705, p. 2 [Ref. 9].)

13.8.2 International and National Signaling Networks

The worldwide signaling network is structured into two functionally independent levels: international and national as shown in Figure 13.7. Such a structure allows a clear division of responsibility for signaling network management and permits numbering plans of signaling points of the international network and the different national networks to be independent of one another.

An SP including an STP, may be assigned to one of three categories:

1. National signaling point (NSP) (an STP), which belongs to the national signaling network (e.g., NSP_1) and is identified by a signaling point code (OPC or DPC) according to the national numbering plan for signaling points.
2. International signaling point (ISP) (an STP), which belongs to the international signaling network (e.g., ISP_3) and is identified by a signaling point code (OPC or DPC) according to the international numbering plan for signaling points.
3. A node that functions both as an international signaling point (STP) and a national signaling point (STP), and therefore belongs to both the international signaling network and a national signaling network and accordingly is identified by a specific signaling point code (OPC or DPC) in each of the signaling networks.[3]

[3]OPC and DPC are discussed in Section 13.7.

If discrimination between international and national signaling point codes is necessary at a signaling point, the network indicator is used.

13.9 SIGNALING PERFORMANCE: MESSAGE TRANSFER PART

13.9.1 Basic Performance Parameters

ITU-T Rec. Q.706 (Ref. 10) breaks down SS No. 7 performance into three parameter groups:

1. Message delay;
2. Signaling traffic load; and
3. Error rate.

Consider the following parameters and values:

Availability. The unavailability of a signaling route set should not exceed 10 min per year.

Undetected Errors. Not more than 1 in 10^{10} of all signal unit errors will go undetected in the message transfer part.

Lost Messages. Not more than 1 in 10^7 messages will be lost due to failure of the message transfer part.

Messages out of Sequence. Not more than 1 in 10^{10} messages will be delivered out of sequence to the user part due to failure in the message transfer part. This includes message duplication.

13.9.2 Traffic Characteristics

Labeling Potential. There are 16,384 identifiable signaling points.

Loading Potential. Loading potential is restricted by the following four factors:

1. Queuing delay;
2. Security requirements (redundancy with changeover);
3. Capacity of sequence numbering (127 unacknowledged signal units); and
4. Signaling channels using bit rates under 64 kbps.

13.9.3 Transmission Parameters

The message transfer part operates satisfactorily with the following error performance:

- Long-term error rate on the signaling data links of less than 1×10^{-6}; and
- Medium-term error rate of less than 1×10^{-4}.

13.9.4 Signaling Link Delays over Terrestrial and Satellite Links

Data channel propagation time depends on data rate (i.e., this reduces transmission time, thus transmitting a data message at 64 kbps requires half the time compared to 32 kbps),

Table 13.2 Calculated Terrestrial Transmission Delays for Various Call Distances

	Delay Terrestrial (ms)		
Arc Length (km)	Wire	Fiber	Radio
500	2.4	2.50	1.7
1,000	4.8	5.0	3.3
2,000	9.6	10.0	16.6
5,000	24.0	25.0	16.5
10,000	48.0	50.0	33.0
15,000	72.0	75.0	49.5
17,737	85.1	88.7	58.5
20,000	96.0	100.0	66.0
25,000	120.0	125.0	82.5

the distance between nodes, repeater spacing, and the delays in the repeaters and in switches. Data rate (in bps) and repeater delays depend on the type of medium used to transmit messages.[4] The velocity of propagation of the medium is a most important parameter. Table 13.2 provides information of delays for three types of transmission media and for various call distances.

Although propagation delay in most circumstances is the greatest contributor to overall delay, processing delays must also be considered. These are a function of the storage requirements and processing times in SPs, STPs, number of SPs/STPs, signaling link loading and message length mix. (Ref. 11). Table 13.3 provides data on maximum overall signaling delays.

13.10 NUMBERING PLAN FOR INTERNATIONAL SIGNALING POINT CODES

The number plan described in ITU-T Rec. Q.708 (Ref. 12) has no direct relationship with telephone, data, or ISDN numbering. A 14-bit binary code is used for identification

Table 13.3 Maximum Overall Signaling Delays

		Delay (ms)[a]; Message Type	
Country Size	Percent of Connections	Simple (e.g., Answer)	Processing Intensive (e.g., IAM)
Large-size to	50%	1170	1800
Large-size	95%	1450	2220
Large-size to	50%	1170	1800
Average-size	95%	1450	2220
Average-size to	50%	1170	1800
Average-size	95%	1470	2240

[a]The values given in the table are mean values.

Source: ITU-T Rec. Q.709, Table 5/Q.709, p. 5 (Ref. 11).

[4]Remember that the velocity of propagation is a function of the type of transmission medium involved.

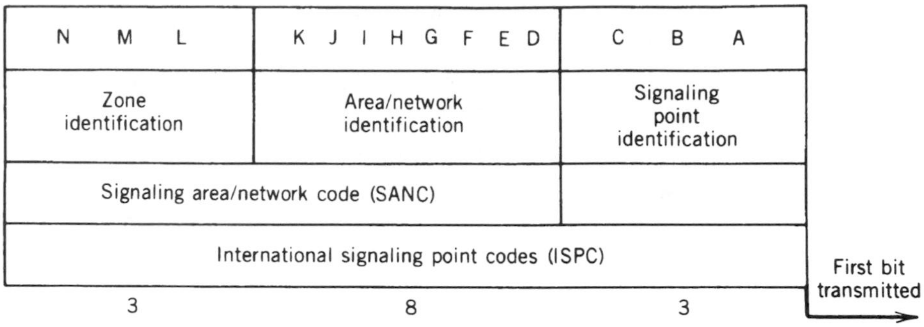

N M L	K J I H G F E D	C B A
Zone identification	Area/network identification	Signaling point identification
Signaling area/network code (SANC)		
International signaling point codes (ISPC)		
3	8	3

First bit transmitted →

Figure 13.8 Format for international signaling point code (ISPC). (From ITU-T Rec. Q.708, Figure 1/Q.708, p. 1 [Ref. 12].)

of signaling points. An international signaling point code (ISPC) is assigned to each signaling point in the international signaling network. The breakdown of these 14 bits into fields is shown in Figure 13.8. The assignment of signaling network codes is administered by the ITU Telecommunication Standardization Sector (previously CCITT).

All ISPCs consist of three identical subfields, as shown in Figure 13.8. The world geographical zone is identified by the N-M-L field consisting of 3 bits. A geographical area or network in a specific zone is identified by the 8-bit field K through D.[5] The 3-bit subfield C-B-A identifies a signaling point in a specific geographical area or network. The combination of the first and second subfields is called a *signaling area/network code* (SANC).

Each country (or geographical area) is assigned at least one SANC. Two of the zone identifications, namely, 1 and 0 codes, are reserved for future allocation.

The ISPC system provides for $6 \times 256 \times 8$ (12,288) ISPCs. If a country or geographical area should require more than 8 international signaling points, one or more additional signaling area/network code(s) would be assigned to it by the ITU-T organization.

A list of SANCs and their corresponding countries can be found in Annex A to ITU-T Rec. Q.708 (Ref. 12). The first number of the code identifies the zone. For example, zone 2 is Europe and zone 3 is North America and its environs.

13.11 SIGNALING CONNECTION CONTROL PART (SCCP)

13.11.1 Introduction

The signaling connection control part (SCCP) provides additional functions to the message transfer part (MTP) for both connectionless and connection-oriented network services to transfer circuit-related and noncircuit-related signaling information between switches and specialized centers in telecommunication networks (such as for management and maintenance purposes) via a Signaling System No. 7 network.

Turn back to Figure 13.3 to see where the SCCP appears in a functional block diagram of an SS No. 7 terminal. It is situated above the MTP in level 4 with the user parts. The MTP is transparent and remains unchanged when SCCP services are incorporated in an SS No. 7 terminal. However, from an OSI perspective, the SCCP carries out the network layer function.

[5]Note here that the alphabet is running backwards, thus K, J, I, H, G . . . D.

The overall objectives of the SCCP are to provide the means for:

- Logical signaling connections within the Signal System No. 7 network; and
- A transfer capability for network service signaling data units (NSDUs) with or without the use of logical signaling connections.

Functions of the SCCP are also used for the transfer of circuit-related and call-related signaling information of the ISDN user part (ISUP) with or without setup of end-to-end logical signaling connections.

13.11.2 Services Provided by the SCCP

The overall set of services is grouped into:

- Connection-oriented services; and
- Connectionless services.

Four classes of service are provided by the SCCP protocol, two for connectionless services and two for connection-oriented services. The four classes are:

0 Basic connectionless class;
1 Sequenced connectionless class;
2 Basic connection-oriented class; and
3 Flow control connection-oriented class.

For connection-oriented services, a distinction has to be made between temporary signaling connections and permanent signaling connections.

Temporary signaling connection establishment is initiated and controlled by the SCCP user. Temporary signaling connections are comparable with dialed telephone connections.

Permanent signaling connections are established and controlled by the local or remote O&M function or by the management function of the node and they are provided for the SCCP user on a semipermanent basis.[6] They can be compared with leased telephone lines.

13.11.3 Peer-to-Peer Communication

The SCCP protocol facilitates the exchange of information between two peers of the SCCP. The protocol provides the means for:

- Setup of logical signaling connection;
- Release of logical signaling connections; and
- Transfer of data with or without logical signaling connections.

13.11.4 Connection-Oriented Functions: Temporary Signaling Connections

13.11.4.1 Connection Establishment. The following are the principal functions used in the connection establishment phase by the SCCP to set up a signaling connection.

[6] O&M stands for operations and maintenance.

- Setup of a signaling connection;
- Establishment of the optimum size of NPDUs (network protocol data units);
- Mapping network address onto signaling relations;
- Selecting operational functions during data-transfer phase (e.g., layer service selection);
- Providing means to distinguish network connections; and
- Transporting user data (within the request).

13.11.4.2 Data-Transfer Phase. The data-transfer phase functions provide the means of a two-way simultaneous transport of messages between two end-points of a signaling connection. The principal data transport phase functions are listed as follows. These are used or not used in accordance with the result of the selection function performed in the connection-establishment phase.

- Segmenting/reassembling;
- Flow control;
- Connection identification;
- NSDU delimiting (M-bit);
- Expedited data;
- Missequence detection;
- Reset;
- Receipt confirmation; and
- Others.

13.11.4.3 Connection Release Functions. Release functions disconnect the signaling connection regardless of the current phase of the connection. The release may be performed by an upper-layer stimulus or by maintenance of the SCCP itself. The release can start at each end of the connection (symmetric procedure). Of course, the principal function of this phase is disconnection.

13.11.5 Structure of the SCCP

The basic structure of the SCCP is illustrated in Figure 13.9. It consists of four functional blocks as follows:

1. *SCCP Connection-Oriented Control.* This controls the establishment and release of signaling connections for data transfer on signaling connections.
2. *SCCP Connectionless Control.* This provides the connectionless transfer of data units.
3. *SCCP Management.* This functional block provides the capability, in addition to the signal route management and lower control functions of the MTP, to handle the congestion or failure of either the SCCP user or signaling route to the SCCP user.
4. *SCCP Routing.* On receipt of the message from the MTP or from the functions listed previously, SCCP routing either forwards the message to the MTP for transfer or passes the message to the functions listed. A message whose called party address is a local user is passed to functions 1, 2, or 3, whereas one destined for a remote user is forwarded to the MTP for transfer to the distant SCCP. (Ref. 13)

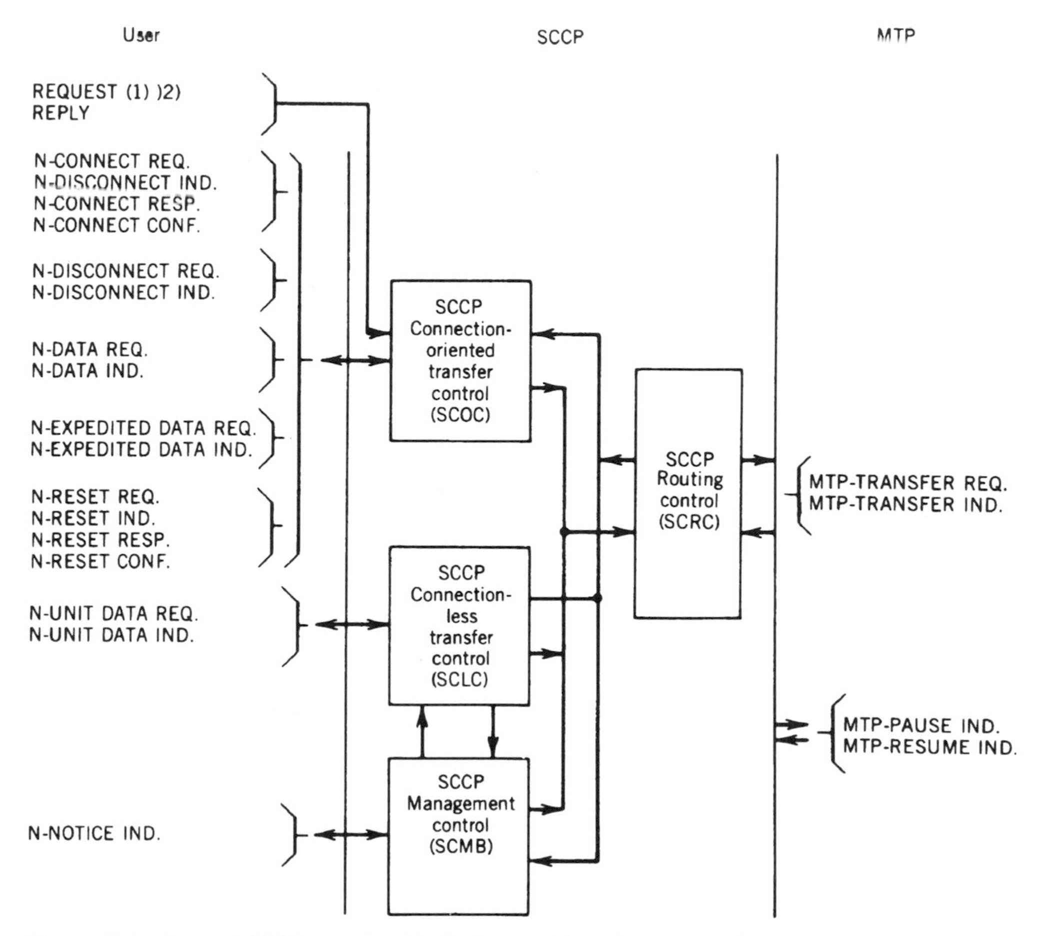

Figure 13.9 General SCCP overview block diagram. Note the listing under user, left side; we find a listing of SCCP primitives. For a discussion of primitives consult Ref. 18. (From ITU-T Rec. Q.714, Figure 1/Q.714, p. 307, [Ref. 15].)

13.12 USER PARTS

13.12.1 Introduction

SS No. 7 user parts, along with the routing label, carry out the basic signaling functions. Turn again to Figure 13.5. There are two fields in the figure we will now discuss: the SIO (service information octet) and the SIF (signaling information field). In the paragraphs that follow we briefly cover one of the user parts, the TUP (telephone user part). As shown in Figure 13.10, the user part, OSI layer 4, is contained in the signaling information field to the left of the routing label. ITU-T Rec. Q.723 (Ref. 14) deals with the sequence of three sectors (fields and subfields of the standard basic message signal unit shown in Figure 13.5).

Turning now to Figure 13.10, we have from right to left, the SIO, the routing label, and the user information subfields (after the routing label in the SIF). The SIO is an octet in length made up of two subfields: the service indicator (4 bits) and the subservice field (4 bits). The service indicator, being 4 bits long, has 16 bit combinations with the following meanings (read from right to left):

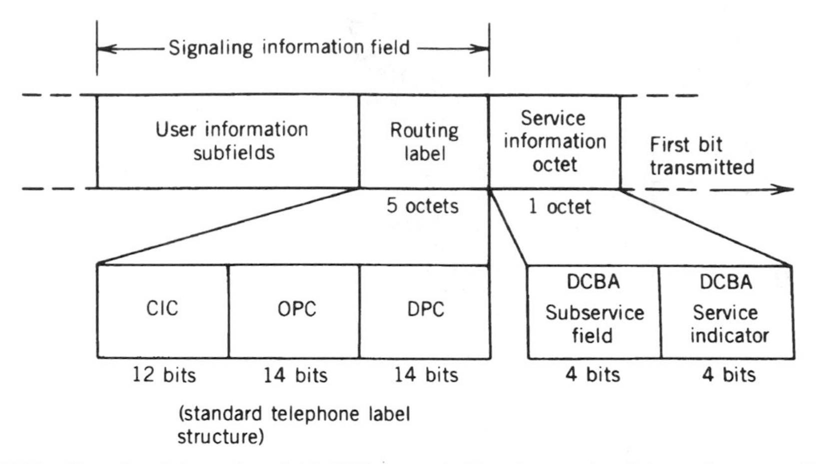

Figure 13.10 Signaling information field (SIF) preceded by the service information octet (SIO). The sequence runs from right to left with the least significant bit transmitted first. DPC = destination point code. OPC = originating point code. CIC = circuit identification code.

BITS DCBA	MEANING
0000	Signaling network management message
0001	Signaling network testing and maintenance
0010	Spare
0011	SCCP
0100	Telephone user part
0101	ISDN user part
0110	Data user part (call- and circuit-related message)
0111	Data user part (facility registration and cancellation)
Remainder (8 sequences)	Spare

The SIO directs the signaling message to the proper layer 4 entity, whether SCCP or user part. This is called *message distribution.*

The subservice indicator contains the network bits C and D and two spare bits, A and B. The network indicator is used by signaling message-handling functions determining the relevant version of the user part. If the network indicator is set at 00 or 01, the two spare bits, coded 00, are available for possible future needs. If these two bits are coded 10 or 11, the two spare bits are for national use, such as message priority as an optional flow procedure. The network indicator provides discrimination between international and national usage (bits D and C).

The routing label forms part of every signaling message:

- To select the proper signaling route; and
- To identify the particular transaction by the user part (the call) to which the message pertains.

The label format is shown in Figure 13.10). The DPC is the destination point code (14 bits), which indicates the signaling point for which the message is intended. The originating point code (OPC) indicates the source signaling point. The circuit identifica-

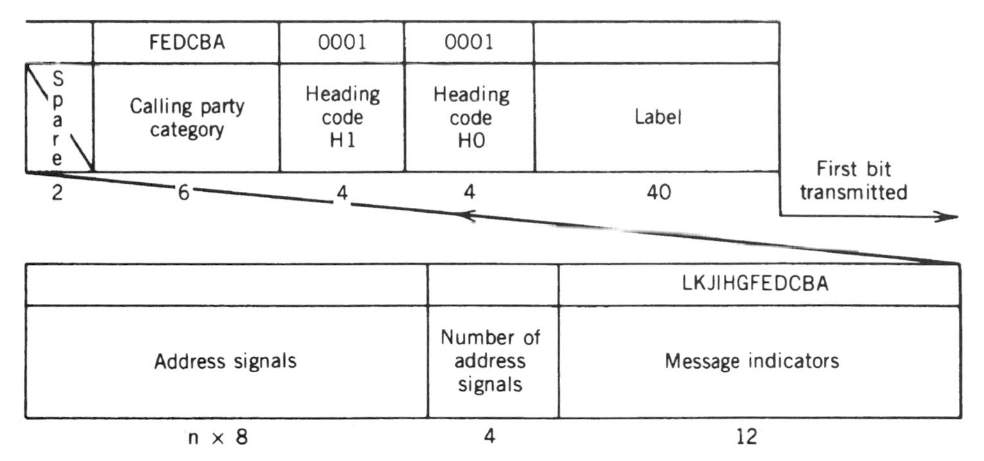

Figure 13.11 Initial address message format. (From CCITT Rec. Q.723, Figure 3/Q.723, p. 23 [Ref. 14].)

tion code (CIC) indicates the one circuit (speech circuit in the TUP case) among those directly interconnecting the destination and originating points.

For the OPC and DPC, unambiguous identification of signaling points is carried out by means of an allocated code. Separate code plans are used for the international and national networks. The CIC, as shown in Figure 13.10, is applicable only to the TUP. CCITT Rec. Q.704 shows a signaling link selection (SLS) field following (to the left) the OPC. The SLS is 4 bits long and is used for load sharing. The ISDN user part address structure is capable of handling E.164 addresses in the calling and called number and is also capable of redirecting address information elements.

13.12.2 Telephone User Part (TUP)

The core of the signaling information is carried in the SIF (see Figure 13.10). The TUP label was described briefly in Section 13.7.2.1. Several signal message formats and codes are described in the following paragraphs. These follow the TUP label.

One typical message of the TUP is the initial address message (IAM). Its format is shown in Figure 13.11. A brief description is given of each subfield, providing further insight of how SS No. 7 operates.

Common to all signaling messages are the subfields H0 and H1. These are the heading codes, each consisting of 4 bits, giving 16 code possibilities in pure binary coding. H0 identifies the specific message group to follow. "Message group" means the type of message. Some samples of message groups are:

MESSAGE GROUP TYPE	H0 CODE
Forward address messages	0001
Forward setup messages	0010
Backward setup messages	0100
Unsuccessful backward setup messages	0101
Call supervision messages	0110
Node-to-node messages	1001

H1 contains a signal code or identifies the format of more complex messages. For instance, there are four types of address message identified by H0 = 0001, and H1 identifies the type of message, such as:

ADDRESS MESSAGE TYPE	H0	H1
Initial address message	0001	0001
IAM with additional information	0001	0010
Subsequent address message	0001	0011
Subsequent address message with signal unit	0001	0100

Moving from right to left in Figure 13.11, after H1 we have the calling party subfield consisting of 6 bits. It identifies the language of the operator (Spanish, English, Russian, etc.). For example, an English-speaking operator is coded 000010. It also differentiates the calling subscriber from one with priority, a data call, or a test call. A data call is coded 001100 and a test call is coded 001101. Fifty of the 64 possible code groups are spare.

Continuing to the left in Figure 13.11, 2 bits are spare for international allocation. Then there is the message indicator, where the first 2 bits, B and A, give the nature of the address. This is information given in the forward direction indicating whether the associated address or line identity is an international, national (significant), or subscriber number. A subscriber number is coded 00, an international number is coded 11, and a national (significant) number is coded 10.

Bits D and C are the circuit indicator. The code 00 in this location indicates that there is no satellite circuit in the connection. Remember that the number of space satellite relays in a speech telephone connection is limited to one relay link through a satellite because of propagation delay.

Bits F and E are significant for common-channel signaling systems such as CCIS, CCS No. 6, and SS No. 7. The associated voice channel operates on a separate circuit. Does this selected circuit for the call have continuity? The bit sequence FE is coded:

BITS F AND E	MEANING
00	Continuity check not required
01	Continuity check required on this circuit
10	Continuity check performed on previous circuit
11	Spare

Bit G gives echo suppressor information. When coded 0 it indicates that the outgoing half-echo suppressor is not included, and when coded 1 it indicates that the outgoing half-echo suppressor is included. Bit I is the redirected call indicator. Bit J is the all-digital path required indicator. Bit K tells whether any path may be used or whether only SS No. 7-controlled paths may be used. Bit L is spare.

The next subfield has 4 bits and gives the number of address signals contained in the initial address message. The last subfield contains address signals where each digit is coded by a 4-bit group as follows:

CODE	DIGIT	CODE	DIGIT
0000	0	1000	8
0001	1	1001	9
0010	2	1010	Spare
0011	3	1011	Code 11
0100	4	1100	Code 12
0101	5	1101	Spare
0110	6	1110	Spare
0111	7	1111	ST

The most significant address signal is sent first. Subsequent address signals are sent in successive 4-bit fields. As shown in Figure 13.11, the subfield contains *n* octets. A filler code of 0000 is sent to fill out the last octet, if needed. Recall in Chapter 7 that the ST signal is the "end of pulsing" signal and is often used on semiautomatic circuits.

Besides the initial address message, there is the subsequent address message used when all address digits are not contained in the IAM. The subsequent address message is an abbreviated version of the IAM. There is a third type of address message, the initial address message with additional information. This is an extended IAM providing such additional information as network capability, user facility data, additional routing information, called and calling address, and closed user group (CUG). There is also the forward setup message, which is sent after the address messages and contains further information for call setup.

CCITT SS No. 7 is rich with backward information messages. In this group are backward setup request; successful backward setup information message group, which includes charging information; unsuccessful backward setup information message group, which contains information on unsuccessful call setup; call supervision message group; circuit supervision message group; and the node-to-node message group (CCITT Recs. Q.722 and Q.723 (Refs. 14, 16).

Label capacity for the telephone user part is given in CCITT Rec. Q.725 (Ref. 17) as 16,384 signaling points and up to 4096 speech circuits for each signaling point.

REVIEW EXERCISES

1. What is the principal rationale for developing and implementing Signaling System No. 7?

2. Describe SS No. 7 and its relationship with OSI. Why does it truncate at OSI layer 4?

3. Give the two primary "parts" of SS No. 7. Briefly describe each part.

4. OSI layer 4 is subdivided into two sublayers. What are they?

5. Layer 2 carries out the functions of the signaling link. Name four of the five functions of layer 2.

6. What are the two basic categories of functions of layer 3?

7. What is a signaling relation?

8. How are signaling points defined (identified)?

9. What does a routing label do?
10. What are the two methods of error correction in SS No. 7?
11. What are the three types of signal units used in SS No. 7?
12. Discuss forward and backward sequence numbers.
13. The routing label is analogous to what in our present telephone system? Name the three basic pieces of information that the routing label provides.
14. Define labeling potential.
15. What is the function of an STP? Differentiate STP with signaling point.
16. Why do we wish to limit the number of STPs in a specific relation?
17. What are the three measures of performance of SS No. 7?
18. Regarding user parts, what is the function of the SIO? Of the network indicator?
19. What is the purpose of circuit continuity?
20. With the TUP, address signals are sent digit-by-digit embedded in the last subfield of the SIF. How are they represented?
21. What are the two overall set of services of the SCCP?
22. What is the purpose of the SCCP?

REFERENCES

1. *Specifications of Signaling System No. 7* (Q.700 Series), Fascicle VI.6, CCITT Yellow Books, VIIth Plenary Assembly, Geneva, 1980.
2. W. C. Roehr, Jr., "Signaling System No. 7," in *Tutorial: Integrated Services Digital Network (ISDN)*, W. Stallings, ed., IEEE Computer Society, Washington, DC, 1985.
3. *Introduction to CCITT Signaling System No. 7*, ITU-Rec. Q.700, ITU, Helsinki, 1993.
4. *Functional Description of the Message Transfer Part (MTP) of Signaling System No. 7*, ITU-T Rec. Q.701, ITU, Helsinki, 1993.
5. R. L. Freeman, *Reference Manual for Telecommunication Engineering*, 2nd ed., Wiley, New York, 1994.
6. *BOC Notes on the LEC Networks—1994*, Bellcore Special Report SR-TSV-002275, Issue 2, Piscataway, NJ, April 1994.
7. *Signaling System No. 7—Signaling Link*, ITU-T Rec. Q.703, ITU, Helsinki, 1993.
8. *Signaling System No. 7—Signaling Network Functions and Messages*, ITU-T Rec. Q.704, ITU, Geneva, 1996.
9. *Signaling System No. 7—Signaling Network Structure*, ITU-T Rec. Q.705, ITU, Geneva, 1993.
10. *Signaling System No. 7—Message Transfer Part Signaling Performance*, ITU-T Rec. Q.706, ITU, Geneva, 1993.
11. *Signaling System No. 7—Hypothetical Signaling Reference Connection*, ITU-T Rec. Q.709, Helsinki, 1993.
12. *Signaling System No. 7—Numbering of International Signaling Point Codes*, ITU-T Rec. Q.708, Helsinki, 1993.

13. *Signaling System No. 7—Functional Description of the Signaling Connection Control Part (SCCP)*, ITU-T Rec. Q.711, ITU, Helsinki, 1993.
14. *Formats and Codes (Telephone User Part)*, CCITT Rec. Q.723, Fascicle VI.8, IXth Plenary Assembly, Melbourne, 1988.
15. *Signaling System No. 7—Signaling Connection Control Part Procedures*, ITU-T Rec. Q.714, ITU, Geneva, 1996.
16. *Signaling System No. 7—General Function of Telephone Messages and Signals*, CCITT Rec. Q.722, Fascicle VI.8, IXth Plenary Assembly, Melbourne, 1988.
17. *Signaling System No. 7—Signaling Performance in the Telephone Application*, ITU-T Rec. Q.725, ITU, Helsinki, 1993.
18. Roger L. Freeman, *Telecommunication System Engineering*, 3rd ed., Wiley, New York, 1996.

14

TELEVISION TRANSMISSION

14.1 BACKGROUND AND OBJECTIVES

Television was developed prior to World War II. However, it did not have any notable market penetration until some years after the war. This was monochrome television. Color television began to come on the market about 1960. The next step in television evolvement was high-definition television (HDTV), and 1998 is considered to be the year when HDTV was launched.

Interfacing standards for television have had a rather unfortunate background. North America, Japan, and much of Latin America follow one standard called NTSC (National Television Systems Committee). The remainder of the world follows a wide variation in standards. For example, there are three different color television standards: NTSC, PAL (phase alternation line), and SECAM (sequential color and memory). These are discussed in Section 14.5.

The television signal, no matter what standard it follows, is a complex analog signal. It is a bandwidth hog, requiring anywhere from 4 MHz to 8 MHz for the video, color subcarrier, and aural (audio) channel(s).

The objectives of this chapter are severalfold. The first is to provide the reader with a clear understanding of how TV works. The second goal is to describe how television is transmitted and distributed over long distances. However, the radio-broadcast of television is not included and is left to other texts. Cable television is covered in Chapter 15. The third goal of the chapter is to provide an overview of digital television, and we cover several generic methods of digitizing original analog television signals.

Our interest in television transmission derives from its impact on the larger telecommunications environment. For example, the PSTN and other carriers are called upon to transport broadcast-quality TV, or to develop and transport a subbroadcast quality TV signal called *conference television*. Conference television is used in the industrial and office environment to facilitate "meetings at a distance." One or more people at location X can meet with one or more people at location Y where attendees at each site can see and hear those at the other site. It can save money on business meetings for travel and lodging.

Current television signals are composed of three parts: (1) the video signal, which is monochrome, (2) the audio subcarrier, and (3) a color subcarrier. We first describe the video signal.

14.2 APPRECIATION OF VIDEO TRANSMISSION

A video transmission system must deal with four factors when transmitting images of moving objects:

1. Perception of the distribution of luminance or simply the distribution of light and shade;
2. Perception of depth or a three-dimensional perspective;
3. Perception of motion relating to the first two factors above; and
4. Perception of color (hues and tints).

Monochrome TV deals with the first three factors. Color TV includes all four factors.

A video transmission system must convert these three (or four) factors into electrical equivalents. The first three factors are integrated to an equivalent electric current or voltage whose amplitude is varied with time. Essentially, at any one moment it must integrate luminance from a scene in the three dimensions (i.e., width, height, and depth) as a function of time. And time itself is still another variable, for the scene is changing in time.

The process of integration of visual intelligence is carried out by *scanning*. The horizontal detail of a scene is transmitted continuously and the vertical detail discontinuously. The vertical dimension is assigned discrete values that become the fundamental limiting factor in a video transmission system.

The scanning process consists of taking a horizontal strip across the image on which discrete square elements called pels or pixels (picture elements) are scanned from left to right. When the right-hand end is reached, another, lower, horizontal strip is explored, and so on, until the whole image has been scanned. Luminance values are translated on each scanning interval into voltage and current variations and are transmitted over the system. The concept of scanning by this means is illustrated in Figure 14.1.

The National Television Systems Committee (U.S.) (NTSC) practice divides an image into 525 horizontal scanning lines.[1] It is the number of scanning lines that determines the vertical detail or resolution of a picture.

When discussing picture resolution, the *aspect ratio* is the width-to-height ratio of the video image (see Section 14.2.1).[2] The aspect ratio used almost universally is 4 : 3.[3] In other words, a TV image 12 in. wide would necessarily be 9 in. high. Thus an image divided into 525 (491) vertical elements would then have 700 (652) horizontal elements to maintain an aspect ratio of 4 : 3. The numbers in parentheses represent the practical maximum active lines and elements. Therefore, the total number of elements approaches something on the order of 250,000. We reach this number because, in practice, the vertical detail reproduced is 64–87% of the active scanning lines. A good halftone engraving may have as many as 14,400 elements per square inch, compared to approximately 3000 elements per square inch for a 9-in. by 12-in. TV image.

Motion is another variable factor that must be transmitted. The sensation of continuous motion, standard TV video practice, is transmitted to the viewer by a successive display of still pictures at a regular rate similar to the method used in motion pictures. The regulate rate of display is called the *frame rate*. A frame rate of 25 frames per second will give the viewer a sense of motion, but on the other hand he/she will be disturbed by luminance flicker (bloom and decay), or the sensation that still pictures are "flicking" on screen one after the other. To avoid any sort of luminance flicker sensation, the image is divided into two closely interwoven (interleaving) parts, and each part is presented in succession at a rate of 60 frames per second, even though *complete* pic-

[1]With most European TV systems this value is 625 lines.

[2]*Resolution* deals with the degree to which closely spaced objects in an image can be distinguished one from another.

[3]High-definition television (HDTV) is a notable exception.

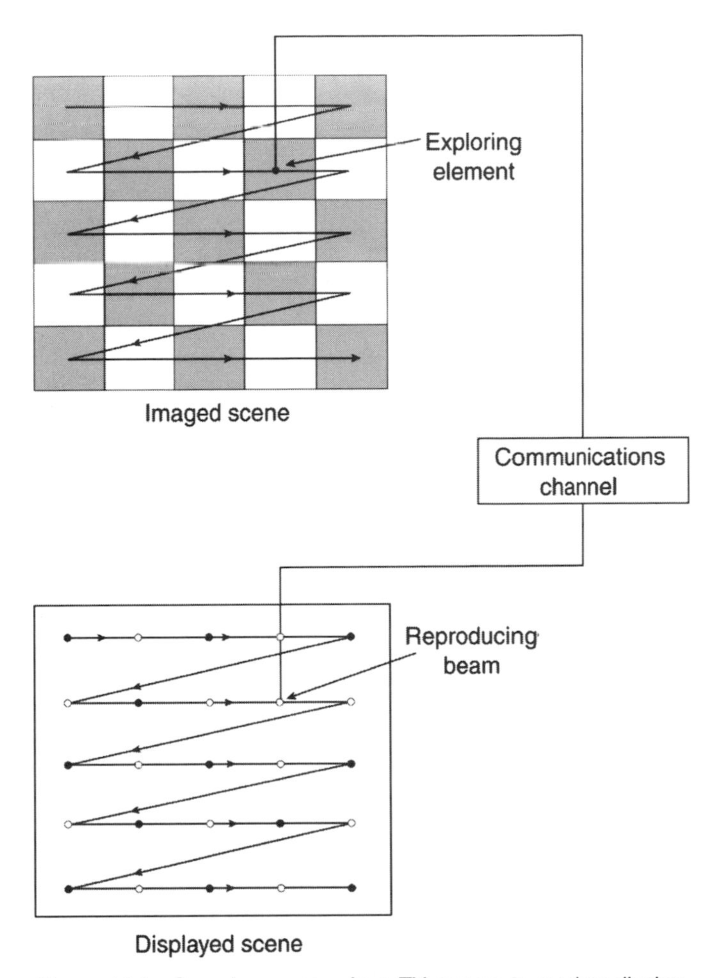

Figure 14.1 Scanning process from TV camera to receiver display.

tures are still built up at a 30 frame-per-second rate. It should be noted that interleaving improves resolution as well as apparent persistence of the cathode ray tube (CRT) by tending to reinforce the scanning spots. It has been found convenient to equate flicker frequency to power line frequency. Hence in North American practice, where power line frequency is 60 Hz, the flicker is 60 frames per second. In Europe it is 50 frames per second to correspond to the 50-Hz line frequency used there.

Following North American practice, some other important parameters derive from the previous paragraphs:

1. A field period is $\frac{1}{60}$ s. This is the time required to scan a full picture on every horizontal line.
2. The second scan covers the lines not scanned on the first period, offset one-half horizontal line.
3. Thus $\frac{1}{30}$ s is required to scan all lines on a complete picture.
4. The transmit time of exploring and reproducing scanning elements or spots along each scanning line is $\frac{1}{15{,}750}$ s (525 lines in $\frac{1}{30}$ s) = 63.5 μs.
5. Consider that about 16% of the 63.5 μs is consumed in flyback and synchro-

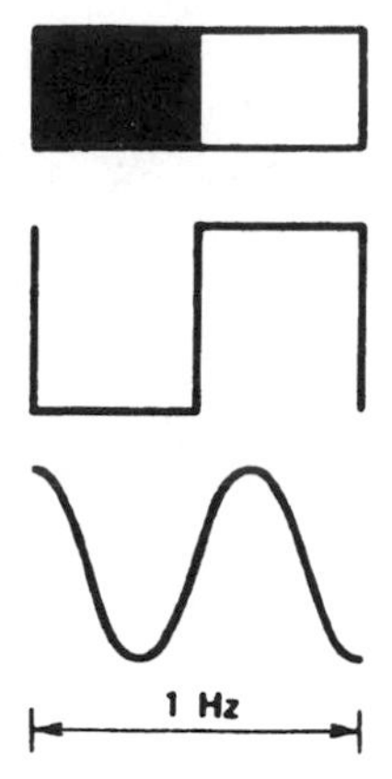

Figure 14.2 Development of a sinusoid wave from the scan of adjacent squares.

nization. Accordingly, only about 53.3 μs are left per line of picture to transmit information.

What will be the bandwidth necessary to transmit images so described? Consider the worst case, where each scanning line is made up of alternate black-and-white squares, each the size of the scanning element. There would be 652 such elements. Scan the picture, and a square wave will result, with a positive-going square for white and a negative for black. If we let a pair of adjacent square waves be equivalent to a sinusoid (see Figure 14.2), then the baseband required to transmit the image will have an upper cutoff frequency of about 6.36 MHz, providing that there is no degradation in the intervening transmission system. The lower limit will be dc or zero frequency.

14.2.1 Additional Definitions

14.2.1.1 Picture Element (pixel or pel). By definition, "The smallest area of a television picture capable of being delineated by an electrical signal passed through the system or part thereof" (Ref. 1), a picture element has four important properties:

1. P_v, the vertical height of the picture element;
2. P_h, the horizontal length of the picture element;
3. P_a, the *aspect ratio* of the picture element; and
4. N_p, the total number of picture elements in an entire picture.

The value of N_p is often used to compare TV systems.

In digital TV, a picture consists of a series of digital values that represent the points along the scanning path of an image. The digital values represent discrete points and we call these pixels (pels).

The resolution of a digital image is determined by its pixel counts, horizontal and vertical. A typical computer picture image might have 640 × 480 pixels.

14.2.1.2 Aspect Ratio. This is the ratio of the frame width to the frame height. This ratio is defined by the active picture. For standard NTSC television and PAL television and computers the aspect ratio is 4 : 3 (1.33 : 1). Wide-screen movies have a 16 : 9 aspect ratio, and HDTV is expected to also use a 16 : 9 aspect ratio.

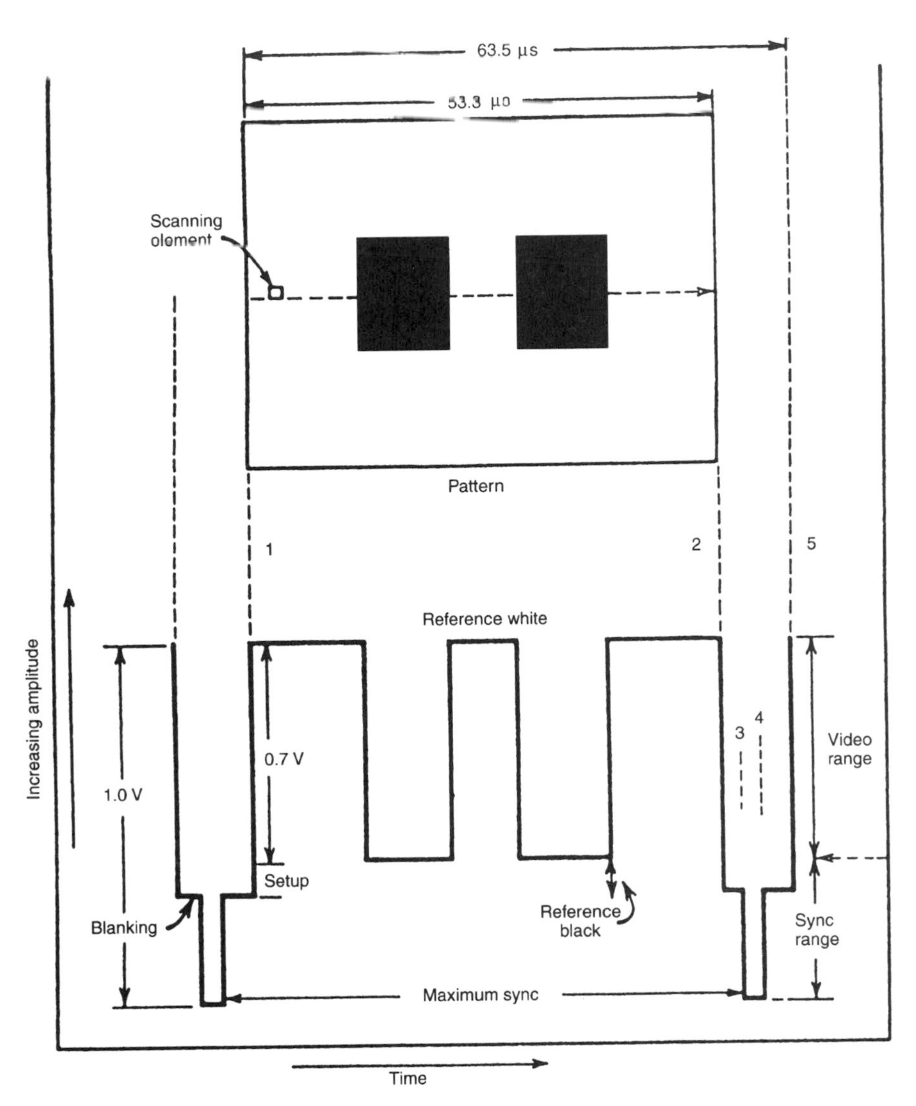

Figure 14.3 Breakdown in time of a scan line.

14.3 COMPOSITE SIGNAL

The word *composite* is confusing in the TV industry. On one hand, composite may mean the combination of the full video signal plus the audio subcarrier; the meaning here is narrower. Composite in this case deals with the transmission of video information as well as the necessary synchronizing information.

Consider Figure 14.3. An image that is made up of two black squares is scanned. The total time for the scan line is 63.5 μsec, of which 53.3 μsec are available for the transmission of actual video information and 10.2 μsec are required for synchronization and flyback.[4]

During the retrace time or flyback it is essential that no video information be trans-

[4]*Flyback* is defined by the IEEE (Ref. 1) as "the rapid return of a beam in a cathode-ray tube in the direction opposite to that of scanning." Flyback is shown in Figure 14.1, where the beam moves left returning to the left side of the screen.

mitted. To accomplish this, a blanking pulse carries the signal voltage into the reference black region. Beyond this region in amplitude is the *blacker than black* region, which is allocated to the synchronizing pulses. The blanking level (pulse) is shown in Figure 14.3.

The maximum signal excursion of a composite video signal is 1.0 V. This 1.0 V is a video/TV reference and is always taken as a peak-to-peak measurement. The 1.0 V may be reached at maximum synchronizing voltage and is measured between synchronizing "tips."

Of the 1.0-V peak, 0.25 V is allotted for the synchronizing pulses and 0.05 V for the setup, leaving 0.7 V to transmit video information. Therefore the video signal varies from 0.7 V for the white-through-gray tonal region to 0 V for black. The best way to describe the actual video portion of a composite signal is to call it a succession of rapid nonrepeated transients.

The synchronizing portion of a composite signal is exact and well defined. A TV/video receiver has two separate scanning generators to control the position of the reproducing spot. These generators are called the horizontal and vertical scanning generators. The horizontal one moves the spot in the *X* or horizontal direction, and the vertical in the *Y* direction. Both generators control the position of the spot on the receiver and must, in turn, be controlled from the camera (transmitter) synchronizing generator to keep the receiver in step (synchronization).

The horizontal scanning generator in the video receiver is synchronized with the camera synchronizing generator at the end of each scanning line by means of horizontal synchronizing pulses. These are the synchronizing pulses shown in Figure 14.3, and they have the same polarity as the blanking pulses.

When discussing synchronization and blanking, we often refer to certain time intervals. These are described as follows:

- The time at the horizontal blanking pulse, 2–5 in Figure 14.3, is called the *horizontal synchronizing interval*;
- The interval 2–3 in Figure 14.3 is called the *front porch*;
- The interval 4–5 is called the *back porch.*

The intervals are important because they provide isolation for overshoots of video at the end of scanning lines. Figure 14.4 illustrates the horizontal synchronizing pulses and corresponding porches.

The vertical scanning generator in the video/TV receiver is synchronized with the camera (transmitter) synchronizing generator at the end of each field by means of vertical synchronizing pulses. The time interval between successive fields is called the vertical interval. The vertical synchronizing pulse is built up during this interval. The scanning generators are fed by differentiation circuits. Differentiation for the horizontal scan has a relatively short time constant (*RC*) and that for the vertical a comparatively long time constant. Thus the long-duration vertical synchronization may be separated from the comparatively short-duration horizontal synchronization. This method of separation of synchronization, known as *waveform separation*, is standard in North America.

In the composite video signal (North American standards) the horizontal synchronization has a repetition rate of 15,750 frames per second, and the vertical synchronization has a repetition rate of 60 frames per second (Refs. 2, 3).

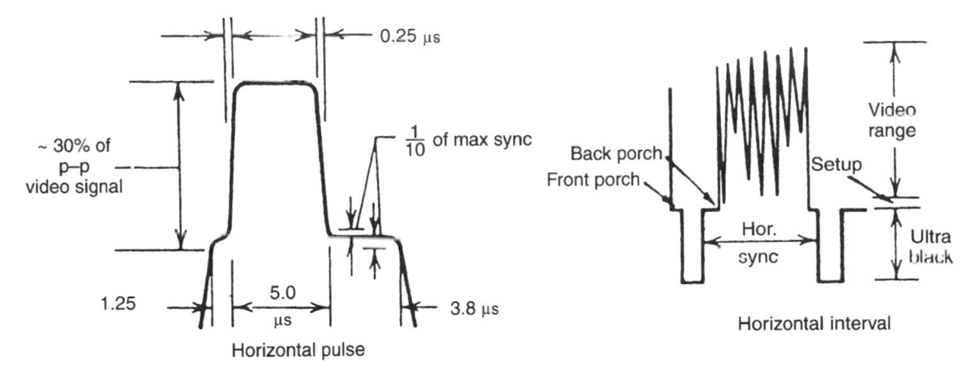

Figure 14.4 Sync pulses and porches.

14.4 CRITICAL VIDEO PARAMETERS

14.4.1 General

Raw video baseband transmission requires excellent frequency response, in particular, from dc to 15 kHz and extending to 4.2 MHz for North American systems and to 5 MHz for European systems. Equalization is extremely important. Few point-to-point circuits are transmitted at baseband because transformers are used for line coupling, which deteriorate low-frequency response and make phase equalization very difficult.

To avoid low-frequency deterioration, cable circuits transmitting video have resorted to the use of carrier techniques and frequency inversion using vestigial sideband (VSB) modulation. However, if raw video baseband is transmitted, care must be taken in preserving its dc component (Ref. 4).

14.4.2 Transmission Standard Level

Standard power levels have developed from what is roughly considered to be the input level to an ordinary TV receiver for a noise-free image. This is 1 mV across 75 Ω. With this as a reference, TV levels are given in dBmV. For RF and carrier systems carrying video, the measurement refers to rms voltage. For raw video it is 0.707 of instantaneous peak voltage, usually taken on synchronizing tips.

The signal-to-noise ratio is normally expressed for video transmission as

$$\frac{S}{N} = \frac{\text{peak signal (dBmV)}}{\text{rms noise (dBmV)}}. \tag{14.1}$$

The Television Allocation Systems Organization (TASO) picture ratings (4-MHz bandwidth) are related to the signal-to-noise ratio (RF) as follows (Ref. 5):

1. Excellent (no perceptible snow) 45 dB
2. Fine (snow just perceptible) 35 dB
3. Passable (snow definitively perceptible but not objectionable) 29 dB
4. Marginal (snow somewhat objectionable) 25 dB

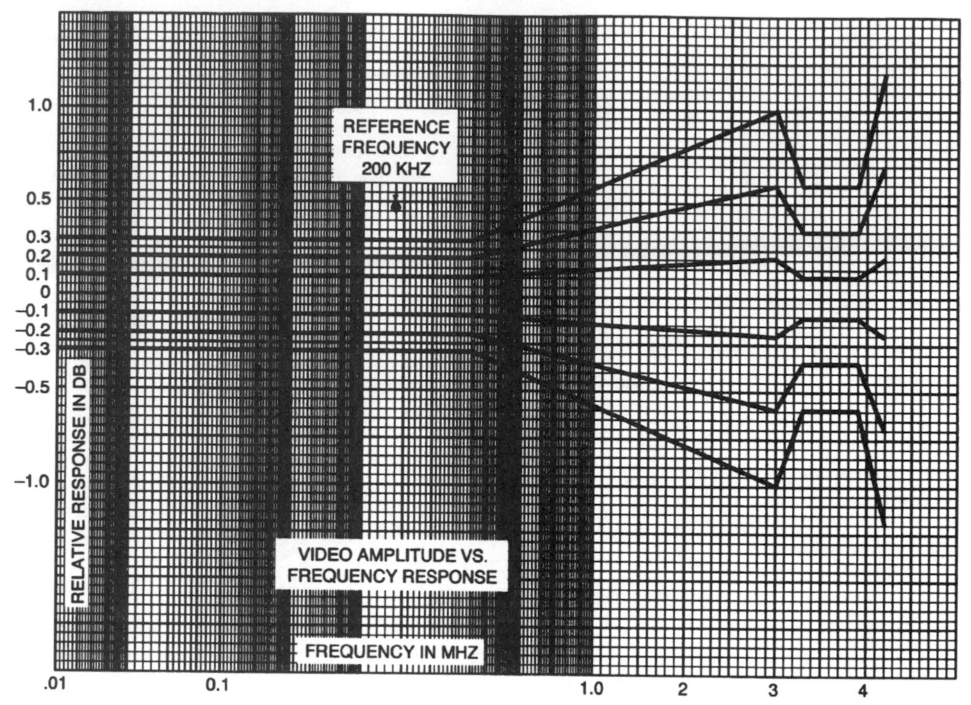

Figure 14.5 Video amplitude–frequency response. (From Ref. 6, EIA-250C. Courtesy of Electronic Industries Association/Telecommunication Industry Association. Reprinted with permission.)

14.4.3 Other Parameters

For black-and-white video systems there are four critical transmission parameters:

1. Amplitude-frequency response (see Figure 14.5).
2. EDD (group delay);
3. Transient response; and
4. Noise (thermal, intermodulation distortion [IM], crosstalk, and impulse).

Color transmission requires consideration of two additional parameters:

5. Differential gain; and
6. Differential phase.

A description of amplitude–frequency response (attenuation distortion) may be found in Section 3.3.1. Because video transmission involves such wide bandwidths compared to the voice channel and because of the very nature of video itself, both phase and amplitude requirements are much more stringent.

Transient response is the ability of a system to "follow" sudden, impulsive changes in signal waveform. It usually can be said that if the amplitude–frequency and phase characteristics are kept within design limits, the transient response will be sufficiently good.

Noise and signal-to-noise ratio are primary parameters for video transmission. Of course, noise is an impairment and is described in Section 3.3.3. Signal-to-noise ratio is the principal measure of video signal quality. Signal-to-noise ratio is defined in Section 3.2.1.

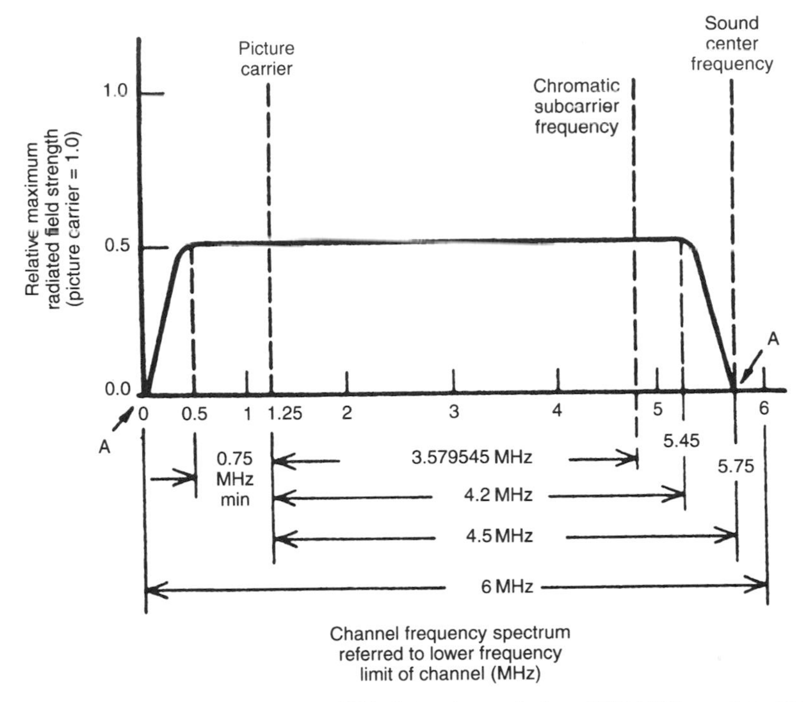

Figure 14.6 RF amplitude characteristics of TV picture transmission, NTSC/US practice. Field strength at points A shall not exceed 20 dB below the picture carrier. Drawing not to scale.

Differential gain is a parameter especially critical for the transmission of color television. It describes how system gain varies as the video signal level varies (i.e., as it traverses the extremes from black to white). *Differential phase* is another parameter that is also critical for the transmission of color television. It is any variation in phase of the color subcarrier as a result of changes in luminance level.[5] Ideally, variation in the luminance level should produce no changes in either amplitude or phase of the color subcarrier.

14.5 VIDEO TRANSMISSION STANDARDS (CRITERIA FOR BROADCASTERS)

The following outlines video transmission standards from the point of view of broadcasters (i.e., as emitted from TV transmitters). Figure 14.6 illustrates the components of the emitted wave (North American practice).

Tables 14.1*a* and 14.1*b* give a capsule summary of some national standards as taken from ITU-R Rec. BT.470-3 (Ref. 7).

14.5.1 Color Transmission

Three color transmission standards exist:

1. NTSC—National Television System Committee (North America, Japan, and many Latin American countries);

[5]Note that the color subcarrier carries its color information by phase modulation.

Table 14.1a Television Standards, NTSC/US

Channel width (see Figure 14.6)	
(Transmission)	6 MHz
Video	4.2 MHz
Aural	± 25 kHz
Picture carrier location	1.25 MHz above lower boundary of channel
Modulation	AM composite picture and synchronizing signal on visual carrier together with FM audio signal on audio carrier
Scanning lines	525 per frame, interlaced 2 : 1
Scanning sequence	Horizontally from left to right, vertically from top to bottom
Horizontal scanning frequency	15,750 Hz for monochrome, or 2/455 × chrominance subcarrier = 15,734.264 ± 0.044 Hz for NTSC color transmission
Vertical scanning frequency	60 Hz for monochrome, or 2/525 × horizontal scanning frequency for color = 59.95 Hz
Blanking level	Transmitted at 75 ± 25% of peak carrier level
Reference black level	Black level is separated from blanking level by 7.5 ± 2.5% of video range from blanking level to reference white level
Reference white level	Luminance signal of reference white is 12.5 ± 2.5% of peak carrier
Peak-to-peak variation	Total permissible peak-to-peak variation in one frame due to all causes is less than 5%
Polarity of transmission	Negative; a decrease in initial light intensity causes an increase in radiated power
Transmitter brightness response	For monochrome TV, RF output varies in an inverse logarithmic relation to brightness of scene
Aural transmitter power	Maximum radiated power is 20% (minimum 10%) of peak visual transmitter power

Source: Refs. 7, 8.

2. SECAM—Sequential color and memory (Europe); and
3. PAL—Phase alternation line (Europe).

The systems are similar in that they separate the luminance and chrominance information and transmit the chrominance information in the form of two color difference signals, which modulate a color subcarrier transmitted within the video band of the luminance signal. The systems vary in the processing of chrominance information.

In the NTSC system, the color difference signals I and Q amplitude-modulate subcarriers that are displaced in phase by $\pi/2$, giving a suppressed carrier output. A burst of the subcarrier frequency is transmitted during the horizontal back porch to synchronize the color demodulator.

In the PAL system, the phase of the subcarrier is changed from line to line, which requires the transmission of a switching signal as well as a color burst.

Table 14.1*b* Basic European TV Standard

Channel width	
(Transmission)	7 or 8 MHz
Video	5, 5.5, and 6 MHz
Aural	FM, ±50 kHz
Picture carrier location	1.25 MHz above lower boundary of channel
Note: VSB transmission is used, similar to North American practice.	
Modulation	AM composite picture and synchronizing signal on visual carrier together with FM audio signal on audio carrier
Scanning lines	625 per frame, interlaced 2 : 1
Scanning sequence	Horizontally from left to right, vertically from top to bottom
Horizontal scanning frequency	15,625 Hz ± 0.1%
Vertical scanning frequency	50 Hz
Blanking level	Transmitted at 75 ± 2.5% of peak carrier level
Reference black level	Black level is separated from blanking by 3–6.5% of peak carrier
Peak white level as a percentage of peak carrier	10–12.5%
Polarity of transmission	Negative; a decrease in initial light intensity causes an increase in radiated power
Aural transmitter power	Maximum radiated power is 20% of peak visual power

Source: Refs. 7, 8.

In the SECAM system, the color subcarrier is frequency modulated alternately by the color difference signals. This is accomplished by an electronic line-to-line switch. The switching information is transmitted as a line-switching signal.

14.5.2 Standardized Transmission Parameters (Point-to-Point TV)

These parameters are provided in Table 14.2*a* for the basic electrical interface at video frequencies (baseband) and Table 14.2*b* for the interface at IF for radio systems. Signal-to-weighted noise ratio at output of final receiver is 53 dB.

Table 14.2*a* Interconnection at Video Frequencies (Baseband)

Impedance	75 Ω (unbalanced) or 124 Ω (resistive)
Return loss	No less than 30 dB
Nominal signal amplitude	1 V peak-to-peak (monochrome)
Nominal signal amplitude	1.25 V peak-to-peak, maximum (composite color)
Polarity	Black-to-white transitions, positive going

Table 14.2b Interconnection at Intermediate Frequency (IF)

Impedance	75 Ω unbalanced
Input level	0.3 V rms
Output level	0.5 V rms
IF up to 1 GHz (transmitter frequency)	35 MHz
IF above 1 GHz	70 MHz

14.6 METHODS OF PROGRAM CHANNEL TRANSMISSION

A *program channel* carries the accompanying audio. If feeding a stereo system, there will, of course, be two audio channels. It may also imply what are called *cue* and *coordination* channels. A cue channel is used by a program director or producer to manage her/his people at the distant end. It is just one more audio channel. A coordination channel is a service channel among TV technicians.

Composite transmission normally is used on TV broadcast and community antenna television (CATV or cable TV) systems. Video and audio carriers are "combined" before being fed to the radiating antenna for broadcast. The audio subcarrier is illustrated in Figure 14.6 in the composite mode.

For point-to-point transmission on coaxial cable, radiolink, and earth station systems, the audio program channel is generally transmitted separately from its companion video providing the following advantages:

- Allows for individual channel level control;
- Provides greater control over crosstalk;
- Increases guardband between video and audio;
- Saves separation at broadcast transmitter;
- Leaves TV studio at separate channel; and
- Permits individual program channel preemphasis.

14.7 TRANSMISSION OF VIDEO OVER LOS MICROWAVE

Video/TV transmission over line-of-sight (LOS) microwave has two basic applications:

1. For studio-to-transmitter link, connecting a TV studio to its broadcast transmitter; and
2. To extend CATV systems to increase local programming content.

As covered earlier in the chapter, video transmission requires special consideration.

The following paragraphs summarize the special considerations a planner must take into account for video transmission over radiolinks.

Raw video baseband modulates the radiolink transmitter. The aural channel is transmitted on a subcarrier well above the video portion. The overall subcarriers are themselves frequency modulated. Recommended subcarrier frequencies may be found in CCIR Rec. 402-2 (Ref. 9) and Rep. 289-4 (Ref. 10).

14.7.1 Bandwidth of the Baseband and Baseband Response

One of the most important specifications in any radiolink system transmitting video is frequency response. A system with cascaded hops should have essentially a flat bandpass in each hop. For example, if a single hop is 3 dB down at 6 MHz in the resulting baseband, a system of five such hops would be 15 dB down. A good single hop should be ±0.25 dB or less out to 8 MHz. The most critical area in the baseband for video frequency response is in the low-frequency area of 15 kHz and below. Cascaded radiolink systems used in transmitting video must consider response down to 10 Hz.

Modern radiolink equipment used to transport video operates in the 2-GHz band and above. The 525-line video requires a baseband in excess of 4.2 MHz plus available baseband above the video for the aural channel. Desirable characteristics for 525-line video then would be a baseband at least 6 MHz wide. For 625-line TV, 8 MHz would be required, assuming that the aural channel would follow the channelization recommended by CCIR Rec. 402-2 (Ref. 9).

14.7.2 Preemphasis

Preemphasis is commonly used on wideband radio systems that employ FM modulation. After demodulation in an FM receiver, thermal noise has a ramplike characteristic with increasing noise per unit bandwidth toward band edges and decreasing noise toward band center. Preemphasis and its companion, deemphasis, makes thermal noise more uniform across the demodulated baseband. Preemphasis-deemphasis characteristics for television are described in CCIR Rec. 405-1 (Ref. 11).

14.7.3 Differential Gain

Differential gain is the difference in gain of the radio relay system as measured by a low-amplitude, high-frequency (chrominance) signal at any two levels of a low-frequency (luminance) signal on which it is superimposed. It is expressed in percentage of maximum gain. Differential gain shall not exceed the amounts indicated below at any value of APL (average picture level) between 10% and 90%:

- Short haul 2%
- Medium haul 5%
- Satellite 4%
- Long haul 8%
- End-to-end 10%

Based on ANSI/EIA/TIA-250C (Ref. 6). Also see CCIR Rec. 567-3 (Ref. 12).

14.7.4 Differential Phase

Differential phase is the difference in phase shift through the radio relay system exhibited by a low-amplitude, high-frequency (chrominance) signal at any two levels of a low-frequency (luminance) signal on which it is superimposed. Differential phase is expressed as the maximum phase change between any two levels. Differential phase, expressed in degrees of the high-frequency sine wave, shall not exceed the amounts indicated below at any value of APL (average picture level) between 10% and 90%:

- Short haul 0.5°
- Medium haul 1.3°
- Satellite 1.5°
- Long haul 2.5°
- End-to-end 3.0°

Based on ANSI/EIA/TIA-250C (Ref. 6).

14.7.5 Signal-to-Noise Ratio (10 kHz to 5 MHz)

The video signal-to-noise ratio is the ratio of the total luminance signal level (100 IRE units) to the weighted rms noise level. The noise referred to is predominantly thermal noise in the 10 kHz to 5.0 MHz range. Synchronizing signals are not included in the measurement. The EIA states that there is a difference of less than 1 dB between 525-line systems and 625-line systems.

As stated in the ANSI/EIA/TIA 250C standard (Ref. 6), the signal-to-noise ratio shall not be less than the following:

- Short haul 67 dB
- Medium haul 60 dB
- Satellite 56 dB
- Long haul 54 dB
- End-to-end 54 dB

and, for the low-frequency range (0–10 kHz), the signal-to-noise ratio shall not be less than the following:

- Short haul 53 dB
- Medium haul 48 dB
- Satellite 50 dB
- Long haul 44 dB
- End-to-end 43 dB

14.7.6 Continuity Pilot

A continuity pilot tone is inserted above the TV baseband at 8.5 MHz. This is a constant amplitude signal used to actuate an automatic gain control circuit in the far-end FM receiver to maintain signal level fairly uniform. It may also be used to actuate an alarm when the pilot is lost. The normal TV baseband signal cannot be used for this purpose because of its widely varying amplitude.

14.8 TV TRANSMISSION BY SATELLITE RELAY

TV satellite relay is widely used throughout the world. In fact, it is estimated that better than three-quarters of the transponder bandwidth over North America is dedicated to television relay. These satellite facilities serve CATV headends, hotel/motel TV service, for TV broadcasters providing country, continental and worldwide coverage for networks, TV news coverage, and so on.[6]

[6]*Headend* is where CATV program derives, its point of origin (see Section 15.2).

Table 14.3 Satellite Relay TV Performance

Parameters	Space Segment	Terrestrial Link[a]	End-to-End Values
Nominal impedance	75 Ω		
Return loss	30 dB		
Nonuseful dc component	0.5 V		
Nominal signal amplitude	1V		
Insertion gain	0 ± 0.25 dB	0 ± 0.3 dB	0 ± 0.5 dB
Insertion gain variation (1s)	± 0.1 dB	± 0.2 dB	± 0.3 dB
Insertion gain variation (1h)	± 0.25 dB	± 0.3 dB	± 0.5 dB
Signal-to-continuous random noise	53 dB[b]	58 dB	51 dB
Signal-to-periodic noise (0–1 kHz)	50 dB	45 dB	39 dB
Signal-to-periodic noise (1 kHz–6 MHz)	55 dB	60 dB	53 dB
Signal-to-impulse noise	25 dB	25 dB	25 DB[c]
Crosstalk between channels (undistorted)	58 dB	64 dB	56 dB
Crosstalk between channels (undifferentiated)	50 dB	56 dB	48 dB
Luminance nonlinear distortion	10%	2%	12%
Chrominance nonlinear distortion (amplitude)	3.5%	2%	5%
Chrominance nonlinear distortion (phase)	4°	2°	6°
Differential gain (x or y)	10%	5%	13%
Differential phase (x or y)	3°	2°	5°
Chrominance–luminance intermodulation	± 4.5%	± 2%	± 5%
Gain–frequency characteristic (0.15–6 MHz)	± 0.5 dB	± 0.5 dB[d]	± 1.0 dB[d]
Delay–frequency characteristic (0.15–6 MHz)	± 50 ns	± 50 ns[d]	± 105 ns[d]

[a]Connecting earth station to national technical control center.

[b]In cases where the receive earth station is colocated with the broadcaster's premises, a relaxation of up to 3 dB in video signal-to-weighted-noise ratio may be permissible. In this context, the term *colocated* is intended to represent the situation where the noise contribution of the local connection is negligible.

[c]Law of addition not specified in CCIR. Rec. 567.

[d]Highest frequency: 5 MHz.

Source: CCIR Rep. 965-1, Ref. 13.

As discussed in Section 9.3, a satellite transponder is nothing more than an RF repeater. This means that the TV signal degrades but little due to the transmission medium. Whereas with LOS microwave, the signal will pass through many repeaters, each degrading the signal somewhat. With the satellite there is one repeater, possibly two for worldwide coverage. However, care must be taken with the satellite delay problem if programming is interactive when two satellites in tandem relay a TV signal. Table 14.3 summarizes TV performance through satellite relay.

14.9 DIGITAL TELEVISION

14.9.1 Introduction

Up to this point in the chapter we have covered the baseband TV signal and radio relaying of that signal via LOS microwave and satellite, both analog systems using frequency modulation. The PSTN is evolving into a fully digital network. It is estimated that by the year 2005 the world's entire PSTN will be digital.

Now convert a TV signal to a digital format. Let's assume it is PCM (Chapter 6) with those three steps required to develop a digital PCM signal from an analog counterpart. We will remember those steps are sampling, quantization, and coding. The bit rate of such a digital signal will be in the order of 80 Mbps–160 Mbps. If we were to assume one bit per Hz of bandwidth, the signal would require in the range of 80 MHz to 160 MHz for just one TV channel! This is a viable alternative on extremely wideband fiber-optic systems. However, for narrower bandwidth systems, such as radio and coaxial cable, compression techniques are necessary.

Television, by its very nature, is highly redundant. What we mean here is that we are repeating much of the same information over and over again. Thus a TV signal is a natural candidate for digital signal processing to remove much of the redundancy, with the potential to greatly reduce the required bit rate after compression.

In this section we will discuss several digitizing techniques for TV video and provide an overview of some bit rate reduction methods, including a brief introduction to a derivative of MPEG-2.[7] The last section reviews several approaches to the development of a conference television signal.

14.9.2 Basic Digital Television

14.9.2.1 Two Coding Schemes. There are two distinct digital coding methods for color television: component and composite coding. For our discussion here, there are four components that make up a color video signal. These are R for red, G for green, B for blue, and Y for luminance. The output signals of a TV camera are converted by a linear matrix into luminance (Y) and two color difference signals R-Y and B-Y.

With the component method of transmission, these signals are individually digitized by an analog-to-digital (A/D) converter. The resulting digital bit streams are then combined with overhead and timing by means of a multiplexer for transmission over a single medium such as specially conditioned wirepair or coaxial cable.

Composite coding, as the term implies, directly codes the entire video baseband. The derived bit stream has a notably lower bit rate than that for component coding.

CCIR Rep. 646-4 (Ref. 14) compares the two coding techniques. The advantages of separate-component coding are the following:

- The input to the circuit is provided in separate component form by the signal sources (in the studio).
- The component coding is adopted generally for studios, and the inherent advantages of component signals for studios must be preserved over the transmission link in order to allow downstream processing at a receiving studio.
- The country receiving the signals via an international circuit uses a color system different from that used in the source country.
- The transmission path is entirely digital, which fits in with the trend toward all-digital systems that is expected to continue.

The advantages of transmitting in the composite form are the following:

[7]MPEG stands for Motion Picture Experts Group.

- The input to the circuit is provided in the composite form by the signal sources (at the studio).
- The color system used by the receiving country, in the case of an international circuit, is the same as that used by the source country.
- The transmission path consists of mixed analog and digital sections.

14.9.2.2 Development of a PCM Signal from an Analog TV Signal. As we discussed in Chapter 6, there are three stages in the development of a PCM signal format from its analog counterpart. These are sampling, quantization, and coding. These same three stages are used in the development of a PCM signal from its equivalent analog video signal. There is one difference, though—no companding is used; quantization is linear.

The calculation of the sampling rate is based on the color subcarrier frequency called f_{sc}. For NTSC television the color subcarrier is at 3.58 MHz. In some cases the sampling rate is three times this frequency ($3f_{sc}$), in other cases four times the color subcarrier frequency ($4f_{sc}$). For PAL television, the color subcarrier is 4.43 MHz. Based on 8-bit PCM words, the bit rates are $3 \times 3.58 \times 10^6 \times 8 = 85.92$ Mbps and $4 \times 3.58 \times 10^6 \times 8 = 114.56$ Mbps. In the case of PAL transmission system using $4f_{sc}$, the uncompressed bit rate for the video is $4 \times 4.43 \times 10^6 \times 8 = 141.76$ Mbps. These values are for composite coding.

In the case of component coding there are two separate digital channels. The luminance channel is 108 Mbps and the color-difference channel is 54 Mbps (for both NTSC and PAL systems).

In any case, linear quantization is employed, and an S/D (signal-to-distortion ratio) of better than 48 dB can be achieved with 8-bit coding. The coding is pure binary or two's complement encoding.

14.9.3 Bit Rate Reduction and Compression Techniques

In Section 14.9.2 raw video transmitted in a digital format required from 82 Mbps to 162 Mbps per video channel (no audio). Leaving aside studio-to-transmitter links (STL) and CATV supertrunks, it is incumbent on the transmission engineer to reduce these bit rates without sacrificing picture quality if there are any long-distance requirements involved.

CCIR Rep. 646-4 (Ref. 14) covers three basic bit rate reduction methods:

1. Removal of horizontal and vertical blanking intervals;
2. Reduction of sampling frequency; and
3. Reduction of the number of bits per sample.

We will only cover this last method.

14.9.3.1 Reduction of the Number of Bits per Sample. There are three methods that may be employed for bit rate reduction of digital television by reducing the number of bits per sample. These may be used singly or in combination:

- Predictive coding, sometimes called differential PCM;
- Entropy coding; and
- Transform coding.

Differential PCM, according to CCIR, has so far emerged as the most popular method. The prediction process required can be classified into two groups. The first one is called *intraframe* or *intrafield*, and is based only on the reduction of spatial redundancy. The second group is called *interframe* or *interfield*, and is based on the reduction of temporal redundancy as well as spatial redundancy.

14.9.3.2 Specific Bit Rate Reduction Techniques. The following specific bit rate reduction techniques are based on Refs. 14 and 15.

Intraframe coding. Intraframe coding techniques provide compression by removing redundant information within each video frame. These techniques rely on the fact that images typically contain a great deal of similar information; for example, a one-color background wall may occupy a large part of each frame. By taking advantage of this redundancy, the amount of data necessary to accurately reproduce each frame may be reduced.

Interframe coding. Interframe coding is a technique that adds the dimension of time to compression by taking advantage of the similarity between adjacent frames. Only those portions of the picture that have changed since the previous picture frame are communicated. Interframe coding systems do not transmit detailed information if it has not changed from one frame to the next. The result is a significant increase in transmission efficiency.

Intraframe and interframe coding used in combination. Intraframe and interframe coding used together provide a powerful compression technique. This is achieved by applying intraframe coding techniques to the image changes that occur from frame to frame. That is, by subtracting image elements between adjacent frames, a new image remains that contains only the differences between the frames. Intraframe coding, which removes similar information within a frame, is applied to this image to provide further reduction in redundancy.

Motion compensation coding. To improve image quality at low transmission rates, a specific type of interframe coding motion compensation is commonly used. Motion compensation applies the fact that most changes between frames of a video sequence occur because objects move. By focusing on the motion that has occurred between frames, motion compensation coding significantly reduces the amount of data that must be transmitted.

Motion compensation coding compares each frame with the preceding frame to determine the motion that has occurred between the two frames. It compensates for this motion by describing the magnitude and direction of an object's movement (e.g., a head moving right). Rather than completely regenerating any object that moves, motion compensation coding simply commands that the existing object be moved to a new location.

Once the motion compensation techniques estimate and compensate for the motion that takes place from frame-to-frame, the differences between frames are smaller. As a result, there is less image information to be transmitted. Intraframe coding techniques are applied to this remaining image information.

14.9.4 Overview of the MPEG-2 Compression Technique

This section is based on the ATSC (Advanced Television Systems Committee) version of MPEG-2, which is used primarily for terrestrial broadcasting and cable TV. The objective of the ATSC standard (Refs. 16, 17) is to specify a system for the transmission

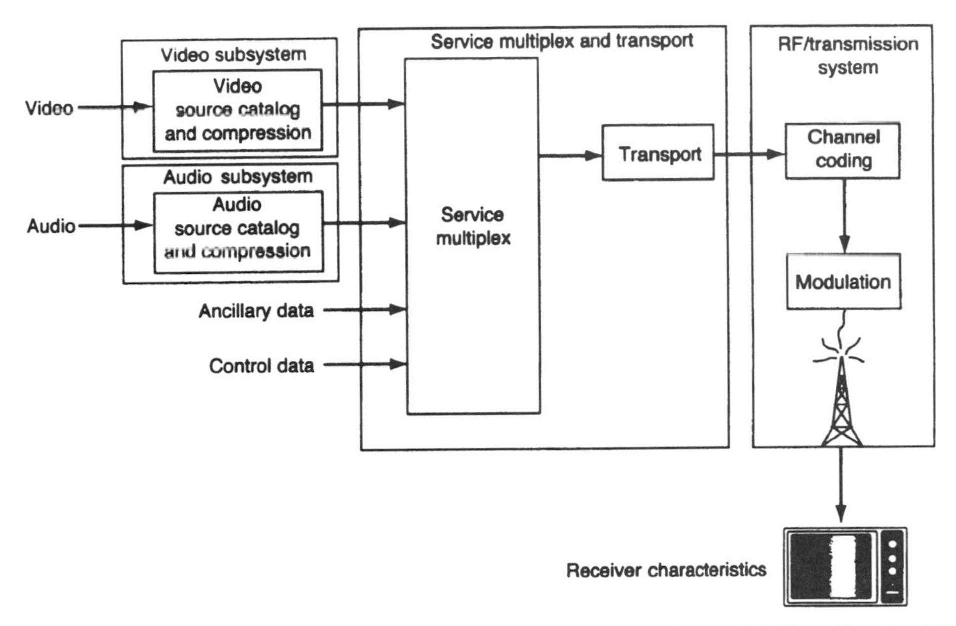

Figure 14.7 Block diagram of the digital terrestrial television broadcasting model. (Based on the ITU-R Task Group 11.3 model. From Ref. 18, Figure 4.1. Reprinted with permission.)

of high-quality video, audio and ancillary services over a single 6-MHz channel.[8] The ATSC system delivers 19 Mbps of throughput on a 6-MHz broadcasting channel and 38 Mbps on a 6-MHZ CATV channel. The video source, which is encoded, can have a resolution as much as five times better than conventional NTSC television. This means that a bit rate reduction factor of 50 or higher is required. To do this the system must be efficient in utilizing the channel capacity by exploiting complex video and audio reduction technology. The objective is to represent the video, audio, and data sources with as few bits as possible while preserving the level of quality required for a given application.

A block diagram of the ATSC system is shown in Figure 14.7. This system model consists of three subsystems:

1. Source coding and compression;
2. Service multiplex and transport; and
3. RF/transmission subsystem.

Of course, *source coding and compression* refers to bit rate reduction methods (data compression), which are appropriate for the video, audio, and ancillary digital data bit streams. The ancillary data include control data, conditional access control data, and data associated with the program audio and video services, such as closed captioning. Ancillary data can also refer to independent program services. The digital television system uses MPEG-2 video stream syntax for coding of the video and Digital Audio Compression Standard, called AC-3, for the coding of the audio.

Service multiplex and transport refers to dividing the bit stream into packets of infor-

[8]The reader will recall that the conventional 525-line NTSC television signal, when radiated, is assigned a 6-MHz channel (see Figure 14.6).

mation, the unique identification of each packet or packet type, and appropriate methods of multiplexing video bit stream packets, audio bit stream packets, and ancillary data bit stream packets into a single data stream. A prime consideration in the system transport design was interoperability among the digital media, such as terrestrial broadcasting, cable distribution, satellite distribution, recording media, and computer interfaces. MPEG-2 transport stream syntax was developed for applications where channel bandwidth is limited and the requirement for efficient channel transport was overriding. Another aspect of the design was interoperability with ATM transport systems (see Chapter 18).

RF/transmission deals with channel coding and modulation. The input to the channel coder is the multiplexed data stream from the service multiplex unit. The coder adds overhead to be used by the far-end receiver to reconstruct the data from the received signal. At the receiver we can expect that this signal has been corrupted by channel impairments. The resulting bit stream out of the coder modulates the transmitted signal. One of two modes can be used by the modulator: 8-VSB for the terrestrial broadcast mode and 16-VSB for the high data rate mode.

14.9.4.1 Video Compression. The ATSC standard is based on a specific subset of MPEG-2 algorithmic elements and supports its Main Profile. The Main Profile includes three types of frames for prediction (I-frames, P-frames, and B-frames), and an organization of luminance and chrominance samples (designated 4 : 2 : 0) within the frame. The Main Profile is limited to compressed data of no more than 80 Mbps. Figure 14.8 is a simplified block diagram of signal flow for the ATSC system.

Video preprocessing converts the analog input signals to digital samples in such a form needed for later compression. The analog input signals are red (R), green (G), and blue (B).

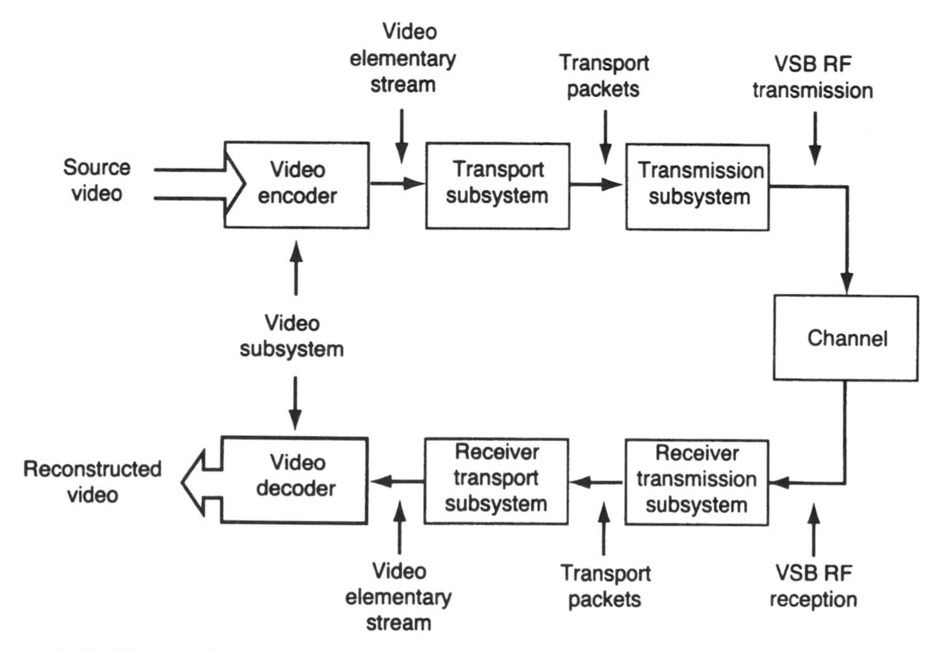

Figure 14.8 Video coding in relation to the ATSC system. (From Ref. 18, Figure 5.1. Reprinted with permission.)

Table 14.4 ATSC Compression Formats

Vertical Lines	Pixels	Aspect Ratio	Picture Rate
1080	1920	16:9	60I, 30P, 24P
720	1280	16:9	60P, 30P, 24P
480	704	16:9 and 4:3	60P, 60I, 30P, 24P
480	640	4:3	60P, 60I, 30P, 24P

Source: Ref. 18, Table 5.1. (Reprinted with permission.)

Table 14.4 lists the compression formats covered by the ATSC standard. The following explains some of the items in the table. *Vertical lines* refers to the number of active lines in the picture. *Pixels* are the number of pixels during the active line. *Apect ratio*, of course, refers to the picture aspect ratio. *Picture rate* gives the number of frames or fields per second. Regarding picture rate values, P refers to progressive scanning and I refers to interlaced scanning. It should be noted that both the 60.00-Hz and 59.95-Hz (i.e., $60 \times 1000/1001$) picture rates are allowed. Dual rates are permitted at 30 Hz and 24 Hz.

Sampling rates. Three active line formats are considered: 1080, 720, and 483. Table 14.5 summarizes the sampling rates.

For the 480-line format, there may be 704 or 640 pixels in an active line. If the input is based on ITU-R Rec. BT 601-4 (Ref. 19), it will have 483 active lines with 720 pixels in each active line. Only 480 of the 483 active lines are used for encoding. Only 704 of the 720 pixels are used for encoding: the first eight and the last eight are dropped. The 480-line, 640-pixel format corresponds only to the IBM VGA graphics format and may be used with ITU-R Rec. BT601-4 (Ref. 19) sources by employing appropriate resampling techniques.

Sampling precision is based on the 8-bit sample.

Colorimetry means the combination of color primaries, transfer characteristics, and matrix coefficients. The standard accepts colorimetry that conforms to SMPTE.[9] Video inputs corresponding to ITU-R Rec. BT 601-4 may have SMPTE 274M or 170M colorimetry.

The input video consists of the RGB components that are matrixed into luminance (Y) and chrominance (Cb and Cr) components using a linear transformation by means of a 3×3 matrix. Of course, the luminance carries picture intensity information (black-and-white) and the chrominance components contain the color. There is a high degree of correlation of the original RGB components, whereas the resulting Y, Cb, and Cr have less correlation and can be coded efficiently.

In the coding process, advantage is taken of the differences in the ways humans perceive luminance and chrominance. The human visual system is less sensitive to the high frequencies in the chrominance components than to the high frequencies in the

Table 14.5 Sampling Rate Summary

Line Format	Total Lines per Frame	Total Samples per Line	Sampling Frequency	Frame Rate
1080 line	1125	2200	74.25 MHz	30.00 fps[a]
720 line	750	1650	74.25 MHz	60.00 fps
480 line (704 pixels)	525	858	13.5 MHz	59.94 Hz field rate

[a]fps = frames per second.

[9]SMPTE stands for Society of Motion Picture and Television Engineers.

luminance component. To exploit these characteristics the chrominance components are low-pass filtered and subsampled by a factor of 2 along with the horizontal and vertical dimensions, thus producing chrominance components that are one-fourth the spatial resolution of the luminance components.

14.10 CONFERENCE TELEVISION

14.10.1 Introduction

Video conferencing (conference television) systems have seen phenomenal growth since 1990. Many of the world's corporations have branches and subsidiaries that are widely dispersed. Rather than pay travel expenses to send executives to periodic meetings at one central location, video conferencing is used, saving money on the travel budget.

The video and telecommunications technology has matured in the intervening period to make video conferencing cost effective. Among these developments we include:

- Video compression techniques;
- Eroding cost of digital processing; and
- Arrival of the all-digital network.

Proprietary video conferencing systems normally use lower line rates than conventional broadcast TV. Whereas conventional broadcast TV systems have line rates at 525/480 lines (NTSC countries) or 625/580 for PAL/SECAM countries, proprietary video conferencing systems use either 256/240 or 352/288 lines. For the common applications of conference television (e.g., meetings and demonstrations), the reduced resolution is basically unnoticeable.

One of the compression schemes widely used for video conference systems is based on ITU-T Rec. H.261 (Ref. 20), titled "Video Codec for Audiovisual Services at pX64 kbps," which is described in the following.

14.10.2 pX64 kbps Codec

The pX64 codec has been designed for use with some of the common ISDN data rates, specifically the B-channel (64 kbps), H_0 channel (384 kbps), and the H_{11}/H_{12} channels (1.536/1.920 Mbps) for the equivalent DS1/E1 data rates. A functional block diagram of the codec (coder/decoder) is shown in Figure 14.9. However, the pX64 system uses standard line rate (i.e., 525/625 lines) rather than the reduced line rate structure mentioned earlier. The reduced line rate techniques are employed in proprietary systems produced by such firms as Compression Labs Inc. and Picture Tel.

One of the most popular data rates for conference television is 384 kbps, which is six DS0 or E0 channels. However, it is not unusual to find numerous systems operating at 64/56 kbps.

14.10.2.1 pX64 Compression Overview

Sampling frequency. Pictures are sampled at an integer multiple of the video line rate. A sampling clock and network clock are asynchronous.

Source coding algorithm. Compression is based on interpicture prediction to utilize temporal redundancy, and transform coding of the remaining signal to reduce spatial redundancy. The decoder has motion compensation capability, allowing optional incor-

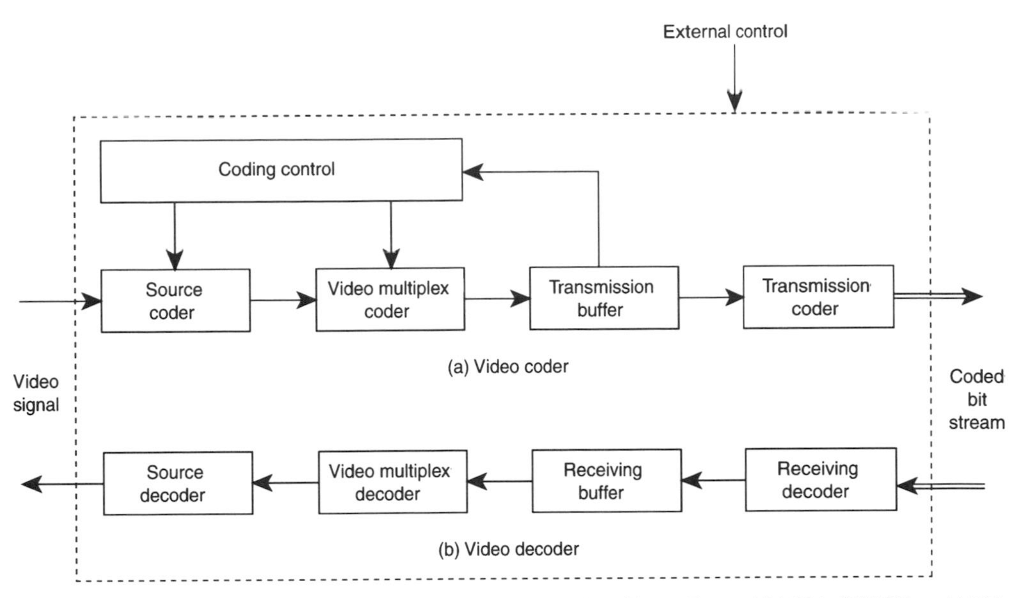

Figure 14.9 Functional block diagram of the pX64 video codec. (From Figure 1/H.261, ITU-T Rec. H.261 [Ref. 20].)

poration of this technique in the coder. There is optional forward error correction available based on the BCH (511,493) code.[10] The codec can support multipoint operation.

14.10.2.2 Source Coder. The coder operates on noninterleaved pictures occurring 30,000/1001 (approximately 29.97) times per second. The tolerance on the picture frequency is ±50 ppm. As in Section 14.9.4, pictures are coded as one luminance and two color difference components (Y, Cb, and Cr). Reference should be made to IT4-R Rec. BT.601 (Ref. 19) for their components and codes representing their sampled values. For example:

Black = 16
White = 235
Zero color difference = 128
Peak color difference = 16 and 240.

The values given are nominal values and the coding algorithm functions with input values of 1 through 254. Two picture scanning formats have been specified.

For the first format (CIF), the luminance structure is 352 pels per line, 288 lines per picture in an orthogonal arrangement. The color-difference components are sampled at 176 pels per line, 144 lines per picture, orthogonal. Figure 14.10 shows the color-difference samples being sited such that the block boundaries coincide with luminance block boundaries. The picture area covered by these numbers of pels and lines has an aspect ratio of 4 : 3 and corresponds to the active portion of the local standard video input.

It should be noted that the number of pels per line is compatible with sampling the active portions of the luminance and color-difference signals from 525- or 625-

[10]FEC coding. Consult Ref. 21 for a discussion of FEC.

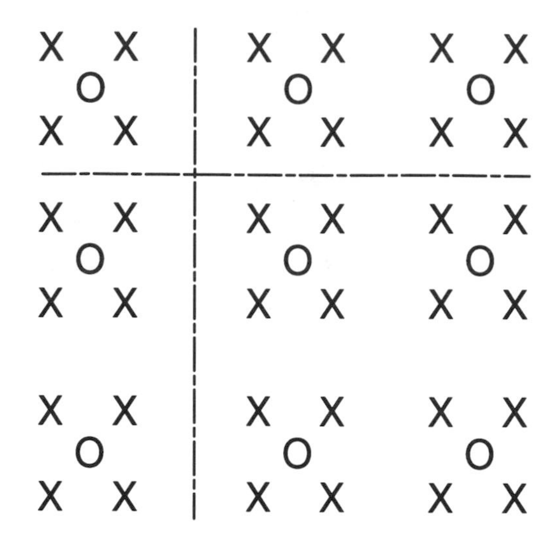

Figure 14.10 Positioning of luminance and chrominance samples. (From Figure 2/H.261, ITU-T Rec. H.261, [Ref. 20].)

line sources at 6.75 MHz or 3.375 MHz, respectively. These frequencies have a simple relationship with those in ITU-R Rec. BT.601 (Ref. 19).

The second format, called quarter-CIF or QCIF, has half the number of pels and half the number of lines stated of the CIF format. All codecs must be able to operate using QCIF.

A means is provided to restrict the maximum picture rate of encoders by having at least 0, 1, 2, or 3 nontransmitted pictures between transmitted pictures. Both CIF/QCIF and this minimum number of nontransmitted frames shall be selectable externally.

Video source coding algorithm. A block diagram of the coder is illustrated in Figure 14.11. The principal functions are prediction, block transformation, and quantization. The picture error (INTER mode) or the input picture (INTRA mode) is subdivided into 8-pel-by-8-pel line blocks, which are segmented as transmitted or nontransmitted. Furthermore, four luminance blocks and two spatially corresponding color-difference blocks are combined to form a macroblock. Transmitted blocks are transformed and the resulting coefficients are quantized and then variable length coded.

Motion compensation is optional. The decoder will accept one vector per macroblock. The components, both horizontal and vertical, of these motion vectors have integer values not exceeding ±15. The vector is used for all four luminance blocks in the macroblock. The motion vector for both color-difference blocks is derived by halving the component values of the macroblock vector and truncating the magnitude parts toward zero to yield integer components.

A positive value of the horizontal or vertical component of the motion vector signifies that the prediction is formed from pels in the previous picture, which are spatially to

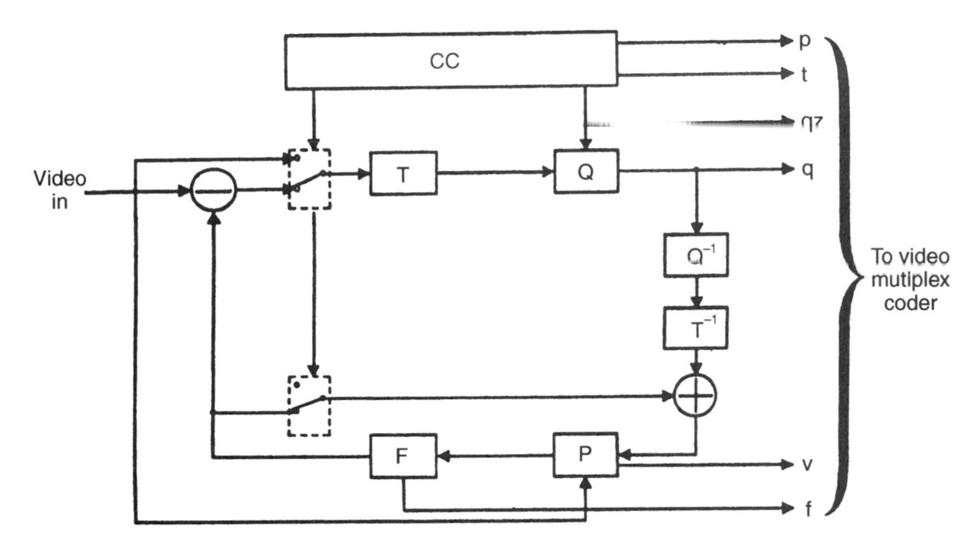

T Transform
Q Quantizer
P Picture memory with motion compensated variable decay
F Loop filter
CC Coding control
p Flag for INTRA/INTER
t Flag for transmitted or not
qz Quantizer indication
q Quantizing index for transform coefficients
v Motion vector
f Switching on/off of the loop filter

Figure 14.11 Functional block diagram of the source coder. (From Figure 3/H.261, ITU-T Rec. H.261 [Ref. 20].)

the right or below the pels being predicted. Motion vectors are restricted such that all pels referenced by them are within the coded picture area.

Loop filter. A two-dimensional spatial filter may be used in the prediction process. The filter operates on pels within a predicted 8-by-8 block. It is separable into one-dimensional horizontal and vertical functions. Both are nonrecursive carrying coefficients of $\frac{1}{4}$, $\frac{1}{2}$, $\frac{1}{4}$ except at block edges, where one of the taps would fall outside the block. In this case the one-dimensional filter is changed to have coefficients of 0, 1, 0. There is rounding to 8-bit integer values at the two-dimensional filter output and full arithmetic precision is retained. Rounding upward is used where values whose fractional part is one-half. The filter is switched on/off for all six blocks in a macroblock according to the macroblock type. There are ten types of macroblocks such as INTRA, INTER, INTER + MC (motion compensation), and INTER + MC + FIL (filter).

Discrete cosine transform. The transmitted blocks are first processed by a separable two-dimensional discrete cosine transform, which is 8 by 8 in size. There is an output range of the inverse transform from −256 to +255 after clipping to be represented by 9 bits.

Quantization. There are 31 quantizers for all other coefficients except the INTRA dc coefficient, which has just 1. The decision levels are not defined in CCIT-Rec. H.261. The INTRA dc coefficient is nominally the transform value linearly quantized with a step size of 8 and no dead-zone. The other 31 quantizers are nominally linear but with a central dead-zone around zero and with a step size of an even value in the range of 2–62.

Clipping and reconstructed picture. Clipping functions are inserted to prevent quantization distortion of transform coefficient amplitudes causing arithmetic overflow in the encoder and decoder loops. The clipping function is applied to the reconstructed picture. This picture is formed by summing the prediction and the prediction error as modified by the coding process. When resulting pel values are less than 0 or greater than 255, the clipper changes them to 0 and 255, respectively.

Coding control. To control the rate of generation of coded video data, several parameters may be varied. These parameters include processing prior to source coder, the quantizer, block significance criterion, and temporal subsampling. When invoked, temporal subsampling is performed by discarding complete pictures.

Forced updating. Forced updating is achieved by forcing the use of the INTRA mode of the coding algorithm. Recommendation H.261 does not define the update pattern. For the control of accumulation of inverse transform mismatch error, a macroblock should be updated forcibly at least once per every 132 times it is transmitted.

14.10.2.3 Video Multiplex Coder. The video multiplex is arranged in a hierarchical structure with four layers. From top to bottom these layers are:

1. Picture;
2. Group of blocks (GOB);
3. Macroblock; and
4. Block.

For further description of these layers, consult CCITT Rec. H.261 (Ref. 20).

14.11 BRIEF OVERVIEW OF FRAME TRANSPORT FOR VIDEO CONFERENCING

This section briefly reviews one method of transporting on the PSTN digital network the video conferencing signals developed in Section 14.10. It is based on ITU-T Rec. H.221 (Ref. 22).

14.11.1 Basic Principle

An overall transmission channel of 64 kbps–1920 kbps is dynamically subdivided into lower rates suitable for transport of audio, video, and data for telematic purposes. The transmission channel is derived by synchronizing and ordering transmission over from one to six B-channels, from one to five H_0 channels or an H_{11} or H_{12} channel.[11] The initial connection is the first connection established and it carries the initial channel in each direction. The additional connections carry the necessary additional channels. The total rate of transmitted information is called the *transfer rate*. The transfer rate can be less than the capacity of the overall transmission channel.

A single 64-kbps channel is structured into octets transmitted at an 8-kHz rate.[12] Each bit position of the octets may be regarded as a subchannel of 8 kbps. A frame

[11] Note the use of ISDN terminology for channel types (e.g., B-channels). Turn back to Section 12.4.2.1 for a description of ISDN user channels.

[12] This is the standard Nyquist sampling rate discussed in Chapter 6.

Bit number								
1	2	3	4	5	6	7	8(SC)	
Sub-channel #1	Sub-channel #2	Sub-channel #3	Sub-channel #4	Sub-channel #5	Sub-channel #6	Sub-channel #7	FAS BAS ECS #8	1 : 8 (FAS) 9 : 16 (BAS) 17 : 24 (ECS) 25 • • • 80 Octet number

FAS Frame alignment signal
BAS Bit-rate allocation signal
ECS Encryption control signal

Figure 14.12 Frame structure for a 64-kbps channel (i.e., a B-channel). (From Figure 1/H.221, ITU-T Rec. H.221 [Ref. 22].)

reflecting this concept is illustrated in Figure 14.12. The service channel (SC) resides in the eighth subchannel. It carries the frame alignment signal (FAS), a bit-rate allocation signal (BAS), and an encryption control signal (ECS).

We can regard an H_0, H_{11}, H_{12} channel as consisting of a number of 64 kbps time slots (TS). The lowest numbered time slot is structured exactly as just described for a 64-kbps channel, whereas the other time slots have no such structure. All channels have a frame structure in the case of multiple B and H_0 channels; the initial channel controls most functions across the overall transmission, while the frame structure in the additional channels is used for synchronization, channel numbering, and related controls. The term *I-channel* is applied to the initial or only B-channel, to time slot one of the initial or only H_0 channels, and to time slot one on H_{11} and H_{12} channels.

REVIEW EXERCISES

1. What four factors must be dealt with by a color video transmission system transmitting images of moving objects?

2. Describe *scanning*, horizontally and vertically.

3. Define a pel or pixel (besides the translation of the acronym).

4. If the *aspect ratio* of a television system is 4 : 3 and the width of a television screen is 12 inches, what is its height?

5. NTSC divides a television image into how many horizontal lines? European systems?

6. How do we achieve a sensation of motion in TV? Relate this to frame rate and flicker.

7. In North American practice, the time to scan a line is 63.5 μsec. This time interval consists of two segments. What are they?

8. What is the standard maximum voltage excursion of a video signal? Just what are we measuring here?

9. Give two definitions of a *composite signal.*

10. At a TV receiver, about what S/N is required for an excellent picture?

11. If we were to measure S/N, we would measure S and we would measure N. In common TV practice, what measurement units are used?

12. What type of modulation is used to transmit the video? The audio? The color subcarrier?

13. On a TV transport system, end-to-end S/N is often specified at 54 dB. Then why is the TV receiver specified at 45 dB? Explain the difference.

14. Regarding TV transport, what is a *program channel*?

15. To digitize a TV signal, what type of generic coding is nearly always used?

16. To digitize a TV signal by PCM, calculate the sampling rate for at least two systems.

17. Give the three basic bit rate reduction techniques suggested by CCIR.

18. Give three ways of reducing the number of bits per sample.

19. Discuss *intraframe coding* regarding redundancy.

20. What is the voltage value of 0 dBm V? When using dBmV, we should state another parameter as well, which we do not have to do when using dBs in the power domain. What is that parameter?

21. How does *differential PCM* bring about bit rate reduction?

22. What are the two broadcast-quality bit rates that can be derived from the ATSC (MPEG-2) coding system?

23. What is one popular bit rate for conference television? Another?

REFERENCES

1. *IEEE Standard Dictionary of Electrical and Electronics Terms*, 6th ed., IEEE Std. 100-1996, IEEE, New York, 1996.
2. *Fundamentals of Television Transmission*, Bell System Practices, Section AB 96.100, American Telephone & Telegraph Co., New York, 1954.
3. *Television Systems Descriptive Information—General Television Signal Analysis*, Bell System Practices, Section 318-015-100, No. 3, American Telephone & Telegraph Co., New York, 1963.

4. A. F. Inglis and Arch C. Luther, *Video Engineering*, 2nd ed., McGraw-Hill, New York, 1996.
5. K. Simons, *Technical Handbook for CATV Systems*, 3rd ed., General Instrument–Jerrold Electronics Corp., Hatboro, PA, 1980.
6. *Electrical Performance for Television Transmission Systems*, ANSI/EIA/TIA-250C, EIA/TIA Washington, DC, 1990.
7. *Television Systems*, ITU-R Rec. BT.470-3, 1994 BT Series, ITU, Geneva, 1994.
8. *Reference Data for Engineers: Radio, Electronics, Computer & Communications*, 8th ed, Sams–Prentice-Hall, Carmel, IN, 1993.
9. *The Preferred Characteristics of a Single Sound Channel Transmitted with a Television Signal on an Analogue Radio-Relay System*, CCIR Rec. 402-2, Part 1, Vol. IX, ITU, Geneva, 1990.
10. *The Preferred Characteristics of Simultaneous Transmission of Television and a Maximum of Four Sound Channels on Analogue Radio-Relay Systems*, CCIR Rep. 289-4, Part 1, Vol. IX, CCIR, Dubrovnik, 1986.
11. *Pre-emphasis Characteristics for Frequency Modulation Radio-Relay Systems for Television*, CCIR Rec. 405-1, Vol. IX, Part 1, ITU, Geneva, 1990.
12. *Transmission Performance of Television Circuits Designed for Use in International Connections*, CCIR Rec. 567-3, Vol. XII, ITU, Geneva, 1990.
13. *Transmission Performance of Television Circuits over Systems in the Fixed Satellite Service*, CCIR Rep. 965-1, Annex to Vol. XII, ITU, Geneva, 1990.
14. *Digital or Mixed Analogue-and-Digital Transmission of Television Signals*, CCIR Rep. 646-4, Annex to Vol. XII, ITU, Geneva, 1990.
15. *Digital Transmission of Component-Coded Television Signals at 30-34 Mbps and 45 Mbps*, CCIR Rep. 1235, Annex to Vol. XII, ITU, Geneva, 1990.
16. *A Compilation of Advanced Television Systems Committee Standards*, Advanced Television Systems Committee (ATSC), Washington, DC, 1996.
17. *ATSC Digital Television Standard*, Doc. A/53, ATSC, Washington, DC, 1995.
18. *Guide to the Use of the ATSC Digital Transmission Standard*, Doc. A/54, ATSC, Washington, DC, 1995.
19. *Encoding Parameters of Digital Television for Studios*, ITU-R Rec. BT.601-4, 1994 BT Series Volume, ITU, Geneva, 1994.
20. *Video Codec for Audiovisual Services at pX64 kbps*, CCITT Rec. H.261, ITU, Geneva, 1990.
21. R. L. Freeman, *Radio System Design for Telecommunications*, 2nd ed., Wiley, New York, 1997.
22. *Frame Structure of a 64 to 1920 kbps Channel in Audiovisual Teleservices*, ITU-T Rec. H.221, ITU, Geneva, 1993.

15

COMMUNITY ANTENNA TELEVISION (CABLE TELEVISION)

15.1 OBJECTIVE AND SCOPE

The principal thrust of community antenna television (CATV) is entertainment. However, since the early 1990s CATV has taken on some new dimensions. It is a broadband medium, providing up to nearly 1 GHz of bandwidth at customer premises. It was originally a unidirectional system, from the point of origin, which we will call the *headend*, toward customer premises, providing from 20 to some 100 television channels.

A CATV headend inserts signals into its transmission system for delivery to residences and offices (see Figure 15.1) from off-the-air local stations, satellite, LOS microwave, and locally generated programs. In older configurations, transmission systems consisted of coaxial cable with wideband amplifiers spaced at uniform distances. New configurations consist of a headend feeding fiber optic trunks up to hubs, which convert the signal format to traditional coaxial cable for the last mile or last 100 feet.

CATV does have the capability of being a two-way system, and many CATV operators are implementing such capability. The CATV band is actually the radio frequency band topping off at about 1000 MHz (see Figure 2.6, from about 1,000,000 downwards in the figure). In this case the radio signals (e.g., the TV channels) are transported in a coaxial cable medium. What a CATV operator might do is to use the band from 50 MHz to 1 GHz for transmission outwards from the headend for TV channels and "other services" and the band from about 5 MHz to 50 MHz to be employed from the user back to the headend. This band will transport the *other services* in a full duplex mode. These other services include POTS (plain old telephone service) and data.

In this chapter we will describe conventional CATV and the concept of supertrunks including HFC (hybrid fiber-coax) systems. We will involve the reader with wideband amplifiers in tandem and the special impairments that we would expect to encounter in a CATV system. System layout, hubs, and last-mile or last 100-ft considerations will also be covered. There will also be a brief discussion of the conversion to a digital system, using some of the compression concepts covered in Chapter 14. Employing these compression techniques, we can expect to see cable television systems delivering up to 500 TV channels to customers.

15.2 EVOLUTION OF CATV

15.2.1 Beginnings

Broadcast television, as we know it, was in its infancy around 1948. Fringe area problems were much more acute in that period. By fringe area, we mean areas with poor or scanty signal coverage. A few TV users in fringe areas found that if they raised their antennas high enough and improved antenna gain characteristics, an excellent picture could be received. These users were the envy of the neighborhood. Several of these people who were familiar with RF signal transmission employed signal splitters so that their neighbors could share the excellent picture service. It was soon found there was a limit on how much signal splitting could be done before signal levels got so low that they were snowy or unusable.

Remember that each time a signal splitter (even power split) is added, neglecting insertion losses, the TV signal drops 3 dB. Then someone got the bright idea of amplifying the signal before splitting. Now some real problems arose. One-channel amplification worked fine, but two channels from two antennas with signal combining became difficult. Now we are dealing with comparatively broadband amplifiers. Among the impairments we can expect from broadband amplifiers and their connected transmission lines (coaxial cable) are the following:

- Poor frequency response. Some part of the received band had notably lower levels than other parts. This is particularly true as the frequency increases. In other words, there was fairly severe amplitude distortion. Thus equalization became necessary.
- The mixing of two or more RF signals in the system caused intermodulation products and "beats" (harmonics), which degraded reception.
- When these TV signals carried modulation, cross-modulation (Xm) products degraded or impaired reception.

Several small companies were formed to sell these "improved" television reception services. Some of the technicians working for these companies undertook ways of curing the ills of broadband amplifiers.

These were coaxial cable systems, where a headend with a high tower received signals from several local television broadcasting stations, amplified the broadband signals, and distributed the results to CATV subscribers. A subscriber's TV set was connected to the distribution system, and the signal received looked just the same as if it were taken off the air with its own antenna. In fringe areas, signal quality, however, was much better than own-antenna quality. The key to everything was that no changes were required in the user's TV set. It was just an extension of her/his TV set antenna. This simple concept is shown in Figure 15.2.

Note in Figure 15.2 that home A is in the shadow of a mountain ridge and receives a weakened diffracted signal off the ridge and a reflected signal off the lake. Here is the typical multipath scenario resulting in ghosts in A's TV screen. The picture is also snowy, meaning noisy, as a result of poor carrier-to-noise ratio. Home B extended the antenna height to be in line-of-sight of the TV transmitting antenna. Its antenna is of higher gain; thus it is more discriminating against unwanted reflected and diffracted signals. Home B has an excellent picture without ghosts. Home B shares its fine signal with home A by use of a 3-dB power split (P) and a length of coaxial cable.

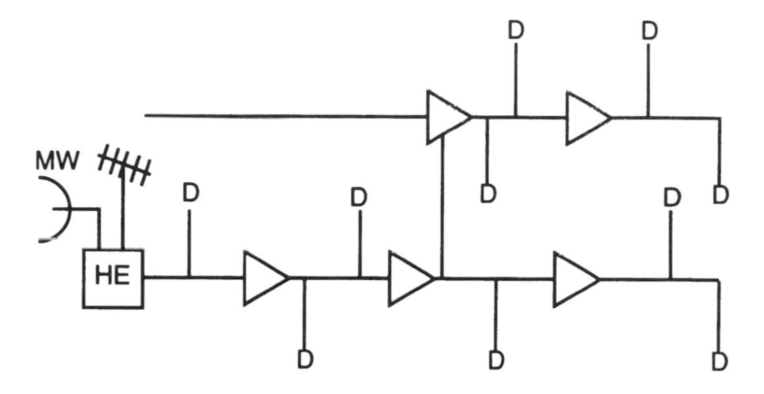

Figure 15.1 An early CATV distribution system.

15.2.2 Early System Layouts

Figure 15.1 illustrates an early CATV distribution system (ca. 1968). Taps and couplers (power splits) are not shown.[1] These systems provided from 5 to 12 TV channels. An LOS microwave system might bring in channels from distant cities. We had direct experience with an Atlantic City, NJ, system where channels were brought in by microwave from Philadelphia and New York City. A 12-channel system was derived and occupied the entire assigned VHF band (i.e., channels 2–13).

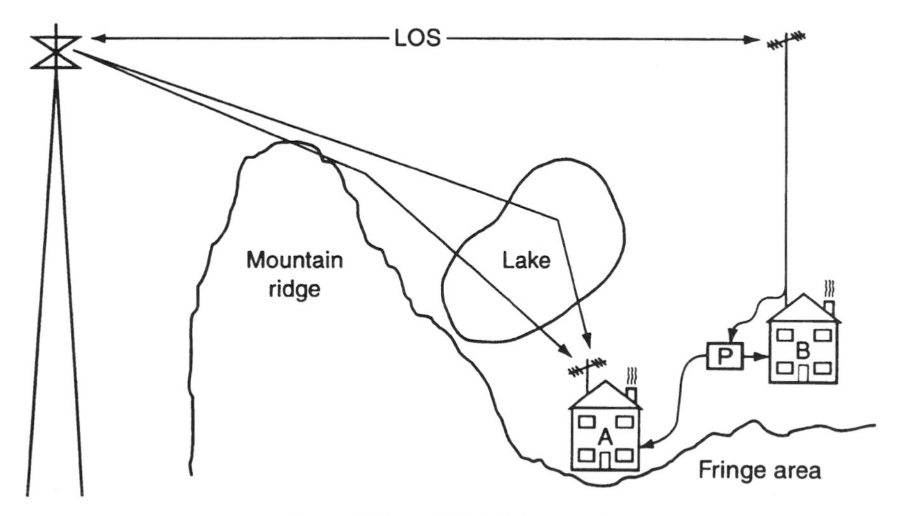

Figure 15.2 CATV initial concept.

[1]A coupler or power split is a passive device that divides up the incoming power in various ratios. A 3-dB split divides the power in half. A 10-dB split divides off one-tenth of the power, leaving approximately nine-tenths of the power on the main route.

As UHF TV stations began to appear, a new problem arose for the CATV operator. It was incumbent on that operator to keep the bandwidth as narrow as possible. One approach was to convert UHF channels to vacant VHF channel allocations at the headend.

Satellite reception at the headend doubled or tripled the number of channels that could be available to the CATV subscriber. Each satellite has the potential of adding 24 channels to the system. Note how the usable cable bandwidth is "broadened" as channels are added. We assume contiguous channels across the band, starting at 55 MHz. For 30 channels, we have 55–270 MHz; for 35 channels, 55–300 MHz; for 40 channels, 55–330 MHz; for 62 channels, 55–450 MHz; and for 78 channels, 55–550 MHz. These numbers of channels were beyond the capability of many TV sets of the day. Set-top converters were provided that converted all channels to a common channel, an unoccupied channel, usually channel 2, 3, or 4, to which the home TV set is tuned. This approach is still very prevalent today.

In the next section we discuss CATV transmission impairments and measures of system performance. In Section 15.4, hybrid-coaxial systems are addressed. The fiber replaced coaxial cable trunks, which made a major stride toward better performance, greater system extension, and improved reliability/availability.

15.3 SYSTEM IMPAIRMENTS AND PERFORMANCE MEASURES

15.3.1 Overview

A CATV headend places multiple TV and FM (from 30 to 125) carriers on a broadband coaxial cable trunk and distribution system. The objective is to deliver a signal-to-noise ratio (S/N) of from 42 dB to 45 dB at a subscriber's TV set. From previous chapters we would expect such impairments as the accumulation of thermal and intermodulation noise. We find that CATV technicians use the term *beat* to mean intermodulation (IM) products. For example, there is triple beat distortion, defined by Grant (Ref. 1) as "spurious signals generated when three or more carriers are passed through a nonlinear circuit (such as a wideband amplifier)." The spurious signals are sum and difference products of any three carriers, sometimes referred to as "beats." Triple-beat distortion is calculated as a voltage addition.

The wider the system bandwidth is and the more RF carriers transported on that system, the more intermodulation distortion, "triple beats," and cross-modulation we can expect. We can also assume combinations of all of the above, such as *composite triple beat* (CTB), which represents the pile up of beats at or near a single frequency.

Grant (Ref. 1) draws a dividing line at 21 TV channels. On a system with 21 channels or fewer, one must expect Xm to predominate. Above 21 channels, CTB will predominate.

15.3.2 dBmV and Its Applications

We define 0 dBmV as 1 mV across 75 Ω impedance. Note that 75 Ω is the standard impedance of CATV, of coaxial cable, and TV sets. From Appendix A, the electrical power law, we have:

$$P_{\mathrm{w}} = E^2/R \tag{15.1}$$

where P_w is the power in W, E the voltage in V, and R the impedance, 75 Ω. Substituting the values from the preceding, then:

$$P_w = (0.001)^2/75$$
$$0 \text{ dBmV} = 0.0133 \times 10^{-6} \text{ W} \qquad \text{or } 0.0133\ \mu\text{W}$$

By definition, then, 0.0133 W = +60 dBmV.

If 0 dBmV = 0.0133×10^{-6} W and 0 dBm = 0.001 W, and gain in dB = $10 \log(P_1/P_2)$, or, in this case, $10 \log[0.001/(0.0133 \times 10^{-6})]$, then 0 dBm = +48.76 dBmV.

Remember that, when working with dB in the voltage domain, we are working with the E^2/R relationship, where $R = 75\ \Omega$. With this in mind the definition of dBmV is

$$\text{dBmV} = 20 \log\left(\frac{\text{voltage in mV}}{1 \text{ mV}}\right). \tag{15.2}$$

If a signal level is 1 V at a certain point in a circuit, what is the level in dBmV?

$$\text{dBmV} = 20 \log(1000/1) = +60 \text{ dBmV}.$$

If we are given a signal level of +6 dBmV, to what voltage level does this correspond?

$$+6 \text{ dBmV} = 20 \log(X_{\text{mV}}/1 \text{ mV}).$$

Divide through by 20:

$$6/20 = \log(X_{\text{mV}}/1 \text{ mV})$$
$$\text{antilog}(6/20) = X_{\text{mV}}$$
$$X_{\text{mV}} = 1.995 \text{ mV, or 2 mV, or 0.002 volt}$$

These signal voltages are rms (root mean square) volts. For peak voltage, divide by 0.707. If you are given peak signal voltage and wish the rms value, multiply by 0.707.

15.3.3 Thermal Noise in CATV Systems

We remember from Section 3.3.3 that thermal noise is the most common type of noise encountered in telecommunication systems. In most cases, it is thermal noise that sets the sensitivity of a system, its lowest operating threshold. In the case of a CATV system, the lowest noise levels permissible are set by the thermal noise level—at the antenna output terminals, at repeater (amplifier) inputs, or at a subscriber's TV set—without producing snowy pictures.

Consider the following, remembering we are in the voltage domain: Any resistor or source that looks resistive over the band of interest, including antennas, amplifiers, and long runs of coaxial cable, generates thermal noise. In the case of a resistor, the thermal noise level can be calculated based on Figure 15.3.

To calculate the noise voltage, e_n, use the following formula:

$$e_n = (4RBk)^{1/2}, \tag{15.3}$$

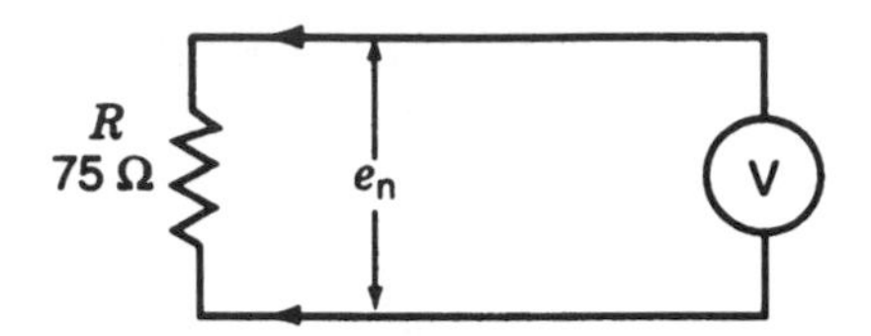

Figure 15.3 Resistor model for thermal noise voltage, e_n.

where e_n = rms noise voltage;
R = resistance in ohms (Ω);
B = bandwidth (Hz) of the measuring device (electronic voltmeter, V); and
k = a constant equal to 40×10^{-16} at standard room temperature.[2]

Let the bandwidth, B, of an NTSC TV signal be rounded to 4 MHz. The open circuit noise voltage for a 75-Ω resistor is

$$e_n = (4 \times 75 \times 4 \times 10^{-16})^{1/2}$$
$$= 2.2 \mu\text{V rms}.$$

Figure 15.4 shows a 2.2-μV noise generating source (resistor) connected to a 75-Ω (noiseless) load. Only half of the voltage (1.1 μV) is delivered to the load. Thus the noise input to 75 Ω is 1.1 μV or −59 dBmV. This is the basic noise level, the minimum that will exist in any part of a 75-Ω CATV system. The value −59 dBmV will be used repeatedly in the following text (Ref. 2). The noise figure of a typical CATV amplifier ranges between 7 dB and 9 dB (Ref. 3).

15.3.4 Signal-to-Noise (S/N) Ratio versus Carrier-to-Noise (C/N) Ratio in CATV Systems

We have been using S/N and C/N many times in previous chapters. In CATV systems S/N has a slightly different definition as follows (Ref. 2):

> This relationship is expressed by the "signal-to-noise ratio," which is the difference between the signal level measured in dBmV, and the noise level, also measured in dBmV, both levels being measured at the same point in the system.

S/N can be related to C/N on CATV systems as:

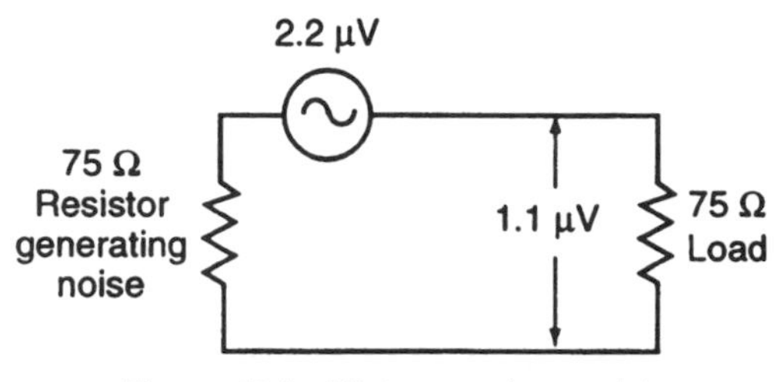

Figure 15.4 Minimum noise model.

[2]This value can be derived from Boltzmann's constant (Chapter 3) at room temperature (68°F or 290 K).

$$C/N = S/N + 4.1 \text{ dB}. \tag{15.4}$$

This is based on Carson (Ref. 4), where the premise is "noise just perceptible" by a population of TV viewers, with an NTSC 4.2-MHz TV signal. Adding noise weighting improvement (6.8 dB), we find:[3]

$$S/N = C/N + 2.7 \text{ dB}. \tag{15.5}$$

It should be noted that S/N is measured where the signal level is peak-to-peak and the noise level is rms.[4] For C/N measurement, both the carrier and the noise levels are rms. These values are based on a VSB-AM (vestigial sideband, amplitude modulation) with an 87.5% modulation index.

The values for S/N should be compared with those derived by the Television Allocations Study Organization (TASO) and published in their report to the U.S. FCC in 1959. Their ratings, corrected for a 4-MHZ bandwidth, instead of the 6-MHZ bandwidth that was used previously, are shown in Section 14.4.2.

Once a tolerable noise level is determined, the levels required in a CATV system can be specified. If the desired S/N has been set at 43 dB at a subscriber TV set, the minimum signal level required at the first amplifier would be −9 dBmV + 43 dB or −16 dBmV, considering thermal noise only. Actual levels would be quite a bit higher because of the noise generated by subsequent amplifiers in cascade.

It has been found that the optimum gain of a CATV amplifier is about 22 dB. When the gain is increased, IM/Xm products become excessive. For gains below this value, thermal noise increases, and system length is shortened or the number of amplifiers must be increased—neither of which is desirable.

There is another rule-of-thumb of which we should be cognizant. Every time the gain of an amplifier is increased 1 dB, IM products and "beats" increase their levels by 2 dB. And the converse is true: every time gain is decreased 1 dB, IM products and beat levels are decreased by 2 dB.

With most CATV systems, coaxial cable trunk amplifiers are identical. This, of course, eases noise calculations. We can calculate the noise level at the output of one trunk amplifier. This is

$$N_V = -59 \text{ dBmV} + NF_{dB}, \tag{15.6}$$

where NF is the noise figure of the amplifier in dB.

In the case of two amplifiers in cascade (tandem), the noise level (voltage) is

$$N_V = -59 \text{ dBmV} + NF_{dB} + 3 \text{ dB}. \tag{15.7}$$

If we have *M* identical amplifiers in cascade, the noise level (voltage) at the output of the last amplifier is

[3]*Weighting* (IEEE, Ref. 10): The artificial adjustment of measurements in order to account for factors that in normal use of the device, would otherwise be different from the conditions during measurement, In the case of TV, the lower baseband frequencies (i.e., from 20 Hz to 15 kHz) are much more sensitive to noise than the higher frequencies (i.e., >15 kHz).

[4]*Peak-to-peak* voltage refers, in this case, to the measurement of voltage over its maximum excursion, which is the voltage of the "sync tips." (See Figures 14.3 and 14.4.)

$$N_V = -59 \text{ dBmV} + NF_{dB} + 10 \log M. \tag{15.8}$$

This assumes that all system noise is generated by the amplifiers, and none is generated by the intervening sections of coaxial cable.

Example 1. A CATV system has 30 amplifiers in tandem; each amplifier has a noise figure of 7 dB. Assume that the input of the first amplifier is terminated in 75 Ω resistive. What is the thermal noise level (voltage) at the last amplifier output?

Use Eq. (15.8):

$$\begin{aligned} N_V &= -59 \text{ dBmV} + 7 \text{ dB} + 10 \log 30 \\ &= -59 \text{ dBmV} + 7 \text{ dB} + 14.77 \text{ dB} \\ &= -37.23 \text{ dBmV}. \end{aligned}$$

For carrier-to-noise ratio (C/N) calculations, we can use the following procedures. To calculate the C/N at the output of one amplifier,

$$C/N = 59 \text{ dBmV} - NF_{dB} + \text{input level (dBmV)}. \tag{15.9}$$

Example 2. If the input level of a CATV amplifier is +5 dBmV and its noise figure is 7 dB, what is the C/N at the amplifier output?

Use Eq. (15.9):

$$\begin{aligned} C/N &= 59 \text{ dBmV} - 7 \text{ dB} + 5 \text{ dBmV} \\ &= 57 \text{ dB}. \end{aligned}$$

With N cascaded amplifiers, we can calculate the C/N at the output of the last amplifier, assuming all the amplifiers are identical, by the following equation:

$$C/N_L = C/N(\text{single amplifier}) - 10 \log N. \tag{15.10}$$

Example 3. Determine the C/N at the output of the last amplifier with a cascade (in tandem) of 20 amplitifers, where the C/N of a single amplifier is 62 dB.

Use Eq. (15.10):

$$\begin{aligned} C/N_L &= 62 \text{ dB} - 10 \log 20 \\ &= 62 \text{ dB} - 13.0 \text{ dB} \\ &= 49 \text{ dB}. \end{aligned}$$

15.3.5 Problem of Cross-Modulation (Xm)

Many specifications for TV picture quality are based on the judgment of a population of viewers. One example was the TASO ratings for picture quality given earlier. In the case of cross-modulation (cross-mod or Xm) and CTB (composite triple beat), acceptable levels are −51 dB for Xm and −52 dB for CTB. These are good guideline values (Ref. 1).

Xm is a form of third-order distortion so typical of a broadband, multicarrier system. It varies with the operating level of an amplifier in question and the number of TV channels being transported. Xm is derived from the amplifier manufacturer specifications.

The manufacturer will specify a value for Xm (in dB) for several numbers of channels and for a particular level. The level in the specification may not be the operating level of a particular system. To calculate Xm for an amplifier to be used in a given system, using manufacturer's specifications, the following formula applies:

$$Xm_a = Xm_{spec} + 2(OL_{oper} - OL_{spec}), \tag{15.11}$$

where Xm_a = Xm for the amplifier in question;
Xm_{spec} = Xm specified by the manufacturer of the amplifier;
OL_{oper} = desired operating output signal level (dBmV); and
OL_{spec} = manufacturer's specified output signal level.

We spot the "2" multiplying factor and relate it to our earlier comments, namely, when we increase the operating level 1 dB, third-order products increase 2 dB, and the contrary applies for reducing signal level. As we said, Xm is a form of third-order product.

Example 1. Suppose a manufacturer tells us that for an Xm of −57 dB for a 35-channel system, the operating level should be +50.5 dBmV. We want a longer system and use an operating level of +45 dBmV. What Xm can we expect under these conditions?
Use Eq. (15.11):

$$\begin{aligned} Xm_a &= -57 \text{ dB} + 2(+45 \text{ dBmV} - 50.5 \text{ dBmV}) \\ &= -68 \text{ dB}. \end{aligned}$$

CATV trunk systems have numerous identical amplifiers. To calculate Xm for N amplifiers in cascade (tandem), our approach is similar to that of thermal noise, namely,

$$Xm_{sys} = Xm_a + 20 \log N, \tag{15.12}$$

where N = number of identical amplifiers in cascade;
Xm_a = Xm for one amplifier; and
Xm_{sys} = Xm value at the end of the cascade.

Example 2. A certain CATV trunk system has 23 amplifiers in cascade where Xm_a is −88 dB. What is Xm_{sys}?
Use Eq. (15.12):

$$\begin{aligned} Xm_{sys} &= -88 \text{ dB} + 20 \log 23 \\ &= -88 + 27 \\ &= -61 \text{ dB} \end{aligned}$$

15.3.6 Gains and Levels for CATV Amplifiers

Setting both gain and level settings for CATV broadband amplifiers is like walking a tightrope. If levels are set too low, thermal noise will limit system length (i.e., number of amplifiers in cascade). If levels are set too high, system length will be limited by excessive CTB and Xm. On trunk amplifiers available gain is between 22 dB and 26 dB (Ref. 1). Feeder amplifiers will usually operate at higher gains, trunk systems at lower gains. Feeder amplifiers usually operate in the range of 26–32-dB gain with output levels in the range of +47 dBmV. Trunk amplifiers have gains of 21–23 dB, with output levels in the range of +32 dBmV. If we wish to extend the length of the trunk plant, we should turn to using lower loss cable. Using fiber optics in the trunk plant is even a better alternative (see Section 15.4).

The gains and levels of feeder systems are purposefully higher. This is the part of the system serving customers through taps. These taps are passive and draw power. Running the feeder system at higher levels improves tap efficiency. Because feeder amplifiers run at higher gain and with higher levels, the number of these amplifiers in cascade must be severely limited to meet CTB and cross-modulation requirements at the end user.

15.3.7 Underlying Coaxial Cable System

The coaxial cable employed in the CATV plant has a nominal characteristic impedance (Z_0) of 75 Ω. A typical response curve for such cable ($\frac{7}{8}$-in., air dielectric) is illustrated in Figure 15.5. The frequency response of coaxial cable is called *tilt* in the CATV industry. This, of course, refers to its exponential increase in loss as frequency increases.

For 0.5-in. cable, the loss per 100 ft at 50 MHz is 0.52 dB; for 550 MHz, 1.85 dB. Such cable systems require equalization. The objective is to have a comparatively "flat" frequency response across the entire system. An equalizer is a network that presents a mirror image of the frequency response curve, introducing more loss at the lower frequencies and less loss at the higher frequencies. These equalizers are often incorporated with an amplifier.

Equalizers are usually specified for a certain length of coaxial cable, where length is

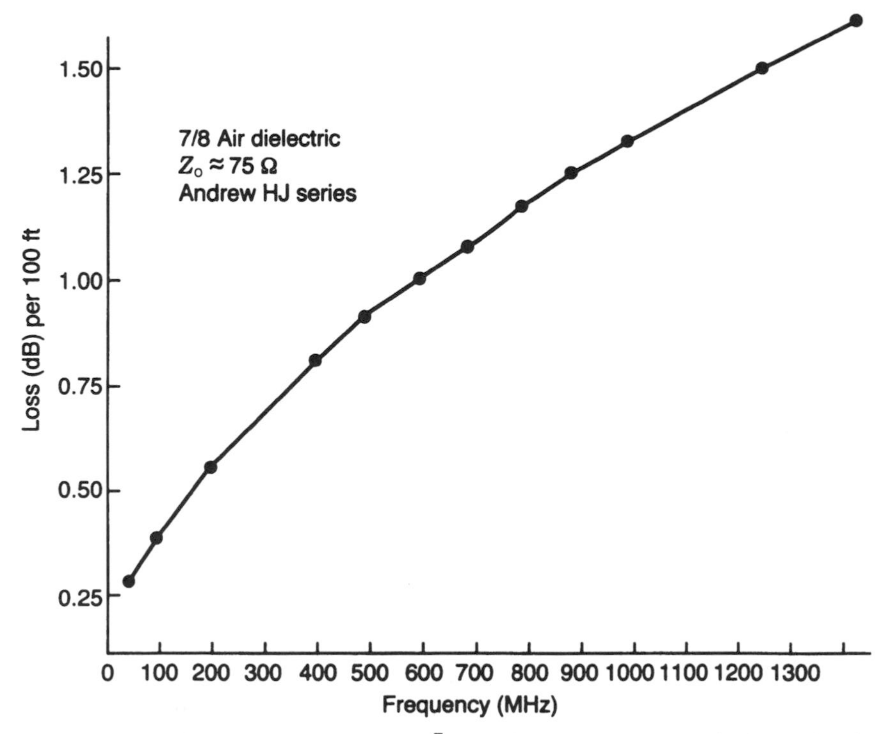

Figure 15.5 Attenuation–frequency response for $\frac{7}{8}$-in coaxial cable, $Z_0 \approx 75\ \Omega$, Andrew HJ series Helix. (Courtesy of Andrew Corp.)

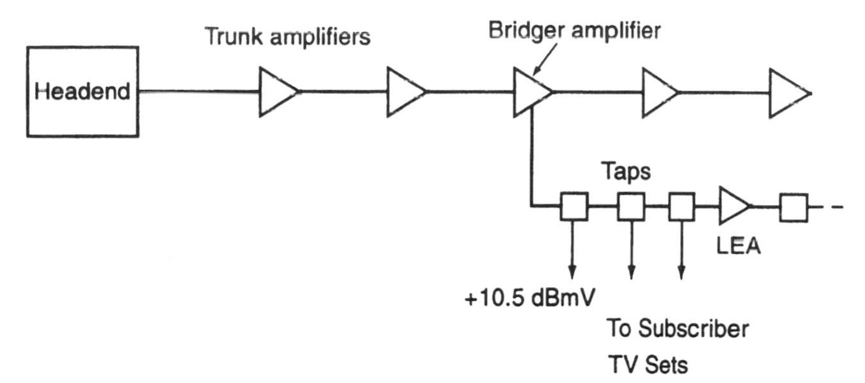

Figure 15.6 A simplified layout of a CATV system showing its basic elements. The objective is to provide +10.5 dBmV signal level at the drops (tap outputs). LEA = line extender amplifier.

measured in dB at the highest frequency of interest. Grant (Ref. 1) describes a 13-dB equalizer for a 300-MHz system, which is a corrective unit for a length of coaxial cable having 13-dB loss at 300 MHz. This would be equivalent to approximately 1000 ft of $\frac{1}{2}$-in. coaxial cable. Such a length of cable would have 5.45-dB loss at 54 MHz and 13-dB loss at 300 MHz. The equalizer would probably present a loss of 0.5 dB at 300 MHz and 8.1 dB at 54 MHz.

15.3.8 Taps

A *tap* is similar to a directional coupler. It is a device inserted into a coaxial cable which diverts a predetermined amount of its RF energy to one or more tap outputs for the purpose of feeding a TV signal into subscriber drop cables. The remaining balance of the signal energy is passed on down the distribution system to the next tap or distribution amplifier. The concept of the tap and its related distribution system is shown in Figure 15.6.

Taps are available to feed 2, 4, or 8 service drops from any one unit. Many different types of taps are available to serve different signal levels that appear along a CATV cable system. Commonly, taps are available in 3-dB increments. For two-port taps, the following tap losses may be encountered: 4, 8, 11, 14, 17, 20, and 23 dB. The insertion loss for the lower value tap loss may be on the order of 2.8 dB and, once the tap loss exceeds 26 dB, the insertion is 0.4 dB and remains so as tap values increase. Another important tap parameter is isolation. Generally, the higher the tap loss, the better the isolation. With an 8-dB tap loss, the isolation may only be 23 dB, but with 29-dB tap loss (two-port taps), the isolation can be as high as 44 dB. Isolation in this context is the isolation between the two tap ports to minimize undesired interference from a TV set on one tap to the TV set on the other tap.

For example, a line voltage signal level is +34.8 dBmV entering a tap. The tap insertion loss is 0.4 dB so the level of the signal leaving the tap to the next tap or extender amplifier is +34.4 dBmV. The tap is two-port. We know we want at least a +10.5 dBmV at the port output. Calculate +34.8 dBmV $-$ X dB = +10.5 dBmV. Then $X = 24.3$ dB, which would be the ideal tap loss value. Taps are not available off-the-shelf at that loss value, the nearest value being 23 dB. Thus the output at each tap port will be +34.8 dBmV $-$ 23 dB = 11.8 dBmV.

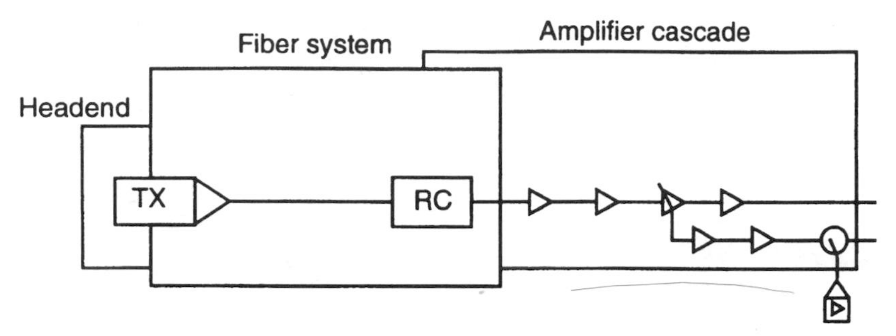

Figure 15.7 A model showing the concept of a hybrid fiber-coaxial cable CATV system. TX = fiber optic transmitter; RC = fiber optic receiver.

15.4 HYBRID FIBER-COAX (HFC) SYSTEMS

The following advantages accrue by replacing the coaxial cable trunk system with optical fiber:

- Reduces the number of amplifiers required per unit distance to reach the furthest subscriber;
- Results in improved C/N and reduced CTB and Xm levels;
- Also results in improved reliability (i.e., by reducing the number of active components); and
- Has the potential to greatly extend a particular CATV serving area.

One disadvantage is that a second fiber link has to be installed for the reverse direction, or a form of WDM is required, when two-way operation is required and/or for the CATV management system (used for monitoring the health of the system, amplifier degradation, or failure).

The concept is illustrated in Figure 15.7. Figure 15.8 shows an HFC system where

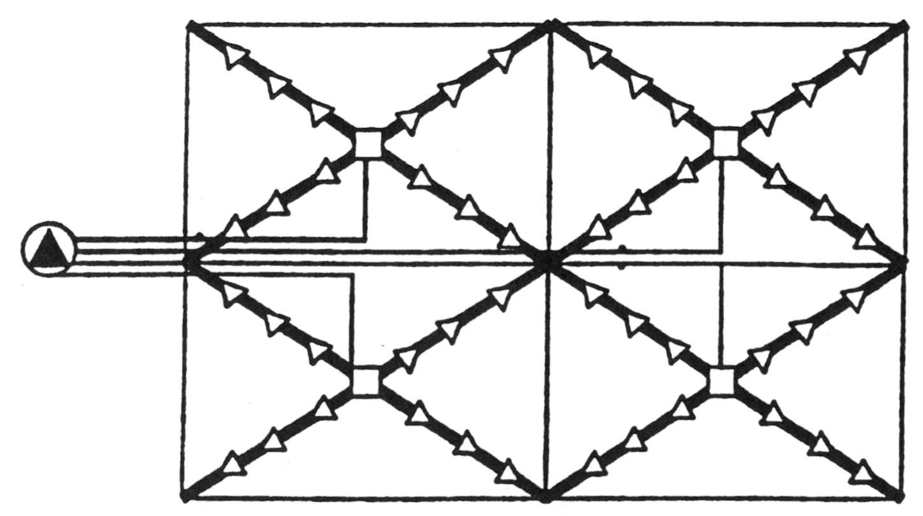

Figure 15.8 HFC system layout for optimal performance (one-way).

there are no more than three amplifiers in tandem to reach any subscriber tap. Also note that with this system layout there cannot be a catastrophic failure. For the loss of an amplifier, only one-sixteenth of the system is affected in the worst case scenario; with the loss of a fiber link, the worst case would be one-sixth of the system.

15.4.1 Design of the Fiber Optic Portion of an HFC System

Before proceeding with this section, it is recommended that the reader turn back to Chapter 9 for a review of the principles of fiber optic transmission.

There are two approaches to fiber optic transmission of analog CATV signals. Both approaches take advantage of the intensity modulation characteristics of the fiber optic source. Instead of digital modulation of the source, amplitude modulation (analog) is employed. The most common method takes the entire CATV spectrum as it would appear on a coaxial cable and uses that as the modulating signal. The second method also uses analog amplitude modulation, but the modulating signal is a grouping of subcarriers that are each frequency modulated. One off-the-shelf system multiplexes in a broad FDM configuration, eight television channels, each on a separate subcarrier. Thus a 48-channel CATV system would require six fibers, each with eight subcarriers (plus 8 or 16 audio subcarriers).

15.4.1.1 Link Budget for an AM System. We will assume a model using a distributed feedback laser (DFB) with an output of +5 dBm coupled to a pigtail. The receiver is a PINFET with a threshold of −5 dBm. This threshold will derive approximately 52-dB S/N in a video channel. The C/N required is about 49.3 dB (see formulas 15.4 and 15.5). This is a very large C/N value and leaves only 10 dB to be allocated to fiber loss, splices, and link margin. If we assign 2 dB for the link margin, only 8 dB is left for the fiber/splices loss. At 1550-nm operation, assuming a conservative 0.4-dB/km fiber/splice loss, the maximum distance from the headend to the coax hub or first fiber optic repeater is only 8/0.4 or 20 km. Of course, if we employ an EDFA (erbium-doped fiber amplifier) with, say, 20 dB gain, the distance can be extended by 20/0.4 or 50 km.

Typical design goals for the video/TV output of the fiber optic trunk are:

C/N (carrier-to-noise ratio) = 58 dB,

Composite second-order (CSO) products = −62 dBc (dB down from the carrier level);

CTB = −65 dBc.

One technique used on an HFC system is to employ optical couplers (a form of power splitter), where one fiber trunk feeds several hubs. A *hub* is a location where the optical signal is converted back to an electrical signal for transmission on coaxial cable. Two applications of optical couplers are illustrated in Figure 15.9. Keep in mind that a signal split not only includes splitting the power but also the insertion loss of the coupler.[5] The values shown in parentheses in the figure give the loss in the split branches (e.g., 5.7 dB, 2 dB).

[5] *Insertion loss* (IEEE, Ref. 10) is the total optical power loss caused by the insertion of an optical component such as a connector, splice, or coupler.

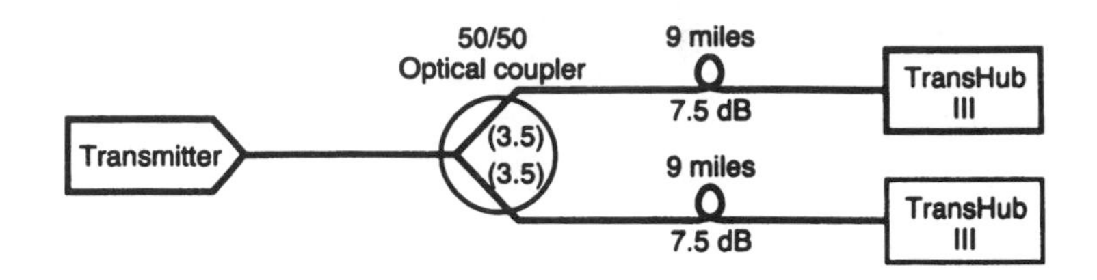

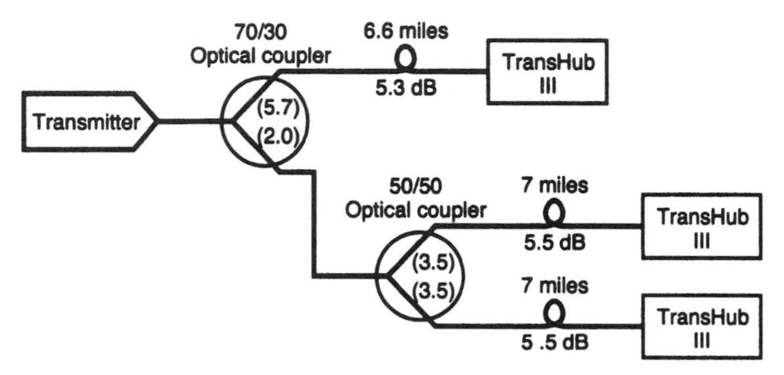

Figure 15.9 Two-way and three-way splits of a light signal transport CATV.

15.4.1.2 FM Systems. FM systems are much more expensive than their AM counterparts but provide improved performance. EIA/TIA-250C (Ref. 5), discussed in Chapter 14, specifies a signal-to-noise ratio of 67 dB for short-haul systems. With an AM fiber-optic system it is impossible to achieve this S/N, whereas a well-designed FM system can conform to EIA/TIA-250C. AM systems are degraded by dispersion on the fiber link; FM systems much less so. FM systems can be extended farther than AM systems. FM systems are available with 8, 16, or 24 channels, depending on the vendor. Of course, channel capacity can be increased by increasing the number of fibers.

Figure 15.10 shows an eight-channel per fiber frequency plan, and Figure 15.11 is a

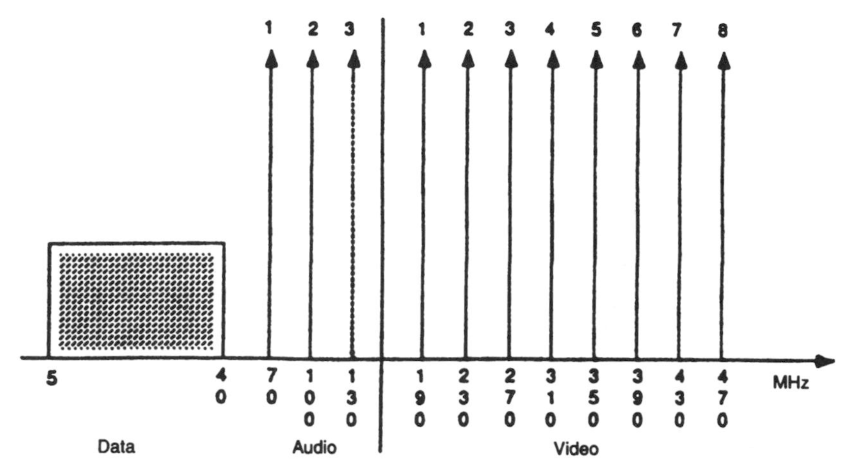

Figure 15.10 Eight-TV-channel frequency plan for an FM system.

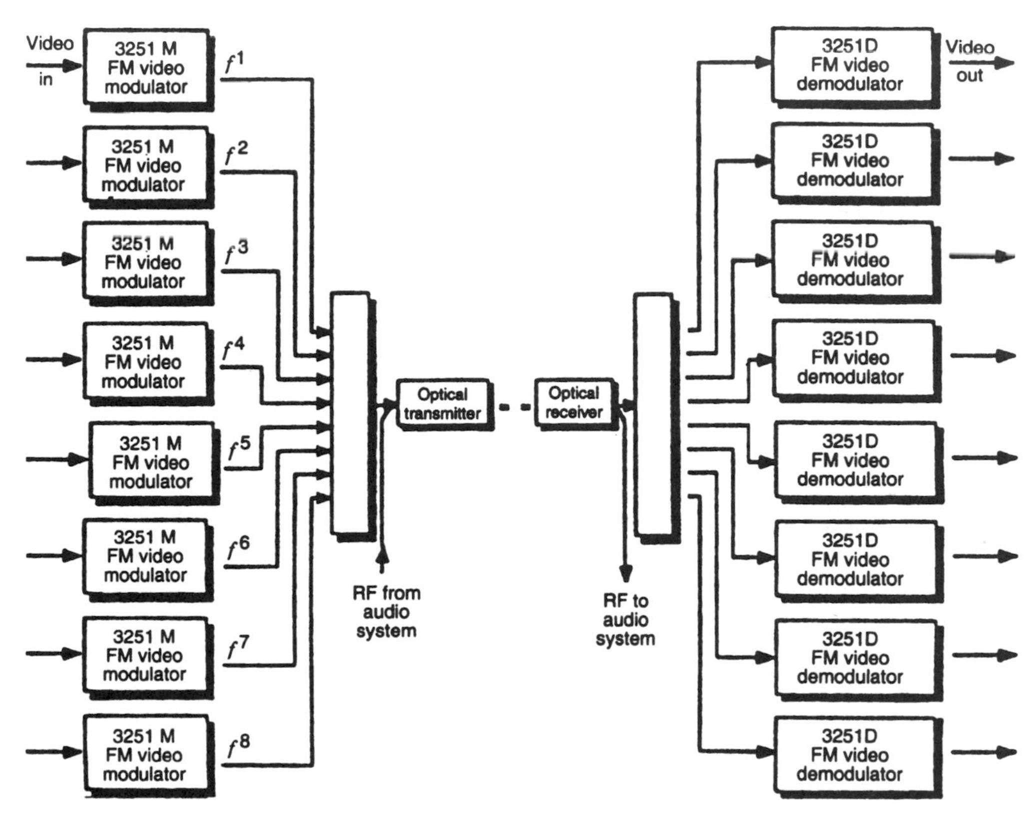

Figure 15.11 FM system model block diagram for the video transmission subsystem. (Courtesy of ADC video systems.)

transmit block diagram for the video portion of the system. Figure 15.12 illustrates a typical FM/fiber hub. Figure 15.13 shows the link performance of an FM system and how we can achieve an S/N of 67 dB and better.

As illustrated in Figure 15.11, at the headend, each video and audio channel must be broken out separately. Each of these channels must FM modulate its own subcarrier (see Figure 15.10). It should be noted that there is a similar but separate system for the associated aural (audio) channels with 30 MHz spacing starting at 70 MHz. These audio channels may be multiplexed before transmission. Each video carrier occupies a 40-MHz slot. These RF carriers, audio and video, are combined in a passive network. The composite RF signal intensity modulates a laser diode source. Figure 15.12 shows a typical fiber/FM hub where this technique is utilized.

Calculation of video S/N for FM system. Given the CNR for a particular FM system, the S/N of a TV video channel may be calculated as follows:

$$\mathrm{SNRw} = K + \mathrm{CNR} + 10\log\frac{B_{\mathrm{IF}}}{B_{\mathrm{F}}} + 20\log\frac{1.6\Delta F}{B_{\mathrm{F}}}, \tag{15.13}$$

where K = a constant (~23.7 dB) made of weighting network, deemphasis, and rms to p-p conversion factors;

CNR = carrier-to-noise ratio in the IF bandwidth;

B_{IF} = IF bandwidth;

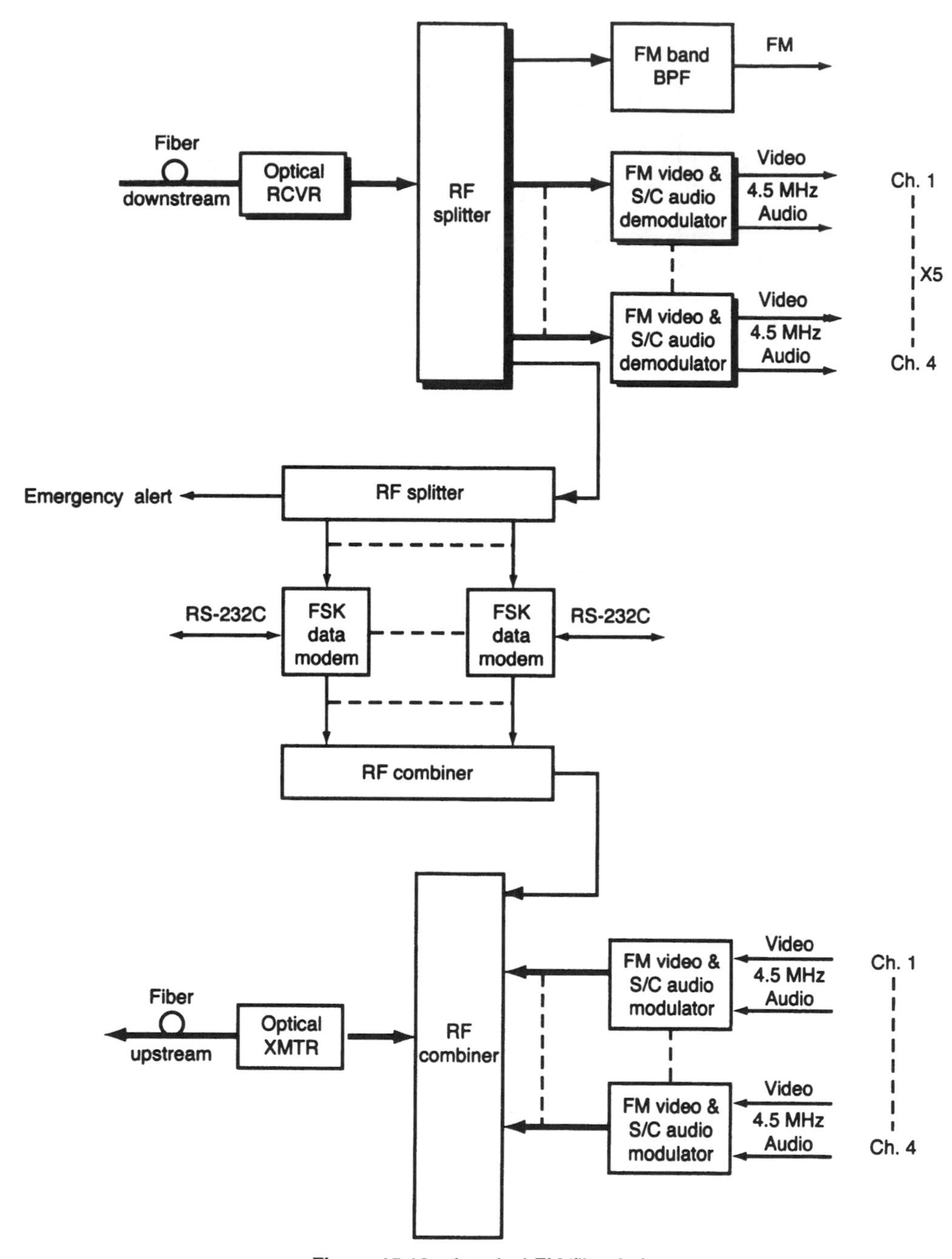

Figure 15.12 A typical FM/fiber hub.

B_F = baseband filter bandwidth; and
ΔF = sync tip-to-peak white (STPW) deviation.

With $\Delta F = 4$ MHz, $B_{IF} = 30$ MHz, and $B_F = 5$ MHz, the SNRw is improved by approximately 34 dB above CNR.

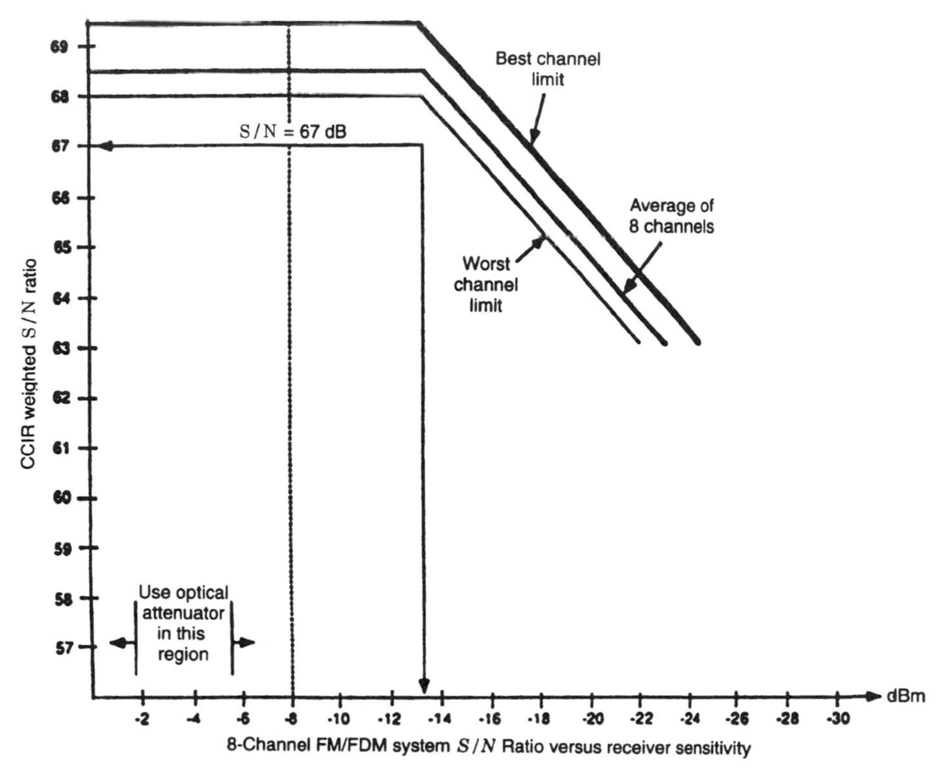

Figure 15.13 Link performance of an FM system. (Courtesy of ADC Video Systems.)

Example. If the C/N on an FM fiber link is 32 dB, what is the S/N for a TV video channel using the given values. Use Eq. (15.14):

$$
\begin{aligned}
S/N &= 23.7\ \text{dB} + 32\ \text{dB} + 10\log(30/5) + 20\log(1.6 \times 4/5) \\
&= 23.7 + 32 + 7.78 + 2.14 \\
&= 65.62\ \text{dB}.
\end{aligned}
$$

Figure 15.13 illustrates the link performance of an FM fiber optic system for video channels. Table 15.1 shows typical link budgets for an HFC AM system.

Table 15.1 Link Budgets for AM Fiber Links

Distance (mi)	Distance (km)	Fiber Loss/km	Total Fiber Loss	Splice Loss/2 km	Total Splice Loss	Total Path Loss	Link Budget	Link Margin
			Mileage, Losses, and Margins—1310 nm					
12.40	19.96	0.5 dB	9.98	0.1 dB	1.00	10.98	13.00	2.02
15.15	24.38	0.4 dB	9.75	0.1 dB	1.22	10.97	13.00	2.03
17.00	27.36	0.35 dB	9.58	0.1 dB	1.37	10.94	13.00	2.06
			Mileage, Losses, and Margins—1550 nm					
22.75	36.61	0.25 dB	9.15	0.1 dB	1.83	10.98	13.00	2.02

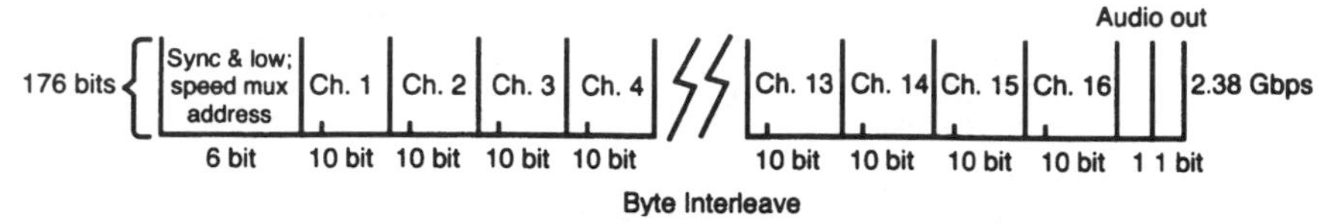

Figure 15.14 Typical frame structure on a single fiber in a CATV trunk. (Courtesy of ADC Video Systems.)

15.5 DIGITAL TRANSMISSION OF CATV SIGNALS

15.5.1 Approaches

There are two approaches to digitally transmitting both audio and video TV signals: transport either raw, uncompressed video or compressed video. Each method has advantages and disadvantages of which some are application-driven. For example, if the objective is digital to the residence/office, compressed TV may be the most advantageous.

15.5.2 Transmission of Uncompressed Video on CATV Trunks

Video, as discussed in Chapter 14, is an analog signal. It is converted to a digital format using techniques similar to the 8-bit PCM covered in Section 6.2. A major difference is in the sampling rate. Broadcast quality TV is generally *oversampled.* Here we mean that the sampling rate is greater than the Nyquist rate. The Nyquist rate, as we remember requires that the sampling rate be two times the highest frequency of interest. In this case, for the video the highest frequency of interest is 4.2 MHz, the video bandwidth. Using the Nyquist rate, the sampling rate would be $4.2 \times 10^6 \times 2$ or 4.2 million samples per second.

One example is the ADC Video Systems scheme, which uses a sampling rate of 13.524×10^6 samples per second. One option is an 8-bit system; another is a 10-bit system. The resulting equivalent bit rates are 108.192 Mbps and 135.24 Mbps, respectively. The 20-kHz audio channel is sampled at 41,880 samples per second and uses 16-bit PCM. The resulting bit rate is 2.68 Mbps for four audio channels (quadraphonic).

ADC Video Systems, of Meriden, CT, multiplexes and frames a 16-channel TV configuration for transmission over a fiber optic trunk in an HFC system. The bit rate on each system is 2.38 Gbps. The frame structure is illustrated in Figure 15.14, and Figure 15.15 is an equipment block diagram for a 16-channel link.

A major advantage of digital transmission is the regeneration capability just as it is in PSTN 8-bit PCM. As a result, there is no noise accumulation on the digital portion of the network. These digital trunks can be extended hundreds of miles or more. The complexity is only marginally greater than an FM system. The 10-bit system can easily provide an S/N ratio at the conversion hub of 67 dB in a video channel and an S/N value of 63 dB with an 8-bit system. With uncompressed video, BER requirements are not very stringent because video contains highly redundant information.

15.5.3 Compressed Video

MPEG compression is widely used today. A common line bit rate for MPEG is 1.544 Mbps. Allowing 1 bit per hertz of bandwidth, BPSK modulation, and a cosine roll-off

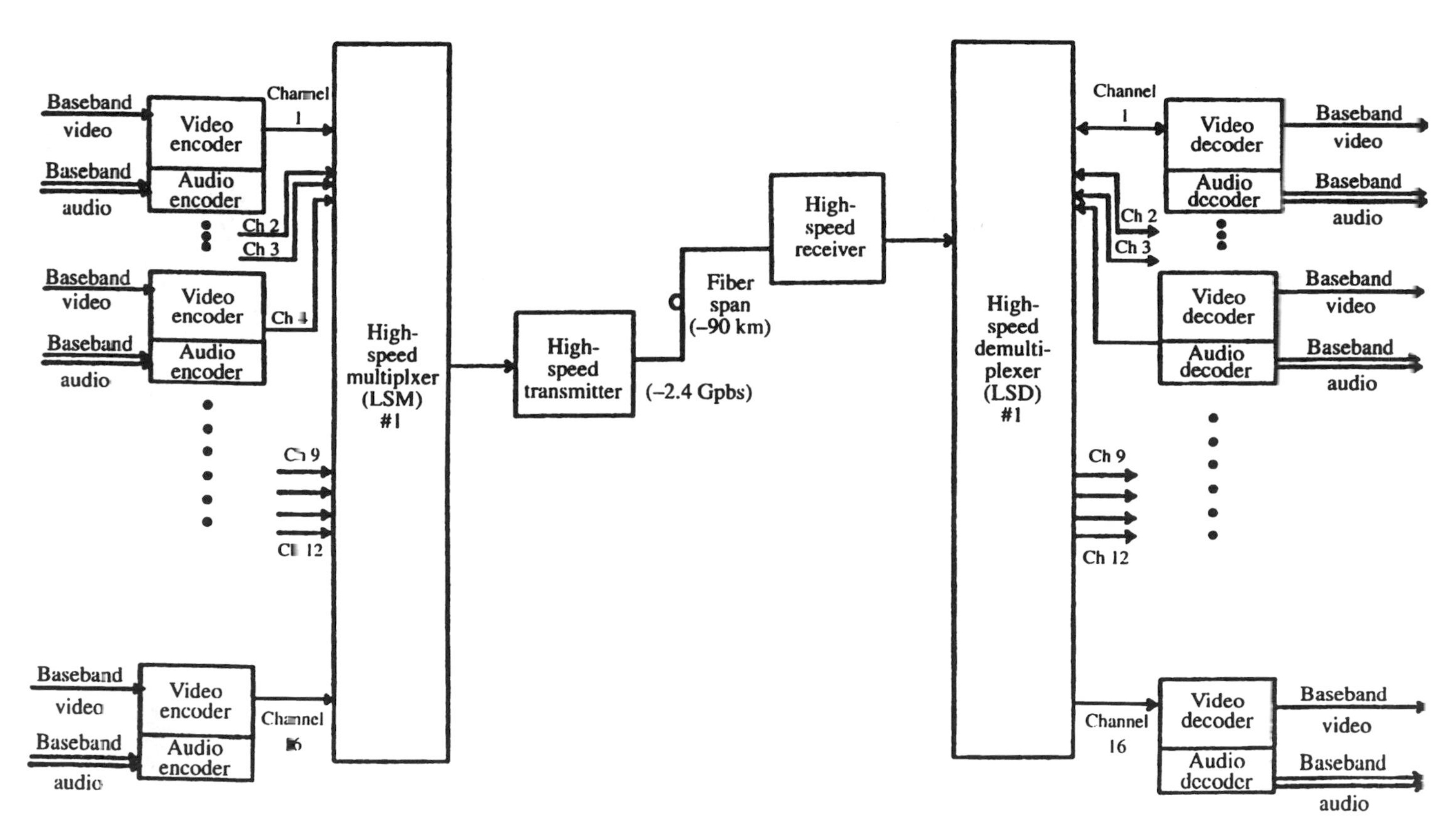

Figure 15.15 Functional block diagram of a 16-channel digital TV system. (Courtesy of ADC Video Systems.)

of 1.4, the 1.544-Mbps TV signal can be effectively transported in a 2-MHz bandwidth. Certainly 1000-MHz coaxial cable systems are within the state of the art. With simple division we can see that 500-channel CATV systems are technically viable. If the modulation scheme utilizes 16-QAM (4 bits/Hz theoretical), three 1.544-Mbps compressed channels can be accommodated in a 6-MHz slot. We select 6 MHz because it is the current RF bandwidth assigned for one NTSC TV channel.

15.6 TWO-WAY CATV SYSTEMS

15.6.1 Introduction

Figures 15.16*a* and 15.16*b* are two views of a CATV system as they might appear on coaxial cable. Of course, with conventional CATV systems, each NTSC television channel is assigned 6 MHz of bandwidth, as shown in Chapter 14, Figure 14.6.

In Figure 15.17*a* only 25 MHz is assigned to upstream services.[6] Not all the bandwidth may be used for voice and data. A small portion should be set aside for upstream telemetry from active CATV equipment in the system (e.g., broadband amplifiers). On the other hand, downstream has 60 MHz set aside for these services. In this day of the Internet, this would be providential, for the majority of the traffic would be downstream.

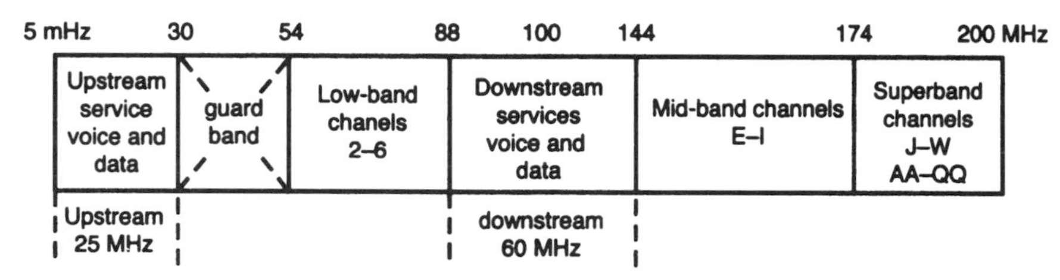

Figure 15.16*a* CATV spectrum based on Ref. 1, showing additional upstream and downstream services. Note the imbalance between upstream and downstream. (Adapted from Ref. 1.)

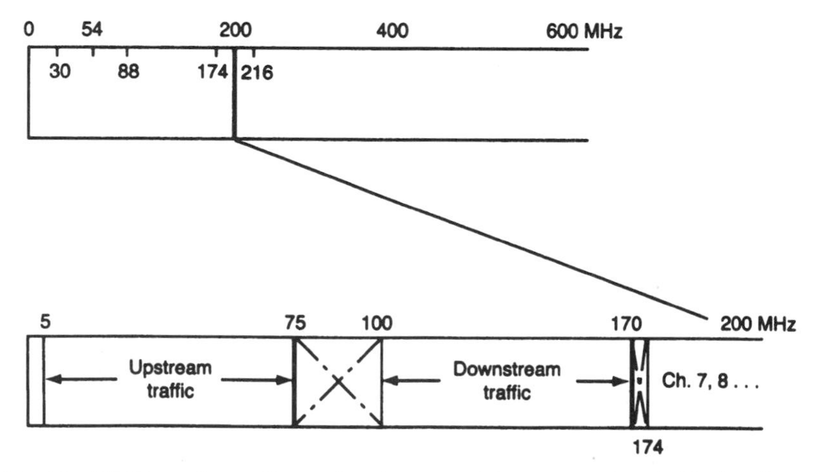

Figure 15.16*b* CATV spectrum with equal upstream and downstream bandwidth for other services.

[6]Remember, *upstream* is the direction from the CATV subscriber to the headend; and *downstream* is in the direction from the headend to the CATV subscriber.

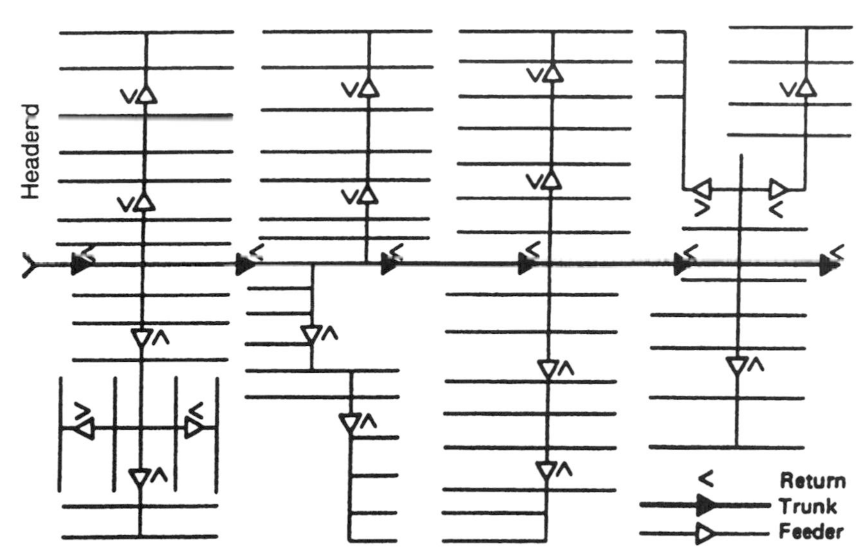

Figure 15.17 Trunk/feeder system layout for two-way operation. (From Ref. 1, reprinted with permission.)

15.6.1.1 Comments on Figures 15.16a and 15.16b. Large guardbands isolate upstream from downstream TV and other services, 24 MHz in Figure 15.16*a* and 25 MHz in Figure 15.16*b*. A small guardband was placed in the frequency slot from 170 MHz to 174 MHz to isolate downstream voice and data signals from conventional CATV television signals.

We assume the voice service will be POTS (plain old telephone service) and that both the data and voice will be digital.

In another approach, downstream voice, data, and special video are assigned the band 550 MHz–750 MHz, which is the highest frequency segment portion of this system (Ref. 6). In this case, we are dealing with a 750-MHz system.

The optical fiber trunk terminates in a node or hub (see Figure 15.12). This is where the conversion from an optical signal to the standard CATV coaxial cable format occurs. Let a node serve four groupings of cable subscribers, each with a coaxial cable with the necessary wideband amplifiers, line extenders, and taps. Typically such subscriber groups would consist of 200–500 terminations (TV sets) each. Assume each termination has upstream service using the band 5–30 MHz (see Figure 15.16*a*). In our example, the node has four incoming 5–30 MHz bands, one for each coaxial cable termination. It then converts each of these bands to a higher-frequency slot 25 MHz wide in a frequency division configuration for backhaul on a return fiber. In one scheme, at the headend, each 25-MHZ slot is demultiplexed and the data and voice traffic is segregated for switching and processing.

Access by data, voice, and special video users of the upstream and downstream assets is another question. There are many ways this can be accomplished. One unique method suggested by a consultant is to steal a page from the AMPS cellular radio specifications (see Chapter 16 for a discussion of AMPS). Because we have twice the bandwidth of an AMPS cellular operator, and because there are no handovers required, there are no shadowing effects and multipath (typical of the cellular environment); thus a much simpler system can be developed. For data communications, CDPD can be applied directly.[7]

[7]CDPD stands for cellular digital packet data.

Keep in mind that each system only serves 500 users as a maximum. Those 500 users are allocated 25 MHz of bandwidth (one-way).

An interesting exercise is to divide 25 MHz by 500. This tells us we can allot each user 50 kHz full-period. By taking advantage of the statistics of calling (usage), we could achieve a bandwidth multiplier of from 4× to 10× by using forms of concentration. However, upstream video, depending on the type of compression, might consume a large portion of this spare bandwidth.

There are many other ways a subscriber can gain access. DAMA techniques, where AMPS cellular is one, are favored. Suppose we were to turn to a digital format using standard 8-bit PCM. Allowing 1 bit/Hz and dividing 25 MHz by 64 kHz, we find only some 390 channels available. Keep in mind that these simple calculations are not accurate if we dig a little further. For instance, how will we distinguish one channel from another unless we somehow keep them in the frequency domain, where each channel is assigned a 64-kHz slot? This could be done by using QPSK modulation, which will leave some spare bandwidth for filter roll-off and guardbands.

One well-thought-out approach is set forth by the IEEE 802.14 (Ref. 7) committee. This is covered in Section 15.7.

Why not bring fiber directly into the home or to the desk in the office? The most convincing argument is economic. A CATV system interfaces with a home/office TV set by means of the set-top box. As discussed earlier, the basic function of this box is to convert incoming CATV channels to a common channel on the TV set, usually either channel 2, 3, or 4. Now we will ask much more of this "box." It is to terminate the fiber in the AM system as well as to carry out channel conversion. The cost of a set-top in 1999 dollars should not exceed $300. With AM fiber to the home, the cost target will be exceeded.

The reason that the driving factor is the set-top box is the multiplier effect. In this case we would be working with multipliers of, say, 500 (subscriber) by >$300. For a total CATV network, we could be working with 100,000 or more customers. Given the two-way and digital options, both highly desirable, the set-top box might exceed $1000 (1997 dollars), even with mass production. This amount is excessive.

15.6.2 Impairments Peculiar to Upstream Service

15.6.2.1 More Thermal Noise Upstream Than Downstream. Figure 15.17 shows a hypothetical layout of amplifiers in a CATV distribution system for two-way operation. In the downstream direction, broadband amplifiers point outward, down trunks, and out distribution cables. In the upstream direction, the broadbnd amplifiers point inward toward the headend and all their thermal noise accumulates and concentrates at the headend. This can account for from 3 dB to 20-dB additional noise upstream at the headend, where the upstream demodulation of voice and data signals takes place. Fortunately, the signal-to-noise ratio requirements for good performance of data and voice are much less stringent than for video, which compensates, to a certain extent, for this additional noise.

15.6.2.2 Ingress Noise. This noise source is peculiar to a CATV system. It basically derives from the residence/office TV sets that terminate the system. Parts 15.31 and 15.35 of the FCC Rules and Regulations govern such unintentional radiators. These rules have not been rigidly enforced.

One problem is that the 75-Ω impedance match between the coaxial cable and the TV set is poor. Thus not only all radiating devices in the TV set, but other radiating devices nearby in residences and office buildings couple back through the TV set into the CATV system in the upstream direction. This type of noise is predominant in the lower frequencies, that band from 5 MHz to 30 MHz that carries the upstream signals. As frequency increases, ingress noise intensity decreases. Fiber optic links in an HFC configuration provide some isolation.

15.7 TWO-WAY VOICE AND DATA OVER CATV SYSTEMS ACCORDING TO THE IEEE 802.14 COMMITTEE STANDARD

15.7.1 General

The narrative in this section is based on a draft edition of IEEE Standard 802.14 dated March 11, 1997 (Ref. 7) and subsequent narrative kindly provided by the chairman of the IEEE 802.14 committee (Ref. 8). The model for the standard is a hybrid fiber-coaxial cable (HFC) system with a service area of 500 households. The actual household number may vary from several hundred to a few thousand, depending on penetration rate.

Two issues limit the depth and completeness of our discussion: (1) reasonable emphasis and inclusion of details versus page count, and (2) because of the draft nature of the reference document, many parameters have not been quantified. Figure 15.18 is a pictorial overview of the 802.14 system.

The IEEE 802.14 specification supports voice, data, video, and file transfer and interactive data services across an international set of networks. These are represented by switched data services such as ATM (asynchronous transfer mode; Chapter 18), variable length data services such as CSMA/CD (Section 11.6.2.1), near constant bit rate services such as MPEG digital video systems (Chapter 14), and very low latency data services such as virtual circuits or STM (synchronous transfer mode; (e.g., E1 and DS1 families covered in Chapter 6). Instantaneous data rates and actual throughput are no longer limited by the protocol, but are, rather, a function of the network traffic engineering and theoretical limit of the media and the modulation schemes that are employed. This

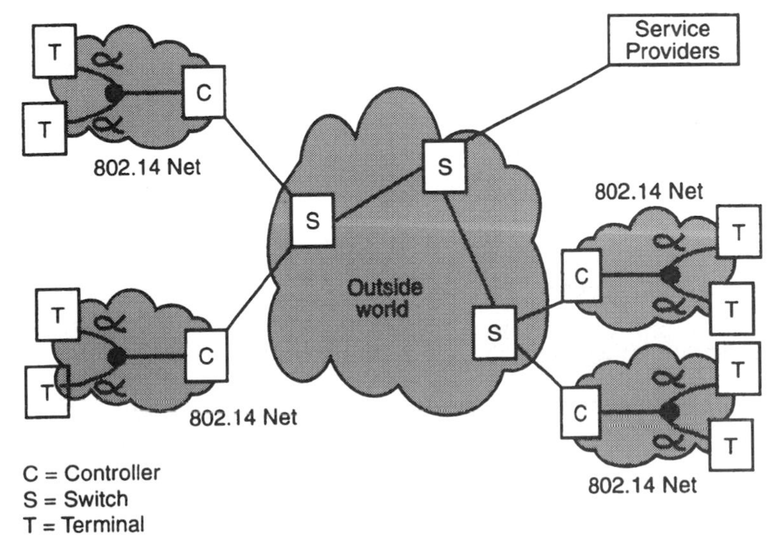

Figure 15.18 A pictorial overview of the IEEE 802.14 networks and their relationship with the outside world. (From IEEE 802.14, Draft R2, p. 14 [Ref. 7].)

generates a wide range of QoS (quality of service) parameters that must be supported simultaneously in order to create a scaleable, multiservice delivery system.

The 802.14 system can be envisioned as having OSI layers 1 and 2. Layer 1, of course, is the physical layer, and layer 2, from a LAN viewpoint, covers the "lower portion" of the data link layer. This portion includes the functions of medium access control (MAC), which is briefly discussed in Section 15.7.2. The "upper portion" of the data link layer is the LLC (logical link control), described in Section 11.5.3. In the following discussion, we will use the acronym PHY for the physical layer.

15.7.2 Overview of the Medium Access Control (MAC)

As overall controller of the actual transmission and reception of information, the MAC must account for the unique physical topology constraints of the network while guaranteeing the required QoS for each type of data to be transported. The network consists of a multicast or broadcast downstream from the headend to individual subscriber groups and multiple allocated and contention upstream channels. The downstream channel consists of a single wideband, high symbol rate channel, composed of six-octet time allocation units. A single unit can be assigned an idle pattern or multiple units can be used to create ATM cells, variable length fragments, or MPEG video streams.

The 802.14 system allows for multiple simultaneous downstream channels as well. The upstream consists of multiple channels divided in time into a series of *minislots*. These represent the smallest orthogonal unit of data allocation. There is enough time in one minislot for the transmission of eight octets of data plus PHY overhead and guard time. Multiple minislots can be concatenated in an upstream channel to create larger packet data units such as ATM cells, variable length fragments, or even MPEG video streams. Because of the varying amount of overhead and guard time required by different physical layers, the number of minislots required for the transmission of any data stream will vary from one upstream PHY to another.

The use of minislots with independent in-channel/in-band control messaging creates a flexible architecture. This flexibility allows one to change traffic flow patterns of the network and to fully integrate multiple channels and time slots.

15.7.3 Overview of the Physical Layer

Similar to the MAC architecture, the physical topology of the hybrid fiber-coax (HFC) plant allows for multicast downstream and multiple converging upstream paths. The 802.14 specification does not specify data types and resident topology. Constraints are defined and resolved while allowing the architecture to adapt on a session-by-session basis. Two distinctly different downstream PHYs are supported. Each type is centered around an existing coding and modulation standard: ITU-T Rec. J.83 (Ref. 9) Annex A/C, which is adopted for European cable systems, and Annex B, which is adopted for North American cable systems. In addition to these standards, 802.14 specifies modulation, coding sequence, scrambling method, symbol rates, synchronization, physical layer timing, message length and formats, transmitter power, and resolution characteristics.

15.7.3.1 Subsplit/Extended Subsplit Frequency Plan. The majority of CATV systems, particularly in North America, are upgraded to subsplit[8] (5 MHz to 30 MHz) or

[8]*Subsplit* is a frequency division scheme that allows bidirectional traffic on a single cable Reverse path signals come to the headend from 5 MHz to 30 MHz, and up to 42 MHz on newer systems. Forward path (downstream) signals go from the headend to end-users from 54 MHz to the upper frequency limit of the system in question.

extended subsplit (5 MHz to 42 MHz) operation. This scenario represents the worst-case design in terms of ingress noise and availability of the reverse channel (upstream) bandwidth. If the design can be deployed in this configuration, the infrastructure upgrade cost for cable systems will be minimal. In the future, the availability of midsplit or highsplit cable plants will enable a physical (PHY) layer with enhanced performance.

15.7.4 Other General Information

15.7.4.1 Frequency Reuse. The assumption is made that the coaxial cable traffic for each service area will be able to use the entire 5–30/5–42-MHz reverse bandwidth. This could be done either by use of a separate return fiber for each service area or by use of a single fiber for several service areas, whose return traffic streams would be combined using block frequency translation at the fiber node.

15.7.4.2 Up to 160-km Round-Trip Cable Distance. The distance coverage of the system can be influenced by several factors such as the fiber optic technology employed and the coaxial cable distribution topology. The limiting factor could be the number of active amplifiers and the resulting noise parameters that must be bounded for an optimal physical design.

15.7.5 Medium Access Control

15.7.5.1 Logical Topology. The logical topology of the CATV plant imposed some significant constraints on the 802.14 protocol design. For example, classic collision detection is impossible to do reliably on the cable plant since a station can only hear transmission by the HC (headend controller) and not by other stations. Even the detection of collisions by the HC is not entirely reliable. The protocol had to take into account the fact that round-trip delay from a station to the HC and back could be as high as 400 μs.

Figure 15.19 shows the elements of MAC topology. Each station has amplifiers in each direction that restrict the data flow unidirectionally. The path from the HC to the stations is familiarly called the *downstream path*. All stations on the network receive the same downstream path. It is incumbent on the station to filter out messages that are not addressed to itself.

The path from the set of user stations to the HC is called the *upstream path*. In the upstream direction any staiton can transmit but only the HC can receive. Diplex amplifiers prevent one station from listening to the transmission of another station. A

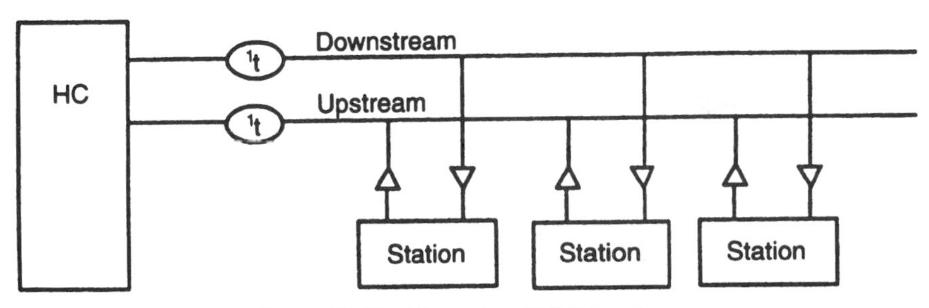

Figure 15.19 Elements of MAC topology.

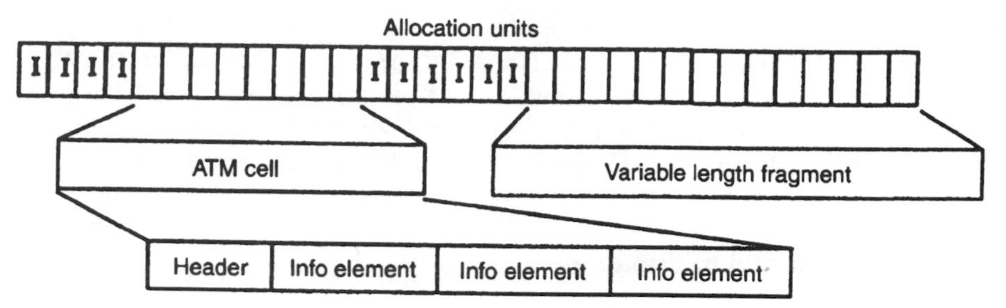

Figure 15.20 Downstream message hierarchy. I indicates and idle pattern. (From IEEE Std. 802.14, Draft R2, Figure M-2, [Ref. 7].)

single network may have several upstream channels to which stations may be assigned. Each station must be capable of changing its upstream channel at the request of the HC. The HC must be able to simultaneously receive all upstream channels within the network.

15.7.5.2 MAC Framing and Synchronization. All stations must be slaved to the master timing source that resides at the HC. To provide a time base to all stations that is synchronized correctly, the HC broadcasts a time-stamped cell to all stations at periodic intervals (~2 ms). The HC can adjust each station's timebase through messages so that all stations are synchronized in time. For any two stations on the network, it is important that, if they both decide to transmit at a given network time, both transmissions will arrive at the HC at the same instant.

15.7.5.3 Channel Hierarchies

15.7.5.3.1 Downstream Hierarchy. The downstream is composed of six-octet allocation units. A single unit can be assigned the idle pattern or multiple units can be used to create ATM cells or variable length fragments. Some of the ATM cells will carry MAC messages in the form of information elements. All MAC messaging is done in ATM cells with certain header values described in the reference publication. The downstream hierarchy is shown in Figure 15.20.

15.7.5.3.2 Upstream Channel Hierarchy. The upstream channel is a multiple access medium. For each upstream channel in the network there is a group of stations that share the assigned bandwidth. Each upstream channel is divided in time into a series of mini (time) slots. A minislot has the time capacity for the transmission of eight octets of data plus PHY overhead and guard time. A PDU (protocol data unit) that only occupies a single minislot is termed a *minipdu*. Minipdus are used primarily for contention opportunities to request bandwidth (bit rate capacity).

15.7.6 Physical Layer Description

15.7.6.1 Overview of the PHY. The PHY of 802.14 supports asymmetrical bidirectional transmission of signals in a CATV HFC network. The network is point-to-multipoint, tree branch access network in the downstream direction, and multipoint-to-point, bus access network in the upstream direction. The downstream transmission originates at the headend node and is transmitted to all end-users located at the tips of the branches in the tree and branch network. An upsteam transmission originates from an end-user

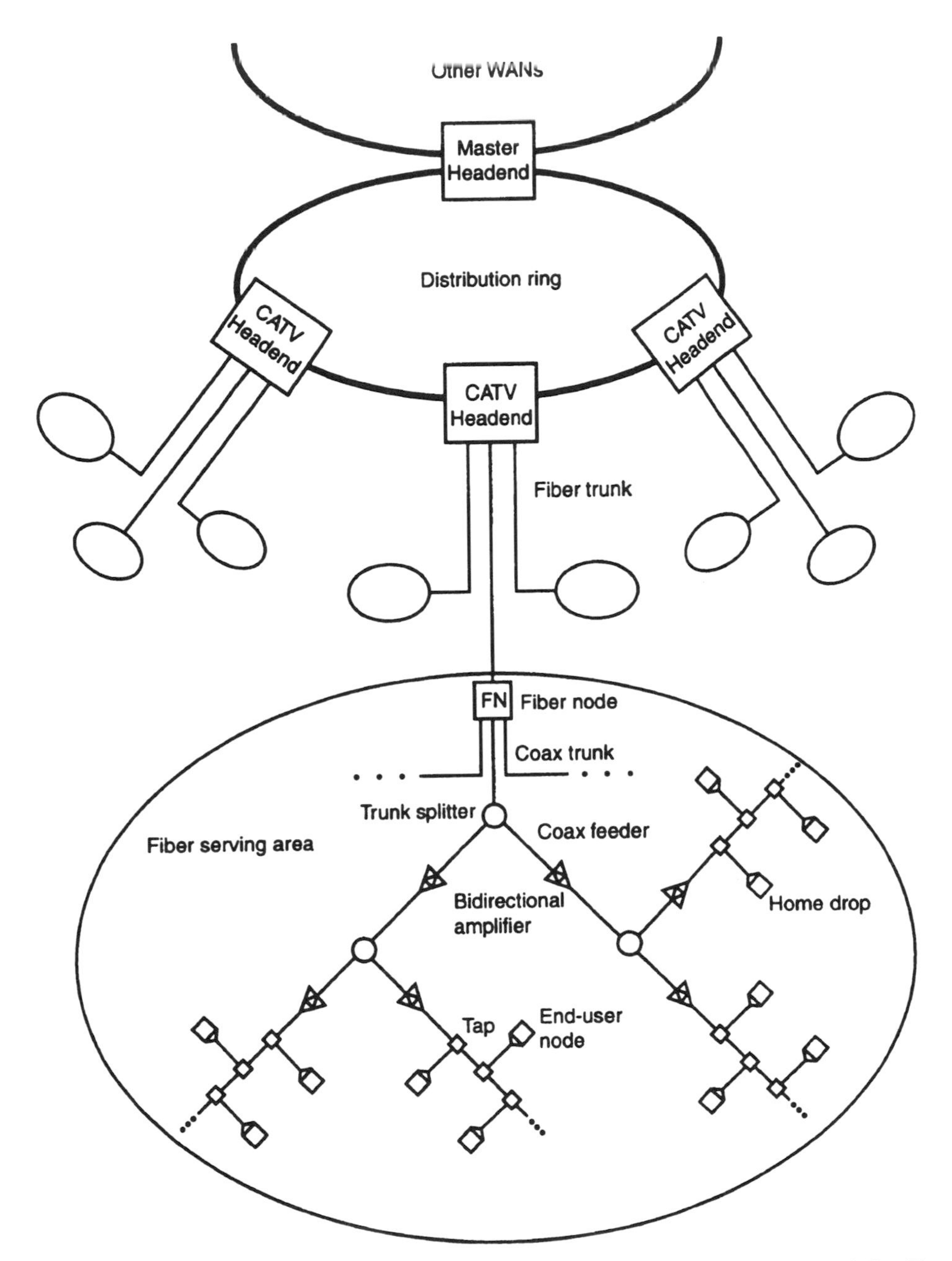

Figure 15.21 A model for HFC CATV serving area topology. (From IEEE Std. 802.14, Draft R2, Figure P-31 [Ref. 7].)

node and reaches the headend node through a multipoint-to-point access network where the access medium is shared by all end-users that are communicating with the same headend.

An example of a CATV HFC network topology is shown in Figure 15.21. In this case we see a multiple of 5–42 MHz upstream channels that are frequency division multiplexed in the fiber node (FN) and that are then transmitted via a single fiber trunk to the CATV headend. This operation is called *frequency stacking*.

15.7.6.2 Downstream Physical Layer Specification. There are two distinctly different PHYs supported by the 802.14 standard. These PHYs are called type A and type B downstream PHYs. The principal difference between the two is the coding method used for FEC (forward error correction; see Section 10.5.3.2). The FEC for type A downstream is based only on RS (Reed-Solomon) block coding. On the other hand, the FEC for type B downstream PHY is based on a concatenated coding method with outer RS coding and inner trellis-coded modulation (TCM). Each downstream PHY type has different modes of operation.

15.7.6.2.1 Downstream Spectrum Allocation. The frequencies from 63 MHz up to the upper frequency limit supported by the CATV cable plant (e.g., 750 MHz) are allocated for downstream transmission. Within this band is a channelized approach (i.e., frequency slots 6 MHz or 8 MHz wide) from the headend to end-user nodes. Standard CATV frequency plans are assumed. The topology model of the system is shown in Figure 15.21.

15.7.6.2.2 Propagation Delay and Delay Variation. The propagation delay for optical fiber is nominally 5 μs/km and for coaxial, 4 μs/km. The propagation delay introduced by the downstream transmission medium should be budgeted such that the total round-trip delay between the heandend and the end-user station should be a maximum TBD (to be determined) milliseconds.[9]

15.7.6.3 Type A Downstream PHY. The Type A downstream PHY supports two modes of operation: 64-QAM and 256-QAM. A block coding approach based on the shortened RS coding is used. A convolutional interleaver mitigates the effects of burst noise.

15.7.6.3.1 Constellations for Type A Downstream PHY. Figure 15.22 illustrates a type A PHY constellation for 64-QAM and 256-QAM waveforms.

15.7.6.4 Type B Downstream PHY. Type B downstream PHY supports two modes of operation: 64-QAM and 256-QAM. As previously mentioned, the coding strategy is different for type B downstream.

15.7.6.5 Downstream Carrier Frequencies. The downstream carrier frequencies are selected in accordance with the following:

$$f_c = (n \times 250 \text{ kHz}) \pm 8 \text{ kHz},$$

where n is an integer such that 63 MHz $\leq f_c \leq$ 803 MHz.

15.7.6.6 Transmitted Signal Levels. The type B downstream PHY is capable of transmitting a signal on the coaxial cable between the ranges of +50 dBmV and +61 dBmV.[10]

[9]If the round-trip (loopback) system extension is 160 km (quoted from above) and seven-eights of the system is optical fiber, then the optical fiber portion is 140 × 5 μs + 40 × 4 μs, for a total of 0.760 ms.

[10]We can get away with calling a voltage a *level* because in this case the impedance is always assumed to be 75 Ω.

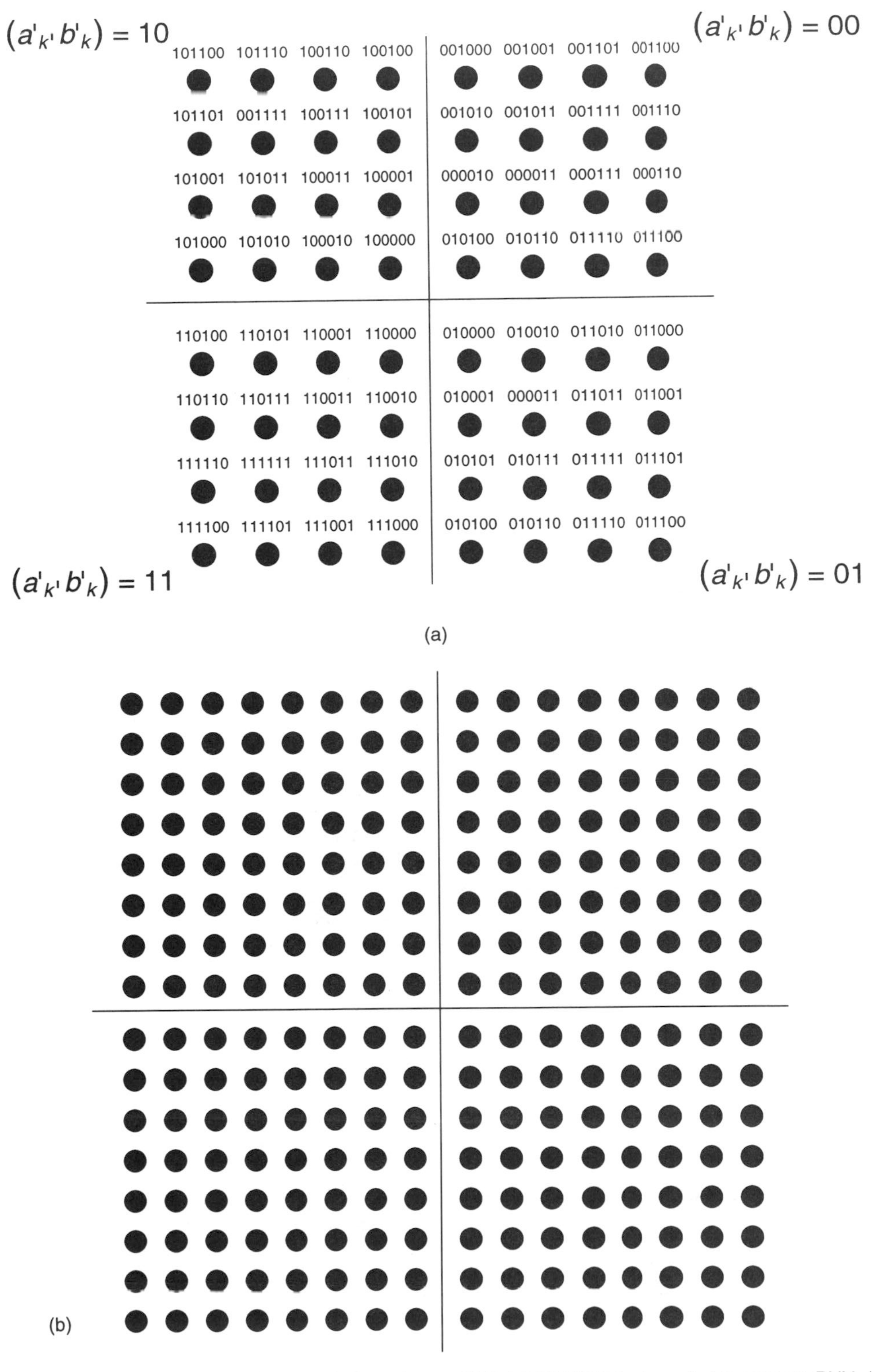

Figure 15.22 Constellations for type A downstream PHY. (*a*) 64-QAM for type A downstream PHY; (*b*) 256-QAM for type A downstream PHY. (From IEEE Std. 802.14, Draft R2, Figure P-37 [Ref. 7].)

15.7.7 Upstream Physical Layer Specification

15.7.7.1 Upstream Spectrum Allocation. The subsplit band (i.e., frequencies between 5 MHz and 42 MHz) is allocated for upstream transmission. In some cable plants, additional frequency spectra for upstream transmission is intended for future use, called *midsplit* and *highsplit* bands. The midsplit extends from 5 MHz to 108 MHz and the highsplit covers the range between 5 MHz and 174 MHz. In some locations, the original subsplit band is modified as 5 MHz to 50 MHz, 5 MHz to 65 MHz, and 5 MHz to 48 MHz in North America, Europe, and Japan, respectively.

15.7.7.2 Upstream Channel Spacing. Channel spacing depends on the modulation rate employed. The minimum channel spacing is:

$$(1 + \alpha) \times R_{S(\min)},$$

where α and $R_{S(\min)}$ denotes spectral roll-off factor and minimum symbol rate, respectively.

15.7.7.3 Carrier Frequencies. Carrier frequencies, f_c, for upstream transmission are selected that:

$$f_c = n \times (32 \text{ kHz}).$$

15.7.7.4 Timing and Synchronization. The headend transmits in the downstream time-stamp messages that are used by a station to establish upstream TDMA synchronization.

15.7.7.4.1 Inaccuracy Tolerance. In order to properly synchronize the upstream transmissions originating from different stations in the TDMA mode, a ranging offset is applied by the station as a delay correction value to the headend time acquired at the station. This process is called *ranging*. The ranging offset is an advancement equal roughly to the round-trip delay of the station from the headend. Upon successful reception of one or more up-stream transmissions from a station, the headend provides the station with a feedback message containing this ranging offset. The accuracy of the ranging offset should be no worse than TBD symbol duration, and resolution thereof is TBD of headend time increment. After the first iteration of ranging, the headend continues to send ranging adjustments, when necessary, to the station. A negative value for

Table 15.2 Standard Data and Modulation Rates, Upstream

Data Rate (Mbit/s) (Mbps)	QPSK Modulation Rate (Mbaud)	16-QAM Modulation Rate (Mbaud)
0.512	0.256	N/A
1.024	0.512	0.256
2.048	1.024	0.512
4.096	2.048	1.024
8.192	4.096	2.048
16.384	N/A	4.096

Source: Table P-14, IEEE 802.14, Draft R2, Ref. 7.

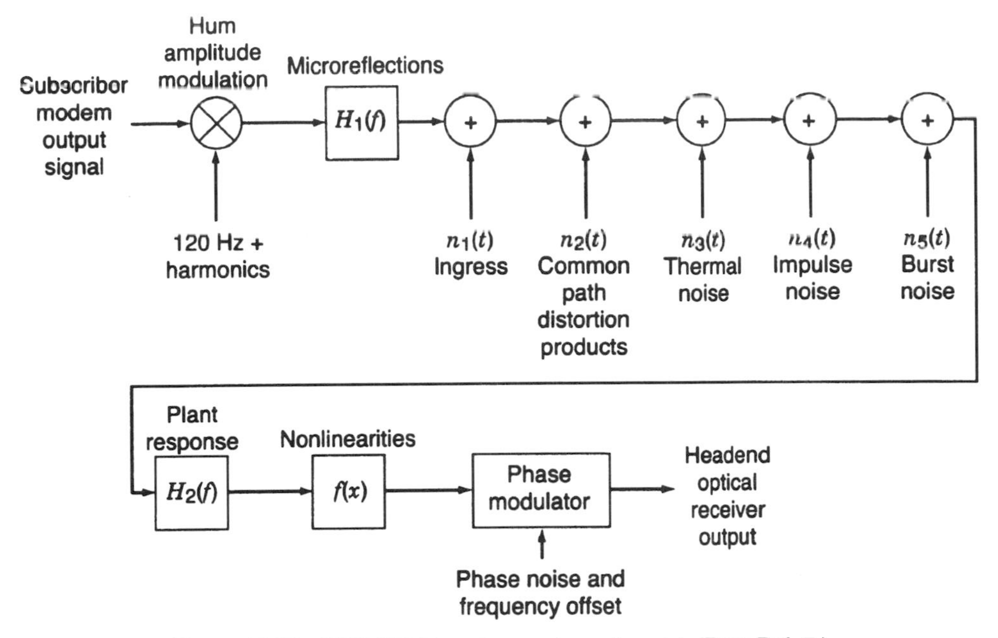

Figure 15.23 IEEE 802.14 upstream channel model. (From Ref. 7.)

the ranging adjustment indicates that the ranging offset at that station is to be decreased, resulting in later times of transmission at the station. The station implements the ranging adjustment with a resolution of at most 1 symbol duration for the symbol rate in use for the given burst. In addition, the accuracy of the station burst transmission timing is TBD ± TBD symbol, relative to the minislot boundaries that are derived at the station based on ideal processing of time-stamp message signals received from the headend.

15.7.7.5 Modulation and Bit Rates. QPSK and 16-QAM are the modulation choices for upstream transmission. Table 15.2 tabulates the upstream data rates and modulation rates. Figure 15.23 is the upstream channel model. The reader should note the numerous upstream channel impairments illustrated in the model.

REVIEW EXERCISES

1. Define a CATV headend. What are its functions?
2. List at least three impairments we can expect from a broadband CATV amplifier (downstream).
3. A signal splitter divides a signal in half, splitting into two equal power levels if the input to a 3-dB splitter were −7 dBm (in the power domain) and the output on each leg would be −10 dBm. Is this a true statement? What is missing here?
4. What was/is the purpose of LOS microwave at a CATV headend?
5. What is the purpose of a set-top converter?
6. What does the term *beat* mean in CATV parlance?
7. Define *composite triple beat.*

8. A signal level is measured at 0.5 V rms. What is the equivalent value in dBmV?

9. What dBmV level can we expect in the CATV minimum noise model?

10. When calculating S/N for TV reception on a CATV system, what is the common value of the noise weighting improvement factor?

11. If the C/N of a CATV system is 40 dB, what is the equivalent S/N?

12. These are ten identical CATV broadband amplifiers in cascade. Each amplifier has a 7-dB noise figure. What is the thermal noise level in dBmV at the output of the tenth amplifier? Use Eq. (15.8).

13. What is an acceptable level down (below wanted signal level) for Xm?

14. A certain CATV system has 22 amplifiers in cascade with an Xm per amplifier of −89 dB. What is Xm_{sys}?

15. Why are levels on feeder systems usually higher than mainline trunk systems?

16. What does *tilt* mean when discussing coaxial cable (CATV parlance)? How do we overcome the tilt?

17. Give three advantages of an HFC CATV system over a straight coaxial cable system.

18. What is a *tap*?

19. Differentiate and give advantages/disadvantages of AM and FM fiber links as part of an HFC system.

20. From a bandwidth viewpoint, why is upstream disadvantaged over downstream?

21. Why is upstream at a disadvantage over downstream from a noise viewpoint?

22. What is *ingress noise*?

23. List at least four telecommunication services that the IEEE 802.14 specification supports.

24. In the IEEE 802.14 system, where does master timing reside and how is the network synchronized?

25. What is the purpose of *ranging*?

26. What are the two types of modulation that may be used on the upstream 802.14 network?

27. List at least four different impairments we might expect to encounter in the 802.14 upstream environment.

REFERENCES

1. W. O. Grant, *Cable Television*, 3rd ed., GWG Associates, Schoharie, NY, 1994.
2. K. Simons, *Technical Handbook for CATV Systems*, 3rd ed., Jerrold Electronics Corp., Hatboro, PA, 1968.
3. E. R. Bartlett, *Cable Television Technology and Operations*, McGraw-Hill, New York, 1990.
4. D. N. Carson, in "CATV Amplifiers: Figure of Merit and the Coefficient System," in *1966*

IEEE International Convention Record, Part I, Wire and Data Communications, pp. 87–97, IEEE, New York, 1966.

5. *Electrical Performance for Television Transmission Systems*, EIA/TIA-250C, Telecommunication Industry Association, Washington, DC, 1990.
6. *Lightwave Buyers' Guide Issue*, Pennwell Publishing Co., Tulsa, OK, 1997.
7. *Multimedia Modem Protocol for Hybrid Fiber-Coax Metropolitan Area Networks*, IEEE Std. 802.14, Draft R2, IEEE, New York, 1997.
8. Private communication, Robert Fuller, Chairman, IEEE 802.14 Committee, April 4, 1997.
9. *Digital Multi-Programme Systems for Television, Sound and Data Services for Cable Distribution*, ITU-T Rec. J.83, ITU, Geneva, Sept. 1995.
10. *The IEEE Standard Dictionary of Electrical and Electronic Terms*, 6th ed., IEEE, New York, 1996.

16

CELLULAR AND PCS RADIO SYSTEMS

16.1 INTRODUCTION

The cellular radio business has expanded explosively since 1980 and continues to expand rapidly. There are several explanations for this popularity. It adds a new dimension to wired PSTN services. In our small spheres of everyday living, we are never away from the telephone, no matter where we are. Outside of industrialized nations, there are long waiting lists for conventional (wired) telephone installations. Go down to the local cellular radio store, and you will have telephone service within the hour. We have found that cellular service augments local telephone service availability. When our local service failed for several days, our cellular telephone worked just fine, although air time was expensive.

Enter PCS (personal communications services). Does it supplement/complement cellular radio or is it a competitor? It is an extension of cellular, certainly in concept. It uses much lower power and has a considerably reduced range. Rappaport (Ref. 1) points out that cellular is hierarchical in nature when connecting to the PSTN; PCS is not. It is hierarchical in that an MTSO (mobile telephone switching office) controls and interfaces up to hundreds of base stations, which connect to mobile users. According to the reference, PCS base stations connect directly to the PSTN. However, a number of PCS strategies have a hierarchy similar to cellular where an MSC (mobile switching center) provides the connectivity to the PSTN. Cellular radio systems operate in the 800-MHz and 900-MHz band; in the United States narrowband PCS operates in the 900-MHz band, and wideband PCS operates in the band 1850-MHz to 1975-MHz. Other PCS operations are specialized, such as the wireless PABX, wireless LAN (WLAN), and wireless local loop (WLL). By WLL we mean a transmission method that will operate in lieu of, supplement, or complement the telephone subscriber loop based on a wire pair.

16.1.1 Background

The earliest radio techniques served a mobile community, namely, ocean vessels. This was followed by vehicular mobile radio including aircraft. Prior to World War II there were police- and ambulance-dispatching systems, followed by growth in the airline industry. However, not until Bell Telephone Laboratories published the famous issue of the *Bell System Technical Journal* devoted entirely to a *new* system called AMPS (advanced mobile phone system), did the cellular idea take hold. It remains the underlying cellular system for the United States and in some Latin-American countries. It uses FM radio, allocating 30 kHz per voice channel.

AMPS set the scene for explosive cellular radio growth and usage. What set AMPS apart from previous mobile radio systems is that it was designed to interface with the PSTN. It was based on an organized scheme of adjoining cells and had a unique capability of handoff when a vehicle moves through one cell to another, or when another cell receives a higher signal level from the vehicle, it will then take over the call. Vehicles can *roam* from one service area to another with appropriate handoffs.

In the late 1980s there was pressure to convert cellular radio from the bandwidth-wasteful AMPS to some sort of digital regime. As the reader reviews this chapter, it will be seen that digital is also bandwidth-wasteful, even more so than analog FM. Various ways are described on how to remedy the situation: first by reducing the bandwidth of a digital voice channel, and second by the access/modulation scheme proposed. Of this latter proposal, two schemes are on the table in North America: a TDMA scheme and a CDMA scheme. They are radically different and competing.

Meanwhile, the Europeans critiqued our approaches and came up with a better mousetrap. It is called Ground System Mobile (GSM) (from the French), and there is some pressure that it be adopted in the United States. GSM is a TDMA scheme, fairly different from the U.S. TIA standard (IS-54C).

As mentioned earlier, PCS is an outgrowth of cellular radio; it uses a cellular concept. The cells, however, are much smaller, under 1-km diameter. RF power is much lower. As with cellular radio, TDMA and CDMA are vying for the national access/modulation method. Unlike the North American popular press, which discriminates between PCS and cellular radio, ITU-R takes a more mature and reasonable view of the affair by placing the two in the same arena. Earlier, CCIR/ITU-R called their conceptual PCS *future public land mobile telecommunication system* (FPLMTS). The name has now changed to UMTS (*universal mobile telecommunication system*). The FPLMTS/UMTS concept breaks down into three terrestrial operational areas: (1) indoor environments (range to 100 m), (2) outdoor environments (100 m to 35 km) for more rural settings, and (3) an intermediate region called outdoor-to-indoor environments, where building penetration is a major theme. They also describe satellite environments.

16.1.2 Scope and Objective

This chapter presents an overview of mobile and personal communications. Much of the discussion deals with cellular radio and extends this thinking inside buildings. The coverage most necessarily includes propagation for the several environments, propagation impairments, methods to mitigate those impairments, access techniques, bandwidth limitations, and ways around this problem. It will cover several mobile radio standards and compare a number of existing and planned systems. The chapter objective is to provide an appreciation of mobile/personal communications. Space limitations force us to confine the discussion to what might loosely be called "land mobile systems."

16.2 BASIC CONCEPTS OF CELLULAR RADIO

Cellular radio systems connect a mobile terminal to another user, usually through the PSTN. The "other user" most commonly is a telephone subscriber of the PSTN. However, the other user may be another mobile terminal. Most of the connectivity is extending "plain old telephone service" (POTS) to mobile users. Data and facsimile services

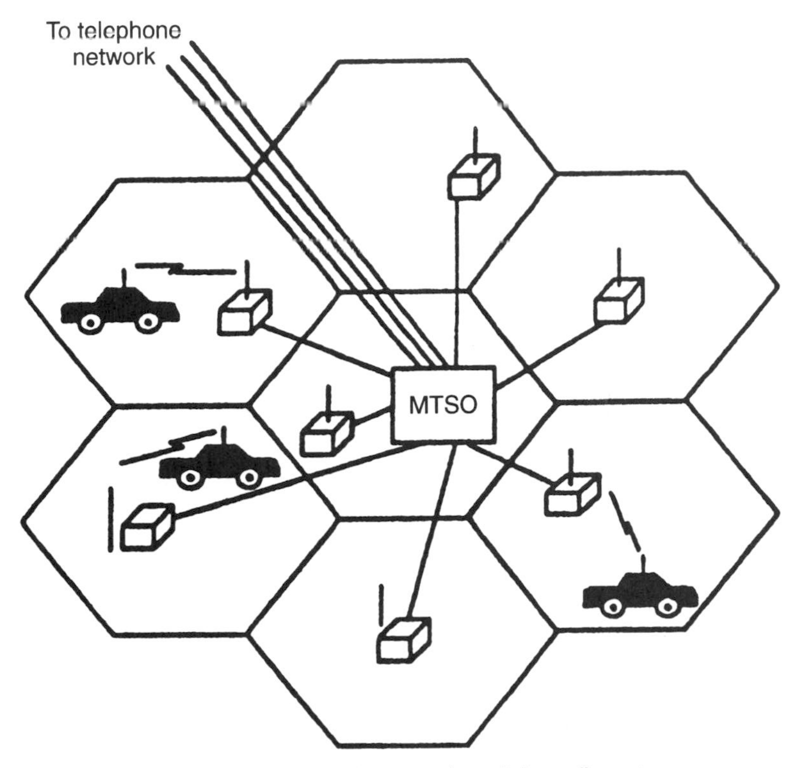

Figure 16.1 Conceptual layout of a cellular radio system.

are in various stages of implementation. Some of the terms used in this section have a strictly North American flavor.

Figure 16.1 illustrates a conceptual layout of a cellular radio system. The heart of the system for a specific serving area is the MTSO. The MTSO is connected by a trunk group to a nearby telephone exchange providing an interface to and connectivity with the PSTN.

The area to be served by a *cellular geographic serving area* (CGSA) is divided into small geographic cells, which ideally are hexagonal.[1] Cells are initially laid out with centers spaced about 4–8 m (6.4–12.8 km) apart. The basic system components are the cell sites, the MTSO, and mobile units. These mobile units may be hand-held or vehicle-mounted terminals.

Each cell has a radio facility housed in a building or shelter. The facility's radio equipment can connect and control any mobile unit within the cell's responsible geographic area. Radio transmitters located at the cell site have a maximum effective radiated power (ERP) of 100 W.[2] Combiners are used to connect multiple transmitters to a common antenna on a radio tower, usually between 50-ft and 300-ft (15-m and 92-m) high. Companion receivers use a separate antenna system mounted on the same tower. The receive antennas are often arranged in a space diversity configuration.

The MTSO provides switching and control functions for a group of cell sites. A method of connectivity is required between the MTSO and the cell site facilities. The

[1]CGSA is a term coined by the U.S. FCC. We do not believe it is used in other countries.

[2]Care must be taken with terminology. In this instance ERP and EIRP are *not* the same. The reference antenna in this case is the dipole, which has a 2.15-dBi gain.

MTSO is an electronic switch and carries out a fairly complex group of processing functions to control communications to and from mobile units as they move between cells as well as to make connections with the PSTN. Besides making connectivity with the public network, the MTSO controls cell site activities and mobile actions through command-and-control data channels. The connectivity between cell sites and the MTSO is often via DS1 on wire pairs or on microwave facilities, the latter being the most common.

A typical cellular mobile unit consists of a control unit, a radio transceiver, and an antenna. The control unit has a telephone handset, a push-button keypad to enter commands into the cellular/telephone network, and audio and visual indications for customer alerting and call progress. The transceiver permits full-duplex transmission and reception between a mobile and cell sites. Its ERP is nominally 6 W. Hand-held terminals combine all functions into one small package that can easily be held in one hand. The ERP of a hand-held is a nominal 0.6 W.

In North America, cellular communication is assigned a 25-MHz band between 824 MHz and 849 MHz for mobile unit-to-base transmission and a similar band between 869 MHz and 894 MHz for transmission from base to mobile.

The first and most widely implemented North American cellular radio system was called AMPS (advanced mobile phone system). The original system description was contained in an entire issue of the *Bell System Technical Journal* (*BSTJ*) of January 1979. The present AMPS is based on 30-kHz channel spacing using frequency modulation. The peak deviation is 12 kHz. The cellular bands are each split into two to permit competition. Thus only 12.5 MHz is allocated to one cellular operator for each direction of transmission. With 30-kHz spacing, this yields 416 channels. However, nominally 21 channels are used for control purposes with the remaining 395 channels available for cellular end-users.

Common practice with AMPS is to assign 10–50 channel frequencies to each cell for mobile traffic. Of course the number of frequencies used depends on the expected traffic load and the blocking probability. Radiated power from a cell site is kept at a relatively low level with just enough antenna height to cover the cell area. This permits frequency reuse of these same channels in nonadjacent cells in the same CGSA with little or no cochannel interference. A well-coordinated frequency reuse plan enables tens of thousands of simultaneous calls over a CGSA.

Figure 16.2 illustrates one frequency reuse method. Here four channel frequency groups are assigned in a way that avoids the same frequency set in adjacent cells. If there were uniform terrain contours, this plan could be applied directly. However, real terrain conditions dictate further geographic separation of cells that use the same frequency set. Reuse plans with 7 or 12 sets of channel frequencies provide more physical separation and are often used depending on the shape of the antenna pattern employed.

With user growth in a particular CGSA, cells may become overloaded. This means that grade of service objectives are not being met due to higher than planned traffic levels during the busy hour (BH; see Section 4.2.1). In these cases, congested cells can be subdivided into smaller cells, each with its own base station, as shown in Figure 16.3. With smaller cells, lower transmitter power and antennas with less height are used, thus permitting greater frequency reuse. These subdivided cells can be split still further for even greater frequency reuse. However, there is a practical limit to cell splitting, often with cells with a 1-mi (1.6-km) radius.

Radio system design for cellular operation differs from that used for LOS microwave operation. For one thing, mobility enters the picture. Path characteristics are constantly changing. Mobile units experience multipath scattering, reflection, and/or diffraction by

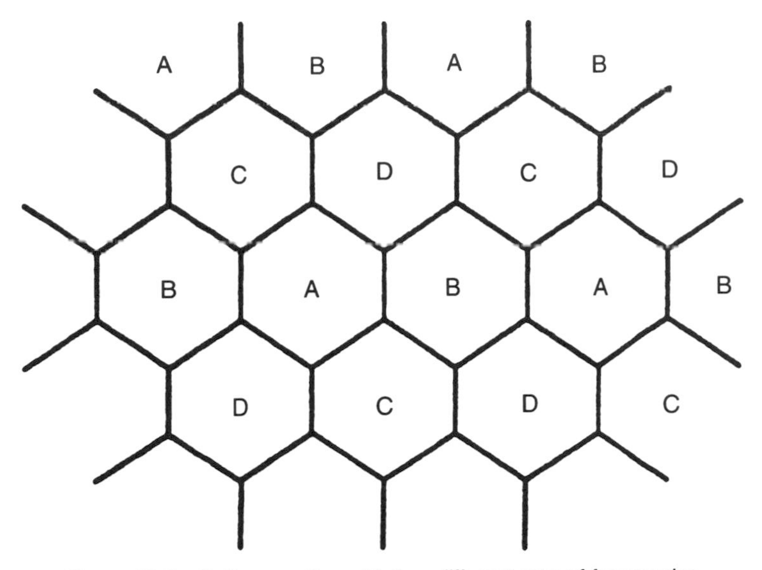

Figure 16.2 Cell separation with four different sets of frequencies.

obstructions and buildings in the vicinity. There is shadowing, often very severe. The resulting received signal under these conditions varies randomly as the sum of many individual waves with changing amplitude, phase, and direction of arrival. The statistical autocorrelation distance is on the order of one-half wavelength (Ref. 2). Space diversity at the base station tends to mitigate these impairments.

In Figure 16.1, the MTSO is connected to each of its cell sites by a voice trunk for each of the radio channels at the site. Also, two data links (AMPS design) connect the

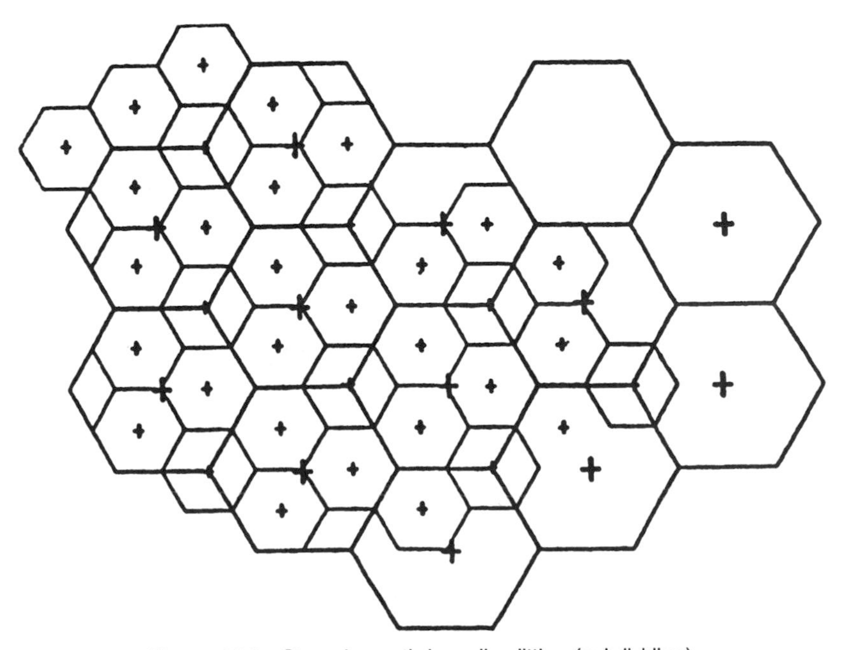

Figure 16.3 Staged growth by cell splitting (subdividing).

MTSO to each cell site. These data links transmit information for processing calls and for controlling mobile units. In addition to its "traffic" radio equipment, each cell site has installed signaling equipment, monitoring equipment, and a *setup* radio to establish calls.

When a mobile unit becomes operational, it automatically selects the setup channel with the highest signal level. It then monitors that setup channel for incoming calls destined for it. When an incoming call is sensed, the mobile terminal in question again samples signal levels of all appropriate setup channels so it can respond through the cell site offering the highest signal level, and then tunes to that channel for response. The responsible MTSO assigns a vacant voice channel to the cell in question, which relays this information via the setup channel to the mobile terminal. The mobile terminal subscriber is then alerted that there is an incoming call. Outgoing calls from mobile terminals are handled in a similar manner.

While a call is in progress, the serving cell site examines the mobile's signal level every few seconds. If the signal level drops below a prescribed level, the system seeks another cell to handle the call. When a more appropriate cell site is found, the MTSO sends a command, relayed by the old cell site, to change frequency for communication with the new cell site. At the same time, the landline subscriber is connected to the new cell site via the MTSO. The periodic monitoring of operating mobile units is known as *locating*, and the act of changing channels is called *handover*. Of course, the functions of locating and handover are to provide subscribers satisfactory service as a mobile unit traverses from cell to cell. When cells are made smaller, handovers are more frequent.

The management and control functions of a cellular system are quite complex. Handover and locating are managed by signaling and supervision techniques, which take place on the setup channel. The setup channel uses a 10-kbps data stream that transmits paging, voice channel designation, and overhead messages to mobile units. In turn, the mobile unit returns page responses, origination messages, and order confirmations.

Both digital messages and continuous supervision tones are transmitted on the voice radio channel. The digital messages are sent as a discontinuous "blank-and-burst" in-band data stream at 10 kbps and include order and handover messages. The mobile unit returns confirmation and messages that contain dialed digits. Continuous positive supervision is provided by an out-of-band 6-kHz tone, which is modulated onto the carrier along with the speech transmission.

Roaming is a term used for a mobile unit that travels such distances that the route covers more than one cellular organization or company. The cellular industry is moving toward technical and tariffing standardization so that a cellular unit can operate anywhere in the United States, Canada, and Mexico.

16.3 RADIO PROPAGATION IN THE MOBILE ENVIRONMENT

16.3.1 Propagation Problem

Line-of-sight microwave and satellite communications covered in Chapter 9 dealt with fixed systems. Such systems are optimized. They are built up and away from obstacles. Sites are selected for best propagation.

This is not so with mobile systems. Motion and a third dimension are additional variables. The end-user terminal often is in motion; or the user is temporarily fixed, but that point can be anywhere within a serving area of interest. Whereas before we dealt with point-to-point, here we deal with point-to-multipoint.

One goal in line-of-sight microwave design was to stretch the distance as much as possible between repeaters by using high towers. In this chapter there are some overriding circumstances where we try to limit coverage extension by reducing tower heights, what we briefly introduced in Section 16.2. Even more important, coverage is area coverage, where shadowing is frequently encountered, Examples are valleys, along streets with high buildings on either side, verdure such as trees, and inside buildings, to name a few typical situations. There are two notable results. Transmission loss increases notably and such an environment is rich with multipath scenarios. Paths can be highly dispersive, as much as 10 μs of delay spread (Ref. 3). If a user is in motion, Doppler shift can be expected.

The radio-frequency bands of interest are UHF (ultra high frequency, the frequency band from 300–3000 MHz), especially around 800 MHz and 900 MHz, and 1700 MHz to 2000 MHz. In certain parts of the world, there is usage in the 400-MHz band.

16.3.2 Propagation Models

We concentrate on cellular operation. There is a fixed station (FS) and mobile stations (MSs) moving through the cell. A cell is the area of responsibility of the fixed station, a cell site. It usually is pictured as a hexagon in shape, although its propagation profile is more like a circle with the fixed station in its center. Cell radii vary from 1 km (0.6 mi) in heavily built-up urban areas to 30 km (19 mi) or somewhat more in rural areas.

16.3.2.1 Path Loss or Transmission Loss. We recall the free space loss (FSL) formula in Section 9.2.3. It simply stated that FSL was a function of the square of the distance and the square of the frequency plus a constant. It is a very useful formula if the strict rules of obstacle clearance are obeyed. Unfortunately, in the cellular situation, it is impossible to obey these rules. Then to what extent must this free space loss formula be modified by the proximity of the earth, the effects of trees, buildings, and hills in, or close to, the transmission path?

There have been a number of models that have been developed that are used as a basis for the calculation of transmission loss, several assumptions are made:

- That we will always use the same frequency band, often 800 MHz or 900 MHz. Thus it is common to drop the frequency term (the 20 log$F$ term) in the FSL formula and include a constant that covers the frequency term. If we wish to use the model for another band, say, 1800 MHz, a scaling factor is added.
- That we will add a term to compensate for the usual great variance between the cell site antenna height when compared with the mobile (or hand-held) antenna height. We often call this the height-gain function, and it tends to give us an advantage. It is often expressed as $-20\ \log(h_T h_R)$ where H_T is the height of the transmit antenna (cell site) and H_R is the height of the receive antenna (on the mobile platform). These are comparative heights. Commonly, the mobile platform antenna height is taken as 6 ft or 3 m.
- That there is a catch-all term for the remainder of the losses, which in some references is expressed as β (in dB);
- That at least three models express the free space loss as just 40 logd_m (d is distance in meters).

16.3.2.2 Okumura Model. Okumura et al. (Ref. 4) carried out a detailed analysis for path predictions around Tokyo for mobile terminals. Hata (Ref. 5) published an empirical formula based on Okumura's results to predict path loss. The Okumura/Hata model is probably one of the most widely applied path loss models in the world for cellular application. The formula and its application follow.

$$\begin{aligned} L_{\mathrm{dB}} &= 69.55 + 26.16\log f - 13.82\log h_{\mathrm{t}} - A(h_{\mathrm{r}}) \\ &\quad + (44.9 - 6.55\log h_{\mathrm{t}})\log d, \end{aligned} \tag{16.1}$$

where r is between 150 MHz and 1500 MHz;
h_{t} is between 30 m and 300 m; and
d is the path distance and is between 1 km and 20 km.

$A(h_{\mathrm{r}})$ is the correction factor for mobile antenna height and is computed as follows:

For a small- or medium-size city,

$$A(h_{\mathrm{r}}) = (1.1\log f - 0.7)h_{\mathrm{r}} - (1.56\log f - 0.8), (\mathrm{dB}) \tag{16.2a}$$

where h_{r} is between 1 and 10 m.

For a large city,

$$A(h_{\mathrm{r}}) = 3.2[\log(11.75h_{\mathrm{r}})]^2 - 4.97(\mathrm{dB}) \tag{16.2b}$$

where ($f \geq 400$ MHz).

Example. Let f = 900 MHz, h_{t} = 40 m, h_{r} = 5 m, and d = 10 km. Calculate $A(h_{\mathrm{r}})$ for a medium-size city.

$$\begin{aligned} A(h_{\mathrm{r}}) &= 12.75 - 3.8 = 8.95 \text{ dB} \\ L_{\mathrm{dB}} &= 69.55 + 72.28 - 22.14 - 8.95 + 34.4 \\ &= 145.15 \text{ dB}. \end{aligned}$$

16.3.2.3 Building Penetration. For a modern multistory office building at 864 MHz and 1728 MHz, transmission loss (L_{dB}) includes a value for clutter loss $L(v)$ and is expressed as follows:

$$L_{\mathrm{dB}} = L(v) + 20\log d + n_{\mathrm{f}}a_{\mathrm{f}} + n_{\mathrm{w}}a_{\mathrm{w}}, \tag{16.3}$$

where the attenuation in dB of the floors and walls was a_{f} and a_{w}, and the number of floors and walls along the line d were n_{f} and n_{w}, respectively. The values of $L(v)$ at 864 MHz and 1728 MHz were 32 dB and 38 dB, with standard deviations of 3 dB and 4 dB, respectively (Ref. 3).

Another source (Ref. 6) provided the following information: At 1650 MHz the floor loss factor was 14 dB, while the wall losses were 3–4 dB for double plasterboard and 7–9 dB for breeze block or brick. The parameter $L(v)$ was 29 dB. When the propagation frequency was 900 MHz, the first floor factor was 12 dB and $L(v)$ was 23 dB. The higher value for $L(v)$ at 1650 MHz was attributed to a reduced antenna aperture at this frequency compared to 900 MHz. For a 100-dB path loss, the base station and mobile terminal distance exceeded 70 m on the same floor, was 30 m for the floor above, and 20 m for the floor above that, when the propagation frequency was 1650 MHz. The corresponding distances at 900 MHz were 70 m, 55 m, and 30 m. Results will vary from building to building, depending on the type of construction of the building, the furniture and equipment it houses, and the number and deployment of the people who populate it.

16.4 IMPAIRMENTS: FADING IN THE MOBILE ENVIRONMENT

16.4.1 Introduction

Fading in the mobile situation is quite different from the static line-of-sight (LOS) microwave situation discussed in Section 9.2.4. In this case radio paths are not optimized as in the LOS environment. The mobile terminal may be fixed throughout a telephone or data call, but is more apt to be in motion. Even the hand-held terminal may well have micromotion. When a terminal is in motion, the path characteristics are constantly changing.

Multipath propagation is the rule. Consider the simplified pictorial model in Figure 16.4. Commonly, multiple rays reach the receive antenna, each with its own delay. The constructive and destructive fading can become quite complex. We must deal with both reflection and diffraction.[3] Energy will arrive at the receive antenna reflected off sides of buildings, towers, streets, and so on. Energy will also arrive diffracted from knife edges (e.g., building corners) and rounded obstacles (e.g., water tanks, hill tops).

Because the same signal arrives over several paths, each with a different electrical length, the phases of each path will be different, resulting in constructive and destructive amplitude fading. Fades of 20 dB are common, and even 30-dB fades can be expected.

On digital systems, the deleterious effects of multipath fading can be even more severe. Consider a digital bit stream to a mobile terminal with a transmission rate of 1000 bps. Assuming NRZ coding, the bit period would be 1 ms (bit period = 1/bit rate). We find the typical multipath delay spread may be on the order of 10 μs. Thus

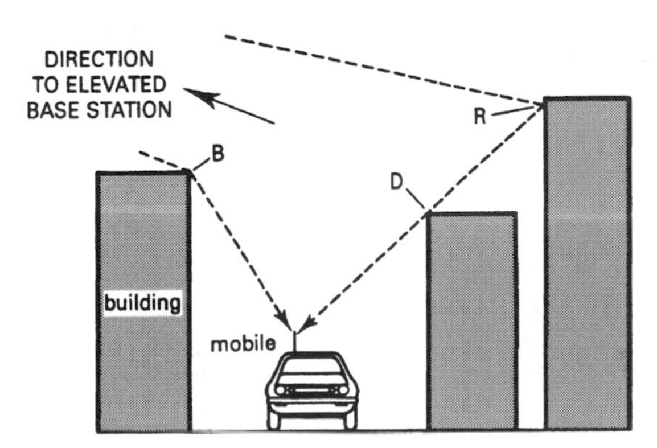

Figure 16.4 Mobile terminal in an urban setting. R = reflection; D = diffraction.

[3]*Diffraction* is defined by the IEEE (Ref. 7) as "The deviation of the direction of energy flow of a wave (ray beam), not attributable to reflection or refraction, when it passes an obstacle, a restricted aperture or other inhomogeneities in a medium.

delayed energy will spill into a subsequent bit (or symbol) for the first 10 μs of the bit period and will have no negative effect on the bit decision. If the bit stream is 64,000 bps, then the bit period is 1/64,000 or 15 μs. Destructive energy from the previous bit (symbol) will spill into the first two-thirds of the bit period, well beyond the midbit sampling point. This is typical intersymbol interference (ISI), and in this case there is a high probability that there will be a bit error. The bottom line is that the destructive potential of ISI increases as the bit rate increases (i.e., as the bit period decreases).

16.4.2 Diversity: A Technique to Mitigate the Effects of Fading and Dispersion

16.4.2.1 Scope. We discuss diversity to reduce the effects of fading and to mitigate dispersion. Diversity was briefly covered in Section 9.2.5, where we dealt with LOS microwave. In that section we discussed frequency and space diversity. In principle, such techniques can be employed either at the base station and/or at the mobile unit, although different problems have to be solved for each. The basic concept behind diversity is that two or more radio paths carrying the same information are relatively uncorrelated, when one path is in a fading condition, often the other path is not undergoing a fade. These separate paths can be developed by having two channels separated in frequency. The two paths can also be separated in space and in time.

When the two (or more) paths are separated in frequency, we call this frequency diversity. However, there must be at least some 2% or greater frequency separation for the paths to be comparatively uncorrelated. This is because, in the cellular situation, we are so short of spectrum, using frequency diversity (i.e., using a separate frequency with redundant information) is essentially out of the question. So it will not be discussed further except for its implicit use in CDMA.

16.4.2.2 Space diversity. Space diversity is commonly employed at cell sites, and two separate receive antennas are required, separated in either the horizontal or vertical plane. Separation of the two antennas vertically is impractical for cellular receiving systems. Horizontal separation, however, is quite practical. The space diversity concept is illustrated in Figure 16.5.

One of the most important factors in space diversity design is antenna separation, to achieve the necessary signal decorrelation. There is a set of empirical rules for the cell site, and another set of rules for the mobile unit. Space diversity antenna separation, shown as distance D in Figure 16.5, varies not only as a function of the correlation

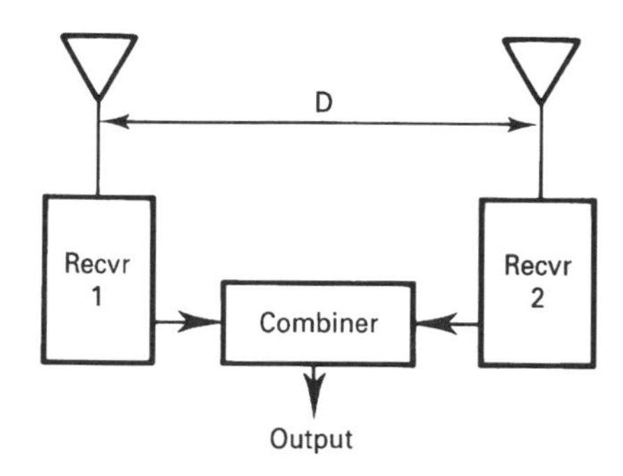

Figure 16.5 The space diversity concept.

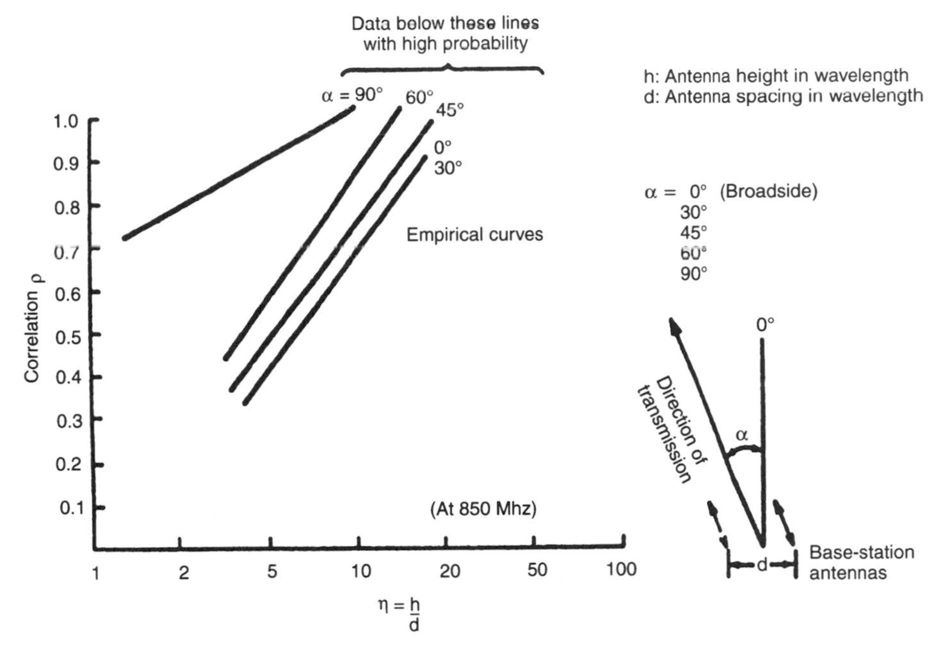

Figure 16.6 Correlation coefficient ρ versus the parameter η for two receive antennas in different orientations. (From Ref. 8, Figure 6.4, reprinted with permission.)

coefficient but also as a function of antenna height, h. The wider the receive antennas are separated, the lower the correlation coefficient and the more uncorrelated the diversity paths are. Sometimes we find that, by lowering the antennas as well as adjusting the distance between them, we can achieve a very low correlation coefficient. However, we might lose some of the height-gain factor.

Lee (Ref. 8) proposes a new parameter η, where

$$\eta = (\text{antenna height})/(\text{antenna separation}) = h/d. \tag{16.4}$$

In Figure 16.6 we relate the correlation coefficient (ρ) with η, where α is the orientation of the antenna regarding the incoming signal from the mobile unit. Lee recommends a value of $\rho = 0.7$. Lower values are unnecessary because of the law of diminishing returns. There is much more fading advantage achieved from $\rho = 1.0$ to $\rho = 0.7$ than from $\rho = 0.7$ to $\rho = 0.1$.

Based on $\rho = 0.7$ and $\eta = 11$, from Figure 16.6 we can calculate antenna separation values (for 850-MHz operation). For example, if $h = 50$ ft (16 m), we can calculate d using formula 16.4:

$$d = h/\eta = 50/11 = 4.5 \text{ ft } (1.36 \text{ m}).$$

For an antenna 120-ft (36.9-m) high, we find that $d = 120/11 = 10.9$ ft (or 3.35 m) (from Ref. 8).

16.4.2.2.1 Space Diversity on a Mobile Platform. Lee (Ref. 8) discusses both vertically separated and horizontally separated antennas on a mobile unit. For the vertical case, 1.5λ is recommended for the vertical separation case and 0.5λ for the horizontal separation case.[4] At 850 MHz, $\lambda = 35.29$ cm. Then $1.5\lambda = 1.36$ ft or 52.9 cm. For 0.5λ, the value is 0.45 ft or 17.64 cm.

16.4.3 Cellular Radio Path Calculations

Consider the path from the fixed cell site to the mobile platform. There are several mobile receiver parameters that must be considered. The first to be derived are signal quality minima from EIA/TIA IS-19B (Ref. 9).

The minimum SINAD (signal + interference + noise and distortion to interference + noise + distortion ratio) is 12 dB. This SINAD equates to a threshold of −116 dBm or 7 μV/m. This assumes a cellular transceiver with an antenna with a net gain of 1 dBd (dB over a dipole). The gross antenna gain is 2.5 dBd with a 1.5-dB transmission line loss. A 1-dBd gain is equivalent to a 3.16-dBi gain (i.e., 0 dBd = 2.15 dBi). Furthermore, this value equates to an isotropic receive level of −119.16 dBm (Ref. 9).

One design goal for a cellular system is to more or less maintain a cell boundary at the 39-dB μ contour (Ref. 10). Note that 39 dBμ = −95 dBm (based on a 50-Ω impedance at 850 MHz). Then at this contour, a mobile terminal would have a 24.16-dB fade margin.

If a cellular transmitter has a 10-w output per channel and an antenna gain of 12 dBi and 2-dB line loss, the EIRP would be +20 dBW or +50 dBm. The maximum path loss to the 39-dBμ contour would be +50 dBm − (−119.16 dBm) or 169 dB.[5]

16.5 CELLULAR RADIO BANDWIDTH DILEMMA

16.5.1 Background and Objectives

The present cellular radio bandwidth assignment in the 800 MHz and 900 MHz bands cannot support the demand for cellular service, especially in urban areas in the United States and Canada. AMPS, widely used in North and South America and elsewhere, requires 30 kHz per voice channel. The system employs FDMA (frequency division multiple access), much like the FDMA/DAMA system described in Section 9.3.5.3. Remember that the analog voice channel is a nominal 4 kHz channel, and 30 kHz is about seven times that value.

The trend is to convert cellular radio to a digital format. Digital transmission, as described in Chapter 6, is notoriously wasteful of bandwidth when compared with the 4-kHz analog channel. We can show that conventional PCM requires 16 times more bandwidth than its 4-kHz analog channel counterpart. In other words, the standard PCM digital voice channel occupies 64 kHz (assuming 1 bit per Hz of bandwidth).

Cellular system designers have taken two approaches to reduce the required bandwidth. First was to use voice compression on the digital voice channel. The second approach was to use more efficient access techniques. We briefly review several techniques of speech compression and then describe two distinctly different schemes for mobile station access to the network. Of course, the real objective is to increase the ratio of users per unit bandwidth when compared with the analog AMPS access method.

[4] Remember that λ is the conventional notation for wavelength. $F\lambda - 3 \times 10^8$ m/s, where F is the frequency in Hz and λ is the wavelength in meters.

[5] The 39-dBμ contour is a threshold for good AMPS operation.

16.5.2 Bit Rate Reduction of the Digital Voice Channel

It became obvious to system designers that conversion to digital cellular required some different technique for coding speech other than conventional PCM, found in the PSTN and described in Chapter 6. The following lists some techniques that have been considered or that have been incorporated in the various systems in North America, Europe, and Japan (Ref. 11):

1. ADPCM (adaptive differential PCM). Good intelligibility and good quality; 32-kbps data transmission over the channel may be questionable;
2. Linear predictive vocoders (voice coders); 2400 bps. Adopted by U.S. Department of Defense. Good intelligibility, poor quality, especially speaker recognition;
3. Subband coding (SBC). Good intelligibility, even down to 4800 bps. Quality suffers below 9600 bps;
4. RELP (residual excited linear predictive) type coder. Good intelligibility down to 4800 bps and fair to good quality. Quality improves as bit rate increases. Good quality at 16 kbps;
5. CELP (codebook-excited linear predictive). Good intelligibility and surprisingly good quality, even down to 4800 bps. At 8 kbps, near-toll quality speech.

16.6 NETWORK ACCESS TECHNIQUES

16.6.1 Introduction

The objective of a cellular radio operation is to provide a service where mobile subscribers can communicate with any subscriber in the PSTN, where any subscriber in the PSTN can communicate with any mobile subscriber, and where mobile subscribers can communicate among themselves via the cellular radio system. In all cases the service is full duplex.

A cellular service company is allotted a radio bandwidth segment to provide this service. Ideally, for full-duplex service, a portion of the bandwidth is assigned for transmission from a cell site to mobile subscriber, and another portion is assigned for transmission from a mobile user to a cell site. Our goal here is to select an "access" method to provide this service given our bandwidth constraints.

We will discuss three generic methods of access: (1) FDMA, (2) TDMA (time division multiple access), and (3) CDMA (code division multiple access). It might be useful for the reader to review our discussion of satellite access in Section 9.3, where we described FMDA and TDMA. However, in this section, the concepts are the same, but some of our constraints and operating parameters are different. It also should be kept in mind that the access technique has an impact on overall cellular bandwidth constraints. TDMA and CDMA are much more efficient, achieving a considerably greater number of users per unit of RF bandwidth than FDMA.

16.6.2 Frequency Division Multiple Access (FDMA)

With FDMA our band of RF frequencies is divided into segments and each segment is available for one user access. Half the contiguous segments are assigned to the cell site for outbound traffic (i.e., to mobile users) and the other half to inbound. A guardband

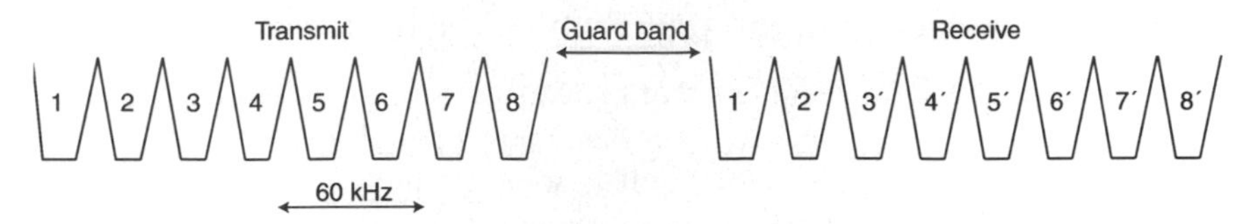

Figure 16.7 A conceptual drawing of FDMA.

is usually provided between outbound and inbound. In North America the guard band at 800 MHz is 20 MHz wide. This FDMA concept is illustrated in Figure 16.7.

Because of our concern to optimize the number of users per unit bandwidth, the key question is the actual width of one user segment. The bandwidth of a user segment is greatly determined by the information bandwidth and the modulation type. With AMPS, the information bandwidth was a single voice channel with a nominal bandwidth of 4 kHz. The modulation is FM and the bandwidth is determined by Carson's rule (Section 9.2). As we pointed out, AMPS is not exactly spectrum conservative (requiring 30 kHz per channel). On the other hand, it has a lot of redeeming features that FM provides, such as noise and interference advantage (FM capture effect).

Another approach to FDMA would be to convert the voice channel to its digital equivalent using CELP (Section 16.5.2), for example, with a transmission rate of 4.8 kbps. Let the modulation be BPSK using a raised cosine filter where the bandwidth would be 1.25% of the bit rate, or just 6 kHz per voice channel. This alone would increase the voice channel capacity five times over AMPS with its 30 kHz per channel. It should be noted that a radio carrier is normally required for each frequency slot.

16.6.3 Time Division Multiple Access (TDMA)

With TDMA we work in the time domain rather than the frequency domain of FDMA. Each user is assigned a time slot rather than a frequency segment and, during the user's turn, the full frequency bandwidth is available for the duration of the user's assigned time slot.

Let's say that there are n users and so there are n time slots. In the case of FDMA, we had n frequency segments and n radio carriers, one for each segment. For the TDMA case, only one carrier is required. Each user gains access to the carrier for $1/n$ of the time and there is generally an ordered sequence of time slot turns. A TDMA frame can be defined as cycling through n users' turns just once.

A typical TDMA frame is illustrated in Figure 16.8. One must realize that TDMA is only practical with a digital system such as PCM or any of those discussed in Section 16.5.2. As we said in Section 9.3.5.2, TDMA is a store-and-burst system.

Incoming user traffic is stored in memory and, when that user's turn comes up, that accumulated traffic is transmitted in a digital burst.

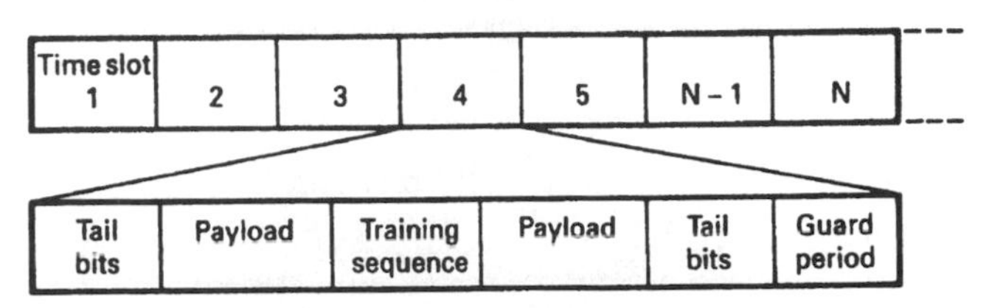

Figure 16.8 A typical TDMA frame.

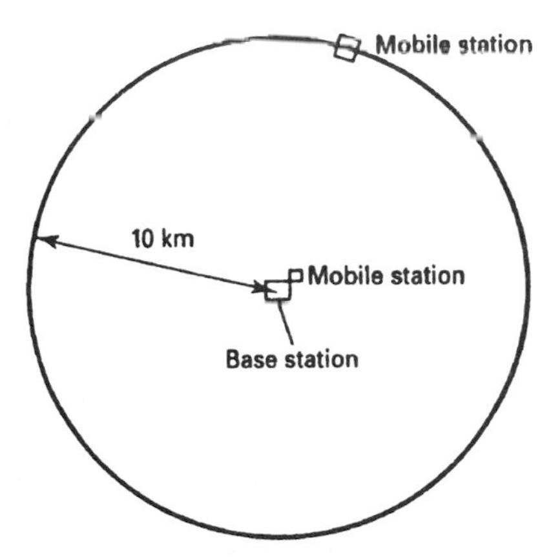

Figure 16.9 A TDMA delay scenario.

Suppose there are ten users. Let each user's bit rate be R, then a user's burst must be at least $10R$. Of course, the burst will be greater than $10R$ to accommodate a certain amount of overhead bits, as shown in Figure 16.8.

We define downlink as outbound, base station to mobile station(s), and uplink as mobile station to base station. Typical frame periods are:

North American IS-54	40 ms for six time slots
European GSM	4.615 ms for eight time slots.

One problem with TDMA, often not appreciated by many, is delay. In particular, this is delay on the uplink. Consider Figure 16.9, where we set up a scenario. A base station receives mobile time slots in a circular pattern and the radius of the circle of responsibility of that base station is 10 km. Let the velocity of a radio wave be 3×10^8 m/s. The time for the wave to traverse 1 km is $1000\ \text{m}/(3 \times 10^8)$ or 3.333 μs. In the uplink frame we have a mobile station right on top of the base station with essentially no delay and another mobile right at 10 km with $10 \times 3.33\ \mu$s or 33.3 μs delay. A GSM time slot is about 576 μs in duration. The terminal at the 10-km range will have its time slot arriving 33.3 μs late compared to the terminal with no delay. A GSM bit period is about 3.69 μs so that the late arrival mutilates about 10 bits and, unless something is done, the last bit of the burst will overlap the next burst (Refs. 3, 12).

Refer now to Figure 16.10, which illustrates GSM burst structures. Note that the access burst has a guard period of 68.25 bit durations or a *slop* of $3.69 \times 68.25\ \mu$s, which will well accommodate the later arrival of the 10-km mobile terminal of only 33.3 μs.

To provide the same long guard period in the other bursts is a waste of valuable "spectrum."[6] The GSM system overcomes this problem by using adaptive frame alignment. When the base station detects a 41-bit random access synchronization sequence with a long guard period, it measures the received signal delay relative to the expected signal from a mobile station with zero range. This delay, called the timing advance, is transmitted to the mobile station using a 6-bit number. As a result, the mobile station advances its time base over the range of 0–63 bits (i.e., in units of 3.69 μs). By this process the TDMA bursts arrive at the base station in their correct time slots and do

[6]We are equating bit rate or bit durations to bandwidth. One could assume 1 bit/Hz as a first-order estimate.

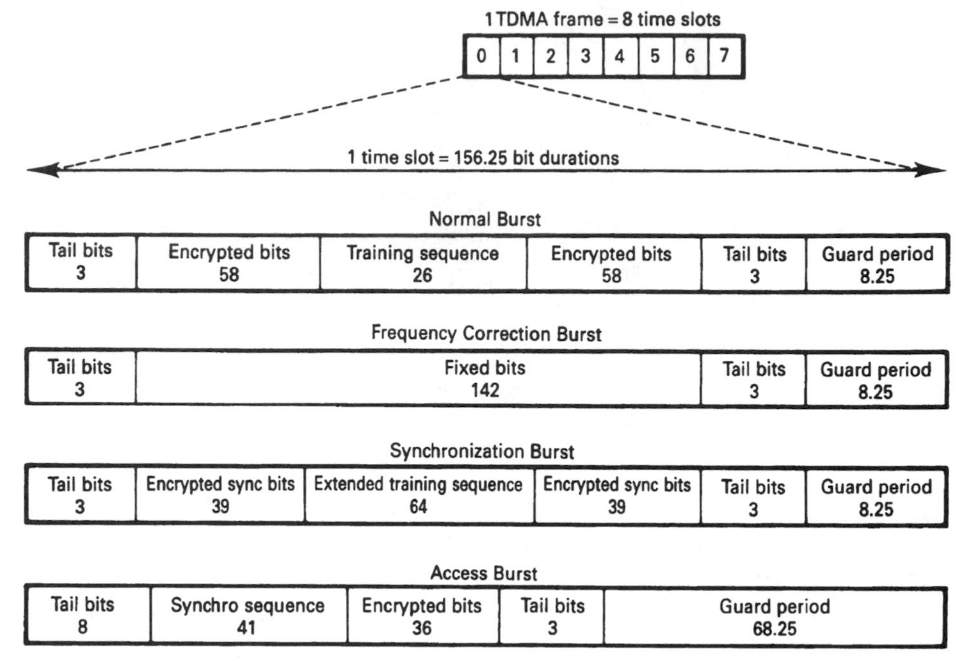

Figure 16.10 GSM frame and burst structures. (From Ref. 3, Figure 8.7. Reprinted with permission.)

not overlap with adjacent ones. As a result, the guard period in all other bursts can be reduced to 8.25×3.69 μs or approximately 30.46 μs, the equivalent of 8.25 bits only. Under normal operations, the base station continuously monitors the signal delay from the mobile station and thus instructs the mobile station to update its time advance parameter. In very large traffic cells there is an option to actively utilize every second time slot only to cope with the larger propagation delays. This is spectrally inefficient but, in large, low-traffic rural cells, admissible (from Ref. 3).

16.6.3.1 Comments on TDMA Efficiency. Multichannel FDMA can operate with a base station power amplifier for every channel, or with a common wideband amplifier for all channels. With the latter, we are setting up a typical generator of intermodulation (IM) products as these carriers mix in a comparatively nonlinear common power amplifier. To reduce the level of IM products, just like in satellite communications discussed in Chapter 9, backoff of the power amplifier is required. This backoff can be in the order of 3–6 dB.

With TDMA (downlink), only one carrier is present on the power amplifier, thus removing most of the causes of IM noise generation. Thus with TDMA, the power amplifier can be operated to full saturation, a distinct advantage. FDMA required some guardband between frequency segments; there are no guardbands with TDMA. However, as we saw previously, a guard time between uplink time slots is required to accommodate the following situations:

- Timing inaccuracies due to clock instabilities;
- Delay spread due to propagation;[7]

[7]*Delay spread* is a variance in delay due to dispersion of emitted signals on delayed paths due to reflection, diffraction/refraction. Lee reports a typical urban delay spread of about 3 μs.

- Transmission delay due to propagation distance (Section 16.6.3); and
- Tails of pulsed signals due to transient response.

The longer guard times are extended, the more inefficient a TDMA system becomes.

16.6.3.2 Advantages of TDMA. The introduction of TDMA results in a much improved transmission system and reduced cost compared to an FDMA counterpart. Assuming a 25-MHZ bandwidth, up to 23.6 times capacity can be achieved with North American TDMA compared to FDMA, typically AMPS (see Ref. 13, Table II.)

A mobile station can exchange system control signals with the base station without interruption of speech (or data) transmission. This facilitates the introduction of new network and user services. The mobile station can also check the signal level from nearby cells by momentarily switching to a new time slot and radio channel. This enables the mobile station to assist with handover operations and thereby improve the continuity of service in response to motion or signal fading conditions. The availability of signal strength information at both the base and mobile stations, together with suitable algorithms in the station controllers, allows further spectrum efficiency through the use of dynamic channel assignment and power control.

The cost of base stations using TDMA can be reduced if radio equipment is shared by several traffic channels. A reduced number of transceivers leads to a reduction of multiplexer complexity. Outside the major metropolitan areas, the required traffic capacity for a base station may, in many cases, be served by one or two transceivers. The saving in the number of transceivers results in a significantly reduced overall cost.

A further advantage of TDMA is increased system flexibility. Different voice and nonvoice services may be assigned a number of time slots appropriate to the service. For example, as more efficient speech CODECs are perfected, increased capacity may be achieved by the assignment of a reduced number of time slots for voice traffic. TDMA also facilitates the introduction of digital data and signaling services as well as the possible later introduction of such further capacity improvements as digital speech interpolation (DSI).

16.6.4 Code Division Multiple Access (CDMA)

CDMA means code division multiple access, which is a form of spread spectrum using direct sequence spreading (see Ref. 14). There is a second class of spread spectrum called *frequency hop*, which is used in the GSM system, but is not an access technique.

Using spread spectrum techniques accomplishes just the opposite of what we were trying to accomplish in Section 9.2.3.5. There bit packing was used to conserve bandwidth by packing as many bits as possible in 1 Hz of bandwidth. With spread spectrum we do the reverse by spreading the information signal over a very wide bandwidth.

Conventional AM requires about twice the bandwidth of the audio information signal with its two sidebands of information (i.e., approximately ±4 kHz).[8] On the other hand, depending on its modulation index, frequency modulation could be considered a type of spread spectrum in that it produces a much wider bandwidth than its transmitted information requires. As with all other spread spectrum systems, a signal-to-noise advantage is gained with FM, depending on its modulation index. For example, with AMPS, a typical FM system, 30 kHz is required to transmit the nominal 4-kHz voice channel.

[8]AM for "toll-quality" telephony.

If we are spreading a voice channel over a very wide frequency band, it would seem that we are defeating the purpose of frequency conservation. With spread spectrum, with its powerful antijam properties, multiple users can transmit on the same frequency with only some minimal interference one to another. This assumes that each user is employing a different key variable (i.e., in essence, using a different time code). At the receiver, the CDMA signals are separated using a correlator that accepts only signal energy from the selected key variable binary sequence (code) used at the transmitter, and then despreads its spectrum. CDMA signals with unmatching codes are not despread and only contribute to the random noise.

CDMA reportedly provides an increase in capacity 15-times that of its analog FM counterpart. It can handle any digital format at the specified input bit rate such as facsimile, data, and paging. In addition, the amount of transmitter power required to overcome interference is comparatively low when utilizing CDMA. This translates into savings on infrastructure (cell site) equipment and longer battery life for hand-held terminals. CDMA also provides so-called soft handoffs from cell site to cell site that make the transition virtually inaudible to the user (Ref. 13).

Dixon (Ref. 15) lists some advantages of the spread spectrum:

1. Selective addressing capability;
2. Code division multiplexing is possible for multiple access;
3. Low-density power spectrum for signal hiding;
4. Message security;
5. Interference rejection.

Of most importance for the cellular user (Ref. 14), "when codes are properly chosen for low cross correlation, minimum interference occurs between users, and receivers set to use different codes are reached only by transmitters sending the correct code. Thus more than one signal can be unambiguously transmitted at the same frequency and at the same time; selective addressing and code-division multiplexing are implemented by the coded modulation format."

Processing gain is probably the most commonly used parameter to describe the performance of a spread spectrum system. It quantifies the signal-to-noise ratio improvement when a spread signal is passed through the appropriate processor. For instance, a certain spread spectrum processor has an input S/N of 12 dB and an output S/N of 20 dB, then its processing gain is 8 dB. Processing gain is expressed by the following:

$$G_{\mathrm{p}} = \frac{\text{spread bandwidth in Hz}}{\text{information bit rate}}. \tag{16.5}$$

More commonly, processing gain is given in a dB value; then

$$G_{\mathrm{p(dB)}} = 10 \log\left(\frac{\text{spread bandwidth in Hz}}{\text{information bit rate}}\right). \tag{16.6}$$

Example. A certain cellular system voice channel information rate is 9.6 kbps and the RF spread bandwidth is 9.6 MHz. What is the processing gain?

$$\begin{aligned} G_{p(dB)} &= 10\log(9.6\times 10^6) - 10\log 9600 \\ &= 69.8 - 39.8(dB) \\ &= 30\ dB \end{aligned}$$

It has been pointed out by Steele (Ref. 3) that the power control problem held back the implementation of CDMA for cellular application. If the standard deviation of the received power from each mobile at the base station is not controlled to an accuracy of approximately ±1 dB relative to the target receive power, the number of users supported by the system can be significantly reduced. Other problems to be overcome were synchronization and sufficient codes available for a large number of mobile users (Ref. 3; see also Ref. 15).

Qualcomm, a North American company, has a CDMA design that overcomes these problems and has fielded a cellular system based on CDMA. It operates at the top of the AMPS band using 1.23 MHz for each uplink and downlink. This is the equivalent of 41 AMPS channels (i.e., 30 kHz × 41 = 1.23 MHz) deriving up to 62 CDMA channels (plus one pilot channel and one synchronization channel) or some 50% capacity increase. The Qualcomm system also operates in the 1.7–1.8-GHz band (Ref. 3). EIA/TIA IS-95 is based on the Qualcomm system. Its processing gain, when using the 9600-bps information rate, is $1.23 \times 10^6/9600$ or about 21 dB.

16.6.4.1 Correlation: Key Concept in Direct Sequence Spread Spectrum. In direct sequence (DS) spread spectrum systems, the *chip rate* is equivalent to the code generator clock rate. Simplistically, a chip can be considered an element of RF energy with a certain recognizable binary phase characteristic. A chip (or chips) is (are) a result of direct sequence spreading by biphase modulating an RF carrier. Being that each chip has a biphase modulated characteristic, we can identify each one with a binary 1 or binary 0.

These chips derive from biphase (PSK) modulating a carrier where the modulation is controlled by a pseudorandom (PN) sequence. If the sequence is long enough, without repeats, it is considered pseudorandom. The sequence is controlled by a key which is unique to our transmitter and its companion far-end receiver. Of course the receiver must be time-aligned and synchronized with its companion transmitter. A block diagram of this operation is shown in Figure 16.11. It is an in-line correlator.

Let us look at an information bit divided into seven chips and coded by a PN sequence + + + − + − − and shown in Figure 16.12*a*. Now replace the in-line correlator with a matched filter. In this case the matched filter is an electrical delay line tapped at delay intervals, which correspond to the chip time duration. Each tap in the delay line feeds into an arithmetic operator matched in sign to each chip in the coded sequence. If each

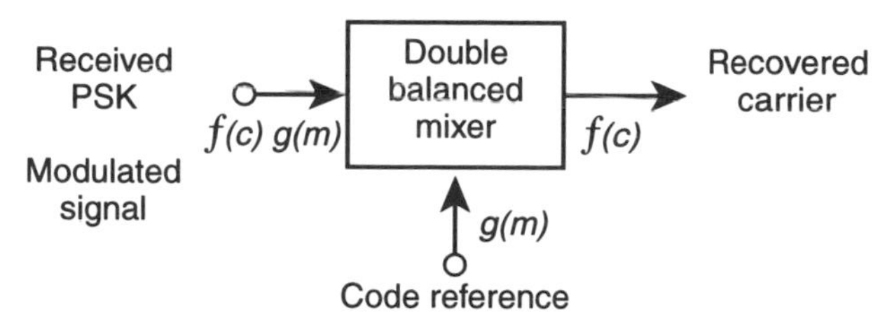

Figure 16.11 In-line correlator.

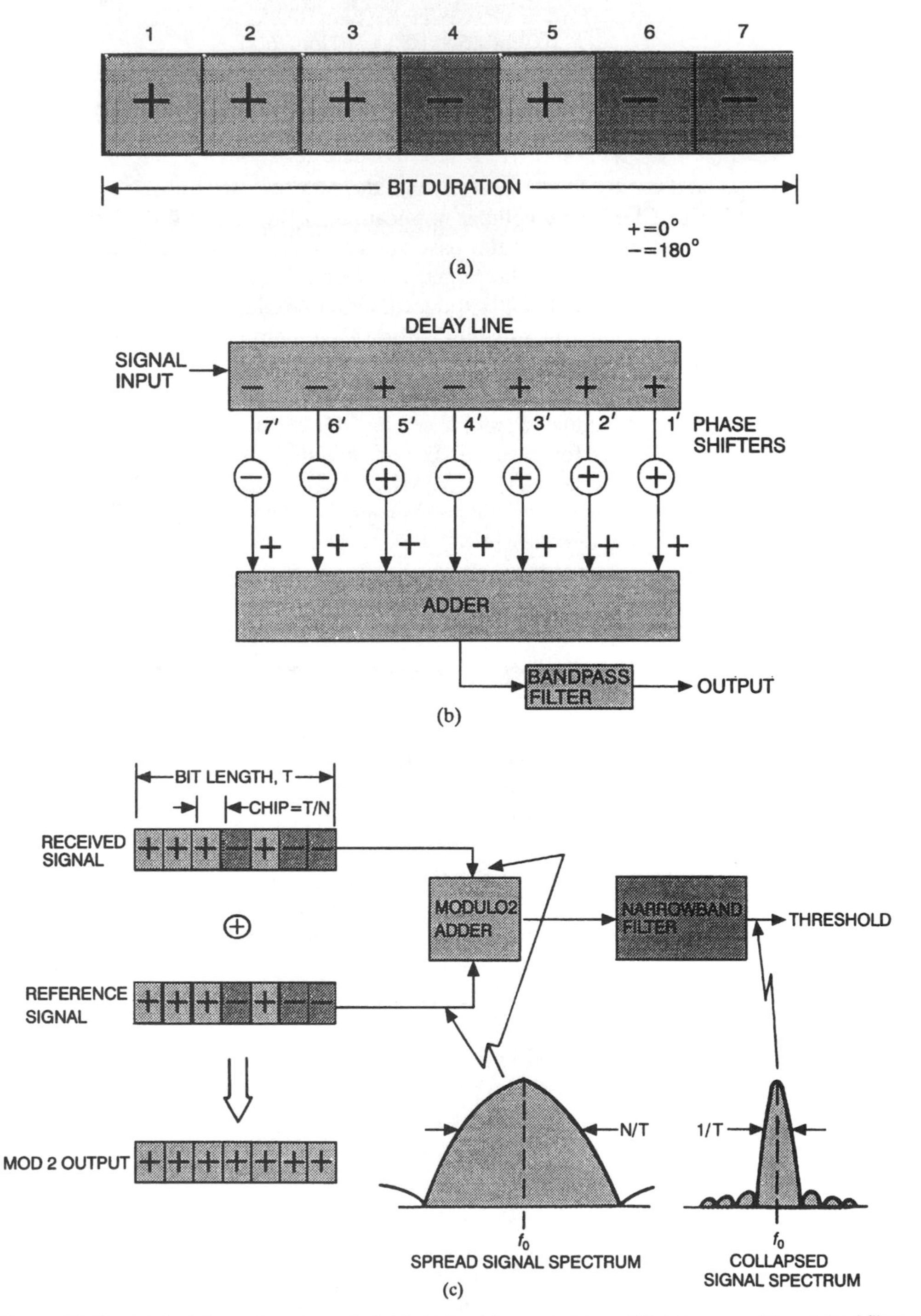

Figure 16.12 (*a*) An information element divided into chips coded by a PN sequence; (*b*) matched filter for 7-chip PN code; (*c*) the correlation process collapses the spread signal spectrum to that of the original bit spectrum. (From Ref. 16. Reprinted with permission.)

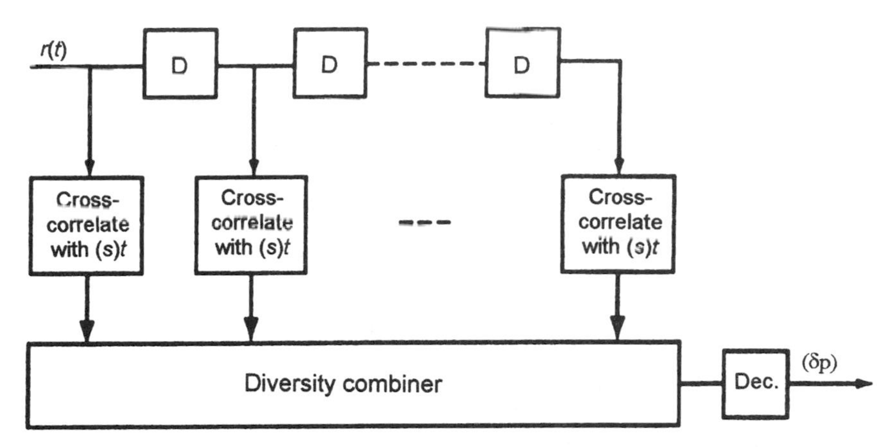

Figure 16.13 A typical RAKE receiver used with direct sequence spread spectrum reception.

delay line tap has the same sign (phase shift) as the chips in the sequence, we have a match. This is illustrated in Figure 16.12*b*. As shown here, the short sequence of seven chips is enhanced with the desired signal seven times. This is the output of the modulo-2 adder, which has an output voltage seven times greater than the input voltage of one chip.

In Figure 16.12*c* we show the correlation process collapsing the spread signal spectrum to that of the original bit spectrum when the receiver reference signal, based on the same key as the transmitter, is synchronized with the arriving signal at the receiver. Of overriding importance is that only the desired signal passes through the matched filter delay line (adder). Other users on the same frequency have a different key and do not correlate. These "other" signals are rejected. Likewise, interference from other sources is spread; there is no correlation and those signals also are rejected.

Direct sequence spread spectrum offers two other major advantages for the system designer. It is more forgiving in a multipath environment than conventional narrowband systems, and no intersymbol interference (ISI) will be generated if the coherent bandwidth is greater than the information symbol bandwidth.

If we use a RAKE receiver, which optimally combines the multipath components as part of the decision process, we do not lose the dispersed multipath energy. Rather, the RAKE receiver turns it into useful energy to help in the decision process in conjunction with an appropriate combiner. Some texts call this implicit diversity or time diversity.

When sufficient spread bandwidth is provided (i.e., where the spread bandwidth is greater or much greater than the correlation bandwidth), we can get two or more independent frequency diversity paths by using a RAKE receiver with an appropriate combiner such as a maximal ratio combiner. Figure 16.13 is a block diagram of a RAKE receiver.

16.7 FREQUENCY REUSE

Because of the limited bandwidth allocated in the 800-MHz band for cellular radio communications, frequency reuse is crucial for its successful operation. A certain level of interference has to be tolerated. The major source of interference is cochannel interference from a "nearby" cell using the same frequency group as the cell of interest. For the 30-kHz bandwidth AMPS system, Ref. 6 suggests that C/I be at least 18 dB. The pri-

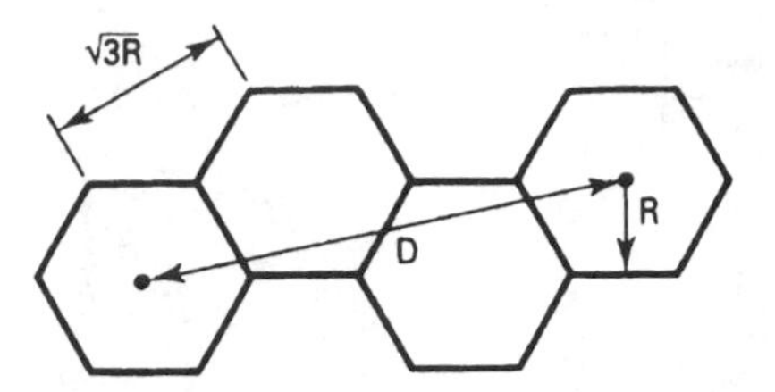

Figure 16.14 Definitions of R and D.

mary isolation derives from the distance between the two cells with the same frequency group. In Figure 16.2 there is only one cell diameter for interference protection.

Refer to Figure 16.14 for the definition of R and D. D is the distance between cell centers of repeating frequency groups and R is the "radius" of a cell. We let:

$$a = D/R. \tag{16.7}$$

The D/R ratio is a basic frequency reuse planning parameter. If we keep the D/R ratio large enough, cochannel interference can be kept to an acceptable level. Lee (Ref. 8) calls a the cochannel reduction factor and relates path loss from the interference source to R^{-4}.

A typical cell in question has six cochannel interferers, one on each side of the hexagon. So there are six equidistant cochannel interference sources. The goal is $C/I \geq$ 18 dB or a numeric of 63.1. So

$$C/I = C/\Sigma I = C/6I = R^{-4}/6D^{-4} = a^4/6 \geq 63.1. \tag{16.8}$$

Then

$$a = 4.4.$$

This means that D must be 4.4 times the value of R. If R is 6 mi (9.6 km) then $D = 4.4 \times 6 = 26.4$ mi (42.25 km).

Lee (Ref. 8) reports that cochannel interference can be reduced by other means such as directional antennas, tilted beam antennas, lowered antenna height, and an appropriately selected site.

One way we can protect a cell that is using the same frequency family as a nearby cell is by keeping that cell base station below line-of-sight of the nearby cell. In other words, we are making our own shadow conditions. Consider a 26.4-mi path. What is the height of earth curvature midpath? From Section 9.2.3.3, $h = 0.667(d/2)^2/1.33 = 87.3$ ft (26.9 m). Providing the cellular base station antennas are kept under 87 ft, the 40-dB/decade rule of Lee holds. It holds so long as we are below line-of-sight conditions.

The total available (one-way) bandwidth is split up into N sets of channel groups. The channels are then allocated to cells, one channel set per cell on a regular pattern, which repeats to fill the number of cells required. As N increases, the distance between channel sets (D) increases, reducing the level of interference. As the number of channel sets (N) increases, the number of channels per cell decreases, reducing the system capacity. Selecting the optimum number of channel sets is a compromise between capacity and quality. Note that only certain values of N lead to regular repeat patterns without gaps. These are $N = 3$, 4, 7, 9, and 12, and then multiples thereof.

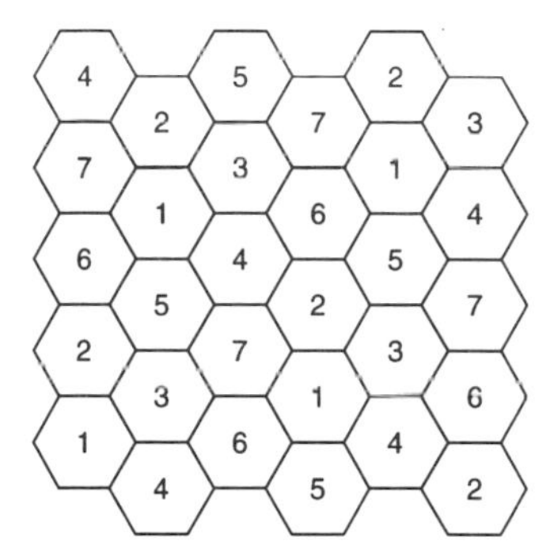

Figure 16.15 A cell layout based on $N = 7$.

Figure 16.15 shows a repeating 7 pattern for frequency reuse. This means that $N = 7$ or there are 7 different frequency sets (or families) for cell assignment. Cell splitting will take place, especially in urban areas, in some point in time because the present cell structure cannot support the busy hour traffic load. Cell splitting, in effect, provides more frequency slots for a given area and relieves the congestion problem. Macario (Ref. 11) reports that cells can be split as far down as 1 km in radius.

Cochannel interference tends to increase with cell splitting. Cell sectorization can reduce the interference level. Figure 16.16 shows a three- and a six-sector plan. Sectorization breaks a cell into three or six parts each with a directional antenna. With a standard cell (using an omnidirectional antenna), cochannel interference enters from six directions. A six-sector plan can essentially reduce the interference to just one direction. A separate channel frequency set is allocated to each sector.

The three-sector plan is often used with a seven-cell repeating pattern (Figure 16.15) resulting in an overall requirement for 21 channel sets. The six-sector plan with its improved cochannel performance and rejection of secondary interferers allows a four-cell repeat plan (Figure 16.2) to be employed. This results in an overall 24-channel set requirement. Sectorization entails a larger number of channel sets and fewer channels per sector. Outwardly it appears that there is less capacity with this approach; however, the ability to use much smaller cells results in a higher capacity operation.

16.8 PERSONAL COMMUNICATIONS SERVICES (PCS)

16.8.1 Defining Personal Communications

Personal communications services (PCS) are wireless. This simply means that they are radio based. The user requires no *tether*. The conventional telephone is connected by a wire pair through to the local serving switch. The wire pair is a tether. We can only walk as far with that telephone handset as the "tether" allows.

Both of the systems we have dealt with in the previous sections of this chapter can be classified as PCS. Cellular radio, particularly with the hand-held terminal, gives the user tetherless telephone communication. Paging systems provided the mobile/ambulatory user a means of being alerted that someone wishes to talk to that person on the telephone or of receiving a short message. The cordless telephone is certainly another example that has extremely wide use around the world with more than 200 million sets. We provide a brief review of cordless telephone sets in the following.

New applications are either on the horizon or going through field tests (1998). One that seems to offer great promise in the office environment is the wireless PABX. It

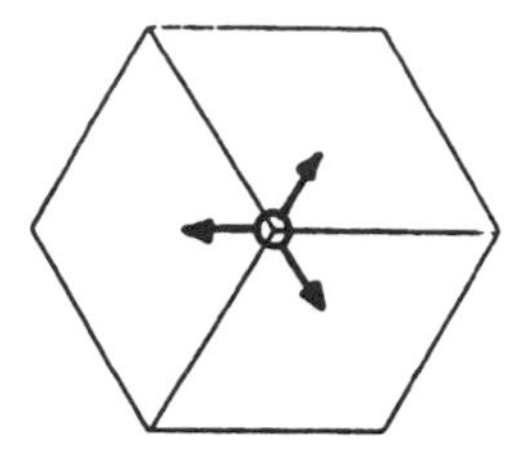

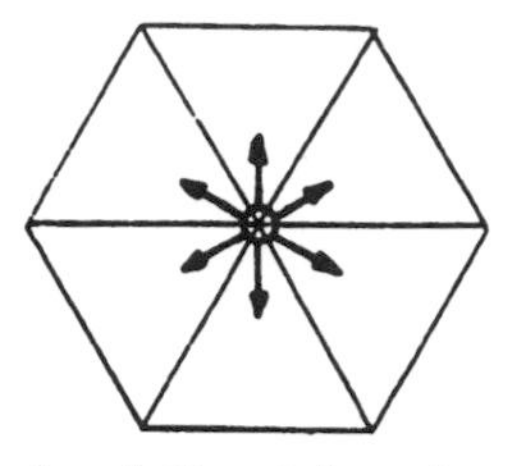

Figure 16.16 Breaking up a cell into three sectors (left) and six sectors (right).

will almost eliminate the telecommunication manager's responsibilities with office rearrangements. Another is the wireless LAN (WLAN).

Developments are expected such that PCS cannot only provide voice communications but facsimile, data, messaging, and possibly video. GSM provides all but video. Cellular digital packet data (CDPD) will permit data services over the cellular system in North America.

Donald Cox (Ref. 17) breaks PCS down into what he calls "high tier" and "low tier." Cellular radio systems are regarded as high-tier PCS, particularly when implemented in the new 1.9-GHz PCS frequency band. Cordless telephones are classified as low tier. Table 16.1 summarizes some of the more prevalent PCS technologies.

16.8.2 Narrowband Microcell Propagation at PCS Distances

The microcells discussed here have a radial range of ≤1 km. One phenomenon is the Fresnel break point, which is illustrated in Figure 16.17. This figure illustrates that signal level varies with distance R as A/R^n, where R is the distance to the receiver. For distances greater than 1 km, n is typically 3.5 to 4. The parameter A describes the effects of environmental features in a highly averaged manner (Ref. 18).

Typical PCS radio paths can be of an LOS nature, particularly near the fixed transmitter where $n = 2$. Such paths may be down the street from the transmitter. The other types of paths are shadowed paths. One type of shadowed path is found in highly urbanized settings, where the signal may be reflected off high-rise buildings (see Figure 16.4). Another is found in more suburban areas, where buildings are often just two stories high.

When a signal at 800 MHz is plotted versus R on a logarithmic scale, as in Figure 16.17, there are distinctly different slopes before and after the Fresnel break point. We call the break distance (from the transmit antenna) R_B. This is the point for which the Fresnel ellipse about the direct ray just touches the ground. Such a model is illustrated in Figure 16.18. The distance R_B is approximated by:

$$R_B = 4h_1h_2/\lambda. \tag{16.9}$$

For $R < R_B$, n is less than 2, and for $R > R_B$, n approaches 4.

It was found that on non-LOS paths in an urban environment with low base station antennas and with users at street level, propagation takes place down streets and around corners rather than over buildings. For these non-LOS paths the signal must turn corners by multiple reflections and diffraction at vertical edges of buildings. Field tests reveal that signal level decreases by about 20 dB when turning a corner.

In the case of propagation inside buildings where the transmitter and receiver are on the same floor, the key factor is the clearance height between the average tops of furniture and the ceiling.

Table 16.1 Wireless PCS Technologies

	High-Power Systems				Low-Power Systems			
	Digital Cellular (HighTier PCS)				Low-Tier PCS		Digital Cordless	
System	IS-54	IS-95(DS)	GSM	DCS-1800	WACS/PACS	Handi-Phone	DECT	CT-2
Multiple access	TDMA/FDMA	CDMA/FDMA	TDMA/FDMA	TDMA/FDMA	TDMA/FDMA	TDMA/FDMA	TDMA/FDMA	FDMA
Frequency band (MHz) Uplink (MHz) Downlink (MHz)	 869–894 824–849 (USA)	 869–894 824–849 (USA)	 935–960 890–915 (Europe)	 1710–1785 1805–1880 (UK)	 Emerging Technology [a] (USA)	1895–1907 (Japan)	1880–1900 (Europe)	864–868 (Europe and Asia)
RF channel spacing Downlink (kHz) Uplink (kHz)	 30 30	 1250 1250	 200 200	 200 200	 300 300	300	1728	100
Modulation	π/4 DQPSK	BPSK/QPSK	GMSK	GMSK	π/4 QPSK	π/4 DQPSK	GFSK	GFSK
Portable txmit power, max./avg.	600 mW/ 200 mW	600 mW	1 W/ 125 mW	1 W/ 125 mW	200 mW/ 25 mW	80 mW/ 10 mW	250 mW/ 10 mW	10 mW/ 6 mW
Speech coding	VSELP	QCELP	RPE-LTP	RPE-LTP	ADPCM	ADPCM	ADPCM	ADPCM
Speech rate (kbps)	7.95	8 (var.)	13	13	32/16/8	32	32	32
Speech channel/RF channel	3	—	8	8	8/16/32	4	12	1
Channel bit rate (kbps) Uplink (kbps) Downlink (kbps)	 48.6 48.6		 270.833 270.833	 270.833 270.833	 384 384	384	1152	72
Channel coding	1/2 rate conv.	1/2 rate fwd 1/3 rate rev.	1/2 rate conv.	1/2 rate conv.	CRC	CRC	CRC (control)	None
Frame (ms)	40	20	4.615	4.615	2.5	5	10	2

[a]Spectrum is 1.85–2.2 GHz allocated by the FCC for emerging technologies; DS is direct sequence.

Source: Ref. 17, Table 1. (Reprinted with permission of the IEEE.)

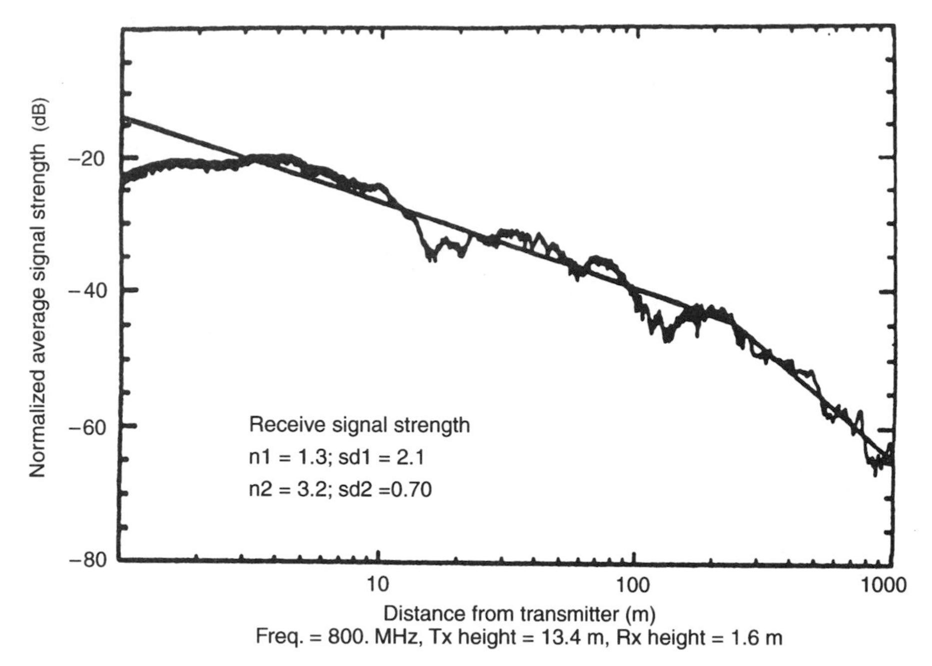

Figure 16.17 Signal variation on a line-of-sight path in a rural environment. (From Ref. 18, Figure 3. Reprinted with permission.)

Bertoni et al. (Ref. 18) call this clearance W. Here building construction consists of drop ceilings of acoustical material supported by metal frames. That space between the drop ceiling and the floor above contains light fixtures, ventilation ducts, pipes, support beams, and so on. Because the acoustical material has a low dielectric constant, the rays incident on the ceiling penetrate the material and are strongly scattered by the irregular structure, rather than undergoing specular reflection. Floor-mounted building furnishings such as desks, cubicle partitions, filing cabinets, and workbenches scatter the rays and prevent them from reaching the floor, except in hallways. Thus it is concluded that propagation takes place in the clear space, W.

Figure 16.19 shows a model of a typical floor layout in an office building. When both the transmitter and receiver are located in the clear space, path loss can be related to the Fresnel ellipse. If the Fresnel ellipse associated with the path lies entirely in the clear space, the path loss has LOS properties ($1/L^2$). Now as the separation between the trans-

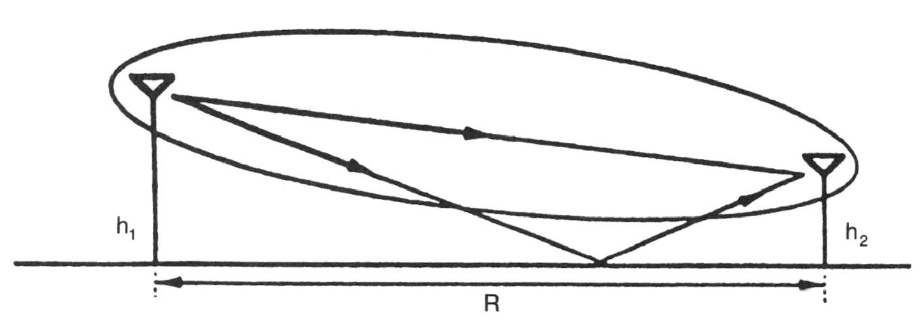

Figure 16.18 Direct and ground-reflected rays, showing the Fresnel ellipse about the direct ray. (From Ref. 18, Figure 18. Reprinted with permission.)

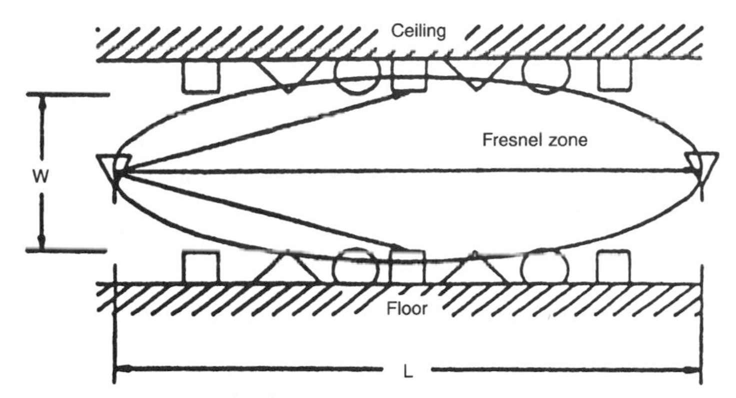

Figure 16.19 Fresnel zone for propagation between transmitter and receiver in clear space between building furnishings and ceiling fixtures. (From Ref. 18, Figure 35. Reprinted with permission.)

mitter and receiver increases, the Fresnel ellipse grows in size so that scatterers lie within it. This is shown in Figure 16.20. Now the path loss become greater than free space.

Bertoni et al. report one measurement program where the scatterers have been simulated using absorbing screens. It was recognized that path loss will be highly dependent on nearby scattering objects. Figure 16.20 was developed from this program. The path loss in excess of free space calculated at 900 MHz and 1800 MHz where $W = 1.5$ m is plotted in Figure 16.20 as a function of path length L. The figure shows that the excess path loss (over LOS) is small at each frequency out to distances of about 20 m to 40 m, respectively, where it increases dramatically.

Propagation between floors of a modern office building can be very complex. If the floors are constructed of reinforced concrete or prefabricated concrete, transmission loss can be 10 dB or more. Floors constructed of concrete poured over steel panels show much greater loss. In this case (Ref. 18), signals may propagate over other paths involving diffraction rather than transmission through the floors. For instance, signals can exit the building through windows and reenter on higher floors by diffraction mechanisms along the face of the building.

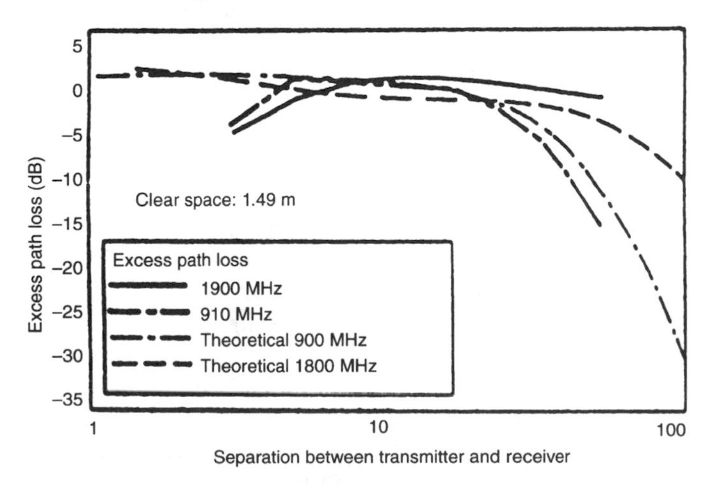

Figure 16.20 Measured and calculated excess path loss at 900 MHz and 1800 MHz for a large office building having head-high cubical partitions, but no floor-to-ceiling partitions. (From Ref. 18, Figure 36. Reprinted with permission.)

16.9 CORDLESS TELEPHONE TECHNOLOGY

16.9.1 Background

Cordless telephones began to become widely used in North America around 1981. Today their popularity is worldwide, with hundreds of millions of units in use. As the technology develops, they will begin to compete with cellular radio systems, where the cordless telephone will operate in microcells.

16.9.2 North American Cordless Telephones

The North American cordless telephone operates in the 50-MHz frequency band with 25 frequency pairs using frequency modulation. Their ERP is on the order of 20 μW. Ref. 19 suggests that this analog technology will continue for some time into the future because of the telephone's low cost. These may be replaced by some form of the wireless local loop (WLL), operating in the 30-GHz or 40-GHz band.

16.9.3 European Cordless Telephones

The first-generation European cordless telephone provided for eight channel pairs near 1.7 MHz (base unit transmit) and 47.5 MHz (handset transmit). Most of these units could only access one or two channel pairs. Some called this "standard" CT0.

This was followed by another analog cordless telephone based on a standard known as CEPT/CT1. CT1 has 40 25-kHz duplex channel pairs operating in the bands 914–915 MHz and 959–960 MHz. There is also a CT1+ in the bands 885–887 MHz and 930–932 MHz, which do not overlap the GSM allocation. CT1 is called a coexistence standard (not a compatible standard), such that cordless telephones from different manufacturers do not interoperate. The present embedded base is about 12.5 million units with some 5 million units expected to be sold in 1998.

Two digital standards have evolved in Europe: the CT2 Common Air Interface and DECT (digital European cordless telephone). In both standards, speech coding uses ADPCM (adaptive differential PCM). The ADPCM speech and control data are modulated onto a carrier at a rate of 72 kbps using Gaussian-filtered FSK (GFSK) and are transmitted in 2-ms frames. One base-to-handset burst and one handset-to-base burst are included in each frame.

The frequency allocation for CT2 consists of 40 FDMA channels with 100-kHz spacing in the band 864–868 MHz. The maximum transmit power is 10 mW, and a two-level power control supports prevention of desensitization of base station receivers. As a by-product, it contributes to frequency reuse. CT2 has a call reestablishment procedure on another frequency after three seconds of unsuccessful attempts on the initial frequency. This gives a certain robustness to the system when in an interference environment. CT2 supports up to 2400 bps of data transmission and higher rates when accessing the 32 kbps underlying bearer channels.

CT2 also is used for wireless pay telephones. When in this service it is called *Telepoint*. CT2 seems to have more penetration in Asia than in Europe.

Canada has its own version of CT2, called CT2+. It is more oriented toward the mobile environment, providing several of the missing mobility functions in CT2. For example, with CT2+, 5 of the 40 carriers are reserved for signaling, where each carrier provides 12 common channel signaling channels (CSCs) using TDMA. These channels support location registration, updating, and paging, and enable Telepoint subscribers to receive calls. The CT2+ band is 944–948 MHz.

DECT takes on more of the cellular flavor than CT2. It uses a picocell concept and TDMA with handover, location registration, and paging. It can be used for Telepoint, radio local loop (RLL), and cordless PABX besides conventional cordless telephony. Its speech coding is similar to CT2, namely, ADPCM. For its initial implementation, 10 carriers have been assigned in the band 1880–1900 MHz.

There are many areas where DECT will suffer interference in the assigned band, particularly from "foreign" mobiles. To help alleviate this problem, DECT uses two strategies: *interference avoidance* and *interference confinement*. The avoidance technique avoids time/frequency slots with a significant level of interference by handover to another slot at the same or another base station. This is very attractive for the uncoordinated operation of base stations because in many interference situations there is no other way around a situation but to change in both the time and frequency domains. The "confinement" concept involves the concentration of interference to a small time–frequency element even at the expense of some system robustness.

Base stations must be synchronized in the DECT system. A control channel carries information about access rights, base station capabilities, and paging messages. The DECT transmission rate is 1152 kbps. As a result of this and a relatively wide bandwidth, either equalization or antenna diversity is typically needed for using DECT in the more dispersive microcells.

Japan has developed the personal handyphone system (PHS). Its frequency allocation is 77 channels, 300 kHz in width, in the band 1895–1918.1 MHz. The upper-half of the band, 1906.1–1918.1 MHz (40 frequencies), is used for public systems. The lower-half of the band, 1895–1906.1 MHz, is reserved for home/office operations. An operational channel is autonomously selected by measuring the field strength and selecting a channel on which it meets certain level requirements. In other words, fully dynamic channel assignment is used. The modulation is $\pi/4$ DQPSK; average transmit power at the handset is 10 mW (80-mW peak power) and no greater than 500 mW (4-W peak power) for the cell site. The PHS frame duration is 5 ms. Its voice coding technique is 32-kbps ADPCM (Ref. 9).

In the United States, digital PCS was based on the wireless access communication system (WACS), which has been modified to an industry standard called PACS (personal access communications services). It is intended for the licensed portion of the new 2-GHz spectrum. Its modulation is $\pi/4$ QPSK with coherent detection. Base stations are envisioned as shoebox-size enclosures mounted on telephone poles, separated by some 600 m. WACS/PACS has an air interface similar to other digital cordless interfaces, except it uses frequency division duplex (FDD) rather than time division duplex (TDD) and more effort has gone into optimizing frequency reuse and the link budget. It has two-branch polarization diversity at both the handset and base station with feedback. This gives it an advantage approaching four-branch receiver diversity. The PACS version has eight time slots and a corresponding reduction in channel bit rate and a slight increase in frame duration over its predecessor, WACS.

Table 16.2 summarizes the characteristics of these several types of digital cordless telephones.

16.10 WIRELESS LANs

Wireless LANs (WLANs), much as their wired counterparts, operate in excess of 1 Mbps. Signal coverage runs from 50 ft to less than 1000 ft. The transmission medium

Table 16.2 Digital Cordless Telephone Interface Summary

	CT2	CT2 +	DECT	PHS	PACS
Region	Europe	Canada	Europe	Japan	United States
Duplexing	TDD		TDD	TDD	FDD
Frequency band (MHz)	864–868	944–948	1800–1900	1895–1918	1850–1910/1930–1990[a]
Carrier spacing (kHz)	100		1728	300	300/300
Number of carriers	40		10	77	16 pairs/10 MHz
Bearer channels/carrier	1		12	4	8/pair
Channel bit rate (kbps)	72		1152	384	384
Modulation	GFSK		GFSK	$\pi/4$ DQPSK	$\pi/4$ QPSK
Speech coding	32 kbps		32 kbps	32 kbps	32 kbps
Average handset transmit power (mW)	5		10	10	25
Peak handset transmit power (mW)	10		250	80	200
Frame duration (ms)	2		10	5	2.5

[a]General allocation to PCS; licensees may use PACS.

Source: Ref. 19, Table 2.

can be radiated light (around 800 nm to 900 nm) or radio frequency, unlicensed. Several of these latter systems use spread spectrum with transmitter outputs of 1 W or less.

WLANs using radiated light do not require FCC licensing, a distinct advantage. They are immune to RF interference but are limited in range by office open spaces because their light signals cannot penetrate walls. Shadowing can also be a problem.

One type of radiated light WLAN uses a directed light beam. These are best suited for fixed terminal installations because the transmitter beams and receivers must be carefully aligned. The advantages for directed beam systems is improved S/N and fewer problems with multipath. One such system is fully compliant with IEEE 802.5 token ring operation offering 4- and 16-Mbps transmission rates.

Spread spectrum WLANs use the 900-MHz, 2-GHz, and 5-GHz industrial, scientific, and medical (ISM) bands. Both direct sequence and frequency hop operation can be used. Directional antennas at the higher frequencies provide considerably longer range than radiated light systems, up to several miles or more. No FCC license is required. A principal user of these higher-frequency bands is microwave ovens with their interference potential. CSMA and CSMA/CD (IEEE 802.3) protocols are often employed.

There is also a standard microwave WLAN (nonspread spectrum) that operates in the band 18–19 GHz. FCC licensing is required.

Building wall penetration loss is high. The basic application is for office open spaces.

16.11 MOBILE SATELLITE COMMUNICATIONS

16.11.1 Background and Scope

This section contains a brief review of satellite PCS/cellular services. It is hard to discern whether these services are PCS or cellular. Most of the active and proposed low earth orbit (LEO) satellite systems discussed here utilize frequency reuse and are based on a cellular concept. By the year 2000 or just after we expect at least two "broadband" satellite systems designed to deliver such services as Internet and various other forms of "high-speed" data. By broadband, we mean bandwidths well in excess of those bands available between 1.5 GHz and 2.6 GHz. Among this group are Teledesic and Celestri.

These systems are more for the fixed environment. Our discussion here will dwell on the narrow-band systems using LEO satellites.

16.11.2 Two Typical LEO Systems

Motorola and a consortium of other entities sponsor IRIDIUM, which is a 66-satellite system in low earth orbit. The second system is GLOBALSTAR, sponsored by Loral and Qualcomm. This system will orbit 48 satellites. Range to these satellites is in the high-700-km above the earth's surface. They will provide truly worldwide coverage. Their access charges may be competitive with terrestrial cellular/PCS systems, but their usage charge is from $0.50 to $4 per minute for telephone service. This is from four to ten times terrestrial usage charge.

16.11.3 Advantages and Disadvantages of LEO Systems

Delay. One-way delay to a GEO satellite is budgeted at 125 ms; one-way up and down is double this value, or 250 ms. Round-trip delay is about 0.5 s. Delay to a typical LEO satellite is 2.67 ms and round-trip delay is 4×2.67 ms or about 10.66 ms. Calls to/from mobile users of such systems may be relayed still again by conventional satellite services. Data services do not have to be so restricted on the use of "handshakes" and stop-and-wait ARQ as with similar services via a GEO system.

Higher Elevation Angles and "Full Earth Coverage." The GEO satellite provides no coverage above about 80° latitude and gives low-angle coverage of many of the world's great population centers because of their comparatively high latitude. Typically, cities in Europe and Canada face this dilemma. LEO satellites, depending on orbital plane spacing, can all provide elevation angles >40°. This is particularly attractive in urban areas with tall buildings. Coverage with GEO systems would only be available on the south side of such buildings in the Northern Hemisphere with a clear shot to the horizon. Properly designed LEO systems will not have such drawbacks. Coverage will be available at any orientation.

Tracking, a Disadvantage of LEO and MEO[9] Satellites. At L-band quasiomnidirectional antennas for the mobile user are fairly easy to design and produce. Although such antennas display only modest gain of several dB, links to a LEO satellite can be easily closed with hand-held terminals. However, large feeder, fixed-earth terminals will require a good tracking capability as LEO satellites pass overhead. Handoff is also required as a LEO satellite disappears over the horizon and another satellite just appears over the opposite horizon. The handoff should be seamless. The quasi-omnidirectional user terminal antennas will not require tracking, and the handoff should not be noticeable to the mobile user.

REVIEW EXERCISES

1. What is the principal drawback in cellular radio, considering its explosive growth over the past decade?

[9] MEO stands for medium earth orbit.

2. Why is transmission loss so much greater on a cellular path compared to a LOS microwave path on the same frequency covering the same distance?
3. What is the function of the MTSO or MSC in a cellular network?
4. What is the channel spacing, in kHz, of the AMPS system? What type of modulation does it employ?
5. Why are cell site antennas limited to just sufficient height to cover cell boundaries?
6. Why do we do *cell splitting*? What is the approximate minimum practical cell diameter (this limits splitting)?
7. When is *handover* necessary?
8. Cellular transmission loss varies with what four factors besides distance and frequency?
9. What are some of the fade ranges (dB) we might expect on a cellular link?
10. If the delay spread on a cellular link is about 10 μsec, up to about what bit rate will there be little deleterious effects due to multipath fading?
11. Space diversity reception is common at cells sites. Antenna separation varies with ________.
12. For effective space diversity operation, there is a law of diminishing returns when we lower the correlation coefficient below *what value*?
13. What is the gain of a standard dipole over a reference isotropic antenna? Differentiate ERP and EIRP.
14. Cellular designers use a field strength contour of ____ dBμ, which is equivalent to ____ dBm.
15. What would the maximum transmission loss to the 39-dBμ contour be if the cell site EIRP is +52 dBm?
16. A cell site antenna has a gain of +14 dBd. What is the equivalent gain in dBi?
17. What are the three generic access techniques that might be considered for digital cellular operation?
18. If we have 10 cellular users on a TDMA frame and the frame duration is 20 ms, what is the maximum burst duration without guard time considerations?
19. What are the two basic elements in digital cellular transmission with which we may improve users per unit bandwidth?
20. What power amplifier advantage do we have in a TDMA system that we do not have in an FDMA system, assuming a common power amplifier for all RF channels?
21. Cellular radio, particularly in urban areas, is gated by heavy interference conditions, especially cochannel from frequency reuse. In light of this, describe how we achieve an interference advantage when using CDMA.
22. A CDMA system has an information bit stream of 4800 bps, which is spread 10 MHz. What is the processing gain?

23. For effective frequency reuse, the value of D/R must be kept large enough. Define D and R. What value of D/R is large enough?

24. In congested urban areas, where cell diameters are small, what measure can we take to reduce C/I?

25. What is the effect of the Fresnel ellipse?

26. What range of transmitter output power can we expect from cordless telephone PCS?

27. Why would it be attractive to use CDMA with a RAKE receiver for PCS systems?

28. Speech coding is less stringent with PCS scenarios, typically 32 kbps. Why?

29. What are the two different transmission media used with WLANs?

30. Give two decided advantages of LEO satellite systems over their GEO counterparts.

REFERENCES

1. T. S. Rappaport, *Wireless Communications: Principles and Practice*, IEEE Press, New York, 1996.
2. *Telecommunications Transmission Engineering*, 3rd ed., Bellcore, Piscataway, NJ, 1989.
3. R. Steele, ed., *Mobile Radio Communications*, IEEE Press, New York, and Pentech Press, London, 1992.
4. Y. Okumura, et al., "Field Strength and Its Variability in VHF and UHF Land Mobile Service," *Rec. Electr. Commun. Lab.*, **16,** Tokyo, 1968.
5. M. Hata, "Empirical Formula for Propagation Loss in Land-Mobile Radio Services," *IEEE Trans. Vehicular Technology*, **VT-20**, 1980.
6. F. C. Owen and C. D. Pudney, "In-Building Propagation at 900 and 1650 MHz for Digital Cordless Telephones," *6th International Conference on Antennas and Propagation*, ICCAP, Pt. 2, Propagation Conf., Pub. No. 301, 1989.
7. *IEEE Standard Dictionary of Electrical and Electronics Terms*, 6th ed., IEEE Std. 100-1996, IEEE, New York, 1996.
8. W. C. Y. Lee, *Mobile Communications Design Fundamentals*, 2nd ed., Wiley, New York, 1993.
9. *Recommended Minimum Standards for 800-MHz Cellular Subscriber Units*, EIA Interim Standard EIA/IS-19B, EIA, Washington, DC, 1988.
10. Cellular Radio Systems, seminar given at the University of Wisconsin—Madison, by A. H. Lamothe, consultant, Leesbury, VA, 1993.
11. R. C. V. Macario, ed., *Personal and Mobile Radio Systems*, IEE/Peter Peregrinus, London, 1991.
12. W. F. Fuhrmann and V. Brass, "Performance Aspects of the GSM System," *Proc. IEEE*, **89**(9), 1984.
13. *Digital Cellular Public Land Mobile Telecommunication Systems* (*DCPLMTS*), CCIR Rep. 1156, Vol. VIII.1, XVIIth Plenary Assembly, Dusseldorf, 1990.
14. M. Engelson and J. Hebert, "Effective Characterization of CDMA Signals," *Wireless Rep.*, London, Jan. 1995.

15. R. C. Dixon, *Spread Spectrum Systems with Commercial Applications*, 3rd ed., Wiley, New York, 1994.
16. C. E. Cook and H. S. Marsh, "An Introduction to Spread Spectrum," *IEEE Communications Magazine*, March 1983.
17. D. C. Cox, "Wireless Personal Communications. What Is It?" *IEEE Personal Communications*, **2**(2), 1995.
18. H. L. Bertoni et al., "UHF Propagation Prediction for Wireless Personal Communication," *Proc. IEEE*, **89**(9), 1994.
19. J. C. Padgett, C. G. Gunter, and T. Hattori, "Overview of Wireless Personal Communications," *IEEE Communications Magazine*, Jan. 1995.

17

ADVANCED BROADBAND DIGITAL TRANSPORT FORMATS

17.1 INTRODUCTION

In the early 1980s fiber optic transmission links burst upon the telecommunication transport scene. The potential bit rate capacity of these new system was so great that there was no underlying digital format to accommodate such transmission rates. The maximum bit rate in the DS1 family of digital formats was DS4 at 274 Mbps, and for the E1 family, E4 at 139 Mbps. These data rates satisified the requirements of the metallic transmission plant, but the evolving fiber optic plant had the promise of much greater capacity, in the multigigabit region.

In the mid-1980s ANSI and Bellcore began to develop a new digital format standard specifically designed for the potential bit rates of fiber optics. The name of this structure is SONET, standing for *synchronous optical network*. As the development of SONET was proceeding, CEPT showed interest in the development of a European standard.[1] In 1986 CCITT stepped in, proposing a singular standard that would accommodate U.S., European, and Japanese hierarchies. This, unfortunately, was not achieved due more to time constraints on the part of U.S. interests. As a result, there are two digital format standards: SONET and the synchronous digital hierarchy (SDH) espoused by CCITT.

It should be pointed out that these formats are optimized for voice operation with 125-μs frames. Both types commonly carry plesiochronous digital hierarchy (PDH) formats such as DS1 and E1, as well as ATM cells.[2]

In the general scheme of things, the interface from one to the other will take place at North American gateways. In other words, international trunks are SDH equipped not SONET equipped. The objective of this chapter is to provide a brief overview of both SONET and SDH standards.

[1] *CEPT* stands for Conference European Post & Telegraph, a European telecommunication standardization agency based in France. In 1990 the name of the agency was changed to ETSI—European Telecommunication Standardization Institute.

[2] Held (Ref. 9) defines *plesiochronous* as "A network with multiple stratum 1 primary reference sources." See Section 6.12.1. In this context, when transporting these PCM formats, the underlying network timing and synchronization must have stratum 1 traceability.

17.2 SONET

17.2.1 Introduction and Background

The SONET standard was developed by the ANSI T1X1 committee with first publication in 1988. The standard defines the features and functionality of a transport system based on the principles of synchronous multiplexing. In essence this means that individual tributary signals may be multiplexed directly into a higher rate SONET signal without intermediate stages of multiplexing.

DS1 and E1 digital hierarchies had rather limited overhead capabilities for network management, control, and monitoring. SONET (and SDH) provide a rich built-in capacity for advanced network management and maintenance capabilities. Nearly 5% of the SONET signal structure is allocated to supporting such management and maintenance procedures and practices.

SONET is capable of transporting all the tributary signals that have been defined for the digital networks in existence today. This means that SONET can be deployed as an overlay to the existing network and, where appropriate, provide enhanced network flexibility by transporting existing signal types. In addition, SONET has the flexibility to readily accommodate the new types of customer service signals such as SMDS (switched multimegabit data service) and ATM (asynchronous transfer mode). Actually, it can carry any octet-based binary format such as TCP/IP, SNA, OSI regimes, X.25, frame relay, and various LAN formats, which have been packaged for long-distance transmission.

17.2.2 Synchronous Signal Structure

SONET is based on a synchronous signal comprised of eight-bit octets, which are organized into a frame structure. The frame can be represented by a two-dimensional map comprising N rows and M columns, where each box so derived contains one octet (or byte). The upper left-hand corner of the rectangular map representing a frame contains an identifiable marker to tell the receiver it is the start of frame.

SONET consists of a basic, first-level, structure called STS-1, which is discussed in the following. The definition of the first level also defines the entire hierarchy of SONET signals because higher-level SONET signals are obtained by synchronously multiplexing the lower-level modules. When lower-level modules are multiplexed together, the result is denoted as STS-N (STS stands for synchronous transport signal), where N is an integer. The resulting format then can be converted to an OC-N (OC stands for optical carrier) or STS-N electrical signal. There is an integer multiple relationship between the rate of the basic module STS-1 and the OC-N electrical signals (i.e., the rate of an OC-N is equal to N times the rate of an STS-1). Only OC-1, -3, -12, -24, -48, and -192 are supported by today's SONET.

17.2.2.1 Basic Building Block Structure. The STS-1 frame is shown in Figure 17.1. STS-1 is the basic module and building block of SONET. It is a specific sequence of 810 octets (6480 bits), which includes various overhead octets and an envelope capacity for transporting payloads.[3] STS-1 is depicted as a 90-column, 9-row structure. With a frame period of 125 μs (i.e., 8000 frames per second). STS-1 has a bit rate of 51.840

[3]The reference publications use the term *byte*, meaning, in this context, and 8-bit sequence. We prefer to use the term *octet*. The reason is that some argue that byte is ambiguous, having conflicting definitions.

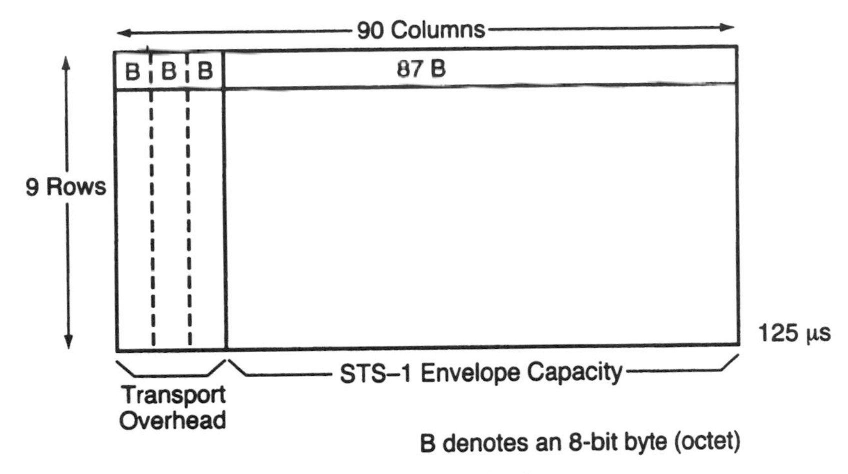

Figure 17.1 The STS-1 frame.

Mbps. Consider Figure 17.1. The order of transmission of octets is row-by-row, from left to right. In each octet of STS-1 the most significant bit is transmitted first.

As illustrated in Figure 17.1, the first three columns of the STS-1 frame contain the transport overhead. These three columns have 27 octets (i.e., 9×3) of which 9 are used for the *section overhead* and 18 octets contain the *line overhead*. The remaining 87 columns make up the STS-1 envelope capacity, as illustrated in Figure 17.2.

The STS-1 synchronous payload envelope (SPE) occupies the STS-1 envelope capacity. The STS-1 SPE consists of 783 octets and is depicted as an 87-column by 9-row structure. In that structure, column 1 contains 9 octets and is designated as the STS path overhead (POH). In the SPE columns 30 and 59 are not used for payload but are designated as *fixed stuff* columns. The 756 octets in the remaining 84 columns are used for the actual STS-1 payload capacity.

Figure 17.3 shows the fixed-stuff columns 30 and 59 inside the SPE. The reference document (Ref. 1) states that the octets in these fixed stuff columns are undefined and

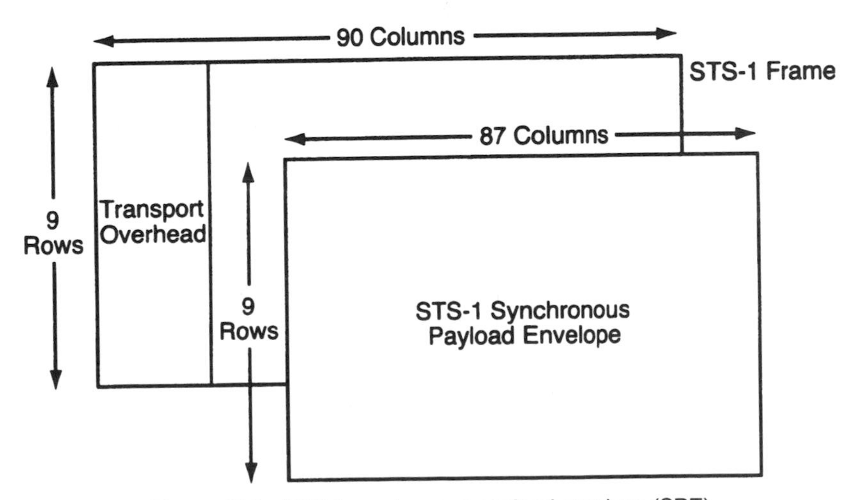

Figure 17.2 STS-1 synchronous payload envelope (SPE).

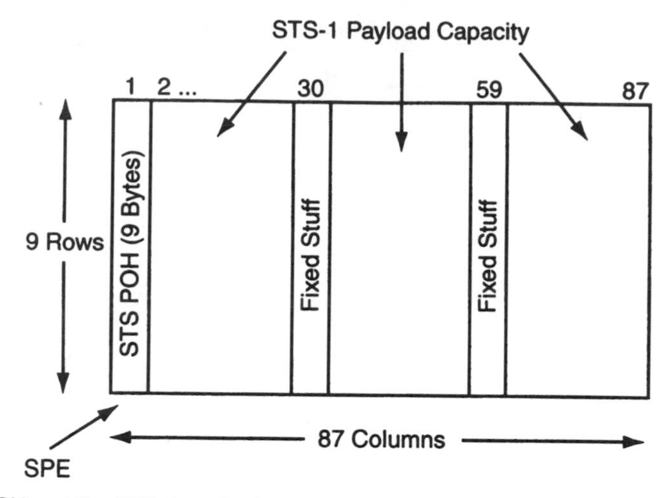

Figure 17.3 POH and the STS-1 payload capacity within the STS-1 SPE. Note that the *net* payload capacity in the STS-1 frame is only 84 columns.

are set to binary 0s. However, the values used to stuff these columns of each STS-1 SPE will produce even parity in the calculation of the STS-1 path BIP-8 (BIP stands for bit interleaved parity).

The STS-1 SPE may begin anywhere in the STS-1 envelope capacity. Typically the SPE begins in one STS-1 frame and ends in the next. This is illustrated in Figure 17.4. However, on occasion, the SPE may be wholly contained in one frame. The *STS payload pointer* resides in the transport overhead. It designates the location of the next octet where the SPE begins. Payload pointers are described in the following paragraphs.

The STS POH is associated with each payload and is used to communicate various pieces of information from the point where the payload is mapped into the STS-1

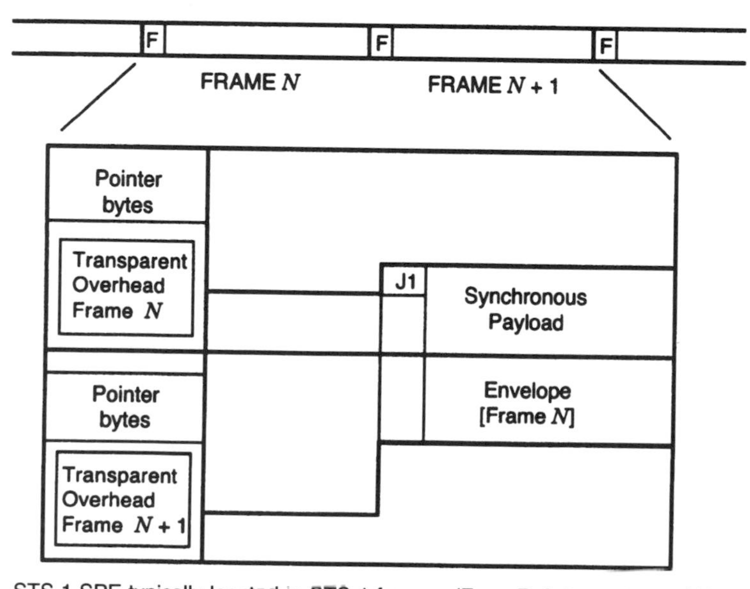

Figure 17.4 STS-1 SPE typically located in STS-1 frames. (From Ref. 2, courtesy of Hewlett-Packard.)

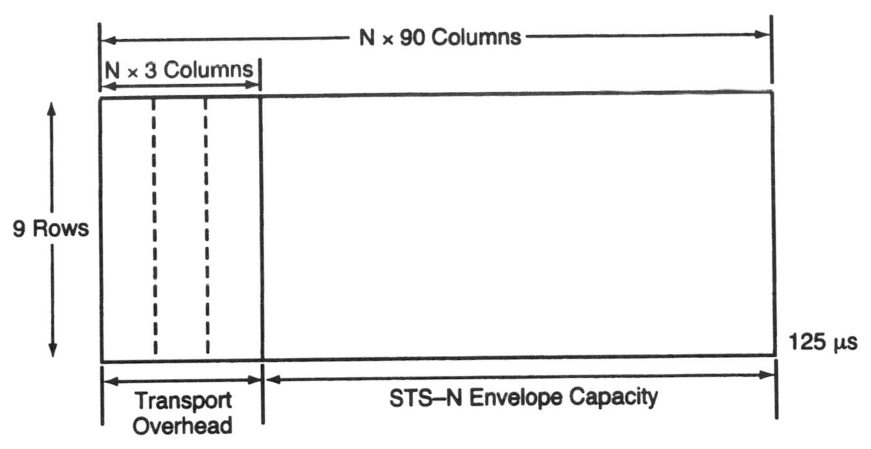

Figure 17.5 STS-N frame.

SPE to the point where it is delivered. Among the pieces of information carried in the POH are alarm and performance data.

17.2.2.2 STS-N Frames. Figure 17.5 illustrates the structure of an STS-N frame. The frame consists of a specific sequence of N × 810 octets. The STS-N frame is formed by octet-interleaved STS-1 and STS-M (<N) modules. The transport overhead of the individual STS-1 and STS-M modules are frame-aligned before interleaving, but the associated STS SPEs are not required to be aligned because each STS-1 has a payload pointer to indicate the location of the SPE or to indicate concatenation.

17.2.2.3 STS Concatenation. Superrate payloads require multiple STS-1 SPEs. FDDI and some B-ISDN payloads fall into this category. Concatenation means the linking together. An STS-Nc module is formed by linking N constituent STS-1s together in a fixed phase alignment. The superrate payload is then mapped into the resulting STS-Nc SPE for transport. Such STS-Nc SPE requires an OC-N or an STS-N electrical signal.[4] Concatenation indicators contained in the second through the Nth STS payload pointer are used to show that the STS-1s of an STS-Nc are linked together.

There are $N \times 783$ octets in an STS-Nc. Such an STS-Nc arrangement is illustrated in Figure 17.6 and is depicted as an $N \times 87$ column by 9-row structure. Because of the linkage, only one set of STS POH is required in the STS-Nc SPE. Here the STS POH always appears in the first of the N STS-1s that make up the STS-Nc (Ref. 3).

Figure 17.7 shows the transport overhead assignment of an OC-3 carrying an STS-3c SPE.

17.2.2.4 Structure of Virtual Tributaries (VTs). The SONET STS-1 SPE with a channel capacity of 50.11 Mbps has been designed specifically to transport a DS3 tributary signal. To accommodate sub-STS-1 rate payloads such as DS1, the VT structure is used. It consists of four sizes: VT1.5 (1.728 Mbps) for DS1 transport, VT2 (2.304 Mbps) for E1 transport, VT3 (3.456 Mbps) for DS1C transport, and VT6 (6.912 Mbps) for DS2 transport. The virtual tributary concept is illustrated in Figure 17.8. The four

[4]"OC-N" stands for optical carrier at the N level. This has the same electrical signal as STS-N and the same bit rate and format structure.

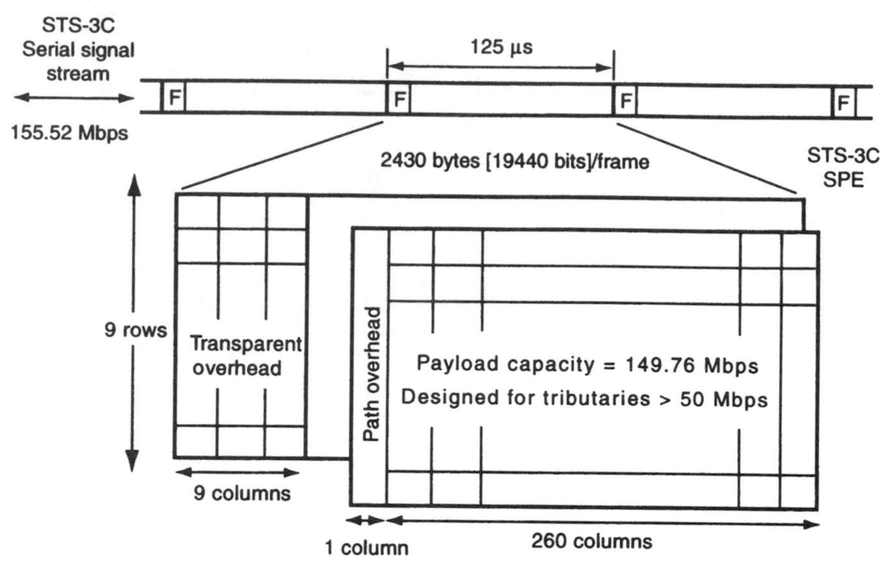

Figure 17.6 STS-3c concatenated SPE. (From Ref. 2, courtesy of Hewlett-Packard.)

VT configurations are shown in Figure 17.9. In the 87-column by 9-row structure of the STS-1 SPE, the VTs occupy 3, 4, 6, and 12 columns, respectively.

There are two VT operating modes: *floating mode* and *locked mode*. The floating mode was designed to minimize network delay and provide efficient cross-connects of transport signals at the VT level within the synchronous network. This is achieved

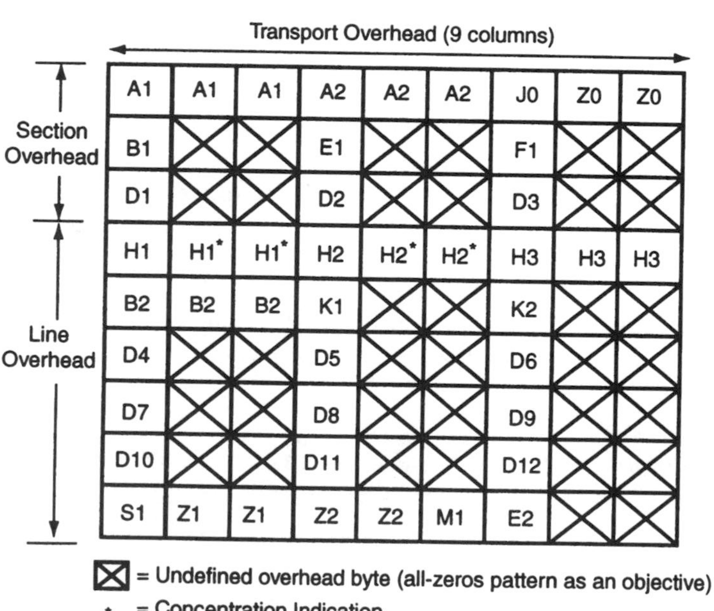

Figure 17.7 Transport overhead assignment showing OC-3 carrying an STS-3c SPE. (From Ref. 1, Figure 3–8, copyright 1994 Bellcore. Reprinted with permission.)

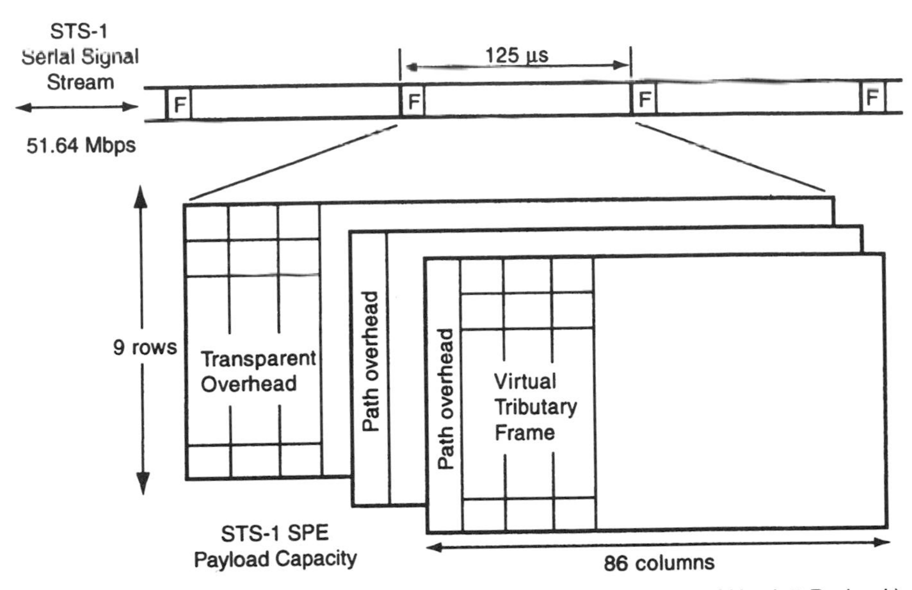

Figure 17.8 The virtual tributary (VT) concept. (From Ref. 2, courtesy of Hewlett-Packard.)

by allowing each VT SPE to float with respect to the STS-1 SPE in order to avoid the use of unwanted slip buffers at each VT cross-connect point.[5] Each VT SPE has its own payload pointer, which accommodates timing synchronization issues associated with the individual VTs. As a result, by allowing a selected VT1.5, for example, to be cross-connected between different transport systems without unwanted network delay, this mode allows a DS1 to be transported effectively across a SONET network.

The locked mode minimizes interface complexity and supports bulk transport of DS1 signals for digital switching applications. This is achieved by *locking* individual VT

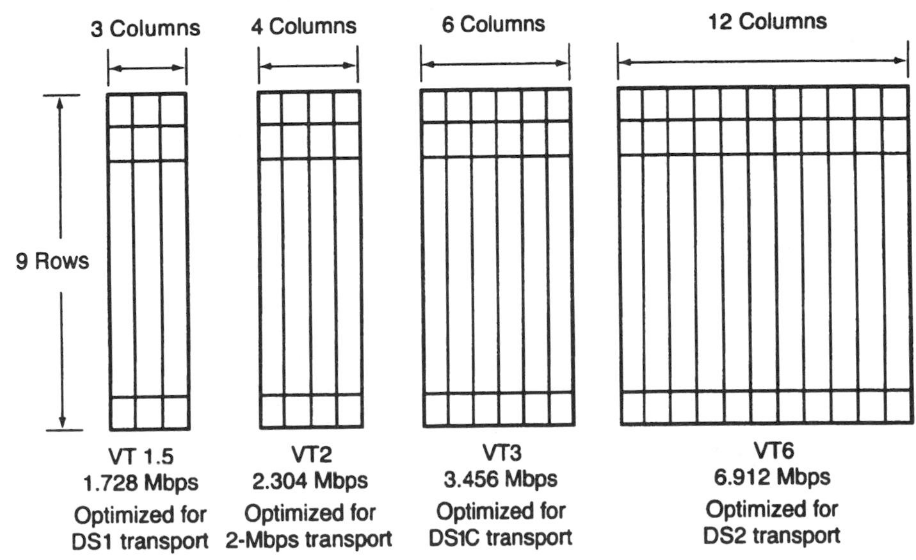

Figure 17.9 The four sizes of virtual tributary frames. (From Ref. 2, courtesy of Hewlett-Packard.)

[5]Slips are discussed in Chapter 6.

SPEs in fixed positions with respect to the STS-1 SPE. In this case, each VT1.5 SPE is not provided with its own payload pointer. With the locked mode it is not possible to route a selected VT1.5 through the SONET network without unwanted network delay caused by having to provide slip buffers to accommodate the timing/synchronization issues.

17.2.2.5 Payload Pointer. The STS payload pointer provides a method for allowing flexible and dynamic alignment of the STS SPE within the STS envelope capacity, independent of the actual contents of the SPE. SONET, by definition, is intended to be synchronous. It derives its timing from the master network clock. (See Section 6.12.2.)

Modern digital networks must make provision for more than one master clock. Examples in the United States are the several interexchange carriers which interface with local exchange carriers, each with their own master clock. Each master clock (stratum 1) operates independently. And each of these master clocks has excellent stability (i.e., better than 1×10^{-11} per month), yet there may be some small variance in time among the clocks. Assuredly they will not be phase-aligned. Likewise, SONET must take into account the loss of master clock or a segment of its timing delivery system. In this case, switches fall back on lower-stability internal clocks. This situation must also be handled by SONET. Therefore synchronous transport must be able to operate effectively under these conditions, where network nodes are operating at slightly different rates.

To accommodate these clock offsets, the SPE can be moved (justified) in the positive or negative direction one octet at a time with respect to the transport frame. This is achieved by recalculating or updating the payload pointer at each SONET network node. In addition to clock offsets, updating the payload pointer also accommodates any other timing phase adjustments required between the input SONET signals and the timing reference at the SONET node. This is what is meant by *dynamic alignment*, where the STS SPE is allowed to float within the STS envelope capacity.

The payload pointer is contained in the H1 and H2 octets in the line overhead (LOH) and designates the location of the octet where the STS SPE begins. These two octets are illustrated in Figure 17.10. Bits 1 through 4 of the pointer word carry the *new data flag*, and bits 7 through 16 carry the pointer value. Bits 5 and 6 are undefined.

Let us discuss bits 7 through 16, the actual pointer value. It is a binary number with a range of 0 to 782. It indicates the offset of the pointer word and the first octet of the STS SPE (i.e., the J1 octet). The transport overhead octets are not counted in the offset. For example, a pointer value of 0 indicates that the STS SPE starts in the octet location that immediately follows the H3 octet, whereas an offset of 87 indicates that it starts immediately after the K2 octet location.

Payload pointer processing introduces a signal impairment known as *payload adjustment jitter*. This impairment appears on a received tributary signal after recovery from a SPE that has been subjected to payload pointer changes. The operation of the network equipment processing the tributary signal immediately downstream is influenced by this excessive jitter. By careful design of the timing distribution for the synchronous network, payload pointer adjustments can be minimized, thus reducing the level of tributary jitter that can be accumulated through synchronous transport.

17.2.2.6 Three Overhead Levels of SONET. The three embedded overhead levels of SONET are:

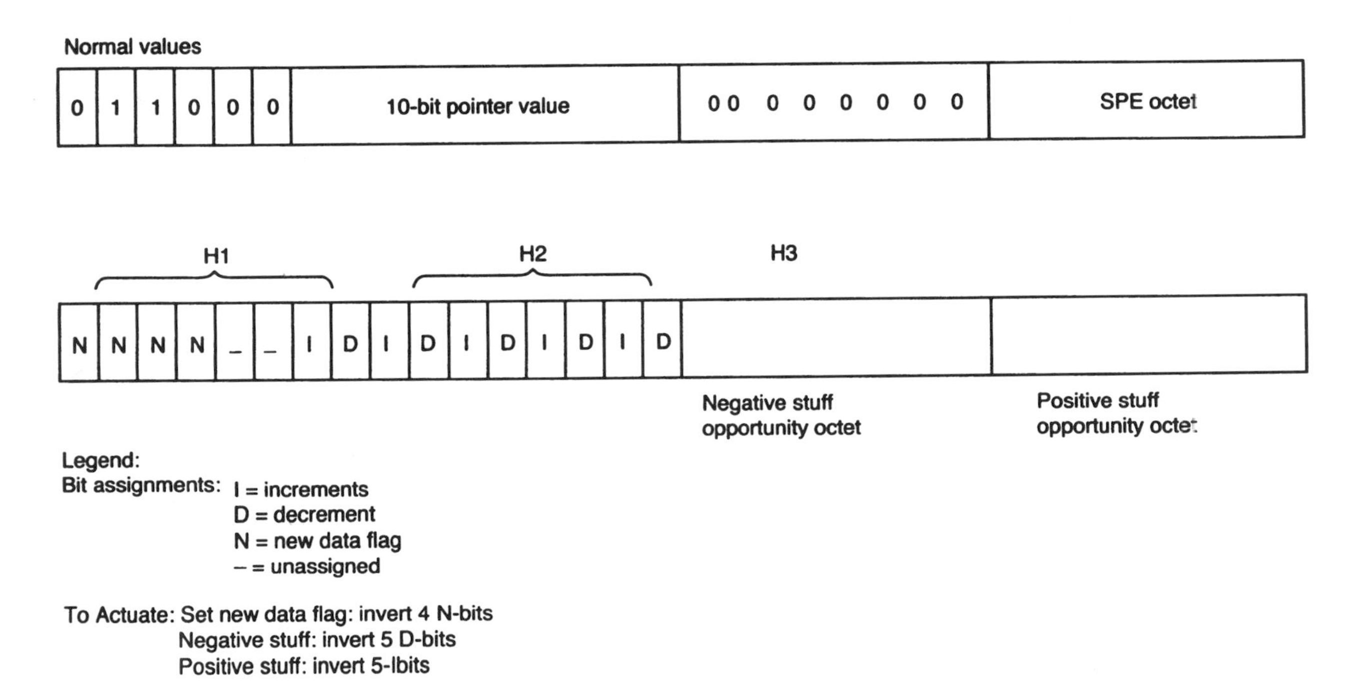

Figure 17.10 STS payload pointer (H1, H2) coding.

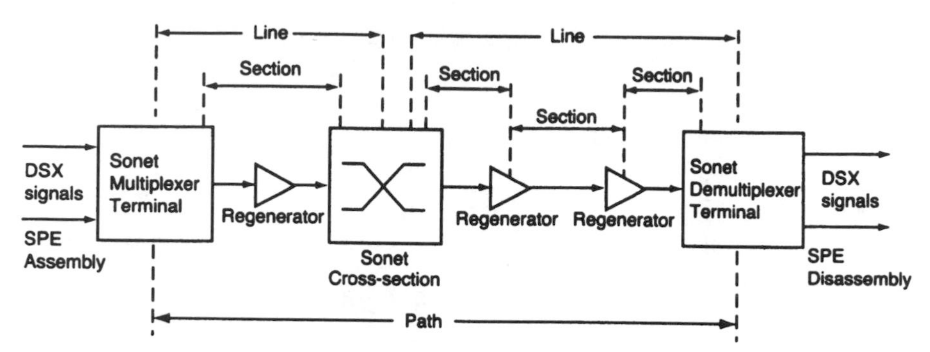

Figure 17.11 SONET section, line, and path definitions.

1. Path (POH);
2. Line (LOH); and
3. Section (SOH).

These overhead levels, represented as spans, are illustrated in Figure 17.11. One important function is to support network operation and maintenance (OAM).

The POH consists of 9 octets and occupies the first column of the SPE, as pointed out previously. It is created and included in the SPE as part of the SPE assembly process. The POH provides the facilities to support and maintain the transport of the SPE between path terminations, where the SPE is assembled and disassembled. Among the POH specific functions are:

- An 8-bit-wide (octet B3) BIP (bit-interleaved parity) check calculated over all bits of the previous SPE. The computed value is placed in the POH of the following frame;
- Alarm and performance information (octet G1);
- A path signal label (octet C2); gives details of SPE structure. It is 8 bits wide, which can identify up to 256 structures (2^8);
- One octet (J1) repeated through 64 frames can develop an alphanumeric message associated with the path. This allows verification of continuity of connection to the source of the path signal at any receiving terminal along the path by monitoring the message string; and
- An orderwire for network operator communications between path equipment (octet F2).

Facilities to support and maintain the transport of the SPE between adjacent nodes are provided by the line and section overhead. These two overhead groups share the first three columns of the STS-1 frame (see Figure 17.1). The SOH occupies the top three rows (total of 9 octets) and the LOH occupies the bottom 6 rows (18 octets).

The line overhead functions include:

- Payload pointer (octets H1, H2, and H3) (each STS-1 in an STS-N frame has its own payload pointer);
- Automatic protection switching control (octets K1 and K2);
- BIP parity check (octet B2);

- 576-kbps data channel (octets D4 through D12); and
- Express orderwire (octet E2).

A *section* is defined in Figure 17.11. Among the section overhead functions are:

- Frame alignment pattern (octets A1, A2);
- STS-1 identification (octet C1): a binary number corresponding to the order of appearance in the STS-N frame, which can be used in the framing and deinterleaving process to determine the position of other signals;
- BIP-8 parity check (octet B1): section error monitoring;
- Data communications channel (octets D1, D2, and D3);
- Local orderwire channel (octet E1); and
- User channel (octet F1).

17.2.2.7 SPE Assembly/Disassembly Process. *Payload mapping* is the process of assembling a tributary signal into an SPE. It is fundamental to SONET operation. The payload capacity provided for each individual tributary signal is always slightly greater than that required by the tributary signal. The mapping process, in essence, is to synchronize the tributary signal with the payload capacity. This is achieved by adding stuffing bits to the bit stream as part of the mapping process.

An example might be a DS3 tributary signal at a nominal rate of 44.736 Mbps to be synchronized with a payload capacity of 49.54 Mbps provided by an STS-1 SPE. The addition of path overhead completes the assembly process of the STS-1 SPE and increases the bit rate of the composite signal to 50.11 Mbps. The SPE assembly process is shown graphically in Figure 17.12. At the terminus or drop point of the network, the original DS3 payload must be recovered, as in our example. The process of SPE disassembly is shown in Figure 17.13. The term used here is *payload demapping*.

The demapping process desynchronizes the tributary signal from the composite SPE signal by stripping off the path overhead and the added stuff bits. In the example, an STS-1 SPE with a mapped DS3 payload arrives at the tributary disassembly location with a signal rate of 50.11 Mbps. The stripping process results in a discontinuous signal representing the transported DS3 signal with an average signal rate of 44.74 Mbps. The timing discontinuities are reduced by means of a desynchronizing phase-locked loop, which then produces a continuous DS3 signal at the required average transmission rate (Refs. 1, 2).

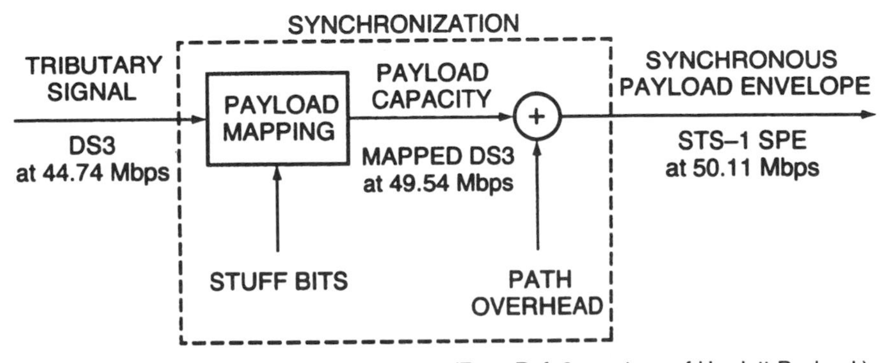

Figure 17.12 The SPE assembly process. (From Ref. 2, courtesy of Hewlett-Packard.)

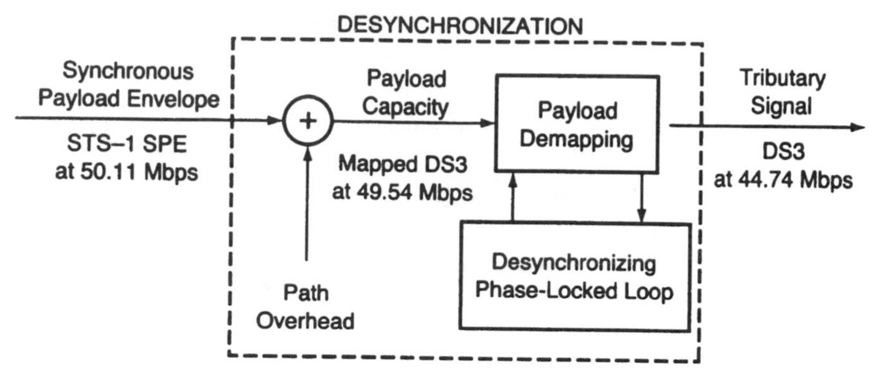

Figure 17.13 The SPE disassembly process. (From Ref. 2, courtesy of Hewlett-Packard.)

17.2.3 Line Rates for Standard SONET Interface Signals

Table 17.1 shows the standard line transmission rates for OC-N and STS-N.

17.2.4 Add–Drop Multiplex (ADM)

The SONET ADM multiplexes one or more DS-*n* signals into the SONET OC-N channel. An ADM can be configured for either the add–drop or terminal mode. In the ADM mode, it can operate when the low-speed DS1 signals terminating at the SONET derive timing from the same or equivalent source as (SONET) (i.e., synchronous) but do not derive timing from asynchronous sources.

Figure 17.14 is an example of an ADM configured in the add–drop mode with DS1 and OC-N interfaces. A SONET ADM interfaces with two full-duplex OC-N signals and one or more full-duplex DS1 signals. It may optionally provide low speed DS1C, DS2, DS3, or OC-M ($M \leq N$) interfaces. There are nonpath-terminating information payloads from each incoming OC-N signal, which are passed the SONET ADM and transmitted by the OC-N interface at the other side.

Timing for transmitted OC-N is derived from either an external synchronization source, an incoming OC-N signal, from each incoming OC-N signals in each direction (called through-timing), or from its local clock, depending on the network application. Each DS1 interface reads data from an incoming OC-N and inserts data into an outgoing OC-N bit stream as required. Figure 17.14 also shows a synchronization interface for local switch application with external timing and an operations interface module (OIM) that provides local technician orderwire, local alarm, and an interface to remote

Table 17.1 Line Rates for Standard SONET Interface Signals

OC-N Level	STS-N Electrical Level	Line Rate (Mbps)
OC-1	STS-1 electrical	51.84
OC-3	STS-3 electrical	155.52
OC-12	STS-12 electrical	622.08
OC-24	STS-24 electrical	1244.16
OC-48	STS-48 electrical	2488.32
OC-192	STS-192 electrical	9953.28

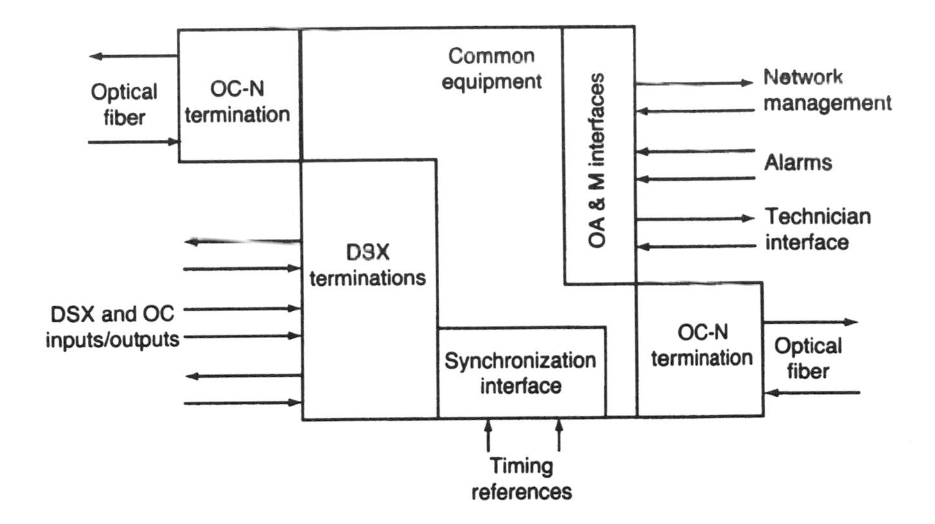

Notes: DSX = DS1–DS3 and OC (optical carrier, SONET)
OA & M = Operations, administration and maintenance

Figure 17.14 SONET ADM add–drop configuration example.

operations systems.[6] A controller is part of each SONET ADM, which maintains and controls the ADM functions, to connect to local or remote technician interfaces, and to connect to required and optional operations links that permit maintenance, provisioning, and testing.

Figure 17.15 shows an example of an ADM in the terminal mode of operation with DS1 interfaces. In this case, the ADM multiplexes up to $N \times$ (28DS1) or equivalent

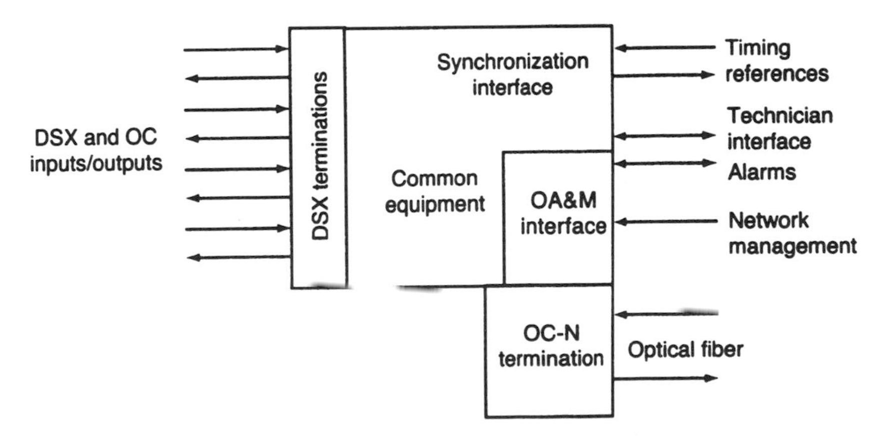

Notes: DSX = DS1–DS3 and OC = optical carrier, SONET
OA & M = Operations, administration and maintenance

Figure 17.15 An ADM in a terminal configuration.

[6]*Orderwire* is a voice or teleprinter circuit for coordinating setup and maintenance activities among technicians.

Table 17.2 SDH Bit Rates with SONET Equivalents

SDH Level[a]	SDH Bit Rate (kbps)	SONET Equivalent Line Rate
1	155,520	STS-3/OC-3
4	622,080	STS-12/OC-12
16	2,488,320	STS-48/OC48

[a]Two other SDH hierarchical levels are under consideration by CCITT (Ref. 5):
Level 8 1,244,160 kbps;
Level 12 1,866,240 kbps.

signals into an OC-N bit stream.[7] Timing for this terminal configuration is taken from either an external synchronization source, the received OC-N signal (called loop timing), or its own local clock, depending on the network application (Ref. 4).

17.3 SYNCHRONOUS DIGITAL HIERARCHY (SDH)

17.3.1 Introduction

SDH was a European/CCITT development, whereas SONET was a North American development. They are very similar. One major difference is their initial line rate. STS-1/OC-1 has an initial line rate of 51.84 Mbps; and SDH level 1 has a bit rate of 155.520 Mbps. These rates are the basic building blocks of each system. SONET's STS-3/OC-3 line rate is the same as SDH STM-1 of 155.520 Mbps.

Another difference is in their basic digital line rates. In North America it is at the DS1 or DS3 lines rates; in SDH countries it is at the 2.048, 34, or 139 Mbps rates (see Chapter 6). This has been resolved in the SONET/SDH environment through the SDH administrative unit (AU) at a 34-Mbps rate. Four such 34-Mbps AUs are "nested" (i.e., joined) to form the SDH STM-1, the 155-Mbps basic building block. There is an AU3 used with SDH to carry a SONET STS-1 or a DS3 signal. In such a way, a nominal 50-Mbps AU3 can be transported on an STM-1 SDH signal.

17.3.2 SDH Standard Bit Rates

The standard SDH bits rates are shown in Table 17.2. ITU-T Rec. G.707 (Ref. 5) states "that the first level digital hierarchy shall be 155,520 kbps ... and ... higher synchronous digital hierarchy rates shall be obtained as integer multiples of the first level bit rate."

17.3.3 Interface and Frame Structure of SDH

Figure 17.16 illustrates the relationship between various multiplexing elements of SDH and shows generic multiplexing structures. Figure 17.17 illustrates one multiplexing example for SDH, where there is direct multiplexing from container-1 using AU-3.

17.3.3.1 Definitions

Synchronous Transport Module (STM). An STM is the information structure used to support section layer connections in the SDH. It is analogous to STS in the SONET regime. STM consists of information payload and section overhead (SOH) information fields organized in a block frame structure that repeats every 125 μs.

[7]This implies a DS3 configuration. It contains 28 DS1s.

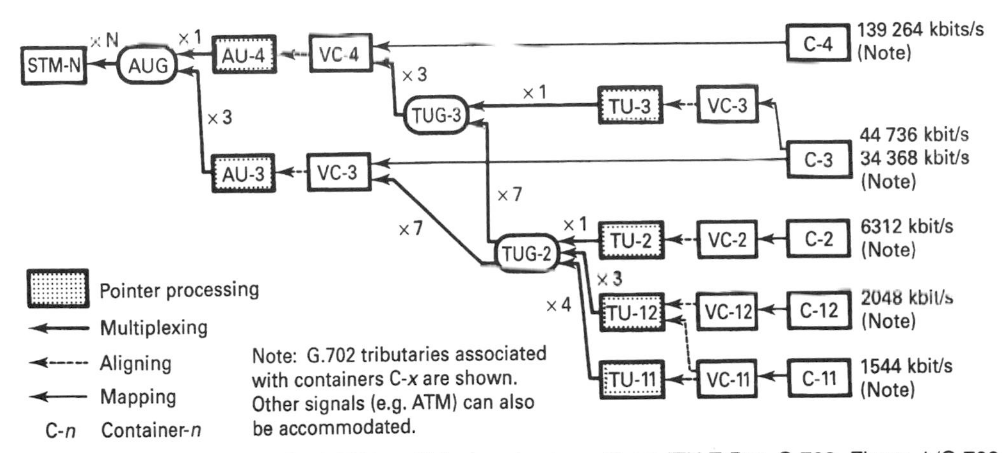

Figure 17.16 Basic generalized SDH multiplexing structure. (From ITU-T Rec. G.709, Figure 1/G.709 [Ref. 7].)

The information is suitably conditioned for serial transmission on selected media at a rate that is synchronized to the network. A basic STM (STM-1) is defined at 155,520 kbps. Higher-capacity STMs are formed at rates equivalent to N times multiples of this basic rate. STM capacities for $N = 4$ and $N = 16$ are defined, and higher values are under consideration by ITU-T. An STM comprises a single administrative unit group (AUG) together with the SOH. STM-N contains N AUGs together with SOH.

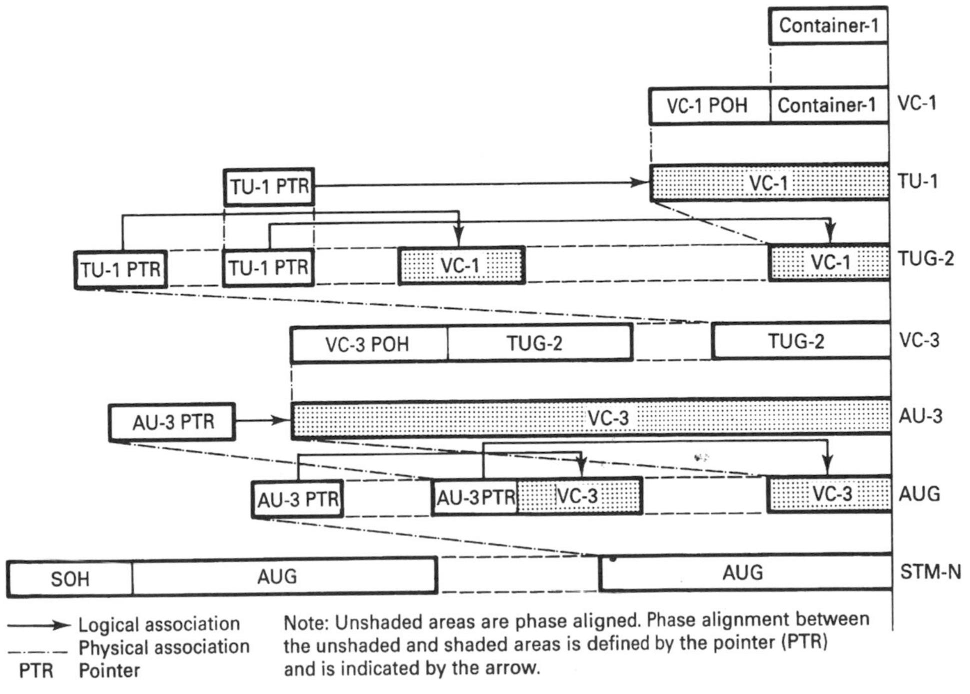

Figure 17.17 SDH multiplexing method directly from container-1 using AU-3. (From ITU-T Rec. G.708, Figure 2-3/G.708 [Ref. 6].)

Container, C-*n* (*n* = 1 to *n* = 4). This element is a defined unit of payload capacity, which is dimensioned to carry any of the bit rates currently defined in Table 17.2, and may also provide capacity for transport of broadband signals that are not yet defined by CCITT (ITU-T Organization) (Ref. 6).

Virtual Container-*n* (VC-*n*). A virtual container is the information structure used to support path layer connection in the SDH. It consists of information payload and POH information fields organized in a block frame that repeats every 125 μs or 500 μs. Alignment information to identify VC-*n* frame start is provided by the server network layer. Two types of virtual container have been identified:

1. *Lower-Order Virtual Container-n, VC-n (n = 1, 2).* This element comprises a single C-*n* (*n* = 1, 2), plus the basic virtual container POH appropriate to that level.
2. *Higher-Order Virtual Container-n, to VC-n (n = 3, 4).* This element comprises a single C-*n* (*n* = 3, 4), an assembly of tributary unit groups (TUG-2s), or an assembly of TU-3s, together with virtual container POH appropriate to that level.

Administrative Unit-*n* (AU-*n*). An administrative unit is the information structure that provides adaptation between the higher-order path layer and the multiplex section. It consists of an information payload (the higher-order virtual container) and an administrative unit pointer, which indicates the offset of the payload frame start relative to the multiplex section frame start. Two administrative units are defined. The AU-4 consists of a VC-4 plus an administrative unit pointer, which indicates the phase alignment of the VC-4 with respect to the STM-N frame. The AU-3 consists of a VC-3 plus an administrative unit pointer, which indicates the phase alignment of the VC-3 with respect to the STM-N frame. In each case the administrative unit pointer location is fixed with respect to the STM-N frame (Ref. 6). One or more administrative units occupying fixed, defined positions in a STM payload is termed an *administrative unit group* (AUG). An AUG consists of a homogeneous assembly of AU-3s or an AU-4.

Tributary Unit-*n* (TU-*n*). A tributary unit is an information structure that provides adaptation between the lower-order path layer and the higher-order path layer. It consists of an information payload (the lower-order virtual container) and a tributary unit pointer, which indicates the offset of the payload frame start relative to the higher-order virtual container frame start. The TU-*n* (*n* = 1, 2, 3) consists of a VC-*n* together with a tributary unit pointer. One or more tributary units occupying fixed, defined positions in a higher-order VC-*n* payload is termed a *tributary unit group* (TUG). TUGs are defined in such a way that mixed-capacity payloads made up of different-size tributary units can be constructed to increase flexibility of the transport network. A TUG-2 consists of a homogeneous assembly of identical TU-1s or a TU-2. A TUG-3 consists of a homogeneous assembly of TUG-2s or a TU-3 (Ref. 6).

Container-*n* (*n* = 1–4). A container is the information structure that forms the network synchronous information payload for a virtual container. For each of the defined virtual containers there is a corresponding container. Adaptation functions have been defined for many common network rates into a limited number of standard containers (Refs. 6 and 7). These include standard E-1/DS-1 rates defined in ITU-T Rec. G.702 (Ref. 8).

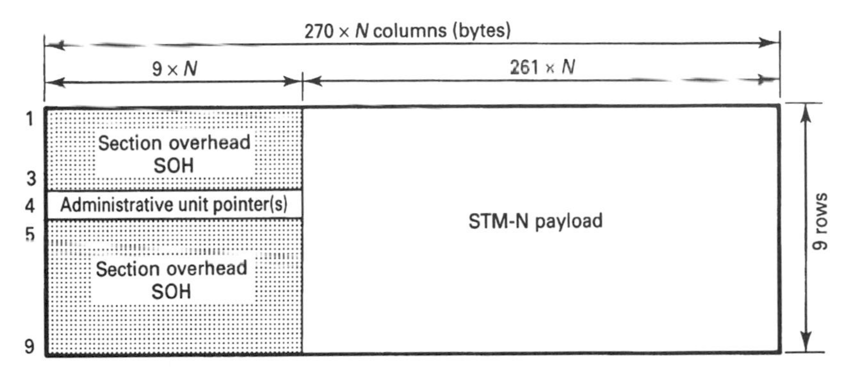

Figure 17.18 STM-N frame structure.

17.3.3.2 Frame Structure. The basic frame structure of SDH is illustrated in Figure 17.18. The three principal areas of the STM-1 frame are section overhead, AU pointers, and STM-1 payload.

17.3.3.2.1 Section Overhead. Section overhead is contained in rows 1–3 and 5–9 of columns $1 - 9 \times N$ of the STM-N shown in Figure 17.18.

17.3.3.2.2 Administrative Unit (AU) Pointers. The AU-*n* pointer (like the SONET pointer) allows flexible and dynamic alignment of the VC-*n* within the AU-*n* frame. Dynamic alignment means that the VC-*n* floats within the AU-*n* frame. Thus the pointer is able to accommodate differences, not only in the phases of the VC-*n* and the SOH, but also in the frame rates.

Row 4 of columns $1 - 9 \times N$ in Figure 17.18 is available for AU pointers. The AU-4 pointer is contained in octets H1, H2, and H3, as shown in Figure 17.19. The three individual AU-3 pointers are contained in three separate H1, H2, and H3 octets, as shown in Figure 17.20.

The pointer contained in H1 and H2 designates the location of the octet where the VC-*n* begins. The two octets (or bytes) allocated to the pointer function can be viewed as one word, as illustrated in Figure 17.21. The last ten bits (bits 7–16) of the pointer word carry the pointer value.

As shown in Figure 17.21, the AU-4 pointer value is a binary number with a range of 0–782, which indicates offset, in three-octet increments, between the pointer and the first octet of the VC-4 (see Figure 17.19). Figure 17.21 also indicates one additional valid point, the concatenation indication. This concatenation is given by the binary sequence 1001 in bit positions 1–4; bits 5 and 6 are unspecified and there are ten 1s in bit positions 7–16. The AU-4 pointer is set to the concatenation indications for AU-4 concatenation.

There are three AU-3s in an AUG, where each AU-3 has its own associated H1, H2, and H3 octets. As detailed in Figure 17.20, the H octets are shown in sequence. The first H1, H2, H3 set refers to the first AU-3, and the second set to the second AU-3, and so on. For the AU-3s, each pointer operates independently. In all cases the AU-*n* pointer octets are not counted in the offset. For example, in an AU-4, the pointer value of 0 indicates that the VC-4 starts in the octet location that immediately follows the last H3 octet, whereas an offset of 87 indicates that the VC-4 starts three octets after the K2 octet (byte) (Ref. 7). Note the similarity to SONET here.

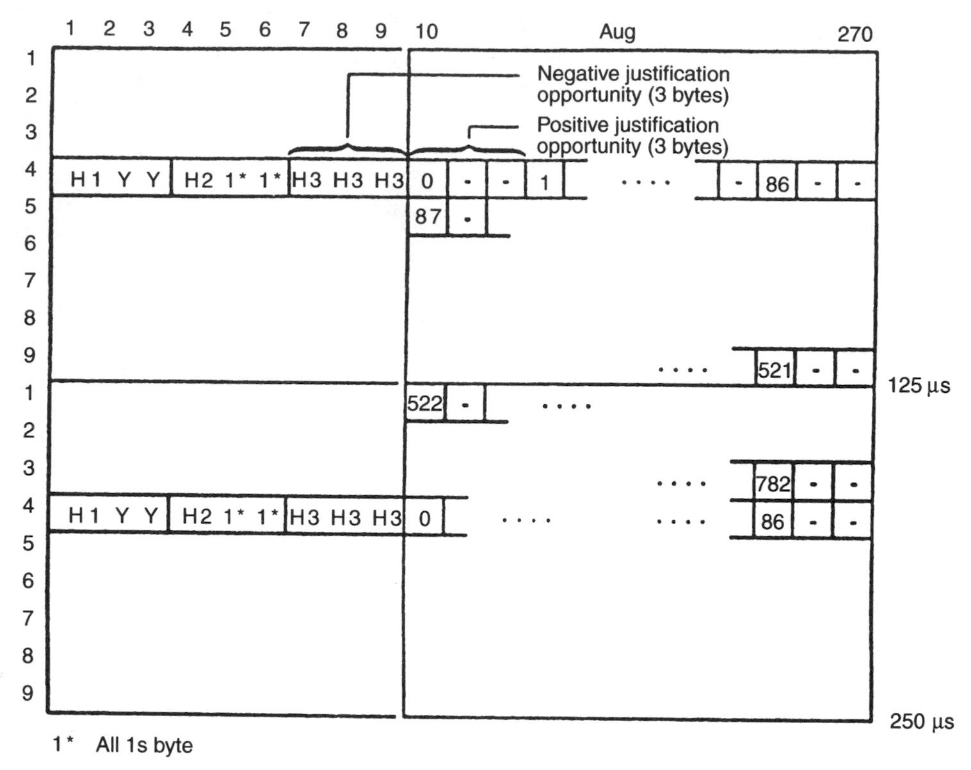

1* All 1s byte

Y 1001SS11 (S bits are unspecified)

Figure 17.19 AU-4 pointer offset numbering. (From ITU-T Rec. G.709, Figure 3-1/G.709 [Ref. 7].)

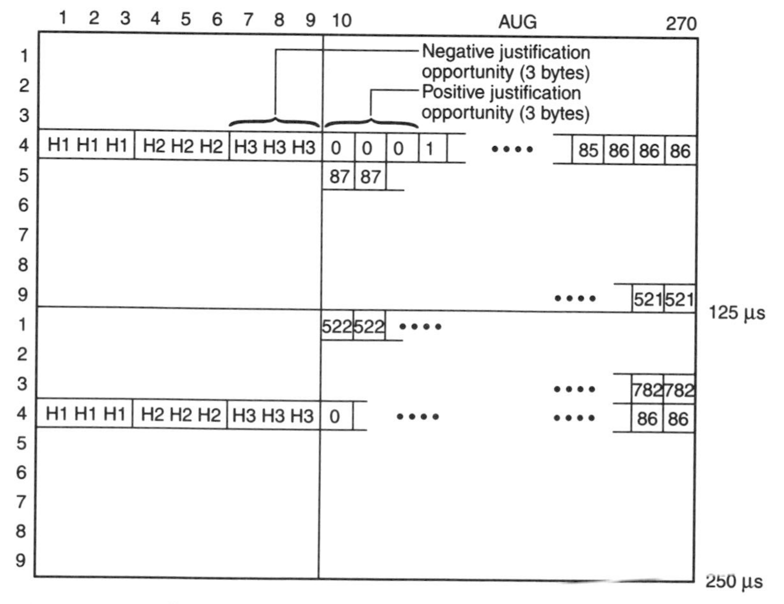

Figure 17.20 AU-3 offset numbering. (From ITU-T Rec. G.709, Figure 3-2/G.709 [Ref. 7].)

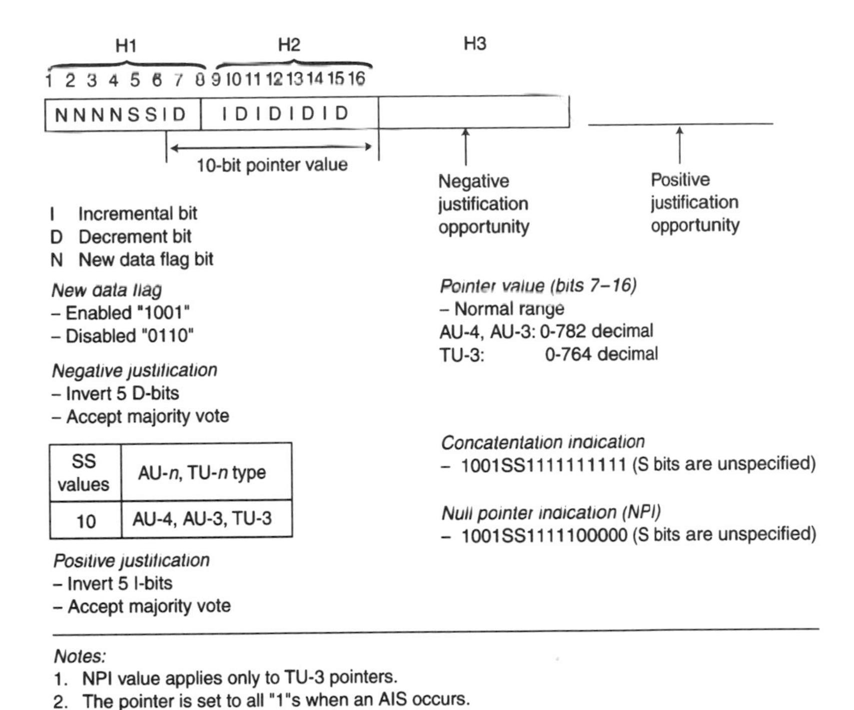

Figure 17.21 AU-*n*/TU-3 pointer (H1, H2, H3) coding. (From ITU-T Rec. G.709, Figure 3-3/G.709 [Ref. 7].)

17.3.3.2.3 Frequency Justification. If there is a frequency offset between the frame rate of the AUG and that of the VC-*n*, then the pointer value will be incremented or decremented as needed, accompanied by a corresponding positive or negative justification octet or octets. Consecutive pointer operations must be separated by at least three frames (i.e., every fourth frame) in which the pointer values remain constant.

If the frame rate of the VC-*n* is too slow with respect to that of the AUG, then the alignment of the VC-*n* must periodically slip back in time and the pointer value must be incremented by one. This operation is indicated by inverting bits 7, 9, 11, 13, and 15 (I-bits) of the pointer word to allow 5-bit majority voting at the receiver. Three positive justification octets appear immediately after the last H3 octet in the AU-4 frame containing the inverted I-bits. Subsequent pointers will contain the new offset.

For AU-3 frames, a positive justification octet appears immediately after the individual H3 octet of the AU-3 frame containing inverted I-bits. Subsequent pointers will contain the new offset.

If the frame rate of the VC-*n* is too fast with respect to that of the AUG, then the alignment of the VC-*n* must periodically be advanced in time and the pointer value must then be decremented by one. This operation is indicated by inverting bits 8, 10, 12, 14, and 16 (D-bits) of the pointer word to allow 5-bit majority voting at the receiver. Three negative justification octets appear in the H3 octets in the AU-4 frame containing inverted D-bits. Subsequent pointers will contain the new offset.

For AU-3 frames, a negative justification octet appears in the individual H3 octet of the AU-3 frame containing inverted D-bits. Subsequent pointers will contain the new offset.

The following summarizes the rules (Ref. 7) for interpreting the AU-*n* pointers:

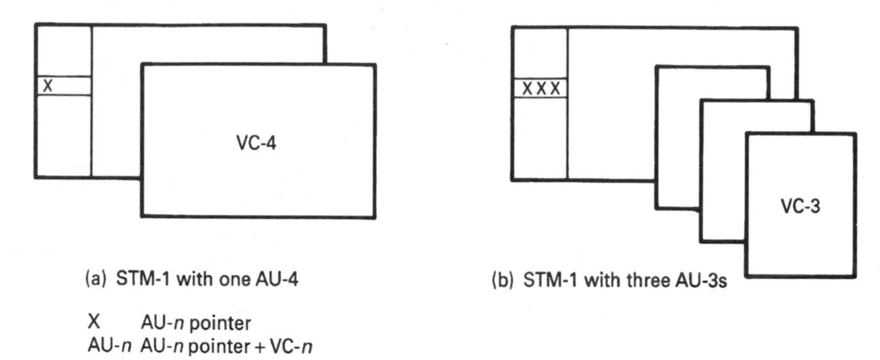

Figure 17.22 Administrative units in an STM-1 frame. (From ITU-T Rec. G.709, Figure 3-2/G.708 [Ref. 6].)

1. During normal operation, the pointer locates the start of the VC-*n* within the AU-*n* frame.
2. Any variation from the current pointer value is ignored unless a consistent new value is received three times consecutively or it is preceded by one; see rules 3, 4, or 5 (below). Any consistent new value received three times consecutively overrides (i.e., takes priority over) rules 3 and 4.
3. If the majority of I-bits of the pointer word are inverted, a positive justification operation is indicated. Subsequent pointer values shall be incremented by one.
4. If the majority of D-bits of the pointer word are inverted, a negative justification operation is indicated. Subsequent pointer values shall be decremented by one.
5. If the NDF (new data flag) is set to 1001, then the coincident pointer value shall replace the current one at the offset indicated by the new pointer values unless the receiver is in a state that corresponds to a loss of pointer.

17.3.3.2.4 Administrative Units in the STM-N. The STM-N payload can support *N* AUGs, where each AUG may consist of one AU-4 or three AU-3s. The VC-*n* associated with each AUG-*n* does not have a fixed phase with respect to the STM-N frame. The location of the first octet in the VC-*n* is indicated by the AU-*n* pointer. The AU-*n* pointer is in a fixed location in the STM-N frame. This is illustrated in Figures 17.17, 17.18, 17.22, and 17.23.

The AU-4 may be used to carry, via the VC-4, a number of TU-*n*'s ($n = 1, 2, 3$) forming a two-stage multiplex. An example of this arrangement is shown in Figure 17.23. The VC-*n* associated with each TU-*n* does not have a fixed phase relationship with respect to the start of the VC-4. The TU-*n* pointer is in a fixed location in the VC-4 and the location of the first octet of the VC-*n* is indicated by the TU-*n* pointer.

The AU-3 may be used to carry, via the VC-3, a number of TU-*n*'s ($n = 1, 2$) forming a two-stage multiplex. An example of this arrangement is shown in Figure 17.22 and 17.23*b*. The VC-*n* associated with each TU-*n* does not have a fixed phase relationship with respect to the start of the VC-3. The TU-*n* pointer is in a fixed location in the VC-3 and the location of the first octet of the VC-*n* is indicated by the TU-*n* pointer (Ref. 7).

17.3.3.3 Interconnection of STM-1s. SDH has been designed to be universal, allowing transport of a large variety of signals, including those specified in ITU-T Rec.

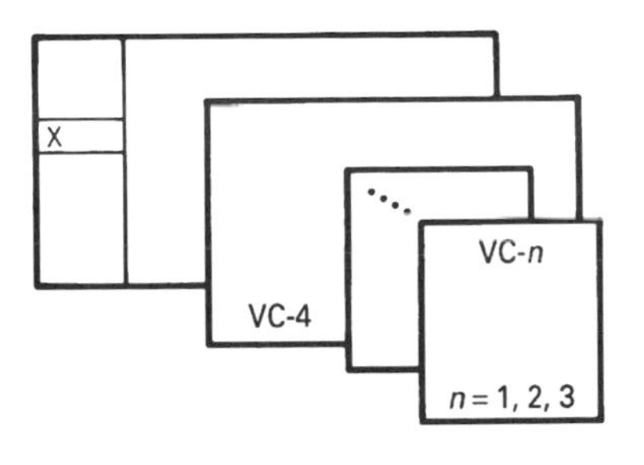

(a) STM-1 with one AU-4 containing TUs

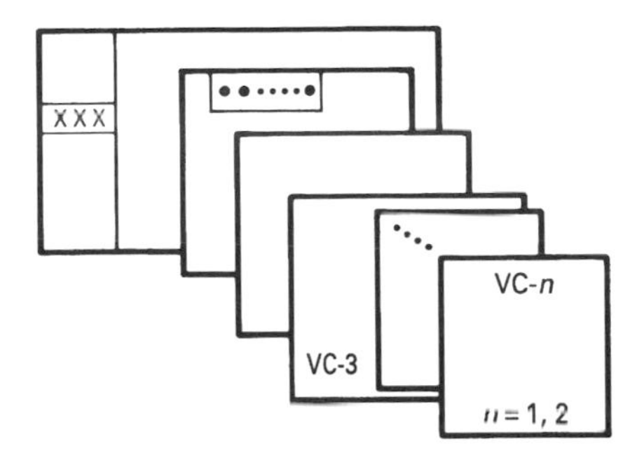

(b) STM-1 with three AU-3s containing TUs

X AU-n pointer
• TU-n pointer
AU-n AU-n pointer + VC-n
TU-n TU-n pointer + VC-n

Figure 17.23 Two-stage multiplex. (From ITU-T G.708, Figure 3-3/G.708 [Ref. 6].)

G.702 (Ref. 8), such as the North American DS1 hierarchy and the European E-1 hierarchy. However, different structures can be used for the transport of virtual containers. The following interconnection rules are used:

1. The rule for interconnecting two AUGs based on two different types of administrative unit, namely, AU-4 and AU-3, is to use the AU-4 structure. Therefore the AUG based on AU-3 is demultiplexed to the TUG-2 or VC-3 level according to the type of payload and is remultiplexed within an AUG via the TUG-3/VC-4/AU-4 route.
2. The rule for interconnecting VC-11s transported via different types of tributary unit, namely, TU-11 and TU-12, is to use the TU-11 structure. VC-11, TU-11, and TU-12 are described in ITU-T Rec. G.709 (Ref. 7).

REVIEW EXERCISES

1. What was the principal driving force for the development of SONET?

2. Being that the frame period of SONET/SDH is 125 μsec, what service are these formats optimized for?

3. Give at least five examples of digital services that can be transported on SONET.

4. Describe in words the basic SONET frame. Show its apparent payload and then its actual payload and explain the differences.

5. How can we calculate the bit rate of STS-3 if we know that the bit rate of STS-1 is 51.84 Mbps?

6. Explain how STS-Nc concatenation operates and its purpose.

7. What are the two primary uses of the *payload pointer*?

8. What network impairment is introduced by the payload pointer? What is its cause?

9. What are the three overhead levels of SONET?

10. Name at least three specific functions of the POH.

11. What is *payload mapping*?
12. If the SONET line rate were OC-48, how could one calculate this bit rate?
13. What is the function of an add–drop multiplexer in the add–drop mode?
14. The SDH STM-1 bit rate equals what SONET line rate?
15. How do we derive higher line rates from the SDH STM-1?
16. What are the two types of pointers used in SDH?
17. When the frame rate of a VC-*n* is too slow with respect to that of the AUG, what happens to the alignment of the VC-*n*? What happens to the pointer value in this example?
18. We change the value of an AU pointer. How many frames must go by before changing it again? Why can we not change it right away, right on the next frame?
19. How many AU-3s or AU-4s may be contained in an AUG?
20. SDH was designed to carry the E1 hierarchy. Can it also carry the DS1 hierarchy?

REFERENCES

1. *Synchronous Optical Network (SONET), Transport Systems, Common Generic Criteria*, Bellcore, GR-253-CORE, Issue 1, Bellcore, Piscataway, NJ, 1994.
2. Introduction to SONET, Seminar, Hewlett-Packard Co., Burlington, MA, Nov. 1993.
3. C. A. Siller and M. Shafi, eds., *SONET/SDH*, IEEE Press, New York, 1996.
4. *SONET Add–Drop Multiplex Equipment (SONET ADM) Generic Criteria*, Bellcore TR-TSY-000496, Issue 2, Bellcore, Piscataway, NJ, 1989.
5. *Synchronous Digital Hierarchy Bit Rates*, ITU-T Rec. G.707, ITU, Helsinki, 1993.
6. *Network Node Interface for the Synchronous Digital Hierarchy*, ITU-T Rec. G.708, ITU, Helsinki, 1993.
7. *Synchronous Multiplexing Structure*, ITU-T Rec. G.709, ITU, Geneva, 1993.
8. *Digital Hierarchy Bit Rates*, CCITT Rec. G.702, Fascicle III.4, IXth Plenary Assembly, Melbourne, 1988.
9. G. Held, *Dictionary of Communications Technology*, Wiley, Chichester, UK, 1995.

18

ASYNCHRONOUS TRANSFER MODE

18.1 EVOLVING TOWARD ATM

Frame relay (Section 12.5) began a march toward an optimized digital format for multimedia transmission (i.e., voice, data, video, and facsimile).[1] There were new concepts in frame relay. Take, for example, the trend toward simplicity where the header was notably shortened. The header was pure overhead, so it was cut back as practically possible. The header also signified processing. By reducing the processing, delivery time of a data frame speeds up.

In the effort to spend up delivery, frames were unacknowledged (at least at the frame relay level); there was no operational error correction scheme. It was unnecessary because it was assumed that the underlying transport system had excellent error performance (BER better than 1×10^{-7}).[2] There was error *detection* for each frame, and a frame found in error was thrown away. Now that was something that we would never do for those of us steeped in old-time data communication. It is assumed that the higher OSI layers would request repeats of the few frames missing (i.e., thrown away). These higher OSI layers (i.e., layer 3 and above) were the customer's responsibility, not the frame relay provider's.

Frame relay also moved into the flow control arena with the BECN and FECN bits and the consolidated link layer management (CLLM). The method of handling flow control has a lot to do with its effectiveness in preventing buffer overflow. It also uses a discard eligibility (DE) bit, which set a type of priority to a frame. If the DE bit were set, the frame would be among the first to be discarded in a time of congestion.

DQDB (distributed queue dual bus) was another antecedent of ATM. It was developed by the IEEE as a simple and unique network access scheme. Even more important, its data transport format is based on the *slot*, which is called a *cell* in ATM. This slot has a format very similar to the ATM cell, which we will discuss at length in this chapter. It even has the same number of octets, 53; 48 of these carry the payload. This is identical to ATM. This "cell" idea may even be found in Bellcore's SMDS (switched multimegabit data service). In each case the slot or cell was 53 octets long. DQDB introduced a comparatively new concept of the HCS (header check sequence) for detecting errors in the header. In neither case was there a capability of detecting errors in the payload. SMDS/DQDB left it to layer 3. Slots or cells carry pieces of messages. The first "piece" of a message is identified by a BOM (beginning of message), all subsequent pieces, but

[1] "Compromise" might be a better term than "optimized."

[2] In North America PSTN BER can be expected to be better than 5×10^{-10}.

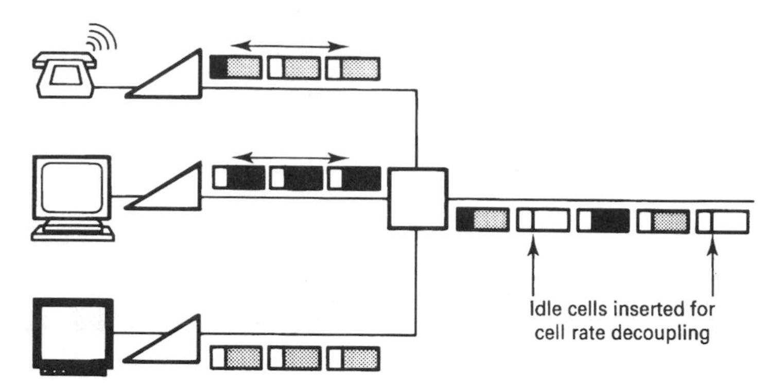

Figure 18.1 An ATM link simultaneously carries a mix of voice, data, and image information. (From Ref. 1, courtesy of Hewlett-Packard Co.)

not the last, are identified with the COM (continuation of message), and the last piece or slot is identified with an EOM (end of message). There is also an MID (message identifier) appended to all pieces of a common message. This is done so that there is no confusion as to which transaction a cell belongs.

18.2 INTRODUCTION TO ATM

ATM is an outgrowth of the several data transmission format systems mentioned in the preceding section, although some may argue this point. Whereas the formats described ostensibly were to satisfy the needs of the data world, ATM (according to some) provides an optimum format or protocol family for data, voice, and image communications, where cells of each medium can be intermixed through the network, as illustrated in Figure 18.1.[3] It would really seem to be more of a compromise from our perspective. Typically, these ATM cells can be transported on SONET, SDH, E1/DS1, and on other digital formats. Cells can also be transported contiguously without an underlying digital network format.

Philosophically, voice and data are worlds apart regarding time sensitivity. Voice cannot wait for long processing and ARQ delays. Most types of data can. So ATM must distinguish the type of service such as constant bit rate (CBR) and variable bit rate (VBR) services. Voice service is typical of constant bit rate or CBR service.

Signaling is another area of major philosophical difference. In data communications, "signaling" is carried out within the header of a data frame (or packet). As a minimum the signaling will have the destination address, and quite often the source address as well. And this signaling information will be repeated over and over again on a long data file that is heavily segmented. On a voice circuit, a connectivity is set up and the destination address, and possibly the source address, is sent just once during call setup. There is also some form of circuit supervision to keep the circuit operational throughout the duration of a telephone call. ATM is a compromise, stealing a little from each of these separate worlds.

Like voice telephony, ATM is fundamentally a connection-oriented telecommunication system. Here we mean that a connection must be established between two stations before data can be transferred between them. An ATM connection specifies the transmission path, allowing ATM cells to self-route through an ATM network. Being

[3]DQDB can also transport voice in its PA (pre-arbitrated) segments.

connection-oriented also allows ATM to specify a guaranteed quality of service (QoS) for each connection.

By contrast, most LAN protocols are connectionless. This means that LAN nodes simply transmit traffic when they need to, without first establishing a specific connection or route with the destination node.

In that ATM uses a connection-oriented protocol, cells are allocated only when the originating end-user requests a connection. They are allocated from an idle-cell pool. This allows ATM to efficiently support a network's aggregate demand by allocating cell capacity on demand based on immediate user need. Indeed it is this concept that lies at the heart of the word *asynchronous* (as in asynchronous transfer mode). An analogy would help. Let's say that New York City is connected to Washington, DC, by a pair of railroad tracks for passenger trains headed south and another pair of tracks for passenger trains headed north. On these two pair of tracks we'd like to accommodate everybody we can when they'd like to ride. The optimum for reaching this goal is to have a continuous train of coupled passenger cars. As the train enters Union Station, it discharges its passengers and connects around directly for the northward run to Pennsylvania Station and loading people bound for New York. Passenger cars are identical in size, and each has the same number of identical seats.

Of course, at 2 A.M. the train will have very few passengers and many empty seats. Probably from 7 A.M. to 9 A.M. the train will be full, no standees allowed, so we'll have to hold potential riders in the waiting room. They'll ride later; those few who try to be standees will be bumped. Others might seek alternate transportation to Washington, DC.

Here we see that the railroad tracks are the transmission medium. Each passenger car is a SONET/SDH frame. The seats in each car are our ATM cells. Each seat can handle a person no bigger than 53 units. Because of critical weight distribution, if a person is not 53 units in size/weight, we'll stick some bricks in the seat to bring the size/weight to 53 units exactly. Those bricks are removed at the destination. All kinds of people ride the train because America is culturally diverse, analogous to the fact that ATM handles all forms of traffic. The empty seats represent idle or unassigned cells. The header information is analogous to the passengers' tickets. Keep in mind that the train can only fill to its maximum capacity of seats. We can imagine the SONET/SDH frame as being full of cells in the payload, some cells busy and some idle/unassigned. At the peak traffic period, all cells will be busy, and some traffic (passengers) may have to be turned away.

We can go even further with this analogy. Both Washington, DC, and New York City attract large groups of tourists, and other groups travel to business meetings or conventions. A tour group has a chief tour guide in the lead seat (cell) and an assistant guide in the last seat (cell). There may be so many in the group that they extend into a second car or may just intermingle with other passengers on the train. The tour guide and assistant tour guide keep an exact count of people on the tour. The lead guide wears a badge that says BOM, all tour members wear badges that say COM, and the assistant tour guide wears a badge that says EOM. Each group has a unique MID (message ID). We also see that service is connection-oriented (Washington, DC, to New York City).

Asynchronous means that we can keep filling seats on the train until we reach its maximum capacity. If we look up the word, it means *nonperiodic*, whereas the familiar E1/DS1 is periodic (i.e., synchronous). One point that seems to get lost in the literature is that the train has a maximum capacity. It is common to read that ATM provides "bandwidth" on demand. The underlying transmission medium has a fixed bandwidth. We believe the statement is supposed to mean "cell rate capacity" on demand. This is true, of course, until we reach the maximum capacity. For example, if the underlying

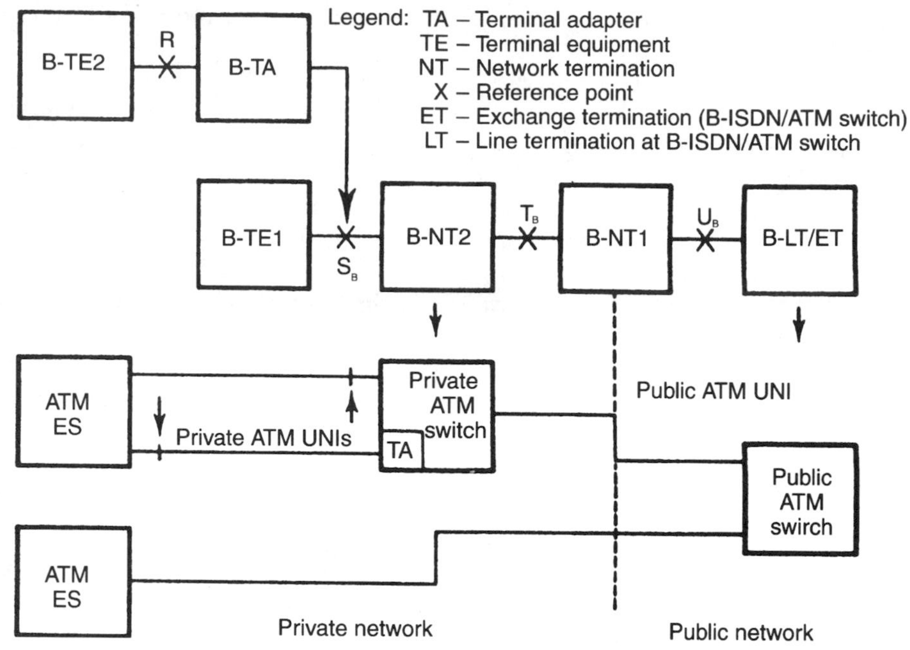

Figure 18.2 ATM reference model and user-network interface configuration. (*Sources:* Refs. 2, 3, and 6.)

transmission format is SONET STS-1, it can provide 84 × 9 octets × 8000/(53) per second or about 114 thousand cells per second of payload (see Chapter 17).

18.3 USER–NETWORK INTERFACE (UNI) AND ARCHITECTURE

ATM is the underlying packet technology of broadband ISDN (B-ISDN). At times in this section, we will use the terms ATM and B-ISDN interchangeably. Figures 18.2 and 18.3 interrelate the two. Figure 18.2 relates the B-ISDN access reference configuration with the ATM user–network interface (UNI). Note the similarities of this figure with Figure 12.15, the reference model for ISDN. The only difference is that the block nomenclature has a "B" placed in front to indicate *broadband*. Figure 18.3 is the traditional CCITT Rec. I.121 (Ref. 2) B-ISDN protocol reference model, showing the extra layer necessary for the several services.

Returning to Figure 18.2, we see that there is an upper part and a lower part. The lower part shows the UNI boundaries. The upper part is the B-ISDN reference configuration with four interface points. The interfaces at the reference points U_B, T_B, and S_B are standardized. These interfaces support all B-ISDN services.

There is only one interface per B-NT1 at the U_B and one at the T_B reference points. The physical media is point-to-point (in each case), in the sense that there is only one receiver in front of one transmitter.

One or more interfaces per NT2 are present at the S_B reference point. The interface at the S_B reference point is point-to-point at the physical layer, in the sense that there is only one receiver in front of one transmitter, and may be point-to-point at other layers.

Consider now the functional groupings in Figure 18.2. B-NT1 includes functions broadly equivalent to OSI layer 1, the physical layer. These functions include:

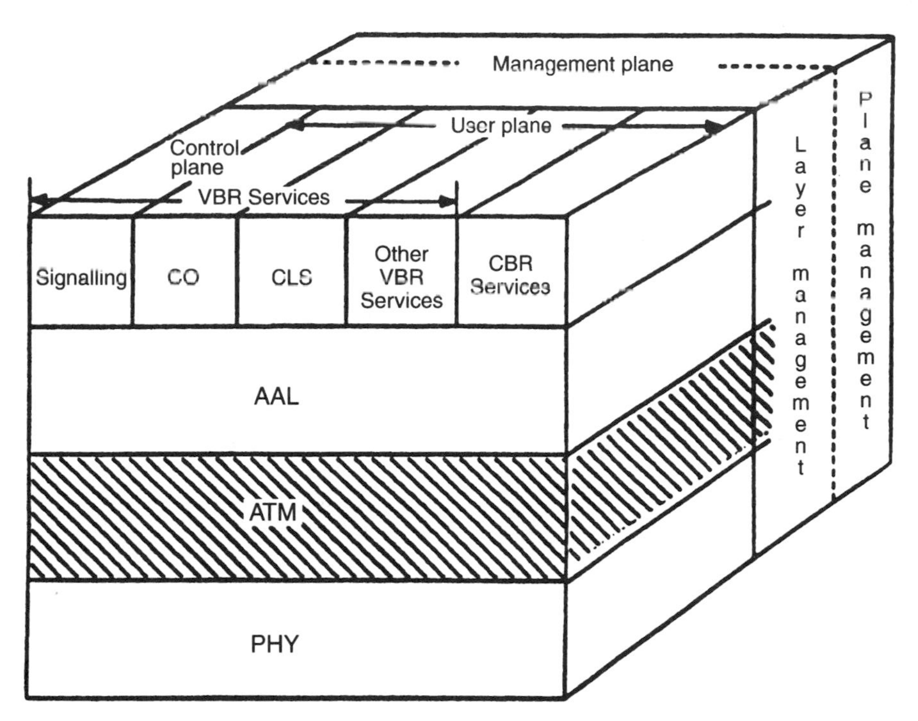

Figure 18.3 B-ISDN reference model. (From Refs. 4, 6, and 7.)

- Line transmission termination;
- Interface handling at T_B and U_B; and
- OAM functions.[4]

The B-NT2 functional group includes functions broadly equivalent to OSI layer 1 and higher OSI layers. The B-NT2 may be concentrated or distributed. In a particular access arrangement, the B-NT2 functions may consist of physical connections. Examples of B-NT2 functions are:

- Adaptation functions for different media and topologies;
- Cell delineation;
- Concentration; buffering;
- Multiplexing and demultiplexing;
- OAM functions;
- Resource allocation; and
- Signaling protocol handling.

The functional group B-TE (TE stands for terminal equipment) also includes functions of OSI layer 1 and higher OSI layers. Some of these functions are:

- User/user and user/machine dialog and protocol;
- Protocol handling for signaling;

[4]OAM stands for operations, administration, and maintenance.

- Connection handling to other equipment;
- Interface termination; and
- OAM functions.

B-TE1 has an interface that complies with the B-ISDN interface. B-TE2, however, has a noncompliant, B-ISDN interface. Compliance refers to ITU-T Recs. I.413 and I.432 as well as ANSI T1.624-1993 (Refs. 4–6).

The terminal adapter (B-TA) converts the B-TE2 interface into a compliant B-ISDN user–network interface.

Four bit rates are specified at the UB, TB, and SB interfaces based on Ref. 5 (ANSI T1.624-1993). These are:

1. 51.840 Mbps (SONET STS-1);
2. 155.520 Mbps (SONET STS-3 and SDH STM-1);
3. 622.080 Mbps (SONET STS-12 and SDH STM-4); and
4. 44.736 Mbps (DS3).

These interfaces are discussed subsequently in this chapter.

The following definitions refer to Figure 18.3:

User Plane (in other literature called the U-plane). The user plane provides for the transfer of user application information. It contains physical layer, ATM layer, and multiple ATM adaptation layers required for different service users such as CBR and VBR service.

Control Plane (in other literature called the C-plane). The control plane protocols deal with call-establishment and call-release and other connection-control functions necessary for providing switched services. The C-plane structure shares the physical and ATM layers with the U-plane, as shown in Figure 18.3. It also includes ATM adaptation layer (AAL) procedures and higher-layer signaling protocols.

Management Plane (in other literature called the M-plane). The management plane provides management functions and the capability to exchange information between the U-plane and the C-plane. The M-plane contains two sections: layer management and plane management. The layer management performs layer-specific management functions, while the plane management performs management and coordination functions related to the complete system.

We will return to Figure 18.3 and B-ISDN/ATM layering and layer descriptions in Section 18.6.

18.4 ATM CELL: KEY TO OPERATION

18.4.1 ATM Cell Structure

As mentioned earlier, the ATM cell consists of 53 octets, 5 of which make up the header and 48 make up the payload or "info" portion of the cell.[5] Figure 18.4 shows an ATM cell stream, delineating the 5-octet header and 48-octet information field of each cell.

[5]Under certain situations, there are 6 octets in the header and 47 octets in the payload.

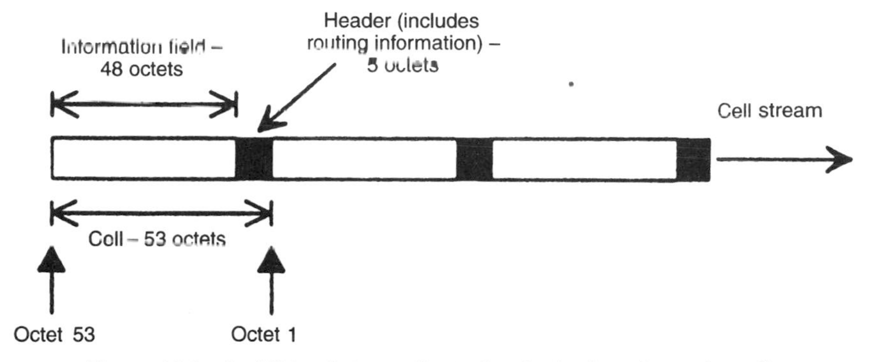

Figure 18.4 An ATM cell stream illustrating the basic makeup of a cell.

Figure 18.5 shows the detailed structure of cell headers at the user–network interface (UNI) (Figure 18.5*a*) and at the network–node interface (NNI) (Figure 18.5*b*).[6]

We digress a moment to discuss why a cell was standardized at 53 octets. The cell header contains only 5 octets. It was shortened as much as possible, designed to contain the minimum address and control functions for a working system. It is obvious that the overhead is nonrevenue-bearing. It is the information field that contains the revenue-bearing payload. For efficiency, we would like the payload to be as long as possible. Yet the ATM designer team was driven to shorten the payload as much as possible. The issue in this case was what is called *packetization delay*. This is the amount of time required to fill a cell at a rate of 64 kbps—that is, the rate required to fill the cell with digitized (PCM) voice samples. According to Ref. 8, the design team was torn between efficiency and packetization delay. One school of though fought for a 64-octet cell, and another argued for a 32-octet cell size. Thus the ITU-T organization opted for a fixed-length 53-octet compromise.

Now let us return to the discussion of the ATM cell and its headers. The left-hand

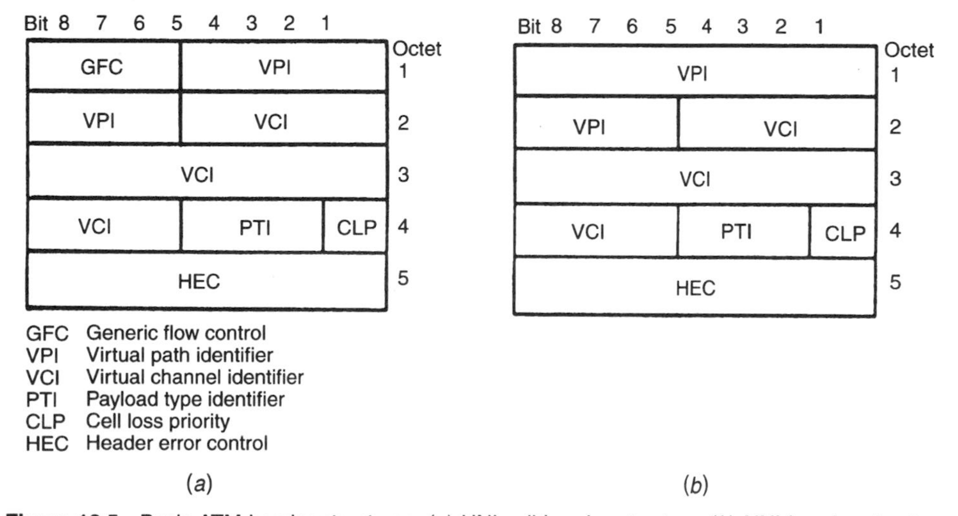

Figure 18.5 Basic ATM header structures. (*a*) UNI cell-header structure; (*b*) NNI header structure.

[6]*NNI* is variously called network–node interface or network–network interface. It is the interface between two ATM network nodes or switches.

side of Figure 18.5 shows the structure of a UNI header, whereas the right-hand side illustrates the NNI header. The only difference is the presence of the GFC (generic flow control) field in the UNI header. The following paragraphs define each header field. By removing the GFC field, the NNI has four additional bits for addressing.

18.4.1.1 GFC–Generic Flow Control. The GFC field contains 4 bits. When the GFC function is not used, the value of this field is 0000. This field has local significance only and can be used to provide standardized local flow control functions on the customer side. In fact, the value encoded in the GFC is not carried end-to-end and will be overwritten by ATM switches (i.e., the NNI interface).

Two modes of operation have been defined for operation of the GFC field. These are *uncontrolled access* and *controlled access*. The uncontrolled access mode of operation is used in the early ATM environment. This mode has no impact on the traffic which a host generates. Each host transmits the GFC field set to all zeros (0000). In order to avoid unwanted interactions between this mode and the controlled access mode, where hosts are expected to modify their transmissions according to the activity of the GFC field, it is required that all CPE (customer premise equipment) and public network equipment monitor the GFC field to ensure that attached equipment is operating in uncontrolled mode. A count of the number of nonzero GFC fields should be measured for nonoverlapping intervals of 30,000 ± 10,000 cell times. If ten or more nonzero values are received within this interval, an error is indicated to layer management (Ref. 3).

18.4.1.2 Routing Field (VPI/VCI). Twenty-four bits are available for routing a cell. There are 8 bits for virtual path identifier (VPI) and 16 bits for virtual channel identifier (VCI). Preassigned combinations of VPI and VCI values are given in Table 18.1. Other preassigned values of VPI and VCI are for further study, according to the ITU-T Organization. The VCI value of zero is not available for user virtual channel identification. The bits within the VPI and VCI fields are used for routing and are allocated using the following rules:

- The allocated bits of the VPI field are contiguous.
- The allocated bits of the VPI field are the least significant bits of the VPI field, beginning at bit 5 of octet 2.
- The allocated bits of the VCI field are contiguous.
- The allocated bits of the VCI field are the least significant bits of the VCI field, beginning at bit 5 of octet 4.

18.4.1.3 Payload Type (PT) Field. Three bits are available for PT identification. Table 18.2 gives the payload type identifier (PTI) coding. The principal purpose of the PTI is to discriminate between user cells (i.e., cells carrying information) and nonuser cells. The first four code groups (000–011) are used to indicate user cells. Within these four, 2 and 3 (010 and 011) are used to indicate that congestion has been experienced. The fifth and sixth code groups (100 and 101) are used for virtual channel connection (VCC) level management functions.

Any congested network element, upon receiving a user data cell, may modify the PTI as follows: Cells received with PTI = 000 or PTI = 010 are transmitted with PTI = 010. Cells received with PTI = 001 or PTI = 011 are transmitted with PTI = 011. Noncongested network elements should not change the PTI.

Table 18.1 Combination of Preassigned VPI, VCI, and CLP Values at the UNI

Use	VPI	VCI	PT	CLP
Meta-signalling (refer to Rec. I.311)	XXXXXXXX[a]	00000000 00000001[e]	0A0	C
General broadcast signaling (refer to Rec. I.311)	XXXXXXXX[a]	00000000 00000010[e]	0AA	C
Point-to-point signaling (refer to Rec. I.311)	XXXXXXXX[a]	00000000 00000101[e]	0AA	C
Segment OAM F4 flow cell (refer to Rec. I.610)	YYYYYYYY[b]	00000000 00000011[d]	0A0	A
End-to-end OAM F4 flow cell (refer to Rec. I.610)	YYYYYYYY[b]	00000000 00000100[d]	0A0	A
Segment OAM F5 flow cell (refer to Rec. I.610)	YYYYYYYY[b]	ZZZZZZZZ ZZZZZZZZ[c]	100	A
End-to-end OAM F5 flow cell (refer to Rec. I.610)	YYYYYYYY[b]	ZZZZZZZZ ZZZZZZZZ[c]	101	A
Resource management cell (refer to Rec. I.371)	YYYYYYYY[b]	ZZZZZZZZ ZZZZZZZZ[c]	110	A
Unassigned cell	00000000	00000000 00000000	BBB	0

The GFC field is available for use with all of these combinations.

A Indicates that the bit may be 0 or 1 and is available for use by the appropriate ATM layer function.

B Indicates the bit as a "don't care" bit.

C Indicates the originating signaling entity shall set the CLP bit to 0. The value may be changed by the network.

[a]XXXXXXXX: Any VPI value. For VPI value equal to 0, the specific VCI value specified is reserved for user signaling with the local exchange. For VPI values other than 0, the specified VCI value is reserved for signaling with other signaling entities (e.g., other users or remote networks).

[b]YYYYYYYY: Any VPI value.

[c]ZZZZZZZZ ZZZZZZZZ: Any VIC value other than 0.

[d]Transparency is not guaranteed for the OAM F4 flows in a user-to-user VP.

[e]The VIC values are preassigned in every VPC at the UNI. The usage of these values depends on the actual signaling configurations. (See ITU-T Rec. I.311.)

Source: ITU-T Rec. I.361, Table 2/I.361, p. 3 (Ref. 9).

Table 18.2 PTI Coding

PTI Coding (Bits 4 3 2)	Interpretation
0 0 0	User data cell, congestion not experienced. ATM-user-to-ATM-user indication = 0
0 0 1	User data cell, congestion not experienced. ATM-user-to-ATM-user indication = 1
0 1 0	User data cell, congestion experienced. ATM-user-to-ATM-user indication = 0
0 1 1	User data cell, congestion experienced. ATM-user-to-ATM-user indication = 1
1 0 0	OAM F5 segment associated cell
1 0 1	OAM F5 end-to-end associated cell
1 1 0	Resource management cell
1 1 1	Reserved for future functions

Source: ITU-T Rec. I.361, p. 4, para. 2.2.4 (Ref. 9).

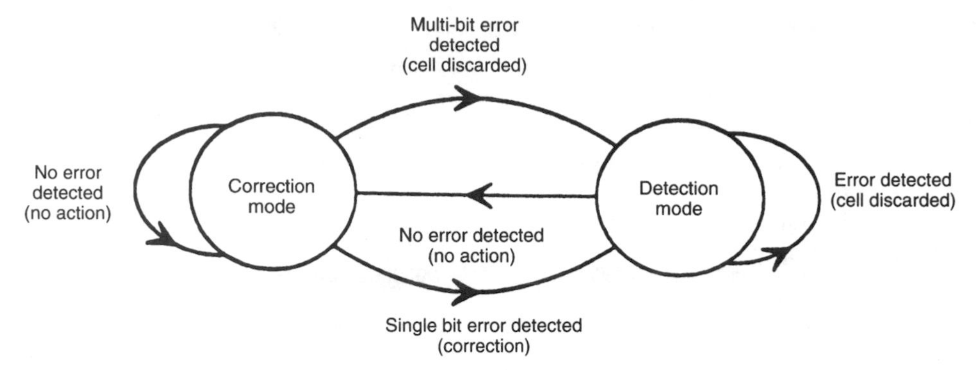

Figure 18.6 HEC: receiver modes of operation. (Based on ITU-T Rec. I.432, Ref. 4.)

18.4.1.4 Cell Loss Priority (CLP) Field. Depending on network conditions, cells where the CLP is set (i.e., CLP value is 1) are subject to discard prior to cells where the CLP is not set (i.e., CLP value is 0). The concept here is identical with that of frame relay and the DE (discard eligibility) bit. ATM switches may tag CLP = 0 cells detected by the UPC (usage parameter control) to be in violation of the traffic contract by changing the CLP bit from 0 to 1.

18.4.1.5 Header Error Control (HEC) Field. The HEC is an 8-bit field and it covers the entire cell header. The code used for this function is capable of either single-bit error correction or multiple-bit error detection. Briefly, the transmitting side computes the HEC field value. The receiver has two modes of operation as shown in Figure 18.6. In the default mode there is the capability of single-bit error correction. Each cell header is examined and, if an error is detected, one of two actions takes place. The action taken depends on the state of the receiver. In the *correction mode*, only single-bit errors can be corrected and the receiver switches to the *detection mode*. In the detection mode, all cells with detected header errors are discarded. When a header is examined and found not to be in error, the receiver switches to the correction mode. The term *no action* in Figure 18.6 means no correction is performed and no cell is discarded.

It should be noted that there is no error protection for the payload of a cell. If an error is found in the header of a cell that cannot be correct, the cell is discarded. The error-protection function provided by the HEC allows for both recovery of single-bit errors and a low probability of delivery of cells with errored headers, under bursty error conditions.

18.4.2 Idle Cells

Idle cells cause no action at a receiving node except for cell delineation including HEC verification. They are inserted and extracted by the physical layer in order to adapt the cell flow rate at the boundary between the ATM layer and the physical layer to the available payload capacity of the transmission media.[7] This is called *cell rate decoupling*. Idle cells are identified by the standardized pattern for the cell header illustrated in Table 18.3. The content of the information field is 01101010 repeated 48 times for an idle cell.

[7]There must be a constant cell flow rate to keep the far-end node synchronized and aligned.

Table 18.3 Header Pattern for Idle Cell Identification

	Octet 1	Octet 2	Octet 3	Octet 4	Octet 5
Header pattern	00000000	00000000	00000000	00000001	HEC = Valid code 01010010

Source: ITU-T Rec. I.432, Table 4/I.432, p. 19 (Ref. 4).

18.5 CELL DELINEATION AND SCRAMBLING

Cell delineation allows identification of the cell boundaries. The cell HEC field achieves cell delineation. Keep in mind that the ATM signal must be self-supporting, in that it has to be transparently transported on every network interface without any constraints from the transmission systems used. Scrambling is used to improve security and robustness of the HEC cell delineation mechanism discussed in the next paragraph. In addition, it helps the randomizing of data in the information field for possible improvement in transmission performance.

Cell delineation is performed by using correlation between the header bits to be protected (the first four octets in the header) and the HEC octet. This octet is produced at the originating end using a generating polynomial covering those first four octets of the cell. The generating polynomial is $X^8 + X^2 + X + 1$. There must be correlation at the receiving end between those first four octets and the HEC octet, which we can call a *remainder*. This is only true, of course, if there is no error in the header. When there is an error, correlation cannot be achieved, and the processor just goes to the next cell.

18.6 ATM LAYERING AND B-ISDN

The B-ISDN reference model was given in Figure 18.3, and its several planes were described. This section provides brief descriptions of the ATM layers and sublayers. Figure 18.7 illustrates B-ISDN/ATM layering and sublayering of the protocol reference model. It identifies functions of the physical layer, the ATM layer, the AAL (ATM adaptation layer), and related sublayers.

18.6.1 Physical Layer

The physical layer consists of two sublayers. The physical medium (PM) sublayer includes only physical medium-dependent functions. The transmission convergence (TC) sublayer performs all functions required to transform a flow of cells into a flow of data units (i.e., bits), which can be transmitted and received over a physical medium. The service data unit (SDU) crossing the boundary between the ATM layer and the physical layer is a flow of valid cells. The ATM layer is unique (independent of the underlying physical layer). The data flow inserted in the transmission system payload is physical medium-independent and self-supported. The physical layer merges the ATM cell flow with the appropriate information for cell delineation, according to the cell delineation mechanism previously described, and carries the operations administration and maintenance (OAM) information relating to this cell flow.

The PM sublayer provides bit transmission capability including bit transfer and bit alignment, as well as line coding and electrical–optical transformation. Of course, the principal function is the generation and reception of waveforms suitable for the medium, the insertion and extraction of bit timing information, and line coding where required.

	Functions	Sublayer	Layer
Layer management	Higher layer functions	Higher layers	
	Convergence	CS	AAL
	Segmentation and reassembly	SAR	
	Generic flow control Cell header generation/extraction Cell VPI/VCI translation Cell multiplex and demultiplex	ATM	
	Cell rate decoupling HEC header sequence generation/verification Cell delineation Transmission frame adaption Transmission frame generation/recovery	TC	Physical layer
	Bit timing Physical medium	PM	

CS Convergence sublayer
PM Physical medium
SAR Segmentation and reassembly sublayer
TC Transmission convergence

Figure 18.7 B-ISDN/ATM functional layering.

The primitives identified at the border between the PM and TC sublayers are a continuous flow of logical bits or symbols with this associated timing information.

18.6.1.1 Transmission Convergence Sublayer Functions. Among the important functions of this sublayer is the generation and recovery of transmission frame. Another function is transmission frame adaptation, which includes the actions necessary to structure the cell flow according to the payload structure of the transmission frame (transmit direction), and to extract this cell flow out of the transmission frame (receive direction). The transmission frame may be a cell equivalent (i.e., no external envelope is added to

Table 18.4 ATM Layer Functions Supported at the UNI

Functions	Parameters
Multiplexing among different ATM connections	VPI/VCI
Cell rate decoupling (unassigned cells)	Preassigned header field values
Cell discrimination based on predefined header field values	Preassigned header field values
Payload type discrimination	PT field
Loss priority indication and selective cell discarding	CLP field, network congestion state
Traffic shaping	Traffic descriptor

Source: Based on Refs. 3 and 8.

the cell flow), an SDH/SONET envelope, an E1/T1 envelope, and so on. In the transmit direction, the HEC sequence is calculated and inserted in the header. In the receive direction, we include cell header verification. Here cell headers are checked for errors and, if possible, header errors are corrected. Cells are discarded where it is determined that headers are errored and are not correctable.

Another transmission convergence function is cell rate decoupling. This involves the insertion and removal of idle cells in order to adapt the rate of valid ATM cells to the payload capacity of the transmission system. In other words, cells must be generated to exactly fill the payload of SDH/SONET, as an example, whether the cells are idle or busy.

Section 18.12.5 gives several examples of transporting cells using the convergence sublayer.

18.6.2 ATM Layer

Table 18.4 shows the ATM layering functions supported at the UNI (U-plane). The ATM layer is completely independent of the physical medium. One important function of this layer is *encapsulation*. This includes cell header generation and extraction. In the transmit direction, the cell header generation function receives a cell information field from a higher layer and generates an appropriate ATM cell header except for the HEC sequence. This function can also include the translation from a service access point (SAP) identifier to a VPI and VCI.

In the receive direction, the cell header extraction function removes the ATM cell header and passes the cell information field to a higher layer. As in the transmit direction, this function can also include a translation of a VPI and VCI into an SAP identifier.

In the case of the NNI, the GFC is applied at the ATM layer. The flow control information is carried in assigned and unassigned cells. Cells carrying this information are generated in the ATM layer.

In a switch the ATM layer determines to where the incoming cells should be forwarded, resets the corresponding connection identifiers for the next link, and forwards the cell. The ATM layer also handles traffic-management functions between ATM nodes on both sides of the UNI (i.e., single VP link segment) while the virtual channel identified by a VCI value = 4 can be used for VP level end-to-end (user $\leftrightarrow$ user) management functions.

What are flows such as "F4 flows"? OAM (operations, administration, and management) flows deal with cells dedicated to fault and performance management of the total system. Consider ATM as a hierarchy of levels, particularly in SDH/SONET, which are the principal bearer formats for ATM. The lowest level where we have F1 flows is the regenerator section (called the *section level* in SONET). This is followed by F2 flows at the digital section level (called the *line level* in SONET). There are the F3 flows for the transmission path (called the *path level* in SONET). ATM adds F4 flows for virtual paths (VPs) and F5 flows for virtual channels (VCs), where multiple VCs are completely contained within a single VP. We discuss VPs and VCs in Section 18.8.

18.6.3 ATM Adaptation Layer (AAL)

The basic purpose of the AAL is to isolate the higher layers from the specific characteristics of the ATM layer by mapping the higher-layer protocols data units (PDUs) into the payload of the ATM cell and vice versa.

18.6.3.1 Sublayering of the AAL. To support services above the AAL, some independent functions are required for the AAL. These functions are organized in two logical sublayers: (1) the convergence sublayer (CS), and (2) the segmentation and reassembly (SAR) sublayer. The primary functions of these layers are:

- SAR—The segmentation of higher-layer information into a size suitable for the information field of an ATM cell. Reassembly of the contents of ATM cell information fields into higher-layer information;
- CS—Here the prime function is to provide the AAL service at the AAL-SAP (service access point). This sublayer is service dependent.

18.6.3.2 Service Classification of the AAL. Service classification is based on the following parameters:

- Timing relation between source and destination (this refers to urgency of traffic): required or not required;
- Bit rate: constant or variable; and
- Connection mode: connection-oriented or connectionless.

When we combine these parameters, four service classes emerge, as shown in Figure 18.8. Examples of the services in the classes shown in Figure 18.8 are as follows:

- Class A: constant bit rate such as uncompressed voice or video;
- Class B: variable bit rate video and audio, connection-oriented synchronous traffic;
- Class C: connection-oriented data transfer, variable bit rate, asynchronous traffic; and
- Class D: connectionless data transfer, asynchronous traffic such as SMDS.

Note that SMDS stands for switched multimegabit data service. It is espoused by Bellcore and is designed primarily for LAN interconnect.

18.6.3.3 AAL Categories or Types. There are five different AAL categories. The simplest is AAL-0. It just transmits cells down a pipe. That pipe is commonly a fiber-optic link. Ideally it would be attractive that the bit rate here be some multiple of 53 × 8 bits or 424 bits. For example, 424 Mbps could handle 100 million cells per second.

Service Parameters	Class A	Class B	Class C	Class D
Timing Compensation	Required		Not Required	
Bit Rate	Constant	Variable		
Connection Mode	Connection-oriented			Connectionless
AAL Types	AAL1	AAL2	AAL3/4 or AAL5	AAL3/4 or AAL5
Examples	DS1, E1, n x 64-kbps emulation	Packet video, audio	Frame relay X.25	IP, SMDS

Figure 18.8 Services classifications of AAL. (Based on Refs. 2, 8, and 10).

18.6.3.3.1 AAL-1. AAL-1 is used to provide transport for synchronous bit streams. Its primary application is to adapt ATM cell transmission to typically E1/DS1 and SDH/SONET circuits. AAL-1 is specifically used for voice communications (POTS; plain old telephone service). AAL-1 robs one octet from the payload and adds it to the header, leaving only a 47-octet payload. This additional octet in the header contains two major fields: (1) sequence number (SN), and (2) sequence number protection (SNP). The principal purpose of these two fields is to check that missequencing of information does not occur, by verifying a 3-bit sequence counter. It also allows for the original clock timing of the data received at the far end of the link. The SAR-PDU format of AAL-1 is shown in Figure 18.9.[8] The 4-bit SN is broken down into a 1-bit CSI (convergence sublayer indicator) and a sequence count. The SNP contains a 3-bit CRC and a parity bit. End-to-end synchronization is an important function for the type of traffic carried on AAL-1. With one mode of operation, clock recovery is via a synchronous residual time stamp (SRTS) and common network clock by means of a 4-bit residual time

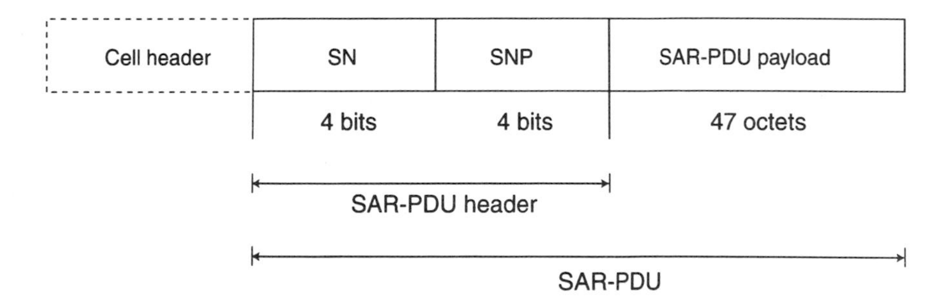

SN Sequence number (4 bits); to detect lost or misinserted cells. A specific value of the sequence number may indicate a special purpose, e.g. the existence of convergence sublayer functions. The exact counting scheme is for further study.

SNP Sequence number protection (4 bits). The SNP field may provide error detection and correction capabilities. The polynomial to be used is for further study.

Figure 18.9 SAR-PDU format for AAL-1. (From ITU-T Rec. I.363, Figure 1/I.363, p. 3 [Ref. 10].)

[8]SAR-PDU stands for segmentation and reassembly, protocol data unit.

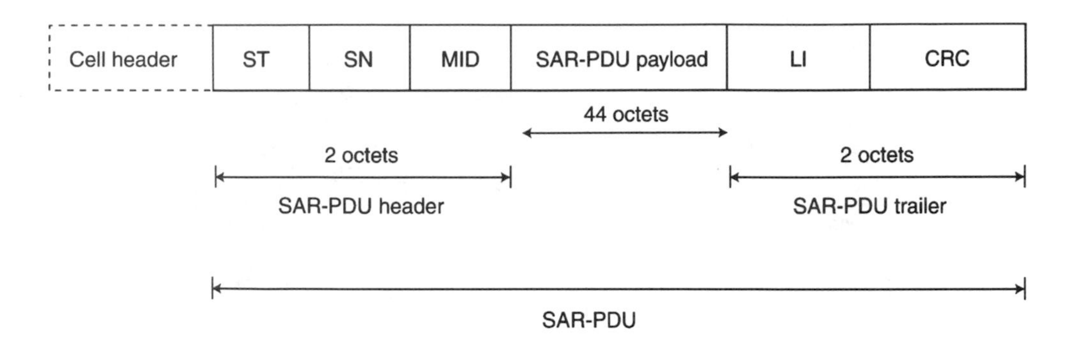

ST Segment type (2 bits)
SN Sequence number (4 bits)
MID Multiplexing identification (10 bits)
LI Length indicator (6 bits)
CRC Cyclic redundancy check code (10 bits)

Figure 18.10 SAR-PDU format for AAL-3/4. (From ITU-T Rec. I.363, Figure 6/I.363, p. 13 [Ref. 10].)

stamp extracted from CSI from cells with odd sequence numbers. The residual time stamp is transmitted over eight cells. It supports DS1, DS3, and E1 digital streams. Another mode of operation is structured data transfer (SDT). SDT supports an octet-structured $n \times$ DS0 service.

18.6.3.3.2 AAL-2. AAL-2 handles the variable bit rate (VBR) scenario such as MPEG (Motion Picture Experts Group) video. It is still in the ITU-T organization definitive stages.

18.6.3.3.3 AAL-3/4. Initially, in ITU-T Rec. I.363 (Ref. 10) there were two separate AALs, one for connection-oriented variable bit rate data services (AAL-3) and one for connectionless service. As the specifications evolved, the same procedures turned out to be necessary for both of these services, and the specifications were merged, to become the AAL-3/4 standard. AAL-3/4 is used for ATM transport of SMDS, CBDS (connectionless broadband data services, an ETSI initiative), IP (Internet protocol), and frame relay. AAL-3/4 has been designed to take variable-length frames/packets and segment them into cells. The segmentation is done in a way that protects the transmitted data from corruption if cells are lost or missequenced. Figure 18.10 shows the cell format of an AAL-3/4 cell. These types of cells have only a 44-octet payload, and additional overhead fields are added to the header and trailer.[9] These carry, for example, the BOM, COM, and EOM indicators [carried in *segment type* (ST)] as well as an MID (multiplexing identifier) so that the original message, as set up in the convergence sublayer PDU (CS_PDU), can be delineated. The header also includes a sequence number for protection against misordered delivery. There is the MID (multiplexing identification) subfield, which is used to identify the CPCS (common part convergence sublayer) connection on a single ATM layer connection. This allows for more than one CPCS connection for a single ATM-layer connection.

The SAR sublayer, therefore, provides the means for the transfer of multiple, vari-

[9]A *trailer* consists of overhead fields added to the end of a data frame or cell. A typical trailer is the CRC parity field appended at the end of a frame.

able length CS-PDUs concurrently over a single ATM layer connection between AAL entities. The SAR_PDU trailer contains a length indicator (LI) to identify how much of the cell payload is filled. The CRC field is a 10-bit sequence used to detect errors across the whole SAR_PDU. A complete CS_PDU message is broken down into one BOM cell, a number of COM cells, and one EOM cell. If an entire message can fit into one cell, it is called a *single segment message* (SSM), where the CS_PDU is 44 or fewer octets long.

AAL-3/4 has several measures to ensure the integrity of the data which has been segmented and transmitted as cells. The contents of the cell are protected by the CRC-10; sequence numbers protect against misordering. Still another measure to ensure against corrupted PDUs being delivered is EOM/BOM protection. If the EOM of one CPCS_PDU and the BOM of the next are dropped for some reason, the resulting cell stream could be interpreted as a valid PDU. To protect against these kinds of errors, the BEtag numeric values in the CPCS_PDU headers and trailers are compared, to ensure that they match. Two modes of service are defined for AAL-3/4:

1. *Message Mode Service.* This provides for the transport of one or more fixed-size AAL service data units in one or more CS-PDUs.
2. *Streaming Mode Service.* Here the AAL service data unit is passed across the AAL interface in one or more AAL interface data units (IDUs). The transfer of these AAL-IDUs across the AAL interface may occur separated in time, and this service provides the transport of variable-length AAL-SDUs. The streaming mode service includes an abort service, by which the discarding of an AAL-SDU partially transferred across the AAL interface can be requested. In other words, in the streaming mode, a single packet is passed to the AAL layer and transmitted in multiple CPCS-PDUs, when and as pieces of the packet are received. Streaming mode may be used in intermediate switches or ATM-to-SMDS routers so they can begin retransmitting a packet being received before the entire packet has arrived. This reduces the latency experienced by the entire packet.

18.6.3.3.4 AAL-5. This type of AAL was designed specifically to carry data traffic typically found in today's LANs. AAL-5 evolved after AAL-3/4, which was found to be too complex and inefficient for LAN traffic. Thus, AAL-5 got the name *SEAL* (simple and efficient AAL layer). Only a small amount of overhead is added to the CPCS_PDU. There is no AAL level cell multiplexing. In AAL-5 all cells belonging to an AAL-5 CPCS_PDU are sent sequentially. To simplify still further, the CPCS_PDUs are padded to become integral multiples of 48 octets, ensuring that there never will be a need to send partially filled cells after segmentation.[10]

18.7 SERVICES: CONNECTION-ORIENTED AND CONNECTIONLESS

The issues such as routing decisions and architectures have a major impact on connection-oriented services, where B-ISDN/ATM end nodes have to maintain or get access to lookup tables, which translate destination addresses into circuit paths. These circuit path lookup tables, which differ at every node, must be maintained in a quasi-real-time fashion. This will have to be done by some kind of routing protocol.

One way to resolve this problem is to make it an internal network problem and use

[10]*Padded* means adding "dummy" octets—octets that do not carry any significance or information.

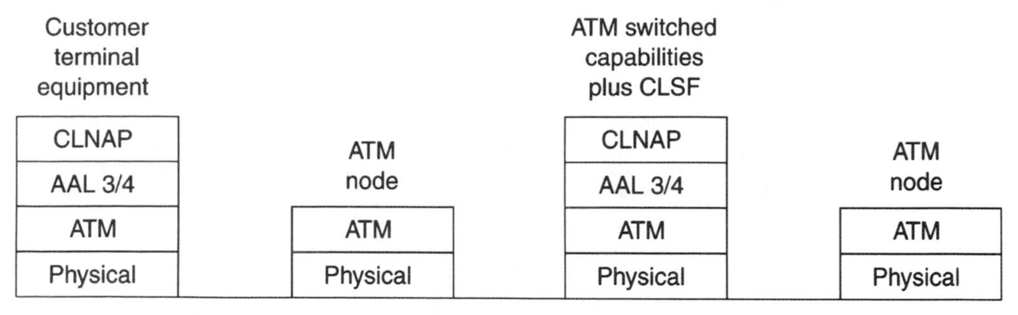

Figure 18.11 General protocol structure for the provision of CL data service in B-ISDN.

a connectionless service as described in ITU-T Rec. I.364 (Ref. 11). We must keep in mind that ATM is basically a connection-oriented service. Here we are going to adapt it to provide a connectionless service.

18.7.1 Functional Architecture

The provision of connectionless data service in the B-ISDN is carried out by means of ATM switches and connectionless service functions (CLSF). ATM switches support the transport of connectionless data units in the B-ISDN between specific functional groups where the CLSF handles the connectionless protocol and provides for the adaptation of the connectionless data units into ATM cells to be transferred in a connection-oriented environment. It should be noted that CLSF functional groups may be located outside of the B-ISDN, in a private connectionless network, with a specialized service provider, or inside the B-ISDN.

The ATM switching is performed by the ATM nodes (ATM switch/cross-connect), which are a functional part of the ATM transport network. The CLSF functional group terminates the B-ISDN connectionless protocol and includes functions for the adaptation of the connectionless protocol to the intrinsically connection-oriented ATM layer protocol. These latter functions are performed by the ATM adaptation layer type 3/4 (AAL-3/4), while the CLSF group terminations are carried out by the services layer above the AAL called the CLNAP (connectionless network access protocol). The connectionless (CL) protocol includes functions such as routing, addressing, and QoS (quality of service) selection. In order to perform the routing of CL data units, the CLSF has to interact with the control/management planes of the underlying ATM network.

The general protocol structure for the provision of CL data service is illustrated in Figure 18.11. Figure 18.12 shows the protocol architecture for supporting connection-

CLNAP user layer
CLNAP
Type 3/4 AAL
ATM
Physical

Figure 18.12 Protocol architecture for supporting connectionless service.

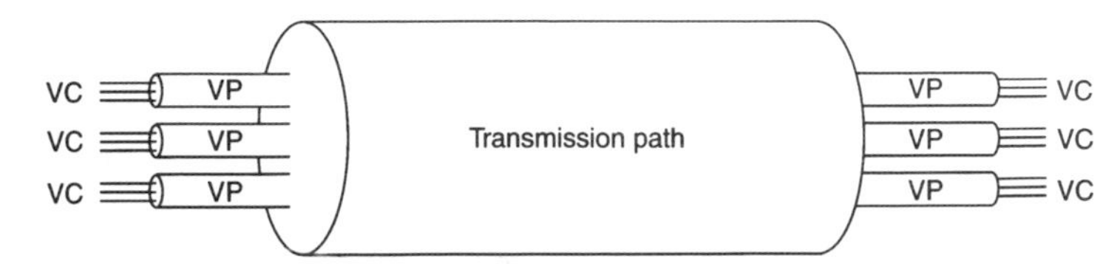

Figure 18.13 Relationship between VC and VP, and the transmission path.

less layer service. The CLNAP layer uses the type 3/4 AAL, unassured service, and includes the necessary functionality for the CL layer service.

The CL service layer provides for transparent transfer of variable-sized data units from a source to one or more destinations in a manner such that lost or corrupted data units are not retransmitted. This transfer is performed using a CL technique, including embedding destination and source addresses into each data unit.

18.8 B-ISDN/ATM ROUTING AND SWITCHING [Ref. 12]

An ATM transmission path supports virtual paths (VPs) and inside those virtual paths are virtual channels (VCs), as illustrated in Figure 18.13. In Section 18.4.1 we described the ATM cell header. Each cell header contains a label that explicitly identifies the VC to which the cell belongs. This label consists of two parts: (1) a virtual channel identifier (VCI), and (2) a virtual path identifier (VPI).

18.8.1 Virtual Channel Level

Virtual channel (VC) is a generic term used to describe a unidirectional communication capability for the transport of ATM cells. A VCI identifies a particular VC link for a given virtual path connection (VPC). A specific value of VCI is assigned each time a VC is switched in the network. A VC link is a unidirectional capability for the transport of ATM cells between two consecutive ATM entities where the VCI value is translated. A VC link is originated or terminated by the assignment or removal of the VCI value.

Routing functions of virtual channels are done at the VC switch/cross-connect.[11] The routing involves translation of the VCI values of the incoming VCI links into the VCI values of the outgoing VC links.

Virtual channel links are concatenated to form a virtual channel connection (VCC). A VCC extends between two VCC endpoints or, in the case of point-to-multipoint arrangements, more than two VCC endpoints. A VCC endpoint is the point where the cell information field is exchanged between the ATM layer and the user of the ATM layer service.

At the VC level, VCCs are provided for the purpose of user–user, user–network, or network–network information transfer. Cell sequence integrity is preserved by the ATM layer for cells belonging to the same VCC.

18.8.2 Virtual Path Level

The virtual path (VP) is a generic term for a bundle of virtual channel links; all the links in a bundle have the same endpoints. A VPI identifies a group of VC links, at

[11] *VC cross-connect* is a network element that connects VC links. It terminates VPCs and translates VCI values, and is directed by management plane functions, not by control plane functions.

a given reference point, that share the same VPC. A specific value of VPI is assigned each time a VP is switched in the network. A VP link is a unidirectional capability for the transport of ATM cells between two consecutive ATM entities where the VPI value is translated. A VP link is originated or terminated by the assignment or removal of the VPI value.

Routing functions for VPs are performed at a VP switch/cross-connect. This routing involves translation of the VPI values of the incoming VP links into the VPI values of the outgoing VP links. VP links are concatenated to form a VPC. A VPC extends two VPC endpoints or, in the case of point-to-multipoint arrangements, more than two VPC endpoints. A VPC endpoint is the point where the VCIs are originated, translated, or terminated. At the VP level, VPCs are provided for the purpose of user–user, user–network, and network–network information transfer.

When VPCs are switched, the VPC supporting the incoming VC links are terminated first and a new outgoing VPC is then created. Cell sequence integrity is preserved by the ATM layer for cells belonging to the same VPC. Thus cell sequence integrity is preserved for each VC link within a VPC.

Figure 18.14 is a representation of a VP and VC switching hierarchy, where the physical layer is the lowest layer composed of, from bottom up, a regenerator section level, a digital section level, and a transmission path level. The ATM layer resides just above the physical layer and is composed of the VP level; just above that is the VC level.

18.9 SIGNALING REQUIREMENTS

18.9.1 Setup and Release of VCCs

The setup and release of VCCs at the user–network interface (UNI) can be performed in various ways:

- Without using signaling procedures. Circuits are set up at subscription with permanent or semipermanent connections;
- By meta-signaling procedures, where a special VCC is used to establish or release a VCC used for signaling. Meta-signaling is a simple protocol used to establish and remove signaling channels. All information interchanges in meta-signaling are carried out via single cell messages;
- User-to-network signaling procedures, such as a signaling VCC to establish or release a VCC used for end-to-end connectivity; and
- User-to-user signaling procedures, such as a signaling VCC to establish or release a VCC within a preestablished VPC between two UNIs.

18.9.2 Signaling Virtual Channels

18.9.2.1 Requirements for Signaling Virtual Channels. For a point-to-point signaling configuration, the requirements for signaling virtual channels are as follows:

- One virtual channel connection in each direction is allocated to each signaling entity. The same VPI/VCI value is used in both directions. A standardized VCI value is used for point-to-point signaling virtual channel (SVC).

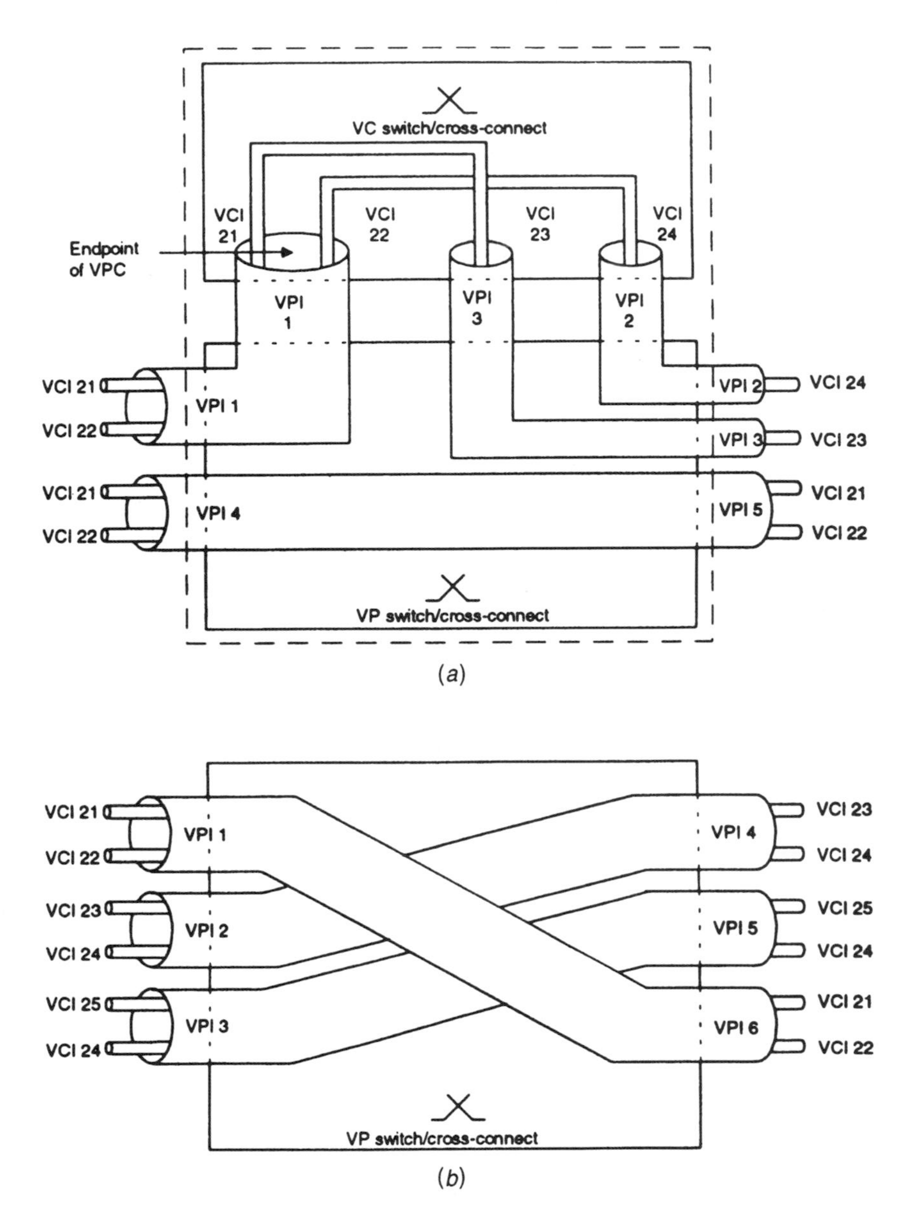

Figure 18.14 Representation of the VP and VC switching hierarchy. (*a*) VC and VP; (*b*) VP switching. (From ITU-T Rec. I.311, Figure 4/I.311, p. 5 [Ref. 12].)

- In general, a signaling entity can control, by means of associated point-to-point SVCs, user-VCs belonging to any of the VPs terminated in the same network element.
- As a network option, the user-VCs controlled by a signaling entity can be constrained such that each controlled user-VC is in either upstream or downstream VPs containing the point-to-point SVCs of the signaling entity.

For point-to-multipoint signaling configurations, the requirements for signaling virtual channels are as follows:

- *Point-to-Point Signaling Virtual Channels.* For point-to-point signaling, one virtual channel connection in each direction is allocated to each signaling entity. The same VPI/VCI value is used in both directions.
- *General Broadcast Signaling Virtual Channel.* The general broadcast signaling virtual channel (GBSVC) may be used for call offering in all cases. In cases where the "point" does not implement service profiles or where "the multipoints" do not support service profile identification, the GBSVC is used for call offering. The specific VCI value for general broadcast signaling is reserved per VP at the UNI. Only when meta-signaling is used in a VP is the GBSVC activated in the VP.
- *Selective Broadcast Signaling Virtual Channels.* Instead of the GBSVC, a virtual channel connection for selective broadcast signaling (SBS) can be used for call offering, in cases where a specific service profile is used. No other uses for SBSVCs are foreseen.

18.10 QUALITY OF SERVICE (QoS)

18.10.1 ATM Quality of Service Review

A basic performance measure for any digital data communication system is bit error rate (BER). Well-designed fiber-optic links will predominate now and into the foreseeable future. We may expect BERs from such links on the order of 1×10^{-12} and with end-to-end performance better than 5×10^{-10} (Ref. 14). Thus other performance issues may dominate the scene. These may be called ATM unique QoS items, namely:

- Cell transfer delay;
- Cell delay variation;
- Cell loss ratio;
- Mean cell transfer delay;
- Cell error ratio;
- Severely errored cell block ratio; and
- Cell misinsertion rate.

18.10.2 Selected QoS Parameter Descriptions

18.10.2.1 Cell Transfer Delay. In addition to the normal delay through network elements and transmission paths, extra delay is added to an ATM network at an ATM switch. The cause of the delay at this point is the statistical asynchronous multiplexing. Because of this, two cells can be directed toward the same output of an ATM switch or cross-connect, resulting in output contention.

The result is that one cell or more is held in a buffer until the next available opportunity to continue transmission. We can see that the second cell will suffer additional delay. The delay of a cell will depend upon the amount of traffic within a switch and thus the probability of contention.

The asynchronous path of each ATM cell also contributes to cell delay. Cells can be delayed one or many cell periods, depending on traffic intensity, switch sizing, and the transmission path taken through the network.

18.10.2.2 Cell Delay Variation (CDV). By definition, ATM traffic is asynchronous, magnifying transmission delay. Delay is also inconsistent across the network. It can be a function of time (i.e., a moment in time), network design/switch design (such as buffer size), and traffic characteristics at that moment in time. The result is cell delay variation (CDV).

CDV can have several deleterious effects. The dispersion effect, or spreading out, of cell interarrival times can impact signaling functions or the reassembly of cell user data. Another effect is called *clumping*. This occurs when the interarrival times between transmitted cells shorten. One can imagine how this could affect the instantaneous network capacity and how it can impact other services using the network.

There are two performance parameters associated with cell delay variation: 1-point cell delay variation (1-point CDV) and 2-point cell delay variation (2-point CDV).

The 1-point CDV describes variability in the pattern of cell arrival events observed at a single boundary with reference to the negotiated peak rate $1/T$ as defined in ITU-T Rec. I.371 (Ref. 13). The 2-point CDV describes variability in the pattern of cell arrival events as observed at the output of a connection portion (MP_1).

18.10.2.3 Cell Loss Ratio. Cell loss may not be uncommon in an ATM network. There are two basic causes of cell loss: (1) error in cell header or (2) network congestion.

Cells with header errors are automatically discarded. This prevents misrouting of errored cells, as well as the possibility of privacy and security breaches.

Switch buffer overflow can also cause cell loss. It is in these buffers that cells are held in prioritized queues. If there is congestion, cells in a queue may be discarded selectively in accordance with their level of priority. Here enters the CLP (cell loss priority) bit, discussed in Section 18.4. Cells with this bit set to 1 are discarded in preference to other, more critical cells. In this way, buffer fill can be reduced to prevent overflow (Ref. 1).

Cell loss ratio is defined for an ATM connection as:

$$\text{Lost cells/Total transmitted cells.}$$

Lost and transmitted cells counted in severely errored cell blocks should be excluded from the cell population in computing cell loss ratio (Ref. 3).

18.11 TRAFFIC CONTROL AND CONGESTION CONTROL

The following functions form a framework for managing and controlling traffic and congestion in ATM networks and are to be used in appropriate combinations from the point of view of ITU-T Rec. I.371 (Ref. 13):

1. *Network Resource Management (NRM).* Provision is used to allocate network resources in order to separate traffic flows in accordance with service characteristics.
2. *Connection Admission Control (CAC).* This is defined as a set of actions taken by the network during the call setup phase or during the call renegotiation phase in order to establish whether a VC or VP connection request can be accepted or rejected, or whether a request for reallocation can be accommodated. Routing is part of CAC actions.

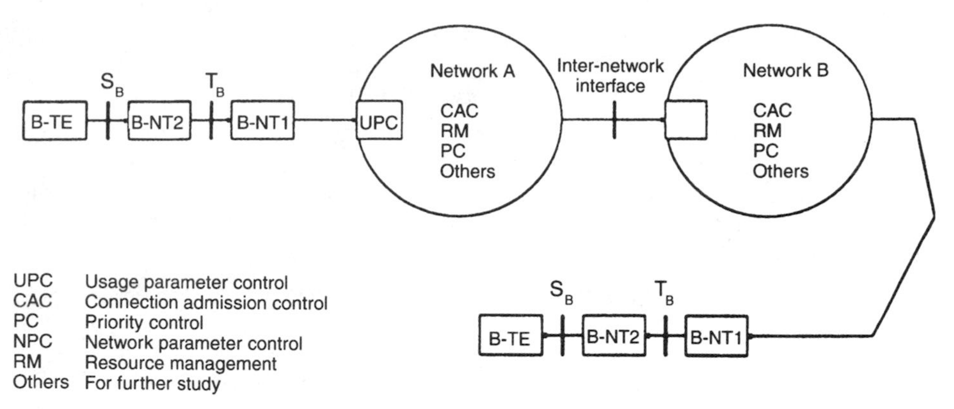

Notes:
1 NPC may apply as well at some intra-network NNI's.
2 The arrows are indicating the direction of the cell flow.

Figure 18.15 Reference configuration for traffic control and congestion control. (From ITU-T Rec. I.371, Figure 1/I.371, p. 3 [Ref. 13].)

3. *Feedback Controls.* These are a set of actions taken by the network and by users to regulate the traffic submitted on ATM connections according to the state of network elements.
4. *Usage/Network Parameter Control (UPC/NPC).* This is a set of actions taken by the network to monitor and control traffic, in terms of traffic offered and validity of the ATM connection, at the user access and network access, respectively. Their main purpose is to protect network resources from malicious as well as unintentional misbehavior, which can affect the QoS of other already established connections, by detecting violations of negotiated parameters and taking appropriate actions.
5. *Priority Control.* The user may generate different priority traffic flows by using the CLP. A congested network element may selectively discard cells with low priority, if necessary, to protect as far as possible the network performance for cells with higher priority (Ref. 13).

Figure 18.15 is a reference configuration for traffic and congestion control on a B-ISDN/ATM network.

18.12 TRANSPORTING ATM CELLS

18.12.1 In the DS3 Frame

One of the most popular high-speed digital transmission systems in North America is DS3 operating at a nominal transmission rate of 45 Mbps. It is also being widely implemented for transport of SMDS. The system used to map ATM cells into the DS3 format is the same that is used for SMDS.

To map ATM cells into a DS3 bit stream, the physical layer convergence protocol (PLCP) is employed. A DS3 PLCP frame is shown in Figure 18.16.

There are 12 cells in a PLCP frame. Each cell is preceded by a 2-octet framing pattern (A1, A2) to enable the receiver to synchronize to cells. After the framing pattern there is an indicator consisting of one of 12 fixed bit patterns used to identify the cell location

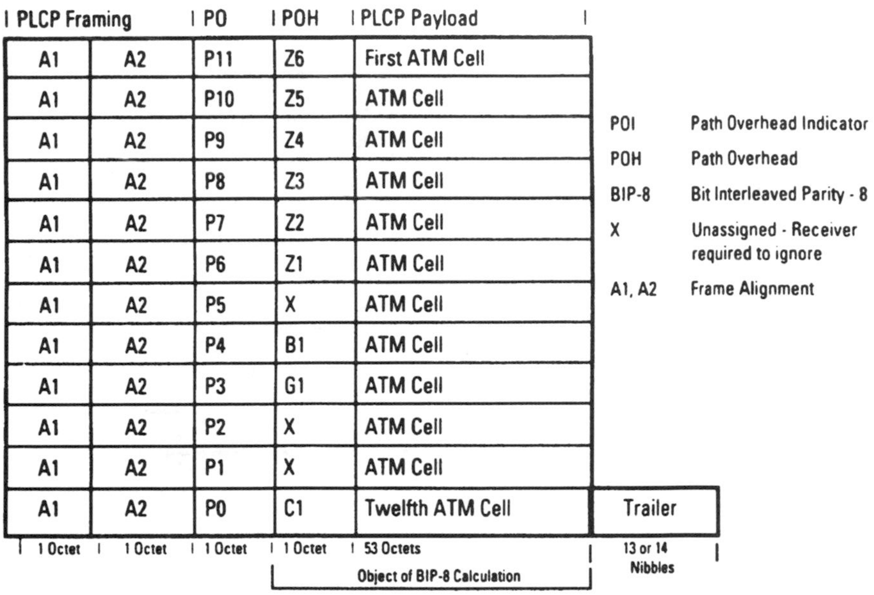

PLCP Framing		PO	POH	PLCP Payload	
A1	A2	P11	Z6	First ATM Cell	
A1	A2	P10	Z5	ATM Cell	
A1	A2	P9	Z4	ATM Cell	
A1	A2	P8	Z3	ATM Cell	
A1	A2	P7	Z2	ATM Cell	
A1	A2	P6	Z1	ATM Cell	
A1	A2	P5	X	ATM Cell	
A1	A2	P4	B1	ATM Cell	
A1	A2	P3	G1	ATM Cell	
A1	A2	P2	X	ATM Cell	
A1	A2	P1	X	ATM Cell	
A1	A2	P0	C1	Twelfth ATM Cell	Trailer
1 Octet	1 Octet	1 Octet	1 Octet	53 Octets	13 or 14 Nibbles

Object of BIP-8 Calculation

POI Path Overhead Indicator
POH Path Overhead
BIP-8 Bit Interleaved Parity - 8
X Unassigned - Receiver required to ignore
A1, A2 Frame Alignment

Figure 18.16 Format of DS3 PLCP frame. (From Ref. 1, courtesy of Hewlett-Packard.)

within the frame (POI). This is followed by an octet of overhead information used for path management. The entire frame is then padded with either 13 nibbles or 14 nibbles (1 nibble = 4 bits) of trailer to bring the transmission rate up to the exact DS3 bit rate. The DS3 frame, as we are aware, has a 125-μs duration.

DS3 has to contend with network slips (added/dropped frames to accommodate synchronization alignment). Thus PLCP is padded with a variable number of stuff (justification) bits to accommodate possible timing slips. The C1 overhead octet indicates the length of padding. The BIP (bit-interleaved parity) checks the payload and overhead functions for errors and performance degradation. This performance information is transmitted in the overhead.

18.12.2 DS1 Mapping

One approach to mapping ATM cells into a DS1 frame is to use a similar procedure as used with the DS3 PLCP. In this case only 10 cells are bundled into a frame, and two of the Z overhead octets are removed. The padding of the frame is set at 6 octets. The entire frame takes 3 ms to transmit and spans many DS1 ESF (extended superframe) frames. This mapping is illustrated in Figure 18.17. The L2_PDU is terminology used with SMDS. It is the upper-level frame from which ATM cells derive through its segmentation.

One must also consider the arithmetic of the situation. Each DS1 time slot is 8 bits long or 1 octet in length. By definition, there are 24 octets in a DS1 frame. This, of course, leads to a second method of transporting ATM cells in DS1, by directly mapping ATM cells into DS1, octet-for-octet (time slot). This is done by groups of 53 octets (1 cell) and would, by necessity, cross DS1 frame boundaries to transport a complete cell.

1	1	1	1	←53 Octets→
A1	A2	P9	Z4	L2_PDU
A1	A2	P8	Z3	L2_PDU
A1	A2	P7	Z2	L2_PDU
A1	A2	P6	Z1	L2_PDU
A1	A2	P5	F1	L2_PDU
A1	A2	P4	B1	L2_PDU
A1	A2	P3	G1	L2_PDU
A1	A2	P2	M2	L2_PDU
A1	A2	P1	M1	L2_PDU
A1	A2	P0	C1	L2_PDU

OH Byte	Function
A1, A2	Framing Bytes
P9-P0	Path Overhead Identifier Bytes
PLCP Path Overhead Bytes	
Z4-Z1	Growth Bytes
F1	PLCP Path User Channel
B1	BIP-8
G1	PLCP Status
M2-M1	SMDS Control Information
C1	Cycle/Stuff Counter Byte

Trailer = 6 Octets

3 msec

Figure 18.17 DS1 mapping with PLCP. (From Ref. 1, courtesy of Hewlett-Packard.)

18.12.3 E1 Mapping

E1 PCM has a 2.048-Mbps transmission rate. An E1 frame has 256 bits representing 32 channels or time slots, 30 of which carry traffic. Time slots (TS) 0 and 16 are reserved. TS 0 is used for synchronization and TS 16 for signaling. The E1 frame is illustrated in Figure 18.18. The sequences of bits from bit 9 to bit 128 and from bit 137 to bit 256 may be used for ATM cell mapping. ATM cells can also be directly mapped into special E3 and E4 frames. The first has 530 octets available for cells (i.e., exactly 10 cells) and the second has 2160 octets (not evenly divisible).

18.12.4 Mapping ATM Cells into SDH

18.12.4.1 At the STM-1 Rate (155.520 Mbps). SDH was described in Section 17.2. Figure 18.19 illustrates the mapping procedure. The ATM cell stream is first mapped

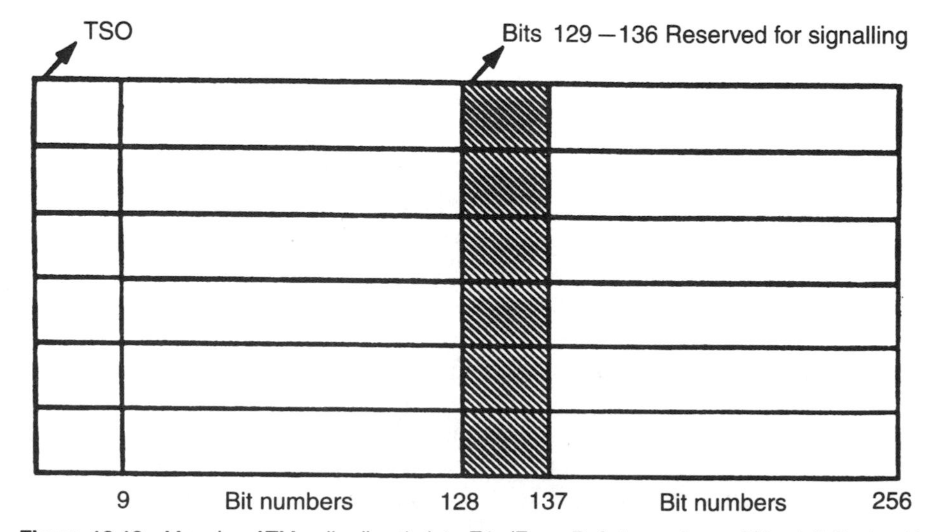

Figure 18.18 Mapping ATM cells directly into E1. (From Ref. 1, courtesy of Hewlett-Packard.)

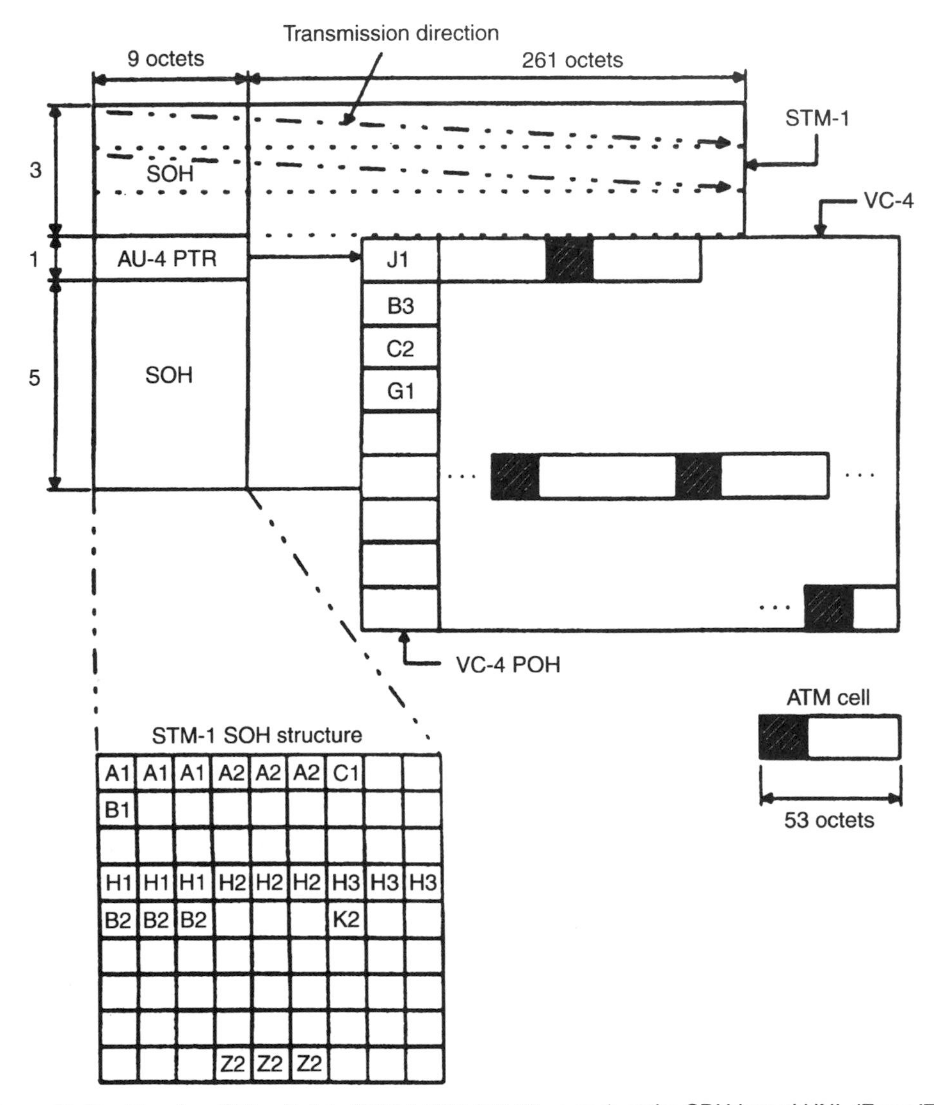

Figure 18.19 Mapping ATM cells into STM-1 (155.520 Mbps rate), at the SDH-based UNI. (From ITU-T Rec. I.432, Figure 8/I.432, p. 13 [Ref. 4].)

into the C-4, which, in turn, is mapped into the VC-4 container along with VC-4 path overhead. The ATM cell boundaries are aligned with STM octet boundaries. Since the C-4 capacity (2340 octets) is not an integer multiple of the cell length (53 octets), a cell may cross C-4 boundaries. The AU-4 pointer (octets H1 and H2 in the SOH) is used for finding the first octet in the VC-4.

18.12.4.2 At the STM-4 Rate (622.080 Mbps). As shown in Figure 18.20, the ATM cell stream is first mapped into the C-4-4c and then packed into the VC-4-4c container along with the VC-4-4c overhead. The ATM cell boundaries are aligned with the STM-4 octet boundaries. The C-4-4c capacity (9360 octets) is not an integer multiple of the cell length (53 octets); thus a cell may cross a C-4-4c boundary. The AU pointers are used for finding the first octet of the VC-4-4c.

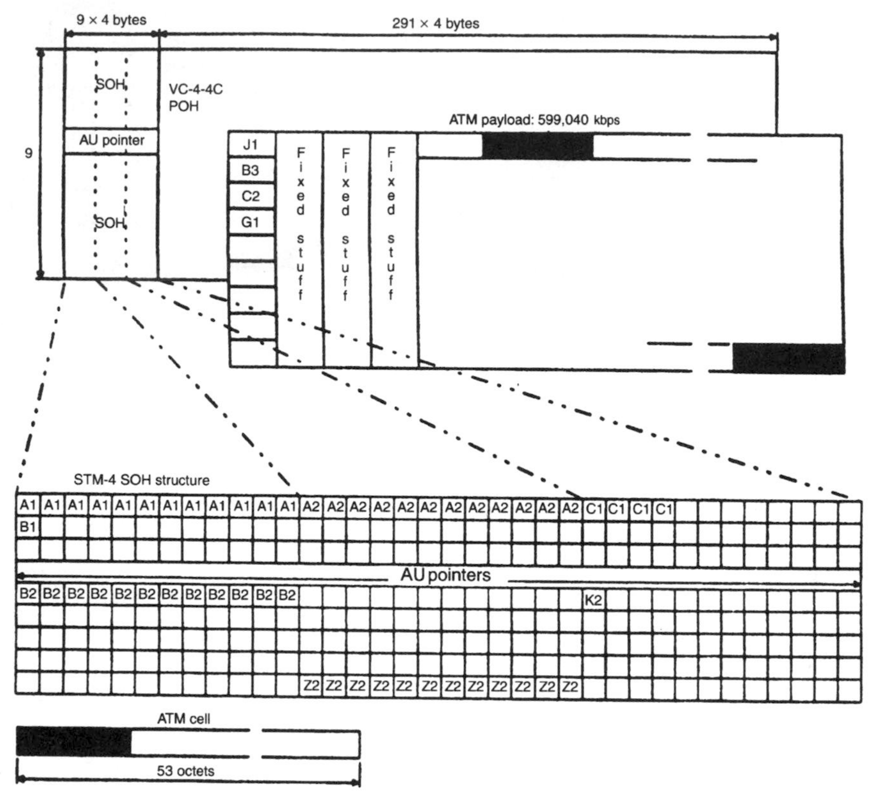

Figure 18.20 Mapping ATM cells into the STM-4 (655.080 Mbps rate) frame structure for the SDH-based UNI. (From ITU-T Rec. I. 432, Figure 10/I.432, p. 15 [Ref. 4].)

18.12.5 Mapping ATM Cells into SONET

ATM cells are mapped directly into the SONET payload (49.54 Mbps). As with SDH, the payload in octets is not an integer multiple of cell length, and thus a cell may cross an STS cell boundary. This mapping concept is shown in Figure 18.21. The H4 pointer

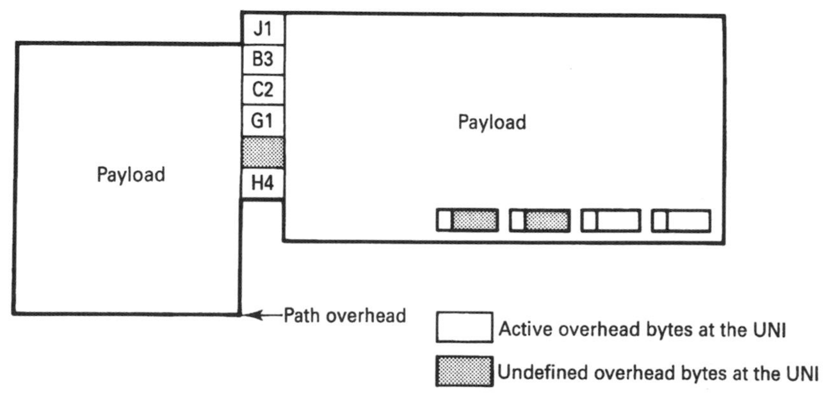

Figure 18.21 Mapping ATM cells directly into a SONET STS-1 frame. (From Ref. 1, courtesy of Hewlett-Packard.)

can indicate where the cells begin inside an STS frame. Another approach is to identify cell headers, and thus the first cell in the frame.

REVIEW EXERCISES

1. What are the two major similarities of frame relay and ATM/B-ISDN?
2. What are some radical differences with frame relay and other data transmission protocols/techniques?
3. ATM offers two basic services. What are they? Relate each service to at least one medium to be transported/switched.
4. Compare signaling philosophy with POTS and data.
5. Leaving aside Bellcore, what are the three standardization bodies for ATM?
6. Describe the ATM cell, its length in octets, the length of the header (and payload). Indicate variants to the lengths in octets.
7. Describe two functions of the HEC field.
8. What is the purpose of the *CLP* bit?
9. Why must there be a constant cell flow on an ATM circuit?
10. What is the principal function of the ATM layer?
11. What are "F4 flows?"
12. What is the principal purpose of the ATM adaptation layer?
13. What is the principal use of AAL-1?
14. What is the principal application of AAL3/4?
15. What does a VPI identify?
16. What service classification parameters is the AAL based on?
17. Where are routing functions of virtual channels done?
18. What happens to cells found with errors in their headers?
19. Name three of the four ways the setup and release of VCCs at the UNI can be performed.
20. Name five of the seven ATM unique quality of service (QoS) items.
21. Cell transfer delay happens at an ATM switch. What is the cause of the delay?
22. Give three causes of CDV (cell delay variation).

22\. Name and explain two effects of CDV.

23\. What is user/network parameter control (UPC/NPC)?

24\. ATM cells are transported on DS1, E1 hierarchies, SONET/SDH hierarchies. With one exception, what is an unfortunate outcome of the 53-octet cell?

REFERENCES

1. *Broadband Testing Technologies*, Hewlett-Packard Co., Burlington, MA, Oct. 1993.
2. *Broadband Aspects of ISDN*, CCITT Rec. I.121, CCITT, Geneva, 1991.
3. *ATM User–Network Interface Specification*, Version 3.0, ATM Forum, PTR Prentice-Hall, Englewood Cliffs, NJ, 1993.
4. *B-ISDN User-Network Interface—Physical Layer Specification*, ITU-T Rec. I.432, ITU, Geneva, March 1993.
5. *Broadband ISDN User–Network Interfaces—Rates and Formats Specifications*, ANSI T1.624-1993, ANSI, New York, 1993.
6. *B-ISDN User–Network Interface*, CCITT Rec. I.413, ITU, Geneva, 1991.
7. *Broadband ISDN–ATM Layer Functionality and Specification*, ANSI T1.627-1993, ANSI, New York, 1993.
8. D. E. McDysan and D. L. Spohn, *ATM Theory and Application*, McGraw-Hill, New York, 1995.
9. *B-ISDN ATM Layer Specification*, ITU-T Rec. I.361, ITU, Geneva, March, 1993.
10. *B-ISDN ATM Adaptation Layer (AAL) Functional Description*, ITU-T Rec. I.363, ITU, Geneva, 1993.
11. *Support of Broadband Connectionless Data Service on B-ISDN*, ITU-T Rec. I.364, ITU, Geneva, March 1993.
12. *B-ISDN General Network Aspects*, ITU-T Rec. I.311, ITU, Geneva, March 1993.
13. *Traffic Control and Congestion Control in B-ISDN*, ITU-T Rec. I.371, ITU, Geneva, March 1993.
14. *Bellcore Notes on the Network*, SR-2275, Issue 3, Bellcore, Piscataway, NJ, Dec. 1997.

APPENDIX A

REVIEW OF FUNDAMENTALS OF ELECTRICITY WITH TELECOMMUNICATION APPLICATIONS

A.1 OBJECTIVE

For a better understanding of this text, a certain basic knowledge of electricity is essential. The objective of the appendix is to cover only the necessary principles of electricity. Each principle is illustrated with one or several applications. Only minimal ac (alternating current) theory is presented. Where possible, these applications favor electrical communications. Analogies to assist in the understanding of a concept are used where appropriate.

We cannot see electricity. We can feel its effects such as electrical shock, the generation of heat, and the buildup and decay of magnetic and electrical fields. For example, a compass will indicate the presence of a magnetic field.

It is assumed that the reader has had at least three years of high school mathematics including algebra, trigonometry, and logarithms. For those who feel that their background in mathematics is insufficient, Appendix B gives an overview of those basic essentials.

A.2 WHAT IS ELECTRICITY?

We define *electricity* as the movement of electrons through a conductor.[1] Certain substances are good conductors and other substances are poor conductors. Poor conductors are called *insulators*. The various materials or substances we have here on earth run the gamut, from superb conductors such as platinum, gold, silver, and copper, to extremely good insulators such as glass, mica, polystyrene, and rubber.

The movement of electrons through a conductor is measured in *amperes*.[2] In fact

[1] *Encylopaedia Brittanica* defines electricity as "The phenomenon associated with positively and negatively charged particles of matter at rest and in motion, individually or in great numbers."

[2] An ampere is a unit of electrical measurement. All units of electrical measurement are named for scientists credited with discovering the particular phenomenon, and often providing a mathematical analysis of that electrical phenomenon. Ampere is one example.

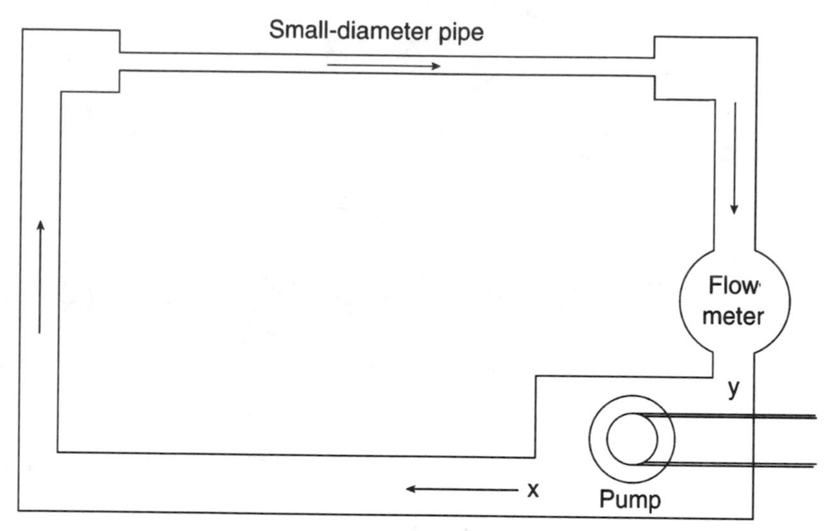

Figure A.1 A water circulating system.

1 ampere is the flow of 6.24×10^{18} electrons per second past a given point in a conductor. Where the flow of electrons is continuous in one direction is called *direct current* (dc). The currents of electrons may also periodically reverse their flow. This is called *alternating current* (ac). Our discussion starts with the traditional laws of direct current. Later we briefly discuss some basic principles of alternating currents.

A.2.1 Electromotive Force (EMF) and Voltage

We require a source of "electric pressure," called *electromotive force* (emf), to establish a flow electrical current. The standard analogy is to use water and its flow through a water circulating system. This is illustrated in Figure A.1. The figure shows the water pressure source as a pump, two different diameter pipes, and some sort of mechanism to measure water flow. Flow can be measured in several ways such as liters or gallons per second or minute. The pump creates a difference in pressure between points x and y. Some will say that the pump sets up a "head of pressure." This will cause water to flow from point x at the output of the pump up the large diameter pipe, across the small diameter pipe, and down through another large diameter pipe in which we installed a flow pressure gauge. The water returns to the low pressure side of the pump, shown as y in the drawing. The amount of water that will flow (i.e., analogy to amperage) depends on the difference in pressure between points x and y and the size of the small pipe. The difference in pressure is analogous to electromotive force.

Figure A.2 is an electrical representation of the circulating water system shown in Figure A.1. The battery (see Section 2.3.1) supplies the electrical pressure or emf. It causes the electricity to flow from the high potential side of the battery, labeled x, to its low potential side, labeled y in Figure A.2. The amount of current that will flow will depend upon the emf and the nature of the resistor, shown between points x and y.

The difference of water pressure may be measured in units such as "difference in head in feet." The emf of the electric circuit, on the other hand, is measured in terms of a unit called a *volt*.

Another important term is *electric potential*. In Figure A.2 we can say that the electric

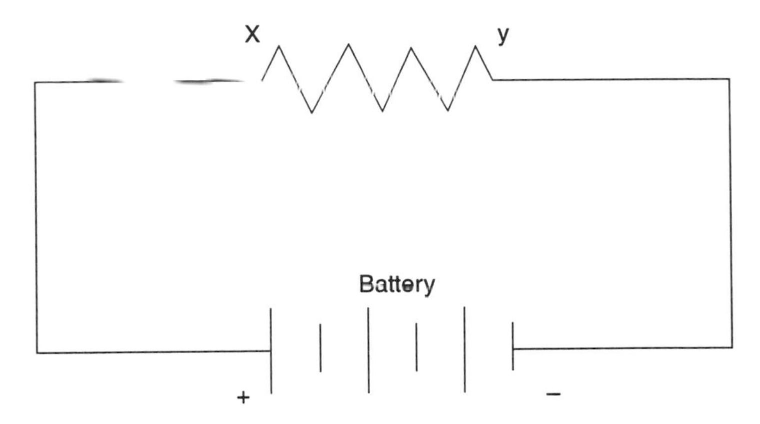

Figure A.2 A simple electric circuit.

potential at the positive terminal of the battery is higher than that of its negative terminal. The difference is the electromotive force of the battery or other source. The potential at point y will be lower than the potential at point x. Thus we say that there is a *potential* between x and y. The potential drop is measured in volts and the magnitude of the drop depends on the resistance of the conductors and the resistor.

A.2.2 Resistance

In Figure A.1, there is no practical unit to measure the "resistance" of the piping system. If the small pipe diameter is yet made smaller, the flow of water will decrease. This pipe resistance, that which reduces the water flow, is analogous to the electrical resistance of Figure A.2. The unit of electrical resistance is well defined and is called the *ohm*.

To review these electrical units:

- The flow of current in a conductors is measured in amperes (A).
- The resistance to this flow is called the ohm (Ω).
- The electrical pressure, called the emf, is measured in volts (V).

A.3 OHM'S LAW

The product of the current (amperes) and resistance (ohms) of an electrical circuit is equal to the voltage (volts). Note how we have carefully stated the units of measure. The law can be stated mathematically by

$$E = IR, \qquad (A.1)$$

where E is the voltage and derives from "emf," I is the current, and R is the resistance. Ohm's law can be stated in other ways by simple algebraic translation of terms. For example:

$$I = E/R, \qquad (A.2)$$

$$R = E/I. \qquad (A.3)$$

Example 1. If a battery has a voltage (emf) of 6 V and is connected to a lamp with a resistance of 200 Ω, what current can be expected in this circuit? Use Eq. (A.2):

$$I = 6/200$$
$$I = 0.03 \text{ A or 30 milliamperes (mA).}$$

Example 2. In a certain electric circuit we measured the current as 2 A and its resistance as 50 Ω. What is the emf (voltage)? Use Eq. (A.1):

$$E = 2 \times 50 = 100 \text{ V.}$$

A.3.1 Voltages and Resistances in a Closed Electric Circuit

Figure A.3 shows three resistances connected in series with a battery source of 48 V. We define a *closed circuit* as one in which current flows; an *open circuit* is a circuit in which no current flows. Figure A.3*a* is a closed circuit (i.e., the switch is in the closed position) and A.3*b* is an open circuit (i.e., the switch is in the open position).

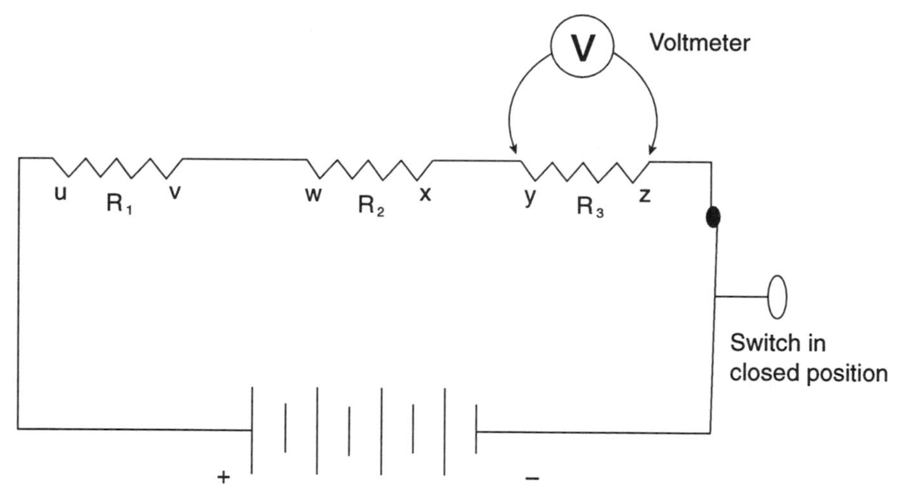

Figure A.3*a* A closed circuit showing three resistances in series with a battery as an emf source. A voltmeter is placed across R_3, which measures the potential difference across that resistance.

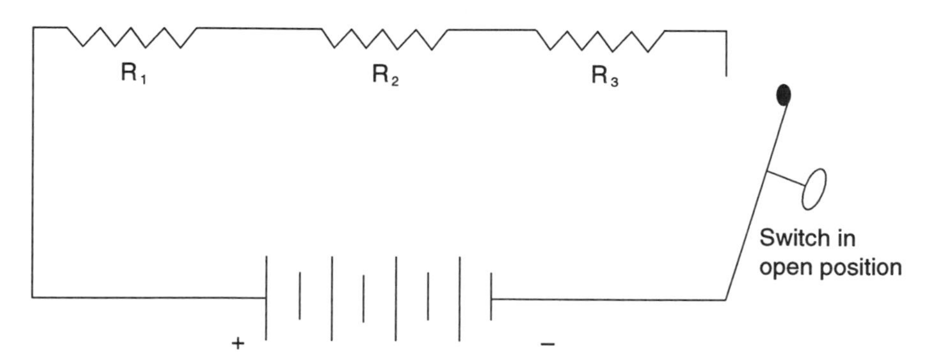

Figure A.3*b* An open circuit with three resistances in series with emf source.

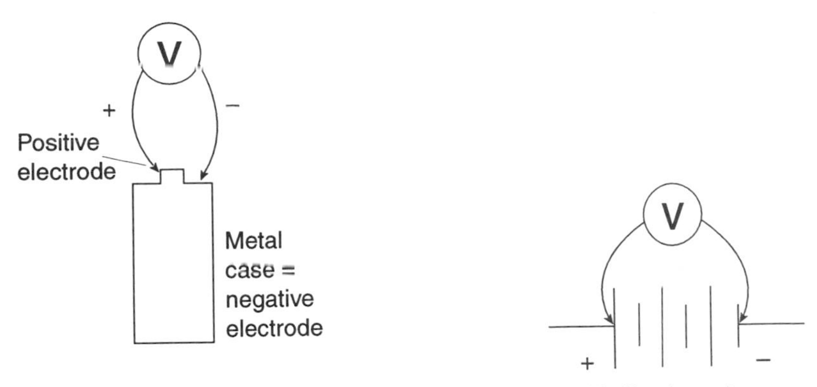

Figure A.4 Measuring the voltage of a battery: (*a*) a typical dry cell; (*b*) Showing voltage measurement using the standard battery drawing symbol. These are open circuit measurements.

Emf is measured with a voltmeter, as shown in Figure A.4, where the open circuit voltage (emf) of a battery is measured. A voltmeter is also used to measure potential drop as shown in Figure A.3*a*. Here the voltmeter is connected across R_3 (across points y and z). In later discussion we will also call this the *IR drop*. Remember Ohm's law: E = IR. Regarding the potential drop across R_3, if we know the resistance of R_3 and we know the current passing through it, we know the potential or IR drop. This is one of the many applications of Ohm's law.

In Figure A.3*a*, if we measure the potential drop (IR drop) in volts across each resistance and sum the values, this will equal the value of the battery emf in the closed circuit condition.

Example of Series Resistances. Four resistors are connected in series. Their resistance values are 250, 375, 136, and 741 Ω. Suppose we were to replace these four with just one resistor. The current through the circuit will be the same for one or for the four resistors in series. What will be the value in ohms of the single replacement resistor?

We know that this resistor must have a value equal to the four of these resistors if there is no change in current. Just sum the resistor values or 250 + 375 + 136 + 741 = 1502 Ω. The rule here is that when we want to calculate the equivalent resistance of resistors in series, we just sum the resistances of each resistor.

A.3.2 Resistance of Conductors

The current we can expect through a conductor with a fixed emf source varies directly with (1) the *resistivity*[3] of the type of conductor, (2) the length of the conductor, and (3) and inversely with its cross-sectional area, a function of its diameter. Unless otherwise specified, all conductors that we will discuss are copper.

Outside plant engineers are faced with the design of the subscriber loop (Section 5.4). This is a wire pair extending from the local serving exchange to the telephone instrument (subset) on the subscriber's premises. A major design constraint is the resistance of the subscriber loop. At some point as the loop is extended in length, there will be so much resistance that its signaling capability is lost. One way we can extend the length and maintain the signaling capability is to increase the wire diameter.

In mainland Europe, wire diameter is given in millimeters. In North America a copper

[3]*Resistivity* is a unit constant used to determine the conductive properties of a material.

Table A.1 American Wire Gauge (AWG) versus Wire Diameter and Resistance

American Wire Gauge	Diameter (mm)	Resistance (Ω/km)[a] at 20°C
11	2.305	4.134
12	2.053	5.210
13	1.828	6.571
14	1.628	8.284
15	1.450	10.45
16	1.291	13.18
17	1.150	16.61
18	1.024	20.95
19	0.9116	26.39
20	0.8118	33.30
21	0.7229	41.99
22	0.6439	52.95
23	0.5733	66.80
24	0.5105	84.22
25	0.4547	106.20
26	0.4049	133.9
27	0.3607	168.9
28	0.3211	212.9
29	0.2859	268.6
30	0.2547	338.6
31	0.2268	426.8
32	0.2019	538.4

[a]These figures must be doubled for loop/km. Remember it has a "go" and "return" path.

wire's diameter is indicated by its gauge. Here we mean American Wire Gauge (AWG), formerly known as Brown & Sharpe (B&S). Table A.1 compares common AWG values with copper wire diameter in mm and the resistance in ohms per kilometer (Ω/km).

A.4 RESISTANCES IN SERIES AND IN PARALLEL, AND KIRCHHOFF'S LAWS

As previously discussed, the total resistance of series resistors is equal to the sum of the individual resistances. If we refer to Figure A.3, the total resistance, R_T is:

$$R_T = R_1 + R_2 + R_3. \tag{A.4}$$

Resistances in parallel are another matter. Consider Figure A.5. Here the current in the circuit divides and each resistor carries its share. Now apply Ohm's law and we find that the current across each resistor must be equal to the potential measured across the particular resistor divided by its value in ohms. In this particular circuit (Figure A.5), the potential across either resistor is equal to the emf supplied by the battery. In reality the battery is supplying two currents, one through resistor w-x and the second through y-z. For any circuit having two resistors in parallel, the current supplied to the combination must be greater than the current supplied to either branch. It follows, then, that we could replace these two resistors with a single resistor in series with the emf supply, and the value of that resistor must be less than the ohmic value of either single resistor in parallel.

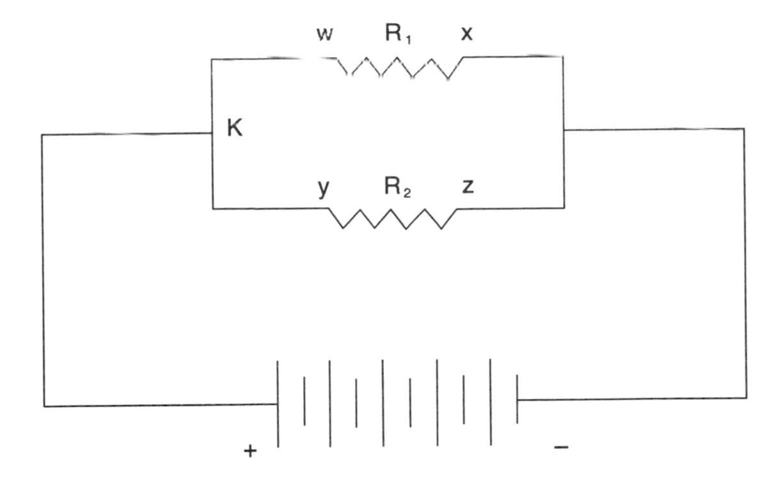

Figure A.5 Model of a circuit with two resistances in parallel and a battery as the emf source.

A.4.1 Kirchhoff's First Law

The first law states that at any point in an electrical circuit there is as much current flowing to the point as away from it. The laws applies no matter how many branches there are in the circuit. Figure A.5 shows point K in the circuit. Let I be the current being supplied by the battery emf source to the combination of the two resistors in parallel; I_1 and I_2 are the currents through the two resistors, respectively. We now can say, based on Kirchhoff's first law, that:

$$I = I_1 + I_2. \tag{A.5}$$

Again consider the circuit in Figure A.5. Let R be the equivalent resistance of the two resistors in parallel. Applying Ohm's law:

$$R = E/I.$$

Substitute Eq. (A.5). We then have:

$$R = E/(I_1 + I_2). \tag{A.6}$$

However, $I_1 = E/R_1$ and $I_2 = E/R_2$. As a consequence:

$$R = E/[(E/R_1) + (E/R_2)]. \tag{A.7}$$

Divide Eq. (A.7) through by E and we now have:

$$R = 1/[(1/R_1) + (1/R_2)]. \tag{A.8}$$

Simplify the compound fraction and we have:

$$R = R_1 \times R_2/(R_1 + R_2). \tag{A.9}$$

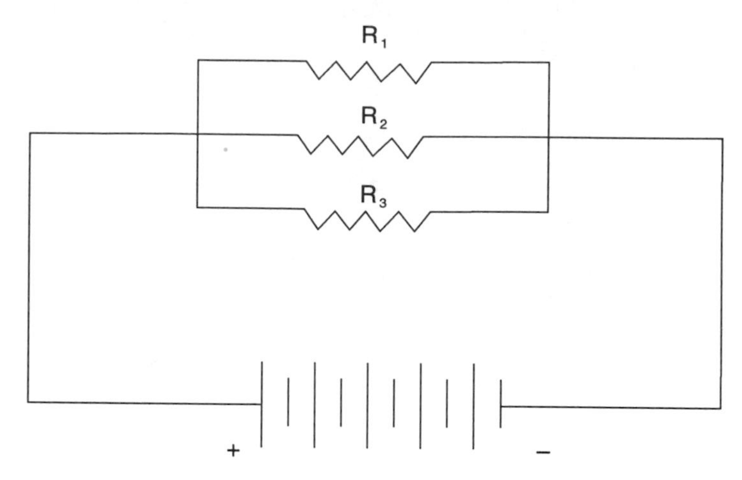

Figure A.6 Model of a circuit with three resistors in parallel.

Example. There are two resistors connected in parallel. Their values are 500 Ω and 700 Ω, respectively. What is the value of the equivalent combined resistance? Use Eq. (A.9).

$$\begin{aligned} R &= 500 \times 700(500 + 700) = 350,000/(500 + 700) = 350,000/1200 \\ &= 291.6\ \Omega. \end{aligned}$$

One self-check for resistances in parallel is that the equivalent combined resistance must be smaller than the value of either resistor.

Figure A.6 shows a group of three resistors in parallel with an emf source that is a battery. We encourage the use of a short-cut when there are more than two resistors in parallel. We introduce a new term, *conductance*. Conductance, G, is the inverse of resistance; stated in an equation:

$$R = 1/G. \qquad \text{(A.10)}$$

The unit of conductance is the *mho*, which the reader will note is "ohm" spelled backwards. To solve the equivalent total resistance problem of Figure A.6, convert each resistance into its equivalent conductance, sum the values, and invert the sum.

Example. The values of the resistances in Figure A.6 are 2000, 2500, and 3000 Ω; the closed circuit emf of the battery is 24 V. What is the value of the current flowing out of the battery? Use Eq. (A.10) and convert each resistance value to its equivalent conductance and sum:

$$\begin{aligned} G &= G_1 + G_2 + G_3 \\ &= 1/2000 + 1/2500 + 1/3000 \\ &= 0.0005 + 0.0004 + 0.00033 \\ &= 0.00123\ \text{mhos} \end{aligned} \qquad \text{(A.11)}$$

Usc the inverse of Eq. (A.11) or R = 1/G; then

$$R = 1/0.00123 = 813\ \Omega.$$

We know the voltage source is 24 V and the circuit resistance is 813 Ω. We now can apply Ohm's law to calculate the current flowing out of the battery. Use Eq. (A.2) or I = E/R

$$I = 24/813 = 0.0295 \text{ A or } 29.5 \text{ mA}.$$

A.4.2 Kirchhoff's Second Law

When current flows through a resistor, there is always a difference in potential when measured across the resistor (i.e., between one end and the other of the resistor). The value in volts of the difference in potential varies with the current flowing through the resistor and the resistance. In fact, as mentioned, the value is the product of the resistance (in ohms) and current (in amperes). This is just the statement of Ohm's law, Eq. (A.1) (E = IR). In our previous discussion, it was called potential difference, voltage drop, or IR drop. This IR drop acts in the opposite direction to, or opposes, the emf which drives the current through the resistor.

Figure A.7 is a model of a circuit with three resistors in series and a battery emf source. For reference, a voltmeter is shown measuring the voltage (the IR drop) across resistor R_2. In such a closed circuit, the sum of the IR drops across the three resistors must be equal to the impressed emf. Let the IR drop across each resistor in Figure A.7 be represented by V_1, V_2, and V_3, respectively. This, then, is how we state Kirchhoff's second law:

$$E = V_1 + V_2 + V_3, \tag{A.12}$$

where E is the impressed emf or closed circuit battery voltage. In the case of Figure A.7, we can write:

$$E = IR_1 + IR_2 + IR_3. \tag{A.13}$$

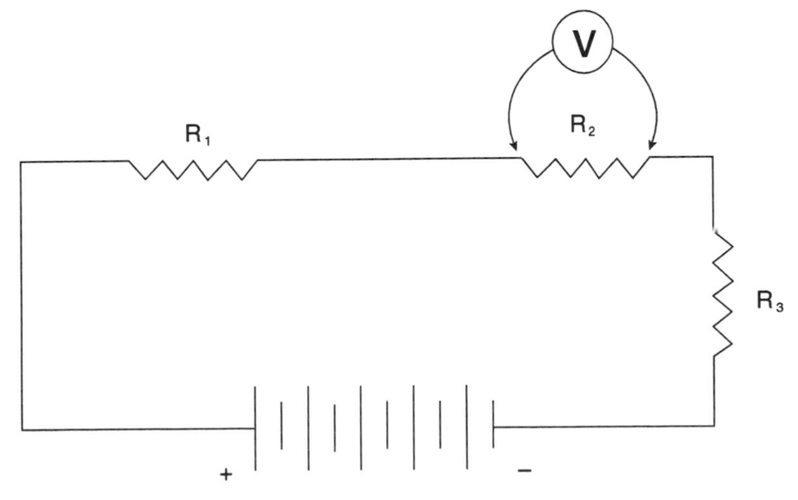

Figure A.7 A model circuit for three resistors in series. A voltmeter measures the IR drop across resistor R_2.

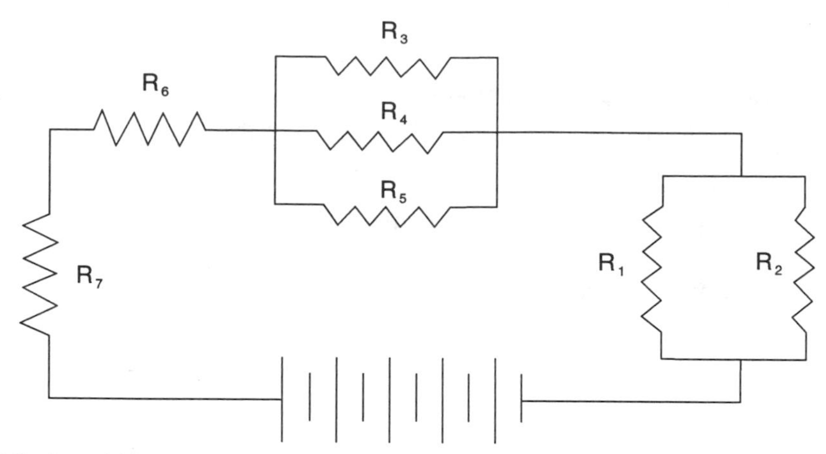

Figure A.8 A model for a circuit with series and parallel resistances with a battery as an emf source.

This equation can be restated as

$$E - IR_1 - IR_2 - IR_3 = 0.$$

A.4.3 Hints on Solving dc Network Problems

Make a network drawing, much like we have in Figures A.1 through A.7. Then assign letters to all unknown values. This should be followed by arrows showing the direction of current flow. With the practical application of Kirchhoff's laws, use correct algebraic signs. There should be one sign (+ or −) given to the electromotive force in the direction of the current flow, and the opposite sign is then given to the IR drops. Often we accept the clockwise direction as positive, and the counterclockwise as negative. For instance, all emfs are labeled positive that tend to make the current flow in the positive direction; and all potential drops are labeled negative, due to this flow of current as well as any emfs tending to make current flow in the opposite direction. When carrying out this exercise, we may find a solution to an equation may be preceded by a minus sign. This merely means that the actual direction of flow of current is opposite to the direction we assumed.

Example Calculation of a Series-Parallel Circuit. Figure A.8 is a model series-parallel circuit. In other words, the circuit has a mix of resistances in series and in parallel. *Hint:* Replace all parallel resistors by their equivalent value first. Thus we end up with a circuit entirely of resistances in series. In other words, first calculate the equivalent resistance for resistors R_1 and R_2; then for R_3, R_4, and R_5. We end up with four resistors in series: the first and second group of parallel resistors, the R_6, and R_7. The total resistance of the circuit is then the sum of these four resistances. Assign the following values to these resistances: $R_1 = 800\ \Omega$; $R_2 = 1200\ \Omega$; $R_3 = 3000\ \Omega$; $R_4 = 5000\ \Omega$; $R_5 = 4000\ \Omega$; $R_6 = 600\ \Omega$; and $R_7 = 900\ \Omega$.

For the first group of parallel resistances we use formula (A.9):

$$\begin{aligned} R &= 800 \times 1200/(800 + 1200) = 960,000/2000 \\ &= 480\ \Omega. \end{aligned}$$

For the second group of parallel resistances use formulas (A.10) and (A.11), as follows:

What is the current flowing out of the 48 V battery?

$$G_3 = 1/3000 = 0.000333; G_4 = 1/5000 = 0.0002; G_5 = 1/4000 = 0.00025.$$

Sum these values using Eq. (A.11) and

$$\begin{aligned} G &= 0.000333 + 0.0002 + 0.00025 \\ &= 0.000783. \end{aligned}$$

Use formula (A.10) to obtain the equivalent resistance

$$\begin{aligned} R &= 1/G = 1/0.000783 \\ &= 1277.14\Omega. \end{aligned}$$

Calculate the total resistance of the circuit with the four resistances, which includes the derived resistances. Use formula (A.4):

$$\begin{aligned} R_T &= 480 + 1277.14 + 600 + 900\ (\Omega) \\ &= 3257.14\ \Omega. \end{aligned}$$

To calculate the current flowing out of the battery, use Ohm's law (Eq. (A.2)):

$$I = E/R = 48/3257.14 = 0.0147 \text{ A or } 14.7 \text{ mA}.$$

A.5 ELECTRIC POWER IN dc CIRCUITS

Batteries store chemical energy. When the battery terminals are connected to supply emf to an electric circuit, the battery chemical energy is converted to electrical energy. This manifests itself as power (work per unit time). The unit of power is the *watt* (W). When we pay our electric bill, we pay for kilowatt-hours of expended electric power.

The resistors in Figures A.2–A.8 dissipate power in the form of heat. In fact, not only are resistors rated for their resistance in ohms, but also for their capability to dissipate heat measured in watts. Let us now examine the electric power dissipated by a resistor or other ohmic device.[4] Let power, expressed in watts, be denoted by the notation P. Then:

$$P = EI. \tag{A.14}$$

Stated in words: *In an electric circuit, if we multiply the electromotive force in volts by the current in amperes, we have an expression for the power in watts* (Eq. (A.14)). The *watt* may, therefore, be defined as the power expended in a circuit having an electromotive force of one volt and a current of one ampere.

There are two variants of Eq. (A.14) by simple substitution of variants of Ohm's law (Eqs. (A.1), (A.2), and A.3)):

$$P = EI = E(E/R) = E^2/R \tag{A.15}$$

and

$$P = EI = IR(I) = I^2R. \tag{A.16}$$

[4]Electrical power, of course, can manifest itself or be useful in other ways besides the generation of heat. For example, it can rotate electric motors (mechanical energy) and it can be used to generate a radio wave.

Example 1. The current flowing through a 1000-Ω resistor is 50 mA. What is the power dissipated by the resistor? Convert mA to A or 0.05 A. Use Eq. (A.16):

$$\begin{aligned} P &= (0.05)^2 \times 1000 \\ &= 2.5 \text{ W}. \end{aligned}$$

Example 2. The potential drop across a 600-Ω resistor is 3 V. What power is being dissipated by the resistor? Use Eq. (A.15):

$$P = E^2/R = 3^2/600 = 0.015 \text{ W or } 15 \text{ mW}.$$

A.6 INTRODUCTION TO ALTERNATING CURRENT CIRCUITS

A.6.1 Magnetism and Magnetic Fields

The phenomenon of magnetism was known by the ancient Greeks when they discovered a type of iron ore called magnetite. Moorish navigators used a needle-shaped piece of this ore as the basis of a crude compass. It was found that when this needle-shaped piece of magnetite was suspended and allowed to move freely, it would always turn to the north. What we are dealing with here are *permanent magnets*. Remember, we used permanent magnets in the telephone subset to set up a magnetic field in its earpiece (Section 5.3.2). Figure A.9 shows an artist's rendition of the magnetic field around a bar magnet, the most common form of permanent magnet.

In the early nineteenth century it was learned that this magnetic property can be artificially induced to a family of iron-related metals by means of an electric current. The magnetic field was maintained while current is flowing; it collapsed when the current stopped.

Magnets, as we know them today, are classed as either *permanent magnets* or *electromagnets*. A typical permanent magnet is a hard steel bar that has been magnetized. It can be magnetized by placing the bar in a magnetic field; the more intense the field, the more intense is the induced magnetism in the steel bar. The influence of a magnetic field can be detected in various ways, such as by a conventional compass or iron filings on paper where the magnet is placed below the paper.

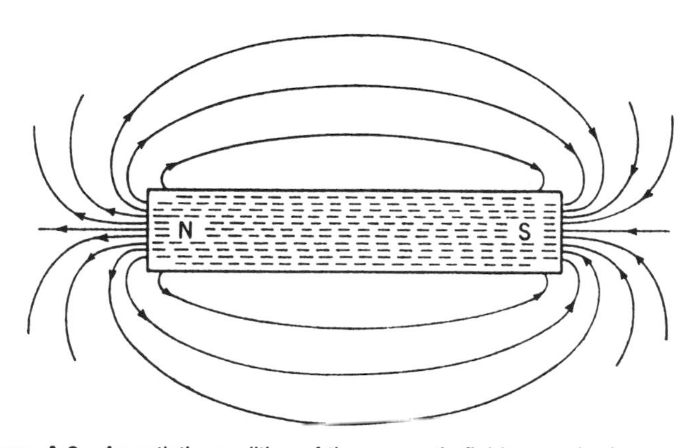

Figure A.9 An artist's rendition of the magnetic field around a bar magnet.

With this latter approach with the iron filings we develop an arrangement of the filings that will appear almost identical with Figure A.9. The lines that develop are commonly known as *lines of magnetic induction*. All of the lines as a group are referred to as the *flux*, which is designated by the symbol ϕ. The flux per unit area is known as the *flux density* and is designated by B. We arbitrarily call one end of the bar magnet the *north pole*, and the opposite end, the *south pole*. Many of us remember playing with magnets when we were children. It was fun to bring in a second bar magnet. It was noted that if its south pole was brought into the vicinity of the first magnet's south pole, the magnets physically repelled each other. On the contrary, when the north pole of one magnet was brought into the vicinity of the south pole of the other, they were attracted.

Suppose the strength of the magnet in Figure A.9 is increased. The magnetic field will be strengthened in proportion, and is conventionally represented by a more congested arrangement of lines of magnetic induction. The force that will be exerted upon a pole of another magnet located at any point in the magnetic field will depend upon the intensity of the field at that point. This field intensity is represented by the notation H.

In the preceding paragraphs we stated that the flux density B is the number of lines of magnetic induction passing through a unit area. By definition, unit flux density is one line of magnetic induction per square centimeter in *air*. Thus, in air, field intensity H and the flux density B have the same numerical value.

A.6.2 Electromagnetism

Any conductor where electrical current flows sets up a magnetic field. When the current ceases, the magnetic field collapses and disappears. If we form that wire conductor into a loop and connect each end to an emf source, current will flow and a magnetic field is set up. Let's say that the resulting magnetic field has an intensity H. Now take the loop and produce a second turn. The resulting magnetic field now has an intensity of 2H; with three turns, 3H, and so forth.

Figure A.10 shows a spool on which we can support the looped wire turns. Some will call this spool arrangement a *solenoid*. If a solenoid is very long when compared with its diameter, the field intensity in the air inside the solenoid is directly proportional to the product of the number of turns and the current and inversely proportional to the length of the solenoid. This can be expressed by the following relationship:

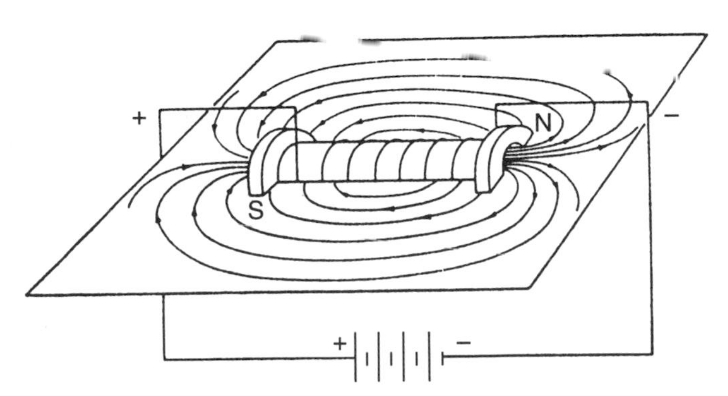

Figure A.10 A magnetic field surrounding a solenoid with an air-core.

$$H = k(NI)/l, \tag{A.17}$$

where l is the length of the solenoid, N is the number of turns, and I is the current. Parameter k adjusts the equation to the type of unit system used. When I is in amperes and l is in meters, then $k = 1$. H is defined as *ampere-turns per meter*. Of course, H is the field intensity.

A.7 INDUCTANCE AND CAPACITANCE

A.7.1 What Happens When We Close a Switch on an Inductive Circuit?

Figure A.11 shows a simple circuit consisting of an air-core inductance, a battery emf supply, and a switch. The resistance across the coil or inductance is 10 Ω and the battery supply is 24 V. We close the switch, and calculate the current flowing in the circuit. Using Ohm's law:

$$I = E/R = 24/10 = 2.4\ \text{A}.$$

Off hand, one would say that at the very moment the switch is closed, the 2.4 A are developed. That is not true. It takes a finite time to build from the zero value of current when the switch is open to the 2.4-A value when the switch is closed. This is a change of state, from off to on. For every electrical change of state there is a finite amount of time required to fully reach the changed-state condition. Imagine a steam-locomotive starting off. It takes considerable time to bring the railroad train to full speed.

When there is an inductance in a circuit, such as in Figure A.11, the buildup is slowed still further because a counter emf is being generated in the coil (L). This is due to the dynamics of the buildup. As current starts to flow in the circuit once the switch is closed, a magnetic field is developing in the coil windings. By definition, this field generates a second field just in the windings themselves. This second magnetic field generates a voltage in the opposite direction of the voltage resulting from the primary field due to the current flow in the circuit. This is called *counter emf.* This counter emf continues so long as there is a dynamic condition, so long as there is a *change in current flow.* Once equilibrium is reached and we have a current flow of exactly 2 A, the counter emf disappears.

Now open the switch in Figure A.11. At the very moment of opening the switch we see an electric arc across the switch contacts. When the switch is opened there is

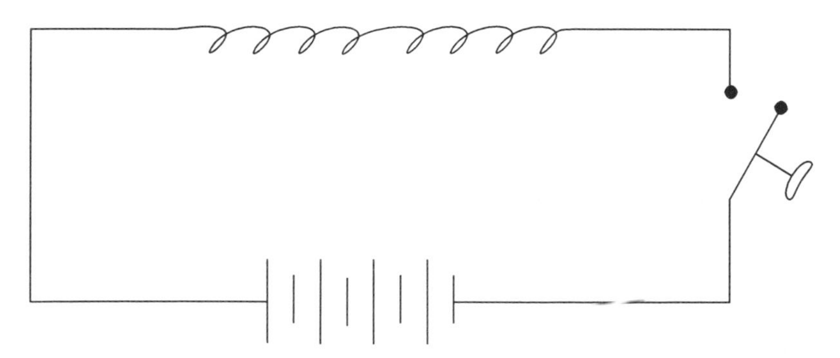

Figure A.11 A simple circuit with an inductance or coil.

no longer support (i.e., no current flows in the circuit) for the magnetic field in coil L. The energy stored in that field must dissipate somewhere. It dissipates back through the circuit again in the opposite direction of original current flow. Again we are faced with the fact that on opening the switch, the current does not drop to zero immediately, but takes a finite amount of time to drop to zero amperes. So a magnetic field of an electromagnet actually stores electricity.

Regarding Figure A.11, we are faced with two conditions: first the buildup of current to its steady-state value, and the decay of electric current to a zero value. It may be said that an electric circuit *reacts* to such current changes.

The magnitude of the induced emf, that reactive effect, is a function of two factors:

1. The first factor is the number of turns of wire in the coil; whether the coil has an iron core; and the properties of the iron core; and
2. The second factor is the rate of change of current in the coil.

These two properties only come into play when the circuit is dynamic—its electrical conditions are changing. There are two such properties that remain latent under steady-state conditions, but come into play when the current attempts to change its value. These are *inductance* and *capacitance*. For analogies, let's consider that inductance is something like inertia in a mechanical device, and capacitance is like elasticity.

Inductance. The property of a circuit which we have called inductance is represented by the symbol L. Its unit of measure is the henry (H). The henry is defined as the inductance of a circuit that will cause an induced emf of 1 V to be set up in the circuit when the current is changing at the rate of 1 A per second. Now the following relationship can be written:

$$E_l = LI/t, \tag{A.18}$$

where E_l is the induced emf; L is the inductance in henrys; I is the current in amperes; and t is time in seconds.

If we have coils in series, then to calculate the total inductance of the series combination, add the inductances such as we add resistances for resistors in series. In a similar manner, if we have inductances in parallel, use the same methodology as though they were resistances in parallel.

Now distinguish between *self-inductance* and *mutual inductance*. Self-inductance is the property of a circuit which creates an emf from a change of current values when the reaction effects are wholly within the circuit itself. If the electromagnetic induction is between the coils or inductors of separate circuits, it is called mutual inductance.

Capacitance. Capacitance was introduced in Section 2.5.1.1, during a discussion of its buildup on wire pair as it was extended. An analogy of capacitance is a tank of compressed air. Air is elastic, and the quantity of air in the tank is a function of the air pressure and capacity of the tank.

Similar to Figure A.11, consider Figure A.12, which shows a capacitor connected across a battery emf supply. In its simplest form, a capacitor may consist of two parallel plates (conductors) separated by an insulator, in this case air. The circuit is equipped with an on–off switch. Close the switch. Unlike the inductance scenario in Figure A.11, there is a surge of current in the circuit. The current is charging the capacitor to a voltage equal to the emf of the battery. As the capacitor becomes charged, the value

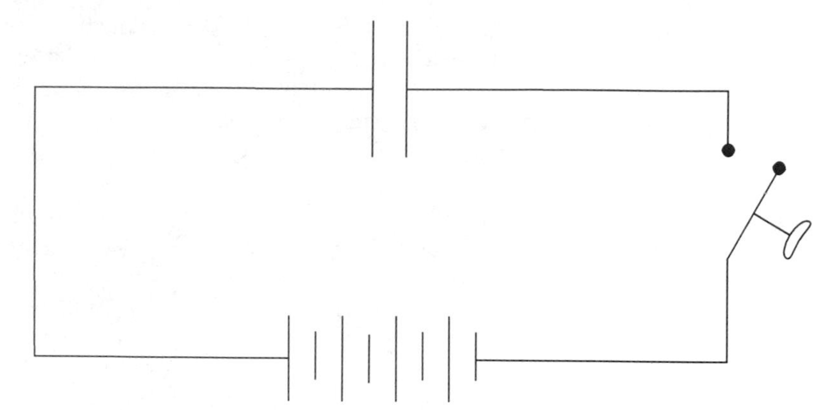

Figure A.12 A circuit model to illustrate capacitance.

of the current decreases until the capacitor is fully charged, when the current becomes zero.

A capacitor is defined as a device consisting of two conductors separated by an insulator (in its simplest configuration). The greater the area of the conductors, the greater the capacitance of the device. (Remember Section 2.5.2.2 where we dealt with a wire pair). Here we have two conductors separated by the insulation on each wire in the pair. Likewise, the longer the wire pair, the greater the capacitance it displays.

The quantity of electricity stored by a capacitor is a function of its capacitance and the emf across its terminals. It is expressed by the relationship:

$$q = \mathrm{EC}, \tag{A.19}$$

where q is the quantity of electricity in coulombs, E is the emf in volts, and C is the capacitance in farads.[5] The capacitance value is a function of the dimensions of the capacitor conductor plates. In the practical world, the farad is too large a unit. As a result we usually measure capacitance in microfarads (μF) or 1×10^{-6} farads, or picofarads (pF) or 10^{-12} farads.

Restating the Eq. (A.19):

$$q = \epsilon_0 \mathrm{EA} \tag{A.20}$$

where ϵ_0 is the permittivity constant, 8.85×10^{-12} farads/meter or 8.85 pF/m; A is the area of the plate; and E is the emf.

Substituting q from Eq. (A.19) in Eq. (A.20) and noting that the capacitance is inversely proportional to the distance, d, between the parallel plates, we have:

$$\mathrm{C} = \epsilon_0(\mathrm{A/d}) \tag{A.21}$$

This tells us that capacitance is a function of geometry; it is directly proportional to the area of the capacitor plates and inversely proportional to the distance between the plates. This assumes that the insulator between the plates is air.

[5]*Coulomb* (IEEE, Ref. 2) is the unit of electric charge in SI units. The coulomb is the quantity of electric charge that passes any cross section of a conductor in 1 second when the current is maintained at 1 A.

Table A.2 Some Properties of Dielectrics

Material	Dielectric Constant (κ)[a]	Dielectric Strength (kV/mm)
Air (1 atm)	1.00054	3
Polystyrene	2.6	24
Paper	3.5	16
Transformer oil	4.5	
Pyrex	4.7	14
Ruby mica	5.4	
Porcelain	6.5	
Silicon	12	
Germanium	16	
Ethanol	25	
Water (20°C)	80.4	
Water (25°C)	78.5	
Titania ceramic	130	
Strontium titanate	310	8

For a vacuum, κ = unity.

[a]Measured at room temperature, except for the water.

Source: Fundamentals of Physics—Extended, 4th ed., p. 751, Table 27-2. (Ref. 3, reprinted with permission.)

Suppose the insulator between the plates is some other material. Table A.2 lists some of the typical materials used as insulators and their *dielectric constant*. The capacitance *increases* by a numerical factor κ, which is the dielectric constant. The dielectric constant of a vacuum is unity by definition. Because air is nearly "empty" space, its measured dielectric constant is only slightly greater than unity. The difference is insignificant.

Let L represent some geometric dimensions as a function of length. For a parallel plate capacitor with an air dielectric (insulator), L = A/d (from Eq. (A.21)). Capacitance (C) can be related to dielectric constant, the geometrical property L, and a constant by:

$$C = \epsilon_0 \kappa L \tag{A.22}$$

where ϵ_0 is the permittivity constant, 8.85 pf/m, and κ is the dielectric constant from Table A.2.

Another effect of the introduction of a dielectric (insulator) including air is to limit the potential difference between the conductors to some maximum voltage value. If this value is substantially exceeded, the dielectric material will break down and form a conducting path between the plates. Every dielectric material has a characteristic *dielectric strength*, which is the maximum value of the electric field that it can tolerate without breakdown. Several dielectric strengths are listed in Table A.2.

Capacitors in Series and in Parallel. When there is a combination of capacitors in a circuit, sometimes we can replace the combination with a single capacitor with an equivalent capacitance value. Just as resistances and inductances can be in series and in parallel, we can have capacitors in series and in parallel. Figure A.13 shows several capacitors in parallel. As we see in the figure, each capacitor has the same potential difference across its plates (i.e., the battery emf). By direct inference from the figure we can see the equivalent capacitance of the three capacitors is the sum of each of the individual capacitances or:

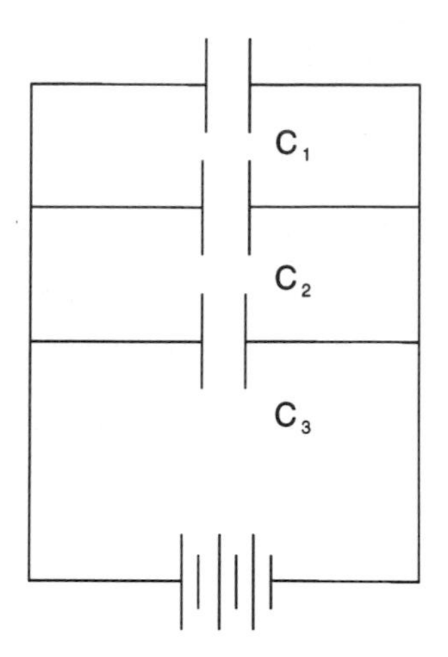

Figure A.13 Capacitors in parallel.

$$C_{eq} = C_1 + C_2 + C_3. \quad \text{(A.23)}$$

Suppose the values of the capacitors were 2, 3, and 4 nF, respectively. What would the equivalent capacitance be? From Eq. (A.23), the value would be 9 nF.

Capacitors in series (see Fig. A.14) are handled in a similar manner as resistances in parallel. The total equivalent capacitance is:

$$1/C_{eq} = 1/C_1 + 1/C_2 + 1/C_3. \quad \text{(A.24)}$$

Suppose, again, the values of the three capacitors were 2, 3, and 4 nF. What would the equivalent capacitance be?

$$\begin{aligned} 1/C_{eq} &= 1/2 + 1/3 + 1/4 \\ &= 6/12 + 4/12 + 3/12 = 13/12, \\ C_{eq} &= 12/13 = 0.923 \text{ nF}. \end{aligned}$$

A.7.2 RC Circuits and the Time Constant

An RC circuit is illustrated in Figure A.15. The capacitor in the figure remains uncharged until the switch is closed. To charge it we throw the switch S to a closed position. The

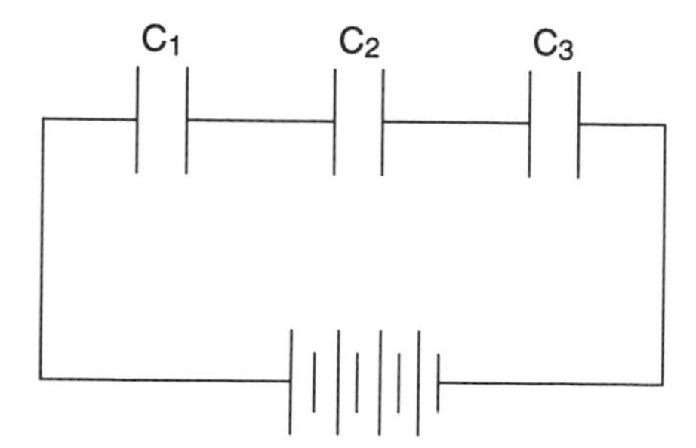

Figure A.14 Capacitors in series.

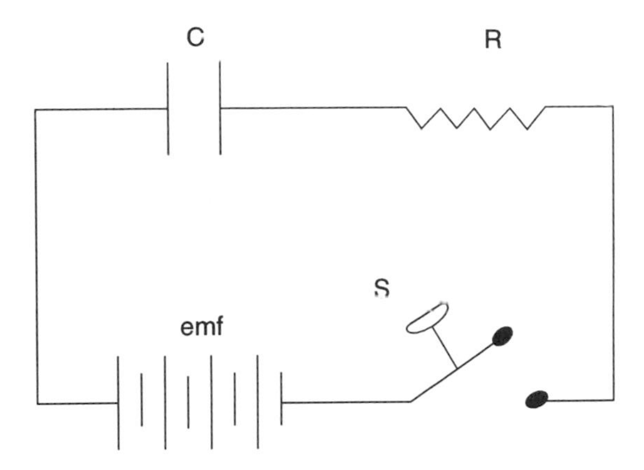

Figure A.15 An RC circuit.

circuit consists of a resistor (R) and capacitor (C) in series. When we close the switch, there is a surge of current whose intensity decreases with time as the charge of the capacitor builds up.

Figure A.16 shows the change in the value of the current with time after the switch is closed. There is a corresponding change in the voltage drop across the capacitor C. This is illustrated with curve E. The instantaneous value of current, I, is determined solely by the value of resistor, R. The total voltage drop is then across R and the drop across C is zero. As the capacitor begins to charge, however, the voltage drop across C gradually builds up, the current decreases, and the voltage drop across R decreases correspondingly. When the capacitor reaches full charge and the current has fallen to a negligible value, the total voltage drop is now across the capacitor. Remember that at all times the sum of the voltage drops across R and C must be equal and opposite to the impressed voltage.

Where does time come into the picture (i.e., time constant)? It takes time for an RC circuit (or an RL circuit) to reach a steady-state value. This time, we find, is a function of the product of R and C. Actually, if we multiply the value of R in ohms by the value of C in farads, it is equal to time in seconds. In a series RC circuit as illustrated in Figure A.15, the product RC is known as the *time constant* of the circuit. By definition, it is the time required to charge the capacitor to 63% of its final voltage. If we plot the curve for voltage (E) in Figure A.16, we have an exponential function of time. This is expressed mathematically as:

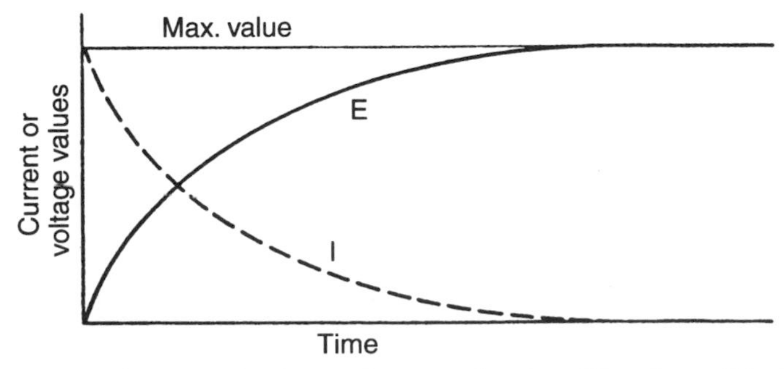

Figure A.16 Current and voltage values as a function of time for an RC circuit.

$$E_i = E_{max}[1 - \epsilon^{-t/RC}], \tag{A.25}$$

where E_i is the instantaneous voltage across the capacitor at any time t, E_{max} is the final voltage, and ϵ is the natural number. This is the base of the natural logarithm or the Naperian logarithm.[6] Its value is 2.718+. When t equals RC, the term in parentheses equals:

$$1 - \epsilon^{-1} = 1 - 1/\epsilon = 1 - 2.718 = 0.63.$$

It should be noted that a similar relationship can be written for an RL circuit where the exponential buildup curve is current rather than voltage. However, here the exponent is L/R rather than RC. Again, it is the time required for the current to build up to 63% of its final value.

RC circuits have wide application in the telecommunications field. Because of the precision with which resistors, capacitors, and inductors may be built, they enable the circuit designer to readily control the timing of current pulses to better than 1-ms accuracy. For such practical applications, the designer considers that currents and voltages in RC and RL circuits reach their final value in a time equal to five times their time constant [i.e., $5 \times RC$ or $5 \times (L/R)$]. It can be shown that at 5 × time constant the current or voltage has reached 99.33+% of its final value.

A.8 ALTERNATING CURRENTS

Alternating current (ac) is a current where the source emf is alternating, having a simple and convenient waveform, namely, the sine wave. Such a sine wave is illustrated in Figure A.17. Any sine wave can be characterized by its frequency, phase, and amplitude. These were introduced in Section 2.3.3, and much of that information is repeated here for direct continuity.

Note in the figure that half a cycle of the sine wave is designated by π and a full cycle by 2π. Remember that there are 2π radians in a circle or 360 degrees, and 1 π

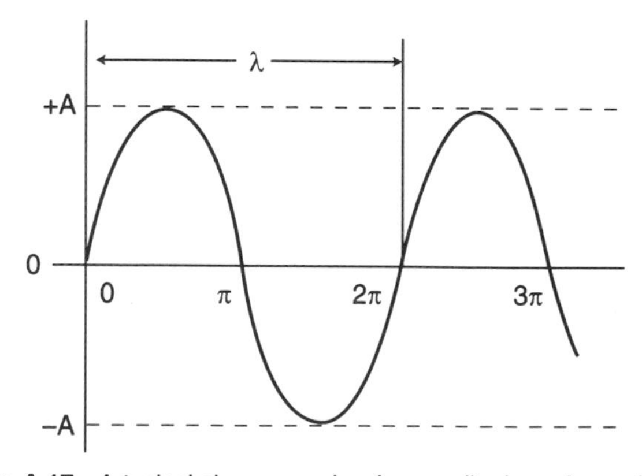

Figure A.17 A typical sine wave showing amplitude and wavelength.

[6]In this text, unless otherwise specified, logarithms are to the base ten. These are called *common logarithms*. Under certain circumstances we will also use logarithms to the base 2 (binary system of notation). In the calculus, nearly all logarithms used are to the natural or Naperian base. This is named in honor of a Scottsman named Napier. However, Napier did not use the base ϵ but used a base approximately equal to ϵ^{-1}.

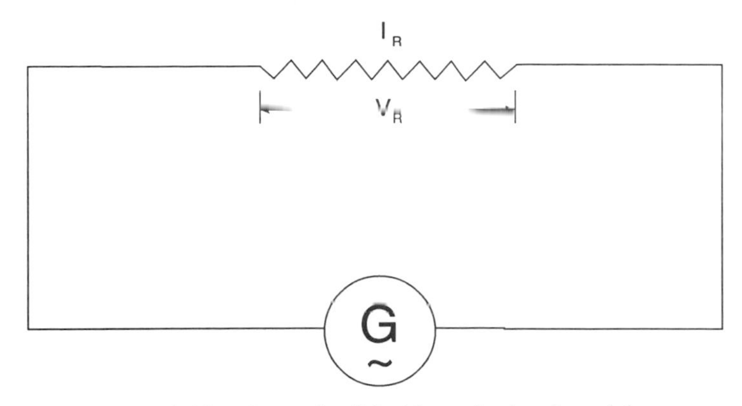

Figure A.18*a* An ac circuit that is predominantly resistive.

radian is 180 degrees. Therefore 1 radian is $180°/\pi$ or 57.29582791 …°. That is an unwieldy number and is not exact. That is why we like to stick to using π, and the value is exact.

In Figure A.17 the maximum amplitude is +A and −A; the common notation for wavelength is λ. To convert wavelength to frequency and vice versa we use the formula given in Section 2.3.3, or

$$F\lambda = 3 \times 10^8 \text{ m/s (meters/second)}. \tag{A.26}$$

We recognize the constant on the right-hand side of the equation as an accepted estimate of the velocity of light (or a radio wave). The *period* of a sine wave is the length of time it takes to execute 1 cycle (1 Hz). We then relate frequency, F, to period, T, with:

$$F = 1/T, \tag{A.27}$$

where F is in Hz and T is in seconds (s).

Voltage and current may be out of phase one with the other. The voltage may lead the current or lag the current depending on circuit characteristics, as we will see.

When an AC circuit is resistive—there is negligible capacitance and inductance—the voltage and current in the circuit will be in-phase. Figure A.18*a* is a typical ac resistive circuit. The phase relations between voltage (V_R) and current (I_R) is illustrated in Figure A.18*b*.

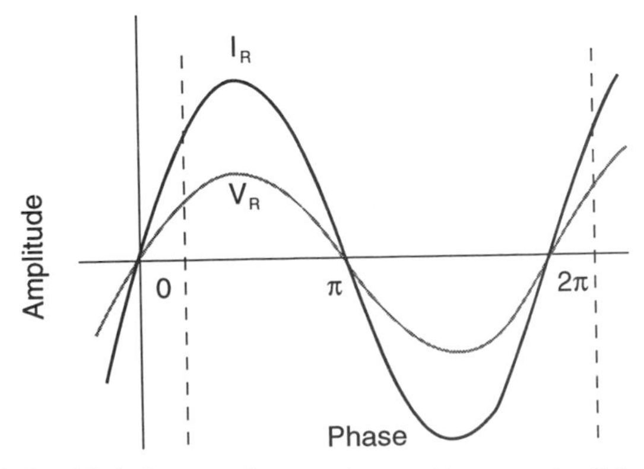

Figure A.18*b* Phase relationship between voltage and current in an ac circuit that is resistive. Here the voltage and current remain in phase.

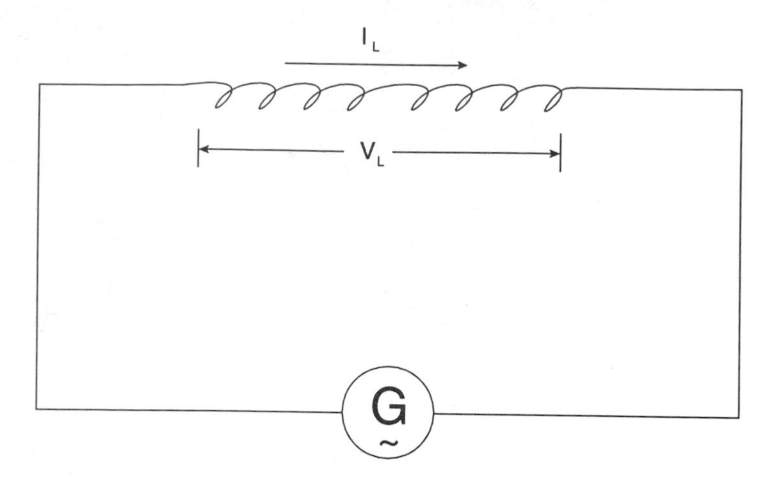

Figure A.19*a* A simple ac inductive circuit. Here the current (I_L) lags the potential difference (voltage V_L) by 90°. G is the ac emf source.

A simple inductive circuit is shown in Figure A.19*a*, and the current and voltage phase relationships of this circuit are shown in Figure A.19*b*. In this circuit, because it is predominantly inductive, the potential difference (voltage V_L) across the inductor leads the current by 90°.

Figure A.20*a* is a simple ac capacitive circuit. In this case the current (I_C) leads the voltage (V_C) by 90°. This is illustrated in Figure A.20*b*.

The magnitudes of current and voltage at some moment in time are usually analyzed using vectors. Vector analysis is beyond the scope of this appendix.

Effective Emf and Current Values. In a practical sense, an arbitrary standard has been adopted so that only the value of current or voltage need be given to define it, its position in time being understood by the convention adopted. One approach is to state the maximum value of voltage or current. However, this approach has some disadvantages. Another and often more useful approach is the average value over a complete half-cycle (i.e., over π radians). For a sine wave this value equals 0.636 of the maximum value.

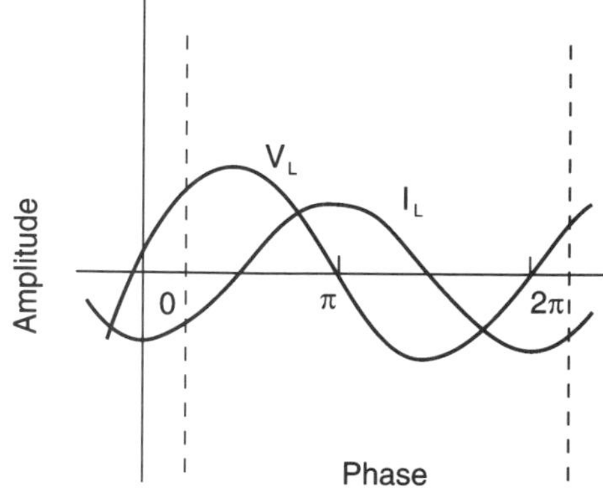

Figure A.19*b* Phase relationship between current (I_L) and potential difference (voltage V_L) in an ac inductive circuit.

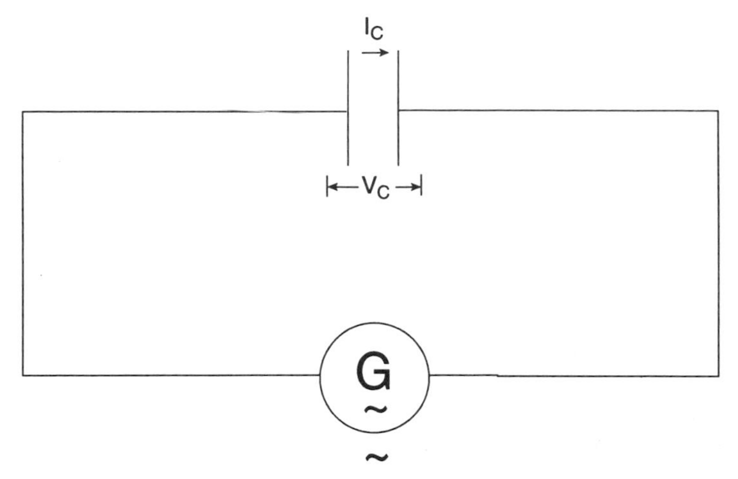

Figure A.20*a* A simple ac capacitive circuit.

One of the most applicable values is based on the heating effect of a given value of alternating current in a resistor that will be exactly the same as the heating value of direct current in the same resistor. This eliminates the disadvantage of thinking that the effects of alternating and direct currents are different. This is known as the *effective value* and is equal to the square root of the average of the squares of the instantaneous values over 1 cycle (2π radians). This results in 0.707 times the maximum value or:

$$I = 0.707(I_{max}) \tag{A.28a}$$

$$E = 0.707(E_{max}), \tag{A.28b}$$

where E and I without subscripts are effective values. Unless otherwise stated, ac voltages and currents are always given in terms of their effective values.

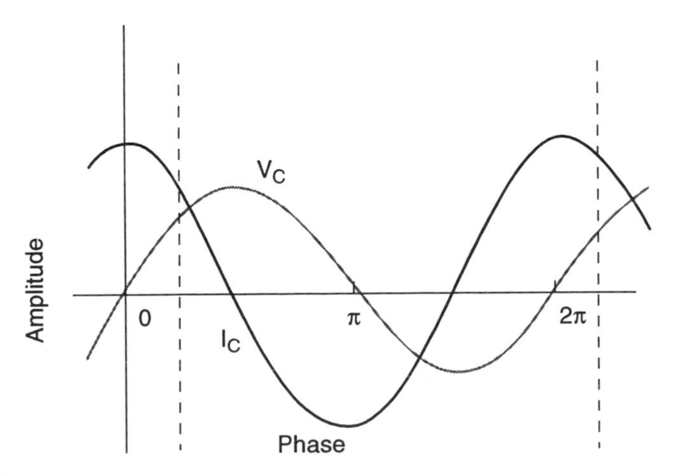

Figure A.20*b* The phase relationship between current (I_C) and emf (voltage, V_C) for an ac capacitive circuit.

A.8.1 Calculating Power in ac Circuits

Equation (A.14) provided an expression for power in a dc circuit. The problem with ac circuits is that the voltage and current are constantly changing their values as a function of time. At any *moment* in time the power generated or dissipated by a circuit is:

$$P = EI^7 \tag{A.29}$$

where P is expressed in W, E in V, and I in A. Also, Eqs. (A.15) and (A.16), Ohm's law variants of Eq. (A.14), are also valid here.

Across a resistive circuit, such as in Figure A.18*a*, where by definition, the ac voltage and current are in phase, Eq. (A.29) expresses the power. Here, voltage and current are effective values.

Now if the circuit that the ac generator looks into is inductive (Figure A.19*a*) or capacitive (Figure A.20*a*), the calculation of power is somewhat more complicated. The problem is that the voltage and current are out of phase one with the other, as illustrated in Figures A.19*b* and A.20*b*. The true power in such circuits will be less than the power calculated with Eq. (A.29) if the circuit were purely resistive. The power in such circumstances can be calculated by applying the *power factor*. Equation (A.29) now becomes:

$$P = EI\cos\theta, \tag{A.30}$$

where θ is the phase angle, the angle that voltage leads or lags current. Earlier we were expressing phase angle in radians (see Figure A.17 and its discussion). Again, there are 2π radians in 360° or π radians in 180°. It follows that 1 radian = $180°/\pi$, where π can be approximated by 3.14159, or 1 radian is 57.296°.

We can look up the value of $\cos\theta$ with our scientific calculator, given the value of θ, which will vary between 0° and 90°. Between these two values, $\cos\theta$ will vary between 0 and 1. Note that when the power factor has a value of 1, $\cos\theta$ is 1 and $\theta°$. This tells us that the voltage and current are completely in phase under these circumstances.

This leads to a discussion of impedance, which in most texts and reference books is expressed by the letter *Z*. In numerous places in our text we have used the notation Z_0. This is the *characteristic impedance*, which is the impedance we expect a circuit or device port to display. For example, we can expect the characteristic impedance of coaxial able to be 75 Ω, of a subscriber loop to be either 600 Ω or 900 Ω, and so forth.

A.8.2 Ohm's Law Applied to Alternating Current Circuits

We can freely use simple Ohm's law relationships (Eqs. (A.14)–(A.16)), when ac current and voltage are completely in phase. For example: R = E/I, where R is expressed in Ω, E in V, and I in A. Otherwise, we have to use the following variants:

$$Z = E/I, \tag{A.31}$$

where Z is expressed in Ω.

One must resort to the use of Eq. (A.31) if an ac circuit is *reactive*. A circuit is reactive when we have to take into account the effects of capacitance and/or inductance in the

[7]This equation is identical to Eq. (A.14).

circuit to calculate the effective value of Z. Under these circumstances, Z is calculated at a specified frequency. Our goal here is to reduce to a common expression in ohms a circuit's resistance in ohms, its inductance expressed in henrys, and its capacitance expressed in microfarads. Once we do this, a particular circuit or branch can be handled simply as though it were a direct current circuit.

We define *reactance* as the effect of opposing the flow of current in an ac circuit due to its capacitance and/or inductance. There are two types of reactance: *inductive reactance* and *capacitive reactance.*

Inductive Reactance. As we learned earlier, the value of current in an inductive circuit not only varies with the inductance but also with the rate of change of current magnitude. This, of course, is frequency (f). Now we can write an expression for a circuit's inductive reactance, which we will call X_L. It is measured in ohms.

$$X_L = 2\pi fL, \tag{A.32}$$

where L is the circuit's inductance in henrys.

Example. Figure A.21 shows a simple inductive circuit where the frequency of the emf source is 1020 Hz at 20 V, the inductance is 3.2 H. Calculate the effective current through the inductance. There is negligible resistance in the circuit. Use Eq. (A.32).

$$\begin{aligned} X_L &= 2 \times 3.14159 \times 1020 \times 3.2 \\ &= 20,508.3\Omega; \\ I &= 20/20,508.3 = 0.000975 \text{ A or } 0.975 \text{ mA}. \end{aligned}$$

In this circuit the voltage lags the current by 90°.

Capacitive Reactance. Capacitive reactive has the opposite behavior of inductive reactance. In this case, the current lags the voltage by 90°. Also, as the frequency increases, the capacitive reactance decreases, whereas with inductive reactance, as the frequency increases, the reactance increases. The following is an expression to calculate capacitive reactance:

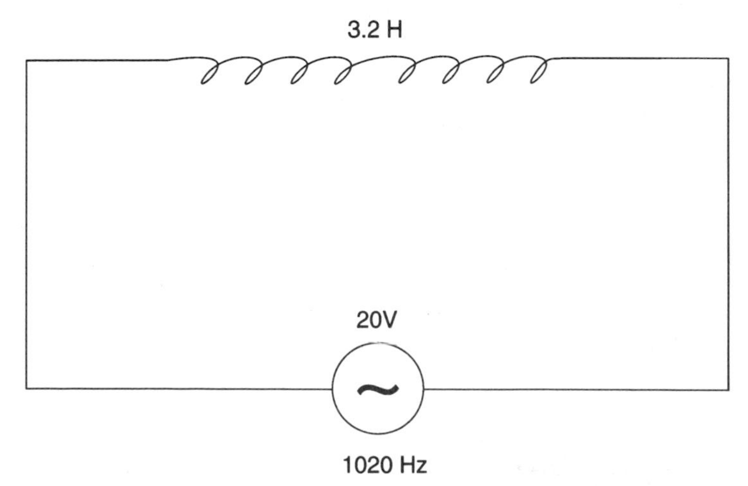

Figure A.21 An ac circuit with inductance only.

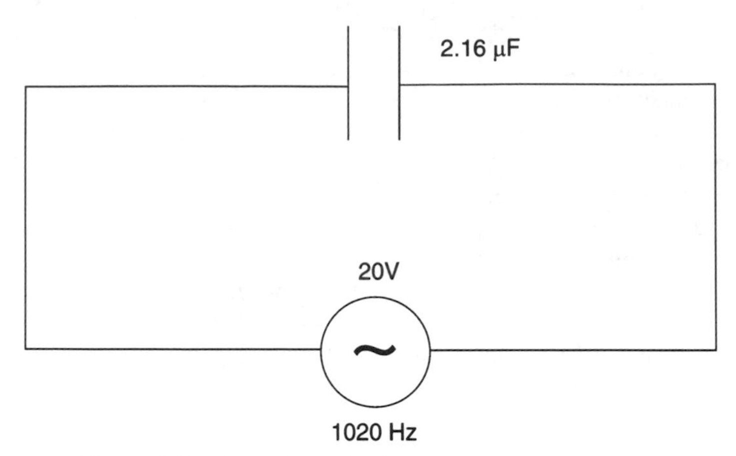

Figure A.22 A simple ac circuit displaying capacitance only.

$$X_C = -1/2\pi fC \tag{A.33a}$$

where f is in Hz and C is in farads.

The more customary capacitance unit is the microfarad (μF). When we use this unit of capacitance, the formula in Eq. (A.33*a*) becomes:

$$X_C = -1 \times 10^6/2\pi fC\ \Omega. \tag{A.33b}$$

Example. Figure A.22 illustrates a capacitive reactance circuit with a standard capacitor of 2.16 μF and an emf of 20 V at 1020 Hz. Calculate the current in amperes flowing in the circuit. Use formula (A.33*b*):

$$\begin{aligned} X_C &= -1 \times 10^6/2 \times 3.14159 \times 1020 \times 2.16 \\ &= -72.24\ \Omega, \\ I &= E/X_C \\ I &= -20/72.24 = -0.277\ \text{A} \qquad \text{(minus sign means leading current).} \end{aligned}$$

Circuits with Combined Inductive and Capacitive Reactance. To calculate the combined or total reactance when an inductance and a capacitance are in series, the following formula is applicable:

$$X = X_L + X_C \quad \text{and} \tag{A.34a}$$

$$X = 2\pi fL - 1 \times 10^6/2\pi fC. \tag{A.34b}$$

A word about signs: If the calculated value of X is positive, inductive reactance predominates, and if negative, capacitive reactance predominates.

Example. Figure A.23 illustrates an example of a circuit with capacitance and inductance in series. The inductance value is 400 mH and the capacitance is 500 nF. The source emf is 20 V. The frequency is 1020 Hz. Calculate the current in the circuit. Assume the resistance is negligible.

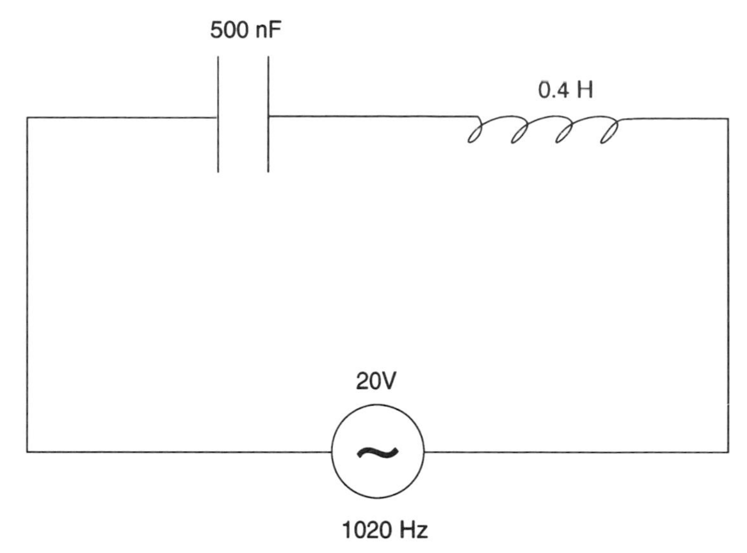

Figure A.23 A simple ac circuit with a coil (inductance) and capacitor in series.

Calculate the combined reactance X of the circuit using formula (A.34). Convert units for capacitance to microfarads, and the units of inductance to henrys. That is 0.5 μf and 0.4 H.

$$\begin{aligned} X &= 2 \times 3.14159 \times 1020 \times 0.4 - 1 \times 10^6/2 \times 3.14159 \times 1020 \times 0.5 \\ &= 2563.537 - 1,000,000/3204.42 \\ &= 2563.537 - 312.068 \\ &= 2251.469\ \Omega. \end{aligned}$$

Apply Ohm's law variant to calculate current:

$$\begin{aligned} I &= E/X = 20/2251.469 \\ &= 0.00888 \text{ A or } 8.88 \text{ mA}. \end{aligned}$$

A.8.3 Calculating Impedance

When we calculate impedance (Z), we must take into account resistance. All circuits are resistive, even though in some cases there is only a minuscule amount of resistance. We first examine the two reactive possibilities; that is, a circuit with inductive reactance and then a circuit with capacitive reactance. For the case with inductive reactance:

$$Z = (R^2 + X_L^2)^{1/2}. \quad \text{(A.35}a\text{)}$$

Substituting:

$$Z = [R^2 + (2\pi fL)^2]^{1/2}. \quad \text{(A.35}b\text{)}$$

For the case with capacitive reactance:

$$Z = (R^2 + X_C^2)^{1/2}. \tag{A.36a}$$

Substituting:

$$Z = [R^2 + (1,000,000/2\pi fC)^2]^{1/2}. \tag{A.36b}$$

We can also state that impedance:

$$Z = (R^2 + X^2).$$

Substituting:

$$Z = [R^2 + (2\pi fL - 1,000,000/2\pi fC)^2]^{1/2}. \tag{A.37}$$

Example. Figure A.24 illustrates a simple ac circuit consisting of resistance (100 Ω), capacitance (2.16 μF), and inductance (400 mH) in series. The frequency is 1020 Hz and the ac emf supply is 20 V. Calculate its impedance, and then calculate the current in the circuit. Use formula (A.37):

$$\begin{aligned} Z &= [10,000 + (2 \times 3.14159 \times 1020 \times 0.4 - 1,000,000/2 \times 3.14159 \times 1020 \times 2.16)^2]^{1/2} \\ &= [10,000 + (2563.53 - 1,000,000/13843.1)^2]^{1/2} \\ &= 2493.3\ \Omega. \end{aligned}$$

To calculate the current in the circuit, again use the variant of Ohm's law:

$$\begin{aligned} I &= E/2493.3 \\ &= 20/2493.3 = 0.00802 \text{ A or } 8.02 \text{ mA}. \end{aligned}$$

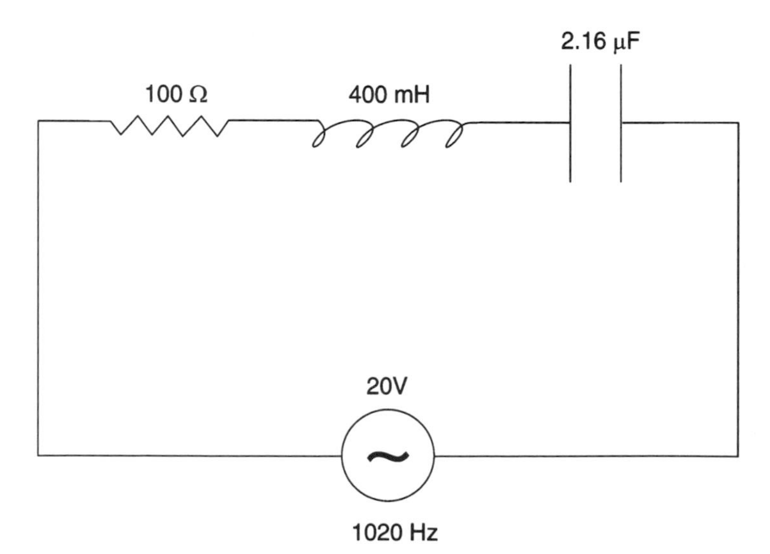

Figure A.24 A simple ac circuit with resistance, capacitance, and inductance in series.

A.9 RESISTANCE IN ac CIRCUITS

In certain situations, ac resistance varies quite widely from the equivalent dc resistance, given the same circuit. For example, the resistance of a coil wound on an iron core where the magnetizing effect demonstrates hysteresis ("the holding back") and resulting *eddy currents* add to the dc resistance. We find that these effects are a function of frequency, the higher the frequency (f), the greater are these effects. Certain comparable losses may occur in the dielectric materials of capacitors, which may have the effect of increasing the apparent resistance of the circuit.

Still more important is the phenomenon called *skin effect*. As we increase frequency of an ac current being transported by wire means, a magnetic field is set up around the wire, penetrating somewhat into the wire itself. Counter currents are set up in the wire as a result of the magnetic field, and as frequency increases field penetration decreases, and the magnitude of the counter currents increases. The net effect is to force the current in the wire to flow nearer the surface of the wire instead of being evenly distributed across the cross section of the wire. Because the actual current flow is now through a smaller area, the apparent ac resistance is considerably greater than its effective dc resistance. When working in the radio frequency (RF) domain, this resistance may be very much greater than dc resistance. However, when working with power line frequencies (i.e., 60 Hz in North America and 50 Hz in many other parts of the world), skin effect is nearly insignificant.

A.10 RESONANCE [Refs. 1 and 4]

As we are aware, the value of inductive reactance and the value of capacitive reactance depend on frequency. When the frequency (f) is increased, inductive reactance increases, and capacitive reaction decreases. We can say that at some frequency, the negative reactance X_C becomes equal and opposite in value to X_L. As a result the reactive component becomes zero, and this is where there is resonance. To determine this frequency we set the combined reactive component equal to zero or:

$$2\pi fL - 1,000,000/2\pi fC = 0.$$

The resonant frequency, f_r, is

$$f_r = 1,000/2\pi(LC)^{1/2}. \tag{A.38}$$

To determine the resonant frequency of a series circuit all we have to know is the value of capacitance and inductance of the circuit.

Example. If the inductance of a particular circuit is 20 mH and the capacitance is 50 nf, what is the resonant frequency? Apply formula (A.38):

$$\begin{aligned} f_r &= 1000/2 \times 3.14159(0.02 \times 50 \times 10^{-9})^{1/2} \\ &= 5,032,991 \text{ Hz or } 5,032,991 \text{ MHz}. \end{aligned}$$

It should be noted that the units of capacitance have been changed to microfarads and of inductance to henrys. These are the units of magnitude in Eq. (A.38).

REFERENCES

1. *Principles of Electricity Applied to Telephone and Telegraph Work*, American Telephone & Telegraph Co., New York, 1961.
2. *IEEE Standard Dictionary of Electrical and Electronics Terms*, 6th ed., IEEE Std. 100-1996, IEEE, New York, 1996.
3. D. Halliday et al., *Fundamentals of Physics—Extended*, 4th ed., Wiley, New York, 1993.
4. H. C. Ohanian, *Physics*, W. W. Norton & Company, New York, 1985.

APPENDIX B

REVIEW OF MATHEMATICS FOR TELECOMMUNICATION APPLICATIONS

B.1 OBJECTIVE AND SCOPE

To derive the full benefit of this text, the reader should have a basic knowledge of algebra, logarithms, and some essentials of trigonometry. To that end we have developed this appendix. The objective is to "bring the reader along," and not to provide an exhaustive primer on basic mathematics. There are four subsections:

1. Introduction;
2. Introductory Algebra;
3. Logarithms to the Base 10; and
4. Essentials of Trigonometry.

B.2 INTRODUCTION

B.2.1 Symbols and Notation

A symbol is commonly used in algebra to represent a quantity. Symbols are also used to indicate a mathematical operations such as $+$, $-$, $\times$, and $\div$. A symbol also may be used to designate an absolute constant. For instance, the speed of light is often denoted by the letter c. We frequently reach into the Greek alphabet, for example, π, which is used to calculate the circumference or area of a circle given its diameter or radius.

Let us say that *notation* is a specific symbol or specific symbols used in a particular procedure or equation. For example, λ is nearly universally used for wavelength; F or f for frequency. Z is used for impedance, and Z_0 for characteristic impedance.

As we said, a symbol in algebra represents a specific quantity. The letter x is the unknown, or x = the *unknown* (quantity). If there is a second unknown, we are apt to call it y. These rules are never hard and fast. For instance, if we are dealing with a geometrical figure with height, length, and width, we would probably use H for height, L for length, and W for width; and we would assign r for radius and d for diameter when dealing with something circular. Angles are often represented by the symbols α and β, and θ is also widely used. These are just more examples of utilizing the Greek alphabet as well as the Roman.

Subscripts. A subscript tells us something about a symbol. For example, P_{dBW} would probably mean "the power expressed in dBW." We might discuss the velocities of two cars. The velocity of car 1 may be expressed as V_1, and the velocity of car 2 as V_2. The use of the subscript allows us to distinguish between the two cars. Subscripts are used widely in the text.

Independent and Dependent Variables. In the selection of symbols to solve a particular problem, we must distinguish between constants and variables specific to the problem at hand. Consider the volume of a circular cone where we keep its height constant (h). The formula for its volume is $V = \pi r^2 h/3$, where r is the radius of the base circle. The height, h, of course, is constant for this problem, and just for this problem alone. It is referred to as a *parameter*. The radius, *r*, is the *independent variable* and *V* is the *dependent variable*.

B.2.2 Function Concept

We remember the equation for power: $P = I^2R$. Let R be 75 Ω; then $P = 75I^2$. When the value of I is known, we can calculate the power. For example, when I = 2 A, the power will be:

$$P = 75 \times 2 \times 2 = 300 \text{ W}.$$

That is to say, that the variable *P* depends on the variable *I*.

In general, we can say that if a variable y depends on another variable x so that for each value of x (in a suitable set), a corresponding value of y is determined. The variable y then is called a function of x. In symbols this is written as:

$$y = f(x).$$

Read this as "*y* is the function of *x*" or just "*y* equals *f* of *x*." Remember here that *f* is not a qualitative symbol but an operative symbol.

Take the conversion formula from absolute temperature to centrigrade: T = 273 + C. Here T = f(C). Another example is noise power. It is a function of absolute temperature and bandwidth. Thus:

$$P_n = -228.6 \text{ dBW} + 10 \log T + 10 \log B_{\text{Hz}}.$$

This is Eq. (9.12) from Chapter 9 of the text. Note the use of an absolute constant, in this case Boltzmann's constant. P_n is the independent variable; T, the noise temperature in kelvins, and B, the bandwidth in Hz, are dependent variables. We could set the bandwidth at 1 MHz. Then $P_n = f(T)$.

B.2.3 Using the Sigma Notation

The Greek letter Σ (capital sigma) indicates summation. The *index of summation* limits the number of items to be summed or added up. The letter *i* is often used for this purpose. Values for *i* are placed above and below sigma. The initial value is placed below, and the final value is placed above the letter sigma. For example, we could have

$$\Sigma w_i = w_1 + w_2 + w_3 + w_4 \ldots + w_9 + w_{10}.$$

Here the initial value is 1 and the final value is 10.

B.3 INTRODUCTORY ALGEBRA

B.3.1 Review of the Laws of Signs

- If we multiply a +factor by another +factor, the product will have a + sign.
- If we divide a +factor by another +factor, the quotient will have a + sign.
- If we multiply a −factor by a +factor, the product will have a − sign (i.e., will be negative).
- Likewise, if we multiply a +factor by a −factor, the product will have a − sign.
- If we divide a +factor by a −factor, the quotient will have a − sign.
- Likewise, if we divide a −factor by a +factor, the quotient will have a −sign.
- If we divide a −factor by a −factor, the quotient will have a + sign.

In other words, $+ \times + = +$; $+/+ = +$; $- \times + = -$; $+ \times - = -$; $-/+ = -$; $+/- = -$; $- \times - = +$; and $-/- = +$.

B.3.2 Conventions with Factors and Parentheses

Two symbols placed together imply multiplication. Each symbol or symbol grouping is called a *factor*. Examples:

xy means x multiplied by y;

$x(y + 1)$ means x multiplied by the quantity $(y + 1)$;

abc means a multiplied by b that is then multiplied by c.

The use of parentheses is vital in algebra. If a parentheses pair has a sign in front of it, that sign is operative on each term inside the parentheses. If the sign is just a +, then consider that each term inside the parentheses is multiplied by +1. Here we just have a flow through.

Example. Simplify $3X + (7X + Y - 10) = 3X + 7X + Y - 10 = 10X + Y - 10$.

Suppose there is a minus sign in front of the parentheses. Assume it is a −1. For the terms inside the parentheses, we change each sign and add.

Consider nearly the same example: Simplify $3X - (7X + Y - 10) = 3X - 7X - Y + 10 = -4X - Y + 10$.

One more example: Simplify $4.5K - (-27.3 + 2.5K) = 4.5K + 27.3 - 2.5K = 2K + 27.3$. We carry this one step further by placing a factor in front of the parentheses.[1]

Example. Simplify: $4X - 5Y(X^2 + 2XY + 10) - 36 = 4X - 5X^2Y - 10XY^2 - 50Y - 36$.

Another example: Simplify: $4Q - 21(5Q^2/3 - 3Q/6 - 4) = 4Q - 105Q^2/3 + 63Q/6 + 84 = 4Q - 35Q^2 + 21Q/2 + 84 = 29Q/2 - 35Q^2 + 84$.

An algebraic expression may have parentheses inside brackets. The rule is to clear the "outside" first. In other words, clear the brackets. Then clear the parentheses in that order.

[1]A *factor* is a value, a symbol that we multiply by. To factor an expression is to break the expression up into components and when these components are multiplied together, we derive the original expression. For example 12 and 5 are factors of 60 because $12 \times 5 = 60$.

Example. Simplify: $(r - 3R) - [(2R - r) + 1] = (r - 3R) - (2R - r) - 1 = r - 3R - 2R + r - 1 = 2r - 5R - 1$.

Fractions. We review the adding and subtracting of fractions in arithmetic. Add 3/4 + 5/8 + 5/12. Remember that we look for the least common denominator (LCD). In this case it is 24. The reason: 24 is divisible by 4, 8, and 12. The three terms are now converted to 24ths or:

$$18/24 + 15/24 + 10/24.$$

Now add 18, 15, and 10, which is 43/24 or 1 19/24.

Another example: Add 1/6 + 1/7 + 1/8. The best we can do here for the LCD is the product of $6 \times 7 \times 8 = 336$. Convert each fraction to 336 or 56/336 + 48/336 + 42/336. Now add 56 + 48 + 42 = 146/336. By inspection we see we can divide the numerator and denominator by 2 and the answer is 73/168.

Subtraction of fractions follows the same procedure, just follow the rules of signs.

Example: Calculate 1/4 – 1/5 + 2/3.

Procedure. Multiply the denominators together for the LCD, which is 60. Convert each fraction to 60ths or 15/60 – 12/60 + 40/60. We now have a common denominator so we can add the numerators: 15 – 12 + 40 = 43/60.

We apply the same procedure when we add/subtract algebraic symbols.

Example. Simplify $1/(X - 4) - 1/(X - 5)$.

Procedure. The LCD is the product $(X-4)(X-5)$. We then have $[(X-5)-(X-4)]/(X-4)(X-5) = -1/(X^2 - 9X + 20)$.

Then multiply the two factors $(X - 4)$ and $(X - 5)$. There are two approaches. Do "long" multiplication as we would do with arithmetic:

$$\begin{array}{rr} & X - 4 \\ & \times\, X - 5 \\ \hline & -5X + 20 \\ & X^2 \;\; - 4X \\ \hline \text{sum} & X^2 \;\; -9X + 20. \end{array}$$

The second approach: $(X - 4) \times (X - 5)$; multiply the leftmost terms (the Xs) and we get X^2; multiply the right sides (–4 and –5) and we get +20; then multiply the means together ($-4X$) and the extremes together ($-5X$) and add $(-4X + -5X) = -9X$. Place together in descending order:

$$X^2 - 9X + 20.$$

There is a grouping we should recognize by inspection in the generic type of $(X^2 - K^2)$, which factors into $(X - K)(X + K)$. Here are several examples: $(X^2 - 1)$, which factors into $(X + 1)(X - 1)$ or $X^2 - 64$, which factors into $(X + 8)(X - 8)$.

Adding and Subtracting Exponents. An *exponent* is a number at the right of and above a symbol. The value assigned to the symbol with this number is called a *power* of the symbol, although *power* is sometimes used in the same sense as *exponent*. If the exponent is a positive integer and X denotes the symbol, then X^n means X if $n = 1$. When $n > 1$ we would have $3^1 = 3$, $3^2 = 9$, $3^3 = 27$, and so on. Note that $X^0 = 1$ if X itself is not zero.

Rules: When we multiply, we add exponents; when we divide, we subtract exponents. For the preceding zero example, we can think of it as $X^2/X^2 = X^0 = 1$. This addition and subtraction can be carried out so long as there is a common base. In this case it was X. For example, $2^3 \times 2^2 = 2^5$. Another example: $X^7/X^5 = X^2$. Because it is division, we subtracted exponents; it had the common base "X."

A negative exponent indicates, in addition to the operations indicated by the numerical value of the exponent, that the quantity is to be made a reciprocal.

Example 1: $X^{-2} = 1/16$.
Example 2: $X^{-3} = 1/X^3$.

Further, when addition is involved, and the numbers have a common exponent, we can just add the base numbers. For example, $3.1 \times 10^{-10} + 1.9 \times 10^{-10} = 5.0 \times 10^{-10}$. We cannot do this if there is not a common base and exponent. The power of ten is used widely throughout the text.

If the exponent is a simple fraction such as $\frac{1}{2}$ or $\frac{1}{3}$, then we are dealing with a root of the symbol or base number. For example, $9^{1/2} = 3$; or $(X^2 + 2X + 1)^{1/2} = X + 1$.

Carry this one step further. Suppose we have $x^{2/3}$. First we square X and then take the cube root of the result. The generalized case is $X^{p/q} = (X^p)^{1/q}$. In other words, we first take the pth power and from that result we take the qth root. These calculations are particularly easy to do with a scientific calculator using the X^y function, where y can even be a decimal such as -3.7.

B.3.3 Simple Linear Algebraic Equations

An equation is a statement of equality between two expressions. Equations are of two types: *identities* and *conditional equations* (or usually simply *equations*). A *conditional equation* is true only for certain values of the variables involved, for example, $x + 2 = 5$ is a true statement only when $x = 3$; and $xy + y - 3 = 0$ is true when $x = 2$ and $y = 1$, and for many other pairs of values of x and y; but for still others it is false.

Equation (9.20) in Chapter 9 of the text is an *identity*. It states:

$$\mathrm{G/T} = \mathrm{G} - 10 \log T,$$

where G is gain and T is noise temperature. Actually the right-hand side of the equation is just a restatement of the left-hand side; it does not tell us anything new. In many cases, identities such as this one are very useful in analysis.

There are various rules for equations. An equation has a right-hand side and a left-hand side. Maintaining equality is paramount. For example, if we add some value to the left side, we must add the same value to the right side. Likewise, if we divide the (entire) left side by a value, we must divide the (entire) right by the same value. As we might imagine, we must carry out similar procedures for subtraction and multiplication.

A linear equation is of the following form: $AX + B = 0$ $(A \neq 0)$. This is an equation with one unknown, X. A will be a fixed quantity, a number; so will B. However, in the parentheses it states that A may not be 0.

Let us practice with some examples. In each case calculate the value of X.

$$X + 5 = 7$$

Clue: We want to have X alone on the left side. To do this we subtract 5 from the left side, but following the rules, we must also subtract 5 from the right side. Thus:

$$X + 5 - 5 = 7 - 5 \quad \text{or}$$
$$X = 2.$$

Another example is

$$3X + 7 = 31.$$

Again, we want X alone. But first we must settle for $3X$. Subtract 7 from both sides of the equation.

$$3X + 7 - 7 = 31 - 7;$$
$$3X = 24.$$

Again, we want X alone. To do this we can divide by 3 (each side).

$$3X/3 = 24/3;$$
$$X = 8.$$

Still another example:

$$z^2 + 1 = 65 \text{ (solve for } z\text{)}.$$

Subtract 1 from each side to get z^2 alone.

$$z^2 + 1 - 1 = 65 - 1;$$
$$z^2 = 64.$$

Take the square root of each side. Thus

$$z = 8.$$

More Complex Equations

Solve for R.

$$0.25(0.54R + 2.45) = 0.24(2.3R - 1.75)$$
$$0.135R + 0.6125 = 0.552R - 0.42$$
$$-0.552R + 0.135R = -0.42 - 0.6125$$
$$-0.417R = -1.0325$$
$$R = 2.476.$$

Another example: Solve for x.

$$\begin{aligned}(x+4)(x-3) &= (x-9)(x-2)\\ x^2-12 &= x^2-11x+18\\ 12x &= 30\\ x &= 2.5.\end{aligned}$$

B.3.4. Quadratic Equations

Quadratic equations will have one term with a square (e.g., X^2) and they take the form:

$$Ax^2 + Bx + C = 0 \qquad (A \neq 0)$$

where A, B, and C are constants (e.g., numbers). A quadratic equation should always be set to 0 before a solution is attempted. For instance, if we have an equation that is $2x^2 + 3x = -21$, convert this equation to: $2x^2 + 3x + 21 = 0$.

We will discuss two methods of solving a quadratic equation: by factoring and by the quadratic formula.

Factoring. Suppose we have the simple relation: $x^2 - 1 = 0$. We remember from the preceding that this factors into $(x-1)(x+1) = 0$. This being the case, at least one of the factors must equal 0. If this is not understood, realize that there is no other way for the equation statement to be true. Keep in mind that anything multiplied by 0 will be 0. So there are two solutions to the equation:

$$x - 1 = 0, \text{thus } x = 1 \qquad \text{or } x + 1 = 0 \text{ and } x = -1$$

Proof that these are correct answers is by substituting them in the equation. Solve for x in this example:

$$x^2 - 100x + 2400 = 0.$$

This factors into: $(x-40)(x-60) = 0$. We now have two factors: $x - 40$ and $x - 60$, whose product is 0. This means that we must have either: $x - 40 = 0$, where $x = 40$ or $x - 60 = 0$ and in this case $x = 60$. We can check our results by substitution that either of these values satisfies the equation.

Another example: Solve for x in $(x-3)(x-2) = 12$. Multiply the factors: $x^2 - 5x + 6$, then $x^2 - 5x + 6 = 12$. Subtract 12 from both sides of the equation so that we set the left-hand side equal to 0. Thus:

$$\begin{aligned}&x^2 - 5x - 6 = 0 \text{ factors into } (x-6)(x+1) = 0\\ &\text{Then } x - 6 = 0, x = 6 \text{ or } x + 1 = 0, x = -1.\end{aligned}$$

Quadratic Formula. This formula may be used on the conventional quadratic equation in the generic form of

$$Ax^2 + Bx + C = 0 \qquad (A \neq 0).$$

The value of x is solved by simply manipulating the constants A, B, and C. The quadratic formula is stated as follows:

$$x = [-B \pm (B^2 - 4AC)^{1/2}]/2A$$

or, rewritten with the radical sign:

$$x = \frac{-B\sqrt{\pm B^2 - 4AC}}{2A}.$$

Just as we did with the factoring method, the quadratic formula will produce two roots (two answers): one with the plus before the radical sign and one with the minus before the radical sign.

Example 1. Solve for x: $3x^2 - 2x - 5 = 0$. Here $A = 3$, $B = -2$ and $C = -5$. Apply the quadratic formula.

$$x = (+2 \pm 4 + 60)/6$$

The first possibility is $(+2 + 64)/6 = 10/6$; the second possibility is $(+2 - 8)/6 = -1$. The quadratic formula will not handle the square root of a negative number. The square root of a negative number can usually be factored down to $(-1)^{1/2}$, which, by definition, is the imaginary number i, and is beyond the scope of this appendix.

Example 2. Solve for E: $E^2 - 3E - 2 = 0$. $A = 1$, $B = -3$ and $C = -2$. Thus $E = (+3 \pm 9 + 8)/2$. The first possibility is $(3 + 4.123)/2$ and the second possibility is $(3 - 4.123)/2$. Thus $E = 3.562$ or -0.562.

B.3.5 Solving Two Simultaneous Linear Equations with Two Unknowns

There are two methods of solving two simultaneous equations:

1. The graphical method, where both equations are plotted and the intersection of the line derived is the common solution; and
2. The algebraic solution.

We will concentrate on the algebraic solution. There are two approaches to solving two simultaneous equations by the algebraic solution:

1. Elimination; and
2. Substitution.

Elimination Method. With this method we manipulate one of the equations such that when the two equations are either added or subtracted, one of the unknowns is eliminated. We then solve for the other unknown. The solution is then substituted in one of the original equations and we solve for the other unknown.

Example:

$$2x + 3y - 8 = 0$$
$$4x - 5y + 6 = 0.$$

Multiply each term by 2 in the upper equation, and we derive the following new equation:

$$4x + 6y - 16 = 0.$$

Place the second equation directly below this new equation, and subtract.

$$4x + 6y - 16 = 0$$
$$4x - 5y + 6 = 0.$$

If we subtract the lower equation from the upper, we eliminate the $4x$ term. Now solve for y.

$$+11y - 22 = 0$$
$$11y = 22$$
$$y = 2.$$

Substitute $y = 2$ in the original upper equation. Then $2x + 6 - 8 = 0$, $2x = 2$ and $x = 1$. So the solution of these equations is $x = 1$ and $y = 2$. Check the solutions by substituting these values into the two original simultaneous equations.

Example:

$$3x - 2y - 5 = 0$$
$$6x + y + 12 = 0.$$

There are several possibilities to eliminate one of the unknowns. This time let us multiply each term in the lower equation by two and we get:

$$12x + 2y + 24 = 0.$$

Place this new equation below the original upper equation:

$$3x - 2y - 5 = 0$$
$$12x + 2y + 24 = 0.$$

Add the two equations and we get $15x + 19 = 0$. Solve for x.

$$15x = -19$$
$$x = -19/15.$$

Substitute this value in the upper equation and solve for y.

$$3(-19/15) - 2y - 5 = 0$$
$$-57/15 - 75/15 = 2y$$
$$2y = -132/15 \text{ and } y = -66/15 \text{ or } -22/5.$$

Substitution Method. Select one of the two simultaneous equations and solve for one of the unknowns in terms of the other. **Example:** (repeating the first example from above)

$$2x + 3y - 8 = 0$$
$$4x - 5y + 6 = 0.$$

We can select either equation. Select the first equation. Then:

$$2x = 8 - 3y$$
$$x = (8 - 3y)/2.$$

Substitute this value for x in the second equation.

$$4(8 - 3y)/2 - 5y + 6 = 0$$
$$16 - 6y - 5y + 6 = 0$$
$$-11y = -22$$
$$y = 2.$$

B.4 LOGARITHMS TO THE BASE 10

B.4.1 Definition of Logarithm

If b is a positive number different from 1, the *logarithm* of the number y, written $\log_a y$, is defined as follows: *if* $ax = y$, *then* x *is a logarithm of* y *to the base* a, *and we write*:

$$\log_a y = x.$$

This shows, therefore, that *a logarithm is an exponent—the exponent to which the base is raised to yield the number*. The expression $\log_a y$ is read: "logarithm of y to the base a." The two equations $a^x = y$ and $\log_a y = x$ are two different ways of expressing the relationship between the numbers x, y, and a. The first equation is in the exponential form and the second is in the logarithmic form. Thus $2^6 = 64$ is equivalent to $\log_2 64 = 6$. Likewise, the statement $\log_{16}(1/4) = -1/2$ implies $16^{-1/2} = 1/4$. These concepts should be thoroughly understood before proceeding.

B.4.2 Laws of Logarithms

In Section B.3 we discussed the laws of adding and subtracting exponents. From these laws we can derive the laws of logarithms. Let us say that the generalized base of a logarithm is a, which is positive, and that x and y are real numbers. Here we mean they are not imaginary numbers (i.e., based on the square root of -1).

Note that a can be any positive number. However, we concentrate on $a = 10$, that is, on logarithms to the base 10. The scientific calculator should be used to obtain the logarithm by using the "log" button. There will also probably be an "ln" button. This button is used to obtain logarithms to the natural base, where $a = 2.71828183+$.

Law 1. *The logarithm of the product of two numbers equals the sum of the logarithms of the factors*. That is:

$$\log_a xy = \log_a x + \log_a y.$$

Law 2. *The logarithm of the quotient of two numbers equals the logarithm of the dividend minus the logarithm of the divisor*. That is:

Table B.1 Selected Powers of Ten

Power of 10	Number	Logarithm of Number	Value of Logarithm
10^4	10,000	log10,000	4
10^3	1000	log1000	3
10^2	100	log100	2
10^1	10	log10	1
10^0	1	log1	0
10^{-1}	0.1	log0.1	−1 or 9 − 10
10^{-2}	0.01	log0.01	−2 or 8 − 10
10^{-3}	0.001	log0.001	−3 or 7 − 10
10^{-4}	0.0001	log0.0001	−4 or 6 − 10

$$\log_a x/y = \log_a x - \log_a y.$$

Law 3. *The logarithm of the nth power of a number equals n times the logarithm of the number.* That is:

$$\log_a x^n = n \log_a x.$$

Law 4. *The logarithm of the pth root of a number is equal to the logarithm of the number divided by n.* That is:

$$\log(x)^{1/p} = 1/p \log_a x.$$

Remember that if $x = 1$, $\log_a x = 0$. Here is an exercise. Express $\log_{10}(38)^{1/2}(60)/(29)^3 = 1/2 \log_{10}(38) + \log_{10}(60) - 3 \log_{10}(29)$.

The logarithm of a number has two components: its *characteristic* and its *mantissa*. The characteristic is an integer and the mantissa is a decimal. If the number in question is 10 or its multiple, the logarithm has a characteristic only, and its mantissa is .000000++. Consider Table B.1 containing selected the powers of 10.

A scientific calculator gives both the characteristic and the mantissa when a number is entered and we press the "log" button. On most calculators, just enter the number then press the "log" button. Its logarithm (to the base 10) will then appear in the display.

Table B.2 gives ten numbers and their equivalent logarithms. One of the logarithms is blatantly in error. Identify it.

Deriving one type of logarithm may be tricky. This group consists of decimals, in other words, numbers less than one (< 1). First check column 4 of Table B.1.

Find the logarithm of 0.00783. In scientific notation we can express the number as 7.83×10^{-3}. Therefore, $\log 0.00783 = \log(7.83 \times 10^{-3}) = \log(7.83) + \log(10^{-3})$.

Suppose we are given the logarithm of a number, assuming the base is 10, and we wish to find the number that generated that logarithm. This is shown on the calculator, usually on the second level, $\log^{-1}$. Sometimes in the literature it is indicated as the *antilog*.

Example. Given the logarithm, find its corresponding number. Use a scientific calculator.

Table B.2 Selected Numbers and their Logarithms

Number	Logarithm to the Base 10
763	2.8825
47	1.6721
14142	2.9637
0.112	−0.9508
167667.2	5.2244
0.000343	−3.4647
3.14159	0.4971
5.616×10^{-4}	−3.2506
1/767	−2.8848
10^{24}	24.0

$$\log^{-1} 1.3010 = 19.9986.$$

Enter the number in the calculator so it appears on the display. Press 2nd or "shift." Press the "log" button. Now 19.9986 appears on the display. The particular calculator we are using indicates that this is the 10^x function, which uses the same button as the "log" function, but on the "shift" or second level of the calculator.

$$\log^{-1} 2.8710 = 743$$
$$\log^{-1} -1.50445 = 0.0313$$
$$\log^{-1} 1.1139 = 13.$$

There are a number of mathematical calculations that are either very difficult to do without the application of logarithms, or nearly impossible. One such operation is to calculate the cube root (or 4th, 5th to the *n*th root).

Example. Calculate the cube root of 9751.

Enter 9751 in the calculator and press the "log" button to get the logarithm of the number. Divide the logarithm by 3 and get the antilog of the result.

$$\log 9751 = 3.9890$$
$$3.9890/3 = 1.32968$$
$$\text{antilog} 1.32968 = 21.364.$$

B.5 ESSENTIALS OF TRIGONOMETRY

B.5.1 Definitions of Trigonometric Functions

Figure B.1 is a right triangle. Remember that a right triangle has one angle that is 90° by definition.

The basic trigonometric functions are defined as follows:

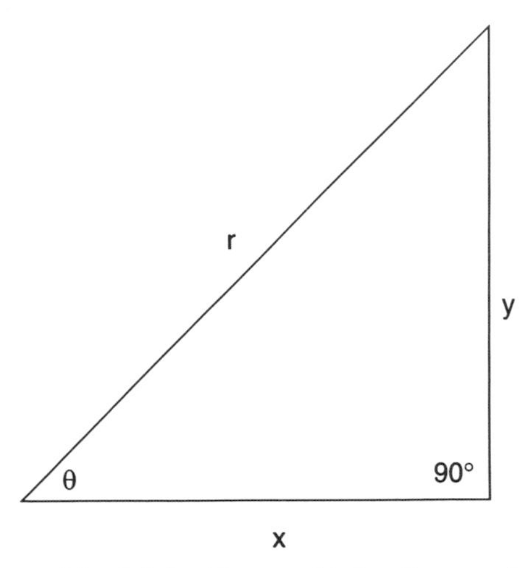

Figure B.1 A right triangle used for defining trigonometric functions; r = hypotenuse; y = opposite side (from the angle θ), and x = the adjacent side.

$$\begin{array}{ll} \text{sine } \theta = y/r & \text{cosine } \theta = x/r \\ \text{tangent } \theta = y/x & \text{cotangent } \theta = x/y \\ \text{secant } \theta = r/x & \text{cosecant } \theta = r/y \end{array}$$

The following are some initial trigonometric relationships:

$$\sin\theta = 1/\operatorname{cosec}\theta \quad \cos\theta = 1/\sec\theta \quad \tan\theta = 1/\cot\theta$$

The triangle in Figure B.1 has a 90° angle and two acute angles. We denominated one of the acute angles θ. Now let us call those two acute angles, A and B. First rule: angle A + angle B + 90° = 180°. In fact, this rule will hold for any triangle; it does not necessarily have to be a right triangle.

The following are useful relationships for a right triangle with acute angles A and B:

$$\begin{array}{ll} \sin A = \cos B & \cot A = \tan B \\ \cos A = \sin B & \tan A = \cot B \\ \sec A = \csc B & \csc A = \sec B \end{array}$$

Using these simple definitions, we can derive: $\sin\theta/\cos\theta = \tan\theta$. Likewise, $\cos\theta/\sin\theta = \cot\theta$.

From Figure B.1 we may remember the Pythagorean relationship:

$$x^2 + y^2 = r^2$$

or, the square of the hypotenuse = sum of the squares of the other two sides. We divide all terms by r^2 and we have:

$$x^2/r^2 + y^2/r^2 = 1.$$

From the basic trigonometric functions we can substitute:

$$\sin^2 \theta + \cos^2 \theta = 1.$$

If both terms on the left-hand side of the preceding equation are divided by $\cos^2 \theta$, we then have:

$$1 + \tan^2 \theta = \sec^2 \theta.$$

In a similar manner, if we divide the two left-hand terms by $\sin^2 \theta$, we have:

$$1 + \cot^2 \theta = \csc^2 \theta.$$

The preceding relationships are called *fundamental trigonometric identities.*

There are three common acute angles that are used repeatedly in geometry and trigonometry. These are 30°, 45°, and 60°. If we know any one of these angles in a right triangle, the other angles are also known. Remember the sum of the three angles is 180°. Of course, with a scientific calculator, the trigonometric functions are easy to obtain. With many scientific calculators, there will be only buttons for three trigonometric functions: sin, cos, and tan. Using the relationships previously provided, sec, csc, and cot can be simply calculated using the inverse key ($1/X$ button).

The following are some algebraic exercises using trigonometric functions. Sometimes we will be given an angle measured in radians, generally related to π. There are 2π radians in 360°; π radians in 180°, $\pi/2$ radians = 90°, $\pi/4$ radians = 45°, and so forth.

Evaluate:

$$\sin 30^\circ + 3\tan 60^\circ - \cot 45^\circ = ?$$

$$0.50 + 5.196 - 1.0 = 4.696.$$

Evaluate:

$$2\tan \pi/6 - 3\sec\pi/3 + 4\csc\pi/4 = ?$$

$$\pi/6 = 30^\circ \qquad \pi/3 = 60^\circ \qquad \text{and } \pi/4 = 45^\circ$$

$$1.1547 - 6 + 5.6568 = 0.8115.$$

B.5.2 Trigonometric Function Values for Angles Greater than 90°

The standard graph based on rectangular coordinates is broken up into four quadrants around the origin, as illustrated in Figure B.2. Some of the trigonometric function values will be negative for angles greater than 90°. Guidance for the assignment of either positive or negative values may be found in Figure B.3. Here only positive values are shown. For all other quadrants, negative values must be assigned.

Most scientific calculators will assign the proper sign for angles greater than 90°. Most trig function tables and older scientific calculators required that the angle be ≤ 90°. When such a situation arises, we subtract the angle from either 180° or 360°, and we assign the proper sign based on Figure B.3.

Examples and Discussion. Obtain values of the indicated trigonometric functions and angles.

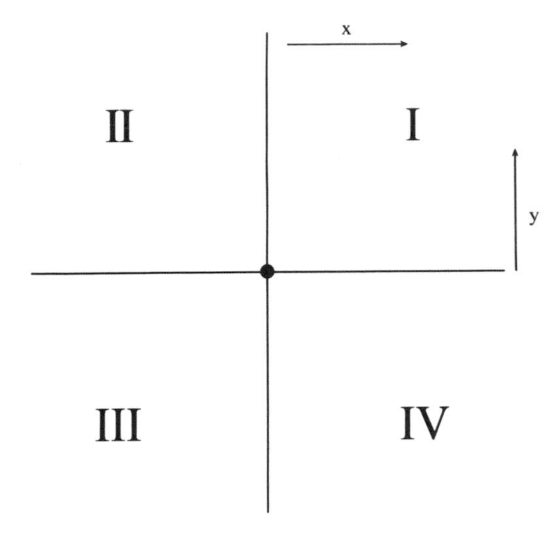

x = abscissa
y = ordinate
center = origin

Figure B.2 A typical graph for plotting with rectangular coordinates showing the four quadrants.

$$\cos 120° = -0.5 \text{ or } \cos(180 - 120°) = \cos 60° = 0.5.$$

Apply proper sign from Figure B.3 or $\cos 120° = -0.5$.

$$\tan 200° = 0.3640 \qquad \text{or } \tan(200 - 180°) = \tan 20° = 0.3640;$$

the sign is positive from Figure B.3.

Given the Trigonometric Function Value, Find the Equivalent Angle. Find the value for θ for the following:

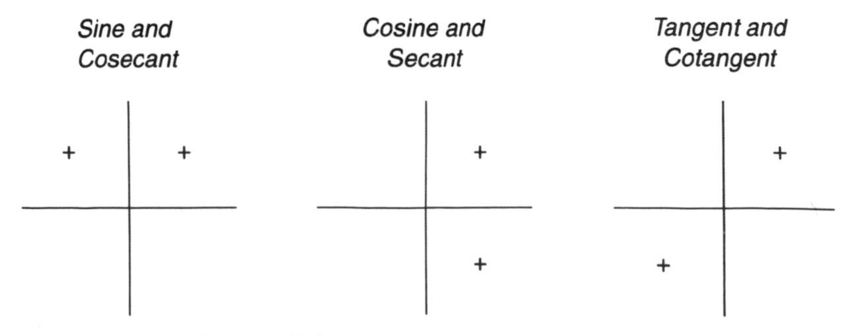

Figure B.3 Quadrant sign diagram. Only positive signs are shown. All other quadrants require negative values.

$$\sin\theta = 0.2952.$$

Enter 0.2952 in the calculator display, press 2nd or "shift" and then the sin key, and we get 17.17°. On many calculators, just above the sin button or key one will find $\sin^{-1}$. This means that given the sin of the angle, find the angle. The reader should realize there is also a valid value in the second quadrant, namely, 180 – 17.17° = 162.83°.

$$\cos\theta = -0.8654.$$

Enter −0.8654 on the calculator display. Press 2nd or "shift" and then depress the cos key or button. We get $\theta = 149.93°$.

$$\tan\theta = 1.6055.$$

Find the third quadrant value (see Figure B.3).

$$\theta = 58.08° \text{ in the first quadrant, } 180 + 58.08° = 238.08°, \text{ in the third quadrant.}$$

Appendix B is based on the author's experience and Refs. 1, 2, and 3.

REFERENCES

1. G. James and R. C. James, *Mathematics Dictionary*, 3rd ed., D. Van Nostrand & Co., Princeton, NJ, 1968.
2. W. R. Van Voorhis and E. E. Haskins, *Basic Mathematics for Engineers and Science*, Prentice-Hall, Englewood Cliffs, NJ, 1952.
3. L. A. Kline and J. Clark, *Explorations in College Algebra*, Wiley, New York, 1998.

APPENDIX C

LEARNING DECIBELS AND THEIR APPLICATIONS

C.1 LEARNING DECIBEL BASICS

When working in the several disciplines of telecommunications, a clear understanding of the decibel (dB) is mandatory. The objective of this appendix is to facilitate this understanding and to encourage the reader to take advantage of this useful tool.

The decibel relates to a ratio of two electrical quantities such as watts, volts, and amperes. If we pass a signal through some device, it will suffer a loss or achieve a gain. Such a device may be an attenuator, amplifier, mixer, transmission line, antenna, subscriber loop, trunk, or a telephone switch, among others. To simplify matters, let's call this generic device a *network*, which has an *input* port and an *output* port, as shown:

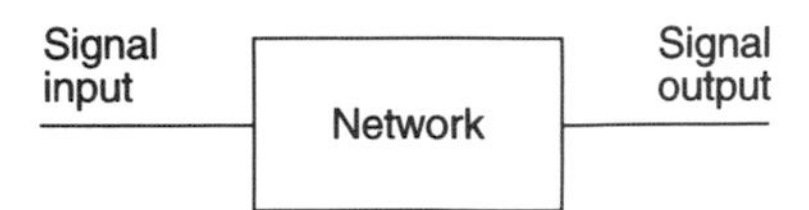

The input and output can be characterized by a signal level, which can be measured in either watts (W), amperes (A), or volts (V). The decibel is a useful tool to compare input-to-output levels or vice versa. Certainly we can say that if the output level is greater than the input level, the device displays a gain. The signal has been amplified. If the output has a lower level than the input, the network displays a loss.

In our discussion we will indicate a gain with a positive sign (+) such as +3 dB, +11 dB, +37 dB; and a loss with a negative sign (−): −3 dB, −11 dB, −43 dB.

At the outset it will be more convenient to use the same unit at the output of a network as at the input, such as watts. If we use watts, for example, it is watts or any of its metric derivatives. Remember:

$$\begin{aligned} 1\ \text{W} &= 1000\ \text{milliwatts (mW)} \\ 1\ \text{W} &= 1,000,000\ (1 \times 10^6)\ \text{microwatts } (\mu\text{W}) \\ 1\ \text{W} &= 0.001\ \text{kilowatts (kW)} \\ 1000\ \text{mW} &= 1\ \text{W} \\ 1\ \text{kW} &= 1000\ \text{W} \end{aligned}$$

We will start off in the power domain (watts are in the power domain, so are milliwatts; volts and amperes are not). We will deal with volts and amperes later. Again, the decibel expresses a ratio. In the power domain (e.g., level is measured in watts or milliwatts), the decibel value of such a ratio is 10 × logarithm of the ratio.

Consider this network:

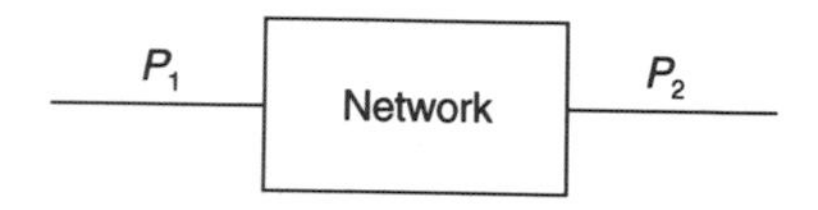

We are concerned about the ratio of P_1/P_2 or vice versa. Algebraically we express the decibel by this formula:

$$\text{dB value} = 10\log(P_1/P_2) \text{ or } 10\log(P_2/P_1). \tag{C1.1}$$

Some readers may feel apprehensive about logarithms. The logarithm (log) used here is to the number base 10. A logarithm is an exponent. In our case it is the exponent of the number 10 such as:

$10^0 = 1$	the log is 0
$10^1 = 10$	the log is 1
$10^2 = 100$	the log is 2
$10^3 = 1000$	the log is 3
$10^4 = 10,000$	the log is 4, etc.

For numbers less than 1, we use decimal values, so

$10^0 = 1$	the log is 0
$10^{-1} = 0.1$	the log is -1
$10^{-2} = 0.01$	the log is -2
$10^{-3} = 0.001$	the log is -3
$10^{-4} = 0.0001$	the log is -4, etc.

Let us now express the decibel values of the same numbers:

$10^0 = 1$	$\log = 0$	dB value $= 10\log 1 = 10 \times 0 = 0$ dB
$10^1 = 10$	$\log = 1$	dB value $= 10\log 10 = 10 \times 1 = 10$ dB
$10^2 = 100$	$\log = 2$	dB value $= 10\log 100 = 10 \times 2 = 20$ dB
$10^3 = 1000$	$\log = 3$	dB value $= 10\log 1000 = 10 \times 3 = 30$ dB
$10^4 = 10000$	$\log = 4$	dB value $= 10\log 10000 = 10 \times 4 = 40$ dB etc.
$10^{-1} = 0.1$	$\log = -1$	dB value $= 10\log .1 = 10 \times -1 = -10$ dB
$10^{-2} = 0.01$	$\log = -2$	dB value $= 10\log .01 = 10 \times -2 = -20$ dB
$10^{-3} = 0.001$	$\log = -3$	dB value $= 10\log .001 = 10 \times -3 = -30$ dB
$10^{-4} = 0.0001$	$\log = -4$	dB value $= 10\log .0001 = 10 \times -4 = -40$ dB, etc.

We now have learned how to handle power ratios of 10, 100, 1000, and so on, and 0.1, 0.01, 0.001, and so on. These, of course, lead to dB values of +10 dB, +20 dB, and +30 dB; −10 dB, −20 dB, −30 dB, and so on. The next step we will take is to learn to derive dB values for power ratios that lie in between 1 and 10, 10 and 100, 0.1 and 0.01, and so on.

One excellent recourse is the scientific calculator. Here we apply a formula (C1.1). For example, let us deal with the following situation:

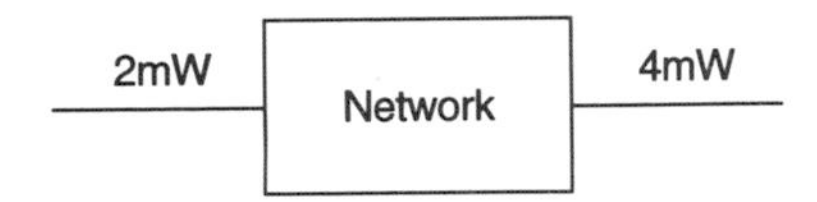

Because the output of this network is greater than the input, the network has a gain. Keep in mind we are in the power domain; we are dealing with mW. Thus:

$$\text{dB value} = 10 \log 4/2 = 10 \log 2 = 10 \times 0.3010 = +3.01 \text{ dB}.$$

We usually round-off this dB value to +3 dB. If we were to do this on our scientific calculator, we enter 2 and press the log button. The value 0.3010−−− appears on the display. We then multiply (×) this value by 10, arriving at the +3.010 dB value.

This relationship should be memorized. The amplifying network has a 3-dB gain because the output power was double the input power (i.e., the output is twice as great as the input).

For the immediately following discussion, we are going to show that under many situations a scientific calculator is not needed and one can carry out these calculations in his or her head. We learned the 3-dB rule. We learned the +10, +20, +30 dB; −10, −20, −30 (etc.) rules. One should be aware that with the 3-dB rule, there is a small error that occurs two places to the right of the decimal point. It is so small that it is hard to measure.

With the 3-dB rule, multiples of 3 are easy. If we have power ratios of 2, 4, and 8, we know that the equivalent (approximate) dB values are +3 dB, +6 dB, and +9 dB, respectively. Let us take the +9 dB as an example problem. A network has an input of 6 mW and a gain of +9 dB. What power level in mW would we expect to measure at the output port?

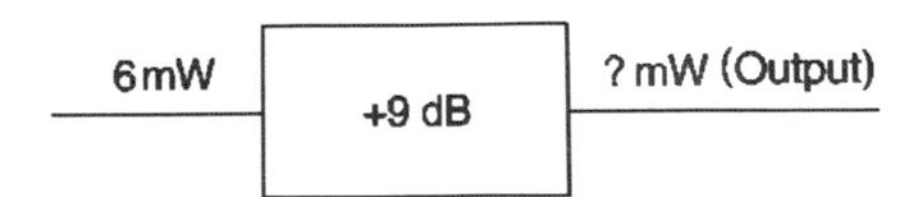

One thing that is convenient about dBs is that when we have networks in series, each with a loss or gain given in dB, we can simply sum the values algebraically. Likewise, we can do the converse: We can break down a network into hypothetical networks in series, so long as the algebraic sum in dB of the gain/loss of each network making up the whole is the same as that of the original network. We have a good example with the preceding network displaying a gain of +9-dB. Obviously 3 × 3 = 9. We break down the +9-dB network into three networks in series, each with a gain of +3 dB. This is shown in the following diagram:

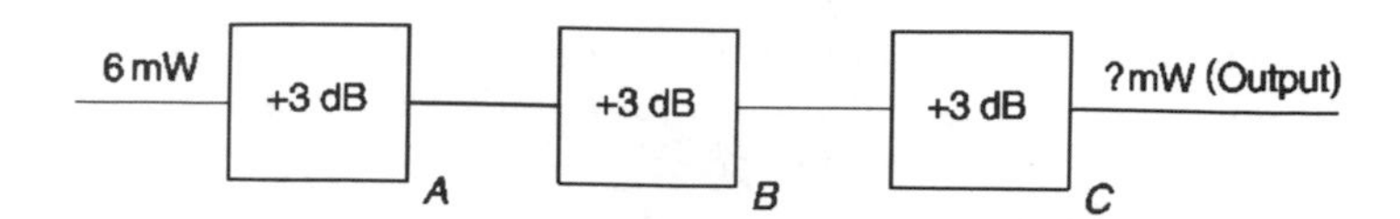

We should be able to do this now by inspection. Remember that +3 dB is double the power; the power at the output of a network with +3-dB gain has 2× the power level at the input. Obviously the output of the first network is 12 mW (point A above). The input to the second network is now 12 mW and this network again doubles the power. The power level at point B, the output of the second network, is 24 mW. The third network—double the power still again. The power level at point C is 48mW.

Thus we see that a network with an input of 6 mW and a 9-dB gain, will have an output of 48 mW. It multiplied the input by 8 times (8 × 6 = 48). That is what a 9-dB gain does. Let us remember: +3 dB is a two-times multiplier; +6 dB is a four-times multiplier, and +9 dB is an eight-times multiplier.

Let us carry this thinking one step still further. We now know how to handle 3 dB, whether + or −, and 10 dB (+ or −), and all the multiples of 10 such as 100,000 and 0.000001. Here is a simple network. Let us see what we can do with it.

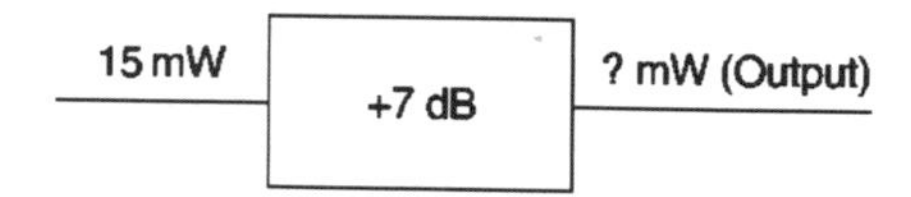

We can break this down into two networks using dB values that are familiar to us:

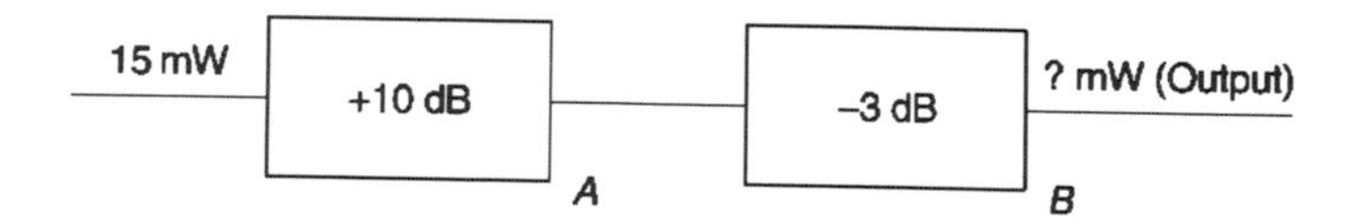

If we algebraically sum the +10 dB and the −3 dB of the two networks in series shown above, the result is +7 dB, which is the gain of the network in question. We have just restated it another way. Let us see what we have here. The first network multiplies its input by 10 times (+10 dB). The result is 15 × 10 or 150 mW. This is the value of the level at A. The second network has a 3-dB loss, which drops its input level in half. The input is 150 mW and the output of the second network is 150 × 0.5, or 75 mW.

This thinking can be applied to nearly all dB values except those ending with a 2, 5, or 8. Even these values can be computed without a calculator, but with some increase in error. We encourage the use of a scientific calculator, which can provide much more accurate results, from 5 to 8 decimal places.

Consider the following problem:

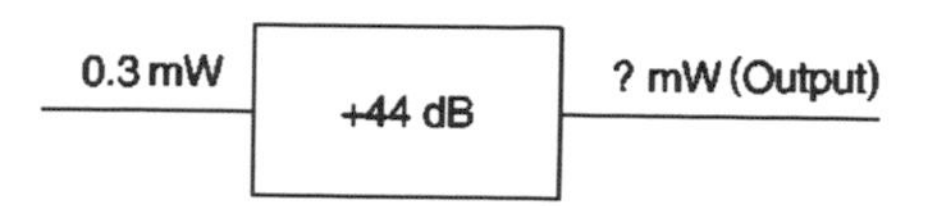

This can be broken down as follows:

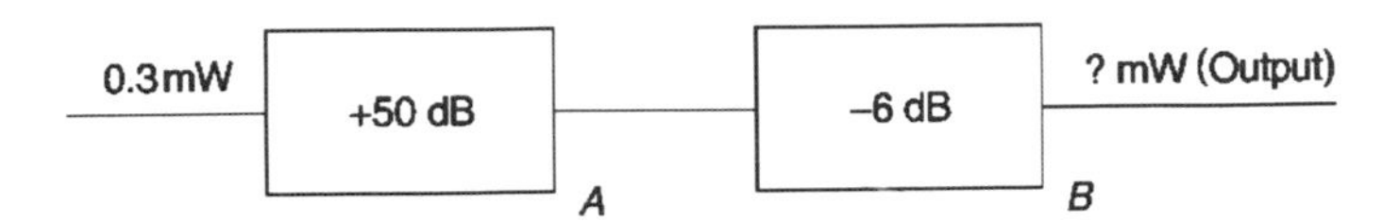

Remember that +50 dB is a multiplier of 10^5 and −6 dB is a loss that drops the power to one quarter of the input to that second network. Now the input to the first network is 0.3 mW and so the output of the first network (A) is 0.3 mW × 100,000 or 30,000 mW (30 w). The output of the second network (B) is one-quarter of that value (i.e., −6 dB), or 7500 mW.

Now we will do a practice problem for a number of networks in series, each with its own gain or loss given in dB. The idea is to show how we can combine these several networks into an equivalent single network regarding gain or loss. We are often faced with such a problem in the real world. Remember, we add the dB values in each network algebraically.

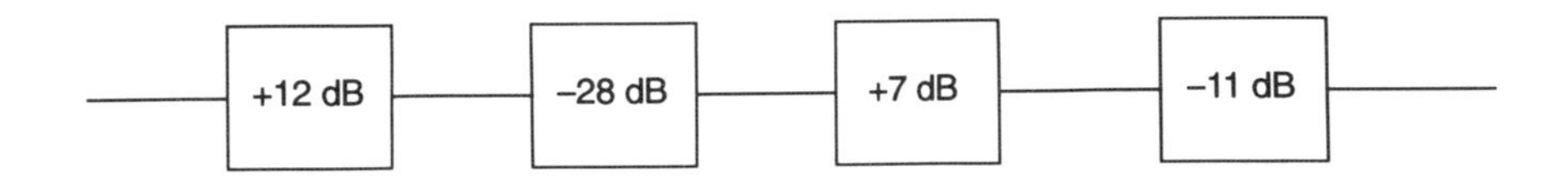

Look what happens when we combine these four networks into one equivalent network. We just sum: +12 − 28 + 7 − 11 = −20, and −20 dB is a number we can readily handle. Thus the equivalent network looks like the following:

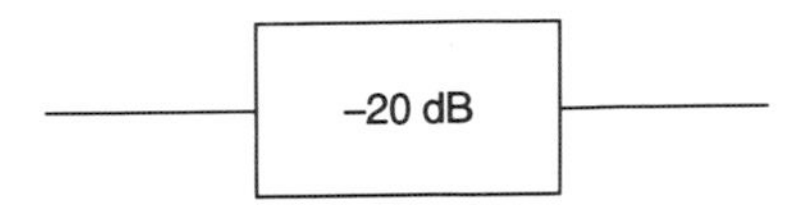

To see really how well you can handle dBs, the instructor might pose a difficult problem with several networks in series. The output power of the last network will be given and the instructor will ask the input power to the first network. Let us try one like that so the instructor will not stump us.

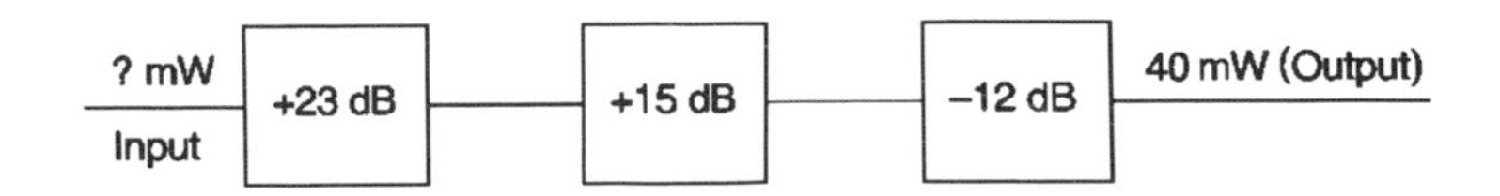

First sum the values to have an equivalent single network: +23 + 15 − 12 = +26 dB. Thus:

? mW | +26 dB | 40 mW (Output)
Input

We first must learn to ask ourselves: *Is the input greater or smaller than the output*? This network has gain, thus we know that the input must be smaller than the output. By how much? It is smaller by 26 dB. What is the numeric value of 26 dB? Remember, 20 dB is 100; 23 dB is 200, and 26 dB is 400. So the input is 1/400 of the output or 40/400 (mW) = 0.1 mW.

C.2 dBm and dBW

These are the first derived decibel units that we will learn. They are probably the most important. The dBm is also a ratio. It is a decibel value related to one milliwatt (1 mW). The dBW is a decibel value related to one watt (1 W). Remember the little m in dBm refers to milliwatt and the big W in dBW refers to watt.

The values dBm and dBW are measures of *real* levels. But first we should write the familiar dB formulas for dBm and dBW:

$$\text{Value (dBm)} = 10 \log P_1/(1 \text{ mW})$$
$$\text{Value (dBW)} = 10 \log P_1/(1 \text{ W}).$$

Here are a few good relationships to fix in our memories:

$$1 \text{ mW} = 0 \text{ dBm} \quad \text{(by definition)}$$
$$1 \text{ W} = 0 \text{ dBW} \quad \text{(by definition)}$$
$$+30 \text{ dBm} = 0 \text{ dBW} = 1 \text{ W}$$
$$-30 \text{ dBW} = 0 \text{ dBm} = 1 \text{ mW}.$$

Who will hazard a guess what +3 dBm is in mW? Of course, it is 3 dB greater than 0 dBm. Therefore it must be 2 mW. Of course, +6 dBm is 4 mW, and −3 dBm is half of zero dBm or 0.5 mW. A table is often helpful for the powers of ten:

$$1 \text{ mW} = 10^0 \text{ mW} = 0 \text{ dBm}$$
$$10 \text{ mW} = 10^1 \text{ mW} = +10 \text{ dBm}$$
$$100 \text{ mW} = 10^2 \text{ mW} = +20 \text{ dBm}$$
$$1000 \text{ mW} = 10^3 \text{ mW} = +30 \text{ dBm} = 0 \text{ dBW}$$
$$10 \text{ W} = 10^4 \text{ mW} = +40 \text{ dBm} = +10 \text{ dBW(etc.)}.$$

Likewise:

$$0.1 \text{ mW} = 10^{-1} = -10 \text{ dBm}$$
$$0.01 \text{ mW} = 10^{-2} = -20 \text{ dBm}$$
$$0.001 \text{ mW} = 10^{-3} = -30 \text{ dBm}$$
$$0.0001 \text{ mW} = 10^{-4} = -40 \text{ dBm}.$$

Once we have a grasp of dBm and dBW, we will find it easier to work problems with networks in series. We now will give some examples.

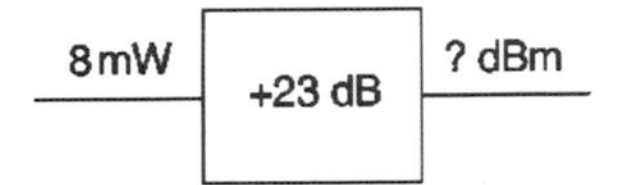

First we convert the input, 8 mW to dBm. Look how simple it is: 2 mW = +3 dBm, 4 mW = +6 dBm, and 8 mW = +9 dBm. Now watch this! To get the answer, the power level at the output is +9 dBm +23 dB = +32 dBm.

Another problem will be helpful. In this case the unknown will be the input to a network.

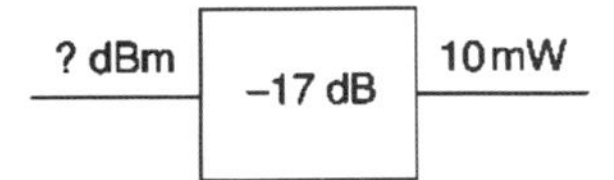

In each case like this we ask ourselves, is the output greater than the input? Because the network is lossy, the input is 17 dB greater than the output. Convert the output to dBm. It is +10 dBm. The input is 17 dB greater, or +27 dBm. We should also be able to say: "that's half a watt." Remember, +30 dBm = 1 W = 0 dBW. Then +27 dBm ("3 dB down") is half that value.

Several exercises are in order. The answers appear after the four exercises.

Exercise 1*a*.

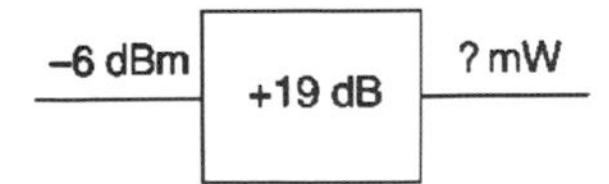

Exercise 1*b*.

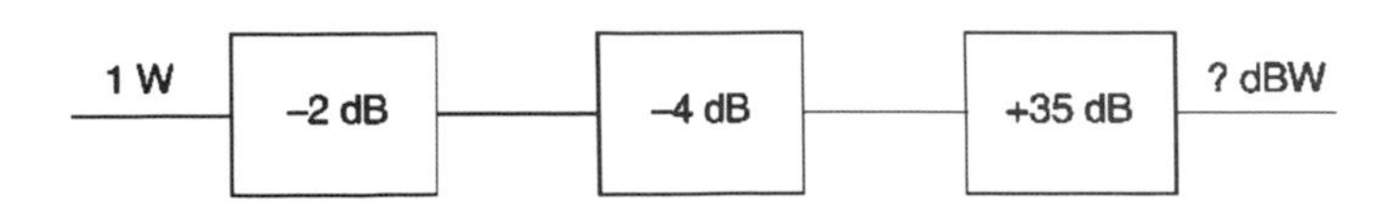

Exercise 1*c*.

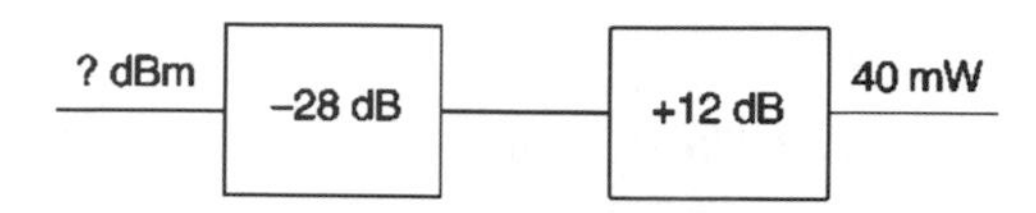

Exercise 1*d*.

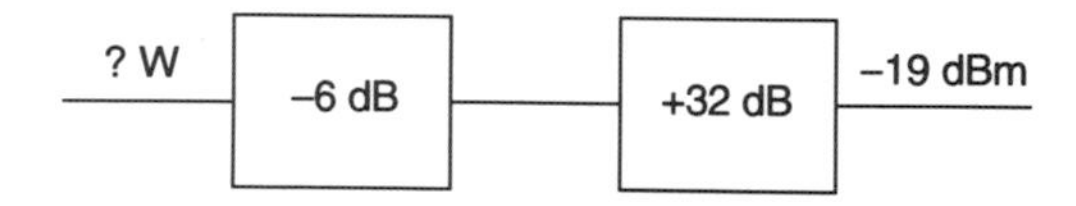

(*Answers*: 1*a*: +13 dBm = 20 mW; 1*b*: +29 dBW, 1*c*: +32 dBm, and 1*d*: +7 dBm = 0.005 W).

C.3 VOLUME UNIT (VU)

The VU is the conventional unit for measurement of speech level. A VU can be related to a dBm only with a sinusoidal tone (a simple tone of one frequency) between 35 Hz and 10,000 Hz. The following relationship will be helpful:

Power level in dBm = VU − 1.4 dB (for complex audio signals).

A complex audio signal is an audio signal composed of many sine waves (sinusoidal tones) or, if you will, many tones and their harmonics.

One might ask: If the level reading on a broadcaster's program channel is −11 VU, what would the equivalent be in dBm? Reading in VU − 1.4 dB = reading in dBm. Thus the answer is −11 VU − 1.4 dB = −12.4 dBm.

C.4 USING DECIBELS WITH SIGNAL CURRENTS AND VOLTAGES

The dB is based upon a power ratio, as discussed. We can also relate decibels to signal voltages and to signal currents. The case for signal currents is treated first. We are dealing, of course, with gains and losses for a device or several devices (called networks) that are inserted in a circuit. Follow the thinking behind this series of equations:

$$\text{Gain/Loss}_{\text{dB}} = 10 \log P_1/P_2 = 10 \log I_1^2 R_1/I_2^2 R_2 = 20 \log I_1/I_2 = 10 \log R_1/R_2.$$

If we let $R_1 = R_2$, then the term 10 log $R_1/R_2 = 0$. (*Hint:* The log of 1 = 0.)

Remember from Ohm's law that E = IR, and from the power law P_w = EI. Thus $P_w = I^2R = E^2/R$.

To calculate gain or loss in dB when in the voltage/current domain we derive the following two formulas from the reasoning just shown:

$$\text{Gain/Loss}_{\text{dB}} = 20 \log E_1/E_2 = 20 \log I_1/I_2.$$

We see, in this case, that we multiply the log by the factor 20 rather than the factor 10 as we did in the power domain (i.e., 20 log vs. 10 log) because we really are dealing with power. Power is the function of the square of the signal voltage (E^2/R) or signal current (I^2R). We use traditional notation for voltage and current. Voltage is measured in volts (E); current, in amperes (I).

We must impress on the reader two important points: (1) equations as written are only valid when $R_1 = R_2$, and (2) validity holds only for terminations in pure resistance (there are no reactive components).

Consider these network examples:

Current (I):

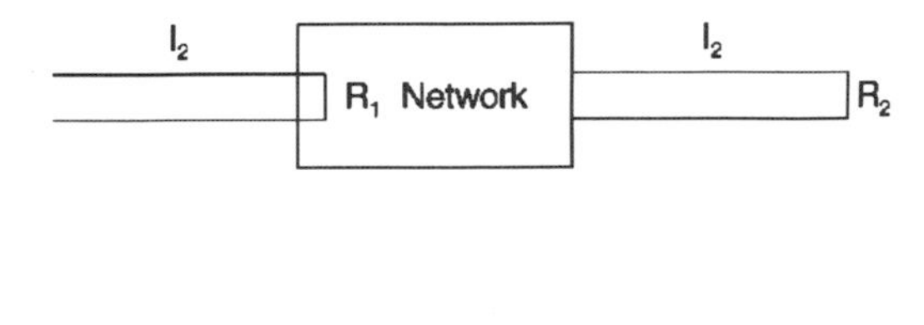

Voltage (E):

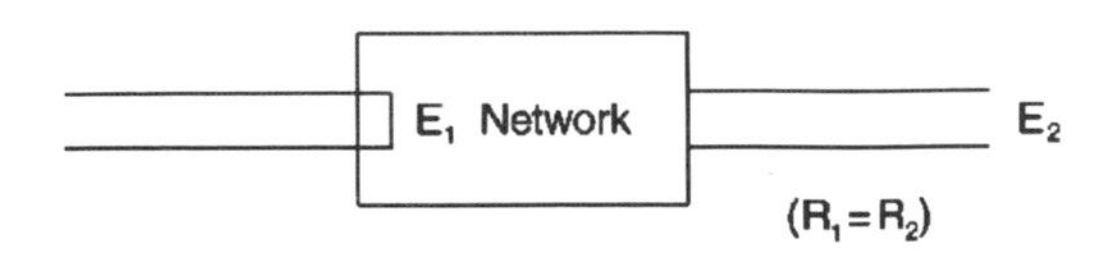

E_1 and E_2 are signal voltage drops across R_1 and R_2, respectively. The incisive reader will tell us that signals at the input are really terminated in an impedance (Z), which should equal the characteristic impedance, Z_0 (specified impedance). Such an impedance could be 600 Ω, for example. An impedance usually has a reactive component. Our argument is only valid if, somehow, we can eliminate the reactive component. The validity only holds true for a pure resistance. About the closest thing we can find to a "pure" resistance is a carbon resistor.

Turning back to our discussion, the input in the two cases cited may not be under our control, and there may be some reactive component. The output can be under our control. We can terminate the output port with a pure resistor, whose ohmic value equals the characteristic impedance. Our purpose for this discussion is to warn of possible small errors when reading input voltage or current.

Let's discuss the calculation of decibels dealing with a gain or loss by an example. A certain network with equal impedances at its input and output ports displays a signal voltage of 10 V at the input and 100 V at the output. The impedances are entirely resistive. What is the gain of the network?

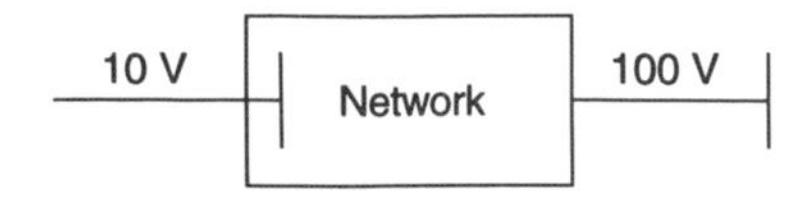

$$\text{Gain}_{\text{dB}} = 20 \log 100/10 = 20 \text{ dB}.$$

A similar network has a signal output of 40 V and a loss of 6 dB. What is the input signal voltage? (Equal impedances assumed.)

$$-6 \text{ dB} = 20 \log 40/X.$$

We shortcut this procedure by remembering our 10 log values. With a voltage or amperage relationship, the dB value is double (20 is twice the magnitude of 10). The value of X is 80 V. Whereas in our 10 log regime 3 dB doubled (or halved), here 6 dB doubles or halves.

A more straightforward way of carrying out this procedure will be suggested in the next section. X can be directly calculated.

C.5 CALCULATING A NUMERIC VALUE GIVEN A dB VALUE

The essence of the problem of calculating a numeric value given a dB value can be stated as such: If we are given the logarithm of a number, what is the number? To express this, two types of notation are given in the literature as follows:

1A) $\log^{-1} 0.3010 = 2$	2A) $\log^{-1} 2 = 100$
1B) antilog(0.3010) = 2	2B) antilog2 = 100.

In the case of example 1, the logarithm is 0.3010, which corresponds to the number 2. If we were to take the log (base 10) of 2, the result is 0.3010. In example 2, the log of 100 is 2 or, if you will, 2 is the logarithm of 100.

For our direct application we may be given a decibel value and be required to convert to its equivalent numeric value. If we turn to our introductory comments, when dealing in the power domain, we know that if we are given a decibel value of 20 dB, we are working with a power gain or loss of 100; 23 dB, 200; 30 dB, 1000; 37 dB 5000, and so on.

A scientific calculator is particularly valuable when we are not working directly with multiples of 10. For instance, enter the logarithm of a number onto the calculator keypad and the calculator can output the equivalent numeric value. Many hand-held scientific calculators use the same button for the log as for the antilog. Usually one can access the antilog function by first pressing the "2nd" button, something analogous to upper case on a keyboard. Often printed directly above the log button is "10^x."

On most calculators we first enter the logarithm on the numerical keypad, being sure to use the proper signs (+ or −). Press the "2nd" button; then press the "log" button. After a short processing interval, the equivalent number is shown on the display.

Let us get to the crux of the matter. We are interested in dBs. Let us suppose we are given 13 dB and we are asked to find its numeric equivalent (power domain). This calculation is expressed by the following formula:

$$\log^{-1}(13/10) = 20.$$

Let us use a calculator and compute the following equivalent numeric values when given dB values:

1. −21.5 dB. Divide by 10 and we have −2.15. Enter this on the keyboard with the negative sign. Press 2nd F (function) to access the upper case, which is the same as the log button but marked right above "10^x." Press = button and the value 0.00708 appears on the display.
2. +26.8 dB. Enter this number on the keypad and divide by 10; press =. Press 2nd F; press log button (10^x) and press =. The equivalent numeric value appears on the display. It is 478.63.

When working in the voltage or current domain, we divide the dB value by 20 rather than 10. Remember we are carrying out the reverse process that we used calculating a dB value when given a number (numeric) (i.e., the result of dividing the two numbers making up the ratio). This is expressed by the following formula:

$$\log^{-1}(\text{dB value}/20) = \text{equivalent numeric.}$$

Consider this example. Convert 26 dB (voltage domain) to its equivalent numeric value. Enter 26 on the keypad and divide by 20. The result is 1.3. Press 2nd F button and press the log button (10^x) and press =. The value 19.952 appears on the display. The reader probably did this in his or her head and arrived at a value of 20.

Try the following six example problems, first in the power domain, and then in the voltage/amperage domain. The correct answers appear just below.

1. −6 dB ________, ________.
2. +66 dB ________, ________.
3. −22 dB ________, ________.
4. +17 dB ________, ________.
5. −27 dB ________, ________.
6. +8.7 dB ________, ________.

Answers: 1: 0.251; 0.501. 2: 3,981,071.7; 1995.26. 3: 0.006309; 0.07943. 4: 50.118; 7.07945. 5: 0.001995; 0.044668. 6: 7.413; 2.7227.

C.5.1 Calculating Watt and Milliwatt Values When Given dBW and dBm Values

We will find that the process of calculating numeric values in watts and milliwatts is very similar to calculating the numeric value of a ratio when given the equivalent value in decibels. Likewise, the greater portion of these conversions can be carried out without a calculator to a first-order estimation. In the case where the dB value is 10 or a multiple thereof, the value will be exact.

Remember: 0 dBm = 1 mW; 0 dBW = 1 W by definition. Further, lest we forget: +3 dBm is twice as large as the equivalent 0 dBm value, thus where 0 dBm = 1 mW, +3 dBm = 2 mW.

Also, +10 dBm numeric value is 10 times the equivalent 0 dBm value (i.e., it is 10 dB larger). So +10 dBm = 10 mW; −10 dBm = 0.1 mW; −20 dBm = 0.01 mW.

In addition, −17 dBm is twice the numeric magnitude of −20 dBm. So −17 dBm = 0.02 mW, and so forth.

Try calculating the numeric equivalents of these dBm and dBW values without using a calculator.

1. +13 dBm ________ mW.
2. −13 dBm ________ mW.
3. +44 dBm ________ dBW, ________ W.
4. −21 dBm ________ mW.

5. +27 dBW __________ W.
6. −14 dBW __________ mW.
7. −11 dBm __________ mW.
8. +47 dBW __________ kW.

Answers: 1: 20 mW. 2: 0.05 mW. 3: +14 dBW, 25 W. 4: 0.008 mW. 5: 500 W. 6: 40 mW. 7: 0.08 mW. 8: 50 kW.

C.6 ADDITION OF dBs AND DERIVED UNITS

Suppose we have a combiner, a device that combines signals from two or more sources. This combiner has two signal inputs: +3 dBm and +6 dBm. Our combiner is an ideal combiner in that it displays no insertion loss. In other words, there is no deleterious effect on the combining action, it is "lossless." What we want to find out is the output of the combiner in dBm. It is not +9 dBm. The problem is shown diagrammatically as:

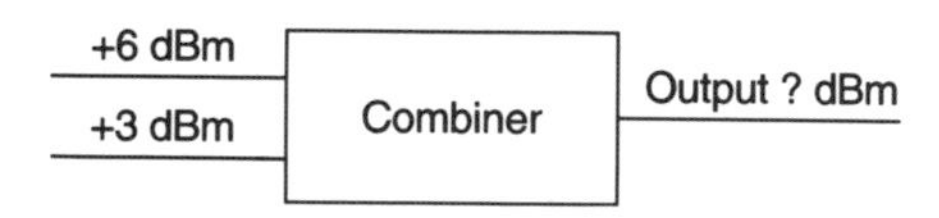

Some texts provide a nomogram to solve such a problem. We believe the following method is more accurate and, with the advent of affordable scientific calculators, easier. It is simple: Convert the input values to their respective numeric values in mW; add and convert the sum to its equivalent value in dBm.

The +3 and +6 dBm values are so familiar that we convert them by inspection, namely 2 and 4 mW. The sum is 6 mW. Now we take 10 log 6 to convert back to dBm again and the answer is +7.78 dBm. Remembering that there is an error when we work "3s" (3, 6, 9, 1, 4 and 7 values), we recalculated using a scientific calculator throughout. The answer was +7.76 dBm showing a 0.02-dB error.

On occasion, we will have to combine a large number of input/outputs where each is of the same level. This is commonly done with frequency division multiplex equipment or with multitone telegraphy or data.

Suppose we have an FDM group (12 voice channel inputs), where each input was −16 dBm. What is the composite output? This is stated as:

$$\begin{aligned}\text{Composite power}_{\text{dBm}} &= -16\text{ dBm} + 10\log 12\\ &= -16\text{ dBm} + 10.79\text{ dB}\\ &= -5.21\text{ dBm.}\end{aligned}$$

The problem of adding two or more inputs in a combiner is pretty straightforward if we keep in the power domain. If we delve into the voltage or current domain with equivalent dB values, such as dBmV (which we cover in Section 15.3.2), we recommend returning to the power domain if at all possible. If we do not, we can open Pandora's box, because of the phase relationship(s) of the inputs. In the next section we will carry out some interesting exercises in power addition.

C.7 dB APPLIED TO THE VOICE CHANNEL

The decibel is used to quantify gains and losses across a telecommunication network. The most common and ubiquitous end-to-end highway across that network is the voice channel (VF channel). A voice channel conjures up in our minds an analog channel, something our ear can hear. The transmit part (mouthpiece) of a telephone converts acoustic energy emanating from a human mouth to electrical energy, an analog signal. At the distant end of that circuit an audio equivalent of that analog energy is delivered to the receiver (earpiece) of the telephone subset with which we are communicating. This must also hold true for the all-digital network.

When dealing with the voice channel, there are a number of special aspects to be considered by the transmission engineer. In this section we will talk about these aspects regarding frequency response across a well-defined voice channel. We will be required to use dBs, dB-derived units, and numeric units.

The basic voice channel is that inclusive band of frequencies where loss with regard to frequency drops 10 dB relative to a reference frequency.[1] There are two slightly different definitions of the voice channel, North American and CCITT:

North America: 200 Hz to 3300 Hz (reference frequency, 1000 Hz);
CCITT: 300 Hz to 3400 Hz (reference frequency, 800 Hz).

We sometimes call this the nominal 4-kHz voice channel; some others call it a 3-kHz channel. (Note: There is a 3-kHz channel, to further confuse the issue; it is used on HF radio and some old undersea cable systems.)

To introduce the subject of a "flat" voice channel and a "weighted" voice channel we first must discuss some voice channel transmission impairments. These are *noise* and *amplitude distortion.*

We all know what noise is. It annoys the listener. At times it can be so disruptive that intelligent information cannot be exchanged or the telephone circuit drops out and we get a dial tone. So we want to talk about how much noise will annoy the average listener.

Amplitude distortion is the same as frequency response. We define amplitude distortion as the variation of level (amplitude) with frequency across a frequency passband or band of interest. We often quantify amplitude distortion as a variation of level when compared to the level (amplitude) at the reference frequency. The two common voice channel reference frequencies are noted in the preceding list.

To further describe amplitude distortion, let us consider a hypothetical example. At a test board (a place where we can electrically access a voice channel) in New York we have an audio signal generator available, which we will use to insert audio tones at different frequencies. At a similar test board in Chicago we will measure the level of these frequencies in dBm. The audio tones inserted in New York are all inserted at a level of −16 dBm, one at a time. In Chicago we measure these levels in dBm. We find the level at 1000 Hz to be +7 dBm, our reference frequency. We measure the 500-Hz tone at +3 dBm; 1200-Hz tone at +8 dBm; 2000-Hz tone +5 dBm, and the 2800-Hz tone at 0 dBm. Any variation of level from the 1000-Hz reference value we may call *amplitude distortion.* At 2800 Hz there was 7 dB variation. Of course, we can expect some of the worst-case excursion at band edges, which is usually brought about by filters or other devices that act like filters.

[1]This value applies when looking toward the subscriber from the local serving exchange. Looking into the network from the local serving exchange the value drops to 3 dB.

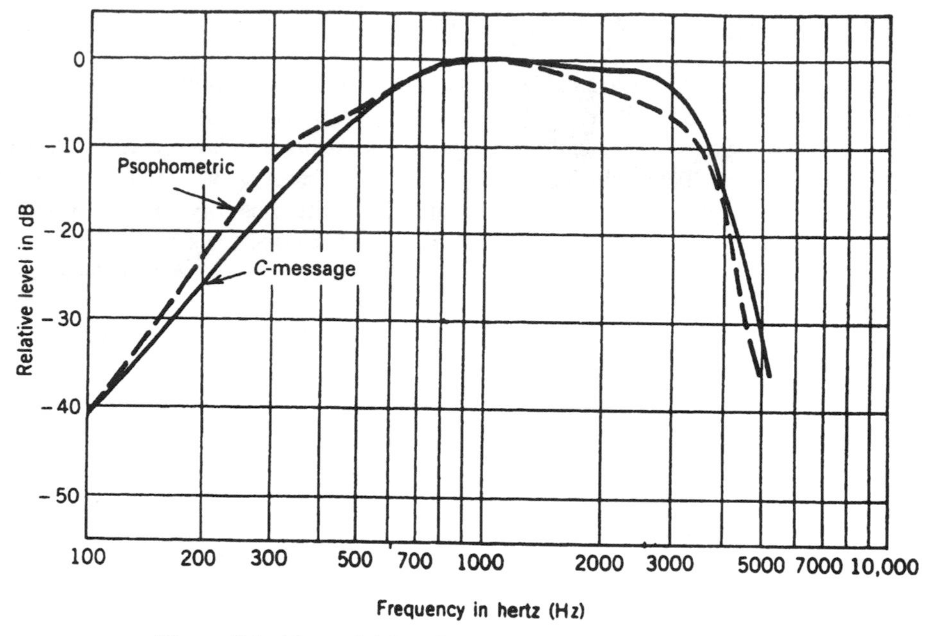

Figure C.1 Line weightings for telephone (voice) channel noise.

The human ear is a filter, as is the telephone receiver (earpiece). The two are in tandem, as we would expect. For the telephone listener, noise is an annoyance. Interestingly we find that noise annoys a listener more near the reference frequencies of a voice channel than at other frequencies. When using the North American 500-type telephone set with average listeners, a simple 0-dBm tone at 1000 Hz causes a certain level of annoyance. To cause the same level of annoyance, a 300-Hz tone would have to be at a level of about +17 dBm; a 400-Hz tone at about +11.5 dBm; a 600-Hz tone at about +4 dBm, and a 3000-Hz tone also at about +4 dBm for equal annoyance levels for a population of average listeners.

The question arose of why should the transmission engineer be penalized in design of a system for noise of equal level across the voice channel? We therefore have "shaped" the voice channel as a function of frequency and "annoyance." This shaping is called a *weighting curve*.

For the voice channel we will be dealing with two types of weighting: (1) C-message, used in North America, and (2) psophometric weighting as recommended by CCITT. Figure C.1 shows these weighting curves.

Weighting networks have been developed to simulate the corresponding response of C-message and psophometric weighting. Now we want to distinguish between *flat* response and *weighted* response. Of course, the curves in Figure C.1 show weighted response. Flat response, regarding a voice channel, has a low-pass response down 3 dB at 3 kHz and rolls off at 12 dB per octave. An octave means twice the frequency, so that it would be down 15 dB at 6 kHz and 27 dB at 12 kHz, and so on.

The term *flat* means equal response across a band of frequencies. Suppose a flat network has a loss of 3 dB. We insert a broad spectrum uniform signal at the input to the network. In the laboratory we generally use "white noise." White noise is a signal that contains components of all frequencies inside a certain passband. We now measure the output of our network at discrete frequencies and at whatever frequency we measure the output, the level is always the same. Figure C.1 shows frequency responses that are decidedly not flat.

We return now to the problem of noise in the voice channel. If the voice channel is to be used for speech telephony, which most of them are, then we should take into account the annoyance factor of noise to the human ear. Remember, when we measure noise in a voice channel, we look at the entire channel. Our noise measurement device reads the noise integrated across the channel. As we said, certain frequency components (around 800 Hz or 1000 Hz) are more annoying to the listener than other frequency components. It is because of this that we have developed a set of noise measurement units that are *weighted*. There are two such units in use today:

1. C-message weighting, which uses the unit dBrnC; and
2. Psophometric weighting, which more commonly uses the numeric unit, the picowatt (pWp) psophometrically weighted.

One interesting point that should be remembered is that the lowest discernible signal that can be heard by a human being is −90 dBm (800 or 1000 Hz).

Another point is that it was decided that all weighted (dB derived) noise units should be positive (i.e., not use a negative sign). First, remember these relationships:

$$1 \text{ W} = 10^{12} \text{ pW} = 10^{9} \text{ mW}$$
$$1 \text{ pW} = 1 \times 10^{-12} \text{ W} = 1 \times 10^{-9} \text{ mW} = -90 \text{ dBm}.$$

A weighted channel has less noise power than an unweighted channel if the two channels have identical characteristics. C-message weighting has about 2 dB *less* noise than a flat channel; a psophometric weighted channel has 2.5 dB *less* noise than a flat channel.

Figure C.2 may help clarify the concept of noise weighting and the noise advantage it can provide. The figure shows the C-message weighting curve. Idealized flat response is the heavy straight line at the arbitrary 0-dB point going right and left from 200 Hz to 3300 Hz. The hatched area between that line and the C-message response curve we may call the *noise advantage* (our terminology). There is approximately 2-dB advantage for C-message weight over flat response. If it were psophometric weighting, there would be a 2.5-dB advantage.

The dBrnC is the weighted noise measurement unit used in North America. The following are useful relationships:

$$0 \text{ dBrnC} = -92 \text{ dBm (with white noise loading of entire voice channel).}$$

Think about this:

$$0 \text{ dBrnC} = -90 \text{ dBm } (1000 - \text{Hz toned}).$$

Figures C.1 and C.2 show the rationale.

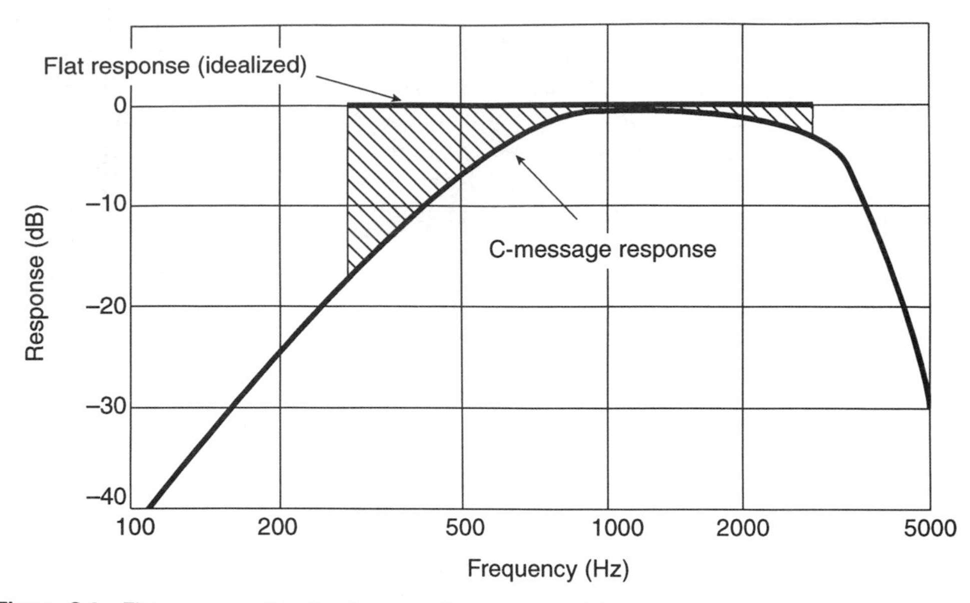

Figure C.2 Flat response (idealized) versus C-message weighting. The hatched area shows how we arrive at approximately a 2-dB noise advantage for C-message weighting. We can only take advantage of C-message improvement for speech telephony. For data transmission we must use flat response.

$$\text{Value in } (-)\text{ dBm} = 10 \log(\text{pW} \times 10^{-9}).$$
$$\text{Value in pWp} = \text{value in pW} \times 0.56.$$
$$-90 \text{ dBm} = -2 \text{ dBrnC and thus} - 92 \text{ dBm} = 0 \text{ dBrnc(white noise loading)}$$
$$-92.5 \text{ dBmp} = -90 \text{ dBm(flat, white noise)}$$
$$1 \text{ pWp} = -90 \text{ dBmp}$$
$$\text{value in dBm} = 10 \log(\text{value in pWp} \times 10^{-9}) + 2.5 \text{ dB}$$
$$\text{dBrnC} = 10(\log \text{ pWp} \times 10^{-9}) - 0.5 \text{ dB} + 90 \text{ dB}$$
$$\text{value in pW} \times 0.56 = \text{value in pWp}$$
$$\text{value in pWp}/0.56 = \text{value in pW}$$

Table C.1 summarizes some of the relationships we have covered for flat and weighted noise units.

Example 1. A hypothetical reference circuit shall accumulate no more than 10,000 pWp of noise. What are the equivalent values in dBrnC, dBm, and dBmp?

$$\begin{aligned}\text{dBrnC} &= 10(\log 10{,}000 \times 10^{-9}) - 0.5 \text{ dB} + 90 \text{ dB}\\ &= 39.5 \text{ dBrnC}.\\ (-)\text{ dBm} &= 10 \log(10{,}000 \times 10^{-9}) + 2.5 \text{ dB}\\ &= -47.5 \text{ dBm}.\end{aligned}$$

Table C.1 Comparison of Various Noise Units

Noise Unit	Total Power of 0 dBm: 1000 Hz	Total Power of 0 dBm: White Noise 0 kHz to 3 kHz	Wideband White Noise of −4.8 dBm/kHz
dBrnc	90.0 dBrnc	88.0 dBrnc	88.4 dBrnc
dBrn 3 kHz FLAT	90.0 dBrn	88.8 dBrn	90.3 dBrn
dBrn 15 kHz FLAT	90.0 dBrn	90.0 dBrn	97.3 dBrn
Psophometric voltage (600 Ω)	870 mV	582 mV	604 mV
pWp	1.26×10^9 pWp	5.62×10^8 pWp	6.03×10^8 pWp
dBp	91.0 dbp	87.5 dBp	87.8 dBp

Source: Based on Table 4.2, p. 60, Ref. 1.

$$\begin{aligned} \text{dBmp} &= 10 \log 10,000 \text{ pWp} \times 10^{-9} \\ &= -50 \text{ dBmp}. \end{aligned}$$

Example 2. We measure noise in the voice channel at 37 dBrnc. What is the equivalent noise in pWp?

$$\begin{aligned} 37 \text{ dBrnC} &= 10(\log X \times 10^{-9}) - 0.5 + 90 \text{ dB}. \\ -52.5 &= 10(\log X \times 10^{-9}). \\ -5.25 &= \log X \times 10^{-9}. \\ \text{antilog}(-5.25) &= 5623 \times 10^{-9}. \\ X &= 5623 \text{ pWp}. \end{aligned}$$

Carry out the following exercises. The answers follow.

1. −83 dBmp = ? pWp
2. 47,000 pWp = ? dBmp
3. −47 dBm = ? dBmp
4. 33 dBrnC = ? dBmp
5. 20,000 pWp = ? dBrnC
6. 50,000 pWp = ? dBm
7. 2000 pW = ? pWp
8. 4000 pWp = ? dBrnC.

Answers: 1: 5 pWp. 2: −43.28 dBmp. 3: −49.5 dBmp. 4: 2238 pWp = −56.5 dBmp = −54 dBm. 5: 42.5 dBrnC. 6: −43 dBmp = −40.5 dBm. 7: 1120 pWp. 8: 35.5 dBrnC.

C.8 INSERTION LOSS AND INSERTION GAIN

When dealing with the broad field of telecommunication engineering, we will often encounter the terms *insertion loss* and *insertion gain*. These terms give us important information about a two-port network in place in a circuit. Two-port just means we have an input (port) and an output (port). A major characteristic of this device is that it will present a loss in the circuit or it will present a gain. Losses and gains are expressed in dB. In the following we show a simple circuit terminated in its characteristic impedance, Z_0.

$Z_0 = 600\ \Omega$

We now insert into this same circuit a two-port network as follows:

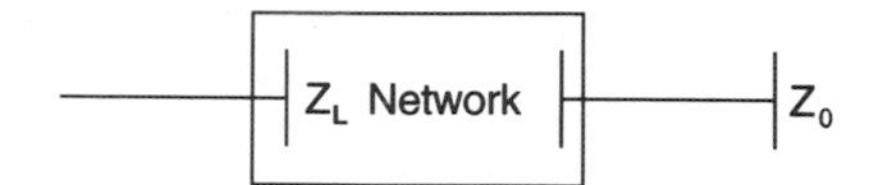

First for the case of insertion loss: Let us suppose the device is an attenuator, a length of waveguide, a mixer with loss, or any other lossy device. Suppose we are delivering power p_2 to the load Z_L with the network in place and power p_0 with the network removed. The ratio expressed in dB of p_0 to p_2 is called the insertion loss of the network:

$$\text{Insertion loss}_{\text{dB}} = 10 \log(p_0/p_2).$$

If Z_L equals Z_0, we can easily express insertion loss as a voltage ratio:

$$\text{Insertion loss}_{\text{dB}} = 20 \log(E_0/E_2).$$

If the network were one that furnished gain, such as an amplifier, we would invert the ratio and write:

$$\text{Insertion gain}_{\text{dB}} = 10 \log(p_2/p_0)$$

or, for the case of voltage,

$$\text{Insertion gain}_{\text{dB}} = 20 \log(E_2/E_0).$$

This may seem to the reader somewhat redundant to our introductory explanation of dBs. The purpose of this section is to instill the concepts of *insertion loss* and *insertion gain*. If we say that that waveguide section had an insertion loss of 3.4 dB, we know that the power would drop 3.4 dB from the input to the output of that waveguide section. If we said that the LNA (low noise amplifier) had an insertion gain of 30 dB, we would expect the output to have a power 30 dB greater than the input.

C.9 RETURN LOSS

Return loss is an important concept that sometimes confuses the student, particularly when dealing with the telephone network. We must remember that we achieve a maximum power transfer in an electronic circuit when the output impedance of a device (network) is exactly equal to the impedance of the device or transmission line connected to the output port. Return loss tells us how well these impedances match; how close they are to being equal in value (ohms) to each other.

Consider the following network's output port and its termination. The characteristic impedance (Z_0) of the output of the network is 600 Ω.

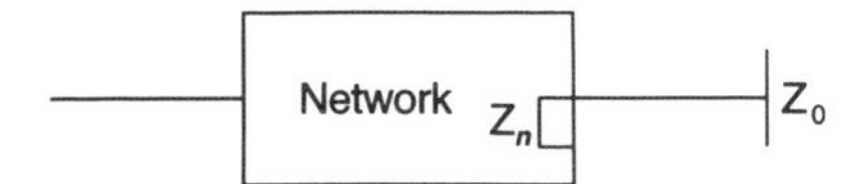

We have terminated this network in its characteristic impedance (Z_0). Let us assume for this example that it is 600 Ω. How well does the network's output port match its characteristic impedance? *Return loss* tells us this. Using the notation in the preceding example, return loss is expressed by the following formula:

$$\text{Return loss}_{\text{dB}} = 20\log(Z_n + Z_0)/(Z_n - Z_0).$$

First let us suppose that Z_n is exactly 600 Ω. If we substitute that in the equation, what do we get? We have then in the denominator 0. Anything divided by zero is infinity. Here we have the ideal case, an infinite return loss; a perfect match.

Suppose Z_n were 700 Ω. What would the return loss be? We would then have:

$$\begin{aligned}\text{Return loss}_{\text{dB}} &= 20\log(700 + 600)/(700 - 600)\\ &= 20\log(1300/100) = 20\log 13\\ &= 22.28\text{ dB.}\end{aligned}$$

Good return loss values are in the range of 25 dB to 35 dB. In the case of the telephone network hybrid, the average return loss is in the order of 11 dB.

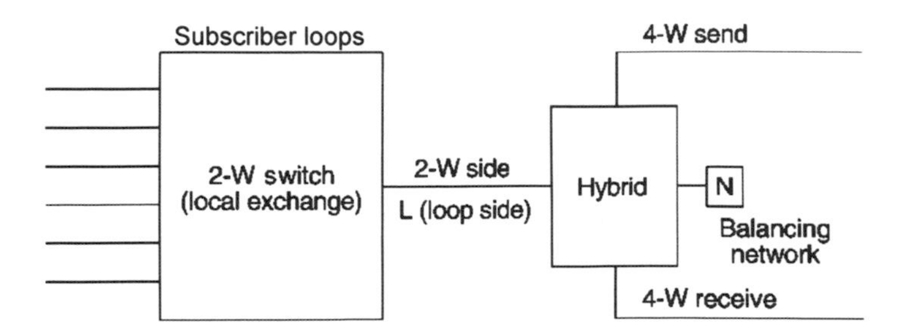

This diagram is the special situation of the 2-wire/4-wire conversion using the hybrid transformer, a 4-port device. Let us assume that the subscriber loop/local exchange characteristic impedance is 600 Ω. We usually can manage to maintain good impedance match with the 4-wire trunks, likewise for the balancing network, often called a *compromise network*. However, the 2-wire side of the hybrid can be switched into very short, short, medium, and long loops, where the impedance can vary greatly.

We will set up the equation for return loss assuming that at this moment in time it is through connected to a short loop with an impedance of 450 Ω; the impedance of the balancing network is 600 Ω, which is Z_0. We now calculate the return loss in this situation:

$$\begin{aligned}\text{Return loss dB} &= 20\log(600 + 450)/(600 - 450)\\ &= 20\log(1050/150) = 20\log 2.333\\ &= 7.36\text{ dB.}\end{aligned}$$

This is a fairly typical case. The mean return loss in North America for this situation is again 11 dB. With the advent of an all-digital network to the subscriber, we should see return losses in excess of 30 dB or possibly we will be able to do away with the hybrid all together.

C.10 RELATIVE POWER LEVEL: dBm0, pWp0, etc.

C.10.1 Definition of Relative Power Level

CCITT defines relative power level as the ratio, generally expressed in dB, between the power of a signal at a point in a transmission channel and the same power at another point in the channel chosen as a reference point, generally at the origin of the channel. Unless otherwise specified (CCITT Recs. G.101, 223), the relative power level is the ratio of the power of a sinusoidal test signal (800 Hz or 1000 Hz) at a point in the channel to the power of that reference signal at the transmission reference point.

C.10.2 Definition of Transmission Reference Point

In its old transmission plan, the CCITT had defined the zero relative level point as being the two-wire origin of a long-distance (toll) circuit. This is point 0 of Figure C.3*a*.

In the currently recommended transmission plan the relative level is −3.5 dBr at the virtual switching point on the transmitting side of a four-wire international circuit. This is point V in Figure C.3*b*. The transmission reference point or zero relative level point (point T in Figure C.3*b*) is a virtual two-wire point which would be connected to V through a hybrid transformer having a loss of 3.5 dB. The conventional load used for computation of noise on multichannel carrier systems corresponds to an absolute mean power level of −15 dBm at point T.

The 0 TLP (zero test level point) is an important concept. It remains with us even in the age of the all-digital network. The concept seems difficult. It derives from the fact that a telephone network has a loss plan. Thus signal levels will vary at different points in a network, depending on the intervening losses. We quote from an older edition (1st

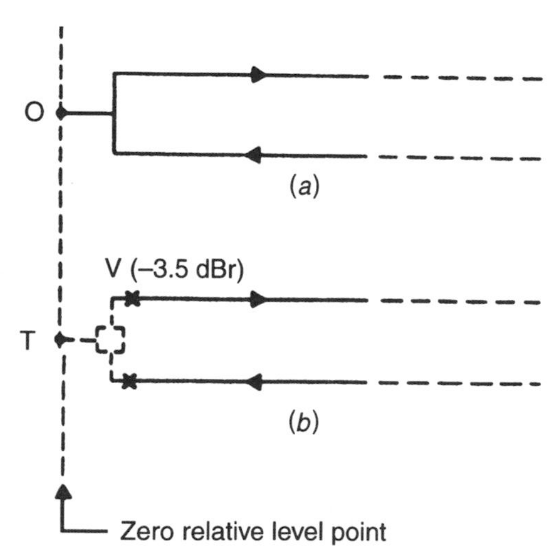

Figure C.3 The zero relative level point.

ed.) of *Transmission Systems for Telecommunications* (Bell Telephone Laboratories, New York, 1959, Vol. I, pp. 2–3):

> In order to specify the amplitudes of signals or interference, it is convenient to define them at some reference point in the system. The amplitudes at any other physical location can be related to this reference point if we know the loss or gain (in dB) between them. In the local plant, for example, it is customary to make measurements at the jacks of the outgoing trunk test panel, or (if one does not wish to include office effects) at the main frame. For a particular set of measurements, one of these points might be taken as a reference point, and signal or noise magnitudes at some other point in the plant predicted from a knowledge of the gains or losses involved.

In toll telephone practice, it is customary to define the toll transmitting switchboard as the reference point or "zero transmission level" point. To put this in the form of a definition:

> The transmission level at any point in a transmission system is the ratio of the power of a test signal at that point to the test signal power applied at some point in the system chosen as a reference point. This ratio is expressed in decibels. In toll systems, the transmitting toll switchboard is usually taken as the zero level or reference point.

Frequently the specification of transmission level is confused with some absolute measure of power at some point in the system. Let us make this perfectly clear. When we speak of −9-dB transmission level point (often abbreviated "the −9 level"), we simply mean that the signal power at such a point is 9 dB below whatever signal power exists at the zero level point. The transmission level does not specify the absolute power in dBm or in any other such power units. It is relative only. It should also be noted that, although the reference power at the transmitting toll switchboard will be at an audio frequency, the corresponding signal power at any given point in a broadband carrier system may be at some carrier frequency. We can, nevertheless, measure or compute this signal power and specify its transmission level in accordance with the definition we have quoted. The transmission level at some particular point in a carrier system will often be a function of the carrier frequency associated with a particular channel.

Using this concept, the magnitude of a signal, a test tone, or an interference (level) can be specified as having a given power at a designated level point. For example, in the past many long toll systems had 9-dB loss from the transmitting to the receiving switchboard. In other words, the receiving switchboard was then commonly at the −9-dB transmission level. Since noise measurements on toll telephone systems were usually made at the receiving switchboard, noise objectives were frequently given in terms of allowable noise at the −9-dB transmission level. Modern practice calls for keeping loss from the transmitting terminals to the receiving terminals as low as possible, as part of a general effort to improve message channel quality. As a result, the level at the receiving switchboard, which will vary from circuit to circuit, may run as high as −4 dB or −6 dB. Because of this, requirements are most conveniently given in terms of the interference that would be measured at zero level. If we know the transmission level at the receiving switchboard, it is easy to translate this requirement into usable terms. If we say, some tone is found to be −20 dBm at the zero level and we want to know what it would be at the receiving switchboard at −6 level, the answer is simply −20 − 6 = −26 dBm.

Quoting from the 4th edition of *Transmission Systems for Communications* (Ref. 2):

Expressing signal mangitude in dBm and system level in dB provides a simple method of determining signal magnitude at any point in a system. In particular, the signal magnitude at 0 TLP is S_0 dBm, then the magnitude at a point whose level is L_x dB is

$$S_x = S_0 + L_x$$

The abbreviation dBm0 is commonly used to indicate the signal magnitude in dBm at 0 TLP. Of course, pWp0 takes on the same connotation, but is used as an absolute noise level (weighted).

Digital Level Plan. The concept of transmission level point applies strictly to analog transmission. It has no real meaning in digital transmission, except where the signal is in analog form. Nevertheless, the concept of TLP is a powerful one, which can be retained.

In North America, when there is cutover to an all-digital network, a fixed transmission loss plan will be in place. The toll network will operate, end-to-end, with a 6-dB loss. A digital toll connecting trunk will have a 3-dB loss. There are two toll connecting trunks in a built-up toll connection, by definition. The remaining intervening toll trunks will operate at 0 dB loss/0 dB gain; thus the 6-dB total loss.

By the following, we can see that the 0 TLP concept still hangs on. We quote from *Telecommunication Transmission Engineering*, Vol. 3 (AT&T, New York):

> It is desirable in the fixed loss network to retain the 6-dB loss for test conditions so that all trunks have an EML (expected measured loss) of 6 dB. To accomplish this, the transmitting and receiving test equipment at digital offices (exchanges) must be equipped with 3-dB pads with analog-digital converters. Because of the use of 3-dB test pads, the No. 4 ESS (ATT digital toll exchange) can be considered at −3 TLP even though signals are in digital form. Since the path through the machine (digital switch) is lossless, the −3 TLP applies to the incoming as well as the outgoing side of the machine, a feature unique to digital switching machines.

C.11 dBi

The dBi is used to quantify the gain of an antenna. It stands for dB above (or below) an *isotropic*. If it is above, we will often use the plus (+) sign, and when below an isotropic, we will use a minus (−) sign. An isotropic is an imaginary reference antenna with uniform gain in all three dimensions. Thus, by definition, it has a gain of 1 dB or 0 dB. In this text, and in others dealing with commercial telecommunications, all antennas will have a “positive” gain. In other words, the gain will be greater than an isotropic. For example, parabolic dish antennas can display gains from 15 dBi to over 60 dBi.

C.11.1 dBd

The *dBd* is another dB unit used to measure antenna gain. The abbreviation dBd stands for dB relative to a dipole. This dB unit is widely used in cellular and PCS radio technology. When compared to an isotropic, the dBd unit has a 2.15-dB gain over an isotropic. For example, +2 dBd = +4.15 dBi.

C.12 EIRP

EIRP stands for "effective isotropically radiated power." We use the term to express how much transmitted power is radiated in the desired direction. The unit of measure is dBW or dBm, because we are talking about power.

$$\mathrm{EIRP}_{\mathrm{dBW}} = \mathrm{P}_{\mathrm{t(dBW)}} + \mathrm{L}_{\mathrm{L(dB)}} + \text{antenna gain}_{\mathrm{(dBi)}},$$

where P_t is the output power of the transmitter either in dBm or dBW. L_L is the line loss in dB. That is the transmission line connecting the transmitter to the antenna. The third factor is the antenna gain in dB.

Warning! Most transmitters give the output power in watts. This value must be converted to dBm or dBW.

Example 1. A transmitter has an output of 20 W, the line loss is 2.5 dB, and the antenna has 27-dB gain. What is the EIRP in dBW?

Convert the 20 W to dBW = +13 dBW. Now we simply algebraically add:

$$\begin{aligned}\mathrm{EIRP} &= +13\ \mathrm{dBW} - 2.5\ \mathrm{dB} + 27\ \mathrm{dB} \\ &= +37.5\ \mathrm{dBW}.\end{aligned}$$

Example 2. A transmitter has an output of 500 mW, the line losses are 5.5 dB, and the antenna gain is 39 dB. What is the EIRP in dBm?

Convert the transmitter output to dBm, which = +27 dBm. Now simply algebraically add [Ref. 3]:

$$\begin{aligned}\mathrm{EIRP} &= +27\ \mathrm{dBm} - 5.5\ \mathrm{dB} + 39\ \mathrm{dB} \\ &= 60.5\ \mathrm{dBm}.\end{aligned}$$

REFERENCES

1. *Transmission Systems for Communications*, 5th ed., Bell Telephone Laboratories, Holmdel, NJ, 1982.
2. *Transmission Systems for Communications*, Revised 4th ed., Bell Telephone Laboratories, Merrimack Valley, MA, 1971.
3. R. L. Freeman, *Telecommunications Transmission Handbook*, 4th ed., Wiley, New York, 1998.

APPENDIX D

ACRONYMS AND ABBREVIATIONS

0 TLP	zero test level point
2B1Q	2 binary to 1 quaternary
AAL	ATM adaptation layer
AAR	automatic alternative routing
ABM	asynchronous balanced mode
ABSBH	average busy season busy hour
ac, AC	alternating current
ACK	acknowledge, acknowledgment
A/D	analog-to-digital
ADM	add–drop multiplex
ADPCM	adaptive differential pulse code modulation
ADSL	asymmetric digital subscriber line
AGC	automatic gain control
AIS	alarm indication signal
ALBO	automatic line build-out
AM	amplitude modulation
AMI	alternate mark inversion
AMPS	advanced mobile phone system
ANSI	American National Standards Institute
APD	avalanche photodiode
APL	average picture level
ARM	asynchronous response mode
ARPA	Advanced Research Projects Agency
ARQ	automatic repeat request
ARP	address resolution protocol
ARR	automatic rerouting
ASK	amplitude shift keying
ASCII	American Standard Code for Information Interchange
ATB	all trunks busy
ATM	asynchronous transfer mode
ATSC	Advanced Television System Committee
AT&T	American Telephone & Telegraph (Corp.)
AU	administrative unit, access unit
AUI	attachment unit interface
AUG	administrative unit group

AWG	American Wire Gauge
AWGN	additive white Gaussian noise
B3ZS, B6ZS, B8ZS	binary 3 zero substitution, binary 6 zero substitution, binary 8 zero substitution
BCC	block check count
BCD	binary-coded decimal
BCH	Bose-Chaudhuri-Hocquenghem (a family of block codes)
BECN	backward explicit congestion notification
Bellcore	Bell Communications Research
BER	bit error rate, bit error ratio
BERT	bit error rate test
BFSK	binary frequency shift keying
BH	busy hour
BIP	bit-interleaved parity
B-ISDN	broadband ISDN
bit	binary digit
BITE	built-in test equipment
BITS	building-integrated timing supply
BIU	baseband interface unit
BNZS	binary n-zeros substitution
BOC	Bell Operating Company
BOM	beginning of message
BP	bandpass
BPS	bits per second
BPV	bipolar violation
BPSK	binary phase shift keying
BRA, BRI	basic rate (interface)
BSN	backward sequence number
BSTJ	*Bell System Technical Journal*
BT	bridged tap
BTL	Bell Telephone Laboratories
CAC	connection admission control
CAP	competitive access provider
CATV	community antenna television
CBDS	connectionless broadband data services
CBR	constant bit rate
CCIR	International Consultive Committee for Radio
CCITT	International Consultive Committee for Telephone and Telegraph
ccs, CCS	cent call second
CD	compact disk; collision detection
CDMA	code division multiple access
CDPD	cellular digital packet data
CDV	cell delay variation
CED	called station identification (fax)
CGSA	cellular geographic serving area
CELP	codebook-excited linear predictive (coder)

CEPT	Conference European Post and Telegraph (from the French)
C/I	carrier-to-interference (ratio)
CIR	committed information rate
CL	connectionless
CLLM	consolidated link layer management
CLNAP	connectionless network access protocol
CLP	cell loss priority
CLR	circuit loudness rating
CLSF	connectionless service functions
C/N_0	carrier-to-noise in 1-Hz bandwidth
C/N, CNR	carrier-to-noise ratio
CODEC, codec	coder-decoder
COM	continuation of message
compander	compressor–expander
CONUS	contiguous United States
COT	central office terminal
CPCS	common part convergence sublayer
CPE	customer premises equipment
CPU	central processing unit
CRC, crc	cyclic redundancy check
CRE	corrected reference equivalent
CREG	concentrated range extender with gain
CRT	cathode-ray tube
CS	convergence sublayer
CSA	carrier serving area
CSI	convergence sublayer indicator
CSMA, CSMA/CD	carrier sense multiple access, carrier sense multiple access with collision detection
CSO	composite second-order (products)
CT	cordless telephone
C/T	carrier (level)-to-noise temperature ratio
CTB	composite triple beat
CUG	closed user group
CVSD	continuous variable slope delta modulation
D/A	digital-to-analog
DA	destination address
DAMA	demand assignment multiple access
DARPA, DARPANET	Defense Advanced Research Projects Agency (network)
dB	decibel
dBc	decibels referenced to the carrier level
dBd	dB referenced to a dipole (antenna)
dBi	dB over an isotropic (antenna)
dBm	dB referenced to a milliwatt
dBmP	dBm psophometrically weighted
dBmV	dB referenced to a millivolt
dBm0	dBm referenced to the zero test level point (0 TLP)
DBPSK	differential binary PSK
dBr	decibels above or below "reference"
dBrnC	dB reference noise C-message weighted

dBμ	decibels referenced to a microvolt
dBW	decibels referenced to 1 watt
dc, DC	direct current
DCE	data communications equipment, data circuit-terminating equipment
DCPBH	double channel planar buried heterostructure
DCS	digital cross-connect (system)
DDS	digital data system
DECT	digital European cordless telephone
DFB	distributed feedback (laser)
DL	data link
DLC	digital loop carrier
DLCI	data link connection identifier
DN	directory number
DNHR	dynamic nonhierarchical routing
dNp	decineper
DoD	Department of Defense (U.S.)
DPC	destination point code
DQDB	distributed queue dual bus
DQPSK	differential QPSK
D/R	distance/radius
DS0, DS1, DS1C, DS2	"digital system" 0, 1, 1C, etc., the North American PCM hierarchy
DS	direct sequence
DSAP	destination service access point
DSL	digital subscriber loop, digital subscriber line
DSU	digital service unit
DTE	data terminal equipment
DUP	data user part
EB	errored block
EBCDIC	extended binary coded decimal interchange code
E_b/I_0	energy per bit per interference density ratio
E_b/N_0	energy per bit per noise spectral density ratio
EC	earth curvature; earth coverage
EDD	envelope delay distortion
EDFA	erbium-doped fiber amplifier
EFS	error-free second
EHF	extremely high frequency—the frequency spectrum from 30–300 GHz
EIA	Electronic Industries Association
EIRP	effective (equivalent) isotropically radiated power
EMC	electromagnetic compatibility
EMI	electromagnetic interference
EN	exchange number
EOM	end of message
EOT	end of text
ERL	echo return loss
ERP	effective radiated power
ES	end section, errored second

ESF	extended superframe
ESS	electronic switching system
ETSI	European Telecommunications Standardization Institute
FAS	frame alignment signal
FCC	Federal Communications Commission (U.S.)
FCS	frame check sequence
FDD	frequency division duplex
FDDI	fiber distributed data interface
FDM	frequency division multiplex
FDMA	frequency division multiple access
FEC	forward error correction
FEXT	far-end crosstalk
FISU	fill-in signal unit
FLTR	filter
FM	frequency modulation
FPLMTS	future public land mobile telecommunication system
FPS	frames per second
FRAD	frame relay access device
FSK	frequency shift keying
FSL	free-space loss
FSN	forward sequence number
FTP	file transfer protocol
GBLC	Gaussian band-limited channel
Gbps	gigabits per second
GBSVC	general broadcast signaling virtual channel
GEO	geostationary earth orbit
GFC	generic flow control
GFSK	Gaussian FSK
GHz	gigahertz (Hz $\times$ 10^9)
GMSK	Gaussian minimum shift keying
GMT	Greenwich mean time
GPS	geographical positioning system
GSM	"group system mobile" (from the French; the digital European cellular scheme), also called Global System for Mobile Communications
GSTN	general switched telecommunications network
G/T	gain (antenna)-to-noise temperature ratio
GTE	General Telephone & Electronics
HC	head-end controller
HDLC	high-level data link control
HDTV	high definition television
HEC	header error control
HF	high frequency; also the radio frequency band 3 MHz to 30 MHz
HFC	hybrid fiber coax
HPA	high power amplifier
HU	high usage (route[s])

Hz	hertz
IAM	initial address message
IBM	International Business Machine (Inc.)
ICMP	Internet control message protocol
IDR	intermediate data rate
IEEE	Institute of Electrical and Electronics Engineers
IF	intermediate frequency
InGaAsP	indium gallium arsenide phosphorus
INTELSAT	International Telecommunication Satellite (consortium)
I/O	input/output (device)
IP	Internet protocol
IRL	isotropic receive level
IS	interim standard
ISC	international switching center
ISDN	integrated services digital networks
ISI	intersymbol interference
ISM	industrial, scientific, and medical (band)
ISO	Internatinal Standards Organization
ISUP	ISDN user part
ITT	International Telephone and Telegraph Co.
ITU	International Telecommunication Union
ITU-R	ITU Radiocommunications Bureau
ITU-T	ITU Telecommunications Standardization Sector
IXC	interexchange carrier
kbps	kilobits per second
kft	kilofeet
kHz	kilohertz
km	kilometer(s)
LAN	local area network
LATA	local access and transport area
LAP, LAPB, LAPD	link access protocol; link access protocol, B-channel; link access protocol, D-channel
LBO	line build-out
LCC	lost calls cleared
LCD	lost calls delayed
LCH	lost calls held
LD	laser diode
LEA	line-extender amplifier
LEC	local exchange carrier
LED	light-emitting diode
LEO	low earth orbit
LLC	logical link control
LMDS	local multipoint distribution system
Ln	log to the natural base
LNA	low noise amplifier
LOH	line overhead
LOS	loss of signal; line-of-sight

LR	loudness rating
LRC	longitudinal redundancy check
LRD	long route design
LSB	lower sideband
LSI	large scale integration
LSSU	link status signal unit
m	meter
mA	milliampere(s)
MAC	medium access control
MAN	metropolitan area network
MAU	medium attachment unit
Mbps	megabits per second
MDF	main distribution frame
MF	multifrequency
MFJ	modification of final judgement
MFSK	multilevel or M-ary FSK
MID	message identifier
MII	medium independent interface
MPSK	multilevel or M-ary PSK
MHz	megahertz
MLP	multilink procedure
MLRD	modified long route design
MPEG	Motion Picture Expert's Group
ms	millisecond
MS	mobile station
MSC	mobile switching center
MSU	message signal unit
MTBF	mean time between failures
MTP	message transfer part
MTSO	mobile telephone switching office
mV	millivolt
mW	milliwatt
MW, M/W	microwave
N/A	not applicable
NA	numerical aperture
NACK	negative acknowledgment
NANP	North American Numbering Plan
NDF	new data flag
NEXT	near-end crosstalk
NF	noise figure
nm	nautical mile; nanometer
NMT	network management
NNI	network–network interface or network-node interface
NPA	numbering plan area
NPC	network parameter control
NRM	normal response mode; network resource management
NRZ	nonreturn to zero
ns, nsec	nanosecond(s)

NSDU	network services signaling data unit
NSP	national signaling point
NT	network termination
NTSC	National Television System Committee
OAM, OA&M	operations and maintenance; operations, administration and maintenance
OC	optical carrier (OC-1, OC-3)
OLR	overall loudness rating
OPC	originating point code
ORE	overall reference equivalent
OSI	open system interconnection
PA	power amplifier
PABX	private automatic branch exchange
PACS	personal access communication services
PAD	packet assembler–disassembler
PAL	phase-alternation line
PAM	pulse amplitude modulation
PAR	positive acknowledgment with retransmission
PBX	private branch exchange
PC	personal computer
PCB	printed circuit board
PCI	protocol control information
PCM	pulse code modulation
PCS	personal communication services; physical coding sublayer
PDH	plesiochronous digital hierarchy
PDN	public data network
PDU	protocol data unit
PEL	picture element
p/f	poll-final
pfd	picofarad
PH	packet handling
PHS	personal handyphone system
PHY	refers to the physical layer (OSI layer 1)
PIN	p-intrinsic-n
PIXEL	picture element (same as pel)
PLCP	physical layer convergence protocol
PLS	physical layer signaling
PM	phase modulation; physical medium
PMA	physical medium attachment
PMD	physical layer medium dependent
PN	pseudonoise
POH	path overhead
POP/POT	point of presence, point of termination
POS	point of sale
POTS	plain old telephone service
PRI	primary service (ISDN)
PRS	primary reference source
PSK	phase shift keying

PSN	public switched network
PSPDN	packet-switched public data network
PSPDS	packet-switched data transmission service
PSTN	public switched telecommunication network
PTI	payload type indicator (identifier)
PVC	permanent virtual circuit
pW	picowatt(s)
pWp	picowatt(s) psophometrically weighted
QAM	quadrature amplitude modulation
QoS	quality of service
QPSK	quadrature phase shift keying
RARP	reverse address resolution protocol
RBOC	Regional Bell Operating Company
RC	resistance capacitance (time constant)
RD	resistance design
RE	reference equivalent
RELP	residual excited linear predictive (coder)
RF	radio frequency
RFI	radio frequency interference
RI	route identifier
RL	return loss
RLR	receive loudness rating
rms	root mean square
RNR	receive not ready
RR	receive ready
RRD	revised resistance design
RRE	receive reference equivalent
RS	Reed-Solomon (code); reconciliation sublayer
RSL	receive signal level
RSU	remote subscriber unit
RT	remote terminal
RZ	return to zero
SA	source address
SANC	signaling area/network code
SAP	service access point
SAPI	service access point identifier
SAR	segmentation and reassembly
SBC	sub-band coding
SBS	selective broadcast signaling
SCADA	supervisory control and data acquisition
SCCP	signaling connection control part
SCPC	signal channel per carrier
S/D	signal-to-distortion ratio
SDH	synchronous digital hierarchy
SDLC	synchronous data link control
SDT	structured data transfer
SDU	service data unit

SECAM	sequential color and memory
SF	signal frequency (signaling)
SFD	start of frame delimiter
SINAD	signal+noise+distortion-to-noise+distortion ratio
SLC	signaling link code
SLIC	subscriber line interface card
SLP	single link procedure
SLR	send loudness rating
SMDS	switched multimegabit data service
SMPTE	Society of Motion Picture and Television Engineers
SMT	station management
SN	sequence number
S/N, SNR	signal-to-noise ratio
SNA	system network architecture
SNAP	subnetwork access protocol
SNP	sequence number protection
SOH	section overhead; start of heading
SONET	synchronous optical network
SP	signaling point
SPC	stored program control
SPE	synchronous payload envelope
SREJ	selective reject
SRTS	synchronous residual time stamp
SSAP	source service access point
SSB, SSBSC	single sideband, single sideband suppressed carrier
SSM	single segment message
ST	segment type
STL	studio-to-transmitter link; Standard Telephone Laboratory
STM	synchronous transport module; synchronous transfer mode
STP	shielded twisted pair
STS	synchronous transport signal; space-time-space (switch)
SVC	switched virtual circuit
SXS	step-by-step (switch)
TA	terminal adapter
TASO	Television Allocation Study Organization
TAT	trans-atlantic (cable)
TBD	to be determined
TC	transmission convergence
TCP/IP	transmission control protocol/Internet protocol
TDD	time division duplex
TDM	time division multiplex
TDMA	time division multiple access
TE	terminal equipment
TelCo	telephone company
T&G	trees & growth
THT	token holding timer
THz	terahertz (1×10^{12} Hz)
TIA	Telecommunication Industry Association
TLP	test level point

TPDU	transport protocol data unit
TRE	transmit reference equivalent
TRT	token rotation timer
TS	time slot
TSI	time slot interchanger
TSTS	time space time space (switching)
TU	tributary unit
TUG	tributary unit group
TUP	telephone user part
TV	television
TVT	valid transmission timer
TWT	traveling-wave tube
UHF	ultra-high frequency (300 MHz to 3000 MHz)
ULP	upper layer protocol
UMTS	universal mobile telecommunication system
UNI	user–network interface
UP	user part
UPC	user parameter control
μs	microsecond
UT	universal time
UTC	universal time coordinated (coordinated universal time)
UTP	unshielded twisted pair
μV	microvolt
μW	microwatt
VBR	variable bit rate
VC	virtual container; virtual connection; virtual channel
VCC	virtual channel connection
VCI	virtual channel identifier
VF	voice frequency
VHF	very high frequency (30 MHz to 300 MHz)
VHSIC	very high speed integrated circuit
VLSI	very large scale integration
VP	virtual path
VPC	virtual path connection
VPI	virtual path identifier
VRC	vertical redundancy check
VSB	vestigial sideband
VSWR	voltage standing wave ratio
VT	virtual tributary
WACS	wireless access communication system
WAN	wide area network
WDM	wave(length) division multiplex
WLAN	wireless LAN
WLL	wireless local loop
Xm	cross-modulation

INDEX

Boldface page number denotes in-depth coverage of a topic.
Italic page number denotes a definition of a term.